The Immune-Neuroendocrine Circuitry

History and Progress

Neuroimmune Biology, Volume 3

Neuroimmune Biology

Series Editors

I. Berczi, A. Szentivanyi

The Immune-Neuroendocrine Circuitry
History and Progress

Volume Editors

Istvan Berczi
Andor Szentivanyi

University of Manitoba, Winnipeg, Manitoba, Canada
and
University of South Florida, Tampa, FL, USA

2003
ELSEVIER
AMSTERDAM – BOSTON – LONDON – NEW YORK – OXFORD – PARIS
SAN DIEGO – SAN FRANCISO – SINGAPORE – SYDNEY – TOKYO

ELSEVIER SCIENCE B.V.
Sara Burgerhartstraat 25
P.O. Box 211, 1000 AE Amsterdam, The Netherlands

First edition 2003

Library of Congress Cataloging in Publication Data
A catalog record from the Library of Congress has been applied for.

British Library Cataloguing in Publication Data
A catalogue record from the British Library has been applied for.

ISBN: 0-444-50851-1

♾ The paper used in this publication meets the requirements of ANSI/NISO Z39.48-1992 (Permanence of Paper).
Printed in The Netherlands.

Foreword

The recognition of the importance of the central nervous system in the co-ordination and regulation of bodily functions goes back to pre-historic times. There was also a consensus about a close relationship between mental and physical health. The first scientific efforts to unravel a possible influence of the central nervous system on allergy and asthma were made during the late 19th century and sporadical experiments were done during the first half of the 20th century. Although some investigators reached conclusions of lasting value, the lack of theoretical knowledge and severe methodological limitations posed insurmountable obstacles to the advancement of knowledge in this area. During the past three decades or so our basic understanding of the nervous-, endocrine- and immune systems has advanced significantly. It became apparent first that the nervous- and the endocrine systems are closely linked. The long-held view that hormones and nerves regulate immune function could now also be investigated. Serious work was done in a handful of laboratories to study this problem during the seventies. This was followed by burst of progress during the subsequent decades. Now it is apparent that the nervous-, endocrine- and immune systems form a regulatory network that plays a fundamental role in the growth, development and function of macro-organisms throughout their life cycle.

The science dealing with this area has been termed by us as *Neuroimmune Biology*. This subject is now covered in a book series bearing the same name. This volume is intended to serve as an introduction of the subject matter to the scientific community. The objective is to present a concise overview of the field by covering historical facts, giving basic information on the subject as well as discussing the recent developments and theoretical advancements. In essence this volume sums up what the Neuroimmune Biology book series will cover in detail. We sincerely hope that the reader will find this book informative about this new and exciting area of Biology.

Istvan Berczi and Andor Szentivanyi

Preface

This volume of Neuroimmune Biology (NIB), and the entire series, have several unique features that are worthy of consideration. Remarkably, the fundamental concepts outlined by the ingenious thinkers and experimentalists of early Physiology and Medicine, Claude Bernard, Pavlov, Cannon and Selye, are being extended in these books by facts and concepts more than ever before. For instance, it is discussed that the concept of homeostasis may now be extended to pathophysiological conditions. It is also clear that Pavlov's thesis of conditioning clearly applies to the immune system as it has been presented in relation to the stress response and to the conditioning of immune reactions (NIB Volume 1). The "fight or flight" concept of Cannon and the "general adaptation syndrome" of Selye have now been related to the acute phase response. It is now clear that the neuroimmune regulatory network coordinates physiological and pathophysiological reactions in higher animals. The health, disease, growth and development, the biology of reproduction, aging and lifespan of animals/man depend to a large extent on the function of this systemic regulatory network.

The simplest animals, sponges, show self-recognition, as demonstrated by their ability to re-aggregate with their own kind after dispersion and by the rejection of tissue grafts from a foreign sponge species. This fundamental form of self-recognition has long been thought to be the basic characteristics of the immune system. In actual fact, this recognition is the pre-requisite of evolution from unicellular to multi-cellular organisms. Adhesion molecules serve as the basis of this organization in the absence of the nervous and endocrine systems. These non-diffusible regulatory molecules maintained their dominant role in tissue organization, the regulation of cell growth and differentiation as well as function also in higher animals. However, during evolution a remarkably complex and very efficient system, the Neuroimmune system, has been superimposed on this cell-to-cell mediated regulation. The result is the Neuroimmune Regulatory Network, with its incredible complexity, precision and efficiency. The central nervous system of man and higher animals is more powerful than the best man-made computer and is still not understood in sufficient detail. The endocrine system harbors the secret of life itself, which remains to be elucidated. The immune system is better than the best biochemist, as it is capable of recognizing instantaneously chemical structure and to synthesize specific reagents (antibodies) to it. Much of immune function remains to be clarified.

In recent years much evidence was produced and important conceptual developments occurred with regards to the function of the neuroimmune interaction. Now it is apparent that in addition to the hypothalamus-pituitary-adrenal axis (HPA), the sympathetic nervous system is very much involved in immunoregulation. It seems clear that the HPA axis and the sympathetic nervous system are equal partners in the profound suppression of the adoptive immune system, and in promoting natural immune mechanisms, which is characteristic of the acute phase response (APR). This gives new emphasis to the long-standing, by now historical, observations that the sympathetic nervous system plays a key role in the regulation of immune function and of inflammatory diseases and that the beta-adrenergic receptor is fundamental to this regulatory process. Mutant beta adrenergic receptors have now been linked to inflammatory disease as discussed in the introductory chapters. It is now apparent, that the ultimate center of immunoregulation is the hypothalamus, which has the power to control immune and inflammatory reactions both by neural and endocrine mechanisms. This allows for an overall systemic control as well as for local regulation that can be customized to the specific local requirements of the tissues and cells involved. In health the hypothalamus maintains immune homeostasis and the homeostasis of the entire organism by securing a *homeostatic neuroimmune milieu*. In acute febrile illness the homeostatic milieu has to be reset in the interest of survival. It should be apparent to the reader of the relevant chapters that the essence of the acute phase response is to suppress the adaptive immune system and to mobilize maximally the natural immune resources of the host in a highly organized fashion in the interest of host survival. This *immunoconversion* is achieved by switching to an *allostatic neuroendocrine milieu*, in which immune derived cytokines, the sympathetic nervous system and the HPA axis play fundamental regulatory roles. Again, the ultimate coordinator of APR is the hypothalamus.

The prevailing view today is that high levels of IgE antibodies cause allergy and frequently asthma as well. It is widely held that the excessive levels of IgE antibodies are solely responsible for allergic inflammatory diseases in man. However, as discussed in the introductory chapters, human IgE antibodies could not transfer allergic disease to monkeys and it was not possible to induce asthma of non-immune origin experimentally. These facts indicate that IgE functions only to deliver the initial signal, that triggers mediator release from the appropriate cells, and that the overall reactivity of these cells is regulated by a different mechanism. Indeed in mice and rats histamine production and release from bone marrow cells was clearly shown to be under the control of the hypothalamus [1].

The early, by now historical, are reviewed in the introduction of this volume in order to re-emphasize the emerging importance of the sympathetic nervous system and of the hypothalamus in NIB. In actual fact this historical introduction is also an up to date presentation of a much neglected but potentially very important area. Indeed this history remains highly relevant to our current understanding of NIB.

Neither a scientific, polemic, nor a historical analysis is comparable to this Volume, which describes the origins, the underlying forces and trends, the principal events, findings, the vital participants, their associates, and witnesses, from the standpoint of their perspectives and vision of the immunological revolution of our times. It hopes to extend the reader's perspective beyond the confines of simplistic formulas that have been used many times to explain immune function. These chapters written by the participants as well as the witnesses are based on broad knowledge, insight, and personal experiences in this phase of the development of Neuroimmune Biology. Thus emerges a rare and penetrating view of the immunological revolution.

Istvan Berczi and Andor Szentivanyi

REFERENCES

1. Szentivanyi A, Berczi I, Pitak D, Goldman A. Studies of the hypothalamic regulation of histamine synthesis. In New foundation of Biology, Neuroimmune Biology Volume 1, Berczi I, Gorczynski R, Editors, Elsevier, 2001; pp. 47–55.

ACKNOWLEDGEMENTS

This book would not have been possible without the talent, devotion and hard work of the following individuals: Sharron Ladd has served as an excellent technical editor for much of this book, Mrs. Carol Funk provided valuable editorial assistance and secretarial services and Rhonda Hannibal-Blount assisted us as a devoted and skilful secretary.

Financial and secretarial support was provided for this work by the Departments of Internal Medicine and Anatomy, University of South Florida, Tampa, FL, and by the Department of Immunology, University of Manitoba, Winnipeg, Canada.

List of Contributors

Edward Baral
The University of Manitoba, Department of Medicine, Faculty of Medicine, Winnipeg, Manitoba, Canada, R3E 0W3

Peter J. Barnes
National Heart & Lung Institute, Imperial College, London SW3 SLY, UK

Istvan Berczi
University of Manitoba, Department of Immunology, Faculty of Medicine, Winnipeg, Manitoba, Canada, R3E 0W3

George Chrousos
National Institute of Child Health and Human Development, Reproductive Neuroendocrinology Branch, Bethesda, MD 20892, USA

K. Eddy Gabry
National Institute of Mental Health, Clinical Neuroendocrinology Branch, Bethesda, MD 20892, USA

Philip W. Gold
National Institute of Mental Health, Clinical Neuroendocrinology Branch, Bethesda, MD 20892, USA

Allan Goldman
The University of South Florida, Department of Internal Medicine, College of Medicine, Tampa, Florida 33612, USA

Vladimir N. Kokryakov
Institute of Experimental Medicine, Department of General Pathology and Pathophysiology, 197376, St. Petersburg, Russia

Elena A. Korneva
Institute of Experimental Medicine, Department of General Pathology and Pathophysiology, 197376, St. Petersburg, Russia

Brian J. MacNeil
University of Manitoba, Department of Pathology, Winnipeg, Manitoba, Canada, R3E 0W3

Eva Nagy
University of Manitoba, Department of Immunology, Faculty of Medicine, Winnipeg, Manitoba, Canada, R3E 0W3

Dwight Nance
University of Manitoba, Department of Pathology, Winnipeg, Manitoba, R3E 0W3 , Canada

Harry Nyanteh
The University of South Florida, Department of Internal Medicine, College of Medicine, Tampa, Florida 33612, USA

Edris Sabbadini
The University of Manitoba, Department of Immunology, Faculty of Medicine, Winnipeg, Manitoba, Canada, R3E 0W3

Andor Szentivanyi
The University of South Florida, Department of Internal Medicine, Tampa, Florida 33612, USA

Contents

PART I: THE NERVOUS SYSTEM, RECEPTORS, LIGANDS AND SIGNAL TRANSDUCTION

I. INTRODUCTION

Edited by I. Berczi and A. Szentivanyi

Introduction

ANDOR SZENTIVANYI[1], ISTVAN BERCZI[2], HARRY NYANTEH[3], and ALLAN GOLDMAN[4]

[1,3,4]*Department of Internal Medicine, College of Medicine, The University of South Florida, Tampa, FL 33612, and* [2]*Department of Immunology, Faculty of Medicine, The University of Manitoba, Winnipeg, Manitoba R3E OW3, Canada*

David Wilson Talmage, one of the most important pioneers of the molecular biology of immunoregulation, in discussing what is beyond molecular immunology, poses the following questions:
"**What will we study when we understand all the genes and their products that impact on the immune system?** Surely in a few years we will understand how lymphocytes differentiate and are activated, how they synthesize, secrete, kill, and migrate. **What will we study?** My prediction is that immunology will then have its greatest challenge. **How can we put it together for the benefit of mankind**? **How will we prevent allergies, autoimmunity, transplant rejection, and how will we cure cancer**? The coming era of immunology will be a stage of synthesis – a return to stage 1, the immunology of the whole animal" [1].

This return to stage 1, that is the immunology of the whole animal has already begun as reflected by this book. Indeed, of the various evolving views on immunologic inflammation, immunity and hypersensitivity (allergy), this book represents an irrevocable shift and turnabout in our concepts of immunoregulation as connected with our growing understanding of the immune-neuroendocrine circuitry. This network, powerful enough both conceptually and in ***de facto*** functioning to bring about a radical change in our conceptions of the human immune, endocrine, and nervous systems has already enlisted the minds, hearts and resources of a large number of our leading laboratories in many areas of life sciences on an international scale.

The immune-neuroendocrine circuitry represents an immensely complex, powerful and wide ranging charter of human physiological and pathologic possibilities that among others, is working its way to the creation of a new immunology based on a wastly enlarged vision of immunologic potential in health and disease. This new immunology will be discussed in the framework of the development of a new foundation of biology, most appropriately called *Neuroimmune Biology*.

REFERENCES

1. Talmage DW. Introduction to basic immunology. In Immunological Diseases, Samter M. editor, Boston, Little Brown, 1988; pp. 1–3.

The Immune-Neuroendocrine Circuitry
Edited by I. Berczi and A. Szentivanyi

History

ANDOR SZENTIVANYI[1], ISTVAN BERCZI[2], HARRY NYANTEH[3] and ALLAN GOLDMAN[4]

[1,3,4]Department of Internal Medicine, College of Medicine, The University of South Florida, Tampa, Fl, 33612 and [2]Department of Immunology, Faculty of Medicine, The University of Manitoba, Winnipeg, Manitoba, R3E OW3, Canada

ABSTRACT

Solomonsen and Madsen published the first paper in 1898 on the role of the nervous system in immunological reactions. This was followed by a book on "Vagotonia" in 1910 that explained anaphylaxis and allergy with a pathological increase in the firing of the vagus nerve. Hyperergic inflammatory reactions were observed in denervated tissue during the 1930s. However, such tissue shows hyperreactivity to neurotransmitters that are present in serum or may reach the tissue through local axonal reflexes. Pavlov's students applied successfully the methodology of conditioning to the phenomenon of anaphylaxis. Unfortunately contemporary scientists rejected these findings. During 1949–1952 Szentivanyi and co-workers obtained definitive scientific evidence for the dominant regulatory role of the hypothalamus (tuber cinereum) over anaphylactic reactions in guinea pigs. Other investigators confirmed these original observations by producing evidence that both the hypothalamus and the pituitary gland are key regulators of the immune system.

1. EARLY HISTORY

Contrary to some prevailing impressions that the available literature on the role of the nervous system in immunologic inflammation, immunity and hypersensitivity (allergy, atopy) started to surface somewhere in the 1960's in our country, there is extensive work available on these matters early in this century in the European literature.

Although the earliest investigations can be traced to the work of Solomonsen and Mandsen in 1898 published in Compt. rend de L'Academie des Sciences [1] development of significant interest in allergy research followed only the publication of the **"vagotonia"** book of Eppinger and Hess [2]. Among other fields of medicine, this book suggested that the phenomena of anaphylaxis and diseases of atopic allergy are due to an abnormal increase in the rate of firing of the vagus nerve. In their ideas, these authors indicated a sharp division and functional separation between the various components of the autonomic nervous system, as at that time referred to as the "neurovegetative system". Today, we are aware that the hypothesis of the vagotonia was erroneous and hence responsible for the contradictory results found in their analysis [3–9].

In our current view the functional antagonism between the sympathetic and parasympathetic systems is an apparent one being subordinated to a major synergism-which in turn ruled by organic functions with respect to their tasks and performances. These can be split into a trophotropic, or preparatory, and a sympathetic i.e. ergotropic functional phase. The relationship between the two phases is in a continuous transition. To put it differently, the sympathetic and parasympathetic innervation of organs is functionally by no means a constant: It is permanently shifting and seeking adjustments, as a compromise to functional exigencies. We recognize further that organs innervated by the autonomic nervous system keep up a permanent interaction and extend it to similarly innervated organs.

Another approach to determine the nature of neural influences in immune and anaphylactic processes used *denervation studies* [10–13]. The same technique was generally adopted in these experiments, i.e. observation of hyperergic inflammatory reactions in the *denervated* tissue. However, the validity of the information obtained in such studies was handicapped by two features of the denervated structure: (1) neural influences cannot be entirely eliminated even at the cost of severing spinal and sympathetic strains or performing ganglionectomy since neural influences can be exercised both through neurohumoral action or by local axonal reflexes and (2) partly or completely denervated tissue rapidly develops an increased reactivity to the natural chemical mediators of immune reactions as to other stimulating substances [14–18]. Concurrently with these developments, a group of Pavlov's former students (Speranski, Brucke, Alpern, Rosle, etc.) introduced the conditional reflex methodology, in the search for the analysis of the nature of the role of the nervous system in anaphylactic phenomenon. Unfortunately, both the methodology as well as the interpretation of these studies was completely rejected by other workers [14–18].

2. RECENT HISTORY

One of the commonly shared features in the interpretation of the experimental results of all the foregoing studies was a lack of an attempt to trace back the observed effect to the specific site of the intervention in the brain. It was this consideration that influenced Szentivanyi and his associates between 1949–1952 to identify the central structure responsible for the anti-anaphylactic effect in the hypothalamus, more precisely the tuber cinereum and the anterior part of the guinea pig hypothalamus.

Beginning in the early 1960's and still continuing at the time of this writing Hugo Besedovsky in Europe and Locke, Ader, Felten and their associates in the United States have added significant understanding to the role of neural influences in mammalian immunoregulation.

However the aforementioned analysis of anaphylaxis for the purpose of determining the nature of the atopic base in all human phenomena of atopic allergy was entirely inappropriate for the following reasons. Anaphylaxis is an immunological phenomenon whereas diseases of atopy (bronchial asthma, allergic rhinitis, atopic dermatitis) all have in addition to the immunological abnormality highly increased sensitivity to recognized neurotransmitters and neurohumors of men (acetylcholine, histamine, serotonin etc.). Consequently it became recognized that the *Bordetella Pertussis* treated mouse and rat shows a critically enhanced reaction to these neurohumors and therefore the pertussis sensitized mouse and rat represents a far more appropriate model for the investigation of the nature of atopy in men.

In the presentation of the pertussis model we need to recognize some of the history of these findings. Parfentjev and Goodline [19] were the first to report an increased sensitivity of certain strains of mice to histamine following injection of dead Bordetella pertussis organisms. More recently this hypersensitivity has been shown to apply, also, to serotonin (5-hydroxytryptamine) and to passive anaphylaxis elicited by an unrelated antigen-antibody systems , reaching a maximum to all these stimuli at approximately 5 days following the injection of the pertussis organisms. The hypersensitivity then declines and the mice apparently regain a normal state.

Pertussis-induced hypersensitivity to histamine and serotonin has particular significance to the fields of immunology and allergy for the somewhat paradoxical reason that it appears to be a unique experimental model of an acquired hypersensitivity not due to classical immunologic mechanisms. Unlike antigen-induced hypersensitivity to histamine and serotonin this model has particular significance to the fields of immunology and allergy for the somewhat paradoxical reason that it appears to be a unique experimental model of an acquired hypersensitivity not due to classical immunologic mechanisms. Unlike antigen-induced hypersensitivity, which is mediated through a specific antibody response, the exaggerated reaction to histamine and serotonin is induced by the injection of another substance of undetermined nature. The hypersensitivity is relatively transient; no anamnestic response is apparent and the histamine and serotonin sensitivity is not transferable by lymphoid cells or serum, the classical agents of the immune response.

Other investigators [20,21] have suggested that the pertussis-induced sensitivity might be the result of an impairment in the normal rapid detoxification of histamine and serotonin. In our experiments, however, no change in the excretion pattern of C-14 serotonin or C-14 histamine or their metabolic products could be detected. These results were interpreted to mean that the handling of histamine and serotonin by the body as a whole is not altered in the hypersensitive animals. For this reason, we concluded that the hypersensitivity was the result of a localized hyper-reactivity of some cells to normally nontoxic concentrations of histamine and serotonin in the extracellular fluid. Our attention was therefore directed toward the muscle and nerve cells, which are the normal receptors for these amines.

The urinary elimination pattern of normal mice following the injection of radioactive histamine and serotonin was compared to that of pertussis-treated animals. The absence of gross differences between these two groups suggested that the handling of histamine and serotonin by the body as a whole is not altered in the highly sensitive pertussis-vaccinated mouse. For this reason, it was concluded that the hypersensitivity is the result of a localized hyperactivity of the effector cells to normally non-toxic concentrations of histamine and serotonin. Attempts were made by a variety of neurohumoral agents known to react with the receptors of neuroeffector cells.

It was found that acute exposure to progressively increasing amounts of histamine and serotonin desensitized the pertussis-treated animals to these amines and the desensitization persisted for at least 24 hrs. Cross-protection was also obtained complete with histamine pretreatment against serotonin, and incomplete with serotonin against histamine. However, desensitization attempts with another toxic amine (agmatine), not known to function as a neurohumor and whose toxicity for mice is not modified by pertussis-sensitization, demonstrated that the resistance of either normal or pertussis treated mice to the same amine was not altered, nor was this procedure of any apparent value in protecting the latter animals against histamine.

The agmatine disparity further emphasized the critical significance of neurohumors in the pertussis-induced sensitive state and led to the administration of various autonomic drugs, which were shown, to be protective in the pertussis- treated animals. Another such substance, Dibenzyline (α-adrenergic blocking agent), was also protective in pertussis-sensitized animals, whereas a β-adrenergic blocking agent, dichloroisoproterenol, duplicated the histamine and serotonin hypersensitivity in normal animals.

Based on these observations, the hypothesis was advanced that the pertussis-induced hypersensitivity is the result of a functional imbalance between two types of adrenergic receptors or in the neural pathways leading to them.

Thus the pertussis induced hypersensitivity in mice put into focus, a possible adrenergic abnormality in manifestations of diseases of atopic allergy and the possible nature of the atopic abnormality itself [22]. At this point we will have to raise this discussion in a historical frame of reference. In 1931, F M. Rackeman wrote the following lines:

"*The situation is somewhat analogous to that of a loaded gun. A good deal of knowledge is being obtained about the great variety of triggers (extrinsic and intrinsic causes) which fire the charge but why is the gun loaded? And what constitutes this load*?" [23]

These questions refer to the fact that only a minority of the total population shows some form of allergic reactivity despite that, by and large, identical conditions of antigenic exposure must be presumed to exist for all members of the same population [24,25]. The nature of the atopic abnormality in disorders of atopic allergy, which determines that only a relatively confined segment of any given population shows atopic reactivity (bronchial, cutaneous, nasal, etc.), i.e., affected pathologically, is as yet unexplained, but it has been traditionally approached through immunological concepts.

These concepts derive from an association of a physicochemically and biologically distinct antibody, the reagin, with these disorders, and the production and unusual reactivity of this type of antibody is thought to account for much of the atopic abnormality in these manifestations [26]. However, as has been extensively discussed [27,28], immunological concepts cannot account for many of the significant facts surrounding these disorders. Therefore, based on model experiments and other considerations, an alternative to the classical immunologic concept, which came to be known as the beta-adrenergic theory of the atopic abnormality, has been postulated [29]. This theory regards disorders of atopic allergy not as immunological diseases but as unique patterns of effector reactivities. In addition to the antigen-antibody interactions, atopic episodes are known to be triggered by a large variety of stimuli such as infection, various synthetic and natural chemicals, conditioned reflexes, psychic stimuli, changes in atmospheric pressure, exposure to thermal changes, non-antigenic dust, fumes, and other irritants, etc.

Any molecular interpretation of a susceptibility to such a large variety of unrelated stimuli would appear to necessitate the postulation that the primary lesion be connected with a final common pathway operating through a biologically unusually broad messenger or signalling system. The adenylcyclase 3',5'-AMP system is such a messenger system capable of responding to a wide variety of neural, humoral and hormonal agents subserving homeostasis, especially to beta-adrenergic receptor activation.

In the experimental analysis of the theory both chronologically and in research strategy, five phases can be distinguished. In the wake of the early presentations of the theory (1962–1972), manifestations of reduced asthmatic responsiveness to catecholamines as measured by various systemic parameters of beta-adrenergic reactivity were found [29,30]. Although they were highly compatible with the beta-adrenergic theory, the interpretation of these results was handicapped by the limitations of *in vivo* studies involving complex homeostatic regulations. It was, therefore, of importance that in the second phase of our experimental analysis (1972–1975), the same pattern of reduced beta-adrenergic responsiveness was demonstrable in *in vitro* preparations of isolated cells derived from asthmatic individuals. In such studies, leukocytes and lymphocytes from atopic donors exhibited a reduced cAMP response to beta-adrenergic agonists [31].

However, as the continuing analysis of the beta-adrenergic approach to atopy entered its third phase (1975–1978), methodological and theoretical problems emerged casting doubt on the validity of interpretations of the findings on leukocytes and lymphocytes [32] as well as on the supportive value of these findings for the *de facto* existence of the postulated beta-adrenergic abnormality.

With the commercial availability of radioactive adrenergic ligands [32] our analysis entered its fourth phase (1979–1986). Using ^{3}H-dihydroergo-cryptine (^{3}H-DHE) (alpha receptors) and ^{3}H-dihydroalprenolol (^{3}H-DHA) (beta receptors) a significant and mutually reciprocal shift was demonstrated in the numbers of adrenoceptors from beta to alpha in lung tissues of patients with reversible airways obstruction. The same receptor shift was found in lymphocytes obtained from patients with atopic dermatitis. Furthermore, when saturation curves for the binding of ^{3}H-DHA and ^{3}H-DHE to adrenergically desensitized lymphocytic membranes from patients with asthma or atopic dermatitis were performed, the numbers of both were quantitatively reduced but the original abnormal ratio between alpha and beta receptors was preserved. This indicated that adrenergic desensitization, even when it contributes to the overall beta-adrenergic subsensitivity, is not the cause of the original adrenergic abnormality in lymphocytes of atopic individuals [33,34].

In the fifth phase (I986-present), the emergence of the lymphocyte as a neuroendocrine cell in the past 15 years [35,36] provided a new approach in the continuing analysis of adrenergic mechanisms in atopy. In these years, it was shown that the lymphocyte can both synthesize as well as store virtually all the known neurotransmitters and neurohormones in the mammalian body. In addition, the lymphocytes were shown to contain two sets of adrenergically active proteins: 1) a secretory variant of beta-arrestin and an IL-1β antagonist, both of which downregulate beta$_2$-adrenergic receptors in A549 human lung epithelial cells, and 2) a mixture of IL-1α and IL-1β that upregulates beta$_{2-}$adrenergic receptors in the same cell line [37,38]. In subsequent studies, it was shown that lymphocytes obtained from patients with asthma and atopic dermatitis possess significantly reduced amounts of the adrenergically upregulating components, but normal amounts of the adrenergically downregulating lymphocytic substances [37]. Finally, it has recently been demonstrated that beta-adrenoceptor mediated relaxation is attenuated in cholinergically stimulated airway smooth muscle from asthmatic patients [39]. In agreement with this finding, reduced beta-adrenergic responsiveness has also been reported in airway smooth muscle isolated from antigen-sensitized animals [40]. It may be added that IgE itself can enhance cholinergic neurotransmission in human airways [41].

Both early and more recent reports have demonstrated the involvement of chromosome 5q31–33 in asthmatic individuals, and also in patients with atopic dermatitis (AD). This region contains the β_2-adrenoceptor and steroid receptor genes. In the same region, the IL-4 cluster has been identified . It has been shown that IL-4 enhances B-cell proliferation and the expression of FC receptors for 1gE (FCεR) [42,43].

Only very recently, it was demonstrated that antigen-induced aggregation of FCεRl modulates the activity of GTP-cyclohydrolase I, the key enzyme for synthesis of the cofactor 6-tetrahydro-bi-opterin (6BH$_4$), followed by increased 6BH$_4$ levels in murine mast cells. This effect was specific for the complete FCεRI, and was absent in murine mastocytoma cells lacking the alpha chain [44] The presence of 6BH$_4$ *de novo* synthesis/recycling has been documented in human epidermal keratinocytes [45,46]. These cells also synthesize and degrade both catecholamines and acetylcholine, and express high numbers of 132 adrenergic, muscarinic and nicotinic receptors [45,46] .

There is accumulating evidence for functional polymorphisms in the β_2-adrenoceptor gene linked to atopy, which influence both β_2-adrenoreceptor structure and function. A linkage analysis of 303 children with severe asthma showed a defect on chromosome 5 5q31~33. Point mutations in the 132-adrenoceptor gene have been identified in mild to moderate (Gln^{27}-Glu^{27}) and in severe asthma (Arg^{16}-Gly^{16}). Most recently, a substitution (Ala^{119}-Asp^{119}) has been shown in 9 unrelated patients with AD [48]. The latter mutation leads to defective β_2-adrenoceptor structure and function in both keratinocytes and lymphocytes with a 6-fold decrease in the K_D for agonist binding [48] These results strongly support Szentivanyi's concept as well as the important contribution from Hanifin's group on cAMP/phosphodiesterase malfunction in AD [49]. For further discussion of these considerations see Rocken, Schallreuter, Renz and Szentivanyi [50].

REFERENCES

1. Salomonsen and Mandsen. Compt. Rend. De l'Academie des Sciences. Influence de quelques poisons sur le pouvoir antitoxiquen du sang. 1898; 125, 1229–1233.
2. Appinger and Hess. Die Vagotonie. Röntgenslogische Befunde der Sella turcica bei hypoplastischen weiblichem Genitale. Mitt. D. Geselesch. F. inn. Med. U. Kindehr in Wien Hirschwald. Berlin. 1910; ix, 26.
3. Auer. The prophylactic action of atropia in immediate anaphylaxis of guinea pigs"; third communication. Amer J Physiol. 1910; 26, 439–452.
4. Bijlsma. Hat Einspritzung von Adrenalin einen Einfluss auf die Fähigkeit zur Antikörsperbildung Centralbl. F. Bakteriol. [etc], 1. abt., Jena. Z f Bact 1921;36, 246.
5. Friedberger and Oshikawa. Beziehungen zwischen Antigen und Antikörperbildung (Der Einfluss des parenterden Antigendepots auf die Antikörperbildung). Ibid Z immune Fschg 1922; 33, 306–316.
6. Joachimoglu and Wada. Ueber den einfluss von Atropin und Pilokarpin auf die Antikörperbildung bei Kanichen. Arch F Exper Path U Pharmakol. 1925; xxiv, 248–250.
7. Tonietti. Le fonti storiche della chimica e dell alchimia in Italia del dott. Giovanni Carbonelli. Rassegna di clin Terap [etc.], Roma. 1925; xxiv, 248–250.
8. Bogendorfer L über den Agglutination. Arc Exp Einfluss des Sympaticus zum Zustendekommen der Path Pharmacol 1928; 133, 107–110.
9. Ader R, Felten D, Cohen N. Psychoneuroimmunology II Edition, 1991.
10. Fischer and Kayserling. Die Experimentelle Lymphogene Allergisch-hyperergische Appendicitis. Virchows Arch 1936; 297, 146–176.
11. Fischer and Kayserling. Untersuchungen zur Frage der Beziehungen des Nervensystems zur Allergisch-hyperegischen Entzündung Virchows Arch 1937;299, 253–269.
12. Leupold E. Der Einfluss der Nervendurschschneidung auf die Resorption von Serumelweiss am Kaninschenohr bei normergisschen un d hyperergishen tieren. Frankfurt. Z Path 1938; 52, 392–406.
13. Lasovsky, JM., Wyropahew DN. Sur les observations morphologiquesconcernant le cours de l'inflammation hyperergique dans le tissue musculaire atrophie, par suite de "l'inactivité" en rapport avec le problème de la participation du systhème nerveaux aux reactions allergiques des tissues. Ann D'Anath Path 1938; 15, 357–370.
14. Meltzer and Auer. Studies on the paradoxical pupil-dilatation caused by adrenalin. J Physiol 1904; 11, 28–51.
15. Dale and Gasser. The Pharmacology of Denervated Mammalian Muscle. Pharm and Exper Ther 1926; 28, 287–315.
16. Luco. The sensitization of inhibted Structures By Denervation. Amer J Physiol 1937, 120, 179–183.
17. Bard. Studies on the Cerebral Cortex; Localized control of placing and hopping reactions in the cat and their normal management by small cortical remnants. Arch Neurol and

Psychiat 1943, 30–44.
18. Cannon. The supersensitiviy of denervated structures. New York, 1949.
19. Parfentjev IA, Goodline MA. Histamine Shock in mice sensitized with Hemophilus pertussis vaccine. J Pharmacol & Exper Therap 1948;92, 411–413.
20. Kind LS. Sensitivity of pertussis inoculated mice to serotonin. Proc Soc Exp Biol Med 1957;95, 200–201.
21. Sanyal, RK, West GB. Sensitizing Properties of Haemophilus Pertussis Vaccine in Laboratory Animals Internat Arch Allergy 1959; 14, 241–248.
22. Fischel CW, Szentivanyi A. The absence of epinephrine induced hyperglycemia in pertussis-sensitizide mice. Federation Proc 1962; 21, 271.
23. Rackerman FM. Clinical Allergy, New York, NY. 1931, 31.
24. Szentivanyi, A, Fillip G, Legeza I. A dohany-allergia mint ipari artolom. Orv Hetil 1951;44, 1–15.
25. Sentivanyi A, Fillip G, Legaza I. Investigations ton tobacco sensitivity. Acta Scie Hung Tomus III Faschilicus 1952;2. 175–184.
26. Szentivanyi A, Maurer PH, Janicki BW, editors. Antibodies; structure, synthesis, funcion and immunologic intervention in disease. New York, Plenum Press: 1987; 1–215.
27. Szentivanyi A, Fischel CW. Die Amin-mediatorstoffe der Allergischen Reaktion und die Reaktionsfahigkeit ihrer Erfolgszellen. In: Fillip G, Szentivanui A, editors. Pathogenese und Therapie Allergischer Reaktionen. Stuttgart, Germany: Grundlagenforschung und Klinik: Ferdinand Enke Verlag, 1966; 588–683.
28. Pearlman DS, Szentivanyi A. Excessive reactivity of defence mechanisms – allergy. In: Cooke RE, editor. The biologic basis of pediatric practice. New York: McGraw Hill, 1968; 536–546.
29. Szentivanyi A. The beta-adrenergic theory of the atopic abnormality in bronchial asthma J Allergy Clin Immunol. 1968; 42: 203–232.
30. Szentivanyi A, Katsh S. Townley RG. J Allergy Clin Immunol 1968;41:107–108.
31. Szentivanyi A, Krzanowski JJ, Polson JB. In: Middeleton E. Reed CE, Ellis EP, editors. Allergy Principles and Practive. 3rd edition. ST. Louis, 1988; 461–493.
32. Szentivannyi A. The radioligand binding approach in the study of lymphocytic adrenoceptors and the constitutional basis of atopy. J Allergy Clin Immunol. 1980; 65: 5–11.
33. Szentivanyi A, Heim O, Schultze P. Changes in adrenoceptor densities in membranes of lung tissue and lymphocytes from patients with atopic disease. NY Acad Sci Ann 1979;332:295–298.
34. Szentivanyi A, Heim O, Schultze P, Szentivanyi J. Adrenoceptor binding studies with [(3)H] dihydroalprenolol and [(3)H] dihydroergocryptine on membranes of lymphocytes from patients with atopic disease. Acta Dermato-Venereol 1980;S92:19–21.
35. Szentivanyi A. Adrenergic regulation. In: Weiss EB. Stein M, editors. Bronchial asthma-mechanisms and therapeutics. 3rd edition. Boston, 1993;165–191.
36. Szentivanyi A. In: Weiss EB, Stein M, editors. Bronchial asthma – mechanisms and therapeutics. 3rd edition. Boston, 1993;421–438.
37. Szentivanyi A, Schultze P, Heim O et al. Abstract 68-69. Int J Immunopharmacol 1991;13/6:739–740.
38. Szentvanyi A, Robicsek S, Heim O et al. Abstract 70-84. Int J Immunopharmacol 1991;13/6:740–741
39. Hakonarson G, Herrick DJ, Grunstein MM. Mechanism of impaired beta-adrenoceptor

responsiveness in atopic sensitized airway smooth muscle. Am J Physiol 1995;269:L654–L652.
40. Candell LM, Yun SH, Tran LLP, Ehlert PJ. Differential coupling of subtypes of the muscarinic receptor to adenylate cyclase and phosphoinositide hydrolysis in the longitudinal muscle of the rat ileum. Mol Pharmacol 1990;38:689–697.
41. Ichinose M, Miura M, Tomaki M et al. Incubation with IgE increases cholinergic neurotransmission in human airways in vitro. Am J Respir Crit Care Med 1996;154: 1272–1276.
42. Howard M, Farrar M, Hilfiker M. Identification of a T cell-derived B cell growth factor distinct from interleukin 2. J Exp Med 1982;155:914–923.
43. Hudak SA, Gollnick SO, Conrad H. Kehry MR. Murine B-cell stimulatory factor 1 (interleukin 4) increases expression of the Fc receptor for IgE on mouse B cells. Proc Natl Acad Sci USA 1987;84:4606–4610.
44. Hesslinger C, Ziegler I, Kremmer E. Hultner L. IgE-mediated signal transduction regulates the tetrahydrobiopterin synthesis in mast cells: a model system for posttranslational modulation of GIP-cyclohydrolase I. In: Pfeiderer W, Rokos H, editors. Chemistry and Biology of Pteridines and Folates. Berlin. Vienna: Blackwell Science, 1997;559–564.
45. Schallreuter KU. Epidermal adrenergic signal transduction as part of the neuronal network in the human epidermis. J Invest Dermatol Symposium Proceedings 1997;2:37–40.
46. Grando SA. Biological functions of keratinocyte cholinergic receptors. J Invest Dermatol Symposium Proceedings 1997;2:41–48.
47. Postma DS, Bleecker ER, Amelung PS et al. Genetic susceptibiligy to asthma- Bronchial hyperresponsiveness coinherited with a major gene for atopy. New Eng J Med 1995;333: 894–900.
48. Schallreuter KU. Genetic aspects of atopic Eczema, In: Dermatology at the Millenium. Proceedings of the 19th World Congress of Dermatology Sydney, Australia, Parthenon Publishing: June 1997.
49. Hanifin JM, Chan SC. Role of cyclic nucleotide metabolism in the patho-physiology of atopic eczema. In: Ruzicka T. Ring J. Przybilla B, editors. Handbook of Atopic Eczema Berlin, Heidelberg: Springer-Verlag, 1991;232–234.
50. Rocken M, Schallreuter K, Renz H, Szentivanyi A. What exactly is Atopy? Exp Dermatology 1998;7:97–104.

Edited by I. Berczi and A. Szentivanyi

The discovery of immune-neuroendocrine circuitry – A generation of progress

ANDOR SZENTIVANYI[1], ISTVAN BERCZI[2], HARRY NYANTEH[3], and ALLAN GOLDMAN[4]

[1,3,4]*Department of Internal Medicine, College of Medicine, The University of South Florida, Tampa, FL 33612, and* [2]*Department of Immunology, Faculty of Medicine, The University of Manitoba, Winnipeg, Manitoba R3E OW3, Canada*

ABSTRACT

The recognition of the immunoregulatory role of the hypothalamus in 1951 coincided with the advent of knowledge about some basic aspects of immunology. The nature of antibody diversity occupied the focus of interest in the light of new genetic studies completed. The "instructionist" theories of immune activation have been abandoned on the basis of new knowledge in favour of "selection theories" as advanced by Talmage and Burnet. These theories served as the basis for the elucidation of the genetic and molecular basis of lymphocyte function during immune responses, as we know it today. The recognition that the immune system is highly adaptable through receptor mutation, possesses memory and is capable of responding under *in vitro* conditions led to the conclusion that it is a virtually autonomous system that defends the body from foreign pathogens. This view is still maintained by many immunologists who see no need or use for additional systemic regulatory intervention.

Under these conditions the scientific community at large has ignored the evidence that has accumulated slowly with regards to the interaction of the Neuroendocrine and Immune systems. Neverteless, the significance of this relationship has been re-emphasised in 1966 and it was proposed that immune homeostasis and the homeostasis of the organism are closely linked. Much evidence has accumulated since then to support this hypothesis.

1. INTRODUCTION

The role of hypothalamic influences in the induction and expression of immunologic inflammation, immunity and hypersensitivity was first discovered in the fall of 1951 at the University of Debrecen School of Medicine in Hungary. The rationale behind the decision for a systematic exploration of the hypothalamus was as follows. Historically, the interpretation of the symptoms and the underlying reaction sequence of human asthma were patterned after the anaphylactic guinea pig. However, the range of atopic responsiveness in asthma includes a variety of stimuli that are non- immunologic in nature. Foremost among these is a broad range of pharmacologically active mediators that could be considered as the chemical organizers of the central and peripheral autonomic regulation. Therefore, it was believed that anaphylaxis could not be used as a model for the investigation of the constitutional basis of atopy in asthma. It was postulated that such a model, if it were to be meaningful, must be able to imitate both the immunologic and autonomic abnormalities of the disease. At that early time (1951), none of the currently existing neuro-active agents (i.e. agonists, antagonists) were available that one could use as an experimental tool to induce an autonomic imbalance It was, therefore, concluded that the best chances to develop such a condition could be through various manipulations at the neuroendocrine hypothalamus. Consequently, hypothalamically imbalanced anaphylactic animals were used. These were produced by the electrolytic lesion, and conversely, by the electrical stimulation of various nuclear groupings in the hypothalamus through permanently implanted depth electrodes placed stereotaxically in the hypothalamus. The resulting cumulative evidence obtained on such a model indicated that the hypothalamus has a modulatory influence on all cellular and humoral immune reactivities and that both neural as well as endocrine pathways are required for hypothalamic modulation of immune responses [1,2].

This was the beginning of the publication of a series of articles between 1952 through the present time with a number of associates including Fillipp G, Mess B, Lezega I, Fornet B, Vegh L, Keszthelyi M, Demeny P, Haraszthy A, Szekely J, Heim O, Reiner S, Schultze P, Talmage DW, Radovich J, Weiler RJ, Hofstra D, Blaisdell R, Hruban Z, Ortez RA, Klein TW, Linaweaver WE, Hurst D, TownleyRG, Trapani IL, Katsh S, Halkais DG, Fishel CW, Szentivanyi J, Haberman K, Schwartz ME, Wagner H, Werhheimer E, and more recently Krzanowski JJ, Polson JB, Fitzpatrick DF, Hackney JF, Williams JF, Lowitt S, Anderson W, Robicsek S, Hurt K, Zority J, Calderon E, Abarca C, Lockey RF, Brooks S, and Nyanteh H.

Concurrently with these developments however, an unparalleled expansion of information on the basic aspects of immunology, in general, and on the nature of antibody diversity, in particular, started to occupy the center stage of immunologic interest. Most importantly, the new perceptions surrounding the nature of antibody diversity began to surface in the late 1950s with the first conclusive genetic studies having been completed and a new set of concepts defined. These circumstances led to a total transformation of prevailing immunologic thought ultimately leading to the replacement of "instructionist" theories by the "selection theories" as advanced by Talmage (Spring 1957) and Burnet (Fall 1957).

Three major postulates were implicit in these theories of heritable cellular commitment: (1) the antigen receptor site and the antibody combining site whose synthesis that cell controls are identical and are derived at least partially from the same structural gene; (2) the condition guaranteeing the correspondence of the immunoglobulin synthesized with the antigen is that they are limited to the same cell that is the cell specialized for the synthesis of a single antibody; and (3) the cell specialization stipulated in item 2 is inherited and therefore, clonal (the clonal selection theory of acquired immunity).

Subsequently, it became established that virtually all antibody diversity and specificity encoded, in the immune system can be accounted for in genetic terms and thus the controls of the antibody response. The initiation of the antibody response is dependent on cell adhesion molecules called the major histocompatibility complex (MHC) members of which present antigen to helper T lymphocytes. In turn helper T cells secrete cytokines that are necessary for the stimulation of antibody secretion by B lymphocytes. Different genes appear to code for immune response, suppression and cell interaction.

Another impediment in the timely recognition of the significance of the immune- neuroendocrine circuitry in immune homeostasis was that immune reactions can proceed *in vitro*. This further supported the concept that the immune system is a totally autonomous and self-regulating unit, a view that among others has overlooked the rich neurohormonal *milieu* in which *in vitro* immune responses occur.

Sporadic refutations of the above postulates have occurred and continue to surface in the literature, but the great bulk of the evidence is supportive of the clonal selection theory. The theory's sheer eloquence, however, has probably been most responsible for its dominant role in immunologic thought and its acceptance as dogma since the early 60s. In any case, these concepts and the large body of supportive evidence has so permeated the field that it became difficult if not impossible, to think of immunology outside of this framework [3].

In the late fifties and throughout the sixties, the two conceptual centers of these ideas were under the leadership of D. W. Talmage at the University of Chicago and later at the University of Colorado, and the group at Walter and Eliza Hall Institute in Australia under F. M. Burnet. Because of the close association with Talmage, extending over a period of 10 years, we were very much under the influence of these views and developed reservations against the significance of the above mentioned hypothalamic findings in immunoregulation. Nevertheless, by 1966 in a chapter of a German text [4] through an extensive analysis of our findings, and the dominant immunologic concepts, we came to articulate the following conclusions: (1) the significance of the immuno-pharmacologic mediators of immune manifestations in normal mammalian physiology is that they are the chemical organizers of central and peripheral autonomic action, (2) the preceding suggests the inseparability of the immune system from the neuroendocrine system, (3) such inseparability indicates the ***de facto*** existence of immune-neuroendocrine circuits and the necessity for a bidirectional flow of information between the two systems, (4) one must distinguish between the concepts of autoregulation as one that primarily satisfies the requirements of antibody diversity and specificity in contrast to the more complex requirements in immune homeostasis, (5) in contrast to autoregulation that is always self-contained homeostatic control is always beyond the constituents of one single cell or tissue system; and (6) thus immune homeostasis must represent a far more sophisticated level of control than autoregulation, and is based on immune-neuroendocrine circuits. Indeed, there is extensive evidence to indicate that no bodily system is as simple, sacred or singular as once thought [5]. Instead, as in any good relationship, the separate components strive for sensitivity, synchrony and synergy. Recognition and communication among the immune, endocrine and nervous systems exemplify the formula for harmony and homeostasis.

While the manifold similarities between the immune and nervous systems are fully realized (see below), the immune system has a major additional level of complexity over that of the nervous system. The nervous system with its spectacular masses of much revealing and well-defined projection patterns is well moored in the body in a static web of axons, dendrites and synapses. In contrast the elements of the immune system are in a continuous mobile phase incessantly scouring over and percolating through the body tissues returning through an intricate system of lymphatic channels, and then blending again in the blood. This dynamism is revealed only by scattered concentrations called *lymphoid organs*. These circumstances would appear to indicate that the junctional plasticity of the immune system is far greater than that of the nervous system, and consequently, its regulation must require a more complex and sophisticated level of control. For these reasons, we have raised the question in the above- mentioned text in 1966, whether the immune system is more "intelligent" than the brain?

The neural network of some 10^{11} nerve cells with at least a thousand times that number of interconnections is more complex and seems in many ways more powerful, than even the largest of modern computers. Our present understanding of the nervous system is limited, and one can scarcely judge whether the comparison makes sense. For instance, we do not know how many functionally distinct categories of nerve cells the brain contains; nor can we give even an outline of the neural computations involved in hearing a word or reaching for an object, let alone proving a theorem or writing a poem. And yet paradoxically, while the brain as a whole remains the most intriguing tissue in the body, the properties of the individual nerve cells, or neurons, are understood better than those of almost any other cell type. At the cellular level at least simple and general principles can be discerned as described below.

REFERENCES

1. Filipp G, Szentivanyi A, Mess B. Anaphylaxis and nervous system, Acta Med Hung 1952; 2:163–175.
2. Szentivanyi A, Filipp G, Legeza I. Investigations on Tobacco sensitivity, Acta Med Hung 1952; 2:175–184.
3. Szentivanyi A. The discovery of the immune-neuroendocrine circuits in the fall of 1951. In Interactions among Central Nervous System, Neuroendocrine and Immune systems. Hadden JW, Nistico G, Masek K, editors, Rome, Pythagora Press, 1989; pp 1–5.
4. Szentivanyi A, Fishel CW. Die Amin-mediatorstoffe der allergischen Reaktion und die reaktionfahiegheit ihrer Erfolgeszellen. In Pathogese und Therapie allergisher Reaktionen. Filipp G, Szentivanyi A, editors, Ferdinand Enke Verlag, Stutgart, 1966; pp. 588–683.
5. Schechter Gail. A good relationship; sensitive, synchronized and synergistic, Prog. Neuro-Endocrine Immunol 1989;2:35–55.

PART II: THE NERVOUS SYSTEM – A HISTORICAL PERSPECTIVE

The Immune-Neuroendocrine Circuitry
Edited by I. Berczi and A. Szentivanyi

Altered Effector Responses

ANDOR SZENTIVANYI[1], ISTVAN BERCZI[2], HARRY NYANTEH[3] and ALLAN GOLDMAN[4]

[1,3,4]Department of Internal Medicine, College of Medicine, University of South Florida, 33612, USA and [2]Department of Immunology, Faculty of Medicine, University of Manitoba, Manitoba R3E OW3 Canada

ABSTRACT

During neurotransmission the second unit should always be regarded as an effector cell, regardless of its nervous or non-nervous nature. In the autonomic system there are three types of effector cells: the neurons, the smooth muscle, and exocrine gland cells. These are the principal target cells of the pharmacologic mediators of allergic responses. The mediators of antigen-antibody responses, when viewed from the standpoint of their physiologic function, are among the natural chemical organizers of autonomic action. Denervation or mediator antagonist drugs cause hypersensitivity in the effector cells towards the transmitter in short supply.

The hypothalamus has two reciprocally antagonistic divisions: the anterior hypothalamus, which mediates primarily cholinergic responses, and the posterior hypothalamus, the stimulation of which results largely in adrenergic responses. A balance between these divisions is important in maintaining normal autonomic functions (e.g., blood pressure). The posterior hypothalamus is normally suppressed by inhibitory impulses transmitted from the sinoaortic barorecaptors, which, among others, keep the posterior hypothalamus in check. When histamine or acethylcholine is given, blood pressure falls, the sinoaortic tension decreases, causing a shift to sympathetic activity. This leads to the release of catechols, which tends to correct and limit the blood pressure drop. Catecholamines increase blood pressure, which shifts the hypothalamic balance to the cholinergic side. It is possible to produce imbalance of the hypothalamus by applying histamine, acethylcholine, or catecholamines directly on hypothalamic structures; by the electrolytic removal or electrical stimulation of one of the divisions of the hypothalamus and by many other stimuli.

1. INTRODUCTION

As defined earlier, transmission of the nerve impulse is the process whereby the nerve terminal activates the second unit or transfers a stimulus to it whether the second unit is a neuron or a non-nervous effector cell. For this reason, the *effector cell* may be defined as the cell that produces a response characteristic to the stimulating agent. By conclusion a *posteriori,* therefore, at both types of junctions (synaptic and neuroeffector), the second unit should always be regarded as an effector cell, regardless of its nervous or non-nervous nature. Thus in the autonomic system there are three types of effector cells: the neurons, the smooth muscle, and exocrine gland cells. These are the principal target cells of the pharmacologic mediators of allergic responses.

The identity of the autonomic effectors and principal target cells of allergic mediators immediately suggests that these substances may participate in impulse transmission or perform some other function in neurohumoral integration. Indeed, as discussed extensively elsewhere [1–6], the pharmacologic mediators of antigen-antibody responses, when viewed from the standpoint of their probable physiologic function, are among those substances that are believed to be the natural chemical organizers of autonomic action.

Among the immediate forms of allergy, the atopic group occupies a unique position in that the mechanism of atopy appears to involve a qualitatively and quantitatively abnormal reactivity to the pharmacologic mediators as well as an immunologic basis for their release [3,4,7–18] Since the nature of this abnormal pharmacologic reactivity in atopy is not known, in the ensuing discussions, conditions are described in which similar abnormal responses have been produced surgically or chemically. In these conditions, the biologic effect of the surgical or chemical intervention is fairly well understood. It is hoped therefore that their analysis will be helpful in our attempts to explore the nature of the atopic reactivity.

We shall discuss these chemical and surgical procedures separately and in succession as though the resultant conditions were mechanistically distinct entities. In reality, there is no clear borderline between them, and each contains elements shared by the others. Because of the interrelated effects of the various mediators in the maintenance of normal autonomic regulation, the altered responsiveness in each of these conditions can be explained on the basis of an imbalance of autonomic regulation at the level of the effector cell [1,5].

2. INHIBITION OF SPECIFIC ENZYME NORMALLY INACTIVATING THE MEDIATOR

Although enhancement or prolongation of effector response to any of the mediators may be obtained by inhibition of its specific catabolizing enzyme, some distinction must be made between ACh and the other mediators. Thus suppression of cholinesterase activity results in a rather consistent enhancement of ACh action, regardless of the species and effector response tested or which of the anticholinesterases is employed. In contrast, results of similar procedures concerning catecholamines, histamine, and serotonin vary with all these factors [20–23]. This disparity might in part be attributed to the several different and competing degradative pathways available for each of these amines. Selective blockade of one of them may be compensated by an increase in the catabolic contribution of the alternative pathways [24]. Furthermore, the relative importance of non catabolic processes in inactivating some of the mediators seems to be greater than is the case with ACh.

3. DEPLETION OF NON-JUNCTIONAL MEDIATOR STORES

It is implicit in the foregoing concept of denervation supersensitivity that the lacking transmitter is normally derived from a store (i.e., synaptic vesicles) that stands in a junctional relationship with the effector cell. However, an alteration in sensitivity of the effector to a mediator can also be produced when the action of the mediator is preceded by depletion of its specific, although non-junctional, stores. For example, neuronal histamine stores are insulated from the action of otherwise highly potent chemical histamine liberators, and repeated administration of such agents for more than 30 days fails to remove the neuronal histamine. Mobilization of neuronal histamine can be accomplished only at high concentrations of the liberator obtainable by direct perfusion of the neuron through its own vascular supply [28]. Therefore, parenteral administration of the liberator will presumably result in the release of histamine from non-junctional stores only. When compound 48/80 or n-octylamine is administered parenterally, a hypersensitivity to histamine coincides with the period during which tissue levels of histamine are known to remain low after treatment with 48/80, n-octylamine, or both [29,30]. This suggests an inverse relationship between sensitivity to histamine and tissue concentration of the same amine. Although available information does not permit a detailed comparison between this altered reactivity and that produced by denervation, it seems that, in contrast to denervation, this hypersensitivity may develop more rapidly and is limited to the mediator whose tissue stores were previously depleted.

4. SELECTIVE BLOCKADE OF ONE OR TWO ANTAGONISTIC SETS OF RECEPTORS SPECIFICALLY ACTIVATED BY THE SAME MEDIATOR

If at the time of action of a mediator there is a shift in relative availability of its two functionally antagonistic sets of specific receptors, the normal response to that mediator will be quantitatively or qualitatively altered [31–35].

For example, in the tracheobronchial tract, the normal balance between the alpha-adrenergic and beta-adrenergic receptor systems appears to favor a beta-adrenergic response of the smooth musculature to epinephrine. If, however, the beta-adrenergic receptors are blocked by dichloroisoproterenol, bronchoconstriction occurs instead of the usual relaxation, exemplifying a qualitatively abnormal response to the mediator [10]. In certain segments of the vascular system, where the normal receptor distribution seems to favor adrenergic contracting effects, the same pharmacologic beta blockade results in a qualitatively normal but quantitatively enhanced response to epinephrine, that exaggerated vasoconstriction of the small cutaneous vessels [32].

In the previous examples, the two antagonistic receptors may be components of the same target cell. Similar considerations apply to those situations in which the two receptors are associated with two different cell systems that have a common function with respect to the organism as a whole. For instance, hyperglycemia to epinephrine is the net result of a combination of the following adrenergic effects [36]: (1) hepatic glycogenolysis leading to the release of free glucose into the extracellular space and (2) a simultaneous reduction of glucose uptake by the muscle and other extrahepatic cells.

In the mouse, hepatic glycogenolysis appears to be an alpha-adrenergic function. When the beta-receptors of the mouse are blocked by dichloroisoproterenol, the blood sugar level fails to rise after epinephrine administration and this occurs despite increased hepatic glycogenolysis [35,37]. This illustrates a quantitatively and qualitatively abnormal response to a mediator produced by the selective blockade of one of its receptors.

Selective blockade of one receptor system, however, can modify the target cell response not only to the specific activator of the receptor affected but also to unrelated amines. An example is the dichloroisoproterenol-induced hypersensitivity of mice to both histamine and serotonin. Within a certain dose range, this beta-adrenergic blockade can increase the normal sensitivity of mice to histamine and serotonin about 50-fold and 10 fold, respectively [38]. This non-specificity is similar to pharmacologic or surgical denervation of the salivary glands. Normally the secretion of these glands is stimulated by both adrenergic and cholinergic impulses. When the cholinergic receptors of the cat's submaxillary gland cells are blocked for a long time by chronic atropinization, they become hypersensitive to epinephrine and to a number of other sympathomimetics, as well as to adrenergic nerve stimulation. The same hypersensitivity to adrenergic agents arises after section of the chorda [15].

5. HORMONAL DEPRIVATION AND OVERDOSAGE

The reactivity of the target cell to the allergic mediators may depend on hormonal balance within the body. The most striking illustration of this is the effect of bilateral adrenalectomy in various laboratory species. *Adrenalectomy* increases the sensitivity to a variety of stressful stimuli and pharmacologically active agents, including histamine and serotonin. For example, cats without adrenals are 13 to 60 times as sensitive to histamine as intact animals [39], and in adrenalectomized mice the corresponding figure for serotonin is about 30 [40]. Since in most reported studies, response after medullectomy or total sympathectomy has not been tested, it is not possible to assess the relative role of catechol versus corticosteroid deficit in these conditions. Nevertheless, in general, the sympathectomized animal occupies an intermediate position between the normal and adrenalectomized animal.

The adrenal corticosteroids and catecholamines appear to operate largely as a functional unit. The sites of action and character of target cell responses to these two natural substances are strikingly similar. Corticosteroids seem to be required for target cell responses to catechols, but there is no evidence of interchangeability between these groups. Although the molecular mechanism underlying the catecholamine-corticosteroid relationship remains obscure, it is well established that in the absence of steroids the receptor threshold for catechols may be raised, occasionally to the point of complete unresponsiveness [41]. Therefore corticosteroid deprivation can clinically simulate either an indiscriminate or a selective adrenergic blockade, depending on the types of effector responses tested [38,41]. Conversely, even a pharmacologically established blockade in an intact animal can be overcome with an excess amount of corticosteroid [35]. This seems to imply that the corticosteroids not only support the target cell responses to catechols but under certain conditions are also capable of sensitizing these cells to the action of these amines.

The sensitizing effects of *thyroxin* and *triiodothyronine* to histamine and serotonin do not require separate discussion, since their effects appear to be indirect and mediated through the adrenal steroids. Hyperthyroidism causes a temporary state of cortical deficiency, whereas hypothyroidism results in a temporary accumulation of corticosteroids in tissues [42].

Excess amounts of *insulin* are also known to increase the responses of the target cells to the allergic mediators. The mechanism of this sensitizing effect is not known but might be related to the physiologic action of this hormone, since an increased peripheral uptake of glucose occurs in some conditions in which hypersensitivity to the mediators has been demonstrated [43–46].

Desensitization of receptors can occur when tissues are exposed to persistent stimulation by agonists. Perkins and co-workers have divided this phenomenon into two major categories, that is, “heterologous” and “homologous” desensitization [47]. Heterologous desensitization occurs after exposure of cells to an agonist that produces tissue refractoriness to itself and to a variety of other pharmacologically different agonists. By contrast, homologous or agonist-specific desensitization is a loss of responsiveness to only the particular agonist that induced the desensitization (or specific group of pharmacologically related agonists, e.g., the catecholamines).

The mechanisms by which *desensitization* is produced are complex and variable, depending on the tissue. For beta-adrenergic receptors at least two different mechanisms are known, and these can be typified by the mechanisms found in two of the cell types that have been studied, that is, turkey and frog erythrocytes. In turkey erythrocytes beta-adrenergic agonists cause a 50–60% decrease in catecholamine-stimulated adenylate cyclase activity, which appears to be dependent on *cyclic* AMP *accumulation*. In this system the receptors appear to be "uncoupled", that is, they are unable to form the high-affinity agonist-receptor-N_s ternary complex that is necessary for cyclase activation (Fig. 8). The *"uncoupled" receptor* protein has been shown to be phosphorylated by endogenous kinase activity, and this may account for its uncoupled state.

In the frog erythrocyte system, a different phenomenon has been observed in desensitized cells. In these cells the *inactivated receptors* have been found to be *internalized* in vesicles, which removes them from the cell surface, the N_s, and catalytic units of the adenylate cyclase. The sequestered receptors are functional and capable of being recycled to the cell surface if agonists are removed and protein synthesis is blocked. Whether the receptors are uncoupled during the desensitization process before they are physically sequestered is a question that remains to be answered [48].

6. REFLEXLY AND CENTRALLY INDUCED HYPOTHALAMIC IMBALANCE

There are two reciprocally antagonistic divisions in the hypothalamus: the *anterior hypothalamus*, which mediates primarily *cholinergic* responses, and the posterior hypothalamus, the stimulation of which results largely in *adrenergic* responses [49]. A balance between these antagonistic divisions is thought to be important in maintaining normal autonomic functions (e.g., blood pressure). The activity of the posterior hypothalamus is normally suppressed by inhibitory impulses transmitted from the sinoaortic baroreceptors. Pressure is the physiologic stimulus of the baroreceptors, and because of the constant stimulus of the blood pressure on these pressure-sensitive devices, nerve impulses at low frequency are continually traveling up through their connecting afferent pathways to the medullary and hypothalamic centers. These inhibitory impulses, among others, keep the posterior hypothalamus in check.

When histamine or ACh is given under appropriate conditions, the systemic blood pressure falls, and consequently the sinoaortic tension decreases, causing a reduction in the frequency of inhibitory impulses traveling to the higher centers. This relieves the suppression of the posterior hypothalamus and results in a shift to sympathetic activity. Accordingly a general increase in rate of firing of peripheral adrenergic structures, including the adrenal medullary cells, follows, leading to the release of catechols. In these circumstances, the normal balance of the hypothalamus is shifted to the adrenergic side, which tends to correct and limit the blood pressure drop [50]. Conversely, when catecholamines are given systemically, the blood pressure rises, and the increased rate of upward discharges shifts the hypothalamic balance to the cholinergic side [50]. Although in these examples we are dealing with a reflexly induced hypothalamic imbalance, it is possible to produce imbalance of the hypothalamus by applying histamine, ACh, or catecholamines directly on hypothalamic structures: by the electrolytic removal or electrical stimulation of one of the divisions of the hypothalamus; and by many other stimuli [50–54].

The reflex- or locally-induced hypothalamic imbalance leads not only to an increased excitability of one hypothalamic division but also to simultaneous inhibition of the antagonistic division. Consequently, the action of either a parasympathomimetic stimulus (ACh, histamine) or a sympathomimetic stimulus (catecholamines) may be reduced as a counteracting imbalance induced in the hypothalamus. If, therefore, at the time of action of an exogenously administered or endogenously released allergic mediator the hypothalamus is already in a state of imbalance for any reason, the capacity of the hypothalamus to counteract and control the stimuli will be markedly reduced. In this way the action of the mediator may be profoundly modified both quantitatively and qualitatively even to the complete reversal of its usual pharmacologic activity [50–54].

7. SURGICALLY OR PHARMACOLOGICALLY INDUCED DENERVATION SUPERSENSITIVITY

Small amounts of neurotransmitters are released continually by the nerve terminals during the resting state of the neuron and in much greater amounts after the arrival of an impulse [58,59]. Apparently the function of the resting secretion, which is too small to activate the effector cell, is to keep the sensitivity of the effector at a low level [56,57,60]. When the effector cells are deprived of this resting secretion, it acquires within a few days a hypersensitivity to its specific neurotransmitter. For example, surgical interruption of the final neuron results in degeneration of the distal portion of the axon and cessation of its resting secretion. In the case of an adrenergic nerve ending, this results in a hypersensitivity to epinephrine in the succeeding effector cell. In many instances, a hypersensitivity of the effector ensues also to other natural neurohumors (serotonin, histamine) and to a number of similar as well as unrelated chemicals. When ACh is used as a test substance and the striated muscle as an indicator tissue, the order of magnitude of the level of sensitivity produced is within the range of 20 to 200 times that of the normal level. A less pronounced hypersensitivity develops after the penultimate neuron, that is the presynaptic neuron, is cut. In this case, the effector is deprived only of the transmitter released by the incoming impulses. Regeneration of the neuron or local application of the transmitter restores the normal level of sensitivity. Although the precise molecular mechanism of the hypersensitivity is not understood, it is probably due to the local and prolonged unavailability of the neurotransmitter.

Denervation hypersensitivity can also be produced by pharmacologic means. Drugs having as their main effect the interruption of impulse transmission at any point and that are allowed to act for an extended period can produce a hypersensitivity essentially identical with that caused by surgical denervation [56]. In general, it would appear that any natural constituent or foreign substance capable of blocking the synthesis, storage, release, or action of the transmitter and thereby depriving the effector cell of its specific controlling agent can produce a denervation supersensitivity.

REFERENCES

1. Pearlman DS, Szentivanvi A. Excessive reactivity of defense mechanisms-allergy. In: Cooke RE, editor, Biologic basis of pediatric practice. New York: McGraw-Hill Book Co. 1968.
2. Schayer RW. Induced synthesis of histamine (microcirculatory regulation and the mechanism of action of the adrenal glucocorticoid hormones). Prog Allergy 1963; 7:187.
3. Szentivanyi A. The beta adrenergic theory of the atopic abnormality in bronchial asthma. J Allergy 1968; 42:203.
4. Szentivanyi A. Effect of bacterial products and adrenergic blocking agents on allergic

reactions. In: Samter M, editor, Textbook of immunologic diseases. Boston: Little, Brown & Co.1971.

5. Szentivanyi A, Fishel CW. Effect of bacterial products on responses to the allergic mediators. In: Samter M, editor, Immunological diseases.Boston : Little, Brown & Co 1965.
6. Szentivanyi A, Fishel CW. Neurohumoral concepts of bronchial asthma. Acta Allergol 1974; (Kbh) 29(suppl. 11):26.
7. Baer RL. Atopic dermatitis. New York: New York University Press; 1955.
8. Curr JJ.The effect of antihistamine substrates and other drugs on histamine bronchoconstriction in asthmatic subjects. J Clin Invest 1946; 25:792.
9. Curry JJ. Comparative action of acetyl-beta-methylcholine and histamine on the respiratory tract in normals, patients with hay fever, and subjects with bronchial asthma. J Clin Invest 1947; 26:430.
10. Curry JJ. The action of histamine on the respiratory tract in normal and asthmatic subjects. J Clin Invest 1947; 26:785.
11. Hajos MK. Clinical studies on the role of serotonin in bronchial asthma. Acta Allergol (Kbh) 1962; 17:358.
12. Halpern BN. Histamine and processes of histamine liberation. In: Prigal SJ, editor: Fundamentals of modern allergy. New York: McGraw-Hill Book Co; 1960.
13. Melon J, Lecomte J. Etude comparee des effects de la bradykinine et des reactions anaphylactiques locales chez l'homme. Int Arch Allergy 1962; 21:89.
14. Rose B. Role of histamine in anaphylaxis and allergy. Am I Med 1947; 3:545.
15. Rose B, Rusted I, Fownes A. Intravascular catheterization studies of bronchial asthma. I. Histamine levels in arterial and mixed venous blood of asthmatic patients before and during induced attacks. J Clin Invest 1950; 29:1113.
16. Samter M. Bronchial asthma and sensitivity to histamine. Z Ges Exp Med 1933; 89:24.
17. Scherbel A, Harrison J. Serotonin hypersensitivity in collagen diseases. Bull Rheum Dis 1959; 9:179.
18. Tiffeneau R. Examen pulmonaire de lasthmatique: deductions diagnostiques prognostiques et therapeutiques. Paris, Masson, Cie, Editeurs; 1957.
19. Chokroverty S, Parameswan KC, Co C. Nonspecific esterases in the myoneural junction of human striated muscle. J Histochem Cytochem 1971;19:798.
20. Angelakos ET, Loew ER. Histamine toxicity in mice and rats following treatment with histaminase inhibitors. J Pharmacol Exp Ther 1957;119:444.
21. Johnson GE, Sellers EA. The influence of iproniazid and pyrogallol on the thermogenic effects of noradrenaline, Can J Biochem Physiol 1962;40:631.
22. Vanov S. Effect of monoamine oxidase inhibitors and pyrogallol on the pressor response to adrenaline, noradrenaline, normetanephrtne, and tyramine in the rat. Arch Int Pharmacodyn Ther 1962; 138:51.
23. Wylie DW, Archer S, Arnold A. Augmentation of pharmacological properties of catecholamines by 0-methyl transferase inhibitors. J Pharmacol Exp Ther 1960; 130:239.
24. Schayer RW. Catabolism of physiological quantities of histamine in vivo. Physiol Rev 1959; 39:116.
25. Thesleff S. Effects of motor innervation on the chemical sensitivity of skeletal muscle. Physiol Rev 1960;40:734.
26. Cannon WB. A law of denervation. Am I Med Sci 1939; 198:737.
27. Cannon WB, Rosenblueth A. The supersensitivity of denervated structures. New York:

Macmillan Publishing Co Inc; 1949.
28. West GB. Histamine in nervous tissue. In: Richter D, editor. Metabolism of the nervous system. New York : Pergamon Press Ltd.; 1957.
29. Guth PS, Byron JW, Ambrus JL. Histamine releasers and histamine sensitivity. J Pharm Pharmacol 1961; 13:254.
30. Harvey JA. Modification of histamine sensitivity after 48/80 treatment. J Pharm Pharmacol 1961; 13:61.
31. Castro de la Mata R, Penna M, Aviado DM. Reversal of sympathomimetic bronchodilation by dichloroisoproterenol. J Pharmacol Exp Ther 1962; 135:197.
32. Fishel CW, Szentivanyi A, Talmage DW. Adrenergic factors in pertussis-induced histamine and serotonin hypersensitivity of mice. In: Landy M, Braun W, editors. Bacterial endotoxins. New Brunswick, New Jersey: Rutgers University Press; 1964.
33. Furchgott RF. Receptors for sympathomimetic amines. In:Vane JR, Wolstenholme GEW, O'Connor M, editors. Ciba Foundation Symposium on Adrenergic Mechanisms. Boston: Little, Brown & Co; 1960.
34. Nickerson M. Blockade of the actions of adrenaline and noradrenaline. Pharmacol Rev 1959; 11:443.
35. Szentivanyi A, Fishel CW, Talmage DW. Adrenaline mediation of histamine and serotonin hyperglycemia in normal mice, and the absence of adrenaline-induced hyperglycemia in pertussis-sensitized mice. J Infect Dis. 1963; 113:86.
36. Ellis S. The metabolic effects of epinephrine and related amines. Pharmacol Rev 1956; 8: 485.
37. Fishel CW, Szentivanyi A. The absence of adrenaline-induced hyperglycemia in pertussis-sensitized mice and its relation to histamine and serotonin hypersensitivity. J Allergy 1963; 34:439.
38. Fishel CW, Szentivanyi A, Talmage DW. Sensitization and desensitization of mice to histamine and serotonin by neurohumors. J Immunol 1962; 89:8.
39. Dale JH. Conditions which are conducive to the production of shock by histamine. Br J Exp Pathol 1920; 1:103.
40. Higginbotham RD. Influence of adrenalectomy and cortisol on resistance of mice to histamine, serotonin, anaphylactic and endotoxin shocks. J Allergy 1962; 33:35.
41. Ramey ER, Goldstein MS, The adrenal cortex and the sympathetic nervous system. Physiol Rev 1957; 37:155.
42. Spencer PSJ, West GB. Further observations on the relationship between the thyroid gland and the anaphylactoid reaction in rats. Int Arch Allergy 1962; 20: 321.
43. Adamkiewicz VW, Langlois YL. Sensitization by insulin to the dextran 'anaphylactoid" reaction, Can J Biochem Physiol 1957; 35:251.
44. Adamkiewicz VW, Langlois YL, Poirier LJ. Sensitization by insulin to dextran anaphylactoid inflammation in spinal rats. Am J Physiol 1958;195:635.
45. Sanyal RK. The effects of insulin on hypersensitivity reactions in the rat. Allerg Asthma 1960; 6:3 17.
46. Sanyal RK, Spencer PSJ, West GB. Insulin and hypersensitivity. Nature 1959; 184:2020.
47. Su YF, Cubeddu-Ximenez L, Perkins JP. Regulation of adenosine-3' ,5 '-monophosphate content of human astrocytoma cells: desensitization to catecholamines and prostaglandins. J Cyclic Nucleotide Res 1976; 2:257.
48. Lefkowitz RJ, Stadel JM, Cerione RA, Strubvici B, Caron M. Structure and function of beta-adrenergic receptors: regulation at the molecular level. Adv Cyclic Nucleotide Res

1984;17:19.
49. Hess WR. Diencephalon, autonomic and extrapyramidal functions. New York: Grune & Stratton Inc; 1954.
50. Gellhorn E. Autonomic imbalance and the hypothalamus. Minneapolis: University of Minesota Press;1957.
51. Filipp G, Szentivanyi A. Anaphylaxis and the nervous system: part III Ann Allergy 1958; 16:306.
52. Szentivanyi A, Filipp G. Anaphylaxis and the nervous system: part II. Ann Allergy 1958; 16:143.
53. Szentivanyi A, Szekely J. Wirkung der konstanten Reizung hypo-thalamischer Strukturer durch Tiefenelekroden auf den histaminbedingten und anaphylaktischen Schock des Meerschweinchens. Acta Physiol Acad Sci Hung 11(suppl.):1957;41.
54. Szentivanyi A, Szekely J. Anaphylaxis and the nervous system: part IV. Ann. Allergy 1958;16:389.
55. Emmelin N. Supersensitivity following "Pharmacological denervation,"Pharmcol Rev 1961; 13:17.
56. Trendenlenburg U. Mechanisms of supersensitivity and subsenseitivity to sympathomimetic amines. Pharmacol Rev 1966; 18: 629.
57. Simpson LL. The use of neuropoisons in the study of cholinergic transmission. Annu Rev Pharmacol 1974 14:305.
58. Hubbard JI, Quastel DMJ. Micropharmacology of Vertebrate neuromascular transmission. Annu Rev Pharmacol 13:199.
59. Martin AR. Quantal nature of synaptic transmission. Physiol Rev 1966, 46:51.
60. Kordas M. An attempt at an analysis of the factors determining the time course of the endplate current. I. The effect of Prostigmine and of the ratio of Mg^{2+} to Ca^{2+} J Physiol (London) 1972; 224:317.

Edited by I. Berczi and A. Szentivanyi

Some Evolutionary, Morphoregulatory and Functional Aspects of the Immune-Neuroendocrine Circuitry

ANDOR SZENTIVANYI[1], ISTVAN BERCZI[2], HARRY NYANTEH[3] and ALLAN GOLDMAN[4]

[1,3,4]*Department of Internal Medicine, College of Medicine, University of South Florida, Tampa, Florida. 33612 and* [2]*Department of Immunology, Faculty of Medicine, University of Manitoba, Winnipeg,, Manitoba, R3E OW3 Canada*

ABSTRACT

The immune system is distributed throughout the body and its basic function is local disease prevention and regulation. The nervous system and the thymus epithelium are both develop from the neural crest. Neuronal immunoglobulin gene superfamilies (Ig SFs) and cytokines and their receptors play important roles during the embryonic development of both systems. The neural cell adhesion molecule (N-CAM) belongs to IgSF and plays an important role in the development of the central nervous system by cell-to-cell and cell-to-matrix adhesion and promotion of synapse formation. Extracellular matrix (ECM) proteins have neurite-promoting activity. These ECM proteins are: type IV collagen, fibronectin, laminin, and a newly discovered protein, *neurite outgrowth factor* (NOF), which promotes neurite outgrowth from sympathetic, parasympathetic, sensory, motor, and central neurons. Cytokines, that are shared by the immune and nervous systems also play a role in neural development and function. They include the cilliary neurotropic factor, the leukemia inhibitory factor, oncostatin M, nerve growth factor, brain-derived neurotropic factor, neurotropin-3 and neurotropins 4/5.Factors known to regulate NGF secretion in astrocytes include IL-1, IL-4 and IL-5.The origin of cytokine receptors also represents a classical example of molecular co-evolution in the immune and nervous systems.

Immunomodulation through the T-cell lineage in rodents can be achieved by the manipulation of hemispheric lateralization in the brain. It appears that the right hemisphere controls the inductive influence on T cells of signal emitted by the left hemisphere.

Two phenomena are associated with learning and memory in the nervous system: habituation and potentiation. Habituation defines decreased responsiveness to a stimulus presented repetitively over time and potentiation refers to increased responsiveness to repetitive stimuli. The hippocampus plays a special role in learning. Experiments on hippocampal slices indicate that the biochemical changes in the synapse represent the molecular bases for long-term memory. They are: (1) a monoamine (serotonin) and a glutamate receptor (NMDA receptor) are involved; (2) binding of the neurotransmitter (serotonin, glutamate) by these receptors initiates a cascade of enzymatic reactions; (3) the first step in this cascade is the activation of a G-protein mediated signal transduction that ultimately affects the efficacy of synaptic transmission. In B memory cells the affinity maturation of the antigen receptor takes place that potentiates signal transduction after antigenic stimulation. Other possible connections between neuronal and immunological memory are: the survival gene, bcl-2, plays a significant role in both immune and neuronal memory; so far IL-1α has been found to be present only in a subset of T-memory cells and hippocampal neurons; transgenic mice overproducing bcl-2 have a long-term persistence of immunoglobulin-secreting cells and an extended lifetime for memory B cells. This suggests that the biochemical basis of all memory-forming processes are highly similar.

Lymphocytic IL-1σ is a cell- and species-specific factor, which is capable of increasing β-adrenoceptor concentration and induction of its gene. These observations are highly important both for normal airway physiology as well as for the possible nature of the β-adrenergic dysregulation in asthma by adding an entirely new dimension to the β-adrenergic theory of the atopic abnormality in bronchial asthma.

Stress alters immune reactivity through specific interactions at every level of the neuroimmune regulatory system. This is based on four critical features shared by both systems: (1) they are composed of extraordinarily large numbers of phenotypically distinct cells organized into intricate networks. (2) Cells of both systems synthesize, secrete, and/or release shared effector molecules. (3) Recognition of these effector molecules is realized by the same cellular receptors and second messenger mechanisms of both cell systems. (4) These cellular and molecular determinants make a continuous, multi-lateral flow of information, which is the *sine qua non* of the unique interactions within the immune-neuroendocrine circuitry.

Neuropeptides take on added significance as immunomodulators, since it is now known that lymphoid organs are directly innervated with nerves secreting these agents. From the standpoint of the integration of information in the immune - neuroendocrine circuitry, future studies will have to examine these parallel signaling pathways in isolation.

1. DEVELOPMENTAL INTERRELATIONSHIPS AMONG THE CELLULAR AND HUMORAL COMPONENTS OF THE IMMUNE – NEUROENDOCRINE SYSTEMS

These interrelationships may be briefly stated through a discussion of (1) the cells involved in the synthesis, storage, secretion, and/or release of the effector molecules of immunologic reactivities; (2) neural crest interactions in the development of the immune system;(3) cerebral dominance or lateralization and immune disorders; (4) the ancient superfamily of immune recognition molecules and the neural cell adhesion molecule; (5) neuroactive immunoregulatory cytokines and growth factors;(6) coevolution of cytokine receptor families in the immune and nervous systems;(7) shared molecular mechanisms in the development of immune and neural memories; (8) the unique recognition and communication powers of the immune and nervous system as shared characteristics.

2. THE CELLS INVOLVED IN THE SYNTHESIS, STORAGE, SECRETION, AND/OR RELEASE OF THE EFFECTOR MOLECULES OF IMMUNOLOGIC REACTIVITIES

The functions of the immune system are the properties of cells distributed throughout the body. They include (1) free or circulating cells of the blood, lymph, and intravascular spaces; (2) similar cells collected into units that allow for close interaction with lymph or circulating blood-lymph nodes, spleen, liver, and bone marrow; and (3) two major control organs for the system, the thymus gland and the hypothalamic-pituitary complex. The cells involved in the synthesis, storage, secretion, and/or release of the effector molecules of immunologic inflammation, immunity, and immunologically based hypersensitivity (allergy) represent a continuous spectrum of related cell types specialized in the production and storage of various physiopharmacologically active effector substances in variable proportions, i.e., of cells that have a common developmental origin with differentiation being determined by the specific requirements of the local neurohumoral regulation. Accounting only for those effector molecules for which the cell type has been identified, this incomplete spectrum of cells and effector substances includes macrophages and lymphocytes (interleukins 1-17, interferons, tumor necrosis factors lysosome and complement components, prostaglandins, leukotrienes, acid hydrolases, neutral proteinases, arginase, nucleotide metabolites, various neuro- active immunoregulatory peptides including ACTH, CRF-like activity, bombesin, endorphins, enkephalins, TSH, growth hormone, prolactin, neurotensin, chorionic gonadotropin, VIP, tachykinin neuropeptides including substance P, substance K, neuromedin K, somatostatins, mast cell growth factor, catecholamines, etc.), neutrophil leukocytes (SRS-A, ECF-A, enzymes, PAF and other vascular permeability factors, kinin-generating substances, a complement-activating factor, histamine releasers, a neutrophil inhibitory factor, VIP, 5-HETE, etc.), basophilic leukocytes (histamine, SRS-A, ECF-A, NCF, PAF, SP, SOMs, etc.), murine basophilic leukocytes (the same as in humans plus serotonin), eosinophilic leukocytes (PAF, 8, 15-diHETE, SRS-1, eosinophil peroxidase, major basic protein, etc.), serosal, connective tissue or TC mast cells (histamine, SRS-A, ECF-A, NCF, PAF, VIP, SP, SOMs, etc.), mucosal or T mast cells (histamine, SRS-A, ECF-A, NCF, PAF, VIP, SP, SOMs, etc), "chromaffin-positive" mast cells (dopamine in ruminants; in other species possibly norepinephrine and neuropeptide Y), the so-called P cells (histamine, serotonin), enterochromaffin cells (serotonin), chromaffin cells (epinephrine, norepinephrine, dopamine, neuropeptide Y, IL-1α, etc.), platelets (depending on species, histamine, serotonin, catecholamines, prostaglandins, 12-HETE), neurosecretory cells (histamine, serotonin, catecholamines, acetylcholine, prostaglandins, and other eicosanoids, kinins, the various hypothalamic substances that release or inhibit the release of the anterior pituitary hormones, and the group of neuro- active immunoregulatory peptides including ACTH, bombesin, neurotensin, endorphins, enkephalins, TSH, growth hormone, prolactin, chorionic gonadotropin, VIP, the tachykinin neuropeptides, somatostatins, neuropeptide Y, IL-1α, IL-1β, IL-6, etc.), medullary thymic epithelial cells and the Hassall's corpusles (thymosins and other thymic factors), SIF cells (dopamine), and other nerve cells including essentially all of the effector molecules listed under the neurosecretory cells. For more detailed information on the foregoing cell types and effector molecules see Szentivanyi et al. [1]. Many of these cell types possess different morphologic, physicochemical, and general biologic characteristics. Nevertheless, in passing from one member of this cell spectrum to another, obvious transitions are seen in all of these characteristics. Furthermore, when one surveys their properties and their probable physiologic function in the higher organism, significant cohesive features become apparent that set them apart from other body constituents as a distinct single class of cells that must be included in current concepts of neurosecretion.

There are some workers who postulated that cellular components of the immune-neuroendocrine circuitry and their effector molecules could be viewed as two different divisions of this network. The two major divisions according to these workers may be defined as involved in neurovascular immunology and neuroendocrine immunology. Neurovascular immunology is concerned with immune response-related actions of vasoactive neurotransmitter substances that function as potent, short-lived local "hormones" first identified and studied as mediators of immunologic inflammation and hypersensitivity. They also play important roles in blood flow, vascular permeability, and pain transmission. These soluble effector molecules are simple compounds (i.e., amine mediators), short-chain peptides (i.e., kinins, substance P, etc.), and short-chain lipids (i.e., prostaglandins, leukotrienes) and have a long evolutionary history in biological defense. They use paracrine and synaptic signaling on their effector cells. The second major division, defined above as involved in neuroendocrine immunology, represents all of the immune response-related hypothalamic, pituitary, and other hormones that use endocrine signaling and are primarily immunomodulatory in character. This perhaps convenient but arbitrary functional separation of these two divisions is intrinsically incorrect as discussed by Szentivanyi [2]. In SLE, spontaneous hyper-reactivity of the cells of the immune system leads to the production of pathogenic autoantibodies. A hyperproliferative state of lymphocytes is indicated by the increased expression of bcl-2.

3. NEURAL CREST INTERACTIONS IN THE DEVELOPMENT OF THE IMMUNE SYSTEM

In discussing neural crest interaction in the development of the immune system at first we must briefly characterize the developmental biology of this structure. The neural crest is produced from ectodermal cells which are released from the apical portions of the neural folds at about the time fusion occurs to form the neural tube and a separate overlying ectodermal layer. The basement membrane underlying the neural crest cells breaks down, the cellular characteristics change, and they become separated from the other components of the neural fold. There is a change in relative spatial relationship, or migration, of the neural crest cells to varied associations and destinies [3].

We have known for some time that neural development is conducted in several distinct, consecutive steps but it is only recently that we have begun to understand the nature of the molecular mechanisms, i.e., by established diffusible factors of vertebrate embryonic induction. This is followed by intracellular regulation of a set of transcriptional factors containing, for instance, Hox and Pou domains [4,5]. When the neural tube is formed, neural crest cells i.e., multipotential stem cells, start to differentiate into the various cell types that comprise the nervous system. This is the stage at which neuronal immunoglobulin gene superfamilies (Ig SFs) and cytokine receptors play important roles. Three processes take place at this stage to form nervous system: cell migration, pathway finding, and determination of cell lineage. Recent studies suggest that the molecules involved in these three processes may have originated from the same protein molecule that has also evolved to generate major molecules in the immune system (see discussion of coevoultion of cytokine receptors in the immune and nervous systems below).

The portion of the neural crest pertinent to this discussion is that which is closely associated with developing brain, specifically, the hindbrain. Crest cells in this cranial portion (anterior to the fifth somite) differentiate into mesenchyme, in addition to other connective tissue, muscular and nervous components. Neural crest cells migrate ventrolateral through the bronchial arches and contribute mesenchymal cells to a number of structures. It is this mesenchyme that forms the layers around the epithelial primordial of the thymus [6]. The full significance of the foregoing will be even more appreciated when viewed in the context of three additional considerations (1) the thymus is formed by contributions from different sources which must interact in a precisely timed sequence for proper development; (2) ablation of small portions of neural crest prevents or alters the development of the thymus; (3) formation of the thymus precedes that of the more secondary, peripheral lymphoid tissues reflecting a critical thymic role already in the early development of the immune system. Taken together, development of the immune system is inherently linked to the neural crest and any aberration in this link results in defective immune development such as that seen for instance in the DiGeorge syndrome.

4. THE ANCIENT SUPERFAMILY OF IMMUNE RECOGNITION MOLECULES AND THE NEURAL CELL ADHESION MOLECULE N-CAM

The ancient superfamily of immune recognition molecules and the N-CAM represent another aspect of the interrelationships among the cellular and molecular components of the immune-neuroendocrine circuitry. Most of the glycoproteins that mediate cell-cell recognition or antigen recognition in the immune system contain related structural elements, suggestion that the genes that encode them have a common evolutionary history. Included in this Ig superfamily are antibodies, T-cell receptors, MHC glycoproteins, the CD2, CD4 and CD8 cell-cell adhesion proteins, some of the polypeptide chains of the CD3 complex associated with T-cell receptors and the various Fc receptors on lymphocytes and other white blood cells, all of which contain one or more Ig-like domains (Ig homology units). Each Ig homology unit is usually encoded by a separate exon, and it seems likely that the entire supergene family evolved from a gene coding Thy-1 or β2-microglobulin which may have been involved in mediating cell-cell interactions. Since a Thy-1-like molecule has been isolated from the brain of squids, it is probable that such a primordial gene arose before vertebrates diverged from their invertebrate ancestors some 400 million years ago. New family members presumably arose by exon and gene duplications and similar duplication events probably gave rise to multiple gene segments that encode antibodies and T-cell receptors [7].

An increasing number of cell-surface glycoproteins that mediate Ca^{2+} independent cell-cell adhesion in vertebrates are being discovered to belong to the Ig superfamily. One of these is the so-called N-Cam, which is a large, single-pass transmembrane glycoproteins (about 1000 amino acid residues long). N-CAM is expressed on the surface of nerves cells and glial cells and causes them to stick together by Ca^{2+}-independent mechanisms. When these membrane proteins are purified and inserted into synthetic phospholipid vesicles, the vesicles bind to one another, as well as to cells that have N-Cam on their surface; the binding is blocked if the cells are pretreated with monovalent anti-N-Cam antibodies. Thus N-Cam binds cells together by a homophilic interaction that directly joins two N-Cam molecules [8].

Anti-N-Cam antibodies disrupt the orderly pattern of retinal development in tissue culture and when injected into the developing chick eye, disturb the normal growth pattern of retinal nerve cell axons. These observations suggest that N-CAM plays an important part in the development of the central nervous system by promotion of cell-cell adhesion. In addition the neural crest cells that form the peripheral nervous system have large amounts of N-CAM on their surface when they are associated with the neural tube, lose it while they are migrating and then re-express it when they aggregate to form a ganglion suggesting that N-CAM play a part in the assembly of the ganglion.

There are several forms of N-CAM, each encoded by a distinct mRNA. The different, mRNAs are generated by alternative splicing of an RNA transcript produced from a single large gene. The large extracellular part of the polypeptide chain (~680 amino acid residues) is identical in most forms of N-CAM and is folded into five domains characteristics of antibody molecules. Thus, N-CAM belongs to the same ancient superfamily of recognition proteins to which antibodies belong [9,10].

We have mentioned earlier the guidance provided by the neural crest in the development of the thymus, that is, a central regulatory organ for the immune system. The converse also appears to be true: the immune system has a special role in the development of the nervous system. In this context, the most critical feature role of the immune system is its tremendous polymorphism so that lymphocytes that are produced can recognize enormous numbers of different antigens. For this reason, the immune system is ideally suited to provide markers, or *"anchoring sites"* that enable developing structures to be built up in precisely the correct form. In no organ system is this type of detailed anchorage mechanism as important as in the developing nervous system in which many millions of nerve fibers traverse great distance and establish connection with particular groups of target cells. Marking by means of histocompatibility antigens provides exactly such a system, as indeed the original function of MHC antigens was defined in the 1970s as the general plasma membrane anchorage site of organogenesis-directing proteins [11].

5. EXTRACELLULAR MATRIX PROTEINS WITH NEURITE-PROMOTING ACTIVITY AND THEIR RELATION TO THE IMMUNOGLOBULIN SUPERFAMILY

Nerve cells i.e. neurons, have two special properties that distinguish them from all other cells in the body. First, they conduct bioelectric signals for long distances without any loss of signal strength. Second, they possess special intercellular connection s with each other nerve cells and with innervated tissues such as muscles and glands. These connections determine the types of information a neuron can receive and the range of responses it can yield in return [12]. The function most revealing structural feature of the neuron is the characteristics specialized contact zone that has been presumptively identified as the site of interneuronal communication, i.e. the *synapse*.

This specialized contact zone is composed of presumed proteinaceous material lining the intracellular portions of the pre- and postsynaptic membranes and filling the synaptic cleft between the apposed cell surfaces. Such types of specialized contacts are a general form of the specialized cell contacts seen between many types of cells derived from the embryonic ectoderm, of which the nerve cell is but one. However, the specialized contact between neurons is polarized; that is, the presynaptic terminal intracellular material is composed of interrupted presynaptic dense projections measuring about 500–700 Å in diameter and separated from each other by 300–400 Å. This material may be present only to bind specific presynaptic nerve endings permanently to specific postsynaptic cell sites. Alternatively, the specialized contact zone could assist in the efficiency of transmission of the nerve impulse and could constitute a method for modulating synaptic transmission in terms of discharge frequency [13,14]. For these reasons, this specialized contact zone must be regarded as a unique association of two cells, and therefore, the synapse represents a special *sui generis* neuronal feature (see below).

Neurite promotion from the neurons is one of the most important events in the process of synapse formation. Therefore, this section focuses on extracellular matrix (ECM) proteins with neurite-promoting activity including their receptors, cell adhesion molecules in addition to the above-described NCAM, and their relation to members of the immunoglobulin superfamily.

Although many ECM proteins have been isolated recently, the ECM proteins that exhibit the neurite-promoting activity are type IV collagen, fibronectin, laminin, and a newly discovered protein, the so-called *neurite outgrowth factor* (NOF) which promotes neurite outgrowth from various neurons [15–17], such as sympathetic, parasympathetic, sensory, motor, and central neurons. Laminin, and its family proteins, exhibit the highest activity for promoting neurite outgrowth.

Laminin is a glycoprotein of about 900 kDa, a main component of the basement membrane, and possesses a variety of cellular functions such as neurite outgrowth-promoting [18], cell adhesive [19], growth factor-like [20], calcium-binding, and metastatic activities. Compared with other ECM proteins, the neurite outgrowth-promoting activity is the strongest. Structurally, laminin is composed of three subunits of A chain (440 kDa), B_1 chain (230 kDa), and B_2 (220 kDa) that are assembled by disulfide bonds to form a cruciform structure.

NOF, a glycoprotein of about 720 kDa, has been shown to be present in the extracellular matrix of muscle fibers and the ciliary ganglion. A partial primary structure of NOF is homologous (80%) with laminin B_1 chain, indicating NOF belongs to the laminin family. The neurite outgrowth by NOF is strictly dependent on the development stage. Thus, ciliary neurons until embryonic day 10 can fully extend the neurites in the presence of NOF, but thereafter the neurons rapidly lose this ability.

Study of the receptors for ECM proteins started in 1984 with the work of Tamkun et al. [21], who isolated a fibronectin receptor cDNA and named it *integrin.* Thereafter, integrin has been used as a general term for ECM protein receptors. The cDNA cloning of the NOF receptor indicates that there are at least two mRNAs that encode NOF receptor, which may be produced by alternative splicing occurring in the region that codes a cytoplasmic domain, because C-terminal length differs in obtained cDNA clones. Since no homology could be found with any known nucleic acid sequences, it is suggested that NOF receptor is a novel protein. The receptor is composed of 583 amino acids including a signal sequence. The deduced amino acid sequence revealed that the NOF receptors have five repeats of immunoglobulin Ig-like domains in its extracellular region, indicating that the NOF receptors belonging to the Ig-superfamily. Weak homology in amino acid sequence (20–30%) is observed with other Ig superfamily molecules such as SCI, NCA and VCAM.

In the nervous system, many Ig superfamily proteins are found to participate in the cell attachment [22]. In the vertebrate brain, NCAM [23] SCI [24], LI [25] MAG [26] and PO [27] are the major Ig superfamily proteins [10].

6. NEUTROPOIETIC IMMUNOMODULATORY CYTOKINES AND GROWTH FACTORS

Many immunomodulatory cytokines and growth factors, when studied for neuroactivity, were found to possess such capacity. At the time of this writing, however, there are only a few agents whose neuropoietic characteristic progressed to a point that their separate and individual discussion is both indicated and possible. They include the cilliary neurotropic factor, the leukemia inhibitory factor, and oncostatin M and beta-nerve growth factor.

7. CILIARY NEUROTROPIC FACTOR (CNTF)

Human CNTF is a 200-amino-acid-residue, single-chain polypeptide of 22.7 kDa. It is highly conserved across species lines as amino acid sequence comparisons of human with rat and rabbit CNFT show 83 and 87% identity, respectively [28–30]. To date, CNTF has been localized to Schwann cells and type 1 astrocytes [31–33]. It has demonstrated activity as a survival and differentiation factors for cells of the nervous system.

Originally identified as a tropic factor for 8-day chick embryo ciliary (parasympathetic) ganglion neurons [34]. These cells normally died without tropic supplement(s). CNTF has subsequently been shown to exhibit *in vitro* survival-promoting activity on sensory (dorsal root) ganglion neurons, nonautonomaic motor neurons and sympathetic motor neurons [34–39]. This enhanced neuronal survival activity was also confirmed in *in vivo* studies. CNTF has been proposed to be a rescue factor for damaged or axotomized neurons, which is supported by the observations that (1) motor neuron death that follows axotomy of the rat facial nerve may be reversed by applying rat CNTF to the proximal axonal segment [40] and (2) recombinant human CNTF injected into rat lateral ventricles prevents the death of axotomized medial septal neurons (axons of cholinergic and GABAergic neurons) projecting to the hippocampus [41,42].

CNTF also serves as a neuronal differentiation factor capable of influencing the phenotype of a neuron. Thus, this cytokine can dictate the type of neurotransmitter used, neuropeptide produced and the type of synapses made between two neurons [43,44]. Indeed CNTF has been shown *in vitro* to induce cholinergic properties in otherwise adrenergic motor neurons including the expression of acetylcholine (ACh) as neurotransmitter and substance P (SP) and vasoactive intestinal peptide (VIP) as ACh-associated neuropeptides [45,46]. Although usually associated with the cholinergic phenotype, CNTF has also been shown to promote the adrenergic phenotype in rat fetal locus coeruleus cell cultures [47] and adrenal chromaffin cells [48]. Another function of CNTF is in glial differentiation. In cultures of oligodendroglia type 2 astrocyte progenitors, if applied in the presence of the extracellular matrix, CNTF promotes the development of either oligodendroglia or type astrocytes [49] which is in functional agreement that type 1 astrocytes are known to synthesize CNTF [31,50]. Finally, like IL-6 CNTF induces fever following intravenous injection [51].

8. LEUKEMIA INHIBITORY FACTOR (LIF)

This cytokine was discovered by it's ability to inhibit the proliferation and to induce differentiation of the murine myeloid leukemic cell line, MI, into macrophages. Subsequently LIF was recognized to be a pleiotropic factor. LIF expression has been detected in a number of cell lines. It stimulated T lymphocytes and monocytes [52,53], mast cell [54] Schwann cell [55], astrocytes [56], bone marrow stromal cells and thymic epithelial cells [57,58]. In some of these cell types, synthesis of LIF has been induced by LPS, phorbol ester, calcium ionophore, retinoic acid and TNF-α [57,59–62].

Human LIF cDNA encodes a 202-amino-acid-residue polypeptide that is cleaved to yield a 180- amino-acid-residue mature human LIF. Native human and murine LIFs are highly glycosylated single-chain molecules varying in molecular mass from 38 to 67 kDa.

LIF has multiple effects on both hematopoietic and nerve cell populations. Its activities overlap those of IL-6, IL-11, CNTF and oncostatin M (shared gp130 signal-transducing subunit in their receptor complexes). LIF also shows synergistic activity with IL-3 to increase the production of primitive hematopoietic progenitor [63] and megakaryocyte colonies [57]. In the nervous system, LIF regulates the levels of both muscarinic and substance P receptor mRNAs in sympathetic neurons from noradrenergic to cholinergic in vitro and in vivo [64,65]. It selectively increases CNS cholinergic differentiation in neurons whose projections are peripheral targets [49] and probably plays a role in early neural crest-derived sensory neuron development [66,67]. Moreover, LIF has the potential to rescue sensory and motor neurons from axotomized death through retrograde axonal LIF transport [68,69]. In cultures of O-2A progenitors, LIF induces oligodendroglia formation and in the presence of endothelial cell-derived extracellular matrix it promotes the differentiation of O-2A progenitors into type 2 astrocytes [49].

9. ONCOSTATIN M (OSM)

This substance was originally isolated as a growth inhibitor of A375 melanoma cells from the conditioned medium of phorbol-ester-treated U937 human histiocytic lymphoma cells [70]. In subsequent studies, OSM became recognized as a member of a family cytokines that includes LIF, IL-6, IL-11 and CNTF. These cytokines have similar protein tertiary structures and gene organization, have overlapping biological activities and share gp130 as a signal transducer [71–78]. OSM is produced by activated monocytes/macrophages and T lymphocytes [71], expressed in AIDS-related Kaposi's sarcoma (KS)-derived spindle cells and is an autocrine growth factor for KS [79]. With a molecular mass of approximately 28 kDa, OSM is a single-chain glycoprotein the DNA of which has been cloned [80]. The gene for OSM has been mapped to chromosome 22 at q12 [81,82] within 19-kb of the LIF gene. The two genes are arranged in a head-to-head orientation, suggesting that common regulatory elements may lie within the 19-kb region between the two genes [83].

Additional OSM activities beyond those shared with LIF, IL-6, IL-11 or CNTF include the induction of differentiation of MI murine myeloid leukemia cells [77,78], the stimulation of G-CSF and GM-CSF production by human endothelial cells, and the blockade of IL-6 in those cell lines that express gp130 but lack functional OSM receptors [84].

10. NERVE GROWTH FACTORS (NGF OR BDNF)

Nerve growth factor is a 26-kDa, non-glycosylated, homo dimeric polypeptide originally identified as a neurotropic factor for embryonic sympathetic neurons [85–87]. In addition to NGF, three other NGF- related molecules have been identified: brain-derived neurotropic factor (BDNF), neurotropin-3 (NT-3 and neurotropins 4/5 (NT-4/5) [88–91]. Cells known to synthesize either NGF or NGF mRNA include submandibular duct epithelium, neurons, Schwann cells, fibroblasts, smooth muscle cells, astrocytes CD 4^{+} T cells, macrophages, mast cells and embryonic keratinocytes [92–94]. Factors known to regulate NGF secretion in astrocytes include IL-1, IL-4 and IL-5 [95,96].

Within the nervous system, NGF has a remarkably broad spectrum of documented activities. It includes the blocking of natural cell death in developing sensory and sympathetic neurons, the rescue of basal forebrain cholinergic neurons following axotomy, the induction of proliferation and differentiation of adult chromaffin cells and support provided to cerebellar granule neurons during ontogenesis [97–99]. NGF plays a significant role in hematopoiesis and the immune system. In early hematopoiesis, NGF is as effective, as IL-1 in synergizing with M-CSF (CSF-1) to generate early hematopoietic progenitors with or without pluripotent stem cells and GM-CSF colonies [100]. NGF is chemotactic for neutrophils, enhances threefold the *in vitro* survival of neutrophils, increases by twofold neutrophil phagocytic activity, induces superoxide production [101] and has a significant role in monocyte maturation [102].

In humoral immunity, NGF induces B-cell differentiation and promotes IgM, IgA and most significantly, IgG4 synthesis in mononucleare cells [103,104]. Both B cells and CD $3^{\prime}$ T cells mononuclear proliferate in response to NGF. In the field of allergy, NGF has established effects on cells of the mast cell/basophil lineage. Thus, NGF enhances the activity of T-cell-derived GM-CSF in the promotion of basophil differentiation from basophil progenitors and prepares basophils for histamine release ion response to anti-IgE exposure [105,106]. NGF has been demonstrated to induce mast cell degranulation in a calcium-independent manner with a substantial increase in histamine release in response to antigen stimulation [101]. Also there is a strong correlation between serums NGF levels and systemic lupus erythematosus, an autoimmune disease [107]. Finally, NGF protection of cholinergic neurons arrests and /or improves memory decline and it was shown to reverse age-related cognitive dysfunction in aged rats [108,109].

11. COEVOLUTION OF CYTOKINE RECEPTOR FAMILIES IN THE IMMUNE AND NERVOUS SYSTEM

Established members of neurotropic factors (NTFs) and neuronal differentiation factors (NDFs) belong to two major groups: cytokine and growth factors. The family of cytokines may be divided into three subclasses based on structural homology of their receptors [72,110]: (1) class I receptors :IL-1R, -3R, -4R, -5R, -6R and –7R, granulocyte-macrophage colony-stimulating factor receptor (GMCSFR), erythropoietin receptor (EPR), cholinergic differentiation factor receptor (CDFR), leukemia inhibitory factor receptor (LIFR), ciliary neutropic factor receptor (CNTFR); (2) class II receptors: interferon -α receptor (INF-αR), interferon γ receptor (INF-γR), and tissue factor receptor (TFR; and (3) class III receptors .i.e. those receptors whose primary structures belong to immunoglobulin gene superfamilies (IG SFS) such as IL-1.Growth factors can be classified into gene families based on the homology of their structures: nerve growth factor (NGF) family, fibroblast growth factor (FGF) family, epidermal growth factor (EGF) family, insulin growth factor (IGF) family. etc.

Class I cytokines receptors can be further divided into at least two subtypes, class Ia and class Ib , based on structural differences [111]. Class IA includes IL-2R,-3R, and-5R and GM-CSFR [112] and consists of the common cytokines domain with four conserved cysteine and Trp-Ser-X (variable residues)-Trp-Ser motif near the transmembrane domains. Class Ib contains IL-6R, CNTFR, LIFR and G-CSFR, and has additional elements,i.e. these fibronectin like domains near the transmembrane domain and an Ig- like domain at the N-terminus in the extracellular sites. It is the group of cytokines in class Ib that is of special interest to us in the context of this discussion. As briefly mentioned above, a number of NTFs and NDFs have been identified recently together with their receptors. It was shown that NTFs enhance survival of certain types of neurons, whereas NDFs affect neuronal phenotype without having any effect on neuronal survival. Cholinergic neurons are sensitive to leukemia inhibitory factor (LIF) that plays an important role in the immune system [113]. Ciliary neutropic factor (CNTF) with an NTF activity on ciliary ganglion neurons also possess NDF activity as that of CDF/LIF by inducing cholinergic differentiation of sympathetic neurons in culture [45]. Another member of the class Ib group is IL-6, an established immunoregulatory molecule, which improves the survival of mesencephalic catecholaminergic and septal cholinergics neurons from postnatal rats (2 weeks old) in culture [114] and triggers the association of its receptor (IL-6R) with the signal transducer gp130 [115]. Generation of sensory neurons is stimulated by LIF [116] and LIFR is structurally related to the IL-6 signal transducer gp130 [117]. Furthermore, there is evidence that not only LIF but also CNTF act on neuronal cells via a shared signal pathway that involves the IL-6 signal transducing receptor component [118].

In view of the very similar NTF and NDF activities of all cytokine members of class Ib ,

Yamamori [111] proposed that their receptors are likely to have a closely similar structure. This suggestion was considerably strengthened by findings in a number of laboratories showing that their receptors belong to a subclass of receptors for cytokines with both neuronal and immunoregulatory activities [117,119–122]. This close structural relationship between the nervous and immune systems at the molecular level may be explained by evolutionary mechanisms in agreement with the proposal of Edelman [123] that the structural resemblance of immunoglobulin and neural cell surface molecule may have resulted from a common evolutionary origin and with the analysis of Bazan [110] of the structural design and molecular evolution of the cytokine receptor superfamily. Finally Yamamori and Sarai [111] in an attempt to trace the evolutionary origin of the cytokine receptors, constructed a higher subdominant structure of the receptor for CDF/LIF based on its known primary structure. In their study, the receptor appears to contain immunoglobulin and fibronectin-like domains in addition to common domains of the cytokines receptor, similar to those cell surface molecules of the neural Ig SFs. Taken together, these and other above-mentioned findings indicate that the class Ib cytokine receptor evolved as a consequence of fusion of the genes for a more primitive cytokine receptor in class Ia and for the IgSF, and similarly that a large number of molecules regulating neural development and vertebrate immunoregulation have a common evolutionary background. Thus the origin of cytokine receptors represents a classical example of molecular coevolution in the immune and nervous systems [111].

12. CEREBRAL DOMAINANCE OR LATERALIZATION AND IMMUNE DISORDERS

The recognition of cerebral lateralization grew out of the discovery in the last century of cerebral dominance, that is, the superior capacity of each side of the brain to acquire particular skills. Over the past 120 years, it was believed that hemispheric dominance was based on functional asymmetry-on the differences in function of the two sides of the brain and of special regions within them. In the face of the prevalent belief that cerebral dominance lacked an anatomic correlate, work over the past three decades has conclusively established that cerebral dominance is based on asymmetries of structure. An example, of this is the early detectable asymmetry in the human brain that involved the upper surface of the posterior portion of the left temporal lobe, the *planum temporale*. The larger size of the left *planum temporale* reflects the greater extent of a particular temporoparietal cytoarchitectonic area on the left. There are other asymmetries in the human brain and the same applies to the findings throughout the animal kingdom. In addition to genetics, several other factors in the course of development, both prenatal and postnatal, influence the direction and extent of these structural differences [124].

Association of anomalous cerebral dominance include not only developmental disorders such as dyslexia, autism, stuttering, mental retardation, and learning disorders, as well as some extraordinary musical, mathematic, athletic, and other talents, but also alterations in many bodily systems including the immune system. In these situations, the same influences that modify structural asymmetry in the brain also modify other systems such as the immune system. The suspected molecular mechanisms involved in the influence of structural asymmetry on the development of immune reactivities are discussed in more detail in two larger reviews [2,125]. Here we shall only mention that animal experiments in the past 20 years provided some insight into the nature of the association between anomalous hemispheric dominance and immune reactivities.

Thus, beginning in 1980 and continuing throughout the decade, Renoux and coworkers

[126] showed that immunomodulation of the T-cell lineage in rodents can be a phenomenon of hemispheric lateralization. In 1980, the initial observation yielded data indicating that lesioning the left cerebral neocortex depressed T-cell-mediated responses in mice without affecting B-cell responses. These observations were extended in experiments where animals with a right cortical lesion served as controls for animals with a left cortical ablation. The findings demonstrated a balanced brain asymmetry in which the right hemisphere controls the inductive influence on T cells of signal emitted by the left hemisphere. In addition, most recent studies found that ditiocarb sodium (Imuthiol), an immunostimulant specifically active on the T-cell lineage, can replace the signals emitted by the left neocortex, since mice without a left neocortex were stimulated to increase T-cell-dependent responses by treatment with ditiocarb sodium, whereas the agent did not modify the responses already increased in right decorticates. B-cell-dependent and some macrophage-dependent responses are not affected by either neocortical ablation or ditiocarb sodium. This lateralization of cortical influence on immune function in rodents is likely to be predictive of an even greater influence in humans with more profound and complex cortical functions [126].

13. IMMUNE AND NEURONAL MEMORY

The immune system and the nervous system possess short- or long-term memory, or both. The latter may be defined as the recording of experiences that can modify behavior. This general definition encompasses a broad spectrum of phenomena from the capacity of sensing chemical gradients to cognitive learning in humans.

The clonal selection theory of acquired immunity provides a useful conceptual framework for understanding the cellular basis of immunologic memory. According to this scheme, immunologic memory is generated during the primary immune response because (1) the proliferation of antigen-triggered virgin cells creates a large number of memory cells a process known as clonal expansion; (2) the memory cells have a much longer life span than do virgin cells and recirculate between the blood and secondary lymphoid organs; and (3) each memory cell is able to respond more readily to antigen than does a virgin cell.

One reason, if not the most important reason, for the increased responsiveness of memory B cells is the higher affinity (avidity) of their antibody receptors for the homologous antigen. Thus, with the passage of time after immunization there is a progressive increase in the affinity of antibodies produced against the immunizing antigen. This phenomenon is known as *affinity maturation* and is the result of the accumulation of somatic mutations in variable (V)-region coding sequences after antigenic stimulation of B-lymphocytes. The rate of somatic mutation in these sequences is estimated to be 10^{-3}/nucleotide pair/per cell generation which is about a million times greater than the spontaneous mutation rate in other genes. This process is called *somatic hyper-mutation*. Since B cells are stimulated to proliferate by the binding of antigen, any mutation occurring during the course of an immune response that increase the affinity of a cell surface antibody molecule will cause the preferential proliferation of the B cell making the antibody, especially when antigen concentration decreases with increasing time after immunization. Thus, affinity maturation is the consequence of repeated cycles of somatic hyper-mutation followed by antigen-driven selection in the course of an antibody response.

Research in the field of neuronal memory is still in an early phase primarily because of methodological difficulties and the validity of approaches currently used. The human brain is extraordinarily complex (10^{12} neurons) and intricate (an average neuron may have 10,000 dendrites interacting with other neurons). This dictates the use of reductionist approaches, which always require a correlation with the whole organism to verify the conclusions reached at the molecular level. Such relationship emphasized the importance of relating the biochemical events in single cells to the more complex organisms such as *Aplysia, Drosophila*, rodents, cats, and humans. The fact, however, that adjacent neurons are almost never identical means that the quantity of material needed to perform biochemical analyses necessitates a cell line approach. Both bacteria and neural cell lines provide a homogeneous population of cells that can be studied biochemically. Bacteria detect chemical gradients using a memory obtained by the combination of a fast excitation process and a slow adaptation process. This model system, which has the advantages of extensive genetic and biochemical information, shows no feature of long-term memory.

To study long-term memory, other biologic systems that exhibit two phenomena associated with learning and memory, habituation and potentiation, must be used. Habituation is defined as the decreased responsiveness to a stimulus when it is presented repetitively over time. Potentiation, on the other hand, is defined as the increased responsiveness of a stimulus when that stimulus is presented repetitively over time. In the mammalian brain, the hippocampus plays a special role in learning: when it is destroyed on both sides of the brain , the ability to form new memories is largely lost although previous long-established memories remain. The evidence obtained on hippocampal slices indicates that the biochemical changes in the synapse represent the molecular bases for long-term memory [127,128]. Despite the wealth of information provided by investigations in mammalian brain slices, it became increasingly clear that the study of cultured neural cells is more desirable because only in such a system could one be certain that the complete biochemical pathway, that is, the complete signal transduction pathway from stimulatory input to a behavioral output, could be analyzed. To study memory, however some modifiable behavior needs to be observed. Because neurons communicate with each other chemically through the release of a neurotransmitter, the secretion of neurotransmitters *(output)* evoked by various chemical stimuli *(input)* could be used to monitor the responsiveness of the cell. This experimental system was used to study the input – output properties of a particular neuron. Both habituation and potentiation could be demonstrated in neuronal cell lines indicating that they can serve as good model systems for the memory process, except that they do not possess synaptic connections. In the early phase of these studies the absence of synaptic connections posed a substantial problem for two major reasons: (1) as stated before, current evidence favors (in more organized neural tissue such as brain slices) the idea that the biochemical changes in the synapse are the molecular basis of long-term memory and (2) the synapse is a unique anatomic association of two cells that occurs only in the nervous system and therefore represents a special *sui generis* neuronal feature.

Whether memories, however, are generally recorded in presynaptic changes or in postsynaptic changes, in synaptic chemistry or in synaptic structure, or indeed in synapses at all, are still open questions. Regardless of the validity of any of these questions, it appears that the biochemical features of all memory-forming processes are highly similar (i.e. habituation, potentiation, and associative learning in invertebrates, mammalian brain slices and cultured clonal neural cell lines: the PC 12 cells and HT4 cells) [129–132]. They can be characterized as follows: (1) a monoamine (primarily serotonin) and a glutamate receptor (known as NMDA receptor because it is selectively activated by the artificial glutamate analogue N-methyl-D-aspartate) are involved; (2) binding of the neurotransmitter (serotonin, glutamate) by these receptors initiates a cascade of enzymatic reactions; (3) the first step in this cascade is the activation of a G-protein which may either interact directly with ion channels or control the production of cyclic adenosine monophosphate (AMP) or Ca^{2+}; (4) the two second messengers in turn regulate ion channels directly or activate kinases that phosphorylate various proteins including ion channels; (5) at many synapses both channel-linked and non-channel-linked receptors are present, and responding either to the same or to different neurotransmitters; (6) responses mediated by non-channel-linked receptors (serotonin) have a slow onset and long duration and modulate the efficacy of subsequent synaptic transmission providing the basis for memory formation; (7) channel-linked receptors that allow Ca^{2+} to enter the cell (NMDA receptor) also mediate long-term memory effects and (8) either too much or too little cyclic AMP can interfere with memory formation [133].

In all of these processes, the interaction between serotonin and glutamate and their respective receptor systems can be illustrated by the studies carried out on the HT4 neural cell line [121]. The HT4 cells do not habituate to repetitive membrane depolarization but after exposure of these cells to various neurotransmitters, serotonin has the capacity to potentiate cellular responsiveness. Depending on the strength of the serotonin stimulus, both short and long-term potentiation can be induced. For instance, a 2-min exposure to serotonin results in a transient increase in cellular responsiveness, whereas a 5-min presentation gives rise to a more permanent potentiation, the difference between the two involving the activation of NMDA receptor. Thus the stronger (5 min) serotonin stimulus results in the endogenous release of excitatory amino acids with activation of NMDA receptors. Consistent with this mechanism, long-term secretory potentiation can also be produced with a 2-min stimulus of serotonin only if glutamate or NMDA is given simultaneously.

14. COMPARISON OF IMMUNOLOGIC VERSUS NEURONAL MEMORY MECHANISMS

As we now begin a comparison of immunologic versus neuronal memory, we have to return to an earlier statement that immunologic memory is related to clonal selection and lymphocyte maturation. Although this is correct in cellular terms, in molecular terms the problem of clonal selection and expansion reduces to the issue of affinity maturation of the antibody on the surface of the lymphocyte. In other words, the entire antigen-driven selection of antibody-producing lymphocytes is based on the strength of the antibody-antigen interaction, which depends on both the affinity and the number of binding sites. The affinity of an antibody reflects the strength of binding of an antigenic determinant to a single antigen-binding site and it is independent of the number of sites. However, the total avidity of an antibody for a multivalent antigen, such as a polymer with repeating subunits, is defined as the total binding strength of all of its binding sites together. When a multivalent antigen combines with more than one antigen-binding site on an antibody, the binding strength is greatly increased because all of the antigen-antibody bonds must be broken simultaneously before the antigen and antibody can dissociate. Thus, a typical IgG molecule will bind at least 1000 times more strongly to a multivalent antigen if both antigen-binding sites are engaged than if only one site is involved. For the same reason, if the affinity of the antigen-binding sites in an IgG and an IgM molecule are the same, the IgM molecule (with ten binding sites) will have a much greater avidity for a multivalent antigen than an IgG molecule (with two sites). This difference in avidity, often 10^4-fold or more, is important because antibodies produced early in an immune response usually have much lower affinities than those produced later. Because of its high total avidity, IgM – the major Ig class produced early in immune responses-can function effectively even when each of its binding sites has only a low affinity [134].

In the late 1950s and early 1960s an excellent correlation was shown between the body temperature of rabbits and the affinity and avidity of the antibody produced against the radiolabeled antigen as tested in equilibrium dialysis experiments [135]. The higher the body temperature was, the greater the affinity and avidity were. In the beginning of these studies, commercially available E. coli endotoxin was used, but later the rise in temperature was reproduced by electrical stimulation of the posterior hypothalamus or hippocampus through stereotaxically implanted permanent depth electrodes in studies without endotoxin administration [136]. Discovery of the peptidoglycan and its derivatives as powerful immunologic adjuvants opened up a new window for the consideration of the relationship between immunologic and neuronal memory. The peptidoglycan, which is the basal layer of the bacterial cell wall, is a rigid macromolecule surrounding the cytoplasmic membrane. It is formed by the polymerization of a disaccharide tetrapeptide subunit; in the intact peptidoglycan, disaccharides form linear chains whereas peptides are linked by inter peptide linkages [137]. The recognition of the immunomodulating properties of peptidoglycans and peptidoglycan fragments is the result of the work aimed at identifying the structure responsible for the adjuvant activity of the mycobacteria cells in Freund's adjuvant [138]. Simple active molecules were soon produced by organic synthesis followed by a vast array of analogues and derivatives that can be classified into several categories. The one that is most pertinent to this discussion is the group of 'simple muramyl peptides'. Of these the smallest immuno- active synthetic muramyl peptide I s N- acetylmuramyl-L-alanyl-D-isoglutamine (MDP) [139,140]. This substance has a pyrogenic effect that originally was attributed to its ability to induce the release of endogenous pyrogen from mononuclear phagocytes. However , a direct central nervous system action could not be excluded since MDP was found to be active by the intracerebroventricular route. MDP was shown to cross the blood-brain barrier [141] in rabbits made leukopenic by nitrogen mustard treatment [137]. In addition, it was subsequently shown that MDP can also induce sleep and the somnogenic effect can be separated from its pyrogenic activity. MDP's pyrogenic activity does not affect brain temperature changes that are tightly coupled to sleep states [142]. More importantly, with respect to the direct central nervous system neuronal effects of MDP, this substance is capable of specific binding to serotonin receptors of synaptosomal membranes of brain tissue and competes with serotonin for these binding sites, and the kinetics of serotonin binding to brain homogenates is altered after sleep deprivation [143]. Additional findings on the capacity of MDP to act directly and specifically on central neurons include the following: MDP alters neuronal firing rates in different regions of the brain [144], humoral antibody responses are enhanced by lowering serotonin levels in the brain [145], administration of para-chlorophenylalanine which markedly decreases the level of brain serotonin completely abolishes the MDP-induced rise in body temperature and the somnogenic effect [146]. Finally, it has been established that immunization decreases the concentration of serotonin in the hypothalamus and the hippocampus [147].

Although the foregoing evidence is fragmentary, it does establish a set of future reference points to begin to undertake a more informed comparison of the molecular mechanism involved in immune and neuronal memory. One of the reference points involves a novel proto-oncogene, bcl-2, that by all available information plays a significant role in both immune and neuronal memory. The other possible reference point may be represented by an IL-1α that so far has been found to be present only in a subset of T-memory cells and hippocampal neurons.

15. THE BCL-2 GENE AND IMMUNE MEMORY

The germinal center forms a specialized microenvironment that plays a key role in the induction of antibody synthesis, affinity maturation of B cells, isotype switching, and memory B-cell formation. Furthermore, it is believed that the germinal center is also involved in the maintenance of T-cell memory. The bcl-2 protein, originally discovered as a proto-oncogene, has proved to be unique in being localized to the mitochondria and in blocking programmed cell death rather than affecting proliferation. It was identified at the chromosomal breakpoint of t (14:18)-bearing B-cell lymphomas. In adults, bcl-2 is topographically restricted to progenitor cells and long lived cells in tissues characterized by apoptotic cell death and it is confined to zones of surviving B cells on germinal centers [148,149]. Within the thymus, bcl-2 is present in the surviving mature thymocytes of the medulla but absent from majority of immature cortical thymocytes, most of which die by apoptosis. Transgenic mice that over express bcl-2 in the B cell lineage demonstrate extended cell survival and prolonged immune responses, and conclusively support a role for bcl-2 in B cell memory. Transgenic models that over express bcl-2 in the thymus have expanded the involvement of bcl-2 to multiple apoptotic pathways and indicate its critical role in thymocyte maturation [150].

Another source of information on the role and functioning of bcl-2 is provided by some recent observations on B chronic lymphocytic leukemia (B-CLL). In the natural history of CLL, at least three categories of genes are believed to be involved: (1) genes that are responsible for transforming event(s) in the target cells; (2) Genes that help the progressive accumulation of malignant cells, and (3) genes that cause the progression toward a more aggressive lymphoma. The possibility that the clonal expansion of B-CLL is related to a prolonged life span of monoclonal B-cells rather than to an acceleration of their proliferative activity is now suggested by recent findings on the expression of the bcl-2 gene in B-CLL cells [151]. As discussed earlier the bcl-2 gene product regulates programmed cell death and recent studies indicate that bcl-2 is involved in the selection and maintenance of long-lived memory B cells rescuing them from apoptotic death and leading to their accumulation in the Go phase of the cell. Variant chromosomal translocations have been detected in a small fraction (5–10%) of B-CLL involving bcl-2 and the Ig light chain gene. Despite the low percentage of bcl-2 rearrangements the expression of mRNA and proteins is appreciable in most samples of fresh B-CLL cells in an amount comparable to that observed in Karpas 422 cells which contain a translocation (14; 18).

The bcl-2 gene is translocated into the Ig loci in about 80% of human follicular lymphomas and in 10% of B-type chronic lymphocytic leukemias (B-CLL), resulting in a high level of expression of bcl-2 transcripts and protein in B-CLL cells in their normal equivalent CD5+ B cells. In normal B-cell populations representative of different *in vivo* and *in vitro* stages of activation and proliferation, bcl-2 was found to be expressed in all of 11 cases of CD5+ B-CLL clones, contrasting with the absent expression in normal CD5+ B cells. Activation of B-CLL cells by phorbol esters to Ig M secretion without concomitant DNA synthesis resulted in a rapid but transient down regulation of bcl-2 expression. In contrast, the reduction of bcl-2 at both the mRNA and protein levels was sustained after mitogenic stimulation, suggesting that bcl-2 expression and proliferation are inversely related in these cells. Furthermore, the pattern of bcl-2 expression in B-CLL resembled that of normal tonsillar follicular B-cells in which a high level of expression was found in resting mantle Zone B cells, but not in the proliferating germinal center B-cells. Based on these findings, and the role of bcl-2 in maintaining B-cell memory, Schena et. al. [152] proposed that the phenotype of B-CLL cells corresponds to a mantle zone memory type B cell. Studying bcl-2 protein in expression in high grade B-cell lymphomas derived from lymph node or mucosa-associated lymphoid tissue (MALT), a significant difference has been found in bcl-2 expression between nodal (39/48) and MALT high grade B-cell lymphoma (1/15). Bcl-2 was usually expressed by memory or resting B-cells, most activated B-cells being bcl-2 negative, except in lymph node originated, high grade B-cell lymphomas, which appear to be mainly bcl-2 positive. Presence of bcl-2 protein in nodal large-cell lymphomas seems to be independent of a t(14:18) translocation, only being found in 19 to 28% of these lymphomas. Translocation constitutes a definitive difference between both tumors, suggesting the existence of different molecular genetic characteristics and pathogenesis [153].

Another dimension of the significance of bcl-2 in B-cell memory has to do with the maintenance of the remarkably constant number of lymphocytes in an animal in the face of antigen-driven proliferation and a high rate of B-cell lymphopoiesis. This reflects the relatively brief life span of many newly generated B cells and argues for a well-regulated death mechanism. Even so, a secondary immune response can be generated years after a primary exposure to antigen. Antigen that might restimulate B cells persists for extended periods on follicular dendritic cells in the light zone of germinal centers. Antigen-binding B cells have also been found months after the end of manifest cell division. It is increasingly apparent that the precise signal that enables certain B cells to emerge as long-term surviving memory cells is bcl-2. Consequently, this agent must have a critical role in the maintenance of immune responsiveness. Indeed, transgenic mice overproducing bcl-2 have a long-term persistence of immunoglobulin-secreting cells and an extended lifetime for memory B cells [154]. Systemic lupus erythematosus (SLE) provides another example reflecting the same. In SLE, spontaneous hyper-reactivity of the cells of the immune system leads to the production of pathogenic autoantibodies. A hyperproliferative state of lymphocytes is indicated by the increased expression of bcl-2. In 19 of 24 patients with SLE, an increased concentration of bcl-2 mRNA was found in unstimulated circulating blood lymphocytes. The overexpression of the bcl-2 gene was more pronounced in patients with active SLE suggesting a pathogenic role of increased bcl-2 expression and prolonged survival of autoimmune memory cells in this disease [150].

The foregoing discussion was focused on B-cell memory and its relation to the bcl-2 gene. One of the central questions in T-cell immunity is how the generation of T-cell memory may proceed simultaneously with the maintenance of T-cell homeostasis. Akbar et al. [155] in their recent article present the role of bcl-2 in regulating T-cell memory as a balancing act between cell death and survival. More specifically, these authors have investigated the bcl-2 protein expression by resting and activated mature T-cell populations. Freshly isolated CD45RO+ T cells within CD4+ and CD8+ subsets expressed significantly less bcl-2 than CD45RO- (CD45RA+) T cells. When CD45RA + T cells within both CD4+ and CD8 + subsets were activated *in vitro,* the transition to CD45RO phenotype was associated with a decrease in bcl-2 expression. In these studies, a significant correlation was seen between low bcl-2 expression by activated T cells and their apoptosis in culture. These findings suggest that the primary activation of T cells leads to the expansion of a population that is destined to perish unless rescued by some extrinsic event. Thus, the suicide of CD45RO+ T cells could be prevented by the addition of IL-2 to the culture medium which resulted in a concomitant increase in the bcl-2 expression of these cells. The paradox that the CD45RO+ population contains the primed/memory T-cell pool yet expresses low bcl-2 and is susceptible to apoptosis can be reconciled by the observations that maintenance of T-cell memory may be dependent on the continuous restimulation of T cells which increases their bcl-2 expression [156].

15.1. The bcl-2 gene and neuronal memory

To assess the role of bcl-2 in neuronal apoptosis, Allsopp et al. [157] microinjected a bcl-2 expression vector into neurotropic factor-deprived embryonic neurons. Sensory neurons that depend for survival on one or more members of the nerve growth factor family of neurotropic factors (nerve growth factor, brain-derived neurotropic factor, and neurotropin-3) were rescued by bcl-2, whereas ciliary neurotropic factor (CNTF)-dependent ciliary neurons were not. Sensory neurons, however, became refractory to bcl-2 after exposure to CNTF. These findings indicate that at least two memory (death) pathways operate in neurons that are distinguished by their susceptibility to bcl-2. Abe-Dohmae et al. [158] have developed an analytical method for quantification of the RNA transcripts of the murine bcl-2 gene. The PCR products from bcl-2 alpha and bcl-2 beta mRNA were fluorometrically analyzed and their specific contents calculated by the internal standard method. Both bcl-2 mRNAs in adult mice were transcribed at the highest level in the thymus and at a comparable level in the spleen. Aside from the immune system, the brain gave the most abundant levels of the bcl-2 mRNAs. During development of the brain, the bcl-2 alpha and bcl-2 beta mRNA levels were highest on embryonic day 15 and about two and three times higher than those of adults, respectively. The results suggest that the bcl-2 gene is highly expressed during neurogenesis and functions to regulate development, memory, and survival of neurons in the central nervous system.

There are at least three additional examples of the effects of bcl-2 in neuronal systems. The first example involves glutamate toxicity. At high concentrations (5-10mM), this substance has been shown to kill cells of the pheochromocytoma cell line PC12. Zhong et al. [159] have recently reported that similar concentrations of glutamate also kill immortalized central neural cell lines and that the expression of bcl-2 in these cell lines blocks glutamate neurotoxicity.

The experiences obtained in Sindbis virus (SV) encephalitis could serve as a second example of neuronal activity by bcl-2. As known, encephalitis in mice provides a model for studying age-dependent susceptibility to acute viral encephalitis. The AR 339 strain of SV causes fatal encephalitis in newborn mice but weanling mice recover uneventfully. Weanling mice with normal immune systems clear the virus from neurons through an antibody-mediated mechanism. This does not happen in newborn mice because the infected neurons die soon after they are infected. Death in immature neurons infected with SV occurs by induction of apoptosis. This can be prevented by cellular expression of bcl-2, an inhibitor of apoptosis, which is expressed by mature neurons in culture [160].

A third example of the neural involvement of bcl-2 is provided by the recent study of Novack and Korsmeyer [161]. These workers have assessed bcl-2's pattern of expression during murine embryogenesis. Immuno-histochemical analysis demonstrated that bcl-2 is widely expressed early in mouse fetal development and that this expression becomes restricted with maturation. For this discussion it is important to note that retinal neuroepithelial cells uniformly express bcl-2 until cells begin to differentiate and then display the topographic distribution maintained into adulthood. The wide distribution of bcl-2 in the developing mouse suggests that many immature cells require a death repressor molecule or that bcl-2 may have roles beyond regulating developmental cell death.

15.2. Bcl-X, a bcl-2-related gene

Boise *et* al. [162] reported the isolation of a bcl-2-related gene that can function as a bcl-2-independent regulator of programmed cell death. Alternative splicing results in two distinct bcl-X mRNAs. The protein product of the larger mRNA, bcl-XL, is similar in size and predicted to have structural similarities to bcl-2. When stably transfected into an IL-3-dependent cell line, bcl-XL inhibits cell death on growth factor withdrawal at least as well as bcl-2. The second mRNA species, bcl-XS, encodes a protein that inhibits the ability of bcl-2 to enhance the survival of growth factor-deprived cells. In vivo, bcl-XS mRNA is expressed at high levels in cells that undergo a high rate of turnover, such as developing lymphocytes. In contrast, bcl-XL is found in tissues containing long-lived, postmitotic cells such as adult brain. Together, these data suggest that bcl-X plays an important role in both positive and negative regulation of programmed cell death, as well as cell memory, respectively.

16. ADRENERGICALLY ACTIVE LYMPHOCYTE SUBSTANCES

Three macromolecular fractions with adrenergic activity can be identified in lymphocyte conditioned medium by diethylaminoethanol (DEAE) ion-exchange high-performance liquid chromatography (HPLC), immunoneutralization, molecular mass, sequence analysis, and biologic characterizations. One of these fractions contains a secretory variant of *β-arrestin* and an *IL-1σ antagonist*, both of which downregulate β_2-adrenergic receptors in A549 human lung epithelial cells. The two other fractions represent protein components that upregulate β_2-adrenergic receptors. One of these contains a mixture of IL-1σ and IL-1β, whereas the adrenergically active components(s) of the remaining fraction is currently being characterized. The first question that may be asked in the context of this chapter is whether the adrenergically highly active IL-1σ obtained from the corresponding macromolecular fraction of lymphocyte conditioned medium has a specific receptor on airway cells. Recent studies designed to explore this question used human bronchial epithelial cells isolated and cultured from the normal bronchi of patients undergoing surgery (for standard clinical reasons) essentially as described by Mattoli et. al. [163]. Using this method, 99% of the final cell population contains epithelial cells. The latter were then incubated with IL-1α radiolabeled by a modified chloramine-T method. In addition to binding of specific, single-class IL-1α receptors, the latter were also identified by internalization of the receptor, affinity cross-linking, and sodium dodecyl sulfate-polyacrylamide gel electrophoresis (SDS-PAGE). Using unlabeled IL-1α and ^{3}H-dihydroalprenolol (DHA) for measuring β-adrenergic mRNA with the guanidium thiocyanate method, the IL-1α-induced accumulation of β_2-adrenoceptor mRNA is demonstrable within 2 hr and an increase in β-adrenoceptor concentration within 4 hr. In other words, concentrated IL-1α derived from human T lymphocytes binds to a specific, single-class surface receptor on human bronchial epithelial cells and induces production of β_2 adrenoceptor mRNA via an associated or separate receptor-linked signaling pathway leading to an increase in epithelial β_2-adrenoceptor concentration [164].

Subsequently, it was shown that lymphocytic IL-1α is a cell- and species-specific factor in increasing β-adrenoceptor concentration and induction of its gene. In this study, we used A549 human lung epithelial cells, A431 human epidermoid cells, DDT_1, MF_2 hamster smooth muscle cells, cultured human bronchial epithelial cells and smooth muscle cells, cultured canine tracheal epithelial and tracheal smooth muscle cells. Concentrated human lymphocytic IL-1α was then cocultured with these various cell populations for 24 hr and β_2-adrenoceptors measured radioactively. The originally shown synergistic β-adrenoceptor upregulation between IL-1α and cortisol [165] was present in the A549 and A431 cells, as well as the human bronchial epithelial cells. Northern blot hybridization showed that levels of β_2-adrenoceptors and β_2-adrenoceptor mRNAs increased significantly by both IL-1α and cortisol, whereas Gs-α, Gi-2α, Gi-3α mRNA levels remained unchanged (see below). In all of these situations, the increase in β_2-adrenoceptor mRNAs always preceded the enhanced expression of the receptor. When DDT_1, MF_2 cells, human and canine tracheal smooth muscle cells were used, IL-1α had no effect on either β_2-adrenoceptors or β_2-adrenoceptor mRNAs whereas cortisol remained active [166]. This extraordinary degree of cell and species specificity of the β_2-adrenoceptor upregulating effect of lymphocytic IL-1α makes these observations highly important both for normal airway physiology as well as for the possible nature of the β-adrenergic dysregulation in asthma by adding an entirely new dimension to the β -adrenergic theory of the atopic abnormality in bronchial asthma as postulated 34 years ago by Szentivanyi [167].

17. THE UNIQUE RECOGNITION AND COMMUNICATION POWERS OF THE IMMUNE AND NEUROENDOCRINE SYSTEMS AS SHARED CHARACTERISTICS

Earlier we cited Schechter, pointing out that a good relationship between two biologic systems must be sensitive, synchronized, and synergistic [168]. There are no two biologic systems where such characterization of an ideal relationship would be more valid than in the case of the immune and neuroendocrine systems as reflected by their unique recognition and communication powers as shared characteristics. Indeed, if intelligence is the capacity to adapt to changing circumstances, then the central nervous system and the immune system manifest this ability beyond all others. As pointed out by Geschwind [169], it would be remarkable if each of these supreme examples of rapid and subtle adaptation did not tap the other's almost limitless potential for variation. While at one time the concept existed that stress might alter immune reactivity through some relatively diffuse activation of the endocrine system of the hypothalamus [170–174], we now find that there are specific interactions at every level, as discussed below. The latter are based on four critical features shared by both systems: (1) they are composed of extraordinarily large numbers of phenotypically distinct cells organized into intricate networks. Moreover, the size of this extensive cellular arsenal continuously increases as new sequence information becomes available and enormous numbers of new members of the IgSF surface each year [10]. Within these cell networks, the individual cells can interact either positively or negatively and the response of one cell reverberates through the system by affecting many other cells. (2) Cells of both systems synthesize, secrete, and/or release the same effector molecules. (3) Recognition of these effector molecules is realized by the same cellular receptors and second messenger mechanisms of both cell systems. (4) These cellular and molecular determinants make a continuous, bilateral flow of information, the *sine qua non* of the unique interactions within the immune-neuroendocrine circuitry, possible.

A more amplified view on the basic biochemistry and molecular biology of receptor-effector coupling by G-proteins (i.e., the fundamental mechanism used by hormones, neurotransmitters, and the immunomodulatory cytokines for signal transmission by G-proteins) is presented by Szentivanyi [175,176,177]. Here we shall only mention that about 80% of all known neurohormones, neurotransmitters, immunomodulatory lymphokines, and other autocrine and paracrine factors that regulate cellular interactions in the immune-neuroendocrine circuitry, called "primary" messengers, elicit cellular responses by combining with specific receptors that are coupled to effector functions by G-proteins. Although the primary messengers are many, the number of physicochemically and biologically distinct receptors that mediate their action is even larger. So far about 80 distinct receptors that recognize 40 hormones, neurotransmitters, and so on, can be identified. It is reasonable to assume that the total number of distinct receptors coupled by G-proteins will be 100 to 150. In contrast to receptors, the number of final effector functions regulated by these receptors and the number of G-proteins that provide for receptor-effector coupling are much lower, probably not much more than 15 each.

In mammals, a total of 8 G-proteins have been purified essentially free from each other (Gt. Gs, Gil, Gi2, Gi3, GOI, G02, and Gz/x) and the cDNAs derived from a total of 9 genes encoding G-alpha subunits have been cloned and designated $alpha_s$ $alpha_{il}$, $alpha_{i2}$ $alpha_{i3}$ $alpha_o$, $alpha_{tr}$ $alpha_{tc}$, $alpha_{olf}$, and $alpha_{z/x}$, giving rise to 12 mRNAs because of the fourfold variation in the splicing of the alpha, precursor mRNA. In addition, there is evidence for the existence of at least 7 additional G-alpha genes. Homology cloning has also revealed that there are at least 4 G-beta genes and 3 G-gamma genes.

The G-proteins are heterotrimeric membrane proteins (alpha, beta, gamma; 1:1:1), distinguished by unique alpha subunits, but sharing common beta subunits. Stated in a different way, G-proteins may be viewed as being composed of a unique, but homologous alpha subunit in reversible association with a complex comprised of a beta and a gamma subunit commonly shared by several different G-protein alpha subunits. Thus, the alpha subunit of Gs (the G-protein that stimulates adenylate cyclase) may share beta-gamma complex in common with the alpha subunits of the family of G-proteins that mediate inhibition of adenylate cyclase (GI), or other G-proteins like GO, Gz, and Gt. The alpha subunits bind and hydrolyze GTP, and are often the substrates for NAD+-dependent ADP ribosylation by bacterial toxins (i.e., pertussis, cholera). Activated P-adrenoceptors catalyze the exchange of GTP for bound guanosine diphosphate (GDP) by the alpha subunit of the holoprotein, promoting the dissociation of the GTP-bound alpha subunit from the beta-gamma complex. It is the "free" GTP ligand alpha subunit of a G-protein that regulates the activity of the membrane-bound effector units such as adenylate cyclase.

The primary sequences of several G-protein-linked effectors, including adenylate cyclase and phospholipase C, have been determined and molecular cloning of phospholipase A2 and Ca^{2+} and K^+ channels are in an advanced stage in several laboratories. As mentioned earlier, G-$alpha_s$ mediates the stimulation of adenylate cyclase and has been shown to regulate Ca2+ channel activity. The G-proteins that mediate the inhibition of adenylate cyclase, termed Gi, constitute a family with at least three members, G-$alpha_{i1}$, G-$alpha_{i2}$ and G-$alpha_{i3}$ each the product of a separate gene. Of these, it is G-$alpha_{i2}$ that mediates the inhibition of adenylate cyclase [175].

At the time of writing, receptors for simple substances, such as the amine mediators and short-chain peptides as well as lipids, and for more than 20 different hypothalamopituitary peptides have been identified in the cells of the immune system, essentially in lymphocytes. In addition to the hypothalamopituitary hormones, lymphocytes also express receptors for peptides secreted from neurons together with other neurotransmitters. These neuropeptides take on added significance as immunomodulators, since it is now known that lymphoid organs are directly innervated with nerves secreting these agents. From the standpoint of the integration of information in the immune-neuroendocrine circuitry, future studies will have to examine these parallel signaling pathways in isolation. In other words, it will be necessary to determine how an individual cell completely processes and integrates information from these individual pathways. This is all the more remarkable because the cell is faced with the task of balancing the need to communicate with other cells with the need for growth and maintenance of the differentiated state while preserving adequate flexibility to support regulation, sensitivity, and gain. One early result of such inquiries is the demonstration of cross-regulation (crosstalk) between the various G-protein-mediated signaling pathways. Thus, it was shown that in the cross regulation between α_1- and

β_2-adrenergic receptor- mediated pathways, activation of β_2-adrenergic receptors increased α_1-adrenergic receptor mRNA levels [178]. Conversely, activation of the $G_{1\alpha}$-mediated inhibitory pathway of adenylate cyclase cross-regulates the stimulatory (Gs_α-mediated) β-adrenergic-sensitive adenylate cyclase system by (1) upregulating β_2-adrenergic receptors and enhancing the activation of the stimulatory (Gs_α-mediated) adenylate cyclase pathway and (2) downregulating elements of the inhibitory adenylate cyclase pathway, $G_{1\alpha2}$ and A_1 adenosine receptor binding, respectively [179]. It may be added that cross-regulation is also observed between signaling pathways that do not share the same effectors. Although much more work remains to be done to unravel the complexities of the coordinated regulation of information processing and integration by the cell, it is already possible to state that there is cross-regulation between neurally derived substances and lymphokines.

The foregoing leads us to a general consideration of the "endocrine" hypothalamus and its regulation by cytokines.

REFERENCES

1. Szentivanyi A, Krzanowski JJ, Polson JB, Abarca. CM. The pharmacology of microbial modulation in the induction and expression of immune reactivities. I. The pharmacologically active effector molecules of immunologic inflammation, immunity, and hypersensitivity. Immunopharmacol Rev 1990; 1:159.
2. Szentivanyi A. The immune-neuroendocrine circuitry and its relation to asthma, in: Bronchial Asthma-Mechanisms and Therapeutics (E. B. Weiss and M. Stein, eds.), Little, Brown, Boston, 1993b; pp. 421–438.
3. Goodman CS, Pearson KG. Neuronal development: Cellular approaches in invertebrates. Neurosci Res Prog Bull 1982; 20:777.
4. He X, Rosenfeld MG. Mechanisms of complex transcriptional regulation: Implication for brain development. Neuron 1991; 7:183.
5. Willkinson DC, Krumlauf R. Molecular approaches to the segmentation of the hindbrain. Trends Neurosci 1990;13:335.
6. LeDouarin N. The Neural Crest. London: Cambridge University Press; 1982.
7. Cunningham BA, Hemperley JJ, Murray BA, Prediger EA, Brackenbury R, Edelman GM. Neural cell adhesion molecule: Structure, immunoglobulin-like domains, cell surface modulation and alternative RNA splicing. Science, 1987; 236:799.
8. Milner RJ, Lai C, Suteliffe JG, Bloom FE. Expression of immunoglobulin-like proteins in the nervous system: Properties of the neural protein 1B236/MAG. In: Neuroimmune Networks: Phvsiology and Diseases. Goetzl EJ, Spector NH, editors. Liss New York 1989; pp. 9–15.
9. Williams AF, Barclay A. The immunoglobulin superfamily-Domains for cell surface recognition. Annu Rev Immmunol 1988; 6:381.
10. Brummendorf T, Rathjen FG. Cell adhesion molecules I: Immunoglobulin superfamily. Protein Profile 1994;1:9.
11. Edelman GM: Neural Darwinism. Basic Books. New York, 1987.
12. Szentivanyi A, Fishel CW. Die Amin-Mediatorstoffe der allergisehen Reaktion und die reaktionsfahiegheit ihrer Erfolgeszellen. in: *Pathogenese und Therapie allergischer Reaktionen* (C. Filipp and A. Szentivanyi. eds.). Ferdinand Enke Verlag. Stuttgart. 1966; pp. 588–683.

13. Szentivanyi A, Polson JB, Krzanowski JJ. The altered reactivity of the effector cells to antigenic and pharmacological influences and its relation to cyclic nucleotides. I. Effector reactivities in the efferent loop of the immune response. In: Pathomechanismus und Pathogenese Allergischer Reaktionen. Filipp C, Banachewski E, editors. Munich: Werk-Verlag, 1980; pp. 460–510.
14. Szentivanyi A, Fitzpatrick DF. The altered reactivity of the effector cells to antigenic and pharmacological influences and its relation to cyclic nucleotides. II. Effector reactivities in the efferent loop of the immune response In: Pathomechanismus und Pathogenese Allergischer Reaktionen. Filipp C, Banachewski E, editors. Munich: Werk-Verlag, 1980; pp. 511–580.
15. Kato S, Negishi K, Hayashi Y, Miki N. Enhancement of neurite outgrowth and aspartate-glutamate uptake system in retinal explants cultured with chick gizzard extract J Neurochem 1983; 40:929.
16. Taniura H, Hayashi Y, Miki N. Purification and characterization of an 82-KD membrane protein as a neurite outgrowth factor binding protein: Possible involvement of NOF binding protein in axonal outgrowth in developing retina. J Cell Biol 1991; 112:313
17. Taira E, Takaha N, Miki N: Extracellular matrix proteins with neurite promoting activity and their receptors. Neurosci Res 1993; 17:1.
18. Evercooren AB, Kleinman HK, Ohno S, Marangos P, Schwartz JP, DuBois-Dalq, ME. Nerve growth factor, laminin and fibronectin promote neurite growth in human fetal sensory ganglia cultures. J Neurosci 1982; 8:179.
19. Graf J, Iwamoto Y, Sasaki M, Martin GR, Kleinman HK, Robey FA, Yamada Y. Identification of an amino acid sequence in laminin mediating cell attachment, chemotaxis, and receptor binding. Cell 1987; 48:989.
20. Panayotou G, End P, Aumailly M, Timpl R, Engel J. Domains of laminin with growth factor activity. Cell 1989; 56:93.
21. Tamkun JW, Schwazbauer JE, Hynes, RO. Proc Natl Acad Sci. 1984; (16):5140–4.
22. Salzer JL, Colman DR, Mechanisms of cell adhesion in the nervous system. Role of the immunoglobulin gene superfamily. Dev Neurosci 1989; 11:377.
23. Edelman CM. Morphoregulatory molecules. Biochemistry 1988; 27:3534.
24. Tanaka H, Matsui T, Agata A, Tomura M, Kubot 1, McFarland KC, Kohn B, Lee A, Phillips H, Shelton DL. Molecular cloning and expression of a novel adhesion molecule. Neuron 1991 7:535.
25. Rathjen FC, Schachner M. Immunocytological and biochemical characterization of a new neuronal cell surface component (LI antigen) which is involved in cell adhesions. EMBO J 1984; 3:1.
26. Arquint M, Roder I, Chia LS, Down J, Wilkinson O, Bayley H, Braun P, Dunn R. Molecular cloning and primary structure of myelin-associated glycoprotein. Proc Natl Acad Sci USA 1987 84:600.
27. Sakamoto Y, Kitamura K, Yoshimura K, Nishijima T, Uyemura K. Complete amino acid sequence of PO protein in bovine peripheral nerve myelin. J Biol. Chem. 1987; 262: 4208.
28. Lam A, Fuller F, Miller J, Kloss J, Manthorpe M, Varon S, Cordell B. Sequence and structural organization of the human gene encoding ciliary neurotrophic factor. Gene 1991; 102:271.
29. Lin LFH, Mismer D, Lile JD, Armes LC, Butler ET III, Vannice JL Collins F. Purification, cloning, and expression of ciliary neurotrophic factor (CNTF). Science 1989; 246:1023.

30. Sporeno E, Paonessa G, Salvati AL, Graziani R, Delmastro P, Giliberto C, Toniatti C. Oncostatin M binds directly to gp 130 and behaves as interleukin-6 antagonist on a cell line expressing gp130 but lacking functional oncostatin M receptors. J Biol Chem 1994; 269: 10991.
31. Stockli KA, Lillien LF, Naher-Noe M, Breitfeld C, Hughes RA, Raff MC, Thoenen H, Sendtner H. Regional distribution, developmental changes, and cellular localization of ciliary neurotrophic factor mRNA and protein in the rat brain. J Cell Biol 1991; 115:447.
32. Rende M, Muir D, Ruoslahti E, Hagg T, Varon S, Manthorpe M. Immunolocalization of ciliary neurotrophic factor in adult rat sciatic nerve. GLIA 1992; 5:25.
33. Rudge JS, Morrissey D, Lindsay RM, Pasnikowski E. Regulation of ciliary neutrophic factor in cultured rat hippocampal astrocytes. Eur J Neurosci 1994; 6:218.
34. Barbin G, Manthorpe M, Varon S. Purification of the chick eye ciliary neuronotrophic factor. J Neurochem 1984; 43:1468.
35. Manthorpe M, Skaper SD, Williams LR, Varon S. Purification of adult rat sciatic nerve ciliary neurotrophic factor. Brain Res 1986; 367:282.
36. Martinou JC, Martinou I, Kato AC. Cholinergic differentiation factor (CDF/LIF) promotes survival of isolated rat embryonic motoneurons in vitro. Neuron 1992; 8:737.
37. Magal E, Burnham P, Varon S. Effects of ciliary neurotrophic factor on rat spinal cord neurons in vitro: Survival and expression of choline acetyltransferase and low-affinity nerve growth factor. Dev Brain Res 1991b; 63:141.
38. Forger NG, Roberts SL, Wang V, Breedlove, SM. Ciliary neurorrophic factor maintains motoneurons and their target muscles in developing rats. J Neurosci 1993; 13:4720.
39. Kotzbauer PT, Lampe PA, Estus S, Milbrandt J, Johnson EM Jr. Postnatal development of survival responsiveness in rat sympathetic neurons to leukemia inhibitory factor and ciliary neurotrophic factor. Neuron 1994; 12:763.
40. Sendtner M, Sebmalbruch H, Stockli KA, Carroll P, Kreutzberg CW, Thoenen H. Ciliary neurotrophic factor prevents degeneration of motor neurons in mouse mutant progressive motor neuronopathy. Nature 1992; 358:502.
41. Lams BE, Isacson 0. Loss of transmitter-associated enzyme staining following axotomy does not indicate death of brainstem cholinergic neurons. Brain Res 1988; 475:401.
42. Hagg T, Quon D, Higaki J, Varon S. Ciliary neurotrophic factor prevents neuronal degeneration and promotes low affinity NGF receptor expression in the adult rat CNS. Neuron 1992; 8:145.
43. Landis S. Target regulation of neurotransmitter phenotype. Trends Neurosci 1990; 13: 344.
44. Patterson PH. The emerging neuropoietic cytokine family: First CDF/LIF, CNTF and IL-6: next ONC, MCF. CCSF? Curr Opin Neurobiol 1992; 2:94.
45. Saadat S, Sendtner M, Rohrer H. Ciliary neurotrophic factor induces cholinergic differentiation of rat sympathetic neurons in culture. J Cell Biol 1989; 108:1807.
46. Rao MS, Tyrrell S, Landis SC, Patterson PH. Effects of ciliary neurotrophic factor (CNTF) and depolarization on neuropeptide expression in cultured sympathetic neurons. Dev Biol 1992; 150:281.
47. Louis JC, Magal E, Burnham P, Varon S. Cooperative effectives of ciliary neurotrophic factor and norepinephrine on tyrosine hydroxylase expression in cultured rat locus coeruleus neurons. Dev Biol 1993; 155:1.
48. Tokiwa MA, Gaspar EM, Doering C. CNTF is superior to NCF as a survival enhancement factor for adrenal medulla cells in vitro. Neuroreport 1994; 5:549.

49. Mayer M, Bhakoo K, Noble M. Ciliary neurotrophic factor and leukemia inhibitorv factor promote the generation, maturation and survival of oligodendrocytes in vitro. Dev Biol 1994; 120:143.
50. Hughes SM, Lillien LE, Roff MC, Rohrer H, Sendtner M. Ciliary neurotrophic factor induces type 2 astrocyte differentiation in culture. Nature 1988; 335:70.
51. Shapiro L, Zhang XX, Rupp RC, Wolff SM, Dinarello CA. Ciliary neurotrophic factor is an endogenous pyrogen. Proc Natl Acad Sci USA 1993; 90:8614.
52. Anegon I, Grolleau D. Regulation of HILDA/LIF gene expression in activated human monocytic cells. J Immunol 1991; 147:3973.
53. Gillett NA, Lowe D, Lu L, Chan C, Ferrara N. Leukemia inhibitory factor expression in human carotid plaques: Possible mechanism for inhibition of large vessel endothelial regrowth. Growth Factors 1993; 9:301.
54. Lorenzo JA, Jastrzebski SL, Kalinowski JF, Downie E, Korn JH. Tumor necrosis factor-alpha stimulates production of leukemia inhibitorv factor in human dermal fibroblast cultures. Clin Immunol Immunopathol 1994; 70:260.
55. Marshall, J.S., Gauldie, J., Nielsen, L., and Bienenstock, J, Leukemia inhibitorv factor production by rat mast cells, Eur I. Immunol., 1993; 23:2116.
56. Aloisi F, Rosa S, Testa U, Bonsi P, Russo G, Peschle C, Levi G. Regulation of leukemia inhibitory factor synthesis in cultured human astrocytes, J Immunol 1994; 152:5022.
57. Metcalf D, Hilton D, Nicola JA. Leukemia inhibitory factor can potentiate murine megakaryocyte production in vitro. Blood 1991; 77:2150.
58. Bhatt H, Brunet LJ, Stewart CL. Uterine expression of leukemia inhibitory factor coincides with the onset of blastocyst implantation. Proc Nati Acad Sci USA 1991; 88:11408.
59. Kurzrock R, Estrov Z, Wetzler M, Gutterman JU, Talpaz M. LIF: Not just a leukemia inhibitory factor. Endocrine Rev 1991; 12:208.
60. Gough NM. Molecular genetics of leukemia inhibitory factor (LIF) and its receptor. Growth Factors. 1992; 7:175.
61. Van Vlasselaer P. Leukemia inhibitory factor: A growth factor with pleiotropic effects on bone biology. Prog Growth Factor Res 1992; 4:337.
62. Aggarwal, BB, Gutterman JU. Leukemia Inhibitory Factor in Human Cytokine. Oxford: Blackwell; 1992.
63. Leary AG, Wong GG, Clark SC, Smith AG, Ogawa M. Leukemia inhibitory factor differentiation-inhibiting activity/human interleukin for DA cells augments proliferation of human hematopoietic stem cells. Blood 1990; 75:1960.
64. Bamber BA, Masters BA, Hoyle GW, Brinster RL, Palmiter KD. Leukemia inhibitory factor induces neurotransmitter switching in transgenic mice. Proc Nati Acad Sci USA 1994; 91:7839.
65. Ludlam WH, Zang Z, McCarson KE, Krause JE, Spray DC, Kessler JA. mRNAs encoding muscarinic and substance P receptors in cultured sympathetic neurons are differentially regulated by LIF and CNTF. Dev Biol 1994; 164:528.
66. Qui L, Bernd P, Fukada K. Cholinergic neuronal differentiation factor (CDF)/leukemia inhibitory factor (LIF) binds to specific regions of the developing nervous system in vivo. Dev Biol 1994; 163:518.
67. Murphy M, Reid K, Brown MA, Bartlett PF. Involvement of leukemia inhibitory factor and nerve growth factor in the development of dorsal root ganglion neurons. Development 1993; 117:1173.
68. Cheema SS, Richards L, Murphy M, Bartlett PF. Leukemia inhibitory factor prevents

the death of axotomized sensory neurons in the dorsal root ganglia of the neonatal rat. J Neurosci Res 1994a; 37:213.
69. Cheema SS, Richards L, Murphy M, Bartlett, PF. Leukemia inhibitory factor rescues motoneurons from axotomy-induced cell death. Neuroreport 1994b; 5:989.
70. Zarling JM, Shoyab M, Marquardt H, Hanson MB, Lioubin, MN, Todaro GJ. Oncostatin M: A growth regulator produced by differentiated histiocytic lymophoma cells. Proc Natl Acad Sci 1986; 83:9739.
71. Rose TM, Bruce AG. Oncostatin M is a member of a cytokine family that includes leukemia inhibitory factor, granulocyte colony-stimulating factor and interleukin-6. Proc Natl Acad Sci 1991; 88:8641.
72. Bazan JF. Neuropoietic cytokines in the hematopoietic fold. Neuron 1991; 7:197
73. Nishimoto N, Ogata A, Shima Y, Tani Y, Ogawa H, Nakagawa M, Sugiyama H, Yoshizaki K, Kishimoto T. Oncostatin M. leukemia inhibitory factor, and interleukin-6 induce proliferation of human plasmacytoma cells via the common signal transducer gpl3O. J Exp Med 1994; 179:1343.
74. Zhang XG, Lu JJ, Lu Z-Y, Kasukawa K, Yancopoulos GD, Yurner K, Shoyab M, Taga T, Kishomoto, T, Bataille R, Klein B. Ciliary neurotropic factor, interleukin-11, Leukemia inhibitory factor and oncostatin M are growth factors for Human myeloma cell lines using the interleukin-6 signal transducer gp130. J Exp Med 1994; 179:1337.
75. Hirano T, Matsuda T, Nakajima K. Signal transduction through gpl3O that is shared among the receptors for the interleukin-6 related cytokine subfamily. Stem Cells 1994; 12:262.
76. Gearing DP. The leukemia inhibitory factor and its receptor. Adv Immunol 1993 53:31.
77. Bruce AG, Hoggatt IH, Rose TM Oncostatin M is a differentiation factor for myeloid leukemia cells. J Immunol 1992a; 149:1271.
78. Bruce AG, Linsley PS, Rose TM. Oncostatin M Prog Growth Factor Res 1992b; 4:157.
79. Cai J, Gill PS, Masood P, Chandrasoma P, Jung B, Law RE, Radka SF. Oncostatin M, is an autocrine growth factor in Kaposi's sarcoma. Am J Pathol 1994; 145:74.
80. Malik N, Kallestad JC, Gunderson NL, Austin SD, Neubauer MG, Ochs V, Marquardt H, Zarling JM, Shoyab M, Wei CM. Molecular cloning, sequence analysis and functional expression of a novel growth regulator, Oncostatin M. Mol Cell Biol 1989; 9:2847.
81. Jeffery E, Price V, Gearing DP. Close proximity of the genes for leukemia inhibitory factor and oncostatin M. Cytokine 1993; 5:107.
82. Giovannini M, Selleri L, Hermanson GG, Evans GA. Localization of the human oncostatin M gene (OSM) to chromosome 22q12, distal to the Ewing's sarcoma breakpoint. Cytogenet Cell Genet 1993; 62:32.
83. Kallestad JC, Shoyab M, Linsley PS. Disulfide bond arrangement and identification of regions required for functional activity of oncostatin M. J Biol Chem 1991; 266:8940.
84. Sporeno E, Paonessa C, Salvati AL, Graziani R, Delmastro P, Giliberto C, Toniatti C. Oncostatin M binds directly to gp 130 and behaves as interleukin-6 antagonist on a cell line expressing gp130 but lacking functional oncostatin M. receptors. J Biol Chem 1994; 269: 10991.
85. Levi-Montalcini R, Hamburger V. Selective growth stimulating effects of mouse sarcoma on the sensory and sympathetic nervous system of the chick embryo. J Exp Zool 1951; 116:321.
86. Ullrich A, Gray A, Berman C, Dull TJ. Human beta-nerve growth factor gene sequence highly homologous to that of mouse. Nature l983; 303:821.
87. Iwane M, Kitamura Y, Kaisho Y, Yosbimura K, Shintani A, Sasada R, Nakagawa S,

Kawabara K, Nakahama K, Kakimura A. Production, purification and characterization of biologically active recombinant human nerve growth factor. Biochem Biophys Res Commun 1990; 171:116.

88. Berkemeier LR, Winslow JW, Kaplan DR, Nikolics K, Goeddel OV, Rosenthal A. Neurotrophin-5: A novel neurotrophic factor that activates trkA and trkB. Neuron 1991; 7: 857.

89. Jones KR, Reichardt LF. Molecular cloning of a human gene that is a member of the nerve growth factor family. Proc Natl Acad Sci USA 1990; 87:8060.

90. Rosenthal A, Goeddel DVM, Nguyen T, Lewis M, Shih A, Laramee CF, Nikolics K, Winslow JW. Primary structure and biological activity of a novel human neurotrophic factor. Neuron 1990; 4:767.

91. Ip NY, lbanez CF, Nye SH, McClain J, Jones PF, Gies DR, Belluscio L, LeBeau MM, Espinosa R III, Squinto SP. Mammalian neurotrophin-4: Structure. chromosomal localization, tissue distribution and receptor specificity. Proc Natl Acad Sci USA 1992a; 89: 3060.

92. Bradshaw RA, Blundell TL, Lapatto R, McDonald NQ, Murray-Rust J. Nerve growth factor revisited. Trends Biochem Sci. 1993; 18:48.

93. Missale C, Boroni F, Sigala S, Zanellato A, DalToso R, Balsari A, Spano P. Nerve growth factor directs differentiation of the bipotential cell line CH-3 into the mammotroph phenotype. Endocrinology 1994; 135:290.

94. Meakin SO, Shooter EM. The nerve growth factor family of receptors. Trends Neurosci 1992; 15:323.

95. Carman-Krzan M, Vige X, Wise BC. Regulation by interleukin-l of nerve growth factor secretion and nerve growth factor mRNA expression in rat primary astroglial cultures. J Neurochem 1991; 56:636.

96. Awatsuji H, Furukawa Y, Hirota M, Murakami Y, Nii S, Furukawa S, Hayashi K. Interleukin-4 and -5 as modulators of nerve growth factor synthesis/secretion in astrocytes. Neurosci Res 1993; 34:539.

97. Tischler AS, Riseberg JC, Hardenbrook MA, Cherington V. Nerve growth factor is a potent inducer of proliferation and neuronal differentiation for adult rat chromaffin cells in vitro. J Neurosci 1993;13:1533.

98. Morimoto M, Morita N, Kawata M. The effects of NGF and glucocorticoid on the cytological features of rat chromaffin cells in vitro. Neuroreport 1994;5:954.

99. Muller Y, Duperray F, Caruso F, Clos J. Autocrine regulation of proliferation of cerebellar granule neurons by nerve growth factor. J Neurosci Res 1994; 38:41.

100. Chevalier S, Praloran V, Smith C, MacGrogan D, Ip NY, Yancopoulos GD, Brochet P, Pouplard A, Cascan H. Expression and functionality of the trkA proto-oncogene product NCF receptor in undifferentiated hematopoietic cells. Blood 1994; 83:1479.

101. DeSimone R, Alleva E, Tirassa P, Aloe L. Nerve growth factor released into the bloodstream following intraspecies fighting induces mast cell degranulation in adult male mice. Brain Behav Immun 1990; 4:74.

102. Ehrhard PB, Erb P, Graumann U, Otten U. Expression of nerve growth factor and nerve growth factor receptor tyrosine kinase Trk in activated CD4-positive T-cell clones. Proc Natl Acad Sci USA 1993a; 90:10984.

103. Kimata H, Yoshida A, Ishioka C, Kusunoki T, Hosoi S, Mikawa H. Nerve growth factor specifically induces human IgG4 production. Eur J Immunol 1991; 21:137.

104. Otten U, Ehrhard P, Peck R. Nerve growth factor induces growth and differentiation of

human B lymphocytes. Proc Natl Acad Sci USA 1989; 86:10059.
105. Tsuda T, Wong D, Dolovich J, Bienenstock J, Marshall J, Denburg JA. Synergistic effects of nerve growth factor and granulocyte macrophage colony stimulating factor on human basophilic cell differentiation. Blood; 1991; 77:971.
106. Bischoff SC, Dahinden CA. Effect of nerve growth factor on the release of inflammatory mediators by mature human basophils. Blood;1992 79:2662.
107. Bracci-Laudeiro L, Aloe L, Levi-Montalcini R, Caleazzi M, Schilter D, Scully JL, Otten U. Increased levels of NGF in sera of systemic lupus erythematosus patients. Neuroreport 1993; 4:563.
108. Barinaga M. Neurotrophic factors enter the clinic. Science 1994; 264:772
109. Markowska AL, Koliatsos VE, Breckler SJ, Price DL, Olton DS. Human nerve growth factor improves spatial memory in aged but not young rats. J Neurosci 1994; 14:48 15.
110. Bazan JF. Structural design and molecular evolution of a cytokine receptor superfamily. Proc Natl Acad Sci USA 1990;87:6934.
111. Yamamori T, Sarai A. Coevolution of cytokine receptor families in the immune and nervous systems. Neurosci Res 1992; 16:235–236.
112. Miyajima A, Kitamura T, Harada N, Yokota N, Arai K. Cvtokine receptors and signal transduction. Annu Rev Immunol 1992; 10:295.
113. Yamamori T, Fukada K, Aebersold R, Korsehing S, Fann SM, Patterson PH. The cholinergic neuronal differentiation factor from heart cells is identical to leukemia inhibitory factor. Science 1989; 246:1412.
114. Hama T, Kushima Y, Miyamoto M, Kubota M, Takei N, Hatanaka H. Interleukin-6 improves the survival of mesencephalic catecholaminergic and septal cholinergic neurons from postnatal, two-week-old rats in cultures. Neuroscience 1991; 40:445.
115. Taga T, Higi M, Hirata Y, Yamasaki K, Yasukawa K, Matsuda T, Hirano T, Kishimoto T. Interleukin-6 triggers the association of its receptor with a possible signal transducer gpl3O. Cell 1989; 58:573.
116. Murphy M, Reid K, Hilton DJ, Bartlett PF. Generation of sensory neurons is stimulated by leukemia inhibitory factor. Proc Natl Acad Sci USA 1991; 88:3498.
117. Gearing DP, Cosman D. Homology of the p40 subunit of natural killer cell stimulatory factor with the extracellular domain of the interleukin-6 receptor. Cell, 1991 66:8.
118. Ip NY, Nye SH, Boulton TG, Davis S, Taga T, Li Y, Birren SJ, Yasukawa K, Kishimoto T, Anderson DJ, Stahl N, Yancopoulos GD. CNTF and LIF act on neuronal cells via shared signal pathway that involve the IL-6 signal transducing receptor component gpl3O. Cell, 1992b; 69:1121.
119. Davis S, Aldrich TH, Valenzuela DM, Wong V, Furth ME, Squinto SP, Yancopoulos GD. The receptor for ciliary neurotrophic factor. Science 1991; 253:59.
120. Hall AK, Rao MS. Cytokines and neurokines: Related ligands and related receptors. Trends Neurosci 1992 ;15:35.
121. Jessell TM, Melton DA. Diffusible factors in vertebrate embryonic induction. Cell 1992; 68:257.
122. Patterson PH. The emerging neuropoietic cytokine family: First CDF/LIF, CNTF and IL-6: next ONC, MCF. CCSF? Curr Opin Neurobiol 1992; 2:94.
123. Edelman GM. Topobiology. Sci Am 1989; 260:44.
124. Geschwind N, Galaburda AM. Cerebral lateralization: Biological mechanisms associations and pathology. Parts I-III Arch Neurol 1985; 42:428.
125. Szentivanyi A. Beta-adrenergic subsensitivity in asthma and atopic dermatitis. A status

report. Acta Biomed Hung Am 1991; 1:1.

126. Renoux G, Biziere K, Renoux M, Bardos P, Degenne D. Consequences of bilateral brain neocortical ablation on imuthiol-induced immunostimulation in mice. Ann N. Y. Acad Sci 1987; 496:346.
127. Malinow R, Tsien RW. Presynaptic enhancement shown by whole-cell recordings of long-term potentiation in hippocampal slices. Nature 1990; 346:177.
128. Bekkers JM, Stevens CF. Presynaptic mechanism for long-term potentiation in the hippocampus. Nature 1990; 346:724.
129. McFadden PN, Koshland DE Jr. Habituation in the single cell: Diminished secretion of norepinephrine with repetitive depolarization in PCI2 cells. Proc Natl Acad Sci USA 1990a; 87:203.
130. McFadden PN, Koshland, DE Jr. Parallel pathways for habituation in repetitively stimulated PCI2 cells. Neuron 1990b; 4:615.
131. Morimoto A, Murakami N, Nakamori T, Sakata Y, Watanabe T. Possible involvement of prostaglandin E in development of ACTH response in rats induced by human recombinant interleukin- I. J Physiol 1989; 411:245—256.
132. Morimoto BH, Koshland DE Jr. Excitatory amino acid uptake and N-methyl-n-aspartate mediated secretion in a neural cell line. Proc Natl Acad Sci USA 1990a, 87:3518.
133. Dudai, Y. Neurogenetic dissection of learning and short-term memory in Drosophila. Annu Rev Neurosci 1988; 11:537.
134. Szentivanyi A, Maurer P, Janicki BW, editors. Antibodies: Structure, Synthesis. Function, and Immunologic Intervention in Disease. New York: Plenum Press; 1987a.
135. Szentivanyi A, Filipp G, Anaphylaxis and the nervous system. Part II, Ann Allergy; 1958 16:143.
136. Szentivanyi A, Berczi I, Nyanteh H, Goldman A. Virus Associated Immune And Pharmacologic Mechanisms In Disorders of Respiratoryan Cutaneous Atopy, Neuroommune Biology, Volume 3, Berczi I., Szentivanyi A. Series ED. Elsevier, 2002 (Volume in press).
137. Hadden JW, Szentivanyi A, editors. The Pharmacology of the Reticuloendothelial System. New York: Plenum Press; 1985.
138. Friedman H, Klein TW, Szentivanyi A editors. Immunomodulation by Bacteria and Their Products. New York, Plenum Press, 1981.
139. Szentivanyi A, Middleton E, Williams IF, Friedman H. Effect of microbial agents on the immune network and associated pharmacologic reactivities. In: Allergy: Principles and Practice. Middleton E, Reed CE, Ellis EF editors. St. Louis: Mosby; pp. 211–236. 1983.
140. Klein TW, Specter S, Friedman H, Szentivanyi A, editors. Biological Response Modifiers in Human Oncology and Immunology. New York: Plenum Press, 1983.
141. Krueger JM, Obal F Jr, Johannsen L, Cady AB, Toth L. Endogenous slow-wave sleep substances: A review In: Current Trends in Slow-Wave Sleep Research. Dugsovic C, Wauquier A, editors. New York: Raven Press, 1988; pp. 97–112.
142. Krueger JM, Obal F Jr, Opp M, Johannsen L, Cady AB, Toth L. Immune response modifiers and sleep. In: Interactions Among Central Nervous System, Neuroendocrine and Immune Systems. Hadden JW, Masek K, Nistico G, editors. Rome: Pythagora Press, 1989; pp. 323—350.
143. Fillion MP, Prudhomme N, Haour F, Fillion G, Bonnet M, Lespinats C, Masek K, Flegel M, Corvaia M, Launay JM. Hypothetical role of the serotonergic system in neuroimmunomodulation: Preliminary molecular studies. In: Interactions Among Central

Nervous System. Neuroendocrine and Immune Systems. Hadden JW, Masek K, Nistico G editors. Rome: Pythagora Press, 1989; pp. 235–250.
144. Dougherty PM. Dafny N. Central opioid systems are differentially affected by products of the immune response. Soc Neurosci Abstr 1987; 13:1437.
145. Eremina 0F, Devoino LV. Production of humoral antibodies in rabbits following destruction of the nucleus of the midbrain raphe. Bull Esp Biol Med 1973; 74:258.
146. Masek K, Horak P, Kadlec 0, Flegel M. The interactions between neuroendocrine and immune systems at the receptor level, The possible role of serotonergic system. In: Interactions Among Central Nervous System. Neuroendocrine and Immune Systems. Hadden JW, Masek K, Nistico G, editors. Rome: Pythagora Press, 1989; pp. 225–234.
147. Vekshina N, Magaeva SV. Changes in the serotonin concentration in the limbic structures of the brain during immunization. Bull Exp Biol Med 1974; 77:625.
148. Korsmeyer SJ. Bcl-2: An antidote to programmed cell death. Cancer Surv 1992a; 15:105.
149. Hockenbery DM. The bcl-2 oncogene and apoptosis. Semin Immunol 1992; 4:413 150. Graninger WB. Transcriptional overexpression of the proto-oncogene bcl-2 in patients with lupus erythematosus. Wien Klin Wochenschr 1992; 104:205.
151. Schena M, Gottardi D, Ghia P, Larsson LG, Carlsson M, Nilsson K. The role of bcl-2 in the pathogenesisof B chronic lymphocyte leukemia. Leuk Lvmphoma 1993; 11:173.
152. Schena M, Larsson LC, Cottardi D, Caidano C, Carlsson M, Nilsson K, Caligaris-Cappio F. Growth and differentiation-associated expression of bcl-2 in B-chronic lymphocytic leukemia cells. Blood 1992; 79:2981.
153. Villuendas R, Piris MA, Orradre JL, Mollejo M, Rodriguez R, Morente M. Different bcl-2 protein expression in high-grade B-cell lymphomas derived from lymph node or mucosaassociated lymphoid tissue. Am **J** Pathol 1991;139:989.
154. Nunez G, Hockenbery D, McDonnel TJ, Sorensen CM, Korsmeyer SJ. Bcl-2 maintains B-cell memory. Nature 1991; 353:71.
155. Akbar AN, Salmon M, Savill J, Janossy G. A possible role for bcl-2 in regulating T-cell memory-A "balancing act" between cell death and survival. Immunol Today 1993b; 14: 526.
156. Akbar AN, Borthwick N, Salmon M, Gombert W, Bofill M, Shamsadeen N, Pilling D, Petts S, Grundy JE, Janossy G. The significance of low bcl-2 expression by CD45RO T-cells in normal individuals and patients with acute viral infections. The role of apoptosis in T-cell memory. J Exp Med 1993a; 178:427.
157. Allsopp TE, Wyatt S, Paterson HF, Davies AM. The proto-oncogene bcl-2 can selectively rescue neurotrophic factor-dependent neurons from apoptosis. Cell 1993; 73:295.
158. Abe-Dohmae S, Harada N, Yamada K, Tanaka R. Bbl-2gene is highly expressed during neurogenesis in the central nervous system. Biochem Biophys Res Commun. 1993; 191: 915.
159. Zhong LT, Kane DJ, Bredsen DE. Bcl-2 blocks glutamate toxicity in neural cell lines. Brain Res 1993; 19:353.
160. Griffin DE, Levine B, Tyor WR, Tucker PC, Hardwick JM. Age-dependent susceptibility to fatal encephalitis: Alpha virus infection of neurons. Arch Virol Suppl 1994; 9:31.
161. Novack DV, Korsmeyer SJ. Bcl-2 protein expression during murine development. Am J Patho 1994; 145:61.
162. Boise LH, Gonzalez-Garcia M, Postema CE, Ding L, Lindsten T, Turka LA, Mao X, Nunez C, Thompson CB. Bcl-X, a bcl-2 related gene that functions as a dominant regulator of apoptopic cell death, Cell 1983; 74:597.

163. Mattoli S, Miante S, Calabro F, Mezzetti M, Fasoli A, Allegra L. Bronchial epithelial cells exposed to isocyanates potentiate activation and proliferation of T cells. Am J Physiol 1990; 259:L320.
164. Robicsek S, Szentivanyi A, Calderon EG, Heim O, Schultze P, Wagner H, Lockey RF, Dwornik JJ. Concentrated IL-1α derived from human T-lymphocytes binds to a specific single class surface receptor on human bronchial epithelial cells and induces the production of beta-adrenoceptor mRNA via an associated or separate receptor-linked signalling pathway. J Allergy Clin Immunol 1992; 89:212.
165. Szentendrei T, Nakane T, Lazarj-Wesley E, Virmani M, Kunos G. Regulation of beta-adrenergic receptor gene expression by interleukin-l. Pharmacologist 1991; 33:225.
166. Szentivanyi A, Calderon EC, Heim 0, Schultze P, Wagner H, Zority J, Lockey RF, Dwornik JJ, Robicsek S. Cell- and species-specific dissociation in the beta-adrenoceptor upregulating effects of IL-1α derived from lymphocyte conditioned medium and cortisol, J Allergy Clin. Immunol 1992; 89:274.
167. Szentivanyi A. The beta-adrenergic theory of the atopic abnormality in bronchial asthma. Allergy 1968; 42:203.
168. Schechter C. A good relationship: Sensitive, synchronized and synergistic. Prog Neuro-Endocrine Immunol 1989; 2:35.
169. Geschwind N. Preface, In: Mind and Immunity. Behavioral lmmunology. Locke SE, Hornig-Rohan M editors. Institute for the Advancement of Health. New York, 1983; pp. 1–6.
170. Selye H. A syndrome produced by diverse nocuous agents. Nature 1936; 138:32.
171. Selye H. Morphological changes in the fowl following chronic overdosage with various steroids. J Morphol 1943; 73:401.
172. Selye H. 1946. The general adaptation syndrome and the diseases of adaptation. J. Clin Endocrinol 1946; 6:117.
173. Selye H. Effect of ACTH and cortisone upon an "anaphylactoid" reaction. Can Med Assoc 1949; 61:553.
174. Karady S, Selye H, Brownie JSL. The influence of the alarm reaction on the development of anaphylactic shock. J Immunol 1938; 35:335.
175. Szentivanyi A. Adrenergic regulation In: Bronchial Asthma-Mechanisms and Therapeutics. Weiss EB, Stein M, editors. Boston: Little Brown, 1993a pp.165–191.
176. Lochrie MA, Simon MI. C protein multiplicity in eukaryotic signal transduction systems. Biochemistry 1988; 17:4957.
177. Birnbaumer L, Brown AM. G proteins and the mechanism of action of hormones, neurotransmitters, and autocrine and paracrine regulatory factors. Am Rev Respir Dis 1990; 141:S106.
178. Morris GM, Hadcock JR, Malbon CC. Cross-regulation between G-protein-coupled receptors. Activation of β_2-adrenergic receptors increases α1-adrenergic receptor mRNA levels. J Biol Chem 1991; 266:2233.
179. Hadcock JR, Port JD, Malbon CC. Cross-regulation between G-protein mediated pathways. Activation of the inhibitory pathway of adenylcyclase increases the expression of β_2-adrenergic receptors. J Biol Chem 1991; 266:11915.

Edited by I. Berczi and A. Szentivanyi

Virus Associated Immune and Pharmacologic Mechanisms in Disorders of Respiratory and Cutaneous Atopy

ANDOR SZENTIVANYI[1], ISTVAN BERCZI[2], HARRY NYANTEH[3] and ALLAN GOLDMAN[4]

[1,3,4]Department of Internal Medicine, College of Medicine, University of South Florida, Tampa. Fl. 33612 and [2]Department of Immunology, Faculty of Medicine, University of Mannitoba, Winnepeg, Mannitoba, R3E OW3, Canada

ABSTRACT

Anaphylaxis represents non-atopic immediate hypersensitivity, whereas manifestations of atopic immediate hypersensitivity include bronchial asthma, hay fever, allergic rhinitis, chronic urticaria, and atopic dermatitis. In spite similar antigen exposure, only a minority of the population shows some form of atopic disease. Atopic disease with its spontaneous pattern of familial occurrence cannot be induced at will.

The exact pathogenesis of atopy is yet to be elucidated. Two theories prevail: 1) atopy is a primary disorder of the immune system with sequelae in the various effector tissues; and 2) a concept of atopy as a primary autonomic imbalance, essentially beta adrenergic in character, with sequelae in effector cells, including those engaged in the production of antibodies. The autonomic imbalance is perceived as caused not by some disorder of the autonomic nervous system itself but by a defector functioning of its effector cells. These two concepts are not mutually exclusive. The IgE antibody, which mediates allergic reactions, is essentially identical with atopic reagin in various animal species.

The beta adrenergic theory regards atopic disorders (i.e., perennial and seasonal allergic rhinitis, bronchial asthma, and atopic dermatitis) not as immunologic diseases but as unique patterns of altered reactivities to a broad spectrum of immunologic, psychic, infectious, chemical and physical stimuli. The antigen-antibody interaction is given the same role as that of a broad category of nonspecific stimuli that function only to trigger the same defective homeostatic mechanism in the various effector cells involved in immediate hypersensitivities. Current evidence favors the possibility that there are inherited and/or acquired multiple abnormalities in the receptor – adenylate cyclase – cyclic AMP system of all effector cells that are critical in the organization of immune reactivities.

Atopic abnormality may be 1) acquired by functional receptor regulatory shifts caused by hormonal changes, infection (viral, bacterial, etc), allergic tissue injury or other event; 2) genetically determined; or 3) caused by autoimmune disease. One, two or all three of these effector mechanisms may be operative in a particular disease.

There is an important relationship between asthma and viral respiratory infection. A history of childhood viral respiratory illness is a risk factor for the development of chronic obstructive airway syndromes in later life. Asthmatic attacks occurred only when the infection produced fever, malaise, cough or coryza. The dominant role of fever in these episodes immediately suggests the profound involvement of adrenergic effector mechanisms. The presence of autoantibodies to beta-adrenoceptors in patients correlated well with a reduced beta - and an increased alpha-adrenergic responsiveness. Virus infections can elicit autoantibody formation.

In patients with atopic dermatitis an increased susceptibility and abnormal host response to viral infections in general. Defective cytotoxic T cells, abnormally functioning macrophages and natural killer cells, a reduced production of IFNα in children, and of IFNγ in atopic patients with food allergy has recently been demonstrated. Lymphocytic cyclic AMP-phosphodiesterase, that destroys cyclic AMP, is increased in atopic dermatitis and in allergic respiratory disease of adults, and this increased activity correlated closely with histamine release from basophils. Peripheral blood leukocytes and lymphocytes in atopic dermatitis have frequently demonstrated impaired beta adrenergic reactivity.

Allergic tissue injury may be initiated by antigen-specific IgE antibodies that combine with Fcε receptors on various cell types and trigger mediator release upon encounter with the antigen. Various noxious agents that are capable of triggering asthma are capable of releasing inflammatory mediators from the same target cells. Accounting only for those pharmacologic mediators where the cell-type has been identified, the spectrum of mediator-storing, synthesizing, or transporting cells includes neutrophil leucocytes, basophilic leucocytes eosinophilic leucocytes; mast cells, "chromaffin-positive" mast cells, enterochromaffin cells, chromaffin cells ; platelets, neurosecretory cells and nerve cells that potentially produce all amine-mediators as well as prostaglandins and kinins.

1. INTRODUCTION

This chapter describes some immune and pharmacologic mechanisms associated with viral infections in certain immunologic diseases that belong into the group of the so-called immediate hypersensitivities. The term *immediate hypersensitivity* denotes an immunologic sensitivity to antigens that manifests itself by tissue reactions occurring within minutes after the antigen combines with its appropriate antibody. Such a reaction may occur in any member of a species *(non atopic immediate hypersensitivity)* or only in certain predisposed or hyperreactive members (*atopic immediate hypersensitivity*).

The prototype of the non-atopic immediate hypersensitivity is localized or generalized anaphylaxis, whereas manifestations of atopic immediate hypersensitivity include bronchial asthma, hay fever, allergic rhinitis, chronic urticaria, and atopic dermatitis.

The discussions that follow shall be confined to the *atopic form of immediate hypersensitivities* in general, and to their respiratory and cutaneous manifestations, in particular. Of the latter, bronchial asthma, and its cutaneous equivalent, atopic dermatitis, shall serve as the *"model" immunologic* manifestations for our analysis below.

2. ATOPIC IMMEDIATE HYPERSENSITIVITIES (DISEASES OF ATOPIC ALLERGY)

Only a minority of the population shows some form of atopic disease in spite of the fact that, by and large, identical conditions of antigens must be presumed to exist for all members of the population. The nature of the constitutional basis of atopy, that is, of the underlying determinant for the development of atopic disease, is as yet unexplained.

Many theories of the constitutional basis of atopy have been proposed since Coca and Cooke's original definition. Only two general ideas, however have survived: 1) the perception of atopy as a primary disorder of the immune system with sequelae in the various effector tissues; and 2) a concept of atopy as a primary autonomic imbalance, essentially beta adrenergic in character, with sequelae in effector cells, including those engaged in the production of antibodies. The autonomic imbalance is perceived as caused not by some disorder of the autonomic nervous system itself but by a defector functioning of its effector cells.

These two concepts are not mutually exclusive. In fact, they may be interdependent. Although the immune features of atopic disease can be understood within the framework of a basic adrenergic disorder of various effector cells, many if not most of the nonimmune features of atopic conditions are not readily explicable of the basis of the primary immune abnormality.

3. THE ORIGINAL CONCEPT OF ATOPY

At the time when the original concepts of atopy were being developed, it had long been known that hay fever and asthma often occurred together in the same individual and both of them showed a marked familial tendency. Similarly, it has been recognized that acute and chronic urticaria as well as gastrointestinal manifestations of idiosyncrasy to a specific food were more common in patients with these diseases than in the general population, and a relation to infantile eczema *(Besnier's prurigo, neurodermatitis*) also was observed. Eczema was found to occur more frequently in the children of patients with hay fever or asthma, and individuals who had eczema in infancy showed an usual incidence of hay fever and asthma later in life.

These diseases were, therefore, considered together by Cooke and Vander Veer as a special group of diseases of human sensitization with a hereditary background, and these authors concluded that such "sensitized individuals transmit to their offspring not their own specific sensitization, but an unusual capacity for developing bioplastic reactivities to any foreign proteins". With further progress in determining additional characteristics of "human sensitization" in contrast to those of experimental anaphylaxis in laboratory animals, Coca and Cooke concluded that a clear distinctions must be made between two types of hypersensitivity manifestations: 1) the anaphylactic type of allergic response to abnormal substances 2) the atopic type of response to substances that are generally innocuous. As they stated:

"*This latter sub-group evidently needs a special term by which it may be conveniently designated and this need is satisfactorily met with the term atopy, which was kindly suggested by Professor Edward D. Perry of Columbia University. The Greek word, from which the term was derived, was used in the sense of a strange disease. However, it is not, on that account, necessary to include under the term all strange diseases; the use of the term can be restricted to the hay fever and asthma group.*"

To these, Wise and Sulzberger then added neurodermatitis under the new designation of "atopic dermatitis". Based on the close association of this condition with other atopic manifestations, Wise and Sulzberger concluded that the skin lesions of this disorder were cutaneous analog of hay fever and asthma and suggested that the name atopic dermatitis replace disseminated neurodermatitis.

Several characteristics of the atopic state emerged from these early concepts: atopy was felt to be a hereditary manifestation, subject to a dominant gene, a peculiarly human disorder with a reacting serum element different from classic antibodies and reminiscent of the Wasserman reagin (hence the name, atopic reagin). Atopic antibodies, furthermore, seemed to occur only in humans, many times without any demonstrable prior exposure to incitant substances and induced by agents that often appeared to be nonantigenic (atopens of Grove and Coca)

Over the years, most of these postulated differences between atopy and anaphylaxis were gradually eliminated. Thus, Ishizaka discovered an antibody, which is essentially identical with atopic reagin in various animal species. Moreover, atopic disease was shown by Patterson to occur in animals. Some of the other distinctions between anaphylaxis and atopy were also amenable to various alternative explanations, indicating that these conditions may not be separated by wide and irreconcilable differences, as the originators of the concept of atopy believed it. Nevertheless, some differences remained, and other important new differences remain, and other new differences emerged, making it imperative that a concept of atopy be reformulated.

4. THE REFORMULATED CONCEPT OF ATOPY

Since the 1960's, it has become increasingly evident that in addition to some of the remaining immunologic difference between anaphylaxis and the immediate hypersensitivities of the atopic and non-atopic type. Thus, it appears that in anaphylaxis we are dealing with a normal (physiologic) antibody response to an unnatural exposure to antigen, whereas in atopic allergic an "abnormal" antibody response to natural antigenic exposure seems to be involved. Anaphylactic reactivity of the sensitized individual depends on the release of an amount of pharmacologically active effector molecules sufficient to be toxic for most members of the same species. In contrast, individuals with atopic disease possess a quantitatively and qualitatively abnormal reactivity to otherwise nontoxic concentrations of endogenously released or exogenously administered pharmacologic mediators. Furthermore, the quantitative change consistently is in the direction of a decreased response when beta-adrenergic agents are the agonists and consistently in the direction of an increased response when any one of the other pharmacologically active effector molecules are involved.

Another essential difference between the atopic and non-atopic varieties of immediate hypersensitivities is the major contributory role played by infection in atopy, whereas infection has not been shown to be causally related to anaphylactic allergy anaphylaxis, the Arthus reaction, or serum sickness. Moreover, atopic conditions can be precipitated by a number of unrelated stimuli, whereas only the specific antigen can bring about anaphylaxis. Finally, the latter conditions may be produced artificially, but atopic disease with its spontaneous pattern of familial occurrence cannot be induced at will. Acute human pulmonary anaphylaxis, which can include asthmatic features, for example, has never been reported to lead to the development of bronchial asthma or atopic disease.

In the reformulation of the original concept of atopy by Szentivanyi [1] the essential difference between immediate hypersensitivities of the non-atopic and atopic varieties is that the former conditions are mediated by normal immune and pharmacologic mechanisms, whereas atopy is based on abnormal immune and pharmacologic mechanisms. This difference between anaphylaxis and atopy is regarded as fundamental. In this view, furthermore, it is the altered pharmacologic reactivity that is considered as the uniformly present, single atopic characteristics of pathognomonic significance.

5. THE DEVELOPMENT OF THE BETA-ADRENERGIC APPROACH TO THE STUDY OF THE CONSTITUTIONAL BASIS OF ATOPY

Authentic atopy cannot be produced at will in animals or humans, neither induced directly nor transferred passively. In addition to animal models of anaphylaxis, there are a number of experimental models simulating human atopy as well as isolated systems suitable for studying segmented areas of atopic reactivity. As such, they are useful for the analysis of some of the individual events (i.e., mediator release in the human reaction). Nevertheless, these in vivo and in vitro models are anaphylactic variants and represent immunologically and pharmacologically normal reactivities. Therefore, they cannot be used for the study of the constitutional abnormality in atopy.

6. THE TWO EXPERIMENTAL MODELS FOR THE STUDY OF THE CONSTITUTIONAL BASIS OF ATOPY

The search for a laboratory model was guided by the premise that, if it is to be meaningful, the model must be able to imitate not only the immunologic but also the pharmacologic abnormality of the atopic state. The latter is manifested against substances that, in mammalian physiology, serve as the natural chemical organizers of autonomic action. It seemed likely, therefore, that an abnormal reactivity to these agents could be most effectively produced through some alteration of normal autonomic regulation significant enough to result in an autonomic imbalance.

7. THE HYPOTHALAMICALLY " IMBALANCED" ANAPHYLACTIC GUINEA PIG

The first attempts to establish a more meaningful experimental counterpart of the atopic state were made by Filip and Szentivanyi in the years from 1952 to 1958 during studies of hypothalamically "imbalanced" anaphylactic guinea pigs. Briefly, by electrolytic removal of one hypothalamic division or electric stimulation of the antagonistic division, it was possible to alter profoundly the anaphylactic reactivity of guinea pigs both immunologically and pharmacologically. From both the immunologic and pharmacologic standpoints, the conditions so produced more closely approximated those of the human atopic state than does anaphylaxis. Nevertheless, it was felt that the artificiality of such surgically induced hypothalamic imbalance is far removed from the natural setting, (involving various inherited or acquired factors or both) that may surround the development of an atopic state. In their efforts to discover an accurate representation of those naturally occurring conditions some of which (i.e., infection) may conceivably serve as a developmental background for atopy. Szentivanyi and Fishel in the early 1960's found that the *Bordetella Pertussis* – induced hypersensitive state served as a more appropriate model.

8. THE BORDETELLA PERTUSSIS-INDUCED HYPERSENSITIVE STATE OF MICE AND RATS

Injection of live or killed *Bordetella Pertussis* organisms into certain strains of mice and rats modifies the normal reaction of these animals to a number of various stimuli. The possible applicability of the results of these investigations to atopy is implied by the following principal features of the *B. Pertussis* induced altered responsiveness: 1) hypersensitivity to endogenously released or exogenously administered histamine, serotonin, bradykinin, slow-reacting substance A, some prostaglandins, and at least in two strains, to acetylcholine; 2) hypersensitivity to less specific stimuli, such as cold, changes in atmospheric pressure, and respiratory irritants; 3) in contrast to these increased sensitivities, a reduced beta adrenergic sensitivity to catecholamines and concerning some metabolic parameters, a reversal of normal beta-adrenergic activity; 4) enhanced antibody formation in general (adjuvant activity) and facilitated production in quantity of antibodies of the IgE class; and 5) presence of a marked eosinophilia.

As described by Szentivanyi, the major advance in these experiments that has paved the way for a meaningful analogy to atopic disorders has been the finding that hypersensitivity of the pertussis-sensitized mouse to pharmacologic mediators may be due to an acquired or genetically determined autonomic imbalance caused primarily by a reduced functioning of the adenylate cyclase coupled beta adrenergic receptors and the associated cyclic AMP system.

9. THE BETA-ADRENERGIC THEORY OF ATOPIC DISORDERS: UPDATED FORMULATION AND POSSIBLE EFFECTOR MECHANISMS

The previously discussed considerations and conclusions of the two consecutive series of animal experiments have culminated in the postulation of the original beta adrenergic theory of atopic disorders as published by Szentivanyi in 1968 [ ]. This theory regarded those disorders (i.e., perennial and seasonal allergic rhinitis, bronchial asthma, and atopic dermatitis) not as immunologic diseases but as unique patterns of altered reactivities to a broad spectrum of immunologic, psychic, infectious, chemical and physical stimuli. This view gives to the antigen-antibody interaction the same role as that of a broad category of nonspecific stimuli that function only to trigger the same defective homeostatic mechanism in the various effector cells of the biochemical reaction sequence of immediate hypersensitivities.

Activation of the same defective mechanism by such a broad spectrum of unrelated stimuli is believed to be made possible by the unusual character of the pharmacologic mediators as biological distinct class of natural substances. These mediators, when viewed from the standpoint of their probable physiologic function, are the chemical organizers of autonomic action as well as of immunoregulation, that is, of homeostatic control. Consequently, regardless of the immunologic or non-immunologic nature of the triggering event, its chemical realization would be expected to be brought about by essentially the same mediators.

Homeostatic adjustment to these influences requires, among other things, mobilization of the adrenergic neurotransmitters and their balanced (uninhibited) interaction with their effector systems. The theory postulated that the constitutional basis of atopy lies in the reduced function of the beta adrenergic effector system, irrespective of what the triggering event may chemically be in a particular case (e.g., immunologic, infectious, or psychic). In this situation, the adrenergic neurotransmitters are released in the face of relatively unresponsive beta-adrenergic effector system, and the resultant autonomic imbalance deprives the effector tissues of their normal counter regulatory adjustment. This constellation of mediators and effectors then lead to a unique pattern of quantitatively and qualitatively altered reactivity to the chemical organizers of autonomic action, mostly in response to trivial trauma.

When the theoretical scheme is applied to respiratory or cutaneous atopy, at least six levels of responses critical in these diseases are expected to be influenced by the beta-adrenergic sub sensitivity in question:

1. A reduction in the normal beta adrenergic inhibition of lysosomal enzyme release, chemotaxis, phagocytosis, antibody dependent cellular cytotoxicity, increased expression of FC IgE receptors, and prostaglandin E synthesis to stimulation with histamine-induced suppressor factor, that is effector mechanisms that are known to play an important role in immunologic inflammation.
2. A reduction in lymphocytic beta-adrenergic sensitivity resulting in abnormality decreased (lymphocyte transformation, E-rosette forming cells, T cells, suppressor cell function) and abnormally increased (IgE-producing B cells, Fc receptor-bearing lymphocytes) lymphocyte reactivities.
3. Mast cell mediator release to immunologic or non-immunologic stimuli, ordinarily suppressed by beta-adrenergic stimulation would become sub sensitive to the same, while both cholinergic and alpha-adrenergic enhancement of mediator release would be exaggerated.
4. Beta-adrenergically mediated bronchial smooth muscle dilation is reduced, while cholinergically and alpha- adrenergically mediated constriction is augmented.
5. Increased mucous secretion in response to alpha adrenergic and cholinergic stimulation, while sodium and water fluxes into tracheobronchial secretions in response to beta-adrenergic stimulation would be reduced.
6. The beta adrenergically mediated eosinopenia would be reduced and replaced by eosinophilia.

All these theoretically predictable manifestations do in fact exist, and represent the cardinal features of atopic disease. Similarly, they all point to the most critical of the malfunctioning effector system that is to the adenylate cyclase-coupled beta-adrenergic receptor and the associated cyclic nucleotide complex. It follows, therefore, that the fundamental abnormality common to all atopic persons may lie in an inherited or acquired lesion that causes defective functioning of this intracellular messenger system.

These reduced responses to catecholamines would reflect alterations to any of a number of sites, including 1) changes in the affinity of catecholamines and their receptor sites; 2) decreases in the number or reactivities of beta receptors; 3) "interconversion" of adrenergic receptors from beta to alpha; 4) alterations in the efficiency of coupling of activated receptors to the catalytic units of adenylate cyclase; and 5) reductions in the concentrations of adenylate cyclase. Alternatively, the postulated lesion may occur at a point beyond the cAMP generation step in the biological sequence leading to the adrenergic end response; in a cAMP related pathway; in a complementary interacting or modulating system such as that provided by acetylcholine, histamine, the prostaglandins, leukotrienes, the interleukins, and a large group of lymphokines, monokines, and cytokines; or in an intracellular messenger system with counter regulatory potential, such as that associated with cyclic GMP. The currently available evidence seems to favor the possibility that there are inherited and/ or acquired multiple abnormalities in the receptor – adenylate cyclase-cyclic AMP system of essentially all effector cells that are critical in the organization of immune reactivities.

Progression of the disease process from subclinical to a clinical form conceivably requires the operation of preparatory or triggering factor. The preparatory factor involves the postulated abnormality, and it may be familial (presumably hereditary) or acquired in nature; however, in either case, it must set the stage for a functional imbalance. The triggering event must be appropriate to result in an increase in the rate of firing of adrenergic neurons, or in any conceivably mediator constellations suitable to make the latent abnormality clinically manifest. However, the preparatory and triggering factors need not be separate or unrelated entities. Infection (probably viral), for example, could serve in both capacities

With the exception of non-nucleated erythrocytes, the adenylate cyclase system has been found in all animal cells examined to date. Its ubiquitous character suggests, that the ultimate clinical manifestation of the fundamentally same atopic abnormality will be determined by the type of cell primarily involved, that is, by effector cell system that primarily harbors the postulated abnormality (cells of bronchial tissue versus those of nasal mucosa and skin and the circulating cells of blood).

For the extensive analysis of the experimental evidence supporting the validity of the beta-adrenergic theory of atopic disorders, and its updated formulation, the reader is referred to major reviews [2,3].

10. DEVELOPMENTAL MECHANISMS OF BETA ADRENERGIC SUBSENSITIVITY IN RESPIRATORY AND CUTANEOUS ATOPY

With respect to the apparent central feature of the atopic abnormality, that is the beta-adrenergic subsensitivity of the effector cells that participate in the cellular organization of the atopic response, the question could be raised as to how such an abnormality could develop. At present, at least three major developmental mechanisms can be envisaged. The abnormality may be 1) acquired by functional receptor regulatory shifts caused by hormonal changes, infection (viral, bacterial, etc) allergic tissue injury or other event; 2) genetically determined; or 3) caused by autoimmune disease. In case of a given atopic disorder, one two or all three of these effector mechanisms may be operative. Because of the orientation of this chapter only the role of viral infection, the allergic tissue injury, and auto-immunity due to anti-receptor antibodies will be discussed below.

11. VIRAL INFECTIONS AS A DEVELOPMENTAL MECHANISM OF BETA ADRENERGIC SUBSENSITIVITY IN BRONCHIAL ASTHMA

Upper respiratory tract infection has frequently been shown to precipitate or exacerbate the asthmatic condition and to produce or increase airway hyperreactivity to bronchospastic agents [4]. In an earlier era when whooping cough was common, *Bordetella Pertussis* infection was considered a frequent cause of asthma or recurrent bronchospasm. Even today, there is evidence of that *Hemophilus influenzae* infection is present in the deeper respiratory tract of asthmatic patients [5]. Since this bacterium was shown to produce in animal experiments [6–8] beta adrenergic impairment and/or an increased cyclic GMP level, the relationship between H. influenzae infection and the pharmacologic abnormality in asthma may be an authentic one.

In the meantime, in contrast to bacterial pathogens, viral respiratory infection was shown to have a more significant role in the pathogenesis of asthma. During the Asian Pandemic in the late 1950s severe exacerbations of bronchial asthma were found to be among the more frequent complications of this infection [9]. Since then numerous reports have appeared describing an important relationship between asthma and viral respiratory infection. Recent refinements in epidemiologic methods, microbiologic isolation techniques, and pulmonary function testing have provided new opportunities to characterize the relationship more precisely [10]

A review of the available information indicates that: 1) experiments in animal models [11] and observations in children [12] suggest that the expression of atopy occurs during a period termed "*allergic breakthrough*" which may follow viral infections; 2) a history of childhood viral respiratory illness is a risk factor for the development of chronic obstructive airway syndromes in later life; 3) if such infection lead to obstructive airway disease, the resultant manifestation is likely to be a *"wheezy"* or asthmatic type of obstructive airway disease; 4) viral as opposed to bacterial respiratory pathogens commonly herald the onset of wheezing in childhood and predispose to the development of atopy, although bronchial pharmacologic hyperreactivity after viral illness may also proceed independent of immunologic mechanisms in bronchospasm; and finally 5) no matter what the ultimate effect of respiratory infection in infancy on the subsequent development of asthma may be, there is little doubt that respiratory viruses commonly induce exacerbations of bronchospasm in the older child, and adults with known asthma.

The significance of one or another respiratory virus in causing asthma or its exacerbations appears to depend on the age of the patient. Thus, in preschool age (0–4 years) the predominant agent is respiratory syncytial virus followed by parainfluenza types 1–3, influenza, rhinovirus and corona virus. In the school age (5–16 years), rhinovirus leads the list followed in descending order of incidence by influenza, parainfluenza types 1–3, and respiratory syncytial virus. In adults, the order of relative significance changes again with the dominance of the influenza virus followed by rhinovirus and respiratory syncytial virus [4,10,12].

Another dimension of viral influences involves the problem of bronchial hyperactivity or lability in asthma. It has been recognized for some time that normal subjects exhibit relatively short lived bronchial hyperirritability to inhaled histamine when they have a viral respiratory infection [13] and as stated above asthma may worsen during and after viral respiratory infection. In some of these situations the documented release in airway resistance could be partially or fully blocked by atropine aerosol prompting the hypothesis that damage to the epithelial surface of the upper airways by viral infection exposes and thereby sensitizes the rapidly adapting sensory irritant receptors in the upper airways to various inhaled irritants, causing reflex parasympathetic vagal bronchoconstriction [14].

There are, however, several considerations that discount the significance of a parasympathetically oriented interpretation of virus induced asthmogenecity:

1) Not all viruses can be implicated in this phenomenon. For instance, adenoviruses herpes virus hominis, influenza Type B virus, and enteroviruses do not show a relationship to episodes of asthma [15,16]. Studies of rhinovirus suggest that only a few rhinovirus subtypes are associated with asthmogenicity. This is a finding that is difficult to reconcile with the damaged epithelium hypothesis (assuming equivalence of infection) and raises the question of other possible mechanisms related to the biochemical properties of the virus [17,18].
2) Conversely, influenza Type A virus that is clearly associated with increased asthma [16] affects lung function largely through an effect on small ways [4] whereas the aforementioned rapidly adapting sensory irritant receptors are primarily distributed in the larger upper airways [19].
3) The hypothesis of a cholinergic hyperactivity as a consequence of epithelial damage presupposes the de facto presence of an active viral infection, which is the cause of the respiratory inflammation, leading to the disruption of the airway epithelial barrier and activating of the subepithelial rapidly adapting sensory irritant receptors. Indeed, the characteristic histologic finding in viral respiratory infection is epithelial destruction. Welliver et al. [20] has shown that IgE was bound to exfoliated nasopharyngeal epithelial cells in most patients during the acute phase of infection with respiratory syncytial virus. They also found that a continued presence of cell bound IgE was more common in patients with bronchiolitis or asthma than in those with mild upper respiratory tract infection. In this regard, it has been established in avian species that experimentally induced viral laryngotracheitis results in disruption of

airway epithelium, with resultant increased permeation of horseradish peroxidase [21] and possibly increased uptake of inhaled antigens. Furthermore, Ida and associates [22] have demonstrated that interferon elaborated during viral infections from leukocytes harvested from patients with ragweed allergy may induce histamine release suggesting that atopic patients may experience bronchial hyperreactivity, if specific antigen exposure occurs at the time of viral infection.

Under these circumstances, it is important to mention that evidence is available that actual respiratory infection with influenza Type A virus is not a necessary condition for the development of increased airway irritability in patients with asthma. Thus, administration of killed influenza virus vaccine to asthmatic patients caused a significant increase in bronchial sensitivity to methacholine aerosol, reaching a maximum after one day and persisting for three days [23]. Normal subjects did not develop the heightened response to methacholine, and no evidence of allergy to the vaccine was detected. These investigators suggested that the effect might be explained by an endotoxin-like action of influenza vaccine. Indeed, endotoxin sensitization of human bronchial smooth muscle to alpha-adrenergic agonists has been reported. Phenylephrine-induced contractions were enhanced two to ten times in normal lung and 1000 times in lungs from a patient with chronic bronchitis. Endotoxin also caused a decrease in cyclic AMP of the tissue [24]. These observations are complemented by recent findings obtained through vaccination with a purified LPS preparation from E. coli that resulted in a decreased number of beta-adrenergic receptors in guinea pig lung [25].

4) There are other observations calling into question the requirement of an active viral infection in the production of bronchial hyperactivity as mediated by cholinergically activated irritant receptors. Thus, pulmonary function abnormalities in adults with viral respiratory illness are generally more pronounced seven to ten days after the onset of symptoms, a time when the clinical manifestations and manifestations of viral inflammation are waning [26]. Also, the abnormalities of pulmonary function are generally prolonged well beyond what we now recognize as the period of viral shedding. Furthermore, administration of the antiviral agent, amantadine hydrochloride, although clinically effective, has no effect on the magnitude or duration of airway hyperactivity [27]
5) In one longitudinal study, Minor et al. [16] found that simple colonization of the respiratory tract by virus was not sufficient to provoke asthma: such attacks occurred only when the infection produced symptoms of fever, malaise, cough or coryza. The dominant role of fever in these episodes immediately suggests the profound involvement of adrenergic effector mechanisms.

Additional support for the important effect of viral infections on adrenergic mechanisms in asthmatic patients came from the extensive studies of Busse and his associates [18]. In their experiments human granulocytes served as a convenient *in vitro* model for the systematic study of virus incubation on the pharmacologic agonist response. At first, it was described [18] that an impairment of the inhibitory action of a beta adrenergic agonist (isoproterenol), normally observed on lysosomal enzyme release, occurred in granulocytes taken from asthmatic patients (lysosomal enzymes are known to play a major role in bronchial immunologic inflammation) [28]. The beta adrenergic impairment of lysosomal enzyme release was significantly greater if the cells were obtained during upper respiratory infection. Following these observations, Busse and colleagues have consistently reported virus-induced impairment of several neutrophil mechanisms, which normally mediate inhibition of enzyme release. Impairment occurred after incubation in vitro with live influenza virus [29] and with live rhinovirus 16 (RV 16) [30], a finding also observed following infection of normal subjects with RV 16 [31]. Both studies reported an impairment of the normal functioning of the beta adrenergic, histamine H_2 and PGE receptors, all responsible for inhibition of lysosomal enzyme release through stimulation of adenylate cyclase, resulting in cyclic AMP formation [32]. Interference with granulocyte beta-adrenergic receptor activity by influenza virus has subsequently been confirmed by Lee [33] and extended to include lymphocytes, resulting in decreased inhibition of E-Rosette formation to beta-adrenergic stimulation by isoproterenol. Another important study by Buckner et. al. [17] showed that parainfluenza 3-virus infection in vivo causes a selective blockade of the beta-adrenergic inhibition of antigen induced contraction of isolated airway smooth muscle. These abnormalities moreover are not limited to those with known atopy but may also be demonstrated in cell systems from normal subjects. Viral infections therefore, may have important inhibitory effects on beta-adrenergic responsiveness in the course of bronchoconstriction. This area has recently been extensively reviewed by Norris and Eyre [34].

6) Up to this point an analysis of the evidence against a parasympatheticlly (cholinergically) oriented virus induced asthmogenecity included arguments based on information obtained from the current state of our understanding of some aspects of virology, immunology, immunologic inflammation, and the cyclic nucleotide system. In the discussion below, these arguments will be extended to include some basic principles of neurophysiology and neuropharmacology, which also contradict the basic validity of the hypothesis of Nadel and his associates [14].

For the purpose of establishing an appropriate framework for this discussion, the basic tenets for this hypothesis will be restated as follows. In the past two decades, Nadel, Boushey, Holtzman, and their associates have accumulated evidence indicating that stimulation of rapidly adapting epithelial nerve receptors of the airways by mechanical, chemical and pharmacologic stimuli, reflexly increases the output of acetylcholine by the vagus nerves, causing a reflex bronchoconstriction. In particular it was shown that, although histamine is capable of constricting airway smooth muscle directly, most of its bronchoconstrictor effect *in vivo* is indirect, and due to this reflex mechanism. This is accomplished both by direct stimulation of these epithelial receptors, and also by decreasing their firing threshold to other introduced stimuli. Thus, when histamine is injected into dog bronchial arteries, most of the airway constriction can be blocked by atropine. In otherwise healthy subjects, viral upper respiratory tract infections, through damage to the bronchial epithelium cause transient bronchial hyperactivity to inhaled histamine and citric acid, a phenomenon that is also abolished by anticholinergic drugs. On the basis of this and similar observations as well as the fact that bronchial hyperactivity is associated with a decrease in cough threshold, these workers suggested that airway epithelial damage with sensitization of airway nerve endings causes exaggerated cough and bronchomotor responses. With this background, Nadel, Boushey, Holtzman, and their associates, further postulated that bronchial asthma is a constellation involving two ingredients: release of pharmacologic mediators, and sensitization of airway epithelial nerve receptors providing a positive feedback system for increasing bronchomotor tone. This mechanism in fact probably contributes to the bronchial obstructive process in asthma. The altered pharmacologic reactivity in atopy, however, is not restricted to airway epithelial effectors, but it is a universal atopic trait. In fact, as explained earlier, the altered pharmacologic reactivity is the uniformly present, single atopic characteristic, which by its very nature must be explained by any theory attempting to elucidate the constitutional basis of atopy [3,35, 36]. For the reasons below, cholinergic overactivity cannot serve in this capacity. Using spontaneously breathing, unanesthetized guinea pigs, it has been found that the vagal reflex component in histamine bronchoconstriction is small and probably a consequence rather than a cause of the constriction. In histamine-sensitive and histamine-insensitive strains of guinea pigs it has been demonstrated that the ease of *in vivo* histamine-induced reduction in lung compliance in the guinea pig is inversely related to its *in vitro* tracheal sensitivity to isoproterenol, revealing the primary homeostatic importance of the tracheobronchial beta- adrenergic receptors rather than that of cholinergic control, in determining the sensitivity of this effector tissue to histamine [37–39]. In harmony with these findings are the extensive studies carried out in humans in the past 20 years [40–49]. The most recent findings by O'Byrne et al. [50] further support the conclusions of these long series of observations indicating not only that blockade of the muscarinic cholinergic receptors has only a small effect on the response to inhaled histamine but also the observations that such a blockade elicits only a minor degree of protection against the response to inhaled allergen [46], exercise [44] and inhalation of cold air [48] in subjects with asthma. Taken together, these experiences indicate that the bronchial effect of histamine is exerted not by reflex bronchoconstriction but through stimulation of H_1 receptors on airway smooth muscle. Therefore, hyper-responsiveness to histamine in asthma is not primarily caused by a defect in the parasympathetic nervous supply to the airway.

Furthermore, no reproducible evidence of elevated levels of acetylcholine in tissues or body fluids of atopic individuals is available. This, in fact, is not surprising when the issue of cholinergic overactivity is examined in the broader biologic context of the general nature of cholinergic versus adrenergic control. Thus, the sympathetic system is distributed to effectors throughout the body, whereas the parasympathetic distribution is much more limited. For instance, sympathetic postganglionic fibers also innervate smooth muscles and glands of somatic (non visceral) regions; no comparable distribution has been established for the parasympathetic division. Moreover, the sympathetic fibers ramify to a much greater extent, and their preganglionic terminals make contact with a large number of postganglionic neurons. In general, the ratio of preganglionic to postganglionic axons may be about 1:20 or more. In addition, there is an overlapping of synaptic innervation so that one ganglion cell is supplied by several pre- ganglionic fibers. By contrast, the parasympathetics are more discrete in their action, i.e., there is a closer to a 1:1 relation between pre- and postganglionic neurons [51]. Also the parasympathetic nervous system has no reinforcing mechanism comparable to that of the adrenal medulla for the sympathetic division.

Usually, when any part of the sympathetic nervous system is stimulated, the entire system, or at least major portions of it, is stimulated at the same time, a phenomenon called *mass discharge*. Norepinephrine and epinephrine, therefore, are almost always released by the adrenal medulla at the same time that the different tissues are being stimulated directly by the sympathetic nerves. The two means of stimulation support each other and either can actually substitute for the other. Without any stimulation, however, the normal resting rate of secretion by the adrenal medulla is sufficient to maintain blood pressure almost to normal even if all direct sympathetic pathways to the cardiovascular system are removed. Another important value of the adrenal medulla is the capability of catecholamines to stimulate structures of the body that are not innervated by sympathetic fibers. In contrast, the characteristics of parasympathetic reflexes are discrete. For instance, they usually act only on the heart to increase or decrease its activity, or frequently cause secretion only in the mouth or, in other instances, secretion only by the stomach glands.

Acetylcholine (ACh), i.e., the cholinergic transmitter released by parasympathetic fibers, is almost instantaneously destroyed in the junctional clefts by an unusual enzyme, acetylcholinesterase (true cholinesterase, AChE). The principal evidence for this is the decay time of the end plate current, which is more rapid than diffusion of ACh out of a synaptic cleft would allow. Also, the most recent preparation of AChE hydrolyzed 960 nmoles ACh per mg of protein per hour, thus placing it among the enzymes having the highest turnover number that is known [51]. This powerful destructive capacity is reinforced by a battery of butyryl cholinesterases ("pseudo" or nonspecific cholinesterases), which destroy most of whatever acetylcholine may have escaped into the blood stream. Thus, it is doubtful whether acetylcholine can reach non-innervated cells or is present in the extracellular space in regulatory concentrations for cells with immunologic significance such as the antigen-sensitive lymphocytes. There is no comparable system of rapid destruction for the catecholamines, a fact that accounts in part for the widespread nature of sympathetic action.

Another way to determine whether we are dealing with a primary cholinergic overactivity in atopy is to examine whether there is any evidence for an enhanced guanylate cyclase activity in cells obtained from patients with atopic disease. This is all the more necessary, since as mentioned earlier, cholinergic and alpha-adrenergic agents activate guanylate cyclase, and markedly reduced adenylate cyclase-cyclic AMP responses to beta-adrenergic stimulation have been shown to be present in atopic individuals.

Under these circumstances, it is highly significant that the available evidence shows not an enhanced but a reduced cholinergic responsiveness in lymphocytes of atopic individuals. Thus, it was found that in normal subjects alpha-adrenergic stimulations with norepinephrine plus propranolol, and cholinergic stimulation with acetylcholine evoked significant increases in cyclic GMP formation. In contrast, the lymphocytic guanylate cyclase activity did not show a significant response to the same agents in patients with acute asthma, but the normal guanylate cyclase responsiveness was found to be partially restored in patients in remission [52]. Similarly, Lang, Goel, and Greico[53], in their study on adrenergic and cholinergic responses of peripheral lymphocytes in the "*active*" E rosette assay, demonstrated not only a subsensitivity of T lymphocytes to beta adrenergic but also to cholinergic stimulation in patients with bronchial asthma. In the same experiments, phenylephrine, an alpha-adrenergic agonist, showed no difference between the normal and asthmatic groups in enhancing the "active" E rosette formation. A subsensitive beta adrenergic and cholinergic system with a normal alpha adrenergic effector system may produce a state of relatively enhanced alpha adrenergic activity, a circumstance which may explain some of the findings showing that by giving alpha receptor blockers one can restore beta adrenergic responsiveness toward normal in lymphocytes of asthmatics [54,55]. There are additionally at least three major arguments against cholinergic overactivity as the primary mechanism of atopy. First, neither pulmonary sympathectomy nor pulmonary vagotomy produces any lasting improvement in bronchial asthma. Second, as discussed in detail elsewhere [3,35,36] it is never the excessive presence of a neurohumor, but if anything, it is its prolonged lack that is likely to result in the development of chronic effector hypersensitivities. Consequently, it is inconceivable that cholinergic overactivity could produce a hypersensitivity to acetylcholine or similar mediators of immediate hypersensitivities. On the contrary, cholinergic overactivity would be expected to lead to desensitization of the cholinergic receptors, as has been extensively demonstrated in numerous preparations such as the skeletal muscles of the frog, the hearts of vertebrates and invertebrates, the Renshaw cells, the neurons of mollusks, etc. [56]. This is in harmony with more recent findings obtained in non-obstructed, non-reversibly obstructed, and reversibly obstructed (asthma) patients. Using H^3-quinuclidinyl benzilate, a stereospecific radioligand for muscarinic cholinergic receptors a significant reduction in receptor density was found in the lung preparations of asthmatics, and no difference in the numbers of such receptors in lung specimens derived from the non-reversibly obstructed and non-obstructed groups as shown by Szentivanyi et al.[57] . An important question of course is how can one possibly find a reduction in the number of muscarinic cholinergic receptors in lung membranes derived from patients with reversible obstruction (asthma) in the simultaneous presence of an exquisite bronchial hyperreactivity to cholinergic agents? At the time of this writing, we can only offer two possible interpretations. One is that the bronchial hyperreactivity to cholinergic agents in asthma is not mediated through cholinergic mechanisms, but is basically due to the beta adrenergic abnormality, which is also responsible for the atopic feature of the disease. The second possibility is that the reduction in muscarinic receptor densities is caused by a heightened vagus activity resulting in a cholinergic "downturn" and ultimately producing a pharmacological "denervation supersensitivity" to cholinergic agents [51,58]. Whether one or the other, or both of these interpretations will prove to be correct, it is evident that they represent important evidence against the validity of Nadel's [14] reflex hypothesis of the virally induced mechanisms of bronchoconstriction in asthma. Finally, if the atopic state were to be due to cholinergic overactivity produced by viruses through the postulated reflex mechanisms, then anticholinergic agents should have a far more demonstrable therapeutic effect than what we are able to observe in asthma [40–42,44] let the other atopic conditions were they are useless.

12. VIRAL INFECTION AS A DEVELOPMENTAL MECHANISM OF BETA-ADRENERGIC SUBSENSITIVITY IN ATOPIC DERMATITIS

The most typical viral infection that affects children with atopic dermatitis is Kaposi's herpetic eruption. It is caused by herpes simplex, type 1 or 2, and also the virus coxsackie A16 can mimic perfectly the eruption caused by a herpetic virus. Other viruses associated with this disease include herpes zoster, vaccinia, warts, and molluscum contagiosum. While it has long been known that recurrent viral cutaneous infections are more prevalent in atopic dermatitis, there is now growing evidence for recurrent cold sores and upper respiratory infections in this condition. Serological studies have also revealed that atopic dermatitis patients display significantly higher serum levels of antibodies against Epstein Barr virus (EBV) than their non-atopic controls. It appears therefore, that the increased susceptibility to viral infections is not restricted to dermatotropic viruses but rather reflects an abnormal host response to viral infections in general [59, 60].

Host defense against most viral infections is dependent to a large extent on cell-mediated immune mechanisms, and there is abundant clinical and experimental evidence of defective cell-mediated immunity in atopic dermatitis. In early studies, a reduction in the number of T cells was found which correlated with the severity of the disease, and these findings were later complemented by the demonstration of a defective functioning of these cells. Soon it was also shown that the T cell defect is particularly evident in suppressor/cytotoxic T cell subsets [61–65].

Defective cytotoxic T cells and also the association of abnormally functioning macrophages [66] and natural killer cells [67], appear to have an important role in the impaired host defense against viral infections in atopic dermatitis. Since these cell types produce, or are capable of producing interferon, a deficient production of this agent may at least be partly responsible for the increased susceptibility to viral infections in atopic dermatitis. A reduced production of interferon alpha in children with atopic dermatitis [68] as well as of interferon-gamma in atopic patients with food allergy has recently been demonstrated [60].

Central to the immunologic and other abnormalities (discussed later) is the T cell defect, which to many workers in this field appears to be a primary, inherited feature of atopic disease. In the context of the T cell, therefore, it is important to focus our interest on the gene products, primarily enzymes that affect T cell maturation or function in atopic dermatitis. For the first time in 1982, it has been reported that the activity of lymphocytic cyclic AMP-phosphodiesterase (that is the enzyme that destroys cyclic AMP) is increased in atopic dermatitis as well as allergic respiratory disease of adults [69], and that this increased activity correlated closely with histamine release from basophils [70]. When the same enzymatic activity together with histamine release was investigated in the newborn using umbilical cord blood, the significant elevation of phosphodiesterase activity was reconfirmed in newborns with a positive atopic history in first-degree relatives, compared to newborns with a negative history. In contrast to adults, however, there was no correlation between phosphodiesterase activity and histamine release [71]. Elevation of cyclic AMP phosphodiesterase activity in cord blood leukocytes before the development of clinical manifestations of atopy strongly suggests that increased cyclic AMP phosphodiesterase activity plays a primary role in the pathogenesis of atopic disease. The lack of correlation between phosphodiesterase activity and histamine release in neonates further suggests that elevated cyclic AMP phosphodiesterase activity is a primary, genetically linked defect rather than secondary to *in vivo* desensitization by inflammatory mediators such as histamine and prostaglandin E_1 [72].

These considerations complete the full circle of the core argument of this chapter and guide us back to the primary nature of the constitutional basis of respiratory and cutaneous atopic disease. Following the publication of the original beta-adrenergic theory by Szentivanyi, [1], a series of experiments have been carried out to examine the applicability of this theory to atopic dermatitis. Studies of peripheral blood leukocytes and lymphocytes in atopic dermatitis have frequently demonstrated impaired beta adrenergic reactivity as revealed by a loss of regulatory effects on lysosomal enzyme secretion [73,74], by reduced formation of cyclic AMP to beta adrenergic stimulation [74–76], by decreased affinity of binding for radio-labeled beta-adrenoceptor agonists [77] and by a shift in the numbers of beta adrenergic receptors to alpha adrenergic receptors resulting in an increased ratio of alpha to beta binding sites [4,78]. More recently, Hannifin has made an extensive effort with his group to determine the lymphocyte and monocyte localization of altered adrenergic receptors, cyclic AMP responses, and cyclic AMP phosphodiesterase in atopic dermatitis [79–84].

In these experiments, the numbers and affinities of beta-adrenergic surface receptors on mononuclear leukocyte subpopulations were measured by the binding of propranolol-displaceable ^{3}H-dihydroalprenolol to cell surfaces. Unfractionated atopic mononuclear leukocytes showed reduced numbers of beta adrenergic receptors per cell together with the absence of a normal, lower affinity subpopulation of high affinity beta receptors. This resulted in a linear Scatchard plot of beta adrenergic binding to mononuclear-leukocytes from atopic patients, instead of the biphasic plot seen in normal control cells. These alterations of surface receptors for cyclic AMP-elevating ligands were localized to T cells and monocytes of patients with atopic dermatitis, whereas atopic B-cell receptor numbers and affinities were identical to those of normal B-cells [79,80]. Of the various subpopulations of T-cells, a lymphocyte subset which is activatable by self Ia-antigen (MHC-II) bearing presenting monocytic cells, has been identified as radiosensitive (functionally dependent upon a proliferative step) OKT4$^+$, T29$^+$ helper/inducer T cell [85–87]. A primary abnormality in the numbers and/or in the intracellular regulation of the cyclic AMP system of the radiosensitive, T29$^+$, helper/inducer T-cells generated by the interaction with autologous Ia antigen presenting macrophages may explain many of the characteristic features of immune dysfunction in atopic dermatitis [81].

For instance, soluble mitogen stimulated proliferation is critically dependent on successful macrophage/T-cell interaction, and can be reduced in patients with atopic dermatitis [88,89]. Development of the pool of blood PWM-recruitable B-cells for in vitro antibody production requires induction by a radiosensitive T-cell inducer [85,90,91], and indeed, B-cells from patients with atopic dermatitis demonstrate decreased mitogen- stimulated antibody secretion, even when corrected for number or when normal T-cells are used to provide helper function [92]. T-cells associated with suppressor and cytotoxic functions, such as T-cells with FcIgG receptors, OKT8 cells and histamine H_2–receptor-bearing T-cells often show significantly reduced values in patients with atopic dermatitis [87,92–94], which is in accord with findings indicating that the development of mature suppressor and cytotoxic effector T cells requires induction by the aforementioned radiosensitive T-helper cells [95–97]. It may be added that the development of cytoxic T lymphocytes is known to be dependent upon Ia^+ monocyte stimulation of helper T-cell factors such as interleukin-2 [95], and a decrease in the production of interleukin 2 as well as interferon by these abnormal helper/inducer T-cells, or their altered ability to respond to these signals may explain the reduced natural killer activity in atopic dermatitis [67]. Thus, the aforementioned multiple abnormalities of the cyclic AMP system in the helper/inducer T-cells in question may account for the immune dysfunction in atopic dermatitis. Alternatively, each of the immune abnormalities listed could be due to altered immune signal processing by distal effector cells with their own malfunctioning intracellular cyclic AMP systems as will be further pointed out below.

In closing the discussion of viral infection as one of the major developmental mechanisms of beta-adrenergic subsensitivity in cutaneous atopy, we need to briefly revisit the issue of whether the atopic diathesis increases the susceptibility to viral infections, or the viruses themselves may produce the atopic disposition. As stated above, the demonstration of higher serum levels of antibodies against EBV in atopic dermatitis were interpreted in support of the assumption that defective immune mechanisms rather than cutaneous alterations predispose for increased susceptibility to viral infections. Although the findings of raised EBV antibody titers in atopic dermatitis may in fact reflect on abnormal host response to the virus, it cannot be excluded that the cause and effect relationship is the reverse, so that EBV may play a causative role for the development of atopic dermatitis. Thus, EBV is a B- cell mitogen, which may stimulate IgE antibody formation, and infectious mononucleosis, which is caused by EBV, is associated with raised serum levels of IgE [98]. The report of a case of atopic dermatitis developing soon after an episode of infectious mononucleosis suggests that EBV may in fact occasionally precipitate atopic disease [99]. Essentially the same causative role of viral agents has been described for respiratory atopy above.

13. THE ALLERGIC TISSUE INJURY AS A DEVELOPMENTAL MECHANISM OF BETA ADRENERGIC SUBSENSITIVITY

The allergic tissue injury is another major developmental mechanism of beta-adrenergic subsensitivity. Advances in knowledge of the immune response and immune reactivity achieved since the early 1960's have been accompanied by a more complete understanding of the different pathways of immune tissue injury. Based on this new understanding, the various types of immunopathologic processes have been subdivided by the classification by Coombs and Cell (1962) into the following four basic types:

Type I: Immediate-hypersensitivities
Type II: Cytotoxic tissue injury
Type III: Immune-complex tissue injury
Type IV: Cell-mediated immune tissue injuries

This classification is oversimplified because of the complex interrelationships that exist between the several events that constitute an inflammatory response. Nevertheless, this view represents the closest approximation of the various basic patterns of immune tissue injury, and the classification does not depend on the host species or on the method of antigen exposure. Another valuable feature of the classification is its integrated emphasis on the important central point that in these various patterns of immune injury the tissue damage results from the immune activation of cellular and biochemical mediator systems of the host. The combination of the immune reactants produces only minimal direct effects, but as a trigger mechanism it sets the destructive factors into play.

Because of the subject of this chapter only the general pattern of immunologic tissue injury that occurs in immediate hypersensitivities will be discussed. Furthermore, no distinction will be made between the atopic and non-atopic varieties of immediate hypersensitivities since in both cases the pattern of tissue injury follows the characteristic triphasic reaction sequence of these manifestations.

In these reactivities, following the initiation of antibody production, the cytotropic antibodies (primarily IgE) that are formed disseminate throughout the circulation to become almost selectively and uniquely attached to the cell membranes of basophils in the circulation and mast cells in the tissues. The attachment occurs through a structural area in the Fc part of the antibody molecule to a specific receptor on the basophil or mast cell membrane. Although evidence indicates that subpopulations of monocytes, macrophages, and lymphocytes also express Fc receptors for IgE antibody, of all mammalian cells only basophils and mast cells exhibit an extraordinary binding affinity for this antibody. There is a relative abundance of these IgE molecules bound along the membrane, and they are located close to each other physically. When the IgE becomes attached, the cells are said to be sensitized, and the individual is now in a sensitive state for reactivity on subsequent exposure to the antigen [100].

A second, or subsequent, exposure can occur via many routes, such as inhalation, ingestion, or injection. The antigen must move across membrane and tissue barriers in order to come to the surface of the sensitized cells. When this close encounter occurs with an antigen of sufficient size to react with the antigen-binding sites of two closely adjacent IgE molecules, it produces a "bridging' effect. In this molecular interaction, one antigen molecule combines with two antibody molecules to form a bridge. This bridging brings together two IgE receptor molecules, which results in coformational changes in the receptors, triggering an enzymatic cascade that causes the release of pharmacologically active effector molecules responsible for the clinical symptomatology of immediate hypersensitivities. Accounting only for those pharmacologic mediators where the cell-type has been identified, the spectrum of mediator-storing, synthesizing, or transporting cells, includes the neutrophil leucocyte [slow-reacting substance of anaphylaxis (SRS-A), eosinophil chemotactic factor of anaphylaxis (ECF-A), enzymes, vascular permeability factors, kinin-generating substances, a complement- activating factor, histamine-releasers, and a neutrophil inhibitory factor (NIF)], basophilic leucocyte [histamine, SRS-A, ECF-A, neutrophil chemo- tactic factor (NCF) and platelet-activating factor (PAF)], the murine basophilic leucocyte (histamine, SRS-A, ECF-A, PAF, and serotonin), the eosinophilic leucocyte (histamine, PAF, and possibly SRS-A), the mast cell (histamine, SRS-A, ECF-A, NCF and PAF), the murine mast cell (histamine, SRS-A, ECF-A, PAF, NCF and serotonin), the "chromaffin-positive" mast cell (dopamine in ruminants; in other mammals possibly norepinephrine), the enterochromaffin cell (serotonin), the chromaffin cell (catecholamines), the platelet (depending on species: histamine, serotonin, catecholamines, and prostaglandins), the neurosecretory cell (histamine, serotonin, catecholamines, acetylcholine, and prostaglandins), and the nerve cell (potentially all amine—mediators as well as prostaglandins and kinins) [101]. Collectively, these pharmacologically active agents produce an increase in blood flow, capillary permeability, constriction of smooth muscles, and secretion of mucous glands, that is manifestations that dominate the clinical picture of immediate hypersensitivities and the associated inflammatory responses.

14. REACTIVITIES OF THE MEDIATOR-STORING CELLS TO ANTIGENIC AND PHARMACOLOGIC INFLUENCES AND THEIR RELATIONS TO CYCLIC NUCLEOTIDES

Depending on concentrations and other experimental conditions, pharmacologically active adrenergic agents can both release as well as inhibit the release of allergic mediators. Thus, amphetamine, phenylethylamine, tyramine, and the like-substances that induce sympathomimetic activity indirectly through the endogenous release of catecholamines are capable of liberating histamine. The same can be accomplished by the exogenous administration of catecholamines and of their specific blocking agents. While all these agents elicit non-immunologic histamine release, they render sensitized mast cells incapable of responding to antigen challenge with histamine release [102–104].

Analysis of these seemingly contradictory findings suggests that 1) adrenergic agents interfere with binding or release of histamine because of their catecholamine-like intrinsic activity, and 2) they operate on a cellular system that is antigen activated and thus central to the mechanism of the allergic reaction.

The first conclusion is supported by the residual agonistic activity of the blocking agents employed, since the only common feature of these directly acting adrenergic compounds is their basic catecholamine structure. The second conclusion is based on the observation that methylxanthine also inhibit immunologic release of histamine [104]. Thus, when ragweed antigen was made to interact in vitro with IgE antibody on the surface of leukocytes from ragweed-sensitive human donors, both methylxanthines and catecholamines inhibited histamine release. The significance of these findings is seen in the fact that methylxanthines are competitive inhibitors of the specific phosphodiesterase that inactivates cyclic 3', 5'AMP; and thereby they may induce "adrenergic action" by increasing the intracellular concentration of the compound. Indeed, catecholamines and methylxanthines were found to act synergistically in inhibiting histamine release, and the phosphodiesterase-inhibitory potencies of the various methylxanthines correlated well with their inhibitory effects on histamine release [3]. Of further significance is the fact that the methylxanthines and catecholamines were shown to inhibit only if added to the cells when antigen was present; they had no effect if removed from the environment of the sensitized cells before antigen exposure [104]. The adenylate cyclase system therefore must be considered as a critical regulatory system in allergic histamine release.

In addition to beta-adrenergic agents, allergic release of histamine or of other pharmacologic mediators of immediate hypersensitivity is also inhibited by prostaglandins of the E series, prostacycline, adenosine, and histamine (i.e., by substances that interact with cell membrane receptors that activate adenylate cyclase [105,106]. Inhibition of allergic mediator release by these agents is generally paralleled by an increase in the intracellular concentration of cyclic AMP in the respective cell preparations. Furthermore, since the release-inhibitory activities of these agents are blocked by their specific antagonists, it is presumed that these agents increase cyclic AMP by acting on receptors linked to adenylate cyclase. The mechanism by which cyclic

AMP blocks mediator release is not known, but current evidence suggests that cyclic AMP acts early in the release process, that it is linked to the obligatory inward flux of calcium, and that it related to microtubule function [107]. There are, however, exceptions when changes in cyclic nucleotide levels do not correlate with inhibition of immunologic mediator release [7,108]. The nature of such dissociation between cyclic AMP elevation and inhibition is unclear but may be explained if there are functionally separate intracellular cyclic AMP pools [109] and if a product of the lipoxygenase or some other pathway can block selectively a biochemical sequence linking adenylate cyclase activation to inhibition of mediator release [107]. The effect of changes in intracellular cyclic GMP levels on allergic mediator release has been less extensively studied. In lung tissue, alpha- adrenergic and cholinergic stimulation increase cyclic GMP levels [110 111], and such effects as well as extracellular cyclic GMP derivatives potentiate antigen-induced mediator release [112]. However, cyclic GMP does not enhance release from rat mast cells and has minimal or no effect on immunologic mediator release from basophils [113]. Furthermore, it is not known whether cyclic GMP-induced enhancement of pulmonary mediator release is a direct effect of alpha-adrenergic or cholinergic agents on the mast cells or whether it reflects their actions on other cell types [57]. Nevertheless, in this context it may be added that pertussis or pharmacologically established beta-adrenergic blockade has been reported to cause peritoneal mast cell degranulation in rats and mice, whereas beta-adrenergic stimulation protects these cells against propranolol induced degranulation [114]. As judged by the PCA reaction in guinea pigs, propranolol has the same enhancing effect on immunologic mediator release [115].

15. THE ALLERGIC TISSUE INJURY, BETA ADRENERGIC SUBSENSITIVITY AND BRONCHIAL ASTHMA

Recent studies in patients with extrinsic asthma [116,117] and in animal models of experimental asthma [118,119,120] raised the possibility that the allergic tissue injury itself may result in the development of some forms of beta-adrenergic subsensitivity. In the studies of deVries et al. [116], and Koeter et al. [117], patients with complaints of episodic wheezing after exposure to allergens, specific IgE and skin tests, and an increased bronchial response to histamine inhalation were included. Symptoms of these seven patients were mild and well controlled without a history of respiratory tract infections or acute asthmatic attacks two months prior to the study. No patient was on beta-adrenergic or corticosteroid therapy.

These studies were designed with the assumption that there might be a relationship between the allergic tissue injury and the adrenergic system. Therefore, the latter was studied before and after an inhalational allergen challenge. Two parameters were measured: 1) *in vivo* propranolol threshold to assess bronchial beta-adrenergic reactivities and 2) *in vitro* lymphocytic cAMP production in response to beta-agonist stimulation. The propranolol threshold changed from 1.32 percent before challenge to 0.86 percent the day after. In the same patients the maximal cAMP response of lymphocytes changed from 339 percent above basal level before the challenge to 194 percent after the challenge.

Recently, the development of beta subsensitivity of airways smooth muscle was studied in greyhound dogs in order to determine its relationship to the hyperreactivity of the same airways to aerosols of *Ascaris suum* antigen [119,120]. Using thoracic trachealis smooth muscle, it was found that the airways hyperreactivity was statistically significantly inversely correlated with 1) beta-adrenoceptor density; 2) isoproterenol stimulated cAMP production; and 3) isoproterenol stimulated relaxation. These authors concluded that the beta-adrenergic subsensitivity of airway smooth muscle that is associated with airways hyperreactivity in this canine asthma model is due to a deficiency of beta-adrenoreceptors, since all post receptor beta-adrenergic responses that were measured (cAMP, protein kinase, relaxation) tended to be depressed in animals with airways hyperreactivity.

These studies clearly indicate that the allergic tissue injury may be one of the contributory factors in the development of beta-adrenergic sub- sensitivity in some forms of human asthma, or alternatively the sole factor in some subsets of human asthma. They may not, however, support the interpretations of deVries [116] and Koeter et al. [117] that these findings suggest that 1) in the seven asthmatics they studied the beta-adrenergic subsensitivity was due to endogenous desensitization by catecholamines released in response to the allergic tissue injury; and 2) in all forms of asthma, this same mechanism is responsible for the manifestation of beta- adrenergic subsensitivity.

Several lines of evidence argue against the general applicability of these interpretations to human asthma and other manifestations of atopy. First of all endogenous catecholamines released from neuronal and adrenal medullary catechol stores would be expected to desensitize both alpha- and beta-adrenoceptors more or less evenly. In the foregoing studies alpha-adrenoceptor concentrations or their sensitivities were not measured, but in those studies where they were, this was not the case. Thus, pulmonary homogenates of sensitized guinea pigs that had been exposed chronically to antigen aerosol showed significant increase in alpha-adrenoceptors and a decrease in beta-adrenoceptors [118] reminiscent of the reciprocal adrenoceptor changes observed in a number of other human and animal studies described in the literature [121–125]. This alpha dominance is also reflected by the demonstration that alpha-adrenergic agonists produce bronchoconstriction in asthmatic patients but not in normal subjects [126,127]. Similarly, in vitro studies show alpha-receptor mediated constriction of bronchial smooth muscle from patients with increased airways resistance but not from normal controls [24,128]. In addition, increased alpha-adrenergic receptor mediated responses in vascular and pupillary smooth muscles have been reported in asthmatics [129].

Furthermore, beta-adrenergic subsensitivity in asthma can be shown to occur in the absence of allergic symptoms or beta-adrenergic medication, and under circumstances in which prior or concurrent beta adrenergic medication can be only one contributing factor to defective beta-adrenergic function. This is also reflected by the presence of beta-adrenergic subsensitivity in atopic dermatitis in which beta-adrenergic medication is not used as a therapeutic modality. Nevertheless, endogenous release of catecholamines in response to the allergic tissue injury may contribute to the development of beta-adrenergic subsensitivity through *homologous* desensitization of the beta-adrenergic receptors. At the same time the endogenous release of other pharmacologic mediators (i.e., histamine) in response to the allergic tissue injury may contribute to the beta-adrenergic subsensitivity through *heterologous* desensitization [36]. The differences between these two mechanisms are explained below.

16. THE ALLERGIC TISSUE INJURY, BETA ADRENERGIC SUBSENSITIVITY, AND ATOPIC DERMATITIS

Differences between homologous and heterologous desensitization resulting in beta-adrenergic subsensitivity could be most conveniently explained through the model of Delean and associates [130] originally used for the interpretation of adrenoceptor-adenylate cyclase interactions and based to a large extent on ligand binding experiments. This model envisions three principal components of the system in the plasma membrane, i.e., the receptor (R), a nucleotide-binding regulatory protein (N), and adenylate cyclase (C). The binding of an agonist (A) to the receptor is believed to bring about a change that either promotes or stabilizes the formation of a ternary complex, ARN. The formation of the complex promotes the dislodging of a tightly bound guanosine diphosphate (GDP) molecule from N and its replacement with guanosine triphosphate (GTP), which enables N to activate C, thus stimulating the formation of cAMP. The GTP is subsequently hydrolyzed to GDP by guanosine triphosphatase (GTPase) associated with N, and this leads to inactivation of the cyclase. All three components of the system are then capable of being reactivated by renewed interactions of R with agonist molecules. The formation of the ternary complex ARN is of crucial importance to the functional coupling of the receptor to the cyclase in this hypothetical model. Antagonists occupy the receptor but do not promote or stabilize the formation of the ternary complex and thus do not activate the catalytic moiety of adenylate cyclase. Biochemical experiments indicate that the free form of the receptor (R) has a low affinity for agonists, while the complex of the receptor plus the nucleotide protein (RN) has a high affinity for agonists. Changes in receptor binding properties showing a lack of high affinity of R for agonists are thus believed to indicate that the crucial ARN complex is not formed. Not surprisingly, there is no activation of the cyclase under these circumstances. This phenomenon resulting in the inability of agonist-stimulated beta-receptors to activate adenylate cyclase is referred to as *uncoupling*. As discussed below, uncoupling is one of the mechanisms by which agonist—induced desensitization of beta-adrenergic receptors take place [131].

The loss of a tissue's responsiveness to an agonist caused by repeated exposure to the agonist has been described using a variety of terms including desensitization, tolerance, refractoriness, and tachyphylaxis. Su and colleagues [132] have divided these phenomena into two major categories, heterologous and homologous desensitization. Heterologous desensitization refers to the desensitization that occurs after exposure of cells to a biologically active agent that produces tissue refractoriness to itself and to a variety of other pharmacologically different agonists. By contrast, homologous or agonist—specific desensitization is a loss of responsiveness to only the particular agonist that induced the desensitization (or to a specific group of pharmacologically related agonists, all acting at the same tissue receptor site, e.g., the catecholamines).

The mechanisms by which desensitization is produced are complex and variable depending on the tissue. Isoproterenol-induced desensitization of turkey erythrocytes appears to fall into the category of heterologous desensitization because loss of sensitivity to fluoride ion and 5'guanylylimidodi- phosphate (Gpp(NH)p)* (a less hydrolyzable analog that can substitute for GTP) is also produced in this situation. There is no decrease in receptor number in the isoproterenol-desensitized turkey erythrocyte, but rather an apparent uncoupling of the beta-adrenergic receptor from adenylate cyclase takes place due to impairment of the ability of occupied receptors to form a stable high-affinity ARN complex [133]. Similar refractoriness can be produced by exposure of the cells to 8-bromo-adenosine 3',5'-cyclic monophosphate, a cAMP analog, suggesting that the desensitization to isoproterenol is caused by the agonist-stimulated levels of cAMP within cells. Stadel and co-workers [133] suggested that cAMP-dependent phosphorylation of the nucleotide regulatory protein (N) may be the mechanism of desensitization in this system, and thus no alteration in the receptor per se but rather uncoupling of the receptor from adenylate cyclase due to modification of the nucleotide regulatory protein takes place in this example of heterologous desensitization.

Homologous desensitization of beta-receptors has been observed in a number of different types of cells including frog erythrocytes and astrocytoma cells and is produced by agonists but is blocked by antagonists [134]. Experiments on astrocytoma cells have shown that beta-adrenoceptor desensitization produced by isoproterenol involves at least two steps. The earliest change to occur is uncoupling of the receptor from the cyclase, followed by the second step, which involves a loss of 80 to 95 percent of the assayable beta-receptors on the cell surface [135]. More recent studies on catecholamine-desensitized frog erythrocytes have shown that the down-regulated (unavailable) beta-receptors are sequestered in cytosolic vesicles apart from the guanine regulatory protein and catalytic moiety of the adenylate cyclase which remain in the plasma membrane [136].

The sequestered receptors appear not to be rapidly degraded and, therefore, may be recycled later during recovery of the tissue from desensitization [136,137]. Human neutrophils have been found to undergo desensitization involving both an uncoupling, which is highly analogous to that demonstrated in frog erythrocytes, and also a 40 percent reduction in number of receptors [138].

Drawing on the information available at the present time, one might formulate the following view: heterologous desensitization of beta-adrenoceptors involves uncoupling due to impairment of the ability of receptors to form the high-affinity ARN complex and consequently to activate adenylate cyclase; this impairment is produced by the agonist-stimulated accumulation of cAMP within cells. By contrast, homologous regulation is a multistep process involving early uncoupling of the receptor from the cyclase followed by later internalization of the uncoupled receptors in vesicles.

With this understanding we can return to the foregoing section on the reactivities of the mediator—storing cells to antigenic and pharmacologic influences and their relations to cyclic nucleotides. As stated earlier, in addition to beta adrenergic agents, allergic release of histamine or of many, if not most of other pharmacologic mediators of immediate hypersensitivities, is also inhibited by prostaglandins of the E series, prostacycline, adenosine, and histamine (i.e., by substances that interact with cell membrane receptors that activate adenylate cyclase). Normal physiologic inhibition of allergic mediator release by these agents is kept in check by their feedback effect on their respective receptors through the mechanisms of homologous and heterologous desensitization. In atopic disease, however, where there are multiple abnormalities in the receptor-adenylate cyclase- and cyclic AMP systems, this physiologic balance between inhibition versus enhancement of mediator release would be expected to lead to an exaggerated release reaction to the allergic tissue injury.

Indeed, enhanced "releasability" of histamine from basophils and mast cells has been shown to occur in atopic dermatitis. "Releasability" is defined in this context as the capacity of mediator secreting cells to release preformed or newly synthesized mediators [139–141]. Among the pharmacologic mediators, histamine is the best-studied substance, and the best-established mechanism is the IgE-mediated release reaction. The first study to show enhanced anti-IgE induced histamine releasability from basophils was performed by Lebel et al. [142], which was confirmed by Ring and his associates [130]. During the last few years numerous, more extensive investigations have been carried out that further confirmed the *de facto* existence of altered releasability in atopic dermatitis [71,144–146]. Similarly, increased releasability of histamine was also found to occur in bronchial asthma [147,148].

In a way, in vitro IgE-secretion by peripheral lymphocytes might also be viewed as a form of "*releasability*" too. Therefore, it is important to mention that several authors have provided evidence of increased spontaneous in vitro IgE-secretion in patients with atopic dermatitis

* 5'Guanylylimidodiphosphate is an analog of GTP that contains an imidodiphosphate rather than a pyrophosphate linkage.

[149–151]. A significant positive correlation between serum IgE and *in vitro* IgE secretion has also been demonstrated [144]. In this connection, it should be noted that although much is known about IgE regulation in rodents [150], the mechanisms involved in the regulation of IgE-synthesis in man are not well established. It is generally assumed that isotype specific suppressor and helper T-cells play an important role, but the relevant subpopulation (perhaps FcE-receptor bearing lymphocytes, [152]), is not known at present. Furthermore, the exact pathogenetic role of IgE—reactions is atopic dermatitis is still controversial [153,154].

17. AUTOIMMIUNITY AS A DEVELOPMENTAL MECHANISM OF BETA-ADRENERGIC SUBSENSITIVITY

The concept that an autoantibody interacting with a cell membrane receptor of a hormone or neurotransmitter could cause functional derangements and subsequent disease is now becoming widely accepted, and the number of diseases that may be mediated by antireceptor antibodies is rapidly growing [155,156].

The leading examples of such diseases include myasthenia gravis, involving autoantibodies directed at nicotinic acetylcholine receptors at the neuromuscular end-plates [157–159], Graves' disease involving autoantibodies to the thyrotropin receptor [160,161] and the severe insulin resistance in Type B insulin-resistant diabetes that has been ascribed to autoantibodies to the insulin receptor [162–164]. Thus, the foregoing diseases may be viewed as receptor diseases, and some subsets of asthma and other atopic diseases may ultimately be recognized as legitimate members of this group. Indeed, it has been described that autoantibodies to beta adrenoceptors can be identified in the plasma of some subjects with atopic allergy [165–167]. Although these antibodies appear to be heterogeneous, they share the ability to affect binding of ^{125}I protein A to calf lung membranes, to inhibit stereospecific beta-adrenergic radioligand binding to calf lung beta$_2$ adrenoceptors, and to precipitate solubilized calf lung beta-adrenergic receptors in an indirect immunoprecipitation assay. Furthermore, the presence of autoantibodies to beta$_2$-adrenoceptors in these subjects correlates well with a reduced beta$_2$ and an increased alpha-adrenergic responsiveness.

It may be added that from the currently available material even the three of the 19 apparently normal subjects with circulating antibodies were significantly less responsive to beta-adrenergic stimulation than the remainder of the normal controls [168].

The precise frequency and distribution of these autoantibodies in various subsets of patients with asthma and other atopic disease are currently under investigation in several American and European laboratories, as is the molecular mechanism by which they produce beta-adrenergic subsensitivities. For a general account of the molecular mechanisms that are involved in the development of autoimmunities in general, the reader is referred to an analysis by Szentivanyi and Szentivanyi [100]. In the orientation of this chapter, however, we shall only mention the role of virus infections as a developmental-mechanism of autoimmunity produced by anti- receptor antibodies specifically directed to beta adrenoceptors. Thus, virus infections can elicit autoantibody formation by two mechanisms. First, the viral antigens and autoantigens may become associated to form immunogenic units. Viral antigens stimulating host T-lymphocytes could then function as helper determinants, thereby stimulating B-lymphocyte responses to auto-antigens. Second, some viruses such as the Epstein-Barr virus (EBV) stimulate proliferation of the B-lymphocyte cell line with autoantibody production. There are two ways in which viral and host antigens can form immunogenic units. Host antigens can be incorporated in the envelopes of some viruses, and viral antigens can appear on the surfaces of infected host cells [100]. The viral antigens also may form complexes with and modify histocompatibility antigens or other membrane constituents such as the contractile protein, actin. The modified viral antigens could stimulate T-cell helper effect and elicit autoantibody formation [100].

In humans, infection with viruses such as influenza, measles, varicella, and herpes simplex has often resulted in autoimmune manifestations such as platelet and red cell autoantibodies. The development of cold autoagglutinins after *Mycoplasma pneumoniae* infection probably occurs by a T-cell bypass mechanism. Following infectious mononucleosis, many patients' sera often react against several autoantigens. These include autoantibodies against nuclei, lymphocytes, erythrocytes, and smooth muscle. In addition, cross-reactive heterophile antibodies may be noted following infectious mononucleosis and other infections. The autoantibody is produced by a mechanism similar to that observed in altered self-component with virus or bacteria.

There has been much speculation about the possible involvement of an oncornavirus in the pathogenesis of human systemic lupus erythematosus [169].

REFERENCES

1. Szentivanyi A. The beta adrenergic theory of the atopic abnormality in bronchial asthma. J Allergy 1968; 42:203.
2. Szentivanyi A. Invited Editorial: The radioligand binding approach in the study of lymphocytic adrenoceptors and the constitutional basis of atopy. J Allergy Clin Immunol 1980; 65:5.
3. Szentivanyi A, Fitzpatrick DF. The altered reactivity of the effector cells to antigenic and pharmacological influences and its relation to cyclic nucleotides. II. Effector reactivities in the efferent loop of the immune response, in: Grafelfing bei Munchen. Fillip G, editor. Werk-Verlag (1980).
4. Szentivanyi A, Middleton W, Friedman H. Effect of microbial agents on the immune network and associated pharmacologic reactivities. In: "Allergy: Principles and Practice," Second Edition. Middleton E, Reed CE, Ellis EF, editors. St. Louis: The C. V. Mosby Company, 1983.
5. Hirschman JV. Everett ED. Haemophilus influenze infections in adults: report of nine cases and a review of the literature. Medicine 1979; 58:80.
6. Bourne HR, Lichtenstein LM, Melmon KL. Modulation of inflammation and immunity by cyclic AMP. Science 1974; 184:19.
7. Schreurs AJK, Terpstra CK, Raaijmakers JAN, Nijkamp FP. Effects of vaccination with Haemophilus influenzae on adrenoceptor function of tracheal and parenchymal strips. J Pharmac Exp Ther 1980; 215:69.
8. Terpstra GK, Raaijmakers JAM, Kreukniet J. Comparison of vaccination of mice and rats with Haemophilus influenzae and Bordetella pertussis as models of atopy. Clin Exp Pharmac Physiol 1979;6:139.
9. Roldaan AC, Masural N. Viral respiratory infections in asthmatic children staying in a mountain resort. Eur J Respir Dis 1982; 63:140.
10. Hall WJ, Hall CB. Bacterial and viral infections in etiology and therapy. In: "Bronchial Asthma: Mechanisms and Therapeutics," Second Edition. Weiss EB, Segal M, Stein M, editors. Boston and Toronto: Little, Brown and Company, 1985.
11. Katz DH. The allergic phenotype: manifestation of "allergic breakthrough" and imbalance

in normal "damping" of IgE antibody production. Immunol Rev 1978; 41:77.
12. Frick OL, German DF, Mills J. Development of allergy in children. I. Association with virus infections. J Allergy Clin Immunol 1979; 63:228.
13. Empey DV, Laitinen LA, Jacobs L. Mechanisms of bronchial hyperreactivity in normal subjects after upper respiratory tract infection. Am Rev Respir Dis 1976; 113:131.
14. Nadel JA. Pathophysiology of asthma: neural effects on airway smooth muscle, submucosal glands, breathing and cough. Prog Respir Res 1980; 14:1.
15. McIntosh K, Ellis EF, Hoffman LS. The association of viral and bacterial respiratory infections with exacerbations of wheezing in young asthmatic children. J Pediatr 1973; 82: 578.
16. Minor TE, Dick EC, DeMeo AN. Viruses as precipitants of asthmatic attacks in children, JAMA 1974; 227:292.
17. Buckner CK, Clayton DE, Shoka AA. Parainfluenza 3 infection blocks the ability of a beta-adrenergic receptor agonist to inhibit antigen induced contraction of guinea pig isolated airway smooth muscle. J Clin Invest 1981; 67:376.
18. Busse WW. Decreased granulocyte response to isoproterenol in asthma during upper respiratory infections. Am Rev Respir Dis 1977; 115:783.
19. Mills JE, Sellick M, Widdicombe JC. Epithelial irritant receptors in the lung in breathing. In: "Breathing: Herring-Breuer Centenary Symposium." Porter R , editor. London: Churchill Livingstone, 1970.
20. Welliver RC, Kaul TN, Ogra PL. The appearance of cell-bound IgE in respiratory-tract epitheliuin after respiratory syncytial virus infection. N Eng J Med 1980; 303:1198.
21. Richardson JB, DeNotariis A, Ferguson CC, Boucher RC. Effect of a viral laryngotracheitis on the epithelial barrier of chicken airways. Lab Invest. 1981; 44:144.
22. Ida S. Enhancement of IgE mediated histamine release from human Basophils by viruses: role of interferon. J Exp Med 1977; 145:8.
23. Ouellette JJ, Reed CE. Increased response of asthmatic subjects after influenza vaccine. J Allergy 1965; 38:558.
24. Simonsson BG, Svedmyr N, Skoogh BE, Andersson R, Bergh NP. In vivo and in vitro studies of alpha-receptors in human airways. Potentiation with bacterial endotoxin. Scan J Re Dis 1972; 53:227.
25. Schreurs AJK, Verhoef J, Nijkamp FP. Bacterial endotoxins decrease the number of respiratory beta-drenergic receptors in guinea pig lung. Agents Actions 1983; 13:449.
26. Hall WJ, Douglas RC Jr, Hyde RW, Roth FK, Cross AS, Speers DM. Pulmonary mechanics after uncomplicated influenza A infection. Amer Rev Respir Dis 1976; 113:141.
27. Little JW, Hall WJ, Douglas RG Jr, Hyde RW, Speers DM. Amantadine effect on peripheral airways abnormalities in influenza. Ann Intern Med 1976; 85:177.
28. Henson PM. The immunologic release of constituents from neutrophil leukocytes. I. The role of antibody and complement on nonphagocytosable surfaces or phagocytosable particles. J Immunol 1971; 107:1525.
29. Busse WW, Cooper W, Warshaver DM, Dick EC, Wallow IHL Albrecht R. Impairment of isoproterenol, H_2 histamine, and prostaglandin E_1 response of human granulocytes after incubation in vitro with live influenza vaccines. Am Rev Respir Dis 1979; 119:561.
30. Busse WW, Anderson CL, Dick EC, Warshaver DM. Reduced granulocyte response to isoproterenol, histamine and prostaglandin E_1 after in vitro incubation with rhinovirus 16. Am Rev Respir Dis 1980; 122:641.
31. Bush RK, Busse WW, Flaherty DK, Warshaver D, Dick EC, Reed CE. Effects of

experimental rhinovirus 16 infection on airways and leukocyte function in normal subjects. J Allergy Clin Immunol 1978; 61:80.
32. Ignarro LJ, Colombo C. Enzyme release from polymorphonuclear leukocyte lysosomes: regulation by autonomic drugs and cyclic nucleotides. Science 1973; 180:1181.
33. Lee TP. Virus exposure diminishes beta-adrenergic response in human leukocytes. Res Comm Chem Path Pharmac 1980; 30:469.
34. Norris AA, Eyre P. Pharmacological abnormality in bronchial asthma and the role of respiratory pathogens. Medical Hypotheses: 1982; 8:199.
35. Szentivanyi A, Williams JF. The constitutional basis of atopic disease. Chapter 14, in: "Allergic Diseases of Infancy, Childhood, and Adolescence," Bierman CW, Pearlman DS editors. Philadelphia and London: W. B. Saunders Company, 1980.
36. Szentivanyi A, Polson JB, Szentivanyi J. Adrenergic regulation, in: Bronchial Asthma: Mechanisms and Therapeutics. Weiss GB, Segal MS, Stein M, editors. Boston: Little, Brown and Company, 1985.
37. Holgate JA, Warner BT. Evaluation of antagonists of histamine, 5-hydroxytryptamine, and acetylchohine in the guinea pig. Br J Pharmacol 1960; 15:561.
38. Douglas JS, Dennia W, Ridgway P, Bouhuys A. Airway constriction in guinea pigs: interactions of histamine and autonomic drugs. J Pharmacol Exp Ther 1973, 184:169.
39. Douglas JS, Brink C, Bouhuys A. Actions of histamine and other drugs on guinea pig airway smooth muscle. In: "Lung Cells in Disease." Bouhuys A, editor. Amsterdam: Elsevier Biomedical Press, 1976.
40. Altounyan REC. Variation of drug action on airway obstruction in man. Thorax 1964; 19: 406.
41. Chan-Yeung MNW, Vyas NN, Crzybowski S. Exercise-induced asthma. Am Rev Respir Dis 1971; 104:915.
42. Casterline CL, Evans R III, Ward CV Jr. The effect of atropine and albuterol aerosols on the human bronchial response to histamine. J Allergy Clin Immunol 1976; 58:607.
43. Fish JE, Rosenthal RR, Summer WR, Menkes H, Norman PS, Permutt S. The effect of atropine on acute antigen-mediated airway constriction in subjects with allergic asthma. Am Rev Respir Dis 1977; 115:371.
44. Deal EC Jr, McFadden ER Jr, Ingram RH Jr, Jaeger JJ. Effects of atropine on potentiation of exercise-induced bronchospasm by cold air. J Appl Physiol 1978; 45:238.
45. Woenne R, Kattan M, Orange RP, Levison H. Bronchial hyperreactivity to histamine and methacholine in asthmatic children after inhalation of SCH 1000 and chlorpheniramine maleate. J Allergy Clin Immunol 1978;62:119.
46. Rosenthal RR, Norman PS, Sumner WK, Permutt S. Role of the parasympathetic system in the antigen-induced bronchospasm. J Appl Physiol 1977; 42:600.
47. Bandouvakis J, Cartier A, Roberts R, Ryan G, Hargreave FE. The effect of ipratropium and fenoterol on methacholine- and histamine-induced bronchoconstriction. Br J Dis Chest 1981; 75:295.
48. 0'Bryne PM, Thomson NC, Morris M, Roberts RS, Daniel EE, Hargreave FE. The protective effect of inhaled chlorpheniramine and atropine on bronchoconstriction stimulated by airway cooling. Am Rev Respir Dis 1983; 128:611.
49. Boulet LP, Latimer KM, Roberts RS, Juniper EF, Cockroft 0W, Thomson NC, Daniels E, Hargreave FE. The effect of atropine on allergen-induced increases in bronchial responsiveness to histamine. Am Rev Respir Dis 1984; 130:368.
50. 0'Bryne PM, Thomson NC, Latimer KM, Roberts RS, Morris MM, Daniel EE, and

Hargreave FE. The effect of inhaled hexamethonium bromide and atropine sulphate on airway responsiveness to histamine. J Allergy Clin Immunol 1985; 76:97.
51. Szentivanyi A, Krzanowski JJ, Polson JB. The autonomic nervous system, in: Allergy: Principles and Practice, Second Edition, Middleton E, Reed CE, Ellis EF, editors. St. Louis: The C. V. Mosby Company, (1983).
52. Haddock AM, Patel KR, Alston WC, Kerr JW. Response of lymphocyte guanyl cyclase to propranolol, noradrenaline, thymoxamine, and acetyleholine in extrinsic bronchial asthma. Br Med J 1975; 2:357.
53. Lang P, Goel Z, Grieco MH. Subsensitivity of T lymphocytes to sympathomimeti and cholinergic stimulation in bronchial asthma. J Allergy Clin Immunol 1978; 61:248.
54. Logsdon PJ, Carnright DV, Middleton E, Coffey KG. The effect of phentolamine on adenylate cyclase and on isoproterenol stimulation in leukocytes from asthmatic and nonasthmatic subjects. J Allergy Clin Immunol 1973; 52:148.
55. Alston WC, Patel KR, Kerr JW. Response of leukocyte adenyl cyclase to isoprenaline and effect of alpha-blocking drugs in extrinsic bronchial asthma. Br Med J 1974; 1:90.
56. Michelson MJ, Danilov AF. Cholinergic transmissions, in:"Fundamentals of Biochemical Pharmacology." Bacq ZM, ed. Oxford: Pergamon Press, 1971.
57. Szentivanyi A. "Adrenergic and Chohinergic Receptor Studies in Human Lung and Lymphocytic Membranes and their Relations to Bronchial Hyperreactivity in Asthma." Darien: Patient Care Publications, Inc., 1982.
58. Szentivanyi A. Effect of bacterial products and adrenergic blocking agents on allergic reactions. In: Textbook of Immunological Diseases, Samter M. editor.Boston: Little, Brown and Company, 1971.
59. Bonifazi E, Garofalo L, Pisani V, Meneghini CL. Role of some infectious agents in atopic dermatitis. Acta Derm Venereol (Stockh.) Suppl 1985, 114:98.
60. Strannegard O, Strannegard IL, Rystedt I. Viral infections in atopic dermatitis. Acta Derm Venereol (Stockh.) Suppl 1985;114:121.
61. Strannegard O, Strannegard IL T-lymphocyte numbers and function in human IgE-immediated allergy. Immunol Rev 1978; 41:149.
62. Schuster DL, Pierson D, Bongiovanni B, Levinson AF. Suppressor cell function in atopic dermatitis associated with elevated immunoglobuhin E. J Allergy Clin Immunol 1979; 64: 139.
63. Jensen JR, Cramers M, Thestrup-Pedersen K. Subpopulations of T-lymphocytes and non-specific suppressor cell activity in patients with atopic dermatitis. Clin Exp Immunol 1981; 45:118.
64. Strannegard O, Strannegard IL, Kang K, Cooper KD, Hanifin JM. FcIgG receptor-bearing lymphocytes and monoclonal antibody defined T-cell subsets in atopic dermatitis. Effect of treatmentwith thymopoletin pentapeptide (TP-5). Int Arch Allergy Appl Immunol 1982; 69:328.
65. Kang K, Cooper KD, Vanderbark A, Strannegard IL, Strannegard O, Hanifin JM. Immunoregulation in atopic dermatitis: T-lymphocyte subsets defined by monoclonal antibodies. Semin Dermatol 1983; 2:59.
66. Kragballe K, Herhin T, Jensen JR. Impaired monocyte cytotoxicity in atopic dermatitis. Arch Dermatol Res 1980; 269:21.
67. Jensen JR, Sard TT, Jorgensen AS, Thestrup-Pedersen K. Modulation of natural killer cell activity in patients with atopic dermatitis. J Invest Dermatol 1984; 82:30.
68. Strannegard IL, Strannegard O. Natural killer cells and interferon production in atopic

dermatitis. Acta Derm Venereol (Stockh.) Suppl 1980; 92:48.
69. Grewe S, Chan S, Hanifin JM. Elevated leukocyte cyclic AMP phosphodiesterase in atopic disease: a possible mechanism for cAMP agonist hyporesponsiveness. J Allergy Clin Immunol 1982; 70:452.
70. Butler JM, Chan SC, Stevens SR, Hanifin JM. Increased leukocyte histamine release with elevated cyclic AMP-phosphodi- esterase activity in atopic dermatitis. J Allergy Clin. Immunol 1983; 71:490.
71. McMillan JC, Heskel NS, Hanifin JM. Cyclic AMP-phospho- diesterase activity and histamine release in cord blood leukocyte preparations. Acta Derm Venereol (Stockh.) Suppl 1985; 114:24.
72. Safko MJ, Chan SC, Cooper KD, Hanifin JM. Heterologous densensitization of leukocytes: a possible mechanism of beta adrenergic blockade in atopic dermatitis. J Allergy Clin Immunol 1981; 68:218.
73. Busse WW, Lee, TP. Decreased adrenergic responses in lymphocytes and granulocytes in atopic eczema. J Allergy Clin. Immunol 1976; 58:586.
74. Reed CE, Busse WW, Lee TP. Adrenergic mechanisms and the adenyl cyclase system in atopic dermatitis. J Invest Dermatol 1976; 67:333.
75. Parker CW, Eisen AZ. Altered cyclic AMP metabolism in atopic eczema. Clin Res 1972; 20:418.
76. Parker CW, Kennedy S, Eisen AZ. Leukocyte and lymphocyte cyclic AMP response in atopic eczema. J Invest Dermatol 1977; 68:30.
77. Pochet R, Delepesse G, DeMaubeuge J. Characterization of beta adrenoreceptors on intact circulating lymphocytes from patients with atopic dermatitis. Acta Derm Venereol (Stockholm) 1980; 92:26.
78. Szentivanyi A, Heim O, Schultze P, Szentivanyi J. Adrenoceptor binding studies with [^{3}H]dihydroalprenolol and 11^{3}H]dihydroergocryptine on membranes of lymphocytes from patients with atopic disease. Acta Dermato Venereol Suppl 1980; 92:19.
79. Cooper KD, Chan SC, Hanifin JM. Differential T and B cell beta adrenergic responsiveness implies alternative regulatory mechanisms in atopy. Clin Res 1981; 29:281A.
80. Cooper KD, Chan SC, Hanifin JM. Absence of low affinity beta adrenergic receptors in atopic dermatitis and histamine desensitized normal leukocytes. Clin Res 1982; 30:27A.
81. Cooper KD, Chan SC, Hanifin JM. Lymphocyte and monocyte localization of altered adrenergic receptors, cAMP responses, and cAMP phosphodiesterase in atopic dermatitis. A possible mechanism for abnormal radiosensitive helper T cells in atopic dermatitis. Acta Derm Venereal (Stockh.) Suppl. 1985; 114:41.
82. Holden CA, Austen DR, Chan SC, Hanifin JM. Defective cyclic nucleotide metabolism in atopic monocytes. Clin Res 1984; 32:591A.
83. Holden CA, Chain SC, Hanifin JM. Adenylate cyclase activity in mononuclear leukocytes from patients with atopic dermatitis. Acta Derm Venereol (Stockh.) Suppl 1985; 114:149.
84. Giustina TA, Chan SC, Thiel NL, Baker JV, Hanifin JM. Increased leukocyte sensitivity to PDE inhibitors in atopic dermatitis: tachyphylaxis after theophylline therapy. J Allergy Clin. Immunol 1984; 74:252.
85. Rheinherz EL, Morimoto C, Penta JA, Schlossman S. Subpopulations of the T4 positive inducer cell subset in man. J Immunol 1981, 126:67.
86. Uchiyama T, Nelson D, Fleisher T, Waldmann T. A monoclonal antibody (Anti-T) reactive with activated and functionally mature human T cells. II. Expression of Tac antigen on activated cytoxic killer T cells. J Immunol 1981; 126:1398.

87. Leung DYM, Wood N, Dubey D, Rhodes AR, Geha RS. Cellular basis of defective cell mediated lympholysis in atopic dermatitis. J Immunol 1983; 130:1678.
88. Lobitz WC, Honeyman JF, Winkler MW. Suppressed cell mediated immunity in two adults with atopic dermatitis. Br J Dermatol 1972; 86:317.
89. Elliott ST, Hanifin JM. Lymphocyte response to phytohemagglutinin in atopic dermatitis. Arch Dermatol 1979; 115:1424.
90. Rheinherz EL, Kung P, Goldstein G, Schlossman S. Further characterization of helper/ inducer T cell subsets defined by monoclonal antibody. J Immunol 1979; 123:2894.
91. Stevens R, Macy E, Morrow C, Saxon A. Characterization of a circulating subpopulation of spontaneous antitetanus toxoid antibody production. J Immunol 1980; 122:2498.
92. Leung DYM, Rhodes AR, Geha RS. Enumeration of T cell subsets in atopic dermatitis using monoclonal antibodies. J Allergy Immunol 1981; 67:450.
93. Schuster DL, Bongiovanni BA, Pierson DO, Barbaro JF, Wong DT, Levinson AJ. Selective deficiency of a T cell subpopularion in active atopic dermatitis. J Immunol 1980; 124:1662 141.
94. Cooper KD, Kazmierowski JA, Wuepper KD, Hanifin JM. Inmunoregulation in atopic dermatitis: functional analysis of T-B cell interactions and the enumeration of Fc receptor bearing T cells. J Invest Dermatol 1983; 80:139.
95. Henney CS, Gillis S. Cell mediated cytotoxicity In: Fundamental Immunology, Paul W, editor.New York: Raven Press, 1984.
96. Rheinherz EL, Schlossman S. Regulation of the immune response-inducer and suppressor T lymphocyte subsets in human beings. N Eng J Med 1980; 303:370.
97. Broder S, Uchiyama TT, Muul L, Goldman C, Sharrow S, Poplack D, Waldman T. Activation of leukemic pre-suppressor cells to become suppressor effector cells. N Eng J Med 1981; 304:1382.
98. Nordbring F, Johansson SGO, Espmark A. Raised serum levels of IgE in infectious mononucleosis. Scand J Infect Dis 1972; 4:119.
99. Barnetson R, Hardie RA, Merret TG. Late onset atopic eczema and multiple food allergies after infectious mononucleosis. Br Med J 1981; 283:1086.
100. Szentivanyi A, Szentivanyi J. The pathophysiology of immunologic and related diseases. In: Sodeman's Pathologic Physiology – Mechanisms of Disease. Seventh Edition. SodemanWA, Sodeman TM, editors. Philadelphia and London: W. B. Saunders Company, 1985b.
101. Szentivanyi A, Szentivanyi J. Cellular and molecular foundation of immunity, immunologic inflammation, and hypersensitivity. Component parts and their relation to neurohumoral control mechanisms, in: Sodeman's Pathologic Physiology - Mechanisms of Disease, Seventh Edition, Sodeman WA, Sodeman TM, editors. Philadelphia and London:W.B. Saunders Company, 1985a.
102. Schild HO, Histamine release and anaphylactic shock in isolated lungs of guinea pigs.Q J Exp Physiol 1936; 26:165.
103. Tabachnick JJ, Gulbenkian A, Schobert LJ. Effects of a beta adrenergic blocking agent and isopropylmethylamine on the release of histamine from guinea pig lung during anaphylaxis in vitro. Biochem Pharmacol 1965, 14:1283.
104. Lichtenstein LM, Margolis S. Histamine release in vitro: inhibition by catecholamines and methylxanthines. Science 1968; 161:902.
105. Bourne HR, Lichtenstein LM, Melmon KL. Modulation of inflammation and immunity by cyclic AMP. Science 1974; 184:19.

106. Orange RP, Austen WG, Austen KF. Immunological release of histamine and slow-reacting substance of anaphylaxis from human lung. I. Modulation by agents influencing cellular levels of cyclic 3',5'- monophosphate. J Exp Med 1971; 134:136.
107. Plaut M, Marone G, Thomas LL, Lichtenstein LM. Cyclic nucleotides in immune responses and allergy. Adv Cyclic Nucleotide Res 1980; 12:161.
108. Marone G, Sobotka AK, Liebtenstein LM. Effects of arachidonic acid and its metabolites on antigen-induced histamine release from human basophils in vitro. J Immunol 1979; 123:1669.
109. Szentivanyi A, Fishel CW. Neurohumoral concepts of bronchial asthma. Acta Allergologica 1974; 29 (Suppl.11):26.
110. Krzanowski JJ, Polson JB, Szentivanyi A. Pulmonary patterns of adenosine-3',5' cyclic monophosphate accumulations in response to adrenergic or histamine stimulation in Bordetella pertussis—sensitized mice. Biochem Pharmacol 1976, 25:1631.
111. Polson JB, Krzanowski JJ, Szentivainyi A. Pulmonary guanosine-3' ,5'-monophosphate in sensitized mice: accumulations produced by H-1 receptor activation and evidence of altered phosphodiesterase activity. J Allergy Clin Immunol 1976; 57:265.
112. Kahiner M, Orange RP, Austen KF. Immunologic release of histamine and slow reacting substance of anaphylaxis from human lung. IV. Enhancement by chohinergic and alpha adrenergic stimulation. J Exp Med 1972; 136:556.
113. Gillespie E, Lichtenstein LN. Histamine release from human leukocytes: relationships between cyclic nucleotide, calcium, and antigen concentrations. J Immunol 1975; 115: 1572.
114. Harrison LE, Flier JS, Kahn CR, Jarrett, DB, Muggeo M, Roth J. In: Genetic Control of Autoimmune Disease. Rose NR, Bigazzi PE, Warner NL, editors. New York. North Holland Biomedical Press, 1978.
115. Nakazawa H, Townley RG. The effects of propranolol on the bronchoconstriction by histamine, acetyl-beta-mechylcholine, and anaphylaxis in inbred guinea pigs in vivo. Ann Allergy 1976; 36:77.
116. deVries K, Koeter GH, Gokemeyer JDM. Some aspects of the regulation of the bronchial tree in obstructive lung disease. Eur J Respir Dis 1982; 63(Supp. 121):60.
117. Koeter GH, Meurs H, Kauffman HF, deVries K. The role of the adrenergic system in allergy and bronchial hyperreactivity. Eur J Respir Dis 1982; 63 (Suppl. 121):72.
118. Barnes PJ, Dollery CT, MacDermot J. Increased pulmonary alpha adrenergic and reduced beta- adrenergic receptors in experimental asthma. Nature, 1980; 285:569.
119. Rinard GA, Jensen AD, Puckett AM, Torphy TJ, Minneman KP, Mayer SE. Depressed beta-adrenergic response(cyclic AMP, protein kinase, relaxation) and receptor density intracheal smooth muscle of allergic-asthmatic dogs, 5th International Conference on Cyclic Nucleotides and Protein Phosphorylation, Milan, Italy (1983).
120. Rinard GA, Rubinfeld AR, Burnston LL, Mayer SE. Depressed cyclic AMP levels in airway smooth muscle from asthmatic dogs. Proc Natl Acad Sci USA 1979; 76:1472.
121. Kunos G, Szentivanyi A. Evidence favoring the existence of a single adrenergic receptor. Nature 1968; 217:1077.
122. Szentivanyi, A., and Pek, L., Characteristic changes of vascular adrenergic reactions in diabetes mellitus, Nature, New Biol.1973; 243:276.
123. Szentivanyi A, Kunos G, JuhaszNagy A. Modulator theory of adrenergic receptor mechanism: vessels of the dog hind limb. Amer J Physiol 1970; 218:869.
124. Szentivanyi A, Heim O, Schultze P. Changes in adrenoceptor densities in membranes of

lung tissue and lymphocytes from patients with atopic disease. Ann NY Acad Sci 1979; 332:295.
125. Szentivanyi A, Krzanowski JJ, Polson JB, Anderson W, Evolution of research strategy in the experimental analysis of the beta adrenergic approach to the constitutional basis of atopy, In:"Advances in Allergology and Clinical Immunology. Oehling A, Mathov E, Glazer I, Arbesman C, editors. Oxford: Pergamon Press, 1980.
126. Anthracite RF, Vachon L, Knapp PH. Alpha-adrenergic receptors in the human lung. Psychosom Med 1977; 33:481.
127. Snashall PD, Boother FA, Sterling GM. The effect of alpha-adrenergic stimulation on the airways of normal and asthmatic man. Clin Sci Mol Med 1978; 54:283.
128. Kneussl MP, Richardson JB. Alpha-adrenergic receptors in human and canine tracheal and bronchial smooth muscle. J Appl Physiol 1978, 45:307.
129. Henderson WR, Shelhamer JH, Reingold DB, Smith LJ, Evans R, Kaliner N. Alpha adrenergic hyperreactivity in asthma. Analysis of vascular and pupillary responses. N Eng J Med 1970; 300:642.
130. DeLean A, Stadel JM, Lefkowitz RJ. A ternary complex model explains the agonist-specific binding properties of the adenylate cyclase-coupled beta-adrenergic receptor. J Biol Chem1980;255:7108.
131. Davies AO, Lefkowitz RJ. Agonist-promoted high affinity state of the beta-adrenergic receptor in human neutrophils: modulation by corticosteroids. J Clin Endocrinol Metabol 1981; 53:703.
132. Su YF, Cubeddu-Ximenez L, Perkins JP. Regulation of adenosine-3',5'-monophosphate content of human ascrocytoma cells: desensitization to catecholamines and prostaglandins. J Cyclic Nucleotide Res 1976; 2:257.
133. Stadel JM, DeLean A, Mulhikin-Kilpatrick D, Sawyer DD, Lefkowitz RJ. Catecholamine-induced desensitization in turkey erythrocytes: cAMP mediated impairment of high affinity agonist binding without alteration in receptor number. J Cyc Nucl Res. 1981; 7: 37.
134. Hoffmann BB, Lefkowitz RJ. Radioligand binding studies of adrenergic receptors: new insights into molecular and physiological regulation. Ann Rev Pharmacol Toxicol 1980; 20:581.
135. Su YF, Harden TK, Perkins JP. Catecholamine-specific desensitization of adenylate cyclase: evidence for a multistep process. J Biol Chem 1980, 255:7410.
136. Stadel JM, Strulovici B, Nambi P, Lavin TN, Briggs MM, Caron MG, Lefkowitz RJ. Desensitization of the beta-adrenergic receptor of frog erythrocytes: recovery and characterization of the down-regulated receptors in sequestered vesicles. J Biol Chem 1983; 258:3032.
137. Doss RC, Perkins JP, Harden TK. Recovery of beta-adrenergic receptors following long term exposure of astrocytoma cells to catecholamine: role of protein synthesis. J Biol Chem 1981; 256:12281.
138. Davies AO, Lefkowitz RJ. In vitro desensitization of beta-adrenergic receptors in human neutrophils: attenuation by corticosteroids. J Clin Invest 1983; 71:565.
139. Conroy NC, Adkinson NF, Lichtenscein LN. Measurement of IgE, on human basophils: relation to serum IgE and anti-IgE-induced histamine release. J Immunol 1977; 118: 1317.
140. Ring JD. Reaktionsbereitschaft ("Releasability"): Veranderte reaction- smuster in der freisetzung vasoaktiver medjatoren. Untersuchungen am beispiel der histamin—

freisetzung aus menschlichen basophilen leukozyten. Allergologie 1984; 7:41.
141. Lichtenstein LM, Schleimer RP, Peters SP, Kagey-Sobotka A, Adkinson NF, Adam CK, Schulman HS, MacGlashan DW. Studies with purified human basophils and mast cells. Monogr Allergy 1983; 18:259.
142. Lebel G, Venencie PY, Saurat JH, Soubrane C, Paupe JP, Anti-IgE induced histamine release from basophils in children with atopic dermatitis. Acta Dermato Venerol Suppl 1980,92:57.
143. Ring JD, Mathison DA, Spiegelberg HL. In vitro releasability of histamine and serotonin: studies of atopic patients. J Clin Lab Immunol 1980; 3:85.
144. Ring JD. and Dorsch W, Altered releasability of vasoactive mediator secreting cells in atopic eczema. Acta Dermato Venereol (Stockh) Suppl 1985; 114:9.
145. Butler JM, Chan SC, Stevens SR, Hanifin JM. Increased leukocyte histamine release with elevated cyclic AMP-phosphodi- esterase activity in atopic dermatitis. J Allergy Clin. Immunol 1983; 71:490.
146. Butler JM, Ebertz N, Chan SC, Stevens SR, Sobieszczuk D, Hanifin JM. Basophil histamine release in atopic dermatitis-its relationship to disordered cyclic nucleotide metabolism. Acta Derm Venereol (Stockh) Suppl 1985; 114:55.
147. Findlay SR, Lichtenstein LN. Basophil "releasibility" in patients with intrinsic asthma. J Allergy Clin Immunol 1978; 61:157.
148. Assem ESK, Atallah NA. Increased release of histamine by anti-IgE from leucocytes of asthmatic patients and possible heterogenicity. Clin Allergy 1981; 11:367.
149. Fiser PM Buckley RH. Human IgE biosynthesis; in vitro studies with atopic and normal blood mononuclear cells and subpopulations. J Immunol 1979; 122:1799.
150. Ishizaka K. Regulation of IgE response by IgE-binding factors. Monogr Allergy 1983; 18: 52.
151. Ring, JD, Senner H. In vitro IgE synthesis in patients with atopic eczema. Monogr Allergy 1983;18:242.
152. Spiegelberg HL. Fc receptors for IgE on macrophages and lymphocytes: IgE synthesis. Fed Proc 1983;42:122.
153. Rajka G. Itch and IgE in atopic dermatitis. Acta Derm Venereol (Stockholm) 1980; 92: 38.
154. Wutrich B. "Zur Immunopathologie der Neurodermitis Constitutionalis," Hans Huber Bern 1975.
155. Rose N. The autoimmune diseases. In: Principles of Immunology, Rose N, Milgrom F, Van Oss C, editors. MacMillan Publishing Company, New York, 1978.
156. Svejgaard A, Christy M, Nerup J. HLA and autoimmune diseases with special reference to thegenetics of insulin-dependent diabetes. In: Genetics Control of Autoimmune Diseases, Rose N, Bigazzi P, Warner N, editors. New York: Elsevier/North Holland, 1978.
157. Abramsky O, Aharonov A, Webb C, Fuchs S, Cellular immune response to acetylcholine receptor- rich fraction in patients with myasthenia gravis. Clin Exp Immunol 1975; 19: 11.
158. Drachman DB, Angus CV, Adams RN, Michelson JD, Hoffman CJ. Myasthenic antibodies cross-link acetylcholine receptors to accelerate degradation. N Eng J Med 1978; 298: 1116.
159. Lindstrom JA, Seybold ME, Lennon VA, Whittingham S, Duane DD. Antibody to acetylcholine receptor in myasthenia gravis: prevalence, clinical correlates, and diagnostic value. Neurology 1976; 26:1054.

160. Manley SW, Bourke GR, Hawker KW. The thyrotrophin receptor in guinea pig thyroid homogenate: interaction with the long-acting thyroid stimulator. J Endocrinol 1974; 61: 437.
161. Smith BR, Hall R. Thyroid-stimulating immunoglobulins in Grave's disease. Lancet 1974; 2:427.
162. Flier JS, Kahn CR, Roth J, Bar RS. Antibodies that impair insulin receptor binding in an unusual diabetic syndrome with severe insulin resistance. Science 1975; 190:63.
163. Harrison LE, Flier JS, Kahn CR, Jarrett DB, Muggeo M, Roth J. In: Genetic Control of Autoimmune Disease, Rose N, Bigazzi P, Warner N, editors. New York: North Holland Biomedical Press, 1978.
164. Kahn CR, Flier GS, Bar RS. The syndromes of insulin resistance and acanthosis nigricans: insulin-receptor disorders in man. N Eng J Med 1976, 294:739.
165. Fraser J, Nadeau J, Robertson D, Wood AJJ. Regulation of human leukocyte beta receptors by endogenous catecholamines: relationship of leukocyte beta receptor density to the cardiac sensitivity to isoproterenol. J Clin. Invest 1981; 67:1777.
166. Kahiner M, Shelhamer JH, Davis PB, Smith LJ, Venter JE. Autonomic nervous system abnormalities and allergy. Ann Intern Med 1982; 96:349.
167. Venter JC, Fraser CM, Harrison LC. Auto-antibodies to beta-adrenergic receptors: a possible cause of adrenergic hyporesponsiveness in asthma and allergic rhinitis. Science 1980; 207:1316.
168. Halonen M, Kaliner M. Determinants of autonomic abnormalities in atopy. In: Proceedings of Invited Symposia XI International Congress on Allergy and Immunology. Kerr JW, Ganderton MA, editors. New York: MacMillan Press, 1983.
169. Theofilopoulos AN. Autoimmunity, in: Basic and Clinical Immunology. Fifth Edition. Stites DP, Stobo JD, Fudenberg HH, aWells JW, editors. California: Lange Medical Publications, 1984.

III. RECEPTORS, LIGANDS AND SIGNALING

The Immune-Neuroendocrine Circuitry
Edited by I. Berczi and A. Szentivanyi

Adhesion Molecules

ISTVAN BERCZI and ANDOR SZENTIVANYI

Department of Immunology, Faculty of Medicine, University of Manitoba, Winnipeg, Manitoba, R3E 0W3, Canada; and Department of Internal Medicine, Faculty of Medicine, University of South Florida, Tampa, Florida 33612, USA

ABSTRACT

Adhesion molecules are cell membrane receptors that mediate cell-to-cell and cell-to-matrix communication in macroorganisms. Because the receptors and ligands involved are non-diffusible, adhesion molecules are capable of organ, tissue and cell specific regulation under physiological and pathophysiological conditions. Consequently, this local (positional) regulatory system has the ability to alter/modify/tune systemic regulatory signals and generate signals locally in order to match tissue/cell function with local demands in health and disease. Four major families of adhesion molecules are distinguished. (i) The immunoglobulin superfamily (IgSF): Fc receptors for immunoglobulins, the CD4 and CD8 co-receptors of T cells, major histocompatibility antigens, and the T lymphocyte antigen receptor are important members of this family. (ii) Integrins are transmembrane glycoprotein receptors that play critical roles in matrix-to-cell and cell-to-cell adhesion and signaling. Integrins play a key role in the early development of invertebrates and vertebrates, in the formation of tissue architecture, in cell migration, cell proliferation and differentiation, angiogenesis, inflammation and in tumor metastasis. Beta-1 integrins play crucial roles in lymphocyte adhesion, migration, proliferation and differentiation. Beta-2 integrins mediate leukocyte cell-to-cell and cell-to-matrix adhesions during inflammation and immune responses. (iii) Selectins recognize carbohydrate molecules on pathogenic microorganisms as well as on self-components. They play a role in cellular interactions. Selectins function as cell surface receptors that mediate leukocyte recirculation and natural killer cell activity as well as play a role in inflammation. Patients with the deficient expression of selectin ligand by neutrophils suffer from recurrent infections, persistent leukocytosis and severe growth and mental retardation. (iv) Cadherins are predominantly expressed in the central nervous system but also present in many other tissues. Classical cadherins, desmosomal cadherins, seven-pass transmembrane cadherins and protocadherins are the major families belonging to this superfamily of adhesion molecules. Cell-to-cell signaling by cadherins affects cytoskeletal proteins and the organization of adherens junctions. Seven-pass transmembrane cadherins signal through G-protein coupled receptors, regulate cell polarity and thus determine neuronal morphology. Protocadherins are numerous and several of them display an immunoglobulin-like genetic structure. Nearly all cadherin genes are tightly linked to each other. The level of cadherin expression by cells determines the strength of adhesions and the type determines the specificity of cell interactions. Cadherins regulate epithelial cell shape and differentiation, contribute to central nervous system (CNS) regionalization, morphogenesis, fiber tract formation and to the maintenance of

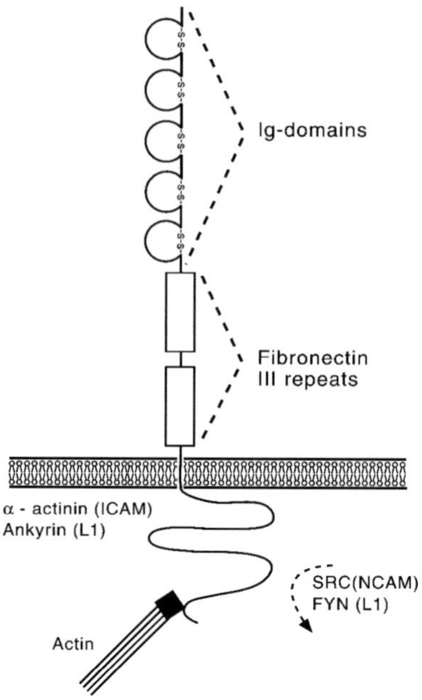

Figure 1. Immunoglobulin Superfamily Adhesion Molecule. Typically these molecules have one or more immunoglobulin like domains that mediate recognition and adhesion and express one or more fibronectin III repeats. The example given here could be representing NCAM, abCAM or fasciclin II of Drosophila L1 that interact with actin in the cytoplasm [Modified from Benson et al. 2000].

functional structures. In reproductive structures, cadherins are responsive to steroid hormones.

1. INTRODUCTION

Adhesion molecules mediate cell-to-cell communication throughout the entire organism. Cell adhesion plays an important role in conception [1] in embryonic development [2], and it regulates cell growth, differentiation and function in every organ and tissue. Adhesion molecules are important regulators under physiological conditions and also play important roles in pathophysiology, including the inflammatory response and immunological reactions [3,4,5,6,7]. The number of diseases where adhesion molecules play a role is rapidly increasing and includes arterosclerosis [8], cancer [9], and pemphigus [10]. Currently, four major groups of adhesion molecules are identified: the immunoglobulin family; integrins; selectins; and cadherins [6].

2. THE IMMUNOGLOBULIN SUPERFAMILY

All members of the immunoglobulin superfamily contain a characteristic loop structure generated by intra-chain disulfide bonds (Ig-like domains) (Figure 1). Within the immune system, antigen receptors of T lymphocytes, the major histocompatibility antigens I and II, and the polymeric immunoglobulin receptor (pIgR) belongs to this category. The Ig superfamily encompasses well over 100 genes and extends across several phylogenic boundaries. This family of adhesion molecules plays a major role in the organization of vertebrates and also in the development of the immune system. The IgSF of proteins is subdivided into sets of A, B and C which are categorized based upon the arrangement of the β-strands of the immunoglobulin fold [11].

2.1. Fc receptors

Fc receptors bind immunoglobulins from the serum which in turn serve as antigen receptors for various forms of leukocytes thereby connecting humoral and cellular immunity. They play important roles in phagocytosis, in antibody-dependent cellular cytotoxicity, and participate in immunoregulation. Fc receptors are capable of inducing cytokine secretion, expression of cell surface receptors and can activate differentiation programs. Fc receptors belong to three large classes of molecules: glycosyl transferases, lectin-like molecules, and receptors that belong to the IgSF. All the receptors that recognize immunoglobulin protein determinants (rather than carbohydrates) belong to the IgSF [11,12,13].

The Fc receptors that are capable of triggering cell activation possess one or several intra-cytoplasmic activation motifs, composed of two YxxL (tyrosine xx lysine) sequences flanking seven variable residues. These sequences are now designated immunoreceptor tyrosine-based activation motifs (ITAM). Activating Fc receptors fall into two types. The first type are multi-chain receptors composed of ligand-binding FcR-α subunit associated with one or two signal transduction subunits in the inter-cytoplasmic domains that contain ITAM. Receptors of the second type consist of two closely related single chain IgG receptors unique to humans and referred to as FcγRIIA and FcγRIIC. These receptors possess a single ITAM that has 12 residues instead of 7 between the two YxxL sequences. Fc receptors that do not trigger cell activation have no ITAM. Inhibitory receptors contain a single YxxL sequence designated immunoreceptor tyrosine-based inhibitory motif (ITIM). Other Fc receptors are involved in the transcytosis of immunoglobulins through epithelia. These are the polymeric IgA and IgM receptors (pIgR) and the neonatal FcR for IgG (FcRN). A human IgG receptor without ITAM, referred to as FcγRIIIB, has no signaling capability (Figure 2) [12,13].

High affinity FcR bind monomeric IgE, IgG and IgA (affinity constants 10^9–$10^{10}M^{-1}$ FcεI, 10^8M^{-1} for FcγR receptors, $5x10^7M^{-1}$ for monocyte Fc-α receptors). Low affinity FcR binds immunoglobulin if aggregated or complexed with multivalent antigens (all IgG receptors). Both high and low affinity receptors trigger cellular responses with equal efficiency. The signal transducing γ-chains associated with Fc receptors (FcRγ) does not belong to the IgSF but is related to the T cell receptor signal transducing ζ-chain (TCRζ). The signal transducing β-chain (FCRβ) present in the FcεRI of mast cells is a tetra-strand molecule that belongs to the CD20 family [12].

The aggregation of Fc receptors induces the phosphorylation of their ITAM which in turn activate tyrosine kinases (src, lyn, lck, hck, syk, Btk, Emt, Tec, Fer). These tyrosine kinases phosphorylate numerous intracellular substrates that include enzyme-adaptor molecules and proteins associated with the cytoskeleton. The Fc receptor induced phosphorylation of tyrosine is modulated by transmembrane and cytoplasmic phosphatases. This may prevent cell activation by the Fc receptor. Cell activation by Fc receptors is followed by increased concentration of intracellular calcium, the activation of protein kinases C and D and the ras pathway which stimulates gene transcription in the nucleus. In addition to activation, the aggregation of Fc receptors may also trigger internalization. Activation may release inflammatory mediators, cytokines, may induce antibody dependent cellular cytotoxicity and may lead to endocytosis and phagocytosis

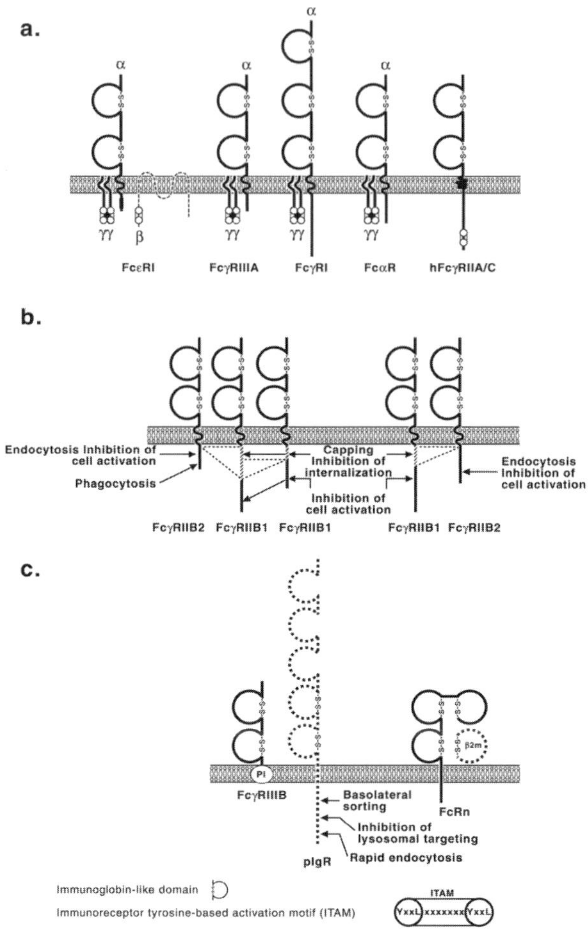

Figure 2. Fc Receptors Belonging to the Immunoglobulin Family. a. *Stimulatory receptors*: These activate immunoreceptor tyrosine based activation motifs (ITAM) b. *Fc receptor γIIB1 and IIB2*: The arrows indicate the approximate locations of sites regulating endocytosis, phagocytosis, capping and cell activation. c. *The polymeric Ig receptor with its intracellular regulatory sites*: Fc receptor IIIB which has no intracellular portion and neonatal Fc receptor (FcRn) [Based on Daëron 1997].

(Table I) [12,13].

FcγRIIB variants constitute a family of single-chain, low affinity IgG receptors that have no ITAM. These receptors cannot activate cells, but rather, they are involved in capping endocytosis and phagocytosis. FcγRIIB triggers calcium responses and oxidative bursts when aggregated [12].

The neonatal Fc receptors are high affinity IgG receptors. They are composed of a ligand-binding α subunit related to MHC Class I molecules and are associated with β-2 microglobulin. These receptors bind IgG ($2–5\times10^7 M^{-1}$). Binding decreases by two orders of magnitude when the pH rises to 7. This enables FcRN to bind maternal IgG with high affinity at acidic pH in ingested milk, and to release them in the newborn blood stream at a neutral pH [12,14].

Table I Fc Receptors.

Receptor	Expression	Motif	Specificity	Function
FcγRI	Moc/Mf, Ne	ITAM	IgG1>IgG3>IgG4	ADCC, phagocytosis
FcγRIIA,B,C	B, Meg, Moc/Mf, Ne, Plt	B:ITIM A,C:ITAM	IgG	Capping, endocytosis, potentiate biological effects, downregulation, Plt aggregation
FCγRIIIA	B, Mas, Moc/Mf, Nκ, T	ITAM	IgG	ADCC, phagocytosis
FCγRIIIB	Ne	NONE	IgG	Clearance of Ag-Ab complex, transport
FcεRI	Ba, Eo, Lan, Mas, Moc	ITAM	IgE	Inflammation, allergy
FcαR	Eo, Moc/Mf, Ne	ITAM	IgA	Mucosal defense
pIgR	Ep		IgA, IgM	Transcytosis of IgA, IgM (secretion)
FcRn	End, Ep, Hep, Trb		IgG, IgM	Immune transfer (absorption), Ig transport

Abbreviations:

Ab	= Antibody	ITIM	= Immunoreceptor tyrosine based inhibitory motif
ADCC	= Antibody-dependent cellular cytotoxicity	Lan	= Langerhans cell
Ag	= Antigen	Mas	= Mast cell
Ba	= Basophil	Meg	= Megakaryocyte
B	= Bone marrow derived lymphocyte	Mf	= Macrophage
End	= Endothelium	Moc	= Monocyte
Ep	= Epithelial cell	Ne	= Neutrophil
FcR	= Fc Receptor	NK	= Natural Killer Cells
FcRn	= Neonatal Fc receptor	pIgR	= Polymeric immunoglobulin receptor
Hep	= Hepatocyte	Plt	= Platelet
Ig	= Immunoglobulin	T	= Thymus-derived lymphocyte
ITAM	= Immunoreceptor tyrosine based activation motif	Trb	= Trophoblast

The polymeric immunoglobulin receptor is a single-chain recognizing polymeric IgA and IgM. These receptors are expressed on plasma membranes on the basal plasmic membrane of epithelial cells of the gastrointestinal, respiratory and urogenital tract. They bind IgA and IgM (approximately 10^9M^{-1} and 10^8M^{-1} respectively), trigger internalization and transfer of the complex to the topical pole of the cell where the extracellular domains of the receptor are cleaved off and released to mucosal surfaces with IgA dimers or secretory IgM. The cleaved domains of pIgR are called secretory components (SC) and protect immunoglobulins from proteolysis. The YxxL-sequence encoded by the third intracytoplasmic axon of FcγRIIB triggers the endocytosis of soluble immune complexes via clathrin-coated pits. Sequences encoded by the first intracytoplasmic axon are necessary and sufficient for capping triggered by FcγRIIB1 and FcγRIIB1' [11,12].

The binding of polymeric immunoglobulin is sufficient to aggregate pIgR, which initiates endocytosis and secretion of immunoglobulin. FcRN and pIgR transport immunoglobulin on opposite directions. FcRN are expressed at the apical pole of epithelial cells whereas pIgR is expressed at the basolateral pole. As a result, FcRN mediates absorption and pIgR secretion.

Most cell types bear more than one type of Fc receptors as well as numerous other receptors that are likely to co-aggregate after cell activation (Table 1). The co-aggregation of FcRN with

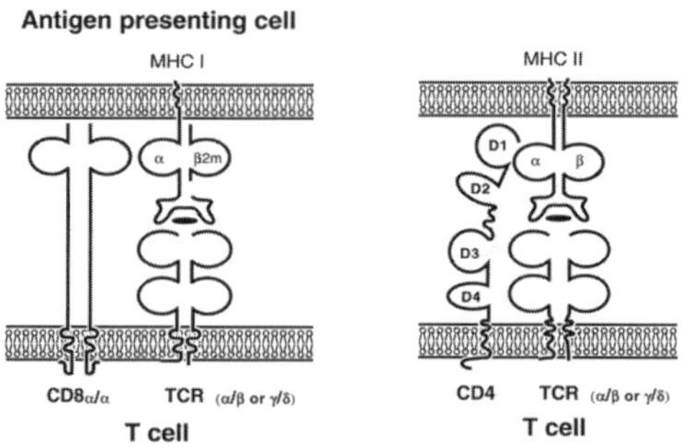

Figure 3. Antigen Presentation by MHC Class I and MHC Class II. Molecules on the surface of antigen presenting cells. The interaction with the T cell receptors and the accessory molecules CD4 and CD8 is illustrated [Modified from Frazer and Capra 1999].

antigen receptors on lymphocytes has a potent negative effect on lymphocyte activation. After aggregation both positive cooperation (synergizm) and negative cooperation (antagonizm) is possible between receptors. For instance, FcγRIIB was demonstrated to inhibit activation via Fc or ITAMs. In FcγRIIB there is a 13 amino acid sequence containing ITIM which is necessary and sufficient for the inhibition of B cell activation. When the tyrosine of the ITIM is phosphorylated, it will bind the phosphatases SHP-1 and SHP-2, which in turn cause dephosphorylation of kinase phosphorylated substrates [12].

2.2. CD-4 and CD-8 molecules

CD-4 and CD-8 molecules serve as "co-receptors" to antigen recognition by T lymphocytes. They participate in TCR/MHC interactions by engaging non-polymorphic regions of the MHC in low affinity interactions. CD-8 molecules bind to Class I MHC proteins while CD-4 molecules interact with Class II. Both CD-8 and CD-4 are glycoproteins and their cytoplasmic region interacts with src-like tyrosine kinases $p56^{lck}$. These molecules boost antigen recognition by over a hundred fold (Figure 3) [11].

2.3. The Major Histocompatibility Complex

The MHC is a set of linked genes, the products of which are involved in the presentation of antigens to lymphocytes. They also mediate self-recognition by T lymphocytes and NK cells. Structurally, MHC belongs to the immunoglobulin family of adhesion molecules and may be divided into class I and class II. MHC class I has three immunoglobulin-like extracellular domains, a transmembrane region and an intra-cytoplasmic tail. MHC-I is linked non-covalently to β-2 microglobulin which is a serum protein. MHC-II molecules consist of two chains inserted in the membrane, α and β, each consisting of two major extracellular domains, α-1 and α-2 and β-1 and β-2 respectively. Both chains have a transmembrane domain and a cytoplasmic sequence. When assembled in the membrane, both MHC-I and MHC-II molecules are non-covalently linked heterodimers consisting of four extracellular domains [15].

The function of MHC molecules is the binding of foreign and self-peptides. Crystallographic studies showed that the amino-terminal domains (α-1 and α-2) of the MHC-I form the floor of

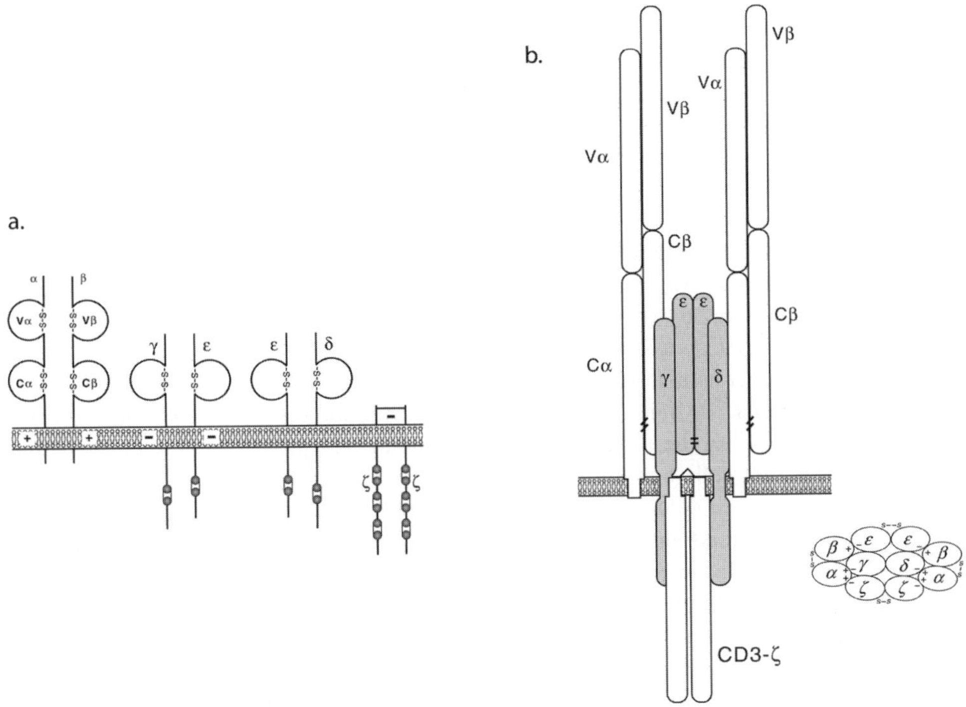

Figure 4. a. T cell receptor: Peptide chains and ITAMs. b. A hypothetical model illustrating the topographical relationship of the various chains in the T cell receptor complex: Left side view, right view from the top [After Terhorst et al. 1996 with modification].

eight strands of anti-parallel β-pleated sheets, and this supports α-helices; one is contributed by the α-1 domain and the other by the α-2 domain which are aligned in an anti-parallel orientation. The α-3 domain is proximal to the membrane has a fold similar to that of immunoglobulin domains and pairs asymmetrically with the other immunoglobulin domain of the molecule contributed by β-2 microglobulin. The peptide-binding site is supported by the β-sheet floor, and the floor is supported by the two immunoglobulin-like domains (Figure 4).

The α-1 and β-1 domains of MHC-II fold very similarly to that of MHC-I. The MHC system regulates the immunoglobulin response to soluble foreign antigens (MHC-II) and also controls reactions to viral and other intra-cellular infectious agents and cancer (MHC-I). The major histocompatibility antigen system of the mouse is called H-2 and of man HLA (for human leukocyte antigen). MHC antigens have been found in all vertebrates so far assessed. In man, Class I HLA alleles are categorized as A, B, C, D, E and G. Class II genes are designated in humans as HLA-D. In the mouse, Class I genes are designated as H-2K, H-2D, H-2L, and Class II genes as H-2Ia. A frequently updated listing of HLA alleles can be found at http://www.icnet.uk/axp/tia [15].

The linked MHC genes of one chromosome are designated as haplotype. Haplotypes generally segregate as a linkage group to children from parents. Individual haplotypes in the mouse are referred to by superscript letters to H2 (eg$H2^b$, $H2^d$, $H2^k$). The binding of peptides by MHC-I or MHC-II molecules is the selective event that permits the antigen presenting cell (APC) to sample its own proteins (MHC-I) or the proteins ingested from the immediate exocellular environment

(MHC-II). The cell-bound MHC-peptide complex is then displayed to T cells. The MHC-I-peptide complex interacts with the receptor of the T lymphocytes and might cause stimulation which might lead to cytolysis of the APC and/or to the secretion of lymphokines by the T cell. MHC Class II molecules bind peptides derived from the degradation of proteins ingested by MHC-II expressing APC and presents them to specific T lymphocytes. The production of antigen fragments from large molecules that are presented by MHC is collectively known as antigen processing. The display of these peptides at the cell surface is known as antigen presentation. Processing involves proteolytic degradation into short peptides. Presentation consists of the binding of the peptide fragments to either MHC-I or MHC-II intra-cellularly which subsequently move to the cell surface for display to the extracellular environment. MHC Class I interacts with CD8 positive T cells where as MHC Class II with CD4-bearing T cells (Figure 3) [15].

Natural killer (NK) cells also recognize MHC antigens for the purposes of identifying self from non-self. NK cells kill tumor cells and virus-infected targets. The expression of the self-MHC Class I molecule by the target cell can in certain cases protect the target by killing by the NK effector. Target cells that are defective in the expression of normal MHC Class I will be lysed by NK cells. The receptors on NK cells may be classified as activating receptors and inhibitory receptors. Structurally, they fall into the immunoglobulin supergene family or to the C-type lectin family.

The MHC is a large region of the genome that spans some four million base pairs on the short arm of chromosome six (between six b21.31 and six b21.32). The mouse MHC occupies a central region of about two centimorgans (cM) of chromosome seventeen (18.0 to 20.0 cM). An updated database for human MHC may be found at http://www.hgmp.mrc.ac.uk [15].

The MHC-III region codes for several complement components for twenty-one hydroxylase which is critical for the biosynthesis of glucocorticoids and for the genes of tumor necrosis factor α and β. MHC antigens show polymorphism which indicates the occurrence of a larger than expected number of genetic variants in the population. Thus, the human MHC-I and II genes show more than 50 alleles at each of the HLA-A-B and –DRB loci. The mechanism of mutation in MHC antigens is a complex phenomenon involving at least two different donor genes.

The biology of the MHC complex has initially been investigated in inbred mice. These animals have been developed by close (brother/sister or parent offspring) breeding. Inbred animals will accept tissue grafts from each other but will reject grafts from a different strain (allogeneic). A difference as little as one amino acid in MHC antigens may be detected by transplantation. Tumor antigens are also detectable by similar transplantation methods. Some diseases which are primarily auto-immune in nature are associated in man by certain MHC antigens. Examples are narcolepsy, ankylosing spondylitis, Reiter's disease, insulin-dependent diabetes, melanosis, rheumatoid arthritis, Hodgkin's disease, hemochromatosis, psoriasis, celiac disease, and multiple sclerosis. Several of the human diseases are associated with MHC Class III genes that control glucocorticoid biosynthesis. The precise mechanism underlying the association of most of the diseases with a particular MHC haplotype is unknown.

MHC molecules are synthesized in the endoplasmic reticulum and destined for surface expression. Expression is controlled by genetic factors and by transcriptional mechanisms (regulated by interferon γ). The rest of the biosynthetic pathway of MHC-I is dependent on cytosolic peptides of the proteasome and the delivery to the ER by the adaptor protein PAP, core-like expression in the ER, transport through the Golgi and arrival to the plasma membrane. A number of persistent viruses have evolved to inhibit the expression of MHC antigens. A disease due to MHC-II deficiency is known as *bare lymphocyte syndrome* [15].

2.4. The T Lymphocyte Antigen Receptors

T cell antigen receptors (TCR) share many similarities both structurally and genetically with B cell antigen receptors (immunoglobulins). An important difference is the MHC-restricted recognition of antigens. This involves the recognition of fragments of antigen (peptides) bound to specific MHC molecules. TCR is known in two distinct forms of heterodimers, α-β or γ-δ, both of which are associated with the non-polymorphic CD-3 polypeptides γ, δ, ε and ζ. The CD-3 complex is fundamental for intracellular signaling. α-β TCR recognize peptide antigens bound to a Class I or Class II MHC molecule. T cells bearing γ-δ TCR are less numerous than the α-β type and the recognition of antigen by these cells may be more similar to that of B lymphocytes. They do not seem to require MHC proteins or other molecules to present antigens, but rather, appear to recognize antigens directly. The T cell receptors consist of 85–90 kD molecular weight disulfide-bounded heterodimer chains that consist of 40–50 kD molecular weight glycosylated subunits referred to as α and β. The TCR genes have antibody-like variables (V), joining (J), and constant (C) regions and do re-arrange in T lymphocytes. All antigen-specific helper or citotoxic cells express TCR α-β heterodimers. The TCR polypeptides exhibit an immunoglobulin-like structure including the "β-barrel" folding structure [16].

The TCR-peptide-MHC complexes have an inherent ability to self-associate (aggregation). Of the CD3 polypeptides, the extracellular domains of γ, δ, and ε chains each consist of a single Ig superfamily domain. The extracellular domain of the ζ chain contains only nine amino acids and one cysteine which is responsible for the disulfide linkage of the ζ homodimer or the alternative association with Fc receptor γ chain. All of the CD-3 polypeptides have a conserved negatively charged amino acid which is complementary to the positive charges seen in the TCR transmembrane region and are necessary for proper assembly. The intra-cellular domains of the γ, δ, ε and ζ chains contain immunoreceptor related tyrosine based activation motif. These domains mediate tyrosine phosphorylation, which is one of the earliest steps in T cell signaling. The TCR gene segments are organized similarly to those of immunoglobulins. The re-combination of V, D and J regions are analogous to the Ig genes and then joined with the C region. Somatic hyper-mutation which is characteristic of antibody maturation has not been consistently detected for TCR genes. No evidence was found for natural selection from either an α-β or γ-δ TCR. Heavy chain switching does not occur in TCR either (Figure 4) [13,16].

γ-δ T cells are the first to appear in the thymus two days before α-β T cells develop in the mouse. But in later weeks α-β T cells predominate. γ-δ T cells preferentially localize in the mucous membranes and in the skin. The antigens recognized by γ-δ T cells do not have to be processed and presented and they need not be proteins either. Recent evidence indicates that γ-δ T lymphocytes mediate natural immune responses [16,17].

3. INTEGRINS

Integrins are heterodimeric (alphabeta) transmembrane glycoprotein receptors that play critical roles in both cell-matrix and cell-cell adhesion (Figure 5). Seventeen α and 8 β subunits forming at least 12 heterodimers are known in mammals. They are also involved in cytoskeletal rearrangements, co-regulation of growth factor activities and activation of signal transduction. Integrins play a key role in the early development of both invertebrates and vertebrates [2]. Integrin function is highly regulated, largely through the formation of specific associations with both structural and regulatory components within cells. These intracellular interactions allow

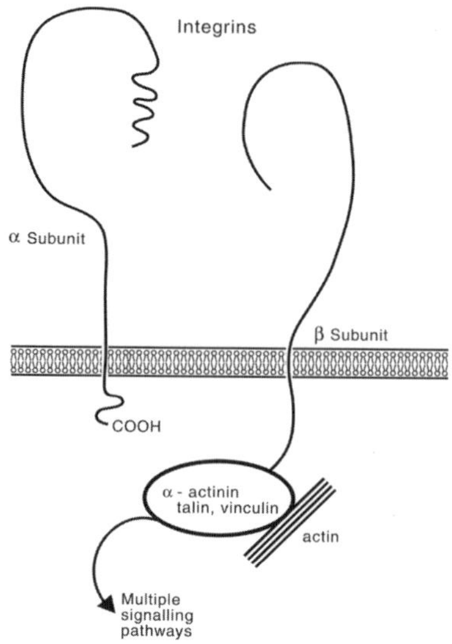

Figure 5. Example of an integrin molecule. The α and β subunits are noncovalently associated, the cytoplasmic domain interacts with actin via talin, α actinin or vinculin and also with numerous cytoplasmic signaling molecules [Modified from Benson et al. 2000].

integrins to affect many biochemical signaling pathways and to influence complex cellular activities [18].

Different integrins can bind to the same ligand and one integrin can bind to several ligands. Integrins frequently recognize the arginine-glycine-asparagine (RGD) sequence motif in their ligands. The clustering of integrins on the cell surface, lateral associations with other proteins and divalent cations contribute to the regulation of integrin ligand affinity. Laminins are trimeric glycoproteins present in the extracellular matrix and are major constituents of basement membranes. Several members of the integrin family serve as laminin receptors on a variety of cell types. This interaction plays an important role in adhesion-mediated events in vertebrates, including embryonic development, tissue architecture, cell migration and tumor cell invasiveness. Cell proliferation and differentiation as well as basement membrane assembly are also regulated by laminin-binding integrins [2,19]. Alpha1beta1 and alpha2beta1 integrins serve as specific cell surface receptors capable of recognizing distinct collagen subtypes. Integrin alpha1beta1 is abundant on smooth muscle cells and alpha2beta1 is present on epithelial cells and platelets. Many cell types, including fibroblasts, osteoblasts, chondrocytes, endothelial cells, and lymphocytes may concomitantly express both of the receptors. The two receptors are connected to distinct signaling pathways and their ligation may lead to opposite cellular responses [20].

Recent studies document a role for integrins and growth factor activated Src family kinases in angiogenesis. Integrin antagonists may prove to be effective inhibitors of angiogenesis during tumor growth, inflammatory disease, and retinopathy [21].

The receptor function of integrins is dependent on the interplay between Mg^{2+} and Ca^{2+}. A unique Mg^{2+}/Mn^{2+} binding site is the MIDAS motif located in the integrin alpha subunit I domain. Divalent cation, bound at this site, coordinates the binding of ligand to the I domain containing integrins. The I-like domain of the integrin beta subunit also has a MIDAS-like motif.

The N-terminal region of the integrin alpha subunit contains three or four 'EF hand' type divalent cation binding motifs. It is certain that most integrins have a high affinity Ca^{2+} binding site which is critical for alphabeta heterodimer formation. The intracellular Ca^{2+} fluxes activate the Ca^{2+} dependent enzyme, calpain, which regulates cluster formation of leukocyte integrins [22].

The cytoplasmic domain of integrins interact with cytoskeletal proteins (α-actinin, talin, filamin) providing a physical link between the extracellular matrix and the cytoskeletal structure. Integrins also participate in various signal transduction processes involving pp125FAK, caveolin, integrin-linked kinase (IAK), and the receptor for activated protein kinase C_1. The cytoplasmic domain also interacts with regulatory proteins such as paxillin, calreticulin, cytohesin, endonexin and ICAP, which are capable of modifying the affinity of integrins for their extracellular ligands [2].

Beta-1 integrins play crucial roles in lymphocyte adhesion, migration, proliferation and differentiation. PLC-gamma, pp125FAK (focal adhesion kinase), pp105, paxillin, p59fyn, p56lck and ERK1/2 are phosphorylated in their tyrosine residues upon engagement of beta1-integrins. Pp105 is a Cas (Crk-associated substrate)-related protein which is predominantly expressed in the cells of lymphoid lineage, hence designated as Cas-L. Cas-L contains a single SH3 domain and multiple SH2 binding sites (YXXP motif), which is suggested to bind SH2 domains of Crk, Nck, and SHPTP2. Pp125FAK binds Cas-L on its SH3 domain and phosphorylates its tyrosine residues upon beta1-integrin stimulation. Cas-L is also involved in the T-cell receptor (TCR)/CD3 signaling pathway as well as the beta1-integrin signaling pathway. Cas-L is transiently phosphorylated following CD3 cross-linking, and binds to Crk and C3G. Cas-L might be the bi-modal docking protein that assembles the signals through beta1-integrins and TCR/CD3. It also participates in a variety of T-cell functions [23]. The Cas proteins (p130Cas, HEF1/Cas-L and Efs/Sin), a family of docking proteins containing multiple interaction domains, are important components of integrin receptor signaling and have been implicated in all integrin-related functions [24].

Members of the beta-2 integrin family are dominantly expressed on leukocytes. They play a major role in leukocyte cell-cell and cell-matrix adhesions during inflammation and other immune responses. Beta-2 integrins are signaling receptors, and are targets of and are functionally affected by intracellular signals. Accordingly, beta-2 integrins (and integrins in general) exhibit two types of signaling: transmission of signals into the cell following binding of ligands or counter-receptors to the integrins (outside-in signaling) and regulation of the avidity and conformation of integrins by signals generated by other receptors within the cell (inside-out signaling). One of the earliest events in beta-2 integrin signaling is activation of non-receptor tyrosine kinases, which in turn trigger downstream activation of various signaling pathways affecting different functional responses of the cell [25].

4. SELECTINS (SEL)

Selectins are carbohydrate recognizing molecules that play a role in cellular interactions and in pathogen recognition. Many of these sugar binding animal lectins belong to the C-type or Ca^{2+}-dependent lectin family. Two subgroups of C-type lectins may be distinguished. *Collectins* are soluble proteins and are involved in pathogen neutralization through the activation of the complement cascade. The macrophage mannose receptor (MNR) mediates phagocytosis of microorganisms. *Selectins* serve as cell surface receptors that mediate cell-to-cell signaling. These receptors mediate leukocyte recirculation and serve as receptors for natural killer (NK)

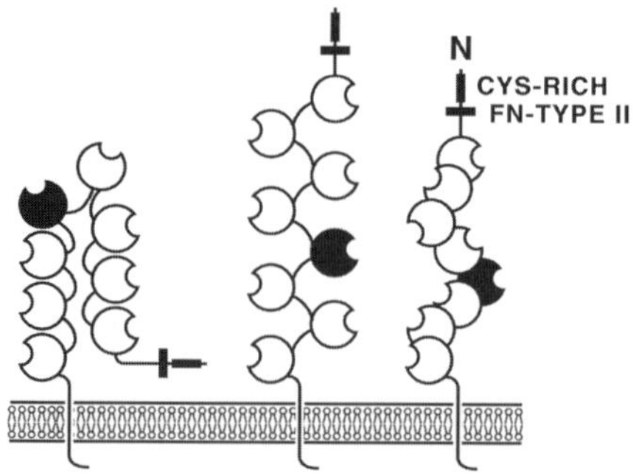

Figure 6. Mannose Receptor. Domain organization is illustrated with three hypothetical possibilities of arrangement. CYS-RICH = cysteine rich domain ; FN-TYPE II = fibronectin type two domain. [Adapted from Weis et al. 1998, with minor modifications]

cells (Figure 6) [26].

The MNR binds mannose, N-acetylglucosamine and fucose, which are uncommon in the terminal position on mammalian polysaccharides but are frequent on the surface of microorganisms. This receptor mediates phagocytosis, which is followed by antigen processing and presentation. Dendritic cells also express this receptor. MNR is a type I transmembrane protein with the extracellular portion containing 8 carbohydrate recognition domains (CRD), a fibronectin type II repeat and a unique N-terminal cysteine-rich domain (Figure 6). More than one domain is required for triggering the endocytosis of microorganisms. High affinity binding by the mannose receptor is achieved through clustering of multiple CRDs that individually display a weak affinity to single sugars. Three other endocytic receptors showing the organization of MNR have been discovered recently. One of these is DEC-205, which is expressed by dendritic cells and T lymphocytes [26]

The SEL family of glycoprotein adhesion molecules (P-selectin, E-selectin, and L-selectin) play a role in the migration of lymphocytes through the high endothelium of postcapillary venules as well as in the transmigration of monocytes and neutrophils to sites of inflammation. Selectins are involved in the pathogenesis of a number of inflammatory disease states. SEL modulate the early adhesive interactions between circulating neutrophils and the endothelium. Both P-selectin and E-selectin can be expressed on the surface of endothelial cells in response to inflammatory mediators. L-selectin is constitutively expressed on the surface of neutrophils at very high levels. Neutrophils also express ligands for the endothelial selectins, that include the carbohydrate sialyl Lewis(x) and the high-affinity ligand P-selectin glycoprotein ligand 1, which facilitate neutrophil-endothelial interactions. Selectins are involved in the pathogenesis of ischemia/reperfusion injury. Highly specific selectin antagonists have been developed for the prevention of ischemic injury [26, 27].

Natural killer cells express surface receptors that are analogous to, but not homologous with, C-type lectins. The binding sites of these molecules do not contain the Ca^{2+}-dependent carbohydrate recognizing domain, but rather form the N-terminal domain of NK receptors (NKD). Some evidence indicates that these receptors recognize carbohydrate moieties in a Ca^{2+}-independent manner. Such receptors are the Ly49 receptors on murine NK cells, which mediate inactivation after recognizing MHC antigens. A second subgroup, NKR-P1, activates killing of target cells. Other related molecules on NK cells are NKG2, CD69 and CD94. These are molecules with similar organization, which are less restricted on NK cells [26].

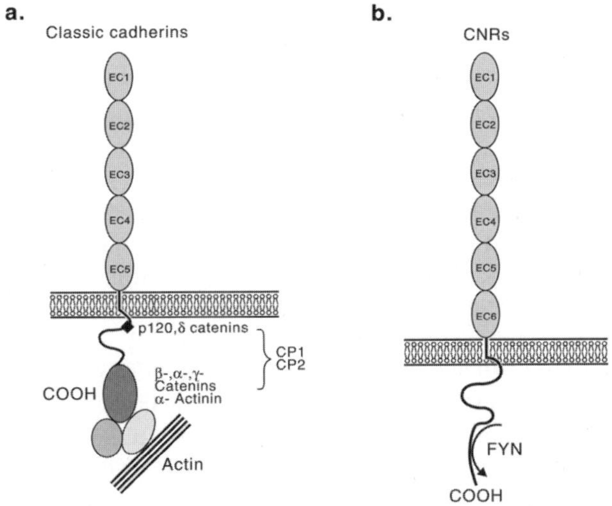

Figure 7. Examples of Cadherin Molecules. The basic structure of classical cadherians and cadherian-like neural receptors (CNRs) is shown. The extracellular domains and intracellular association with actinin and with the tyrosine kinase FYN is shown [Based on Benson et al. 2000].

IL-12 induced selectins during inflammation, which were selectively expressed on Th1 cells generated *in vitro*. Selectin expression on effector cells occurring *in vivo* was less well correlated with the Th subset. [28].

It was shown in genetically engineered mice that selectins not only mediate leukocyte rolling, but also platelet rolling on the vessel wall. Mutant mice also revealed the anti-inflammatory role of platelet P-selectin. P-selectin binding to leukocytes promoted the production of an anti-inflammatory mediators that limited the extent of acute glomerulonephritis. Moreover, soluble P-selectin was shown recently to be shed from both activated platelets and endothelium. There are strong indications for the attenuating effect of soluble P selectin on inflammatory disease progression. It was also discovered in selectin-deficient mice that P- and E-selectins have the crucial role of homing hematopoietic progenitor cells to the bone marrow. The use of selectin inhibitors was evaluated in two major disease processes: cancer and atherosclerosis [8].

Type II leukocyte adhesion deficiency (LAD II) is a rare disorder characterized by recurrent infections, persistent leukocytosis, and severe mental and growth retardation. In this condition, neutrophils are deficient in the expression of selectin ligand activity, and exhibit a diminished ability to roll on endothelium and to traffic to inflammatory sites *in vivo*. Such patients exhibit a deficiency of cell surface fucosylated glycan structures that include the H and Lewis blood group determinants and the sialyl Lewis x epitope. The corresponding fucosyltransferase activities responsible for synthesis of these structures are expressed at normal levels. The molecular defect in LAD II has been localized to the synthesis of GDP-fucose from GDP-mannose [29].

5. CADHERINS (CAD)

A large number of cadherin genes have been identified to date; most of them are expressed in the CNS. Within a single mammalian species, at least 80 cellular receptors belonging to the cadherin superfamily are expressed. The superfamily members include classical cadherins, desmogleins, desmocollins, protocadherins, CNR, Fats, seven-pass transmembrane cadherins and Rat tyrosine kinase. All members are transmembrane proteins and contain the cadherin motif (or EC domain) containing the negatively charged DXD, DRE, and DXNDNAPXF sequence motifs, which are involved in Ca^{2+} binding. The EC domains are tandemly repeated in the extracellular segment of all cadherins and the number of EC domains is considerably varied in different molecular species (Figure 7).

Classical cadherins (CDH1-20) and the *desmosomal cadherin (DSC1-3,DSG1-3)* subfamily are homophilic adhesion molecules. The EC domain plays a crucial role in adhesion interactions by these molecules. The intracellular domains interact with catenin $p120^{ctn}$ and β-catenin. β-catenin binds with α-catenin and this molecular complex associates with vinculin and with other cytoskeletal proteins that results in the organization of adherens junction. The cadherins *Fat* and *Dachsous* play a role in embryonic morphogenesis and differentiation. Some of the frequently used synonyms are epithelial (E,*CDH1*), neural (N, *CDH2*), placental (P,*CDH3*), osteoblast (OB, *CDH11*) and kidney (K, *CDH6*) cadherin [30,31].

Seven-pass transmembrane cadherins (FMI, FMI1, MEGF2) are similar to G-protein-coupled receptors and are present in vertebrates and invertebrates. These cadherins signal through cytoplasmic pathways which are distinct from that of classical cadherins. Recent results suggest that seven-pass transmembrane cadherins regulate cell polarity and therefore play a role in determining neuronal morphology.

Protocadherin family (PCDHα, -β, -γ, -1-68). A large number of cadherin superfamily genes that code for surface molecules in the CNS, belong to this family, identified as protocadherins (Pt) or cadherin-related neuronal receptors (CNR). Interestingly, several of these genes display an unusual genomic structure resembling immunoglobulin or the T cell receptor genes. These molecules contain five to six repeats of the EC domain, yet their cytoplasmic domain shows no similarity to that of classical cadherins. Protocadherins control morphogenic cell migrations during embryonic development. CNR cadherins are associated with the Fyn tyrosine kinase and have been localized in the synaptic membrane.

Nearly all cadherin genes are tightly linked to each other. Two large groups may be distinguished according to genomic structure: classical and desmosomal cadherin genes containing 12–17 exons and CNR genes containing long first exons that encode the extracellular domain the transmembrane region and part of the cytoplasmic tail. The carboxy-terminal cytoplasmic domain of each CNR protein is identical and is coded for by three small exons. The genomic organization of CNR (*Pcdhα* in human) is similar to the organization of the Ig and TCR genes. Wu and Maniatis have designated the amino terminal of large exons as the *variable region,* and the commonly used three exons for the carboxy terminus as the *constant region.* CNR genes give rise to three alternative transcripts. Each single variable exon can potentially be connected to three different carboxy terminal tails This is comparable to immunoglobulin class switching. The human *Pcdh* gene cluster consists of 52 members on chromosome 5q31. Three groups are distinguished: *Pcdhα, Pcdhβ* and *Pcdhγ.* The organization of these genes is similar to that of CNR [31,32,33].

Mice deficient of the enzymatic machinery essential for DNA rearrangements (e.g. for generating Ig and TCR diversity) exhibit a striking increase in the death of young neurons and neuroblasts. This correlates with the differential stage of neurogenesis and suggests that gene rearrangements are important for the development of the CNS. Moreover, a number of cadherin genes map within the potential regions that have been identified as carriers of the susceptibility to schizophrenia [32].

The level of cadherin expression determines the strength of adhesion and the type of cadherin expressed determines the specificity of cell interactions. The strength of interactions is modulated by growth factors and by other signals. Cadherin-mediated adhesion is regulated by (i) rapid cell surface regulation of adhesion; (ii) adherens junction formation associated with major changes in cell states (e.g. epithelial-mesenchymal transition); and (iii) control of biogenesis or turnover of junctional molecules.

E-cadherin complexes with catenins are believed to link cadherins functionally to the actin cytoskeleton. α-Catenin interacts with a number of actin binding proteins, such as α-actinin, vinculin, ZO-1 as well as with actin itself. Tyrosine phosphorylation of β-catenin correlates with the inhibition of cadherin-mediated adhesion, resulting from kinase activation. Cadherin-mediated adhesion involves the small GTP-ases, Rac, Rho, and Cdc42. These enzymes regulate actin-membrane interactions. Small GTP-ases can influence the status of cell contacts at the level of the association of α-catenin with the cadherin complex. Cadherin maintains epithelial cell shape and differentiation by interacting with actin filaments and with the Rho family of small GTP-ases.

In mammals, afadin and shroom are additional proteins that participate in some way in the formation and organization of epithelial cell junctions. Afadin and the Ig-type adhesion molecule nectin, are more abundant in the adherens junction than are E-cadherin and catenins. The functional differences between the junctional and non-junctional forms of cadherin-mediated adhesions is not understood [31,34,35].

In the CNS, approximately 30 cadherins are expressed differentially at almost all stages of development and in the mature brain. Most cadherins are expressed by the embryonic subdivisions of the early embryonic brain (e.g. developing nuclei, cortical layers and regions, fiber tracts, neural circuits and synapses). Each cadherin has a distinct expression pattern. Cadherins contribute to CNS regionalization, morphogenesis and fiber tract formation and to the maintenance of functional structures in the CNS [36].

Cadherins exert multiple functions in reproductive tissues. Cadherin concentrations in reproductive tissues are responsive to steroid hormones. Progesterone induces E-cadherin in mouse uterine epithelium, and modulates the amounts of OB-cadherin, K-cadherin and β-catenin in isolated endometrial stromal cells. Estrogen induces E-cadherin in the uterus and upregulates N-cadherin in mouse testis and in rat granulosa cells. Estrogen also modulates N-cadherin in pituitary somatolactotrophs. A combination of FSH and testosterone modulates N-cadherin-mediated binding of round spermatids to Sertoli cells. Sertoli cells upregulate N-cadherin expression in response to FSH, LH and estrogen. The expression of LH receptors by granulosa cells after stimulation with FSH is dependent on cadherin-mediated adhesion [30].

REFERENCES

1. Kimber SJ, Spanswick C. Blastocyst implantation: the adhesion cascade. Semin Cell Dev Biol 2000;11:77–92.
2. Darribere T, Skalski M, Cousin HL, Gaultier A, Montmory C, Alfandari D. Integrins: regulators of embryogenesis. Biol Cell 2000;92:5–25.

3. Madri JA, Graesser D. Cell migration in the immune system: the evolving inter-related roles of adhesion molecules and proteinases. Dev Immunol 2000;7:103–16.
4. Bochner BS. Road signs guiding leukocytes along the inflammation superhighway. J Allergy Clin Immunol 2000;106(5 Pt 1):817–828.
5. Holmskov UL. Collectins and collectin receptors in innate immunity. APMIS Suppl 2000;100:1–59.
6. Rojas AI, Ahmed AR. Adhesion receptors in health and disease. Crit Rev Oral Biol Med 1999;10:337–58.
7. Zhang P, Summer WR, Bagby GJ, Nelson S. Innate immunity and pulmonary host defense. Immunol Rev 2000;173:39–51.
8. Hartwell DW, Wagner DD. New discoveries with mice mutant in endothelial and platelet selectins. Thromb Haemost 1999;82:850–7.
9. Ohene-Abuakwa Y, Pignatelli M. Adhesion molecules in cancer biology. Adv Exp Med Biology 2000;465:115–26.
10. Hertl M. Humoral and cellular autoimmunity in autoimmune bullous skin disorders. Int Arch Allergy Immunol 2000;122:91–100.
11. Frazer JK, Capra JD. Immunoglobulins : structure and function. In: Paul WE, editor. Fundamental Immunology, 4th Edition. Philadelphia: Lippincott-Raven, 1999;37–74
12. Daëron M. Fc receptor biology. Annu Rev Immunol 1997;15:203–34; functions Matrix Biol 2000;19:319–23.
13. Isakov N. Immunoreceptor tyrosine-based activation motif (ITAM), a unique module linking antigen and Fc receptors to their signaling cascades. J Leukoc Biol 1997;6:6–16.
14. Ghetie V, Ward ES. Multiple roles for the major histocompatibility complex class I- related receptor FcRn. Annu Rev Immunol 2000;18:739–66.
15. Margulies DH. The major histocompatibility complex. In: Paul WE, editor. Fundamental Immunology, 4th Edition. Philadelphia: Lippincott-Raven;1999;263–286.
16. Davis MM, Chien YH. T-cell antigen receptors. In: Paul WE, editor. Fundamental Immunology, 4th Edition. Philadelphia: Lippincott-Raven, 1999;341–66.
17. Berczi I, Gorczynski R, Szentivanyi A, editors. New Foundations of Biology : Neuroimmune Biology. Amsterdam:Elsevier, 2001.
18. Coppolino MG, Dedhar S. Bi-directional signal transduction by integrin receptors. Int J Biochem Cell Biol 2000;32:171–88.
19. Belkin AM, Stepp MA. Integrins as receptors for laminins Microsc Res Tech 2000;51(3): 280–301.
20. Heino J. The collagen receptor integrins have distinct ligand recognition and signaling functions. Matrix Biol 2000;19:319–323.
21. Eliceiri BP, Cheresh DA. Role of alpha v integrins during angiogenesis. Cancer J Sci Am 2000;6 Suppl 3:S245–9.
22. Leitinger B, McDowall A, Stanley P, Hogg N. The regulation of integrin function by Ca(2+). Biochim Biophys Acta 2000;1498:91–98.
23. Iwata S, Ohashi Y, Kamiguchi K, Morimoto C. Beta 1-integrin-mediated cell signaling in T lymphocytes. J Dermatol Sci 2000;23:75–86.
24. O'Neill GM, Fashena SJ, Golemis EA. Integrin signaling: a new Cas(t) of characters enters the stage. Trends Cell Biol 2000;10:111–9.
25. Dib K. Beta 2 integrin signaling in leukocytes. Front Biosci 2000;5:D438–51.
26. Weiss WI, Taylor ME, Drickamer K. The C-type lectin superfamily in the immune system. Immunol Rev 1998;163:19–34.

27. Lefer DJ. Pharmacology of selectin inhibitors in ischemia/reperfusion states. Annu Rev Pharmacol Toxicol 2000;40:283–94.
28. Syrbe U, Siveke J, Hamann A. Th1/Th2 subsets: distinct differences in homing and chemokine receptor expression? Springer Semin Immunopathol 1999;21:263–85.
29. Becker DJ, Lowe JB. Leukocyte adhesion deficiency type II. Biochem Biophys Acta 1999;1455:193–204.
30. Rowlands TM, Symonds JM, Farookhi R, Blaschuk OW. Cadherins: crucial regulators of structure and function in reproductive tissues. Rev Reprod 2000;5:53–61.
31. Nollet F, Kools P, van Roy F. Phylogenetic analysis of the cadherin superfamily allows identification of six major subfamilies besides several solitary members. J Mol Biol 2000;299:551–72.
32. Yagi T, Takeichi M. Cadherin superfamily genes: functions, genomic organization, and neurologic diversity. Genes Dev 2000;14:1169–80.
33. Ranscht B. Cadherins: molecular codes for axon guidance and synapse formation. Int J Dev Neurosci 2000;18:643–51.
34. Gumbiner BM. Regulation of cadherin adhesive activity. J Cell Biol 2000;148:399–404.
35. Braga V. Epithelial cell shape: cadherins and small GTPases. Exp Cell Res 2000;261: 83–90.
36. Redies C. Cadherins in the central nervous system. Prog Neurobiol 2000;61:611–48.

Immunoglobulins

ISTVAN BERCZI and ANDOR SZENTIVANYI

Department of Immunology, Faculty of Medicine, The University of Manitoba, Winnipeg, Manitoba R3E OW3, Canada; and Department of Internal Medicine, Faculty of Medicine, The University of South Florida, Tampa, Florida 33612, USA

ABSTRACT

Immunoglobulins (Ig) are serum proteins produced by bone marrow-derived (B) lymphocytes which function as specific antibodies within the immune system. Five classes of immunoglobulins are known in man and higher animals: IgA, IgD, IgE, IgG and IgM. Human IgA is further divided into 2, and IgG into 4 subclasses. This classification is based on the characteristics of the dimeric heavy (H) chain content, that are unique to each class and are designated as α-, δ-, ε-, γ- and μ-chains, respectively. Each Ig molecule contains two identical light (L) chains, which may belong to 2 major classes: κ or λ. The variable (V) regions of one H and L chain form the antigen combining site, giving rise to 2 sites per Ig molecule. H and L chains also have constant (C) regions. The C region of the H chains mediates biological functions, that include complement fixation, phagocytosis, cytotoxicity, secretion, regulation of cellular functions and transportation across membranes. IgA is secreted to mucosal surfaces as a dimer and IgM exists both as pentamer and hexamer in serum and secretions. The mammalian genome contains multiple genes coding for variable (v), diversity (D), joining (J) and constant (c) regions of immunoglobulins. The germline genes are spliced and joined together to achieve VDJC configuration for H chains and VJC for L chains. During the joining process mutations also occur, which give rise to affinity maturation of antibodies during the immune response.

1. INTRODUCTION

Antibody molecules are defined as globular proteins containing carbohydrates and are commonly referred to as immunoglobulins. Heilderberger and coworkers pioneered the biochemical characterization of antibodies. Tiselius and Kabat were the first to observe that gamma globulins are elevated in the serum of immunized animals. Antibodies isolated from specific precipitates with various antigens are also gamma globulins. The term immunoglobulin is broader than gamma globulin since it includes proteins that extend into the β- and α-2 range of electrophoretic mobility [1].

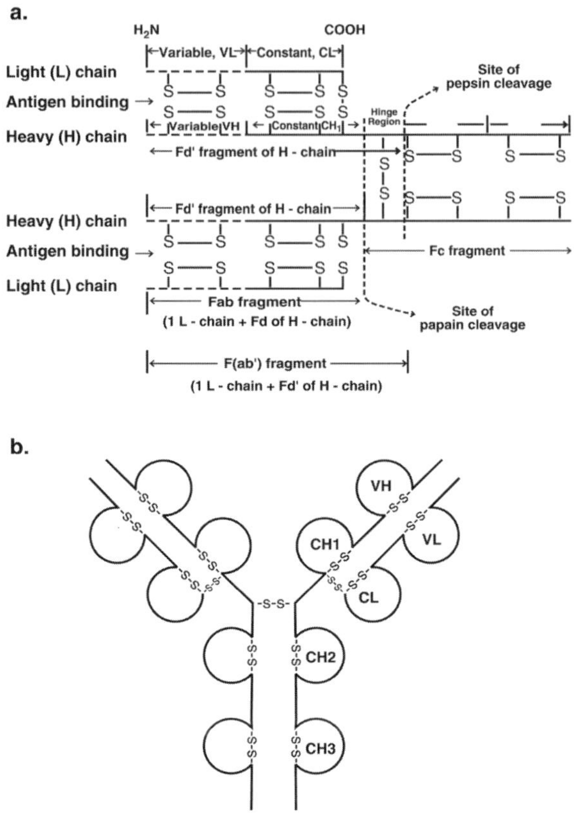

Figure 1. Schematic Representation of Immunoglobulin G. a. *Structure*: (Please see text for detailed explanation.) b. *Domains*: (Please see text for detailed explanation.)

2. IMMUNOGLOBULIN STRUCTURE

Today, it is clear that immunoglobulins belong to the immunoglobulin superfamily of proteins, many of which serve as important cellular receptors within the immune system, in the nervous system and in various other tissues of the body. The basic structure of immunoglobulins is shown in Figure 1. The various members of this family show domain organization consisting of approximately 110 amino acids, that form a loop linked by a disulfide bond. The L chain of immunoglobulins contains two such domains. Light chains belong to two major groups: κ or λ. The H chains contain 4 to 5 domains and determine the class of immunoglobulins. The immunoglobulin heavy chain for IgA is called alpha (α), for IgD it is called delta (δ), IgE, epsilon (ε), IgG, gamma (γ), and for IgM, it is mu (μ). If the IgG molecule is digested with papain, Fab and Fc fragments result. The Fab fragment consists of one L-chain and the Fd fragment of one H-chain. The crystallizable Fc fragment consists of two identical portions of the H-chain. Pepsin cleaves the molecule behind the inter H-chain disulfide bonds which yields an $F(ab)_2$ fragment. The antigen binding sites of the molecules are formed by the V regions of both the H and L chains which show hyper-variability of amino acid sequences. In κ and λ chains three hyper-variable or complementary determining regions (CDR) are known, and these are interspersed

Table I Properties of human immunoglobulins.

	IgG	IgA	IgM	IgD	IgE
Synonyms	$_{\gamma}G, 7S_{\gamma}, \gamma_2, \gamma_{ss}$	$_{\gamma}A, \beta_2A, \gamma_1A$	$_{\gamma}M, 19S_{\gamma}, \beta_2M, \gamma_1M$	$_{\gamma}D$	$_{\gamma}E$, IgND
Physicochemical Properties					
Sedimentation constant	6.5–7.0	7,10,13,15,17	18–20, 30	6.2–6.8	8
Molecular weight	150 kD	160 kD, 400 kD *	950 kD, 1150 kD	175 kD	190 kD
Total carbohydrate (%)	2.9	7.5	11	9	13
Immunochemical Properties					
Heavy-chain classes	γ	α	μ	δ	ϵ
Light-chain types	κ, λ	κ, λ	κ, λ	κ, λ	κ, λ
Molecular formula	$\gamma_2\kappa_2, \gamma_2\lambda_2$	$\alpha_2\kappa_2, \alpha_2\lambda_2$ $(\alpha_2\kappa_2)2$ $(\alpha_2\lambda_2)2$	$(\mu_2\kappa_2)5; (\mu_2\lambda)5$ $(\mu_2\kappa_2)6; (\mu_2\lambda)6$	$\delta_2\kappa_2, \delta_2\lambda_2$	$\epsilon_2\kappa_2, \epsilon_2\lambda_2$
Accessory chains	None	J_1SC*	J_1SC	None	None
Number of subclasses known	4	2	1	1	1
Biological Values					
Serum concentration (mg/ml)	9.5–12.5	1.5–2.6	0.7–1.7	0.04	0.0003
Synthesis rate (mg/kg/day)	20–40	65	3.2–16.9	0.4	0.016
Serum half life (days)	23	6	5	3	2.5
Distribution (% in I.V. pool)	48–62	40	65–100	63–86	
Antibody activity	Yes	Yes	Yes	Yes	Yes
Placental passage	Yes	No	No	No	No
Presence in cerebrospinal fluid	Yes	Yes	No		
Complement fixation	Yes	No	Yes	Yes (alternate)	No

* In exocrine secretions, IgA has a dimeric structure and an attached secretory component (SC) and J chain.
Based on Berczi [4] and Frazer and Capra [2].

among four framework regions [2,3].

The remaining domains of H and L chains show much less variability and are called constant regions. The four chains of the basic immunoglobulin molecules are linked to each other covalently by disulfide bonds, the positions and numbers of which vary somewhat according to the various immunoglobulin classes and sub-classes. The hinge region of the Y shaped immunoglobulin molecule allows for extreme flexibility of the antigen combining sites. The Fc portions of all immunoglobulin molecules perform important biological functions that include complement fixation, transplacental transfer, secretion, binding to cell surface receptors and triggering phagocytosis, target cell killing and regulating immune function. A number of these functions are mediated by carbohydrate moieties on the Fc region rather than the protein sequences (Tables 1, 2) [2,4].

The molecular weight of the heavy chain polypeptide is approximately 55 kD, and the light chains are 25 kD. The Ig domains are characterized by the presence of a structural motif termed the immunoglobulin fold. This structure is referred to as a specialized "β-barrel" which is comprised of 7 polypeptide strands forming anti-parallel "β-pleated" sheets in the folded domain (Figure 2) [5].

The various immunoglobulin classes are referred to as *isotypes*. Five isotypes are present in

Table II Properties of human IgG subclasses.

Property	IgG1	IgG2	IgG3	IgG4
Disulfide linkages	2	4	5–15	2
Molecular Weight	146 kD	146 kD	165 kD	146 kD
Average adult free serum level (mg/ml)	5.9 ± 2.6	3 ± 2.5	0.6 ± 0.55	0.9 ± 0.25
Macrophage binding by FcγR	+	–	+	+
Placental transfer	+	++	+	++
ADCC	+++	+	+++	+
Classical complement activation	+++	++	++++	+
Alternative complement activation	+	+	+	+++

Modified from Frazer and Capra [2].

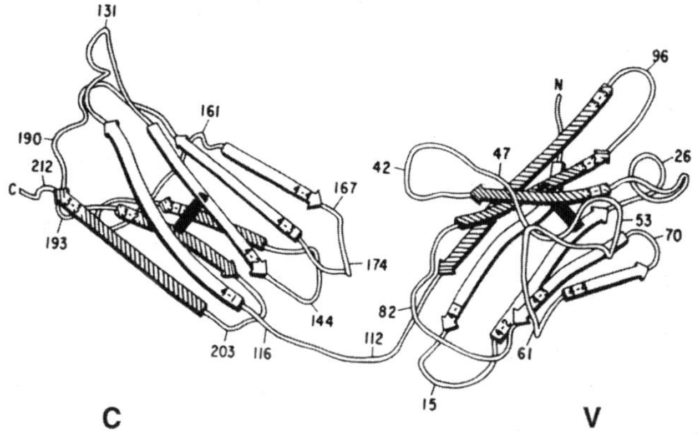

Figure 2. The Immunoglobulin Fold - Ribbon diagram of V and C domains of a light chain.
The β-strands are represented as arrows in the antiparallel β-pleated sheets of each domain. The arrows are shaded in the three-stranded sheets and those in the four stranded sheets are white. The black bars represent intradomain disulfide bonds. Selected amino acids are numbered with position 1 at the N-terminus. Adapted from Edmundson et al. [5].

all mammals, for which immunoglobulin profiles have been determined. Human IgA is divided into -1 and -2 subclasses, and human IgG is separable into four γ subclasses: γ-1, γ-2, γ-3 and γ-4. Murine IgG also has four subclasses: γ-1, γ-2a, γ-2b and γ-3. *Allotypes* are antigenic determinants found on the immunoglobulins of some, but not all, members of a species and are localized in the C region of heavy or light chains. Allotypes are inherited in a typical Mendelian fashion. *Idiotypes* are determinants found in the V regions of antibodies and frequently correlate with antigenic specificity. Idiotypes are considered markers for an individual immunoglobulin molecule [2].

The characteristic 4-chain assembly of the immunoglobulin molecule is referred to as the *primary structure.* This assembly is fundamental to the function of the molecule. In the variable region, diversity sequences are clustered in three regions which were called hyper-variable regions (HVR's) in the past. These highly variable segments of heavy and light chains are thought to form the antigen combining site and are now termed "complimentary determining regions". Other well conserved stretches of the V-region are named "framework regions" (FR) as they provide the platform that supports the CDR. FR's coincide with the β-strands of the immunoglobulin fold while CDR chiefly corresponds to the loops that join β-strands on the C-region-distal and the V-domain. In humans, the heavy chain locus is found on chromosome 14 and the κ and λ loci are on chromosome 2 and 22 respectively. In the mouse, the heavy chain genes are located on chromosome 12, κ on 6, and λ on 16. In humans, 70% of immunoglobulins contain κ chains, while in murine systems approximately 95% of antibodies are in the κ class. The hinge region consists of distinct domains in the μ and ε chains and are represented as shorter specialized segments for α, γ and δ chains that connect the Fab and Fc portions of the molecule. Non-mammalian heavy chain sequences lack extensive hinge structure. Immunoglobulin molecules are synthesized in the rough endoplasmic reticulum (RER) which is followed by secretion [2].

The heavy chains of all immunoglobulin isotypes can exist either as secreted antibodies or as membrane-bound immunoglobulins (mIg). Existence is determined by alternative mRNA splicing at the 3' end. Surface Ig molecules contain hydrophobic terminal sequences which anchor the immunoglobulin to the cell membrane and followed by a short cytoplasmic tail. The terminal sequence of mIg allows interaction with other constituents of the B-cell antigen receptor. This is necessary for signal propagation and eventual activation of the cell by antigens [6].

The immunoglobulin fold is referred to as *secondary structure*. The secondary structure is essential for the positioning of CDR for antigen binding. *Tertiary structure* refers to the fully folded immunoglobulin molecule and is maintained by key core residues in spite of the presence of dramatic differences in the primary sequence. The greatest divergence between the constant domain of immunoglobulins are in those sequences implicated in mediating the different biological effects that distinguish one class or subclass from the other. Under *quaternary structure,* the shared and unique characteristics of a particular immunoglobulin monomer are taken into consideration. The fact that IgM exists as a pentamer or hexamer and IgA as a dimer is referred to as *higher-order immunoglobulin structure* [2].

Immunoglobulin polymerization is mediated by the J chain and secretion is facilitated by the secretory component (SC). The J-chain is a 134 amino acid polypeptide synthesized by antibody-producing plasma cells. It covalently interacts with one or more systems of immunoglobulin monomers that undergo multimerization. There is always one J-chain present per polymer, regardless of the number of monomeric units joined. The J-chain is not always a part of the polymeric IgM molecule but is always present in polymeric IgA [2,7].

The SC or secretory piece was discovered as a polypeptide complex to the Fc of secreted IgA and IgM. It is a portion of the larger trans-membrane polymeric Ig receptor protein (pIgR) which consists of seven immunoglobulin-like domains, five of which are extracellular. The first five domains represent the secretory piece. The polymeric Ig receptor is synthesized in the epithelial cells of the respiratory -, gastrointestinal - and genitourinary tracts and is expressed on the basal lateral aspects of these cells. The pIgR captures IgA and IgM during which covalent disulfide bondage occurs. This is followed by endocytosis via clathrin coated pits. The polymeric Ig receptor is then cleaved between domain five and six and the secretory immunoglobulin is exocytosed at the apex of the cell endomucosal surfaces. It is commonly held that the SC helps to protect secretory immunoglobulins in the harsh mucosal environment [2].

Recent observations indicate that only J-chain-containing IgA and IgM polymers show high affinity for the polymeric Ig receptor. This indicates that the J chain is involved in creating the binding site for pIgR/SC in the Ig polymers by interacting directly with the receptor protein. Therefore, both the J chain and the pIgR/SC are key proteins in the secretion of immunoglobulins [7].

3. IMMUNOGLOBULIN FUNCTION

Collectively, antibodies are able to activate both the classical and alternative pathways of the complement cascades; transcytose to mucosal surfaces to provide protection; transfer maternal humoral immunity to the fetus trans-placentally and by the colostrum; induce phagocytosis, antibody-dependent cellular cytotoxicity by lymphocytes; maintain antiparasitic immune responses by eosinophils; produce inflammation by mast cells and basophils and inactivate directly infectious agents such as viruses. Antibodies also fulfill important regulatory functions that include cell activation, differentiation, induction of anergy and even apoptosis. The surface Ig of memory B-cells is important in antigen recognition and allows them to function as specific antigen-presenting cells (Tables 1, 2) [3,6,8].

Apart from some exceptions, the specific functions of antibodies can be ascribed to individual domains of the molecule. For optimal and exquisitely specific antigen binding, the V-regions of the heavy and light chains have to interact. Most of the binding contacts between the V-domains and antigen take place between amino acids residues in the CDR and the major epitopes of the antigen. Framework residues may also contact the antigen. Antigen-antibody interactions only occur if the binding reaction releases enough free energy to be thermodynamically favored. At the amino acid level tyrosine, tryptophan, serine and asparagine are involved in antigen binding because these residues have lower conformation freedom and less entropy to lose upon binding. In addition, the side chains of these residues foster the chemical interaction, driving changes in enthalpy necessary to promote binding energetically. The antigen antibody interaction involves a variety of forces that include electrostatic bonds between opposite charges, hydrogen bonds, Van der Waals forces and hydrophobic forces. With some specific immunoglobulin-ligand complexes, salt bridges and other forms of interaction may also play a role. Covalent bonds rarely occur between antigen and antibody. For this reason antigen-antibody complexes are readily dissociated by solvents that break the above bonds, such as high salt concentrations, organic solvents and urea [2,9].

Superantigens such as Staphylococcus protein A (SPA) are mitogenic for human B cells. They interact with the V region of the surface immunoglobulin, which is independent of the specificity of the antibody. In general, SPA does not block antigen binding and is able to deliver activation signals to stimulate the differentiation of B cells via V_H3-N coded receptors and also stimulate antibody productions. SPA was shown to interact simultaneously with FR1, CDR2 and FR3 on the V_H region. Another superantigen is the human immunodeficiency virus (HIV) envelope glycoprotein gp-120. Superantigens generally require both V_H and V_L domains for effective binding [2,6].

Fc mediated functions can be broadly classified into three categories: (a) activation of complement, (b) interaction with cells, and (c) transport and compartmentalization of immunoglobulins.

4. IMMUNOGLOBULIN M (IGM)

IgM is the most versatile antibody produced during the primary immune response. It forms the antigen receptor, represents 5–10 % of Ig in the serum and is a major isotype present on mucosal surfaces. Its name comes from the original description of macroglobulin due to its polymeric structure. IgM is present in most vertebrate species and is effective in classical complement fixation. IgM also serves as a secretory immunoglobulin at mucosal surfaces and is secreted into

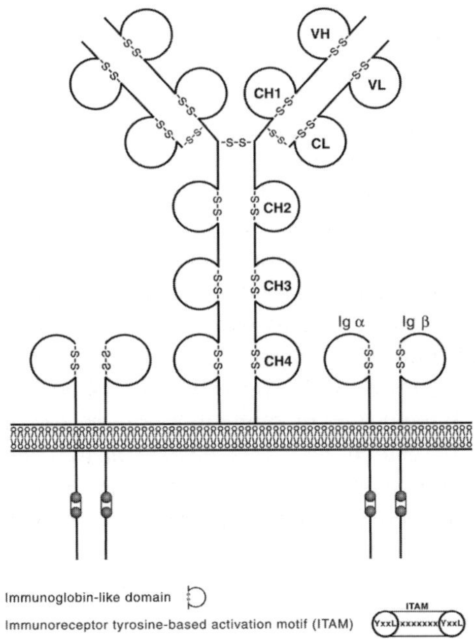

Figure 3. The Antigen Receptor Structure of B-lymphocytes. Antigen recognition is mediated by the binding sites formed by the VH and VL domains of the heavy and light chains of membrane bound immunoglobulin. Initially IgD and IgM serve as membrane receptors, which will change to the immunoglobulin class secreted by the cell after class switching. Intracellular signal transduction is initiated by the receptor-associated Igα ad Igβ chains that contain immunoreceptor tyrosine based activation motifs (ITAM) in their cytoplasmic tail. ITAMs initiate the phosphorylation cascade, leading to cell activation.

breast milk as well [2].

5. IMMUNOGLOBULIN D (IGD)

In serum, IgD constitutes less than 0.5% of total serum Ig. IgD is able to fix complement, however because of its low serum concentrations, this function is not likely to fulfill a significant role in host defense. To date, no specific function unique to IgD has been defined. Research involving IgD knockout mice failed to reveal a definitive function to this immunoglobulin. Membrane-bound IgM and IgD are transcribed together from the same primary RNA message from which they are derived by differential splicing. Memory B cells do not express surface IgD. Similar to IgM, IgD is non-covalently associated in the membrane with Ig-α and Ig-β heterodimers which serve as signaling peptides for the B cell antigen receptor (Figure 3). Ligation of IgD by the specific antigen can independently mediate B cell activation, deletion or anergy [2,10].

6. IMMUNOGLOBULIN G (IGG)

IgG is the predominant immunoglobulin in blood and tissue fluids. High affinity IgG antibodies are produced during the secondary immune response. Electrophoretically, IgG migrates to the γ-range of serum globulins, hence it was designated earlier as gamma globulin. IgG is comprised of four subclasses: IgG 1 – 4 in man. The subclasses activate the classical complement pathway to varying degrees. Much of the biological effect of IgG antibodies is exerted via Fc-γ receptors. A number of such receptors exist, each with specificity for binding a particular IgG subclass. Fc-γ receptors are expressed patterns on various cell types and exert different biologic activities. Macrophages, polymorphonuclear cells and lymphocytes express Fc-γ receptors which trigger many functional effects: including phagocytosis, antibody dependent cell mediated cytotoxicity and modulation of lymphocyte function. This includes antigen presentation, cytokine release, cytokine receptor expression and immunoglobulin secretion. Cerebral Fc-γ receptors with the capacity to bind IgG also exist. IgG Fc receptors mediate the trans-placental movement of maternal antibodies during gestation. Nuclear magnetic resonance analysis of the IgG molecule after binding to one FcγR revealed conformational changes in the Fc region of IgG. This conformational change precludes the binding of two FcγR to one IgG [2,10].

7. IMMUNOGLOBULIN A (IGA)

IgA is the major mucosal immunoglobulin and is present in saliva, mucus, sweat, gastric fluid and tears. In addition, it is present in large amounts in the colostrum and breast milk. The secreted form of IgA is polymeric and includes the J chain and SC. Monomeric IgA is an important component of serum. The secretion rate of IgA is higher than all other immunoglobulins combined. The majority of IgA synthesized is secreted to mucosal surfaces and IgA producing plasma cells reside in the sub-epithelial mucosa. Individuals with IgA deficiency suffer from recurrent mucosal infections. In humans, IgA-1 exists primarily as a monomeric molecule in the serum, whereas, IgA-2 is a polymer when secreted. IgA is believed to protect primarily by binding to pathogens and preventing their uptake across epithelia and by facilitating their expulsion in mucus excretion. It does not induce inflammatory responses which could have detrimental consequences on mucosal surfaces. IgA-1 fixes complement by the alternate cascade, and it optimizes phagocytosis which is mediated by the Fc-α receptor found on macrophages, monocytes and neutrophils. This receptor binds secretory IgA with higher affinity than serum IgA. Eosinophils also possess Fc-αR which mediates anti-parasite immunity [2].

8. IMMUNOGLOBULIN E (IGE)

The serum concentration of IgE is the lowest of all immunoglobulins. The function of IgE is to provide basophils and mast cells with specific antigen receptors. These cells, in turn, act as potent inducers of inflammatory reactions. IgE producing plasma cells are chiefly found in the lungs and skin. Circulating IgE is bound to high affinity Fc-ε receptors on mast cells and basophilic leukocytes, and remains stabily bound for weeks or months. These cells have IgE antibodies of many different specificities on their surface. Multivalent antigen can then cross link the bound IgE which leads to the degranulation of mast cells and basophils as well as the release of pro-inflammatory mediators. These mediators are capable of inducing mucous secretion, coughing, sneezing, vomiting, diarrhea and inflammation. While such a response may serve as a host defense against parasites, the pathological manifestation of such reactions is allergy and anaphylaxis in predisposed individuals. Langerhans cells also express Fc-ε RI. The CD23 surface antigen has been shown to be a low-affinity IgE receptor (Fc-ε RII). This receptor is expressed on monocytes and some follicular B cells. Monocytes are capable of secreting soluble Fc-ε RII [2].

Pooled human intravenous immunoglobulin (IVIg) ameliorates successfully the clinical manifestations of humoral immune deficiencies, haematological disorders, HIV infection and of many other diseases. The mechanism of action of IVIg remains unclear. The mechanisms proposed include Fcγ receptor blockade, accelerated clearance of auto-antibodies, inhibition of the complement cascade, neutralization of superantigens and bacterial toxins, anti-cytokine and anti-idiotype effects. A major factor in IVIG-mediated immunomodulation is the modification of immune/inflammatory activities of monocyte/ macrophages [11].

9. IMMUNOGLOBULIN GENES

The diversity of immunoglobulins is achieved through gene recombination, mutations, class switching and affinity maturation. For the variable region of immunoglobulin heavy chains, three separate sets of genes exist in the germ line. V_H genes number approximately 100; D genes with numbers over 10, and there are a small number of J genes (four in the mouse). In order to assemble a functional variable gene from germ line genes, first a D gene will recombine with a J gene which is followed by recombination of the D J with the V gene. Thus creating a functional variable region which is associated with genes coding for the constant portion of immunoglobulin classes (Figure 4). This recombination takes place through translocation and followed by excision of intervening DNA sections. A large number of V genes are created within the immune system by recombination. Junctional imprecision between D and J, and V and D increases diversity. For light chain variable regions, similar rules exist except that D genes are not involved. Only V and J elements recombine and are associated with constant region genes for κ and λ chains. This adds to the diversity of light chain class switching. Initially, B-lymphocytes form IgM antibodies which may be switched to other immunoglobulin classes by switching C genes while maintaining V region genes and thereby antigen specificity. During this switch, the membrane-bound receptors of the B cell are also switched to the class secreted, for instance from IgM to IgG, A or E receptors. Upstream of the C genes, specialized switch (S) regions are located that contain repetitive sequences with the exception of the δ-C gene. An immunoglobulin class switch is illustrated in Figure 4d [12,13]. The process of immunoglobulin class switch is poorly understood in molecular terms at the present time. Even the nature of the switch recombinase remains elusive. Current results on the subject indicate an unprecedented complexity, and for this reason the process could not be reproduced so far in a cell free systems [14].

Interleukin 4 is capable of inducing a switch from IgG 4 to IgE. Interferon-γ induces a switch to IgG-2 A in the mouse and TGF-β promotes switching to IgA. The CD-4 ligand on T cells and bacterial lipopolysaccharides (LPS) are also known to promote immunoglobulin switching.

B-1/CD5+ B cells mature in the liver during fetal development and produce natural antibodies, which maintain a germ line configuration of immunoglobulin genes. Many of these natural antibodies recognize autoantigens. These autoreactive natural antibodies may fulfill physiological regulatory functions rather than represent pathogenic autoreactivity. Autoantigens play a role in the development of these B lymphocytes [15,16].

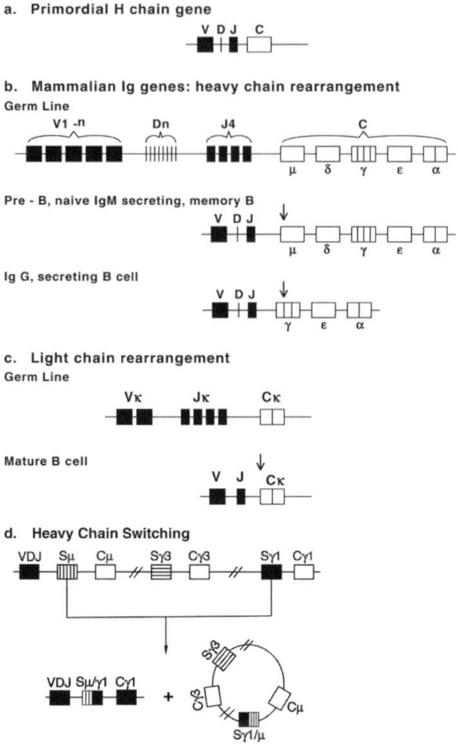

Figure 4. Immunoglobulin Genes and their Rearrangements. a. *A Primordial Form of the Immunoglobulin Gene Present in Sharks*. Sharks generate multiple copies of VDJC sequences. b. *In Mammals Multiple Copies of V, D, J and C Genes Exist Separately in the Germline*. Recombination initially starts with joining D and J, and continued with DJ joining to V and eventually to the C genes. A rearranged VDJ with a full complement of C genes is characteristic of pre-B, naïve IgM secreting B and memory B cells. After class switching some of the C genes are lost due to excision by splicing. c. *Light Chain Rearrangement*. Here D genes are not involved, so a VJC combination is reached after rearrangement. d. *Immunoglobulin Class Switching*. The process by which a B cell is switching from IgM production to IgG1 is illustrated. In this case the severance (S) points are Sμ and Sγ1. After cleavage the VDJCγ1 gene is created and a circular episome containing the excised genes results. Transcription initiating upstream of VDJ results in the formation of IgG1 heavy chain protein.

10. AFFINITY MATURATION

Initially B cells produce low-affinity IgM antibodies while T cell-dependent B cell stimulation and somatic hypermutation is going on in the germinal centers. The mutation rate is one per thousand base pairs per generation which greatly increases antibody diversity. This ensures that the B cells that emerge will bind antigen with high affinity. These cells are rescued from apoptosis by antigen-dependent T cell stimulation, increase the affinity of antibodies produced which in turn leads to "affinity maturation" of the antibody response.

The secondary antibody response is faster and of greater magnitude than is the primary response. The antibodies produced are of higher affinity and of different isotype: typically IgG, rather than IgM. This is the result of V gene somatic hypermutation and heavy chain class switching [17].

REFERENCES

1. Heilderberger M. The precipitin reaction, microbial agglutination, complement fixation and relations between chemical structure and antigen-antibody interaction. In: Szentivanyi A, Friedman H, editors. The Immunological Revolution. Facts and Witnesses. Boca Raton, FL; CRC Press, 1994;1–11.
2. Frazer JK, Capra JD. Immunoglobulins: Structure and function. In: Paul WE, editor. Fundamental Immunology. New York: Lippincott-Raven, 1999;37–74.
3. Szentivanyi A, Maurer PH, Janicki BW. Antibodies: Structure, Synthesis and Immunogic Intervention in Disease. New York: Plenum Press;1987.
4. Berczi I. Humoral factors of immunity. In: Berczi I, editor. Pituitary Function and Immunity. Boca Raton, FL: CRC Press, 1986;11–16.
5. Edmundson AB, Ely KR, Abola EE, Schiffler M, Panagiotopoulos N. Rotational allomerism and divergent evolution of domains in Ig light chains. Biochemistry 1975;14: 3953–3961.
6. Melchers F, Rolink A. B-lymphocyte development and biology. In: Paul WE, editor. Fundamental Immunology, 4th edition. New York: Lippincott-Raven, 1999;183–224.
7. Johansen FE, Braathen R, Brandtzaeg P. Role of J chain in secretory immunoglobulin formation. Scand J Immunol 2000;52:240–8.
8. Berzofsky JA, Berkower IJ, Epstein ES. Antigen-antibody interactions and monoclonal antibodies. In: Paul WE, editor. Fundamental Immunology. New York: Lippincott-Raven, 1999;75–110.
9. Berczi I, Chow D, Bertok L. Neuroimmunoregulation and Natural Immunity. Ann NY Acad Sci 2001;917:248–257.
10. Kato K, Fridman WH, Arata Y, Sautes-Fridman C. A conformational change in the Fc precludes the binding of two Fc gamma receptor molecules to one IgG. Immunol Today 2000;21:310–2.
11. Rhoades CJ, Williams MA, Kelsey SM, Newland AC. Monocyte-macrophage system as targets for immunomodulation by intravenous immunoglobulin. Blood Rev 2000;14: 14–30.
12. Lansford R, Okada A, Chen J, Oltz EM, Blackwell TK, Alt FW, Rathbun G. Mechanism and control of immunoglobulin gene rearrangement. In: Hames, BD, Glover DM, editors. Molecular Immunology, 2nd edition. Oxford University: IRL Press, 1996;1–100.
13. Max EE. Immunoglobulins: Molecular genetics. In: Paul WE, editor. Fundamental Immunology, 4th edition. New York: Lippincott-Raven, 1999;111–182.
14. Zhang K. Immunoglobulin class switch recombination machinery: progress and challenges. Clin Immunol 2000;95:1–8.
15. Hardy RR, Li YS, Allman D, Asano M, Gui M, Hayakawa K. B-cell commitment, development and selection. Immunol Rev 2000;175:23–32.
16. Chow D, Kraut R, Want X.. Natural immune regulation of activated cells. In: Bertczi I, Gorcynski R, editors. New Foundations of Biology, Neuroimmune Biology. Amsterdam: Elsevier, 2001;331–345.
17. Neuberger MS, Ehrenstein MR, Rada C, Sale J, Batista FD, Williams G, Milstein C. Memory in the B-cell compartment: antibody affinity maturation. Philos Trans R Soc Lond B Biol Sci 2000;355:357–60.

Edited by I. Berczi and A. Szentivanyi

Growth and Lactogenic Hormones, Insulin-Like Growth Factor and Insulin

ISTVAN BERCZI and ANDOR SZENTIVANYI

Department of Immunology, Faculty of Medicine, University of Manitoba, Winnipeg, Manitoba, R3E 0W3, Canada; and Department of Internal Medicine, Faculty of Medicine, University of South Florida, Tampa, Florida 33612, USA

ABSTRACT

Prolactin (PRL), growth hormone (GH) and placental lactogenic hormones (PL) belong to the growth and lactogenic hormone (GLH) family. These hormones show molecular heterogeneity and the variant hormones differ in their biological activity. GLH receptors are also heterogeneous and have common features with hemopoietic and lymphocyte cytokine receptors. GLH commonly uses signal transducers and activators of transcription (STAT) family transcription factors in the early phase of induction. However, they also act through other signaling systems.

Growth hormone and prolactin induce insulin-like growth factor-I (IGF-I) in cells of the immune system and in other tissues and organs. The IGF receptors belong to the transmembrane tyrosine kinase family which signal through phospholipase C-protein kinase C pathway.

Placental hormones rather than pituitary GH and PRL support the embryonic development of the immune system. After parturition pituitary GH and PRL in conjunction with IGF-I maintain the bone marrow, thymus and the secondary lymphoid tissue (spleen, lymph nodes, mucosal and cutaneous lymphoid system) in a functional state. In the absence of pituitary GH and PRL, the immune system loses rapidly cellularity which is associated with a profound decrease in immunocompetence. Bone marrow function is also decreased. For the normal function of the bone marrow and the immune system, it is sufficient to have normal levels of either pituitary GH or pituitary PRL. Pituitary GLH in conjunction with IGF-I support all immune functions that involve humoral and cell mediated immune responses, innate immune responses and also the internal immunoregulatory pathways. These may lead to the induction of immunological tolerance, anergy and apoptosis induced by cell-to-cell signaling. On this basis, pituitary GLH may be designated as the hormones of immunocompetence.

Insulin (INS) is a pleiotropic growth factor which interacts with both GH and IGF-I. Insulin is required for normal immune development and function. In general, INS stimulates immune reactions but inhibitory effects have also been observed. Insulin may be a key regulator of immune function in acute illnesses, which are also characterized by insulin resistance.

1. INTRODUCTION

Prolactin (PRL), growth hormone (GH) and placental lactogen (PL) diverted from a common ancestral gene. All of these hormones show molecular heterogeneity and are currently referred to as the GLH family. Experimental evidence indicates that the variant hormones differ in their biological activity [1–7].

2. RECEPTORS AND SIGNALING

PRL receptors have been characterized in various species as short, intermediate and long, and GH receptors as short or long. These receptors share features with hemopoietic and lymphocyte cytokine receptors and may be designated as the GLH-cytokine receptor family. The receptors for erythropoietin (EPO), granulocyte colony stimulating factor (G-CSF), granulocyte-macrophage colony stimulating factor (GM-CSF), and interleukin (IL)-2, -3, -4, -5, -6 and -7 belong to this family [8,9]. The short and long receptors are present in the thymus, spleen, lymph nodes and bone marrow of rats and mice, [10]. Immunization [11] and autoimmune disease [12] lead to the increased expression of PRL receptors (PRLR) by lymphocytes.

The expression of GH receptors (GHR) was analyzed in murine lymphoid organs using flow cytofluorometry and biotinylated bovine GH. GHRs are expressed in all murine hematopoietic tissue in fetuses, newborns, and 3- and 7-week-old animals. All hematopoietic lineages express variable levels of GH receptors in the bone marrow, whereas in the thymus, expression is mainly seen in CD4-, CD8-, CD4+CD8+, and CD8+ subpopulations. Fifty percent of splenocytes and peripheral blood lymphocytes and 20% of lymph node cells are GHR positive. A wider receptor expression was present on B cells and macrophages (approximately 50%) than on T cells (approximately 20%), that included both CD4+ and CD8+ cell subsets. The proportion of GHRs bearing CD4+ and CD8+ splenocytes is increased after T-cell activation with Con A or anti-CD3. All peripheral T cells expressing GHRs also express prolactin receptors [13].

Growth hormone receptors were readily detected in early- and mid-gestational bovine fetal thymocytes, but they were less readily detected in thymocytes from older fetuses. In contrast, GHRs were easily detected in splenocytes at all times. The CD4+ and $CD8^+$ (single positive) thymocyte subsets were positive for GHR [14].

In human tonsil lymphocytes, B cells constitutively expressed GHR and quiescent cells had fewer receptors than activated cells. Activated T cells expressed GHR, but the typical CD4+ CD57+ T cells in germinal centers did not [15]. GH receptor expression on immune cells of non-syndromic short children is inversely related to linear growth and body mass index [16].

Growth hormone induces insulin-like growth factor-I (IGF-I) in the liver as well as all other tissues and organs. IGF-I is an obligate mediator for many of the biological effects of GH. IGF-I is structurally related to insulin and cross reacts with it at the receptor level [17]. Prolactin stimulates synlactin in its target cells, which is analogous to IGF-I [18].

The crosslinking of two receptor molecules by a single hormone molecule (dimerization or oligomerization) is required for effective signal transduction by the growth hormone receptor. Consequently, at high hormone concentrations when each receptor is occupied by a single hormone molecule, crosslinking becomes impossible. In this situation it is visualized that the hormone would serve as its own antagonist [19].

The signaling pathway for PRL and related hormones is not fully resolved. G-proteins, tyrosine kinase and protein kinase C have all been suggested to play a role [20,9,21]. Direct nuclear signaling by PRL and GH has also been proposed [22–26].

Much evidence suggests that phosphorylation of transcription factors by protein kinases plays a major role in signal transduction by the PRL receptor. DNA binding of transcription factors and transactivation can be effected both positively and negatively by these factors. GLH and cytokines commonly use STAT (signal transducers and activators of transcription)-family transcription factors in the early phase of induction. The use of other signaling systems, such as *Src*-family kinases, phospholipase C/protein kinase C and mitogen activated protein (MAP) kinase is variable during GLH signaling. Although initially there are shared protein kinases and STATs, the combination of transacting factors working at a particular promoter in a given cell type is distinct as a rule. This may specify a very different biological effect upon the stimulation of different target cells by the same hormone. Therefore, the specificity of hormone action depends to a large extent on the target cells' repertoire of transacting factors *(*Figure 1*)* [27,28,23,7].

A fair amount of information is available about the function of the seven STAT genes in mammals. STAT1 is activated by IFNα, -β, -γ, GH, IL-2 epidermal growth factor (EGF) and angiotensin. In STAT1 knockout mice an IFN signaling defect and immunodeficiency were observed. STAT3 is activated by multiple cytokines (IL-2, -3, -6, -7, -10, -11, -15, IFNα, -γ, G-CSF, LIF, OSM, CNTF) and some hormones and growth factors (leptin, EGF, PDGF). The receptors of these mediators harbor common STAT3 docking motifs (YxxQ) in their cytoplasmic domain. STAT3-deficient mice exhibit fetal lethality which may be due to defective visceral endoderm function and nutritional insufficiency. In adult tissues the role of STAT3 was studied using the Cre-loxP recombination system. This system allows the deletion of STAT3 in specific tissues. STAT3 deficient T cells showed a severely impaired proliferation to IL-6, as IL-6 was unable to prevent apoptosis in such cells. The response to IL-2 was also suppressed because of impaired IL-2 mediated IL-2Rα induction. Mice with STAT3 deficient macrophages and neutrophils demonstrated the increased production of inflammatory cytokines (e.g. TNFα, IL-1 and IFNγ), and were highly susceptible to endotoxin shock. Aging mutant mice developed chronic enterocolitis. The suppressive effect of IL-10 on cytokine production by macrophages and neutrophils was completely abolished. STAT3 deficiency in the skin led to severely impaired wound healing. In STAT4 deficient mice IL-12-mediated functions are impaired: these include the production of IFNγ, cell proliferation, natural killer (NK) cell activity, and TH1 differentiation. STAT6 deficient mice have defects in IL-4 and IL-13-mediated functions. CD23 and MHC-II expression, class switching to IgE, B- and T cell proliferation, TH2 development and macrophage activation suffer. STAT6 plays a crucial role in allergen-induced airway inflammation. Diverse biological functions for the two STAT5 forms, STAT5a and STAT5b, have been discovered in Stat5-deficient mice. Roles are indicated for one or both STAT5 forms in the immune system, hematopoiesis, sexually dimorphic growth, mammary development, hair growth, deposition of adipose tissue, and pregnancy. Splenocytes of STAT5a null (0) mice are defective in anti-CD3 induced proliferation that can be overcome by high IL-2 doses. In such mice the IL-2 induced IL-2Rα expression by T lymphocytes was impaired. IL-2Rα impairment is also present in STAT5b-null mice and it cannot be corrected by IL-2 treatment. Splenocytes of STAT5b-0 mice show a greatly diminished proliferation in response to IL-2 and IL-15, and this cannot be corrected by high dose IL-2. NK cytotoxicity is greatly diminished, IL-2Rβ expression and the proliferative response and augmentation of cytotoxicity in response to IL-2 and IL-15 are greatly diminished. Bone marrow-derived macrophages of STAT5-0 mice show a defective response to GM-CSF [29,30].

The IGF-I receptors belong to the transmembrane tyrosine kinase family. Signal transduction involves the phosphorylation of enzymes, such as phospholipase C, phosphatydil-inositol-3'kinase, *ras*, guanosine triphosphatase activating protein and *Src* and *Src*-like thyrosine kinases.

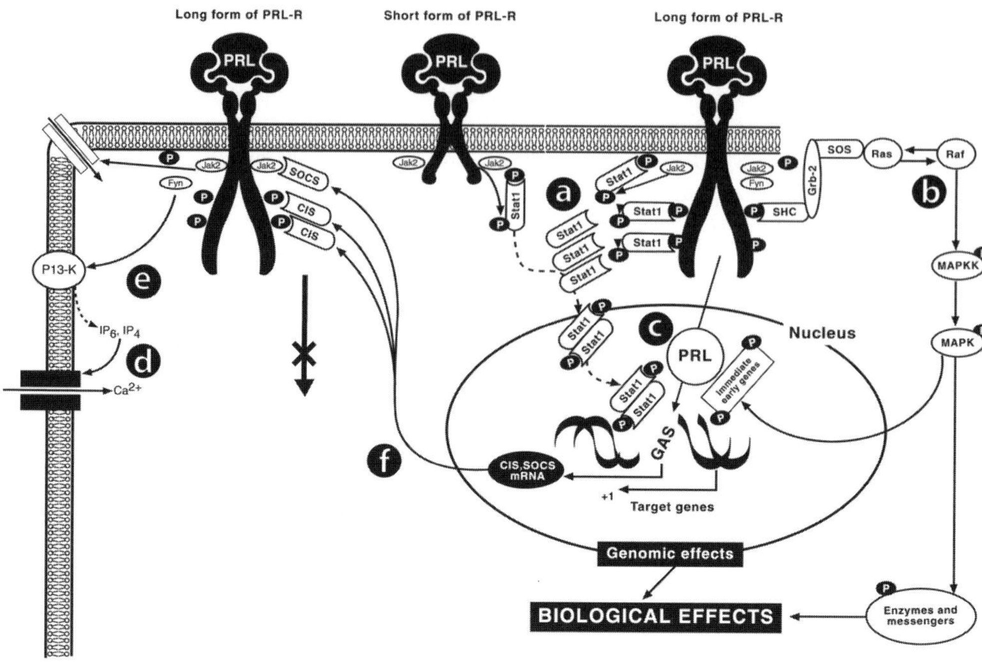

Figure 1. Signal Transduction by Prolactin. a. *The Jak/STAT pathway*: Receptor dimerization is required for the activation of the PRLR. A single PRL molecule contains two binding sites and first site number one interacts with a receptor molecule through the N-terminal extracellular domain number one (ECD1). The ECD1 and ECD2 extracellular domains of PRLR show analogies with the fibronectin type III molecule, which drives the receptor-ligand interactions in cytokine receptors. The initial binding induces the interaction of binding site two on the same PRL molecule with a second PRLR to achieve dimerization. This is necessary for the activation of Janus kinase 2 (Jak2) which is constitutively associated with the intracellular domain of PRLR. In turn, Jak2 kinases trans-phosphorylate each other and phosphorylate (P) the tyrosine residues (Y) of the PRLR itself. Phosphorylation of Jak2 is a key event in the activation of all prolactin receptors. Tyrosine phosphorylation does not occur in the case of the short PRLR although tyrosine residues are present. Signal Transduction: The Jak/STAT pathway that includes STAT1, STAT3, STAT5a and STAT5b, which are members of the signal transducer and activator of transcription (STAT) protein family, is central to signal transduction pathways by PRLR. STATs contain a DNA-binding domain, an SH2-like and SH3-like domain and an N- and C-terminal trans-activating domain. A phosphorylated tyrosine residue of the activated long PRLR interacts with the SH2 domain of STAT while it is docked at the receptor and is phosphorylated by Jak kinase. The phosphorylated STAT will dissociate then from the receptor and form hetero- and homodimers through its phosphotyrosine residues with the SH2 domain of another phosphorylated STAT molecule. The dimers then translocate to the nucleus and activate the DNA-binding motif in the promoter of a target gene termed GAS (γ-interferon activated sequence). Although the tyrosine resides of the short form of PRLR are not phosphorylated, the phosphotyrosine of Jak2 can serve as a docking site for STAT1 which, in turn, contributes to signal transduction by the Jak STAT pathway. b. *The MAPK cascade*: The PRLR is also capable of activating the mitogen-activated protein kinase (MAPK) cascade. This pathway involves the activation of a wide range of transcription factors/immediate early genes by phosphorylation. The phosphotyrosine residues of the activated long PRLR isoforms serve as docking sites for adapter proteins (Shc/Grb2/SOS) which connect the receptor to the Ras/Raf/MAPK cascade. The Jak/STAT and MAPK pathways communicate with each other. c. *Nuclear signaling*: By the direct translocation of PRL into the nucleus. d. *Ion channels*: Box 1 of the intracellular domain of the PRLR is involved in the activation of tyrosine kinase-dependent calcium-sensitive potassium channels through Jak2. The C-terminal of PRLR is involved in the production of the intracellular messenger's inositol 1, 3, 4, 5-tetrakisphosphate (IP_4) and inositol hexa-

Phospholipase C activation generates inositol triphosphate and diacylglycerol which in turn act as second messengers, mobilizing intracellular Ca^{++} and activating protein kinase C [8].

In monocyte cell lines the first exon and promoter sequence for GH expression is analogous to the pituitary promoter for GH. This promoter also contains negative regulatory elements [31].

3. GLH AND THE DEVELOPMENT OF THE IMMUNE SYSTEM

3.1. Embryonic development

The administration of prolactin to hypophysectomized (Hypox) rats results in immune-infiltration of the corpus luteum by monocytes and macrophages that induce regression. The PRL-induced infiltration of monocytes-macrophages is not accompanied by a decrease in luteal tissue progestin, at least in the early stages of luteal regression [32].

In fetuses lacking the pituitary gland, the immune system develops normally [33]. This indicates that during embryonic development placental and perhaps maternal GLH support the development of the hemopoietic and immune systems. Bone marrow function, thymus cellularity and the immune response can be restored in Hypox rats by human placental lactogen (PL) [34,35]. Human PL is mitogenic for the Nb2 rat lymphoma cell line which is dependent on lactogenic hormones for proliferation [2]. PRL and pituitary grafts placed onto the chorioallantoic membrane stimulated the early maturation of thymocytes in decapitated chicken embryos [36]. Neonatal rats treated with anti-GH serum for 8 weeks showed a significant decrease in thymus and spleen weights and cellularities and a diminished antibody response. All these abnormalities were corrected by treatment with bovine GH [37].

Both PRL and its structurally modified isoforms are abundant in early milk and gradually diminish with the progression of lactation. Milk PRL is absorbed by and is biologically active in the neonate. There is evidence to support the hypothesis that milk PRL regulates differentiation and maturation of neonatal neuroendocrine, reproductive, and immune systems [38].

3.2. Bone marrow

Hormones have long been known to play a role in bone marrow function [39]. Jepson and Lowenstein [40] first demonstrated the erythropoietic effect of PRL. Hypox rats develop anemia and exhibit a profound impairment of DNA and RNA synthesis in the bone marrow, which is associated with leukocytopenia and thrombocytopenia. All these deficits are restored to normal levels by syngeneic pituitary grafts (SPG) placed under the kidney capsule or by treatment with purified PRL, GH or human PL [34,35,41–43]. PRL and ACTH stimulated the proliferation of bone marrow progenitors, partly through preprotachykinin-I and neurokinin-1 induction [44]. PRL stimulated the phosphorylation of the PRLR-associated Janus tyrosine kinase (JAK)-2 in rat bone marrow and spleen cells. This lead to the activation and subsequent binding of STAT5b to an interferon regulatory factor-1 (IRF-1) gamma activation sequence (GAS). PRL, at physio-

---

kisphosphate (IP_6) that open voltage-independent calcium channels. e. *Src kinases*: The members of this kinase family, c-src and Fyn, which are involved in the tyrosine phosphorylation of phosphatidyl inositol 3-kinase (PI3K) also play a role. f. *Downregulation*: Jak/STAT pathways are inhibited by SOCS (suppressors of cytokine signaling) which inhibit Jak kinase or by CIS (cytokine inducible SH2-containing protein) which compete with STAT for docking sites on PRL receptors [7,23,28].

logical concentrations (0.01 mg/ml), stimulated the expression of the IRF-1 gene in these normal cells. PRL could therefore affect several aspects of the immune response [45].

Prolactin receptors are present in murine bone marrow B-cell precursors and PRL induced significant expansion of defined B-lineage cell populations in short-term bone marrow cell cultures [46].

A recombinant form of human prolactin (r-hPRL) enhanced cell function, accelerated lymphoid and myeloid reconstitution and promoted hematopoiesis *in vitro* and in animal model systems. These results have been produced and evaluated in a wide variety of preclinical models. The multi-lineage effect of r-hPRL makes it an attractive candidate for clinical situations presenting with immune deficiency or myelosuppression [47].

Human hemopoietic progenitor cells were found to form granulocyte and erythroid colonies if stimulated with PRL in the presence of IL-3, GM-CSF and EPO [48]. Insulin-like growth factor-I mediated the action of GH on the bone marrow, including B lymphocyte growth [49–51]. PRL was shown to regulate the function of bursa of Fabricius in birds [52,53].

3.3. The thymus

The stimulatory effect of GH on thymus growth has long been established [54–56]. GH induces thymus growth in hormone deprived and old animals, which positively correlates with an increase in immunocompetence. GH stimulates the production of thymic hormones and has a direct mitogenic effect on thymocytes *in vitro* [57–60]. Many effects of GH on the thymus are mediated by IGF-I. Hormone production by human and rat thymic epithelial cells was stimulated by GH and by IGF-I [61–63]. In mice with severe combined immunodeficiency, human GH promoted the engraftment of human thymocytes [64]. Syngeneic pituitary grafts, or treatment with GH or PRL restores impaired DNA synthesis and involution of the thymus in Hypox rats, and reverses the profound immunodeficiency present in these animals. In treated rats, thymocyte proliferation resumes and immunocompetence returns to normal [60]. Pituitary grafts increased thymus weight as well as the number of thymocytes in Ames dwarf mice [65]. In thymocytes, TRL stimulated the expression of the Thy-1, LT-34 (CD4), and TL antigens [59,66,67]. PRL stimulated thymic epithelial cells for hormone production. PL selectively increased the growth of the thymus in Snell-Bagg pituitary dwarf mice [68,62].

Human GH (1–100 ng/ml) increased the total number and nucleation of multinucleated cells (MNC) in long-term human bone marrow cultures. Cytological and immunological characterization of MNC revealed several macrophage features [69].

4. GLH AND IMMUNE FUNCTION

4.1. The antibody response

It has been suggested that GH augments the antibody response (Berczi 1986). The impaired antibody response of Hypox rats and of rats immunosuppressed by bromocriptine (BRC) or adrenocorticotropic hormone (ACTH) treatment can be restored by replacement doses of either GH or PRL [34]. The treatment of mice with bovine PRL enhanced the antibody response according to a biphasic dose curve [70] as did SPG both in young and old mice [71]. Bovine PRL also modulated the antibody response in cockerells [72,53].

The immunization of rats with sheep red blood cells (SRBC, a T cell-dependent antigen) induced an increase of hypothalamic thyroid stimulating hormone (TRH) mRNA at 4–24 h. In contrast, after treatment with lipopolysaccharide (LPS), a decrease in TRH mRNA was observed. SRBC increased pituitary TRH receptor mRNA and plasma PRL levels but there was no change in TSH or GH. Unlike the early corticosterone peak induced by LPS, the activation of the HPA suppressive response appeared 5–7 days after SRBC treatment. The treatment of SRBC immunized rats with TRH antisense nucleotide given i.c.v. inhibited the specific antibody response, and caused a concomitant inability to produce the peak in plasma PRL levels. These results indicate, that the T cell-dependent immune response is critically dependent on the early activation of TRH and PRL and that the neuroendocrine changes are profoundly different from those occurring during the T cell-independent and inflammatory responses (LPS model) [73].

The treatment of pituitary dwarf children with GH induced transient changes in the mitogenic response of peripheral B lymphocytes and immunoglobulin secretion [74,75]. Recombinant human GH stimulated human B lymphocytes for proliferation and immunoglobulin synthesis in serum free medium [76]. Furthermore, human GH and IGF-I enhanced the secretion of IgG1, IgG2, IgG3, IgG4, IgA1, IgA2 and IgM, and stimulated the growth of human tonsillar B lymphocytes after activation with *Staphylococcus aureus* Cowan strain I [77]. Human GH and IGF-I induced IgE and IgG4 secretion by normal human mononuclear cells, from which IgE^+ and $IgG4^+$ B lymphocytes were depleted. IgM, IgG1, -2, -3, IgA1 and 2 production was not affected in the same culture system. Purified surface IgE and surface IgG4 negative B cells did not produce immunoglobulin when exposed to GH or IGF-I. However, IgE and IgG4 production was induced when anti-CD40 monoclonal antibody was also added to the system. This treatment did not enhance Ig production by surface IgE^+ and $IgG4^+$ B lymphocytes. It was suggested that GH and IGF-I induced immunoglobulin class switching in this system [78]. Human GH inhibited the differentiation of human B lymphocytes into plasma cells *in vitro* [79].

Epstein Barr virus (EBV) transformed human B lymphocytes produced significant amounts of IGF-I after stimulation with GH. EBV transformed B cells of pygmies produced significantly less IGF-I after a similar stimulation with GH [80]. Hypox rats showed a 30% reduction of IGF-I and a 50% reduction of IGF-II in their spleens, which was reversed by GH treatment [81].

Prolactin stimulated the proliferation of murine B cell hybridomas in a dose-dependent manner and enhanced their proliferation in response to IL-4, IL-5, and IL-6. This resulted in an overall increase in antibody production. Hybridoma cell lines incubated with TGF-beta demonstrated a dose-dependent decrease in proliferation, which could be reversed by the addition of PRL to the cultures [82].

In hypophysectomized rainbow trout (*Oncorhynchus mykiss*) plasma IgM levels were decreased to 30% of those in sham-operated controls 1 or 4 weeks after the operation. Implantation of a cholesterol pellet containing salmon PRL or GH restored plasma IgM levels to normal [83].

Aged female monkeys received treatment with recombinant human GH or IGF-I for 7 weeks. Both treatments induced lymphocyte phenotype changes, which was different in blood, spleen and lymph nodes. In the blood, the percentage of B cells and $CD8^+$ cells rose after IGF therapy and thus caused a fall in the CD4/CD8 ratio. In the spleen, the CD4/CD8 ratio nearly doubled. Combined treatment with GH and IGF-I increased the percentage of T cells in the spleen. Both treatments increased the antibody response to tetanus toxoid [84].

In young GH deficient adult humans, GH treatment for 6 months caused a rapid and transient decrease in circulating immunoglobulins and enhanced circulating immune complexes [85].

4.2. Cell mediated immunity

Physiological (10–20 ng/ml) concentrations of PRL interacted synergistically with granulocyte-macrophage colony stimulating factor (GM-CSF) in the stimulation of dendritic cell (DC) maturation. The effect was similar to that induced by IL-4 on GM-CSF-driven DC maturation. The physiological concentrations of PRL were inhibitory, when applied alone, whereas higher concentrations (80 ng/ml) were stimulatory for DC precursors [86]

In rats, Hypox inhibits contact sensitivity reactions and prolongs the survival of allogeneic skin grafts [87]. If Hypox animals are given SPG or treated with PRL, GH or PL, contact sensitivity reactions are restored. Treatment of normal rats with the dopaminergic agents, BRC or pergolide prevents the development of contact sensitivity reactions to dinitrochlorobenzene. Contact sensitivity can be restored by additional treatment with PRL or GH. Restoration may be antagonized if replacement doses of ACTH are also given [88,34]. Treatment of normal rats with BRC prolonged the survival of allogeneic skin grafts [34]. Combined treatment of rats with dopaminergic agents (e.g. BRC or CQP-201-403) and with the immunosuppressive drug cyclosporine A (CSA), showed synergism in the suppression of the mixed lymphocyte reaction, the local graft versus host reaction, and the rejection of kidney and cardiac allografts [89–91].

In mice given allogeneic skin grafts, pituitary PRL mRNA was significantly increased and PRL bioactivity in the serum was elevated [92]. The T cell dependent induction of macrophage tumoricidal activity in mice by *Listeria monocytogenes* or *Mycobacterium bovis* could be prevented by bromocriptine treatment. Additional treatment with ovine PRL reversed this effect [93]. Serum PRL levels increased significantly and consistently in patients with cardiac allografts during the primary rejection episode, but such an increase became irregular during recurrent rejections [94].

The proliferation of human T cells stimulated by IL-2 or phytohemagglutinin (PHA) was significantly enhanced by PRL. At physiological concentrations PRL also stimulated B, T and NK cells when applied together with mitogens, whereas 5 to 10-fold higher levels inhibited the response of T cells to IL-2 [95,96]. Human peripheral blood NK cells significantly increased the secretion of IFNγ in response to stimulation by PRL. In turn IFNγ stimulated the cytotoxic activity of NK and LAK cells [97]. Bovine PRL, but not bovine GH, increased the expression of IL-2 receptors by $CD4^+$ and $CD8^+$ T lymphocytes in the peripheral lymph nodes when administered to mice. The expression of IL-2 receptors in popliteal lymph nodes of dwarf mice was only 50% of normal after the injection of ConA. Treatment with PRL or GH corrected this deficiency [98]. Bovine GH (bGH) was able to stimulate the proliferation of Con A- or anti-CD3-activated murine T cells. A bell-shaped dose-response curve was observed. LPS activated B cells were not affected by bGH [99]

The 20 kD and the 22 kD variants of human GH (hGH) equally stimulated the proliferation of human Th2 and Th0 cells in the presence of mite antigen. Th1 cells were not affected. The 20 kD GH works exclusively through the hGH receptor. Growth hormone binding protein (GHBP) inhibited Th1 proliferation and GH-stimulated Th2 and Th0 growth. Mite antigen stimulated Th1 cells that produced GH, but not Th2 cells [100].

Human T cell lines derived by transformation with the human T lymphotropic virus (HTLV) responded to GH *in vitro* by proliferation and IGF-I production, which in turn stimulated colony formation by these cells. T cell lines derived from individuals with Laron-type dwarfism did not respond to GH treatment by proliferation and IGF secretion [101]. Murine spleen cells also produced IGF-I in response to GH stimulation [102].

In two dwarf children, acute rejection episodes of renal allografts occurred after treatment with GH [103]. Myositis developed in two boys while on GH therapy. The disease resolved in both patients two months after the cessation of therapy, but returned within three months when therapy was resumed [104]. In 22 adults with growth hormone deficiency the absolute number and relative percentage of NK cells was significantly decreased [105]. In 14 girls with Turner syndrome, GH therapy decreased the CD4/CD8 T cell ratio and increased the number of NK cells [106]. GH therapy increased NK cell mediated cytotoxicity in healthy adults as well as in female patients with impaired GH secretion [107,108].

The experiments on the effect of PRL on NK cell mediated cytotoxicity are controversial, Matera et al. [109] showed that PRL stimulates the growth and cytotoxic activity of purified NK cells. However, there was no effect on NK activity when mixed populations of peripheral blood lymphocytes (PBL) were treated. The lack of reactivity in this case was due to the activation of suppressor cells by PRL. Treatment of PBL or purified NK cells for 4 days with PRL in serum free medium increased target cell killing, but novel cytotoxic cells were not induced on NK susceptible target cells (K562 and U937). However, cytotoxicity against lymphokine activated killer (LAK) target cells (HL60, Jurkat, Daudi, and Sutt-1) was induced in both the NK and T cell populations. PRL had a biphasic (e.g. stimulatory and inhibitory) effect on NK cells with peaks either at 25 or 200 ng/ml, whereas LAK activation occurred only at 200 ng/ml. Physiological concentrations of PRL stimulated the generation of NK and LAK activities when combined with low doses of IL-2. Pathologically high concentrations of PRL reversibly inhibited the generation of LAK cells, whereas IL-2 activated NK cells were stimulated [110].

Human GH and IGF-I bind to CD16+/CD3- NK cells. In GH-deficient patients, there is a decreased NK cell activity. The absolute number and percentage of NK cells is normal. Replacement therapy with GH normalizes NK cell activity in deficient individuals [111]. In young GH deficient adults GH treatment for 6 months increased the CD4/CD8 T cell ratio [85].

4.3. The effect of GLH on phagocytic cells

Human GH, at picomolar concentrations, exerts a chemotactic effect on human monocytes. Somatostatin and its long acting analogue, octreotide, show a similar effect, but micromolar concentrations were required. Somatostatin inhibited the chemotactic effect of GH on monocytes in a dose dependent manner. Interferon and substance P were also inhibitory [112].

The phagocytic activity of monocytes and polymorphonuclear cells was increased significantly in children treated with GH for 6 months. This was also true for long term GH replacement therapy [113]. GH, PRL and growth hormone releasing hormone at very high concentrations enhanced hydrogen peroxide production in human monocytes stimulated with phorbol myristate acetate (PMA). IGF-I had no effect on this system. Monocytes expressed mRNA for GH and PRL receptors and bound specifically radiolabelled GH [114].

Treatment with GH, and to a lesser extent with PRL, significantly increased the number and size of testicular macrophages in long term Hypox rats. After testicular Leydig cells had been killed selectively with ethylenedimethanesulphonate, the clearance of dead cells was largely increased in GH and GH plus PRL treated rats [115].

In neutrophilic leukocytes, GH stimulated lysosomal enzyme production, oxidative metabolism, adhesiveness, modulated chemotaxis, and priming for superoxide production [116,117]. Aged rats exhibit a reduction in superoxide anion secretion and in bactericidal activity of neutrophilic leukocytes. This deficiency could be corrected *in vitro* by treatment with IFN-γ or GH. Neutrophils from aged rats grafted with a syngeneic GH secreting pituitary tumor GH3 could be primed *in vitro* for superoxide secretion by IFN-γ whereas control rats could not [118]. Neutrophils from patients with acromegaly and hyperprolactinemia showed a decrease in chemotactic activity [119].

Growth hormone treatment (100 ng/ml) of human polymorphonuclear leukocytes *in vitro* inhibited apoptosis and up-regulated the production of reactive oxygen intermediates. GH treatment had no effect on apoptosis of monocytes and lymphocytes in the same culture system [120].

4.4. The effect of GLH on cytokine production

IFN-γ production by murine spleen cells and by human peripheral blood mononuclear cells is enhanced by PRL treatment *in vitro* [74,93,121]. Furthermore, interferon-regulatory factor-I gene expression and IFN-γ production were induced by PRL in the Nb2 rat lymphoma cell line and in T lymphocytes. GH stimulated the production of IL-2 by human lymphocytes and IL-1, as well as tumor necrosis factor-α (TNFα) and superoxide anion production by monocytes [122,123,27]. GH activated monocytes for superoxide production, but not for TNF production, cell adherence or killing of *Mycobacterium tuberculosis* [114]. Growth hormone increased the release of IFN-γ from murine splenocytes stimulated with enterotoxin A, and inhibited the release of IL-1α. Prolactin also decreased the release of IL-1α, but had no affect on IFN-γ release under these conditions [124]. In cultures of human peripheral blood lymphocytes (PBL) placental and pituitary GH reduced IL-5 production slightly and stimulated IFN-γ production. PRL also enhanced IFN-γ and could be induced by the PRL-receptor [125].

Murine splenocytes were stimulated with Protein A (PA), Toxic Shock Syndrome Toxin-1 (TSST-1) and Streptolysin S (SLS). In splenocytes stimulated with PA, GH induced a 40% and 50% drop in IL-1α and IFN-γ release respectively, compared to controls, while no change was seen in IL-4 release. The release of IFN-γ by TSST-1-stimulated splenocytes fell by 30%, but no changes were shown in IL-1α and IL-4 release after GH treatment. The release of IL-1α by SLS-stimulated splenocytes increased by 50% in the presence of GH, and no changes were shown in IFN-γ and IL-4 release [126].

Growth hormone treatment of cows reduced plasma levels of TNFα, cortisol, thromboxane-B2 and thromboxane/prostacyclin ratios in response to endotoxin injection [127].

High dose GH (13 IU/m^2/day) given to patients undergoing laparoscopic surgery did not affect the cytokine response to surgery (TNF-α, IL-6). PBL from healthy individuals exposed to GH concentrations between 100 to 1000 ng/ml for 6–72 hrs. GH had no effect on the production of TNF-α, IL-6 and IFN-γ. The incubation of PBL with the GH antagonist B2036 had no effect on the production of these cytokines [128].

In thymocytes from bovine fetuses at approximately mid-gestation, GH treatment for 30 minutes down-regulated c-jun and c-fos mRNA approximately 2- and 2.8-fold, respectively. After 6 hours of treatment, GH increased transcript levels for IL-1α, -β, IL-6, and GM-CSF. These results suggest that GH is intimately involved in lymphocyte function and expression of certain cytokines during a critical period of fetal immune development [14].

Enzymatically cleaved (16K)-PRL, but not full-length PRL, acted to promote the expression of the inducible isoform of nitric oxide synthase (iNOS) and nitric oxide (NO) production by pulmonary fibroblasts and alveolar type II cells. The potency of 16K-PRL was comparable to that of the proinflammatory cytokines IL-1β, IFN-γ, and TNF-α. This effect of 16K-PRL versus PRL occurs through a receptor distinct from known PRL receptors. Moreover, pulmonary fibroblasts express the PRL gene and endogenously produce 16K-PRL. This indicates that this pathway may serve both autocrine and paracrine roles in the regulation of NO production. Therefore, proteolytic cleavage of PRL converts this classical hormone to a potent iNOS inducer, with a possible role in inflammatory and immune processes [129]

4.5. The acute phase response

Human recombinant GH increased the production of the acute phase proteins (APP) α2-macroglobulin, haptoglobin and fibrinogen and the expression of the junB gene in the HepG2 human hepatoma cell line. However, when HepG2 cells were also stimulated by IL-6, additional GH treatment inhibited junB activation and the production of APP [130].

5. IMMUNE-DERIVED GLH

Considerable information is available with regard to lymphocyte derived prolactin (LPRL), which is produced both by B and T lymphocytes, some B lymphoblast cell lines and by some lymphoid tumors [131–134]. The molecular weight of LPRL is varied from 11 to 48 kDa, which suggests that monomeric, dimeric and cleaved forms of LPRL are all produced within the immune system [133–137]. In lymphocytes the regulatory sequences of the PRL gene are analogous to the placental and not the pituitary regulatory sequences. Pituitary regulators of PRL secretion have no effect on hormone production by the IM-9 human B cell line, but dexamethasone is inhibitory [138–141]. B lymphocytes are most involved in the production of PRL. [142,7]

The transcription factor Pit-1, that controls PRL, GH and TSH secretion in the pituitary gland is also expressed in rat thymocytes and splenocytes and in human peripheral blood cells [143,144]. GH and Pit-1/GHF-1 transcripts were detected in thymocytes and splenocytes from bovine fetuses at 60, 90, 120, and 270 days of gestation. The lymphoid GH mRNA was approximately 350 nucleotides larger than in the pituitary. The coding regions as well as 3' untranslated regions of the lymphocyte GH and pituitary transcripts were the same. Analysis of the 5'-untranslated region of the lymphocyte GH mRNA showed that transcription began upstream from the start site in the pituitary gland, suggesting differences in regulation in lymphoid tissue. Fetal thymocytes and splenocytes expressed the 2.5-kb transcript of Pit-1/GHF-1 [145].

A variant form of growth hormone (hGH-V) was considered to be exclusively expressed in human placenta. It was shown that hGH and hGH-V transcripts are simultaneously produced by peripheral blood mononuclear cells (PBMC) in both men and women, including pregnant women. The PBMC of a PIT-1-negative woman expressed only the hGH-V transcript, but not the hGH-N. Therefore, it is likely that hGH-V is involved in immune regulation, as is pituitary GH [146]. The activity of the human PRL upstream extra pituitary promoter can be induced by activators of T-cells, as well as by a cAMP analogue. The signal is transmitted by protein kinase A [147].

In the immune organs of children, almost all components of the growth hormone-insulin-like growth factor axis were expressed, but the levels of expression varied. Growth hormone, IGF-I and IGF binding proteins 1 to 6 were all produced by immune cells. The expression of growth hormone receptors on peripheral blood lymphocytes was age-related [148]. GH is produced in the immunological tissues of humans and the binding of GH to its receptor on lymphocytes stimulates the production of IGF-I. Many of the immune effects of GH are shared by IGF-I [149]. Mitogen and antigen stimulated human Th1 cells, but not Th2 cells, produced GH in tissue culture. Th2 cells were shown to be stimulated for proliferation by mite antigen and GH in the same system, indicating a paracrine relationship between Th1 and Th2 [100]. In addition

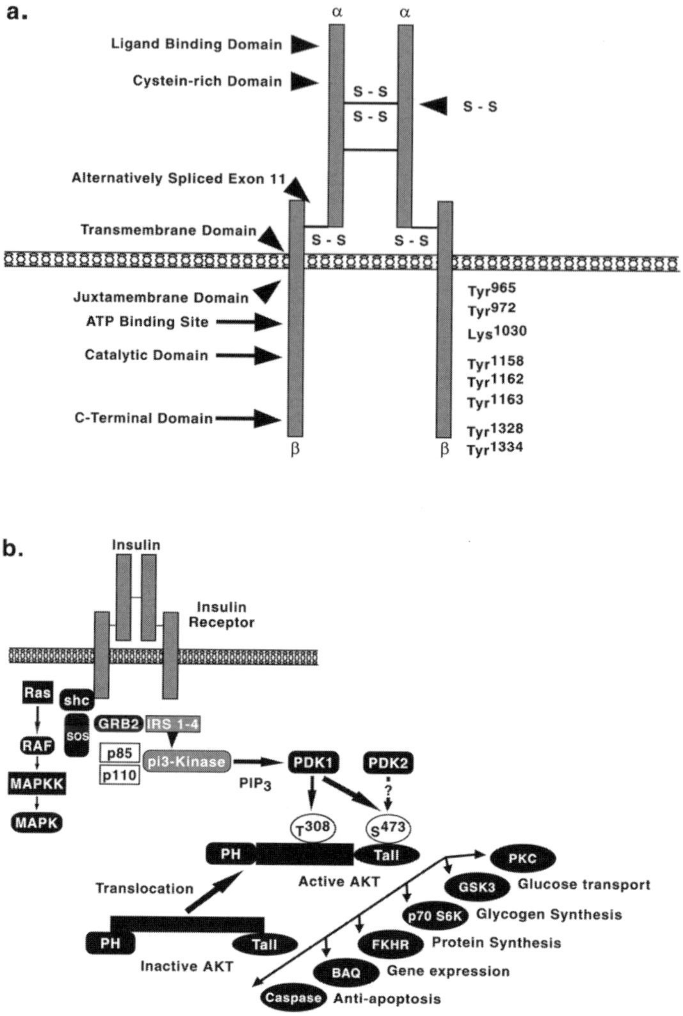

Figure 2. Signal Transduction by the Insulin Receptor. a. *Subunit structure of the insulin receptor*: The insulin receptor is the product of a single gene located on chromosome 19. A single chain polypeptide precursor undergoes post translational cleavage followed by dimerization and export to the plasma membrane. The insulin binding domain is localized at the N-terminus of the α-subunit. The intracellular domain of the β-subunit contains the tyrosine specific protein kinase activity. Insulin binding to the extracellular domain causes a conformational change in the intracellular domain and the receptor undergoes autophosphorylation followed by the binding of ATP. The tyrosine residues that are phosphorylated in the receptor juxtamembrane domain, catalytic loop, and carboxyl-terminal domain are listed in the figure. The variably spliced exon 11 is indicated at the C-terminus of the α-subunit. b. *Insulin signaling pathways*: Multiple signaling pathways are activated. The pathways associated with the activation of insulin receptor substrate (IRS) proteins are shown in detail. Four closely related IRS proteins are known. The enzyme phosphatidyl inositol 3-kinase (PI3-kinase) catalyzes the addition of phosphate on the D3 position of the inositol ring of phosphoinositol leading to the generation of PI3-phosphate (PIP3). This leads to the activation of PIP3-dependent kinases such as PI-dependent protein kinase-1 (PDK1) and PDK2. These enzymes then phosphorylate AKT (a product of the *AKT* protooncogene) on amino acid residues T^{308} and S^{473} leading to its activation. This is followed by the nuclear translocation of AKT resulting in gene

to the role of GH and IGF-I, recent findings indicate a possible role for IGF-II and IGF-binding proteins in immune function [150].

T lymphocytes infiltrating the synovium in patients with rheumatoid arthritis, and to a lesser extent synovial fibroblast cells, produce PRL. PRL enhances the proliferation of synovial cells, augments the production of the proinflammatory cytokines IL-6 and IL-8 and of matrix metalloproteinase, while the production of tissue inhibitor of metalloproteinases was inhibited. The stimulation of synovial cells by PRL induced a rapid translocation of STAT5 to the nucleus. Bromocriptine inhibited the secretion of proinflammatory cytokines and of collagenases by rheumatoid synovial cells. These results indicate that locally produced PRL acts as a paracrine and autocrine activator of the inflammatory response in rheumatoid arthritis [151].

6. EXPERIMENTS IN GENETICALLY ALTERED MICE

Humoral and cell-mediated immune responses were normal in mice with mutations that result in a selective reduction of PRL, GH, IGF-I, and/or thyroid hormone production. The response to T-independent and T-dependent antigens has been examined [152,153]. In PRL receptor deficient mice there were no alterations in thymic or splenic cellularity or in the composition of the lymphocyte subsets present in bone marrow, thymus, spleen and lymph nodes. Lymphocytes from such mice proliferated normally when stimulated by mitogens, cytokines, and allogeneic cells. PRLR negative splenocytes displayed normal NK-mediated cytotoxicity. PRLR null mice were able to generate normal steady-state Ig levels, mounted a normal specific antibody response, eliminated injected allogeneic tumor cells, and controlled *Listeria monocytogenes* infection. These results show that the development and function of the immune system proceed normally in the absence of PRL-mediated signaling [154].

The over expression of IGF-II in transgenic FVB/N mice stimulated only T cell development. IGF-II transgenic Snell-dwarf mice that are deficient in PRL, GH and TSH, show low serum levels of IGF-I. In these mice T cell development was stimulated to the same extent as in FVB/N mice. IGF-II also increased the number of nucleated bone marrow cells, including immature B lymphocytes. Mature B cells were not affected in the spleen [155,156].

7. INSULIN

Insulin is a pleiotropic growth factor, which shows a regulatory interaction with GH and also with IGF-I. Insulin receptors are expressed at low density on resting T and B lymphocytes, but high affinity specific receptors appear on both cell types after activation *(*Figure 2*)* [157]. Both B and T lymphocytes require INS for normal development and function. Insulin was found to have a variable effect on lymphocyte mitogenesis. INS was an obligate growth factor for the human mixed lymphocyte reaction in serum free medium. Activated human T cells respond to

---

expression and protein synthesis. Some of the known cellular targets of AKT are indicated. In addition to binding insulin, the insulin receptor can bind insulin like growth factor – I, and – II. The Ras/mitogen-activated protein kinase (MAPKK) pathway can be activated by insulin through the formation of complexes within the exchange factors SOS (son-of-sevenless) and growth factor receptor binding protein 2 (GRB2). This pathway may play a role in certain tissues to stimulate the actions of insulin on growth and proliferation [173].

INS with chemotaxis [59,158]. Insulin induces an anaphylactic inflammatory promoting factor, potentiates anaphylaxis and enhances fibrinolysis and phagocytosis. Fc receptors of guinea pig macrophages were downregulated and antibody-dependent cytotoxicity was inhibited by INS. Insulin suppressed the production of IL-1 and IFN-γ elicited in murine spleen cells by Staphylococcal entertoxin-A. Glucagon and somatostatin antagonized INS action in lymphoid tissue [59,159,124].

Treating AIDS patients with insulin (0.3 U/kg body weight) for 6 months resulted in marked weight gain, improved metabolic profiles and energy balance. Simultaneously, a marked rise in CD4 T lymphocyte counts and an improvement in the thyroid hormone profile were noted [160]. HIV patients with lipodystrophy (LD) had significantly reduced (twofold) peripheral insulin sensitivity, but normal levels of free fatty acids as well as reduced levels of insulin-like growth factor binding protein-1 (IGFBP-1), relative to the nonlipodystrophy groups. This indicates that the loss of insulin sensitivity was more pronounced in skeletal muscle than in liver or fat. The loss of peripheral fat in the HIV-LD group (34%;$p<0.05$) correlated with the reduced peripheral insulin sensitivity ($p=.0001$). Levels of soluble TNF receptors (sTNFR2) were higher in those with lipodystrophy compared to those without, and sTNFR2 levels strongly correlated with a reduction in insulin sensitivity ($p=0.0001$). Insulin resistance in HIV-LD is distinct from type 2 diabetes and obesity. In this case an inflammatory stimulus is contributing to the development of HIV-associated lipodystrophy [161]. Increased fasting plasma insulin and higher waist-to-hip ratios were found in non-wasting individuals with AIDS, compared with HIV-negative controls [162]

The percentage of γδ TCR+ cells and CD5+ B lymphocytes was found to be significantly ($p< 0.0001$ and $p=0.03$, respectively) higher in diabetic and prediabetic groups of children, than in controls. Young children with insulin-dependent diabetes (IDDM) associated antibodies had a higher risk of developing clinical disease than older individuals. The percentage of both CD5+ B and γδ T lymphocytes was higher in the younger population. The age-dependent decrease for both lymphocyte subpopulations in IDDM-prediabetic patients was less than in healthy controls. These observations suggest a higher capacity of older individuals to express "natural autoimmunity" and would explain, at least in part, the increased risk for antibody positive young children developing IDDM (163). In humans, it has been suggested that genetic differences exist in TNF-receptor 2 (TNFR2) expression. Individuals with high expression are susceptible to obesity, develop insulin resistance and ultimately may develop type II diabetes. This hypothesis is supported by observations in TNFR2 null mice and in obese women. Such individuals exert higher natural immune defenses and are more resistant to starvation compared to subjects with low TNFR2. However, these individuals are unfit for the western high carbohydrate, high fat, low fiber diet [164].

TNF-alpha may cause insulin resistance in skeletal muscle by inhibiting insulin receptor substrate (IRS)-1- and IRS-2-mediated phosphatidyl inositol (PI) 3-kinase activation, p42 mitogen activated protein kinase (MAPK) and p44(MAPK) tyrosine phosphorylation, which results in impaired insulin-stimulated glucose uptake [165]. Interleukin-12 also plays a role in the development of insulin resistance [166]. Insulin resistance and concomitant hyperinsulinemia are present early in the course of renal disease, even in patients with a normal glomerular filtration rate and irrespective of the type of renal disease. This observation is relevant to the high cardiovascular morbidity and mortality of patients with renal disease [167].

The catabolic effect of TNF-alpha on myotubes was mediated by macrophage migration inhibitory factor (MIF), which serves as an autocrine stimulus for glycolysis [168]. MIF plays a central role in host immunity and regulates glucocorticoid effects on the immune and inflammatory systems. MIF is produced by activated immune cells and also by the anterior pituitary gland upon stimulation of the hypothalamic-pituitary-adrenal axis. MIF is also secreted together with insulin from the pancreatic beta-cells and acts as an autocrine factor to stimulate insulin release. Circulating MIF levels are elevated during stress or systemic inflammatory processes [169].

Stress elicited an infectious disease in BALB/c mice infected with the opportunistic pathogen *Mycobacterium avium*. This was associated with a progressive loss of CD4+ T cells. In contrast, in autoimmune diabetes prone NOD mice, stress or infection significantly increased the number of cells in the thymus glands. This finding supports the previously reported resistance of NOD mouse lymphocytes to glucocorticoids [170].

In the human and mouse thymus, the pancreatic hormones preproinsulin, proglucagon, prosomatostatin and propancreatic polypeptide are produced in small quantities. Proglucagon, prosomatostatin and propancreatic polypeptide were expressed in a macrophage population, and preproinsulin expression was restricted to dendritic cells, which are more potent antigen-presenting cells. [171]

Insulin caused marked hypoglycemia, increased the plasma corticosterone level, and induced the fragmentation of thymocyte DNA in fasted but not in fed rats. Glucose or the glucocorticoid receptor antagonist, RU486, inhibited the apoptosis of thymocytes elicited by insulin. This indicates that insulin-induced hypoglycemia causes thymocyte apoptosis by promoting glucocorticoid secretion from the adrenal gland [172].

REFERENCES

1. Nicoll CS. Prolactin and growth hormone: specialists on one hand and mutual mimics on the other. Perspect Biol Med 1982;25:1369–381.
2. Nickel BE, Kardami E, Cattini, PA. The human placental growth hormone variant is mitogenic for rat lymphoma Nb2 cells. Endocrinology 1990; 126: 971–976.
3. Southard JN, Talamantes F. Placental prolactin-like proteins in rodents -variations on a structural theme. Mol Cell Endocrinol 1991;79:C133–140.
4. Berczi I. The immunology of prolactin. Sem Reprod Endocrinol 1992;10:196–219.
5. Lewis UJ. Growth hormone – what is it and what does it do. Trends Endocrinol Metab 1992;3:117–121.
6. Duckworth ML, Robertson MC, Schroedter ML et al. Molecular genetics and biology of the rat placental prolactin family. In: Soares MJ, Handwerger S, Talamantes F, editors. Trophoblast Cells: Pathways for Maternal-Embryonic Communication. New York: Springer, 1993;169–190.
7. Freeman ME, Kanyicska B, Lerant A, Nagy G. Prolactin: structure, function, and regulation of secretion. Physiol Rev 2000;80:1523–631.
8. Aaronson SA. Growth factors and cancer. Science 1991;254:1146–1153.
9. Kishimoto T, TagaT, Akira S. Cytokine signal transduction. Cell 1994;76: 253–262.
10. Touraine P, Leite de Moraes MC, Dardenne M, Kelly PA. Expression of short and long forms of prolactin receptor in murine lymphoid tissues. Mol Cell Endocrinol 1994;104: 183–190.
11. Koh CY, Phillips JT. Prolactin receptor expression by lymphoid tissues in normal and immunized rats. Mol Cell Endocrinol 1993;92:R21–R25.
12. Gagnerault MC, Touraine P, Savino W. Expression of prolactin receptors in murine lymphoid cells in normal and autoimmune situations. J Immunol 1993;150:5673–5681.

13. Corbacho AM, Nava G, Eiserich JP, Noris G, Macotela Y, Struman I, Martinez De La Dardenne M, Mello-Coelho V, Gagnerault MC, Postel-Vinay MC. Growth hormone receptors and immunocompetent cells. Ann N Y Acad Sci 1998;840:510–7.
14. Chen HT, Schuler LA, Schultz RD. Growth hormone receptor and regulation of gene expression in fetal lymphoid cells. Mol Cell Endocrinol 1998;137:21–9.
15. Thellin O, Coumans B, Zorzi W, Barnard R, Hennen G, Heinen E, Igout A. Expression of growth hormone receptors by lymphocyte subpopulations in the human tonsil. Dev Immunol 1998;6:295–304.
16. Valerio G, Bond HM, Badolato R, Petrella A, Di Maio S, Salerno M, Waters MJ, Venuta S, Tenore A. Expression of growth hormone receptor by peripheral blood lymphocytes in children: evaluation in clinical conditions of impaired growth. Clin Endocrinol (Oxf) 1997;47:329–35.
17. Cohick WS, Clemmons DR. The insulin-like growth factors. Annu Rev Physiol 1993;55: 131–153.
18. Nicoll CS, Hebert NJ, Delidow BC, English DE, Russell SM. Prolactin and synlactin: comparative aspects. Prog Clin Biol Res 1990;342:211–8.
19. Wells JA, de Vos AM. Structure and function of human growth hormone : implications for the hematopoietins. Annu Rev. Biophys. Biomol. Struct 1993;115:46–52.
20. Kelly PA, Djiane J, Postel-Vinay MC, Edery M. The prolactin/growth hormone receptor family. Endocrine Rev 1991;12:235–251.
21. Rui H, Lebrun JJ, Kirken RA. JAK2 activation and cell proliferation induced by antibody-mediated prolactin receptor dimerization. Endocrinology 1994;135:1299–1306.
22. Clevenger CV, Sillman AL, Prytowsky MB. Interleukin-2 driven nuclear translocation of prolactin in cloned T-lymphocytes. Endocrinology 1990;127:3151–3159.
23. Clevenger CV, Freier DO, Kline JB. Prolactin receptor signal transduction in cells of the immune system. J Endocrinol 1998;157:187–97.
24. Rao YP, Olson MD, Buckley DJ, Buckley AR. Nuclear co-localization of prolactin and the prolactin receptor in rat Nb2 node lymphoma cells. Endocrinology 1990;133:3062–3065.
25. Lobie PE, Mertani H, Morel G. Receptor-mediated nuclear translocation of growth hormone. J Biol Chem 1994;269:21330–21339.
26. Belkowski SM, Levine JE, Prystowsky MB. Requirement of PI3-kinase activity for the nuclear transport of prolactin in cloned murine T lymphocytes. J Neuroimmun 1999 94;40–7.
27. Horseman ND, Yu-Lee LY. Transcriptional regulation by the helix bundle peptide hormones: growth hormone, prolactin and hematopoietic cytokines. Endocrine Rev 1994;15:627–649.
28. Bole-Feysot C, Goffin V, Edery M, Binart N, Kelly PA. Prolactin (PRL) and its receptor: actions, signal transduction pathways and phenotypes observed in PRL receptor knockout mice. Endocr Rev 1998;19:225–68.
29. Akira S. Functional roles of STAT family proteins: lessons from knockout mice. Stem Cells 1999;17:138–46.
30. Davey HW, Wilkins RJ, Waxman DJ. STAT5 signaling in sexually dimorphic gene expression and growth patterns. Am J Hum Genet 1999;65:959–65.
31. Weigent DA, Vines CR, Long JC, Blalock JE, Elton TS. Characterization of the promoter-directing expression of growth hormone in a monocyte cell line. Neuroimmunomodulation 2000;7:126–34.
32. Port CB, Bowen JM, Keyes PL, Townson DH. Effects of a 3beta-hydroxysteroid

dehydrogenase inhibitor on monocyte-macrophage infiltration into rat corpus luteum and on apoptosis: relationship to the luteolytic action of prolactin. J Reprod Fertil 2000;119: 93–9.
33. Potter EL, Craig JM. Pathology of the fetus and the infant, 3rd Edition. Chicago:Year Book;1975.
34. Nagy E. Berczi I, Friesen HG. Regulation of immunity in rats by lactogenic and growth hormones. Acta Endocrinol 1983;102:351–357.
35. Berczi I, Nagy E. Human placental lactogen is a hemopoeitic hormone. Brit J Haematol 1991;79:355–358.
36. Moreno J, Vicente A, Heijnen I, Zapata AG. Prolactin and early T-cell development in embryonic chicken. Immunol Today 1994;15:524–526.
37. Crilly PR, Johnson PC, Gardner MJ, Flint DJ. Immunoneutralisation of GH in neonatal rats results in defects in lymphatic tissues and the humoral immune response. Endocrine J 1994;2:105–109.
38. Ellis LA, Mastro AM, Picciano MF. Milk-borne prolactin and neonatal development. J Mammary Gland Biol Neoplasia 1996;1:259–69.
39. Crafts RC, Meinecke HA. The anemia of hypophysectomized animals. Ann NY Acad Sci 1959;77:501–517.
40. Jepson JH, Lowenstein L. The effect of testosterone, adrenal steroids and prolactin on erythropoiesis. Acta Haematol 1967:38:292–299.
41. Berczi I, Nagy E. The effect of prolactin and growth hormone on hemolyphopoietic tissue and immune function. In: Berczi I, Kovac K, editors. Hormones and Immunity. Lancaster, England: MTP Press, 1987;145–171.
42. Nagy E, Berczi I. Pituitary dependence of bone marrow function. Brit J Haematol 1989;71: 457–462.
43. Murphy WJ, Durum SK, Anver MR, Longo DL. Immunologic and hematologic effects of neuroendocrine hormones. J Immunol 1992;148:3799–3805.
44. Maloof PB, Joshi DD, Qian J, Gascon P, Singh D, Rameshwar P. Induction of preprotachykinin-I and neurokinin-1 by adrenocorticotropin and prolactin. Implication for neuroendocrine-immune-hematopoietic axis. J Neuroimmunol 2001;112:188–96.
45. Dogusan Z, Book ML, Verdood P, Yu-Lee LY, Hooghe-Peters EL. Prolactin activates interferon regulatory factor-1 expression in normal lympho-hemopoietic cells. Eur Cytokine Netw 2000;11:435–42.
46. Morales P, Carretero MV, Geronimo H, Copin SG, Gaspar ML, Marcos MA, Martin-Perez J. Influence of prolactin on the differentiation of mouse B-lymphoid precursors. Cell Growth Differ 1999;10:583–90.
47. Richards SM, Murphy WJ. Use of human prolactin as a therapeutic protein to potentiate immunorhematopoietic function. J Neuroimmunol 2000;109:56–62.
48. Bellone G, Geuna M, Carbone A. Regulatory action of prolactin on the *in vitro* growth of CD34+ve human hemopoietic progenitor cells. J Cell Physiol 1995;163:221–231.
49. Merchav S, Tatarsky I, Hochberg Z. Enhancement of human granulopoiesis *in vitro* by biosynthetic insulin-like growth factor I/somatomedin C and human growth hormone. J Clin Invest 1988a;81:791–797.
50. Merchav S, Tatarsky I, Hochberg Z. Enhancement of erythropoiesis *in vitro* by human growth hormone is mediated by insulin-like growth factor-I. Brit J Haematol 1988b;70: 267–271.
51. Jardieu P, Clark R, Mortensen D, Dorshkind K. In vivo administration of insulin-like

growth factor-I stimulates primary B lymphopoiesis and enhances lymphocyte recovery after bone marrow transplantation. J Immunol 1994;152:4320–4327.
52. Bhat G, Gupta SK, Maiti BR. Influence of prolactin on mitogenic activity of the bursa of Fabricius of the chick. Gen Comp Endocrinol 1983;52:452–5.
53. Skwarlo-Sonta K, Sotowaska-Brochocka J. Rosolowska-Huszcz D. Effect of prolactin on the diurnal changes in immune parameters and plasma corticosterone in white leghorn chickens. Acta Endocrinol 1987;116:172–178.
54. Frænkel-Conrat H, Li CH. Hormonal effects on the nucleic acid and phospholipid turnover of rat liver and thymus. Endocrinology 1949;44:487–91.
55. Baroni CD, Fabris N, Bertoli G. Effects of hormones on development and function of lymphoid tissues: Synergistic action of thyroxin and somatotropic hormone in pituitary dwarf mice. Immunology 1969;17:303–14.
56. Baroni CD, Pesando PC, Bertoli G. Effects of hormones on development and function of lymphoid tissues II: Delayed development of immunological capacity in pituitary dwarf mice. Immunology 1971;21:455–61.
57. Pandian MR, Talwar GP. Effect of growth hormone on the metabolism of thymus and on the immune response against sheep erythrocytes. J Exp Med 1971;134:1095–113.
58. Kelley KW, Brief S. Westly HJ. GH_3 pituitary adenoma cells can reverse thymic aging in rats. Proc Natl Acad Sci USA 1986;83:5663–5667.
59. Berczi I. Pituitary Function and Immunity. Boca Raton, Florida:CRC Press; 1986.
60. Berczi I, Nagy E, DeToledo SM, Matusik RJ, Friesen HG. Pituitary hormones regulate c-myc and DNA synthesis in lymphoid tissue. J Immunol 1991:146:2201–6.
61. Timsit J, Savino W, Safieh B. Growth hormone and insulin-like growth factor-I stimulate hormonal function and proliferation of thymic epithelial cells. J Clin Endocrinol Metab 1992;75:183–8.
62. Savino W. Mello-Coelho V, Dardenne M. Neuroendocrine control of the thymic microenvironment: role of pituitary hormones. In: Berczi I, Szelenyi J, editors. Advances in Psychoneuroimmunology. New York:Plenum Press, 1994;75–82.
63. Yamada M, Hato F, Kinoshita Y. The indirect participation of growth hormone in the thymocyte proliferation system. Cell Mol Biol Noisy le Grand 1994;40:111–121.
64. Taub DD, Tsarfaty G, Lloyd AR. Growth hormone promotes human T cell adhesion and migration to both human and murine matrix proteins *in vitro* and directly promotes xenogeneic engraftment. J Clin Invest 1994;94:293–300.
65. Esquifino AI, Vallanua MA, Szary A, Yau J, Bartke A. Ectopic pituitary transplants restore immunocompetence in Ames dwarf mice. Acta Endocrinol 1991;125:67–72.
66. Singh U, Owen JJT. Studies on the maturation of the thymus stem cells: The effects of catecholamines, histamines and peptide hormones on the expression of T cell alloantigens. Eur J Immunol 1976;6:59–62.
67. Russell DH, Mills KT, Talamantes RJ, Bern HA. Neonatal administration of prolactin antiserum alters the developmental pattern of T- and B- lymphocytes in the thymus and spleen of BALB/c female mice. Proc Natl Acad Sci USA 1988; 85:7404–7.
68. Arezzini C, DeGori V, Tarli P, Neri P. Weight increase of body and lymphatic tissues in dwarf mice treated with human chorionic somatomammotropin (HCS). Proc Soc Exp Biol Med 1972;141:98–100.
69. Guicheux J, Heymann D, Gouin F, Pilet P, Faivre A, Daculsi G. Growth hormone stimulates multinucleated cell formation in long-term bone marrow cultures. Eur J Cell Biol 1998;7559–65.

70. Spangelo BL, Hall NRS, Ross PC, Goldstein AL. Stimulation of *in vivo* antibody production and concanavalin-A-induced mouse spleen cell mitogenesis by prolactin. Immunopharmacology 1987;14:11–20.
71. Cross RJ, Campbell JL, Markesbery WR, Roszman TL. Transplantation of pituitary grafts fail to restore immune function and to reconstitute the thymus glands of aged mice. Mech Aging Develop 1990;56:11–22.
72. Sotowaska-Brochocka J, Rosolowska-Huszcz D, Skwarlo-Sonta, Gajewska A. Effect of exogenous prolactin on immunity in chickens. Res Vet Science 1984;37:123–125.
73. Castro AG, Hauser TM, Cocks BG, Abrams J, Zurawski S, Churakova T, Zonin F, Robinson D, Tangye SG, Aversa G, Nichols KE, de Vries JE, Lanier LL, O'Garra A. Molecular and functional characterization of mouse signaling lymphocytic activation molecule (SLAM): differential expression and responsiveness in Th1 and Th2 cells J Immunol 1999;163: 5860–70.
74. Rapaport R, Oleske J, Ahdieh H. Suppression of immune function in growth hormone deficient children during treatment with human growth hormone. J Pediat 1986;109: 434–439.
75. Rapaport R, Oleske J, Ahdieh H. Effects of human growth hormone on immune functions: *In vitro* studies on cells of normal and growth hormone-deficient children. Life Sci 1987;41:2319–2324.
76. Yoshida A, Ishioka C, Kimata H, Mikawa H. Recombinant human growth hormone stimulates B-cell immunoglobulin synthesis and proliferation in serum-free medium. Acta Endocrinol 1992;126:524–529.
77. Kimata H, Yoshida A. Effect of growth hormone and insulin-like growth factor-I on immunoglobulin production by and growth of human B cells. J Clin Endocrinol Metab 1994;78:635–641.
78. Kimata H, Fujimoto M. Growth hormone and insulin-like growth factor I induce immunoglobulin IgE and IgG4 production by human B cells. J Exp Med 1994;180: 727–732.
79. Fornari MC, Scolnik MP, Palacios MF. Growth hormone inhibits normal B-cell differentiation and neutrophils chemotaxis *in vitro*. Int J Immunopharmacol 1994;16: 667–673.
80. Merimee TJ, Grant MB, Broder CM. Insulin-like growth factor secretion by human B-lymphocytes: a comparison of cells from normal and pygmy subjects. J Clin Endocrinol Metab 1989;69:978–84.
81. Yakar S, Damene H, Meiden R. Growth hormone (GH) stimulates insulin-like growth factor-I (IGF-I) and IGF-binding protein (IGFBP)-2 gene expression in spleens of juvenile rats. Hormones Metab Res 1994;26:363–366.
82. Richards SM, Garman RD, Keyes L, Kavanagh B, McPherson JM. Prolactin is an antagonist of TGF-beta activity and promotes proliferation of murine B cell hybridomas. Cell Immunol 1998;184:85–91.
83. Yada T, Nagae M, Moriyama S, Azuma T. Effects of prolactin and growth hormone on plasma immunoglobulin M levels of hypophysectomized rainbow trout, Oncorhynchus mykiss. Gen Comp Endocrinol 1999;115:46–52.
84. LeRoith D, Yanowski J, Kaldjian EP, Jaffe ES, LeRoith T, Purdue K, Cooper BC, Pyle R, Adler W. The effects of growth hormone and insulin-like growth factor I on the immune system of aged female monkeys. Endocrinology 1996;137:1071–9.
85. Lebl J, Sediva A, Snajderova M, Pruhova S, Rakosnikova V. Immune system in adults with

childhood-onset growth hormone deficiency: Effect of growth hormone therapy. Endocr Regul 2000;34:169–173.
86. Matera L, Galetto A, Geuna M, Vekemans K, Ricotti E, Contarini M, Moro F, Basso G. Individual and combined effect of granulocyte-macrophage colony-stimulating factor and prolactin on maturation of dendritic cells from blood monocytes under serum-free conditions. Immunology 2000;100:29–36.
87. Nagy E, Berczi I. Immunodeficiency in hypophysectomized rats. Acta Endocrinol 1978;89:530–537.
88. Berczi I, Nagy E, Asa SL, Kovacs K. Pituitary hormones and contact sensitivity in rats. Allergy 1983;38:325–330.
89. Hiestand PC, Gale JM, Melker P. Soft immunosuppression by inibition of prolactin release: synergism with cyclosporine in kidney allograft survival and in the localized graft-versus-host reaction. Transplant Proc 1986a;18:870–2.
90. Hiestand PC, Melker P, Nordmann R. Grieder A, Permongkol C. Prolactin as a modulator of lymphocyte responsiveness provides a possible mechanism of action for cyclosporine. Proc Natl Acad Sci USA 1986b;83:2599–603.
91. Wilner ML, Ettenger RB, Koyle MA, Rosenthal JT. The effect of hypoprolactinemia alone and in combination with cyclosporine on allograft rejection. Transplantation 1990;49: 264–7.
92. Shen, GK, Montgomery DW, Ulrich ED. Upregulation of prolactin gene expression and feedback modulation of lymphocyte proliferation during acute allograft rejection. Surgery 1992;112:387–393.
93. Bernton EW, Meltzer MT, Holaday JW. Suppression of macrophage activation and T-lymphocyte function in hypoprolactinemic mice. Science 1988;239:401–4.
94. Cosson C, Myara I, Guillemain R. Serum prolactin as a rejection marker in heart transplantation. Clin Chem 1989;35:492.
95. Matera L, Cesano A, Bellone G, Oberholtzer E. Modulatory effect of prolactin on the resting and mitogen-induced activity of T, B and NK lymphocytes. Brain Behav Immun 1992;6:409–17.
96. Athreya BH, Pletcher J, Zulian F. Subset-specific effects of sex hormones and pituitary gonadotropins on human lymphocyte proliferation *in vitro*. Clin Immunol Immunopathol 1993;66:201–211.
97. Matera L, Contarini M, Bellone G, Forno B, Biglino A. Up-modulation of interferon-γ mediates the enhancement of spontaneous cytotoxicity in prolactin-activated natural killer cells. Immunology 1999;98:386–92.
98. Gala RR, Shevach EM. Influence of prolactin and growth hormone on the activation of dwarf mouse lymphocytes *in vivo*. Proc Soc Exp Biol Med 1993;204:224–230.
99. Postel-Vinay MC, de Mello Coelho V, Gagnerault MC, Dardenne M. Growth hormone stimulates the proliferation of activated mouse T lymphocytes. Endocrinology 1997;138: 1816–20.
100. Yamashita Y, Hashimoto Y, Honjo M, Yamashita N. The effect of growth hormone on the proliferation of human Th cell clones. Life Sci 2000;66:1929–35.
101. Geffner M, Bersch N, Lippe BM. Growth hormone mediates the growth of T-lymphoblast cell lines via locally generated insulin-like growth factor-I. J Clin Endocrinol Metab 1990;71:464–469.
102. Baxter JB, Blalock JE, Weigent DA. Characterization of immunoreactive insulin-like growth factor-I from leukocytes and its regulation by growth hormone. Endocrinology

1991;129:1727–34.
103. Tyden G, Berg U, Reinholt F. Acute renal graft rejection after treatment with human growth hormone. Lancet 1990;336:1455–1456.
104. Yordam N, Kandemir N, Topaloglu H, Gogus S, Kucukalu T. Myositis associated with growth hormone therapy. J Pediat 1994;125:671 (letter).
105. Span JP, Pieters GF, Smals AG, Koopmans PP, Kloppenborg PW. Number and percentage of NK-cells are decreased in growth hormone-deficient adults. Clin Immunol Immunopathol 1996;78:90–2.
106. Rongen-Westerlaken C, Rijkers GT, Scholtens EJ. Immunologic studies in Turner syndrome before and during treatment with growth hormone. J Pediat 1991;119:268–272.
107. Crist DM, Peake GT, Mackinnon LT. Exogenous growth hormone treatment alters body composition and increases natural killer cell activity in women with impaired endogenous growth hormone secretion. Metabolism 1987;36:1115–7.
108. Crist DM, Kraner JC. Supplemental growth hormone increases the tumor cytotoxic activity of natural killer cells in healthy adults with normal growth hormone secretion. Metabolism 1990;39:1320–1324.
109. Matera L, Cesano A, Muccioli G, Veglia F. Modulatory effect of prolactin on the DNA synthesis rate and NK activity of large granular lymphocytes. Int J Neurosci 1990;51: 265–67.
110. Cesano A, Oberholtzer E, Contarini M. Independent and synergistic effect of interleukin-2 and prolactin on development of T- and NK-derived LAK effectors. Immunopharmacology 1994;28:67–75.
111. Bidlingmaier M, Auernhammer CJ, Feldmeier H, Strasburger CJ. Effects of growth hormone and insulin-like growth factor I binding to natural killer cells. Acta Paediatr Suppl 1997;423:80–1.
112. Wiedermann CJ, Reinisch N, Braunsteiner H. Stimulation of monocyte chemotaxis by human growth hormone and its deactivation by somatostatin. Blood 1993;82:954–60.
113. Manfredi R, Turnietto F, Azzaroli L. Growth hormone (GH) and the immune system : imparied phagocytic function in children with idiopathic GH deficiency is corrected by treatment with biocynthetic GH. J Pediat Endocrinol 1994;7:245–251.
114. Warwick-Davies J, Lowrie DB, Cole PJ. Growth hormone is a human macrophage activating factor priming of human monocytes for enhanced release of H_2O_2. J Immunol 1995;154:1909–1918.
115. Gaytan F, Romero JL, Bellido C. Effects of growth hormone and prolactin on testicular macrophages in long-term hypophysectomized rats. J Reprod Immunol 1994;27:73–84.
116. Rovensky J, Ferencikova J, Vigas M. Effect of growth hormone on the activity of some lysosomal enzymes in neutrophilic polymorphonuclear leukocytes of hypopituitary dwarfs. Int J Tissue React 1985;7:153–59.
117. Wiedermann CJ, Niedermuhlbicher M, Geissler D. Priming of normal human neutrophils by recombinant human growth hormone. Brit J. Hæmatol 1991;78:19–22.
118. Fu YK, Arkins S, Li YM. Reduction in superoxide anion secretion and bactericidal activity of neutrophils from aged rats : reversal by the combination of gamma interferon and growth hormone. Infect Immun 1994;62:1–8.
119. Cecilia-Fornari M, Fernanda-Palacios M, Diez RA, Intebi AD. Decreased chemotaxis of neutrophils in acromegaly and hyperprolactinemia. Eur J Endocrinol 1994;130:463–468.
120. Matsuda T, Saito H, Inoue T, Fukatsu K, Han I, Furukawa S, Ikeda S, Muto T. Growth

hormone inhibits apoptosis and up-regulates reactive oxygen intermediates production by human polymorphonuclear neutrophils. JPEN J Parenter Enteral Nutr 1998; 22:368–74.
121. Cesario TC, Yousefi S, Carandang G. Enhanced yields of gamma interferon in prolactin treated human peripheral blood mononuclear cells. Proc Soc Exp Biol Med 1994;205: 89–95.
122. Edwards CK, Ghiasuddin SM, Yunger LM. *In vivo* administration of recombinant growth hormone or gamma-interferon activates macrophages: Enhanced resistance to experimental Salmonella-typhimurium infection is correlated with generation of reactive oxygen intermediates. Infec Immun 1992;60:2514–2521.
123. Kelley KW, Arkins S, Li YM. Growth hormone, prolactin and insulin-like growth factors: New jobs for old players. Brain Behav Immun 1992;6:317–26.
124. Galdiero M, Donnarumma G, Cipollareo de L'Ero G, Marcatili A, Scarfogliero P, Sommesse L. Effect of growth hormone, prolactin and insulin on the release of the IL-1 alpha, IFN-gamma and IL-4 by staphylococcal enterotoxin A-stimulated splenocytes. Euro Cytokine Netw 1995;6:187–94.
125. Thellin O, Coumans B, Devos S, Hooghe-Peters EL, Zorzi W, Igout A, Heinen E, Hooghe R. Limited effects of placental and pituitary growth hormone on cytokine expression in vitro Eur Cytokine Netw 2000;11:452–5.
126. Galdiero M, Vitiello M, Scarfogliero P, Sommese L. Growth hormone release of interleukin-1 alpha, interferon-gamma and, interleukin-4 from murine splenocytes stimulated with staphylococcal protein A, toxic shock syndrome toxin-1 and streptococcal lysin S. Eur Cytokine Netw 1997;8:83–90.
127. Elsasser TH, Fayer R, Rumsey TS, Hartnell GF. Recombinant bovine somatotropin blunts plasma turmor necrosis factor-alpha, cortisol, and thromboxane-B2 responses to endotoxin *in vivo*. Endocrinology 1994;134:1082–88.
128. Zarkesh-Esfahani SH, Kolstad O, Metcalfe RA, Watson PF, von Laue S, Walters S, Revhaug A, Weetman AP, Ross RJ. High-dose growth hormone does not affect proinflammatory cytokine (tumor necrosis factor-alpha, interleukin-6, and interferon-gamma) release from activated peripheral blood mononuclear cells or after minimal to moderate surgical stress. J Clin Endocrinol Metab 2000;85:3383–90.
129. Corbacho AM, Nava G, Eiserich JP, Noris G, Macotela Y, Struman I, Martinez De La Escalera G, Freeman BA, Clapp C. Proteolytic cleavage confers nitric oxide synthase inducing activity upon prolactin. J Biol Chem 2000;275:13183–86.
130. Derfalvi B, Igaz P, Fulop KA, Szalai C, Falus A. Interleukin-6-induced production of type II acute phase proteins and expression of junB gene are downregulated by human recombinant growth hormone in vitro. Cell Biol Int 2000;24:109–14.
131. Montgomery DW, Zukoski CF, Shah GN. Concavalin A-stimulated murine splenocytes produce a factor with prolactin-like bioactivity and immunoreactivity. Biochem Biophys Res Commun 1987;145:692–698.
132. Hartmann DP, Holaday JW, Bernton DW. Inhibition of lymphocyte proliferation by antibodies to prolactin. FASEB J 1989;3:2194–202.
133. DiMattia GE, Gellersen B, Bohnet HG, Friesen HG. A human B-lymphoblastoid cell line produces prolactin. Endocrinology 1988;122:2508–517.
134. Hatfill SJ, Kirby R, Hanley M. Hyperprolactinemia in acute myeloid leukemia and indication of ectopic expression of human prolactin in blast cells of a patient subtype M4. Leukemia Res 1990;14:57–62.
135. Montgomery DW, LeFevre JA, Ulrich ED. Identification of prolactin-like proteins

synthesized by normal murine lymphocytes. Endocrinology 1990;127:2601–2603.
136. Montomery DW, Shen GK, Ulrich ED. Human thymocytes express a prolactin-like messenger ribonucleic acid and synthesize bioactive prolactin-like proteins. Endocrinology 1992;131:3019–26.
137. Wu H, Devi R, Malarkey WB. Expression and localization of prolactin messenger ribonucleic acid in the human immune system. Endocrinology 1996;137:349–353.
138. DiMattia GE, Gellersen B, Duckworth ML, Friesen HG. Human prolactin gene expression: The use of an alternative noncoding exon in decidua and the IM-9_P3 lymphoblastoid cell line. J Biol Chem 1990;265:16412–421.
139. Gellersen B, DiMattia GE, Friesen HB, Bohnet HG. Regulations of prolactin secretion in the human B-lymphoblastoid cell line IM-9-P3 by dexamethasone but not other regulators of pituitary prolactin secretion. Endocrinology 1989;125:2853–61.
140. Gellersen B, DiMattia GE, Friesen HG, Bohnet HG. Prolactin (PRL) mRNA from human decidua differs from pituitary PRL mRNA, but resembles the IM-9-P3 lymphoblast PRL transcript. Mol Cell Endocrinol 1989;64:127–30.
141. Gellersen B, Kempf R, Telgmann R, DiMattia GE. Nonpituitary human prolactin gene transcription is independent of Pit-1 and differentially controlled in lymphocytes and in endometrial stroma. Mol Endocrinol 1994;8:356–373.
142. Rapaport RG, Bozzola M. Role of B-cells in growth hormone-immune interactions. Acta Paediatr Suppl 1997;423:82–3.
143. Delhase M, Vergani P, Malur A, Hooghe Peters EL, Hooghe RJ. The transcription factor Pit-1/HGF-1 is expressed in hematopoietic and lymphoid tissues. Eur J Immunol 1993;23: 951–955.
144. Hooghe RJ, Hooghe-Peters EL. The developmental biology of the anterior pituitary: A biased review. In: Growth hormone, prolactin and insulin-like growth factor-I as lymphohemopoietic cytokines. Texas: RG Landes Co.,1995;3–34.
145. Chen HT, Schuler LA, Schultz RD. Growth hormone and Pit-1 expression in bovine fetal lymphoid cells. Domest Anim Endocrinol 1997;14:399–407.
146. Melen L, Hennen G, Dullaart RP, Heinen E, Igout A. Both pituitary and placental growth hormone transcripts are expressed in human peripheral blood mononuclear cells (PBMC). Clin Exp Immunol 1997;110:336–40.
147. Reem GH, Ray DW, Davis JR. The human prolactin gene upstream promoter is regulated in lymphoid cells by activators of T-cells and by cAMP. J Mol Endocrinol 1999;22:285–92.
148. Yang Y, Guo L, Ma L, Liu X. Expression of growth hormone and insulin-like growth factor in the immune system of children. Horm Metab Res 1999;31:380–4.
149. Geffner M. Effects of growth hormone and insulin-like growth factor I on T- and B-lymphocytes and immune function. Acta Paediatr Suppl 1997;423:76–9.
150. Buul-Offers SC van, Kooijman R. The role of growth hormone and insulin-like growth factors in the immune system. Cell Mol Life Sci 1998;54:1083–94.
151. Nagafuchi H, Suzuki N, Kaneko A, Asai T, Sakane T. Prolactin locally produced by synovium infiltrating T lymphocytes induces excessive synovial cell functions in patients with rheumatoid arthritis. J Rheumatol 1999;26:1890–900.
152. Foster MP, Jensen ER, Montecino-Rodriguez E, Leathers H, Horseman N, Dorshkind K. Humoral and cell-mediated immunity in mice with genetic deficiencies of prolactin, growth hormone, insulin-like growth factor-I, and thyroid hormone. Clin Immunol 2000;96:140–9.

153. Dorshkind K, Horseman ND. The roles of prolactin, growth hormone, insulin-like growth factor-I, and thyroid hormones in lymphocyte development and function: insights from genetic models of hormone and hormone receptor deficiency. Endocr Rev 2000;21:292–312.
154. Bouchard B, Ormandy CJ, Di Santo JP, Kelly PA. Immune system development and function in prolactin receptor-deficient mice. J Immunol 1999;163:576–82.
155. Kooijman R, Buul-Offers SC van, Scholtens EJ, Schuurman HJ, Brande JL Van den, Zegers ZJM. Cell development in insulin-like growth factor II transgenic mice. J Immunol 1995;154:5736–45.
156. Kooijman R, Buul-Offers SC van, Scholtens LE, Reijnen-Gresnight RG, Zegers BJM. T and B cell development in pituitary deficient insulin-like growth factor-II transgenic dwarf mice. J Endocrinol 1997;155:165–170.
157. Helderman JH, Strom TB. Specific insulin binding site on T and B lymphocytes as a marker of cell activation. Nature 1978;274:62–63.
158. Berman JS, Center DM. Chemotactic activity of porcine insulin for human T lymphocytes *in vitro*. J Immunol 1987;138:2100–103.
159. Berczi I. Hormonal interactions between the pituitary and immune system. In: Grossman CJ, editor. Bilateral Communication Between the Endocrine and Immune Systems. Berlin: Springer, 1994;96–144.
160. Kabadi UM, Reust CS, Kabadi MU. Weight gain, improvement in metabolic profile, and CD4 count with insulin administration in an AIDS patient. AIDS Patient Ce STDS 2000;14:575–9.
161. Mynarcik DC, McNurlan MA, Steigbigel RT, Fuhrer J, Gelato MC. Association of severe insulin resistance with both loss of limb fat and elevated serum tumor necrosis factor receptor levels in HIV lipodystrophy. J Acquir Immune Defic Syndr 2000;25:312–21.
162. Shikuma CM, Waslien C, McKeague J, Baker N, Arakaki M, Cui XW, Souza S, Imrie A, Arakaki R. Fasting hyperinsulinemia and increased waist-to-hip ratios in non-wasting individuals with AIDS. AIDS 1999;13:1359–65.
163. Gyarmati J, Szekeres-Bartho J, Fischer B, Soltesz G. Fetal type lymphocytes in insulin dependent diabetes mellitus. Autoimmunity 1999;30:63–9.
164. Fernandez-Real JM, Ricart W. Insulin resistance and inflammation in an evolutionary perspective: the contribution of cytokine genotype/phenotype to thriftiness. Diabetologia 1999;42:1367–74.
165. del Aguila LF, Claffey KP, Kirwan JP. TNF-alpha impairs insulin signaling and insulin stimulation of glucose uptake in C2C12 muscle cells. Am J Physiol 1999;276:E849–55.
166. Cseh K, Baranyi E, Winkler G. The role of cytokines of the innate and adaptive immune system in the regulation of insulin resistance. Diebetologia 1999;42:497–8.
167. Fliser D, Pacini G, Engelleiter R, Kautzky-Willer A, Prager R, Franek E, Ritz E.Insulin resistance and hyperinsulinemia are already present in patients with incipient renal disease. Kidney Int 1998;53:1343–7.
168. Benigni F, Atsumi T, Calandra T, Metz C, Echtenacher B, Peng T, Bucala R. The proinflammatory mediator macrophage migration inhibitory factor induces glucose catabolism in muscle. J Clin Invest 2000;106:1291–1300.
169. Waeber G, Calandra T, Bonny C, Bucala R. A role for the endocrine and pro-inflammatory mediator MIF in the control of insulin secretion during stress. Diabetes Metab Res Rev 1999;15:47–54.
170. Martins TC, Aguas AP. NOD mice are resistant to depletion of thymic cells caused by

acute stress or infection. Autoimmunity 1999;29:273–80.
171. Throsby M, Pleau J, Dardenne M, Homo-Delarche F. Thymic expression of the pancreatic endocrine hormones. Neuroimmunomodulation 1999;6:108–14.
172. Morishita S, Sato EF, Takahashi K, Manabe M, Inoue M. Insulin-induced hypoglycemia elicits thymocyte apoptosis in the rat. Diabetes Res Clin Pract 1998;40:1–7.
173. Kido Y, Nakae J, Accilli D. The insulin receptor and its cellular targets. J. Clin Endocrinol Metab 2001;86:972–979.

The Immune-Neuroendocrine Circuitry
Edited by I. Berczi and A. Szentivanyi

The Hypothalamus-Pituitary-Adrenal Axis and Opioid Peptides

ISTVAN BERCZI and ANDOR SZENTIVANYI

Department of Immunology, Faculty of Medicine, The University of Manitoba, Winnipeg, Manitoba RE3 0W3, Canada; and Department of Internal Medicine, Faculty of Medicine, The University of South Florida, Tampa, Florida 33612, USA

ABSTRACT

Corticotropin releasing factor (CRF) regulates adrenal corticotrope hormone (ACTH) release, integrates the stress response in the CNS and acts centrally as an immunosuppressive agent by causing sympathetic outflow. Lymphocytes also produce CRF, and it functions within the immune system as an internal regulator. Urocortin (UCN) is a novel endogenous ligand for Type II CRF receptors which is produced within the immune system.

ACTH exerts an immunosuppressive and anti-inflammatory effect mainly by the stimulation of glucocorticoids produced in the adrenal glands. However, lymphocytes do express receptors for ACTH. ACTH is also produced within the immune system and has a direct regulatory effect on immune function. In addition to ACTH, β-endorphin (END) and α-melanocyte stimulating hormone (αMSH) are also derived from the proopiomelanocortin molecule by splicing in the pituitary gland and other tissues including the immune system. Kappa (κ), delta (δ) and mu (μ)-type opioid receptors are present in immune cells. The effect of endogenous opioids on immune function is diverse and may be the result of direct action on the immune system or modulating immune function through the CNS.

Alpha-MSH is a major regulator of fever and inflammation and functions as an antagonist of proinflammatory cytokines. It inhibits fever and inflammation by acting on both the CNS and directly on immune cells. Alpha-MSH is produced within the immune system in the skin and in many other tissues and functions locally as an immunoregulator.

2. CORTICOTROPIN RELEASING FACTOR (CRF)

CRF stimulates the release of ACTH from the pituitary gland and mediates, in part, the cytokine induced ACTH release. CRF integrates the stress response in the central nervous system and also acts centrally as an immunosuppressive agent, which is not mediated by glucocorticoids but rather by the stimulation of sympathetic outflow. Receptors for CRF are present in the immune system and immunocytes have the capacity to produce CRF (1,2)

Within the immune system CRF production increases during inflammatory responses and it functions as an important anti-inflammatory hormone. However, some forms of inflammation actually are enhanced by CRF [3]. CRF reduced lipopolysaccharide (LPS)-induced pulmonary vascular leak in mice which was mediated by glucocorticoids. Leukocyte infiltration was significantly repressed by CRF and a modest prolongation of survival occurred in treated mice. CRF also antagonized the stimulation of human granulocytes by TNF [4,5,6]. In rats CRF treatment led to a decrease in various parameters of immune cell function [7]. Elevated levels of CRF were detected in synovial fluids of rats with experimental arthritis and of patients suffering from rheumatoid arthritis [8]. Studies on the *in vitro* effect of CRF on lymphocyte reactions are contradictory. T lymphocyte proliferation, IL-1, -2 and -6 secretion, NK cell activity have all been affected by CRF [2].

By immunohistochemical analysis, only a few clusters of CRF receptor 1 (CRFR1) positive cells were found in spleens of naive mice at sites typical for granulopoietic islands. A 17-fold increase in the mean number of CRFR1+ cells occurred within hours following a challenge of acute systemic inflammation induced by i.p. administration of LPS. The majority of these cells were mature neutrophils. CRFR1 mediated the suppression of IL-1β secretion by these cells. At later time points a large number of granulocyte-macrophage precursors were strongly labeled with anti-CRFR1 antibody. Western blot analysis of splenic membranes revealed a molecular size of 70 kD for CRFR1. CRFR1 mRNA was not detected in bone marrow and peripheral blood leukocytes from naive mice. These findings indicate that CRFR1 is not expressed constitutively by mature or immature neutrophils, but rather, expression is triggered by inflammatory stimuli [9].

CRFR1-deficient (CRFR1-0) mice show marked impairment of the pituitary-adrenal axis. However, plasma ACTH concentrations of unstressed mutant mice are similar to those in wild-type animals. Arginine vasopressin (AVP) is a major ACTH secretagogue in CRFR1-0 mice under resting conditions. The administration of anti-AVP serum, but not anti-CRF serum, markedly reduced (by 60%) resting plasma ACTH concentrations in null mutants. The administration of turpentine into the hind-limb of CRFR1-0 mice produced a slightly (15–25%) smaller swelling of the limb, but a 10 fold greater rise in plasma IL-6 levels, compared to normal controls. Turpentine-induced local inflammation produced pronounced elevations in the plasma concentrations of both ACTH and corticosterone in both CRFR1-0 and wild-type mice. Anti-CRF and anti-AVP sera only inhibited ACTH secretion in wild-type mice. These data indicate that resting ACTH secretion in CRFR1-0 mice is attributable in part to AVP. Furthermore, mice deficient in CRFR1 are still able to mount a pituitary-adrenal response via mechanisms that do not depend critically on either CRF or AVP action [10].

High levels of CRFR1 are present in multiple regions of the brain, whereas the CRFR2α message was localized in a few structures of the limbic system and positive signals for the CRFR2 β mRNA was observed only in the choroid plexus. LPS stimulated the transcription of the gene encoding the CRFR1 (but not CRFR2) receptor in selective hypothalamic nuclei. High doses of LPS caused a strong expression of CRFR1 mRNA in the parvocellular and magnocellular paraventricular nucleus (PVN) and in the supraoptic nucleus (SON). Low and moderate doses induced a more specific expression of CRFR1 in the parvocellular division of the PVN. The inhibition of cyclo-oxygenase pathways significantly inhibited the expression of CRFR1 in the PVN and SON of rats sacrificed 6 h after a moderate or a low dose of LPS [11].

In rats immunoreactive CRF was present in cells in the marginal zone and red pulp of the spleen, in connective tissue septa and the subcapsular region of the thymus and in the medullary cords and sinuses of the mesenteric lymph nodes. Double-label tests for CRF and markers for specific immunocyte subsets suggested that CRF+ cells in the spleen and thymus are macrophages. CRF+ cells in primary and secondary lymphoid organs reside in compartments that are innervated by sympathetic nerves [12]

The amygdala contains high levels of CRF, and has been implicated as a central site for stress-related autonomic and behavioral response. IFN-α will release AVP from both amygdala and hypothalamus. Guanylate cyclase inhibitors block CRF release by IFN-α, IL-2 and acetylcholine, indicating a role for cGMP. TGFβ1 had no effect on basal release of CRF, nor on the CRF-release induced by IL-2, but it selectively blocked the acetylcholine-induced release in both amygdala and hypothalamus [13].

Urocortin (UCN) is a new mammalian member of the CRF family. UCN is a candidate endogenous ligand for type 2 CRF receptors. In adult male rats UCN mRNA was abundant in the gastrointestinal tract and in the thymus and spleen. LPS induced a 2-fold time-dependent increase in thymic UCN mRNA levels within 6 h. Splenic UCN mRNA levels decreased after LPS administration. UCN mRNA in thymus, but not in the spleen, was significantly increased after ACTH injection. LPS did not influence UCN expression in the thymus of adrenalectomized rats that received corticosterone replacement, despite substantial increases in ACTH. Corticosterone, given s.c., stimulated UCN mRNA similarly to that of LPS. These results indicate that UCN mRNA expression increases within the thymus after immune activation in a corticosterone-dependent manner, and that this is the consequence of HPA axis activation [14].

Intracerebroventricular (i.c.v.) injection of UCN (1 ng) to rats produced a marked decrease in the proliferative response of splenocytes to a mitogen. This suppressive effect of urocortin was abolished by pretreatment with a ganglionic blocking agent (chlorisondamine) or a beta-adrenergic receptor antagonist (propranolol), but not by adrenalectomy. These results indicate that the immunosuppressive effect of UCN is mediated by the sympathetic nervous system [15]

IL-6 activates the HPA axis and increases brain tryptophan and serotonin metabolism. IL-6 and its receptor (IL6R) and their respective mRNAs have been detected in several regions of the brain. Expression of IL-6 has been observed in several CNS disorders, some of which have been associated with disorders of serotonin metabolism. The interactions between IL-6 and brain serotonin are complex and involve CRF and opioid peptides [16] .

Systemic injection of IL-6 to rats induced a specific transcription of c-fos gene in most of the sensorial circumventricular organs, including the organum vasculosum, lamina terminalis, subfornical organ, median eminence, and area postrema, as well as in the central nucleus of the amygdala and bed nucleus of the stria terminalis. Central injection of IL-6 increased cellular level of c-fos mRNA in the ependymal layer and the walls of the ventricles, meninges, nucleus of the solitary tract, and circumventricular organs. The signals for c-fos mRNA were detected 1 h after both treatments and vanished 3 h afterwards. In these experiments IL-6 did not activate the CRF gene in the paraventricular nucleus of the hypothalamus [17]

In addition to secretion by lymphocytes, IL-10 is produced in pituitary, hypothalamic, and neural tissues. In hypothalamic and pituitary tissues IL-10 enhances CRF and ACTH production, respectively. Endogenous IL-10 has the potential to contribute to regulation of glucocorticosteroid production both tonically and following stressors [18].

IL-1α (100 U/ml) increased prostacyclin (PGI2) synthesis in endothelial cells and the production of PGE2 in fibroblasts. The PG response to IL-1α was suppressed by simultaneous exposure to CRF in both cell types. IL-1α enhanced both phospholipase A2 and prostaglandin H synthase activities, and the two effects were completely abrogated by CRF. IL-1β (100 U/ml) was as active as IL-1α in triggering release of PGI2 from endothelial cells and PGE2 from fibroblasts. However, CRF failed to alter the IL-1β-induced PG synthesis in both cell types. This difference in regulation of pro-inflammatory cytokines by CRF attests to the precision and specificity of neuro-immune control of inflammatory processes [19].

LPS induced a dose-dependent increase in leukocyte rolling, adhesion, and emigration in mesenteric venules, as measured by intravital microscopy in rats. This was associated with the upregulation of endothelial ICAM-1 expression. Intracisternal injection of CRF abrogated the increased expression of ICAM-1 and leukocyte recruitment. After i.v. administration, the specific CRF receptor antagonist, astressin, did not modify leukocyte-endothelial cell interactions induced by a high LPS dose, but enhanced leukocyte adhesion induced by a low dose. Blockade of glucocorticoids but not α-MSH receptors reversed the inhibitory action of CRF on leukocyte-endothelial cell interactions during endotoxemia [20]

Locally expressed CRF seems to act on CRF receptors on immune cells in inflamed tissue, and induce the release of opioid peptides. The opioids act on peripheral sensory nerves to inhibit the transmission of painful stimuli [21].

Both CRF and urocortin suppressed experimental autoimmune encephalomyelitis (EAE). Suppression by CRF involved the stimulation of the HPA axis and inhibitory effects on an encephalitogenic T cell line. CRF also suppressed EAE in adrenalectomized rats, where glucocorticoid stimulation plays no role. The encephalitogenicity of a T cell line exposed to CRF in vitro was reduced [22].

During chronic inflammation there is an alteration in the hypothalamic control mechanism, where CRF is no longer the major corticotrophin-releasing factor. This has been noted in a number of immune-mediated disease models, including experimental allergic encephalomyelitis, eosinophilia myalgia syndrome, systemic lupus erythematosus, and leishmaniasis. AVP takes over as the major stimulator of the axis. Arthritic rats are unable to mount a response to acute stressors, but such animals are capable of mounting a response to an acute immune challenge. Central neurotransmitter systems are able to influence the severity of peripheral inflammation. The depletion of serotonin at the time of the development of the inflammatory episode reduces the severity of the inflammation [23].

The immune system of the CRF-transgenic (CRF-TG) mice has profound changes compared to littermate controls, including a marked reduction in both cell number and immune responsiveness. In the spleen CRF-TG mice had a greater percentage of T lymphocytes compared to littermate controls. Adrenalectomy of CRF-TG animals reversed their immunological phenotype and restored immune responsiveness. CRF overexpression leads to profound impairment of lymphocyte development and function mediated via corticosteroids [24].

Corticotropin-releasing factor-binding protein (CRFBP) is predominantly produced in the liver. It is a glycoprotein with a molecular weight of 37 kDa. This binding protein is distributed and expressed independently of the expression of CRFRs. Most CRF in the plasma is bound to CRFBP and therefore, is inactive and unable to bind to its receptor. Other competitives can reverse the binding, liberating CRF. Together with the CRF neuropeptides and CRFR, CRFBP plays a role in the hypothalamic-pituitary-adrenal axis, in immune/inflammatory reactions as an auto/paracrine proinflammatory regulator and in pregnancy [25].

3. ADRENOCORTICOTROPIC HORMONE (ACTH)

ACTH treatment has an antiinflammatory effect and influences leukocyte recirculation in various species of mammals and birds. These effects are due to the stimulation of glucocorticoid production by ACTH in the adrenal gland. In animals ACTH treatment inhibited antibody formation, antibody mediated reactions (anaphylaxis, Arthus type hypersensitivity), cell mediated immunity (graft rejection, tuberculin response) [26].

ACTH antagonized the stimulatory effect of growth hormone on the antibody response, which was first described by Hayashita and Li in 1957 [27]. In Hypox rats replacement doses of ACTH inhibited the restoration of immunocompetence by PRL or GH treatment [28].

Immunocompetent cells synthesize proopiomelanocortin (POMC) mRNA species and generate POMC related peptides, that includes ACTH. Lymphoid cells express specific receptors for these peptides [29,30,31]. Keratinocytes of the epidermis produce ACTH , α-melanocyte stimulating hormone (α-MSH) and immunoregulatory cytokines [32]. Lymphocytes are not able to secrete ACTH in quantities that are sufficient to stimulate adrenal steroidogenesis in Hypox rats [33].

ACTH has a direct effect on lymphocyte proliferation [34,35] and modulates IgM and IgE synthesis *in vitro*. Its effect on IgE is mediated by accessory cells [36,37]. ACTH suppressed both the Ca^{++}-dependent and -independent phagocytosis of murine peritoneal macrophages [38]. ACTH exerts an anti-pyretic effect by acting on the central nervous system [39].

Mice deficient in CRF demonstrate severely impaired glucocorticoid production in response to psychological and metabolic challenge, but near normal responses to stimuli that activate the immune system. Plasma cortisone (CON) levels after bacterial LPS injection in mice deficient in CRF or IL-6 were significantly lower than in wild-type mice but significantly greater than in mice deficient in both CRF and IL-6. In mice deficient in CRF or IL-6 immune activation by the 2C11 antibody to the T cell receptor resulted in a normal CON response, but there was a markedly decreased response in mice deficient in both CRF and IL-6. The relative contribution of IL-6 for modulation of the adrenal response to stress was greater in female than in male mice. IL-6 receptors are present on pituitary corticotrophs and on adrenocortical cells, which explains the ability of IL-6 to bypass CRF in augmentation of adrenal function [40]

Studies in rats revealed that selective central inhibition of epinergic function reduced endotoxin-stimulated ACTH secretion, an effect mainly due to a decrease in CRF activity. Central epinergic mechanisms exert a positive and negative effect on CRF and AVP, respectively. Inhibition of peripheral epinephrine synthesis reduced immune function [41]

Studies in rats indicated that nitric oxide (NO) acts presynaptically to inhibit the release of ACTH secretagogues from nerve terminals in the infundibulum. The acute ACTH response to immune-derived cytokines is believed to primarily depend on events taking place within the median eminence. In this case blockade of NO formation results in exaggerated ACTH release. During exposure to shocks and other nonimmune forms of stress, the increased ACTH secretion is primarily due to the activation of hypothalamic neurons. In this case endogenous NO is required to stimulate the hypothalamic perikarya that manufacture CRF and/or of the afferents to these neurons. Here blockade of NO synthase activity blunts CRF production, and consequently ACTH release [42].

Mature IL-1β derives from a 31-kD precursor (pro-IL-1β) that is processed by IL-1β-converting enzyme (ICE). Plasma levels of CON and ACTH in response to LPS were similar in ICE-deficient mice and wild-type (WT) controls. Plasma ACTH response to restraint or to IL-1β was higher in ICE-0 mice than in WT animals. This hyper-reactivity of the HPA axis in ICE-0 mice was not related to the production of plasma IL-1β or IL-6, which was similar to that of WT mice after IL-1β injection. In ICE-0 mice LPS elicited a smaller increase in plasma IL-1β and IL-6 than in controls. After restraint stress no increase occurred in plasma cytokine levels [43]

In ovariectomized monkeys (5 animals) the i.c.v. infusion of IL-1α caused rapid and profound suppression of lymphocyte mitogen responsiveness for 7 hrs. Mean ACTH and cortisol levels increased significantly. Lymphocyte suppression after i.c.v. IL-1 was much more profound than after i.v. ACTH ($p < 0.01$). These studies demonstrate in the primate that centrally injected IL-1 has a profound suppressive effect on lymphocyte function. There is a significant immunosuppressive message produced by IL-1 that is not accounted for by the associated increases in

ACTH and cortisol [44].

The administration of GM-CSF to Wistar rats (10 μg/kg b.w.) significantly increased the secretion of ACTH and CON. Prolonged administration of GM-CSF increased the ACTH ($P < 0.001$) and CON ($P < 0.001$) concentration in blood. GM-CSF also increased the proliferative activity of corticotrophs ($P < 0.05$), but it did not significantly change the total cell proliferation in the anterior pituitary gland. Cell proliferation in the adrenal cortex was also increased by GM-CSF ($P < 0.001$) [45].

Interleukin-18 is a proinflammatory cytokine and a stimulator of cell-mediated immune responses. Acute stress stimulates the production of IL-18 mRNA in the glucocorticoid (GC)-producing cells of the adrenal cortex. In adult male Sprague-Dawley rats ACTH stimulated IL-18 mRNA and protein in the adrenal gland. Additional treatment with CON did not inhibit this response [46].

Natural rat IFN-αβ and recombinant human IFN-α increased both ACTH and glucocorticoid blood cells after i.v. administration to rats through an indwelling catheter. 300–600 u/g body weight doses were effective with maximum release at 30–60 minutes after injection. The IFN preparations were free of LPS, and the injection and sampling of animals was stress-free [47].

Leukemia inhibitory factor (LIF) increased ACTH secretion, but not PRL or GH, from dispersed rat pituitary cells [Kim and Melmed 1999]. LIF is secreted by anterior pituitary cells and it stimulates both basal and CRF-stimulated ACTH secretion by AtT20 anterior pituitary tumor cells and from dispersed sheep pituitary cells. LIF colocalized with TSH and LH in pituitary cells [48].

Oncostatin M (OSM), a cytokine of IL-6 family, at 1 nM concentration released ACTH from dispersed rat pituitary cells. The release of GH and PRL was not affected [49]

The intravenous administration of LPS to rats increased IL-1β and CRF in the extracellular fluid of the brain within 15 minutes, which was followed by ACTH at 30 minutes. Plasma glucocorticoid levels reached their maximum at 45 minutes and plasma IL-β showed a significant elevation at 30 minutes and reached its maximum at 3 hours. These results indicate that LPS activates the HPA axis by the stimulation of hypothalamic IL-1β production [50].

The neuroendocrine effects of commercial LPS was compared to a native and a radiodetoxified LPS preparation in rats. Plasma ACTH, CON and βEND levels increased dramatically after the administration of commercial and native LPS preparations. Hormone levels in radiodetoxified LPS-treated rats did not rise significantly. These results suggest that radio-detoxification blunts the capacity of LPS to stimulate the cytokines that act on the HPA axis [51].

In male rats subjected to hemorrhagic shock plasma ACTH and CON, adrenal CON and cAMP were significantly decreased 20 h after resuscitation and adrenal morphology was altered. Treatment with the testosterone receptor antagonist, flutamide (25 mg/kg body wt s.c.), improved the CON content, restored cAMP and attenuated adrenal morphological alterations. Flutamide also improved the ACTH-induced CON release and adrenal cAMP response at 20 h after hemorrhage and resuscitation. The diminished CON response to ACTH stimulation in isolated adrenal preparations was also improved with flutamide [52].

In rats restraint stress significantly increased the contents of neuropeptides in thymic, but not splenic, extracts. A single icv injection of IL-1β increased CRF, AVP, ACTH and βEND in the spleens of both sham-operated and adrenalectomised (ADX) rats. IL-1β increased thymic contents of CRF and ACTH in sham-operated rats but not in ADX rats. These results indicate that the stimulation of neuropeptides by IL-1β in the spleen is independent of glucocorticoids, whereas in the thymus stimulation is dependent on circulating glucocorticoids [53].

Anaphylactoid reactions induced in male rats resulted in a rapid and significant activation of

ACTH secretion and induced c-Fos immunoreactivity in CRF-secreting subset of the parvocellular neurosecretory neurons. Magnocellular neurosecretory neurons and autonomic-related projection neurons in the PVN became also c-Fos positive upon challenge [54].

The effect of tetanus toxoid vaccine on HPA axis activity in 10 healthy premenopausal women (aged 28.6 +/- 2.6 year) was examined. Compared to placebo, tetanus toxoid elicited significant increases in plasma ACTH and serum cortisol (COL), with the maximum COL increase of 1.6-fold occurring 4.5 hr after toxoid administration. Urinary free COL increased 1.8-fold in the 8 h after tetanus toxoid administration. There was a significant inverse correlation ($r = 0.87$; $P < 0.005$) between the tetanus toxoid-induced increase in serum COL and the increase in tetanus antibody levels measured 1 month postvaccination [55].

4. BETA-ENDORPHIN (END) AND OTHER OPIOID PEPTIDES (OP)

Opioid receptors of κ-, δ- and μ-types are present on lymphocytes, on monocyte/macrophages and on polymorphonuclear leukocytes [56,57,58,59]. These receptors consist of about 400 amino acids and have the characteristics of 7 transmembrane domain structure of G-protein-coupled receptors. The amino acid homology between the different opioid receptor types is about 60% and there is about 90% identity of the same type of receptors in different animal species. Upon agonist binding, opioid receptors mediate inhibition of adenylyl cyclase [60].

Beta-endorphin is derived in the pituitary gland and in other tissues from the POMC peptide by enzymatic cleavage. CRF and immune derived cytokines, such as IL-1β and TNFα regulate the production and secretion of β-END in the pituitary gland [61,62]. Lymphoid cells also produce β-END, especially in inflamed tissues. Opioids in general and β-END in particular exert an anti-inflammatory effect and downregulate the immune response [63,64]. *In vitro*, both μ and κ opioid agonists exhibit a suppressive effect on the antibody response [65]. The κ opioid agonist drug, U50,488H, suppressed the production of IL-1 and TNFα by mouse peritoneal macrophages at 1 nM and IL-6 at 10 nM concentrations. This suppressive effect was completely reversed by the κ selective antagonist, norbinaltorphimine [66].

The effects of endogenous OP peptides on immune function are diverse and many of the findings are contradictory. Antibody production, NK cell activity, LAK cell activity, cytotoxic T lymphocytes, the mixed lymphocyte reaction, the production of IL-1, -2, -4, -6, IFN-γ, prostaglandins, mast cell degranulation and the activity of neutrophilic leukocytes were all affected by opioid peptides. Opioids are capable of modulating the immune system by acting through the central nervous system [67,68,69,70,71,72,73,74,75,76].

The tetrapeptide of human βEND (MPF: Lys-Lys-Gly-Glu) augments the proliferative response of human lymphocytes to the T-cell mitogen, concanavalin A. Human MPF similarly stimulates rat lymphocytes, but the peak effect is seen at a 100-fold higher dose (10^{-6} M). Rat MPF (Lys-Lys-Gly-Gln) has a peak effect at 10^{-6} M with human lymphocytes, but the peak effect with rat lymphocytes is at a 1000-fold lower concentration. Truncated forms of the MPFs (Gly-Glu, Gly-Gln, Gly, Glu, Gln) and OP (βEND, [Leu] and [Met] ENK) show only weak stimulatory or inhibitory effects. These results indicate that MPF acts on non-opioid receptors located on lymphocytes and that endogenously released MPF is an immunoregulatory mediator [77].

Physiological concentrations of βEND enhanced Con-A stimulated Ca mobilization by murine splenic T cells ($p < 0.01$), which was mediated by the δ opioid receptor [78].

Inducible nitric oxide synthase (iNOS) was expressed and cAMP levels were raised in human peripheral blood monocytes after incubation with βEND. Immunostaining for iNOS decreased, when an inhibitor of protein kinase A (H-89) was also added to the system. The cAMP increase was lowered by naloxone, which also reduced slightly iNOS expression [79].

In order to study the effect of endogenously released transmitters on spontaneous IL-6 secretion, spleen slices were electrically stimulated with 1, 5 10, 50, and 100 Hz. Spontaneous IL-6 secretion was significantly reduced at a frequency of 10 Hz with 10^{-7} M cortisol present ($P < 0.0001$). This inhibition was most probably due to a net inhibitory effect of opioidergic transmission [80].

Opioid peptide-containing immune cells migrate to inflamed sites, where they release βEND which inhibits pain. The initial step of immunocyte recruitment is mediated by selectins. Anti-selectin treatment abolishes peripheral opioid analgesia by blocking the infiltration by immunocytes [81].

Met-enkephalin (MENK) and βEND, at physiological concentrations (10^{-8} and 10^{-6}M) enhanced fMLP, PMA or OZ stimulated chemiluminescence (CL), induced a chemotactic response and up-regulated the expression of CD11b and CD18 by human neutrophils. High concentrations of βEND (10^{-3} M) or MENK (10^{-5} M) decreased CL. Tus OP, like inflammatory cytokines, may prime neutrophils for inflammatory response [82].

Consistently, the opiate receptor antagonist, naloxone, when given alone at the time of transplantation, but not later, accelerates graft rejection and increases splenocyte IL-2 and IFNγ production. Splenocyte βEND concentrations are lower in transplanted animals [83].

βEND was analyzed in MRL lpr/lpr mice, that develop a lupus-like autoimmune disease, and in the obese strain (OS) of chicken with spontaneous autoimmune thyroiditis. In both species hypothalamic concentrations of βEND were significantly lower than in normal controls. In MRL lpr/lpr mice, βEND was already lower at 1 month of age, when no clinical sign of the disease was evident. Low levels of βEND were also present in OS chickens already at the embryonic stage, way before the onset of thyroiditis. A further decrease of βEND was observed in OS chickens at the time of thyroid infiltration by mononuclear cells. Central βEND is known to exert an immunosuppressive effect. In susceptible animals the low hypothalamic βEND concentrations may be one of several factors predisposing for the development of autoimmune disease [84].

Beta-endorphin concentrations in the peripheral blood mononuclear cells of chronic fatigue patients were significantly lower ($p < 0.001$) than in healthy subjects [85].

Plasma βEND levels in severe RA patients were significantly lower (16.1 +/- 6.2 pg/ml) compared to patients with mild RA (45.4 +/- 2.8 pg/ml). The mean serum levels of βEND were also significantly lower in both RA groups than in normal controls (62.1 +/- 5.7 pg/ml). There is an inverse correlation between plasma βEND levels, the rheumatoid disease activity score, and the duration of RA. The depressed plasma βEND level in patients with RA may be used as an indicator of the disease activity [86].

Opioids exert a wide range of biological activities in the body including reproduction and development, normal physiological function and a reaction to stress and infectious disease [87]. Endogenous opioid peptides play multiple roles in the CNS, from development to immune modulation. Specific opioid receptors are uniquely localized in different brain regions and cells. Exogenous opioids may influence neural and glial cell function leading to either protection or damage of the brain. The mechanisms underlying opioid effects in the CNS may be mediated by immune-derived cytokines, beta-chemokines and free radicals (reactive oxygen intermediates produced by microglia and astrocytes). There are many conflicting results in the field of opiate research [88]. The precise mechanisms involved in the centrally mediated immune modulation by opioids are not well-defined and the physiological role of these mechanisms is obscure at the present time. The modulation phenomenon may represent an integral feedback loop within the homeostatic control system of the body. The HPA axis in immune regulation serves to balance immune function in concert with multiple feedback systems. The physiological role of endogenous opioid control may be to monitor and subvert chronic inflammatory processes and autoimmune disorders [89].

Morphine suppresses a variety of immune responses that include natural killer cells, T and B lymphocytes, macrophages and polymorphonuclear leukocytes. These effects have been reported in humans, monkeys and rodents. Phagocytic cell function is down-regulated both *in vivo* and *in vitro.* A direct effect of morphine on NK cells has not been proven. The overall evidence supports the existence of a neuroimmune circuit in which morphine acts to depress the function of all cells in the immune system. However, endogenous opioids have been reported to potentiate immune function in *in vitro* systems [90].

Enkephalins, the opioid antagonist naloxone, and an inhibitor of enkephalin-degrading endopeptides, as thiorphan, modulate the proliferation and differentiation of hematopoietic cells from murine bone marrow in culture. The effect is partially dependent on accessory hematopoietic elements and follows a circadian rhythm. Enkephalins bind to specific opioid receptors on bone marrow cells and the signaling cascade includes diacyl-glycerol, protein kinase C and calcium ions [91].

Cytokines regulate opioid and opioid receptor gene expression in neuronal and immune cells. The μ, δ, and κ opioid receptor promoter regions contain potential cytokine response elements (NF-IL6 binding sites). However, other experiments suggest that there is no significant cytokine stimulation of opioid receptor gene expression in immune cells and that the cytokine response element in opiate receptor genes are not functional [92]. The opioid receptors expressed in the immune system are the same or similar to the ones in the neuroendocrine systems, particularly, δ- and κ- type receptors. Delta receptor agonists appear to function in an autocrine/paracrine manner [93].

The expression of neuronal opiate receptors in the immune system does not completely explain the biological role of opiates in immune function [94]. Studies in rats with opioid receptor agonist drugs revealed that immunomodulation by opioids are mediated by the μ opioid receptor [95]. The μ opioid receptor (OP3) is coupled to G-proteins expressed by human embryonic kidney cells (HEK-293) [96]. The chemotactic response of monkey leukocytes to IL-8, MIP-1β and RANTES was inhibited by μ opiate receptor agonists (morphine, DAMGO, methadone and endomorphine). This inhibition could be reversed by naloxone [97]. The porcine μ opioid receptor shows 96% homology with the human receptor. It is expressed in the nervous system, gut, thymus, Peyer's patches and in other immune tissues and in isolated immune cells [98].

In μ opioid receptor knockout mice (μOPR-O), morphine had no effect on macrophage phagocytosis and the secretion of TNF-α. However, thymic and splenic cell numbers have been reduced and mitogen induced proliferation as well as IL-1 and IL-6 secretion were inhibited. These later results are consistent with the existence of a naloxone-insensitive morphine receptor (non-opioid morphine binding site) present on immune cells [99]. The cell active κ-opioid agonist trans-3-4-dichloro-N-methyl-N-[2- (1-pyrolidinyl) cyclohexyl] benzeneacetamide methanesulfonate (U50,488H) significantly inhibited the transcription of IL-7 and elevated the expression of chemokine receptor (CCR-2) in thymocytes stimulated with superantigen [100]. The priming of microglia with IFN-γ or TNF-α enhanced the production of superoxide N ion, a reactive oxygen intermediate produced during inflammation. The selective κ-opioid receptor ligand (U50,488) inhibited the priming effect of cytokines [101].

Endomorphine-1 (EMO-1) and endomorphine-2 (EMO-2) are opioid tetrapeptides discovered within the nervous system which have a very high specificity for the μ-opioid receptor. EMO-1 and EMO-2 have been also detected in rat spleen and thymus [102]. Leucine-enkephalin elicits the potentiation of immune responses through the OP-1 (δ) receptor and suppression through the OP-2 (κ) receptor. The OP-3 receptor is also required for the immune response since it has a permissive effect for central immunomodulation of endogenous opioid peptides and LENK. In contrast, MENK enhances immune function by OP-1 receptors independent of OP-3 [103].

Opioid receptor subtypes of δ, κ and μ similar to those detected in the brain have been identified in immune cells. The treatment of human lymphocytes with morphine increased the expression of μ opioid receptors [104].

Methionine enkephalin at concentrations ranging from 10^{-7} to 10^{-15} M increased intracellular calcium levels after the ligation of TCR/CD-3 complex in T lymphocytes in a 20 to 40 percent of experiments performed. The levels of intracellular calcium in this particular fraction were lower than control levels prior to activation [105]. The endogenous opioid peptides EMO-1 and EMO-2 inhibited, in a dose-dependent manner, the production of super oxide N ions by PAM-stimulated neutrophils [106]. Pre-proenkephalin mRNA and MENK are present in T lymphocytes and the concentration was higher in the latter cell type. Lymphocytes from PENK-deficient mice could be driven into the TH-2 phenotype by IL-4 treatment alone suggesting that PENK is not required for differentiation [107].

The treatment of mice subjected to hemorrhagic shock by naloxone decreased peritoneal macrophage and splenic immune functions [108]. The secretion of IL-6 in the spleen is inhibited by electrical pulses. This inhibition is mediated by α adrenergic and μ opioidergic endogenous transmitters [109]. Morphine at nanomolar concentrations synergized with LPS and augmented the secretion of both IL-6 and TNF-α. At micromolar concentrations morphine inhibited LPS induced synthesis of IL-6 and TNF-α. Similarly, low concentrations of morphine augmented the activation of NFκB activation by LPS whereas high (micromolar) concentrations were inhibitory. Augmentation was reversible with naloxone, whereas inhibition of NFκB by morphine was not reversed by naloxone indicating the involvement of a non-classical opioid receptor [110].

5. ALPHA-MELANOCYTE STIMULATING HORMONE (AMSH)

The POMC derived α-MSH is a major regulator of fever and inflammation. α-MSH is a cytokine antagonist and inhibits the pyrogenic and proinflammatory effects of IL-1, IL-6, TNF and IFN-γ. It acts within the brain to inhibit fever and peripheral inflammation. However, α-MSH also acts on the periphery, where it clearly exerts an anti-inflammatory effect [111].

α-MSH is produced by spleen cells, by keratinocytes and is also present in the aqueous humour of the eye [112,32,113). α-MSH inhibits thymocyte proliferation, neutrophilia, the production of acute phase proteins by the liver, TNF and IFN-γ production, the induction of prostaglandin E in fibroblasts and contact sensitivity reactions. The immunosuppressive effect of IL-1β, given i.c.v., could be blocked by the simultaneous infusion of α-MSH [114,115,116,117,118,119,120,121]. Alpha-MSH administered to mice either systemically or locally at the site of treatment with contact sensitizing agents resulted in the induction of hapten specific tolerance. Lymphocytes from regional lymph nodes of α-MSH treated mice failed to produce IL-2 five days after sensitization. Tolerance induction by α-MSH was abrogated *in vivo* by the administration of antibodies to IL-10 to the site of sensitization [122,123].

The anti-inflammatory effect of centrally administered α-MSH [1–13] was inhibited in mice by propranolol (β-adrenergic receptor blocker) and by a specific β2-adrenergic receptor antagonist. A marked anti-inflammatory effect was exerted by α-MSH [11–13] when injected i. p. to animals with spinal cord trans-section [124]. α-MSH is involved in adjuvant arthritis, adult respiratory distress syndrome and septic shock. Plasma levels of α-MSH are elevated in patients with acquired immunodeficiency syndrome and in the early stages of myocardial infection. α-MSH was found in the synovial fluid of joints affected by rheumatoid arthritis. The synovial level did not correlate with plasma α-MSH levels [111].

Exogenous αMSH antagonizes the stimulatory effects of IL-1 on the HPA axis. The ACTH response of rats to IL-1β was enhanced by i.c.v. infused αMSH antiserum. The CON response was also significantly enhanced in antiserum treated animals. These results indicate that αMSH plays a physiological role in limiting the HPA response to this inflammatory cytokine [125].

Receptors for α-MSH belong to the G-protein-receptor family. Alpha-MSH shows binding affinities to five different subtypes of G-protein-receptors. The pro-B lymphocyte cell line Ba/F3 express melanacortine-5 (MC-5) receptors and bind α-MSH specifically. Alpha-MSH stimulates Janus kinase 2 (JAK-2), signal transducers and activators of transcription (STAT-1) and tyrosine phosphorylation in both Ba/F3 cells and the human IM9-B lymphocyte line. Alpha-MSH also activated JAK-2 in mouse L-cells transfected with the human MC-5 receptors. Alpha-MSH promoted cell proliferation in these target cells [126]. A low number of melanocortin receptors have been found on unstimulated human monocytes. After activation of monocytes by LPS or PHA the receptors were up regulated for three to five days and could be further augmented by cytokine treatment with IL-2, IFN-γ, IL-4 and IL-10. Monocytes expressed the MC-1 receptor which binds α-MSH and ACTH. α-MSH significantly inhibited the expression of CD-86 on LPS-treated monocytes [127]. Cardiopulmonary bypass results in an inflammatory response which is characterized in part by hyperstimulation of leukocytes. Plasma from such patients stimulated monocytes and granulocytes which were inhibited by pre-treatment with α-MSH (10^{-6} M). If the cells were first exposed to plasma and then treated with α-MSH, there was no effect [128].

Alpha-MSH is an antagonist of pro-inflammatory cytokines (e.g. IL-1, IL-6 and TNF-α). The treatment of human keratinocytes with α-MSH significantly reduced the IL-1β mediated secretion of the C-X-C chemokines IL-8 and Gro-α. Alpha-MSH also downregulated the IL-1β induced activation of NFκB which appears to be a crucial event responsible for the downregulation of cytokine gene transcription [129]. The treatment of human peripheral blood mononuclear cells with α-MSH significantly enhanced the release of IL-10 in a dose-dependent manner. Only the C-terminal tripeptide of α-MSH was required to enhance IL-10 production. A tripeptide from α-MSH strongly induced IL-10 in purified monocytes. T lymphocytes did not produce IL-10 in response to α-MSH [130].

Pro-opiomelanocortin (POMC)-derived neuropeptides such as α-MSH, β-ENB and ACTH are produced in the skin. IL-1 produces an appregulation of cutaneous levels of POMC mRNA, POMC peptides and MSH receptors. Ultraviolet radiation also stimulates all the components of the CRF/POMC system including the expression of the appropriate receptors. The receptors expressed are MC-1, MC-5 and μ-opiate receptors. CRF receptors are also present. The CRF/POMC system of the skin regulates pigmentary, immune and other defense mechanisms in the interest of maintaining skin integrity [131]. Eleven α-MSH is released in the skin after stimulation by epidermal cells including keratinocytes, Langerhans cells and melanocytes as well as immunocompetent cells. Alpha-MSH inhibits the production and activity of proinflammatory cytokines such as IL-1, IL-2, IFN-γ, and downregulates the expression of co-stimulatory molecules (B7) on antigen presenting cells and it stimulates the inhibitory cytokine IL-10. Monocytes express the MC-1 receptor for MSH which is upregulated by mitogens, endotoxins and proinflammatory cytokines. α-MSH is an immunosuppressive neurohormone [132].

Keratinocytes release increased amounts of cytokines following exposure to UV radiation. PMOC-derived peptides are also released after UV exposure including α-MSH. α-MSH is an important immunomodulating and antiinflammatory hormone mainly by acting on antigen-presenting cells and on the vascular endothelium. In mice, α-MSH given intravenously or after topical application inhibited contact hyper-sensitive reactions and induced hapten-specific tolerance. In the epidermis after UV irradiation, α-MSH may mediate UV-induced immunosuppression [133].

POMC is produced in both the immune and neuroendocrine systems and the receptors are also shared by the peptide hormones derived from POMC. This allows for bidirectional communication between the immune and neuroendocrine systems. There is a regulatory interaction between IL-1 and α-MSH which functions as an IL-1 antagonist. POMC peptides exert a local antinociceptive effect and influence hematopoiesis. The deficiency of ACTH receptors on lymphocytes leads to ACTH insensitivity [134,135,136].

The α-MSH derived peptide (1–13) and it's carboxy-terminal tripeptide α-MSH (1–13) have exerted a potent antiinflammatory effect in all major models of inflammation. These include experimental inflammatory bowel disease, CNS ischemia/perfusion injury, and LPS-induced inflammation in the brain. Clearly, these peptides provide a basic regulatory mechanism to the inflammatory process. Alpha-MSH inhibits inflammation by three general mechanisms: (i) inhibition of the production of inflammatory mediators; (ii) inhibition of inflammatory action of mediators; and (iii) inhibition of peripheral host cells [137].

In rats the IL-1β-induced stimulation of ACTH release was significantly enhanced by antagonizing the activity of α-MSH by specific antibodies. Corticosterone production was also increased after i.c.v. application of α-MSH antibodies [124].

REFERENCES

1. Besedovsky HO, del Rey A, Klusman I, Furukawa H, Monge Arditi G, Kabiersch A. Cytokines as modulators of the hypothalamus-pituitary-adrenal axis. J Steroid Biochem Mol Biol 1991;40:613–8.
2. Tsagarakis S, Grossman A. Corticotropin-releasing hormone: interactions with the immune system. Neuroimmunomodulation 1994;1:329–34.
3. Karalis K, Sano H, Redwine J. Autocrine or paracrine inflammatory actions of corticotropin-releasing hormone *in vivo*. Science 1991;254:421–23.
4. Smith EM, Hughes TK, Cadet P, Stefano GB. Corticotropin-releasing factor-induced immunosuppression in human and invertebrate immunocytes. Cell Mol Neurobiol 1992;12:473–81.
5. Kelley DM, Lichtenstein A, Wang J, Taylor AN, Dubinett SM. Corticotropin-releasing factor reduces lipopolysaccharide-induced pulmonary vascular leak. Immunopharmacol Immunotoxicol 1994;16:139–48.
6. Vamvakopoulos NC, Chrousos GP. Hormonal regulation of human corticotropin-releasing hormone gene expression: implications for the stress response and immune/inflammatory reaction. Endocrine Rev 1994;15:409–420.
7. Jain R, Zwickler D, Hollander CS. Corticotropin-releasing factor modulates the immune response to stress in the rat. Endocrinology 1991;128:1329–35.
8. Crofford LJ, Sano H, Karalis K, Friedman TC, Epps HR, Remmers EF, Mathern P, Chrousos GP, Wilder RL. Corticotropin-releasing hormone in synovial fluids and tissues

of patients with rheumatoid arthritis and osteoarthritis. J Immunol 1993;151:1587–96.
9. Radulovic M, Dautzenberg FM, Sydow S, Radulovic J, Spiess J. Corticotropin-releasing factor receptor 1 in mouse spleen: expression after immune stimulation and identification of receptor-bearing cells. J Immunol 1999;162:3013–21.
10. Turnbull AV, Smith GW, Lee S, Vale WW, Lee KF, Rivier C. CRF type I receptor-deficient mice exhibit a pronounced pituitary-adrenal response to local inflammation. Endocrinology 1999;140:1013–7.
11. LaCroix S, Rivest S. Role of cyclo-oxygenase pathways in the stimulatory influence of immune challenge on the transcription of a specific CRF receptor subtype in the rat brain. J Chem Neuroanat 1996;10:53–71.
12. Brouxhon SM, Prasad AV, Joseph SA, Felten DL, Bellinger DL. Localization of corticotropin-releasing factor in primary and secondary lymphoid organs of the rat. Brain Behav Immun 1998;12:107–22.
13. Raber J, Koob GF, Bloom FE. Interferon-alpha and transforming growth factor-beta 1 regulate corticotropin-releasing factor release from the amygdala: comparison with the hypothalamic response. Neurochem Int 1977;30:455–63.
14. Kageyama K, Bradbury MJ, Zhao L, Blount AL, Vale WW. Urocortin messenger ribonucleic acid: tissue distribution in the rat and regulation in thymus by lipopolysaccharide and glucocorticoids. Endocrinology 1999;140:5651–8.
15. Okamoto S, Ishikawa I, Kimura K, Saito M. Potent suppressive effects of urocortin on splenic lymphocyte activity in rats. Neuroreport 1998;9:4035–9.
16. Barkhudaryan N, Dunn AJ. Molecular mechanisms of actions of interleukin-6 on the brain, with special reference to serotonin and the hypothalamo-pituitary-adrenocortical axis. Neurochem Res 1999;24:1169–80.
17. Vallieres L, Lacroix S, Rivest S. Influence of interleukin-6 on neural activity and transcription of the gene encoding corticotrophin-releasing factor in the rat brain: an effect depending upon the route of administration. Eur J Neurosci 1997;9:1461–72.
18. Smith EM, Cadet P, Stefano GB, Opp MR, Hughes TK. IL-10 as a mediator in the HPA axis and brain. J Neuroimmunol 1999;100:140–8.
19. Fleisher-Berkovich S, Danon A IL-1alpha but not IL-1beta-induced prostaglandin synthesis is inhibited by corticotropin-releasing factor. Cytokine 1999;11:239–43.
20. Casadevall M, Saperas E, Panes J, Salas A, Anderson DC, Malagelada JR, Pique JM. Mechanisms underlying the anti-inflammatory actions of central corticotropin-releasing factor. Am J Physiol 1999;276:G1016–26.
21. Schafer M, Mousa SA, Stein. Corticotropin-releasing factor in antinociception and inflammation. Eur J Pharmacol 1997;323:1–10.
22. Poliak S, Mor F, Conlon P, Wong T, Ling N, Rivier J, Vale W, Steinman L. Stress and autoimmunity: the neuropeptides corticotropin-releasing factor and urocortin suppress encephalomyelitis via effects on both the hypothalamic-pituitary-adrenal axis and the immune system. J Immunol 1997;158:5751–6.
23. Harbuz MS, Conde GL, Marti O, Lightman SL, Jessop DS. The hypothalamic-pituitary-adrenal axis in autoimmunity. Ann N Y Acad Sci 1997;823:214–24.
24. Boehme SA, Gaur A, Crowe PD, Liu XJ, Tamraz S, Wong T, Pahuja A, Ling N, Vale W, De Souza EB, Conlon PJ. Immunosuppressive phenotype of corticotropin-releasing factor transgenic mice is reversed by adrenalectomy. Cell Immunol 1997;176:103–12.
25. Menzies R, Phelps C, Wiranowska M, Oliver J, Chen L, Horvath E, Hall N. The effect of interferon-alpha on the pituitary-adrenal axis. J Interferon Cytokine Res. 1996;16:

619–29.
26. Berczi I. Pituitary Function and Immunity. Boca Raton, Florida:CRC Press;1986.
27. Hayashita T, Li CH. Influence of adrenocorticotropic and growth hormone on antibody formation. J Exp Med 1957;105:93.
28. Berczi I, Nagy E. The effect of prolaction and growth hormone on hemolymphopoietic tissue and immune function. In: Berczi I, Kovacs K, editors. Hormones and Immunity. Lancaster, England:MTP Press,1987;145–171.
29. Smith EM, Brosnan P, Meyer WJ, Blalock JE. An ACTH receptor on human mononuclear leukocytes: relation to adrenal ACTH-receptor activity. N Engl J Med 1987;317:1266–69.
30. Oates EL, Allaway GP, Armstrong GR. Human lymphocytes produce pro-opiomelanocortin-related gene transcripts. J Biol Chem 1988;263:10041–44.
31. Clarke BL, Bost KL. Differential expression of functional adrenocorticotropic hormone receptors by subpopulations of lymphocytes. J Immunol 1989;143:464–69.
32. Schauer E, Trautinger F, Kock A. Proopiomelanocortin-derived peptides are synthesized and released by human keratinocytes. J Clin Invest 1994;93:2258–62.
33. Olsen NJ, Nicholson WE, DeBold CR, Orth DN. Lymphocyte-derived adrenocorticotropin is insufficient to stimulate adrenal steroidogenesis in hypophysectomized rats. Endocrinology 1992;130:2113–19.
34. Heijnen CJ, Zijlstra J, Kavelaars A. Modulation of the immune response by POMC-derived peptides. I. Influence on proliferation of human lymphocytes. Brain Behav Immun 1987;1:284–91.
35. Kavelaars A, Ballieux RE, Heijnen C. Modulation of the immune response by proopiomelanocortin derived peptides. II. Influence of adrenocorticotropin hormone on the rise in intracellular free calcium concentration after T-cell activation. Brain Behav Immun 1988;2:57–66.
36. Bost KL, Clarke BL, Xu J. Modulation of IgM secretion and H chain mRNA expression in CH12.LX.C4.5F5 B cells by adrenocorticotropic hormone. J Immunol 1990;145:4326–31.
37. Aebischer I, Stampfli MR, Zurcher A. Neuropeptides are potent modulators of human *in vitro* immunoglobulin E synthesis. Eur J Immunol 1994;24 :1908–13.
38. Ichinose M, Sawada M, Mæno T. Suppression of phagocytosis by adrenocorticotropic hormone in murine peritoneal macrophages. Immunol Lett 1994;42:161–5.
39. Zimmer JZ, Lipton JM. Central and peripheral injections of ACTH (1–24) reduce fever in adrenalectomized rabbits. Peptides 1981;2:413–417.
40. Bethin KE, Vogt SK, Muglia LJ. Interleukin-6 is an essential, corticotropin-releasing hormone-independent stimulator of the adrenal axis during immune system activation. Proc Natl Acad Sci U S A 2000;97:9317–22.
41. Giovambattista A, Chisari AN, Gaillard RC, Spinedi E. Modulatory role of the epinergic system in the neuroendocrine-immune system function. Neuroimmunomodulation 2000;8: 98–106.
42. Rivier C. Role of nitric oxide and carbon monoxide in modulating the ACTH response to immune and nonimmune signals. Neuroimmunomodulation 1998;5:203–13.
43. Liege S, Moze E, Kelley KW, Parnet P, Neveu PJ. Activation of the hypothalamic-pituitary-adrenal axis in IL-1 beta-converting enzyme-deficient mice. Neuroimmunomodulation 2000;7:189–94.
44. Sullivan GM, Canfield SM, Lederman S, Xiao E, Ferin M, Wardlaw SL.

Intracerebroventricular injection of interleukin-1 suppresses peripheral lymphocyte function in the primate. Neuroimmunomodulation 1997;4:12–8.

45. Zylinska K, Mucha S, Komorowski J, Korycka A, Pisarek H, Robak T, Stepien H. Influence of granulocyte-macrophage colony stimulating factor on pituitary-adrenal axis (PAA) in rats in vivo. Pituitary 1999;2:211–6.

46. Conti B, Sugama S, Kim Y, Tinti C, Kim H, Baker H, Volpe B, Attardi B, Joh T. Modulation of IL-18 production in the adrenal cortex following acute ACTH or chronic corticosterone treatment. Neuroimmunomodulation 2000;8:1–7.

47. Kim DS, Melmed S. Stimulatory effect of leukemia inhibitory factor on ACTH secretion of dispersed rat pituitary cells. Endocr Res 1999;25:11–9.

48. Schwartz J, Ray DW, Perez FM. Leukemia inhibitory factor as an intrapituitary mediator of ACTH secretion. Neuroendocrinology 1999;69:34–43.

49. Kim DS, Choi HS, Park YS, Kim TW. Effects of oncostatin M on hormone release of rat pituitary cells in primary culture. J Korean Med Sci 2000;15:323–6.

50. Ma XC, Chen LT, Oliver J, Horvath E, Phelps CP. Cytokine and adrenal axis responses to endotoxin. Brain Res. 2000;861:135–42.

51. Barna I, Bertok L, Koenig JI, Makara GB. Radiodetoxified Lipopolysaccharide Fails to Activate the Hypophyseal-Pituitary-Adrenal Axis in the Rat. Neuroimmunomodulation 2000;8:128–131.

52. Ba ZF, Wang P, Koo DJ, Zhou M, Cioffi WG, Bland KI, Chaudry IH. Testosterone receptor blockade after trauma and hemorrhage attenuates depressed adrenal function. Am J Physiol Regul Integr Comp Physiol 2000;279:R1841–8.

53. Jessop DS, Douthwaite JA, Conde GL, Lightman SL, Dayan CM, Harbuz MS. Effects of Acute Stress or Centrally Injected Interleukin-1beta on Neuropeptide Expression in the Immune System. Stress 1997;2:133–144.

54. Foldes A, Nemethy Z, Szalay O, Kovacs KJ. Anaphylactoid reactions activate hypothalamo-pituitary-adrenocortical axis: comparison with endotoxic reactions. Brain Res Bull 2000;52:573–9.

55. Oken E, Kasper DL, Gleason RE, Adler GK. Tetanus toxoid stimulation of the hypothalamic-pituitary-adrenal axis correlates inversely with the increase in tetanus toxoid antibody titers. J Clin Endocrinol Metab 1998;83:1691–6.

56. Gaveriaux C, Peluso J, Simonin F, Laforet J, Kieffer B. Identification of kappa- and delta-opioid receptor transcripts in immune cells. FEBS Lett 1995;369–272–6.

57. Chuang TK, Killam KR, Chuange LF, Kumg HF, Sheng WS, Chao CC, Yu L, Chuange RY. Mu opioid receptor gene expression in immune cells. Biochem Biophys Res Comm 1995;216:922–30.

58. Sedqi M, Roy S, Ramakrishnan S, Elde R, Loh HH. Complementary DNA cloning of a mu-opioid receptor from rat peritoneal macrophages. Biochem Biophys Res Commun 1995;209:563–74.

59. Wick MJ, Minnerath SR, Roy S, Ramakrishnan S, Loh HH. Differential expression of opioid receptor genes in human lymphoid cell lines and peripheral blood lymphocytes. J Neuroimmunol 1996;64:29–36.

60. Knapp RJ, Malatynska E, Collins N, Fang L, Wang JY, Hruby VJ, Roeske WR, Yamamura HI. Molecular biology and pharmacology of cloned opioid receptors. FASEB J 1995;9: 516–25.

61. Ruzicka BB, Akil H. Differential cellular regulation of pro-opiomelanocortin by interleukin 1β and corticotropin-releasing hormone. Neuroendocrinology 1995;61:136–

51.
62. Sacerdote P, Brini AT, Locatelli L, Radulovic J, Panerai AE. Tumor necrosis factor alpha differentially regulates beta-endorphin concentrations and proopiomelanocortin RNA in the anterior and neurointermediate pituitary *in vivo.* Neuroimmunomodulation 1994;1: 357–60.
63. Panerai AE, Radulovic J, Monastra G, Manfredi B, Locatelli L, Sacerdote P. Beta-endorphin concentrations in brain areas and peritoneal macrophages in rats susceptible and resistant to experiemental allergic encephalomyelitis; a possible relationship between tumor necrosis factor alpha and opioids in the disease. J Neuroimmunol 1994;51:169–75.
64. Stein C, Schafer M, Hassan AHS. Peripheral opioid receptors. Ann Med 1995;27:219–21.
65. Taub DD, Eisenstein TK, Geller EB, Adler MW, Rogers TJ. Immunomodulatory activity of μ- and κ-selective opioid agonists. Proc Natl Acad Sci USA 1991;88:360–64.
66. Alicea C, Belkowski S, Eisenstein TK, Adler MW, Rogers TJ. Inhibition of primary murine macrophage cytokine production *in vitro* following treatment with the κ-opioid agonist U50,488H. J Neuroimmunol 1996;64:83–90.
67. Plotnikoff NP, Faith RE, Murgo AJ, Good RA. Enkephalins and Endorphins: Stress and the Immune System. New York:Plenum Press;1986.
68. Gilmore W, Weiner LP. Beta-Endorphin enhances interleukin-2 (IL-2) production in murine lymphocytes. J Neuroimmunol 1988;18:125–38.
69. Stefano GB. Role of opioid neuropeptides in immunoregulation. Prog Neurobiol 1989;33: 149–59.
70. Weber RJ, Pert A. The periaqueductal gray matter mediates opiate-induced immunosuppression. Science 1989;245:188–190.
71. Apte RN, Durum SK, Oppenheim JJ. Opioids modulate interleukin-1 production and secretion by bone-marrow macrophages. Immunol Lett 1990;24:141–48.
72. Millar DB, Hough CJ, Mazorow DL, Gootenberg JE. Beta-endorphin's modulation of lymphocyte proliferation is dose, donor and time dependent. Brain Behav Immun 1990;4: 232–242.
73. Chiappelli F, Nguyen L, Bullington R, Fahey JL. Beta-Endorphin blunts phosphatidylllinositol formation during *in vitro* activation of isolated human lymphocytes: Preliminary report. Brain Behav Immun 1992;6:1–10.
74. van den Bergh P, Dobber R, Ramlal S. Role of opioid peptides in the regulation of cytokine production by murine CD4+ T cells. Cell Immunol 1994;154:109–122.
75. Hiddinga HJ. Isaak DD, Lewis RV. Enkephalin-containing peptides processed from proenkephalin significantly enhance the antibody-forming cell responses to antigens. J Immunol 1994;152:3748–59.
76. Mazorow DL, Simpkins CO, Millar DB. Beta-endorphin modulates calcium channel activity in human neutrophils. J Neuroimmunol 1994;50:77–83.
77. Owen DL, Morley JS, Ensor DM, Miles JB. The C-terminal tetrapeptide of beta-endorphin (MPF) enhances lymphocyte proliferative responses. Neuropeptides 1998;32:131–9.
78. Sharp BM, McKean DJ, McAllen K, Shahabi NA. Signaling through delta opioid receptors on murine splenic T cells and stably transfected Jurkat cells. Ann N Y Acad Sci 1998;840: 420–4.
79. Aymerich MS, Bengoechea-Alonso MT, Lopez-Zabalza MJ, Santiago E, Lopez-Moratalla. Inducible nitric oxide synthase (iNOS) expression in human monocytes triggered by beta-endorphin through an increase in cAMP. Biochem Biophys Res Commun 1988;245:

717–21.
80. Straub RH, Dorner M, Riedel J, Kubitza M, Van Rooijen N, Lang B, Scholmerich J, Falk W. Tonic neurogenic inhibition of interleukin-6 secretion from murine spleen caused by opioidergic transmission. Am J Physiol 1998;274:R997–1003.
81. Machelska H, Cabot PJ, Mousa SA, Zhang Q, Stein C. Pain control in inflammation governed by selectins Nat Med 1998;4:1425–8.
82. Pasnik J, Tchorzewski H, Baj Z, Luciak M, Tchorzewski M. Priming effect of met-enkephalin and beta-endorphin on chemiluminescence, chemotaxis and CD11b molecule expression on human neutrophils in vitro. Immunol Lett 1999;67:77–83.
83. Sacerdote P, di San Secondo VE, Sirchia G, Manfredi B, Panerai AE. Endogenous opioids modulate allograft rejection time in mice: possible relation with Th1/Th2 cytokines. Clin Exp Immunol 1998;113:465–9.
84. Sacerdote P, Lechner O, Sidman C, Wick G, Panerai AE. Hypothalamic beta-endorphin concentrations are decreased in animals models of autoimmune disease. J Neuroimmunol 2000;102:222.
85. Conti F, Pittoni V, Sacerdote P, Priori R, Meroni PL, Valesini G. Decreased immunoreactive beta-endorphin in mononuclear leucocytes from patients with chronic fatigue syndrome. Clin Exp Rheumatol 1998;16:729–32.
86. Elbeialy A, Elbarbary M, Kamel M. Peripheral beta-endorphin in rheumatoid arthritis: A correlation with the disease activity. Scand J Rheumatol 1997;26:88–91.
87. Olson GA, Olson RD, Kastin AJ. Endogenous opiates. Peptides 1997;18:1651–88.
88. Sheng WS, Hu S, Gekker G, Zhu S, Peterson PK, Chao CC. Immunomodulatory role of opioids in the central nervous system. Arch Immunol Ther Exp 1997;45:359–66.
89. Brinkman WJ, Hall DM, Suo JL, Weber RJ. Centrally-mediated opioid-induced immunosuppression. Elucidation of symphathetic nervous system involvement. Adv Exp Med Biol 1998;437:43–9.
90. Eisenstein TK, Hilburger ME. Opioid modulation of immune responses: effects on phagocyte and lymphoid cell populations. J Neuroimmunol 1998;83:36–44.
91. Boranic M, Krizanac-Bengez L, Gabrilovac J, Marotti T, Breljak D. Enkephalins in hematopoiesis. Biomed Pharmacother 1997;51:29–37.
92. Im HJ, Kang SW, Loh HH. Opioid receptor gene: cytokine response element and the effect of cytokines. Brain Res 1999;829:174–9.
93. Sharp BM, Roy S, Bidlack JM. Evidence for opioid receptors on cells involved in host defense and the immune system. J Neuroimmunol 1998;83:45–56.
94. Madden JJ, Whaley WL, Ketelsen D. Opiate binding sites in the cellular immune system: expression and regulation. J Neuroimmunol 1998;83:57–62.
95. Nelson CJ, Schneider GM, Lysle DT. Involvement of central mu- but not delta- or kappa-opioid receptors in immunomodulation. Brain Behav Immun 2000;14:170–84.
96. Burford NT, Want D, Sadee W. G-protein coupling of mu-opioid receptors (OP3): elevated basal signaling activity. Biochem J 2000;349Pt3:531–7.
97. Choi Y, Chuang LF, Lam KM, Kung HF, Wang JM, Osburn BI, Chuang RY. Inhibition of chemokine-induced chemotaxis of monkey leukocytes by mu-opioid receptor agonists. In Vivo 1999;13:389–96.
98. Pampusch MS, Osiski MA, Brown DR, Murtaugh MP. The porcine mu opioid receptor: molecular cloning nad mRNA distribution in lymphoid tissues. J Neuroimmunol 1998;90: 192–8.
99. Roy S, Barke RA, Loh HH. MU-opioid receptor-knockout mice: role of mu-opioid receptor

in morphine mediated immune functions. Brain Res Mol Brain Res 1998;61:190–4.

100. Zhang L, Rogers TJ. Kappa-opoid regulation of thymocyte IL-7 receptor and C-C chemokine receptor 2 expression. J Immunol 2000;164:5088–93.

101. Hu S, Peterson PK, Chao CC. Kappa-opioid modulation of human microglial cell superoxide anion generation. Biochem Pharmacol 1998;56:285–8.

102. Jessop DS, Major GN, Coventry TL, Kaye SJ, Fulford AJ, Harbuz MS, De Bree FM. Novel opioid peptides endomorphin-1 and endomorphin-2 are present in mammalian immune tissues. J Neuroimmunol 2000;106:53–9.

103. Dimitrijevic M, Stanojevic S, Kova&cbreve;evic-Jovanoic V, Miletic T, Vujic-redzic V, Radulovic J. Modulation of humoral immune responses in the rat by centrally applied met-Enk and opioid receptor antagonists: functional interactions of brain OP(1), OP(2) and OP(3) receptors. Immunopharmacology 2000;49:255–62.

104. Suzuki S, Miyagi T, Chuang TK, Chuang LF, Doi RH, Chuang RY. Biochem Biophys Res Commun 2000;279:621–8.

105. Sorensen AN, Claesson MH. Effect of the opioid methionine enkephalinamide on signal transduction in human T-lymphocytes. Life Sci 1998;62:1251–6.

106. Azuma Y, Want P, Shinohara M, Ohura K. Immunomodulation of the neutrophil respiratory burst by endomorphins 1 and 2. Immunol Lett 2000;75:55–9.

107. Hook S, Camberis M, Prout M, Konig M, Zimmer A, Van Heeke G, Le Gros G. Preproenkephalin is a Th2 cytokine but is not required for TH2 differentiation in vitro. Immunol Cell Biol 1999;77:385–90.

108. Wichmann MW, Zellweger R, Ayala A, Chaudry IH. Effect of naloxone on immune responses after hemorrhagic shock. Crit Care Med 2000;28:184–9.

109. Straub RH, Herrmann M, Berkmiller G, Frauenholz T, Lang B, Scholmerich J, Falk W. Neuronal regulation of interleukin 6 secretion in murine spleen: adrenergic and opioidergic control. J Neurochem 1997;68:1633–9.

110. Roy S, Cain KJ, Chapin RB, Charboneau RG, Barke RA. Morphine modulates NF kappa B activation in macrophages. Biochem Biophys Res Commun 1998;245:392–6.

111. Catania A, Lipton JM. The neuropeptide alpha-melanocyte-stimulating hormone: a key component of neuroimmunomodulation. Neuroimmunomodulation 1994;1:93–99.

112. Jessop DS, Jukes KE, Lightman SL. Release of alpha-melanocyte-stimulating hormone from rat splenocytes *in vitro* is dependent on protein synthesis. Immunol Lett 1994;41: 191–94.

113. Taylor AW, Streilein JW, Cousins SW. Immunoreactive vasoactive intestinal peptide contributes to the immunosuppressive activity of normal aqueous humor. J Immunol 1994;153:1080–86.

114. Glyn-Ballinger JR, Bernardini GL, Lipton JM. Alpha-MSH injected into the septal region reduces fever in rabbits. Peptides 1983;4:199.

115. Murphy MT, Richards DB, Lipton JM. Antipyretic potency of centrally administered alpha-melanocyte stimulating hormone. Science 1983;221:192–93.

116. Cannon JG, Tatro JB, Reichlin S, Dinarello CA. Alpha melanocyte stimulating hormone inhibits immunostimulatory and inflammatory actions of interleukin 1. J Immunol 1986;337:2232–36.

117. Dardenne M, Itoh T, Homo-Delarche F. Presence of glucocorticoid receptors in cultured thymic epithelial cells. Cell Immunol 1986;100:112–118.

118. Robertson B, Dostal K, Daynes RA. Neuropeptide regulation of inflammatory and immunologic responses: The capacity of α-melanocyte-stimulating hormone to inhibit

tumor necrosis factor and IL-1-inducible biologic responses. J Immunol 1988;140:4300–4307.

119. Sundar SK, Becker KJ, Cierpial MA. Intracerebro-ventricular infusion of interleukin 1 rapidly decreases peripheral cellular immune responses. Proc Natl Acad Sci USA 1989;6398–402.
120. Goelst K, Mitchell D, Laburn H. Effects of α-melanocyte stimulating hormone on fever caused by endotoxin in rabbits. J Physiol 1991;441:469–476.
121. Taylor AW, Streilein JW, Cousins SW. Alpha-melanocyte-stimulating hormone suppresses antigen-stimulated T cell production of gamma-interferon. Neuroimmunomodulation 1994;1:188–194.
122. Shimizu T, Streilein JW. Influence of alpha-melanocyte stimulating hormone on induction of contact hypersensitivity and tolerance. J Dermatol Sci 1994;8:187–93.
123. Grabbe S, Bhardwaj RS, Mahnke K, Simon MM, Schwartz T, Luger TA. Alpha-melanocyte-stimulating hormone induces hapten-specific tolerance in mice. J Immunol 1996;156:473–8.
124. Macaluso A, McCoy D, Ceriani G. Antiinflammatory influences of α-MSH molecules: central neurogenic and peripheral actions. J Neurosci 1994;14:2377–82.
125. Papadopoulos AD, Wardlaw SL. Endogenous alpha-MSH modulates the hypothalamic-pituitary-adrenal response to the cytokine interleukin-1beta. J Neuroendocrinol 1999;11: 315–9.
126. Buggy JJ. Binding of α-MSH to its G-protein-coupled receptor on B-lymphocytes activates the Jak/STAT pathway. Biochem J 1998;331(Pt 1):211–6.
127. Bhardwaj R, Becher E. Mahnke K, Hartmeer M, Schwarz T, Scholzen T, Luger TA. Evidence for the differential expression of the functional α-MSH MC-1 on human monocytes. J Immunol 1997;158:3378–84.
128. Billfinger TV, Hughes TK, Rodriguez M, Glass R, Casares F, Stefano GB. Hyperstimulation of leukocytes by plasma from cardiopulmonary bypass patients is diminished by α-MSH pretreatment. Int J Cardiol 1996;53Suppl:S47–53.
129. Brzoska T, Kalden DH, Scholzen T, Luger TA. Molecular basis of the α-MSH/IL-1 antagonism. Ann NY Acad Sci 1999;885:230–8.
130. Bhardwaj RS, Schwarz A, Becher E, Mahnke K, Aragane Y, Schwarz T, Luger TA. Pro-opiomelanocortin-derived peptides induce IL-10 production in human monocytes. J Immunol 1996;156:2517–21.
131. Slominski A, Wortsman J, Luger T, Paus R, Solomon S. Corticotropin releasing hormone and proopiomelanocortin involvement in the cutaneous response to stress. Physiol Rev 2000;80:979–1020.
132. Luger TA, Scholzen T. Brzoska T, Becher E, Slominski A, Paus R. Cutaneous immunomodulation and coordination of skin stress responses by α-MSH. Ann NY Acad Sci 1998;840:381–94.
133. Klecha A, Genaro A, Lysionek A, Caro R, Colluccia A, Cremaschi G. Experiemental evidence pointing to the bidirectional interaction between the immune system and the thyroid axis. Int J Immunopharmacol 2000;20:491–500.
134. Blalock JE. Proopiomelanocortin and the immune-neuroendocrine connection. Ann NY Acad Sci 1999;885:161–72.
135. Luger TA, Kalden D, Scholzen TE, Brzoska T. α-MSH as a mediator of tolerance induction. Pathobiology 1999;67:318–21.
136. Luger TA, Schwarz T, Kalden H, Scholzen T, Schwarz A, Brzoska T. Role of epidermal

cell-derived α-MSH in ultraviolet light mediated local immunosuppresion. Ann NY Acad Sci 1999b; 885:209–16.

137. Lipton JM, Catania A. Mechanisms of antiinflammatory action of the neuroimmunomodulatory peptide α-MSH. Ann NY Acad Sci 1998;840:373–80.

Edited by I. Berczi and A. Szentivanyi

The Hypothalamus-Pituitary-Thyroid Axis

ISTVAN BERCZI and ANDOR SZENTIVANYI

Department of Immunology, Faculty of Medicine, University of Manitoba, Winnipeg, Manitoba, R3E 0W3, Canada; and Department of Internal Medicine, Faculty of Medicine, University of South Florida, Tampa, FL 33612, USA

ABSTRACT

Thyrotropin releasing hormone (TRH) is produced in the hypothalamus and also within the immune system. In the pituitary gland, TRH regulates the release of thyroid stimulating hormone (TSH), prolactin and growth hormone. Within the immune system, TRH has an effect on the development of intestinal T lymphocytes, on the production of interleukin-2 and interferon-γ. TSH also effects immunocytes directly and receptors are detectable at low concentration on B and T lymphocytes, monocytes and NK cells; whereas, high levels were detected on dendritic cells. Human monocytes and lymphocytes synthesize TSH. TSH stimulated IL-2 production by lymphocytes, NK cell cytotoxicity and the phagocytic activity and the production of IL-1β and IL-12 by dendritic cells. Bone marrow cells also express TSH receptors and respond to TSH by increased cytokine production. Thyroxin (T4) and tri-iodothyronine (T3) directly effect lymphocytes and monocytes through nuclear receptors. Lymphocytes are able to convert T4 to T3 which regulates their metabolism, maturation and function. The immunological effects of thyroid hormones are diverse and somewhat controversial. While thyroid deficiency is usually, but not always, associated with immune deficiency, supplemental treatment with T3 yielded mostly negative results. Primary B lymphocyte development is defective in mice deficient in the TSH receptor and thyroid receptor α-negative mice. Other hematopoietic lineages and the development of secondary lymphocytes were normal in these mice.

1. THYROTROPIN RELEASING HORMONE (TRH)

Immune cells have receptors for TRH. Human peripheral blood lymphocytes and tonsil-derived leukocytes showed receptor expression in non-activated and phytohemagglutinin (PHA)-activated T and B cells [1].

TRH treatment of rats significantly increased the proliferative response of spleen cells to Con-A [2]. In neonatally thymectomized (NTX) mice intestinal intraepithelial αβ T lymphocytes did not develop, whereas intraepithelial γδ T cells were normal. The treatment of NTX mice with TRH restored the normal development of intestinal epithelial lymphocytes. TRH stimulated T cell development only in the gut and not in the spleen [3]. In rats immunized with a T-cell-dependent antigen the antibody response was critically dependent on the activation of the TRH-PRL axis [4].

In patients serum IL-2 levels were found to be significantly elevated during standard TRH tests, [5]. Five normoprolactinemic women with irregular menstrual cycles were given intravenous bolus of LHRH/TRH . Increased plasma IFN-γ levels were observed, with the maximum response at 45 min, after treatment. LHRH and TRH, separately and together, significantly enhanced IFN-gamma production *in vitro* by staphylococcal enterotoxin A and concanavalin-A (con-A) activated peripheral blood mononuclear cells from the same patients [6].

Repeated TRH administration in critical illness results in a repetitive increase of TSH, PRL, GH, T4, and T3, without increasing reverse T3. These observations suggest that TRH may be used for reversing the euthyroid sick syndrome, GH deficiency, and immune dysfunction associated with critical illness [7].

2. THYROID STIMULATING HORMONE (TSH)

TSH receptor (TSHR) has been studied by immunoprecipitation and by flow cytometric analysis. Trace amounts of TSHR were precipitated from enriched B cells and T cells populations, whereas high levels were precipitated from the dendritic cell fraction. Flow cytometry revealed that only 2–3% of splenic T cells were TSHR+, whereas 10–20% of CD4+8- and CD4-8+ lymph node T cells expressed the TSHR, which was exclusively associated with $CD45RB^{high}$ cells and was not expressed during or after activation. TSHR was not detectable on cells of the immune system during fetal or neonatal life. Recombinant TSHβ significantly enhanced the phagocytic activity of dendritic cells from adult animals and selectively augmented IL-1β and IL-12 cytokine responses of dendritic cells following phagocytic activation. TSH serves as a novel endocrine mediator associated with professional antigen presenting cells and naive T cells [8].

Other investigators also observed that TSH receptors are present at very low concentrations on human monocytes, NK cells and T and B lymphocytes. Receptor number increases in activated B, but not in T, lymphocytes. TSH moderately increased Ig secretion by activated B cells, but had no effect on lymphocyte proliferation [9, 10]. TSH stimulated the production of IL-2 by PHA activated human blood lymphocytes [5], enhanced the proliferative response of murine spleen cells to IL-2, and significantly increased IL-2 induced NK cell cytotoxicity [11]. TSH also enhanced the expression of MHC-II surface molecules by human thyroid epithelial cells [12].

In human monocytes TRH induced TSH release *in vitro,* which could be totally inhibited by the administration of T3 [5]. Human lymphocytes synthesized the β subunit of TSH and the mRNA for TSHβ subunit was regulated by thyromimetic compounds [13].

The TSHR was stimulated on bone marrow cells, using highly purified or recombinant TSH, or by anti-TSHR receptor antibodies. The synthesis of IL-6, IFNβ, TNFα, TNFβ, TGFβ2, and lymphotoxin-β responses were reproducibly induced by TSH within 2–3 hr of stimulation. TSH stimulation of bone marrow cells caused rapid increases in cAMP levels and induced the phosphorylation of the Jak2 protein kinase. This is a novel G-protein-coupled receptor/cytokine synthesis pathway. These findings clearly demonstrate that TSH can serve as a primary inductive signal of cytokine production by bone marrow cells [14].

The addition of dendritic cells purified from Wistar rats to isolated thyroid follicles in culture clearly inhibited their 3H-thymidine uptake, both under basal conditions and after TSH stimulation. In this system DC also suppressed T3 secretion by TSH stimulated follicles. The DC-induced inhibition of thyroid follicle growth was totally abrogated by anti-IL-1β antibodies. Anti-IL-6 had an effect only on the DC inhibition of non-TSH-stimulated thyrocytes. Anti-TNF-α demonstrated no effect at all. Antibodies to MHC and to adhesion molecules had no effect [15].

3. THYROXIN (T4) AND TRIIODOTHYRONINE (T3)

Lymphocytes and monocytes express nuclear receptors for both T3 and T4. Human lymphocytes are able to convert T4 to T3, which is the biologically active form of the hormone. T3 regulates sodium exchange and glucose uptake in lymphocytes, stimulates thymus growth and hormone production, and promotes erythroid burst forming clones and B cell maturation in the bone marrow. Diverse effects of thyroid hormones were reported in various species on mitogen induced lymphocyte proliferation, antibody formation to various antigens, and on various forms of cell mediated immunity, including graft rejection, delayed type hypersensitivity reaction and NK cell activity. Artificially created thyroid deficiency was usually, but not always, associated with immune deficiency, which could be restored by treatment with T3. Hypothyroidism in man was associated with immunodeficiency. The supplemental treatment of normal animals with T3 yielded mostly negative results [16,17,18,19,20,21,22,23].

The cytotoxic activity of splenic NK cells was low in weanling mice fed a low protein diet, but was normal if the diet contained supplemental T3 (0.2 μg/g diet) [24].

The treatment of Con-A stimulated mononuclear cells with T3 (1–100 nM for 3 days) increased the production of soluble interleukin-2 receptor alpha chain (IL2Rα) significantly ($p<0.05$ and $p<0.01$). The stimulatory effect of T3 on IL2Rα chain expression was greater when ConA (5 μg/ml) and IL-2 (1 U/ml) were used together. Similarly, T3 enhanced the stimulation of IL2Rα chain expression by IL-2 in the YT cells (an IL-2 independent natural killer-like cell line) [25].

The effect of TRH, TSH and thyroxine (T4) was examined on intestinal intraepithelial lymphocytes (IEL). IEL in euthymic mice were not affected by TRH and TSH treatment. T4 administered to adult euthymic mice at 3 or 6 weeks of age caused a dramatic reduction in the numbers of TCRαβ, CD8 αβ IEL. These are the same subsets previously shown to be increased by TRH and TSH in athymic mice. T4 had minimal effect on IEL, when given to euthymic mice >8 weeks of age, where the TCRαβ and CD8 αβ IEL subsets had reached normal levels. This indicates that, T4 is directed to developing but not mature IEL. This possibility was confirmed by T4 treatment of bone marrow radiation chimeras during an active phase of T cell regeneration. All IEL development was temporarily halted at a stage characteristic of immature IEL. T4 had no effect on developing thymocytes or on mature peripheral T cells, in either normal euthymic mice or during hematopoietic reconstitution of radiation chimeras [26].

T3 and T4 levels were increased a few days after the immunization of mice with allogeneic lymphoid cells. Treatment of the animals with T4 was shown to increase alloantibody titers during the early stages of immunization and to enhance lymphocyte proliferation in mixed lymphocyte reactions. Conversely, lowering serum thyroid hormone levels by propylthiouracil treatment inhibited the humoral and cellular alloimmune responses [27].

A marked decrease was found in the functional binding of nuclear thyroid hormone receptors to the DR4 thyroid hormone responsive DNA element in the spleens of rats with adjuvant arthritis (AA) or with AA + adrenalectomy [28].

The number and cytolytic activity of NK cells in a group of relatively healthy nonagenarians and centenarians were correlated with endocrine, metabolic and functional parameters that change during ageing. A group of subjects with the highest number of NK cells and the best cytolytic function also presented a well preserved metabolism of thyroid hormones and vitamin

D and integrity of muscle mass. The NK cell number and/or cytolytic activity of healthy subjects > 90 years old correlated positively with serum levels of vitamin D, while T3, FT4, i-PTH hormones and lean body mass were correlated only with NK cell number. These results stress the importance of nutritional evaluation in the clinical assessment of elderly people [29].

On cultured human thyrocytes (TEC) a biphasic, non-cytotoxic and reversible influence of IL-1was found. IL-1 in moderate to high concentrations inhibited thyroid cell function. TNF and IFN-γ supported this physiological role of IL-1 in the regulation of thyroid cell function. IL-1 induced the release of NO and cGMP from TEC. An inhibitor of nitric oxide synthase did not abolish the IL-1-induced inhibition of the release of thyroglobulin and cAMP. IL-1β inhibited the adenylate cyclase mediated pathways and stimulated the guanylate cyclase mediated pathways in TEC. All the demonstrated IL-1 effects were counteracted by IL-1 receptor antagonist indicating, that the effects were exerted through the activation of specific IL-1 receptors on thyrocytes [30].

4. GENETICALLY ALTERED MURINE MODELS

Mice with a mutation in the TSH receptor coded for by the *hyt* gene exert profound hypothyroidism. The gene coding for the thyroid hormone receptor (TR) α and β have been knocked out from other strains of mice, and the two knockout strains have been cross bred. Immunological studies in these mouse strains with deficient thyroid hormone action revealed a defective primary B cell development in *hyt/hyt* and TR-α mice. Other hematopoietic cell lineages and mature B lymphocytes showed no deficiency [31].

REFERENCES

1. Mellado M, Fernandez-Agullo T, Rodriguez-Frade JM, San Frutos MG, de la Pena P, Martinez-A C, Montoya E, Expression analysis of the thyrotropin-releasing hormone receptor (TRHR) in the immune system using agonist anti-TRHR monoclonal antibodies. FEBS Lett 1999;451:308–14.
2. Raiden S, Polack E, Nahmod V, Labeur M, Holsboer F, Larzt E. TRH receptor on immune cells : *in vitro* and *in vivo* stimulation of human lymphocyte and rat splenocyte DNA synthesis by TRH. J Clin Immunol 1995;15:242–49.
3. Wang J, Klein JR. Thymus-neuroendocrine interactions in extrathymic T cell development. Science 1994;265:1860–62.
4. Perez Castro C, Penalva R, Paez Pereda M, Renner U, Reul JM, Stalla GK, Holsboer F, Arzt E. Early activation of thyrotropin-releasing-hormone and prolactin plays a critical role during a T cell-dependent immune response. Endocrinology 1999;140:690–97.
5. Komorowski J, Zylinska K, Pawlikowski M, Stepien H. Stimulatory effect of thyrotropin (TSH) on interleukin-2 (IL-2) release from human peripheral blood lymphocytes. A dose-response study *in vitro*. Hormones Metab Res 1993;25:598–99.
6. Grasso G, Massai L, De Leo V, Muscettola M. The effect of LHRH and TRH on human interferon-gamma production *in vivo* and *in vitro*. Life Sci 1998;62:2005–14.
7. Van den Berghe G, de Zegher F, Vlasselaers D, Schetz M, Verwaest C, Ferdinande P, Lauwers P. Thyrotropin-releasing hormone in critical illness: from a dopamine-dependent test to a strategy for increasing low serum triiodothyronine, prolactin, and growth hormone

concentrations. Crit Care Med 1996;24:590–5.
8. Bagriacik EU, Klein JR. The thyrotropin (thyroid-stimulating hormone) receptor is expressed on murine dendritic cells and on a subset of CD45RB high lymph node T cells: functional role for thyroid-stimulating hormone during immune activation. J Immunol 2000;164:6158–65.
9. Harbour DV, Leon S, Keating C, Hughes K. Thyrotropin modulates B-cell function through specific bioactive receptors. Prog Neuroendocrinimmunol 1990;3:266–76.
10. Coutelier JP, Kehrl JH, Bellur SS. Binding and functional effects of thyroid stimulating hormone on human immune cells. J Clin Immunol 1990;10:204–10.
11. Provinciali M, Di Stefano G, Fabris N. Improvement in the proliferative capacity and natural killer cell activity of murine spleen lymphocytes by thyrotropin. Int J Immunopharmacol 1992;14:865–70.
12. Todd I, Pujol-Borrell R, Hammond LJ, McNally JM, Feldman M, Bottazzo GF. Enhancement of thyrocyte HLA class II expression by thyroid stimulating hormone. Clin Exp Immunol 1987;69:524–31.
13. Peele ME, Carr FE, Baker Jr JR, Wartofsky L, Burman KD. TSH beta subunit gene expression in human lymphocyte. Am J Med Sci 1993;305:1–7.
14. Whetsell M, Bagriacik EU, Seetharamaiah GS, Prabhakar BS, Klein JR. Neuroendocrine-induced synthesis of bone marrow-derived cytokines with inflammatory immunomodulating properties. Cell Immunol 1999;192:159–66.
15. Simons PJ, Delemarre FG, Drexhage HA. Antigen-presenting dendritic cells as regulators of the growth of thyrocytes: a role of interleukin-1beta and interleukin-6. Endocrinology 1998;139:3148–56.
16. Berczi I. Pituitary Function and Immunity. Boca Raton, Florida:CRC Press;1986.
17. Fabris N, Mocchegiani E, Provinciali M. Pituitary-thyroid axis and immune system: a reciprocal neuroendocrine-immune interaction. Horm Res;1995:43:29–38.
18. Haddad EE, Mashaly MM. *In vivo* effects of TRH, T3 and cGH on antibody production and lymphocytes-T and lymphocytes-B proliferation in immature male chickens. Immunol Invest 1991;20:557–68.
19. Stein-Streilein J, Zakarija M, Papic M, McKenzie JM. Hyperthyroxinemic mice have reduced natural killer cell activity: Evidence for a defective trigger mechanism. J Immunol 1987;139:2502–07.
20. Turaihi K, Khan FA, Baron DN, Dandona P. Effect of short term triiodothyronine administration on human leukocyte Fb (K) influx and Na efflux. J Clin Endocrinol Metab 1987;65:1031–34.
21. Schoenfeld PS, Myers JW, Myers L, LaRocque JC. Suppression of cell-mediated immunity in hypothyroidism. South Med J 1995;88:347–49.
22. Ohashi H, Itoh M. Effects of thyroid hormones on the lymphocyte phenotypes in rats: changes in lymphocyte subsets related to thyroid function. Endocr Regul 1994;18:117–23.
23. Cole NA, Allowvan RH, Rodrigues SL, Purdy CW. Influence of triiodothyronine injections on calf immune response to an infectious bovine rhinotracheitis virus challenge and nitrogen balance of lambs. J Amin Sci 1994;72:1263–73.
24. Ingram KG, Crouch DA, Douez DL, Croy BA, Woodward B. Effects of triiodothyronine supplements on splenic natural killer cells in malnourished weanling mice. Int J Immunopharmacol 1995;17:21–31.
25. Nakanishi K, Taniguchi Y, Onji M. Triiodothyronine enhances expression of the

interleukin-2 receptor alpha chain. Endocr J 1999;46:437–42.
26. Wang J, Klein JR. Hormone regulation of murine T cells: potent tissue-specific immunosuppressive effects of thyroxine targeted to gut T cells. Int Immunol 1996;8: 231–5.
27. Klecha AJ, Genaro AM, Lysionek AE, Caro RA, Coluccia AG, Cremaschi GA. Experimental evidence pointing to the bidirectional interaction between the immune system and the thyroid axis. Int J Immunopharmacol 2000;22:491–500.
28. Brtko J, Jurcovicova J, Schmutzler C, Rokyta R, Svik K, Bobal'ova J. Adjuvant arthritis in the rat is associated with decreased binding of nuclear receptors to thyroid hormone responsive element in spleen extracts. Gen Physiol Biophys 1998;17:341–7.
29. Mariani E, Ravaglia G, Forti P, Meneghetti A, Tarozzi A, Maioli F, Boschi F, Pratelli L, Pizzoferrato A, Piras F, Facchini A. Vitamin D, thyroid hormones and muscle mass influence natural killer (NK) innate immunity in healthy nonagenarians and centenarians. Clin Exp Immunol 1999;117:206.
30. Rasmussen AK. Cytokine actions on the thyroid gland. Dan Med Bull 2000;47:94–114.
31. Dorshkind K, Horseman N. The roles of prolactin, growth hormone, insulin-like growth factor-I and thyroid hormones in lymphocyte development and function: insights from genetic models of hormone and hormone receptor deficiency. Endocrine Rev 2000;21: 292–312.

Edited by I. Berczi and A. Szentivanyi

Nerve Growth Factor, Leptin and Neuropeptides

ISTVAN BERCZI and ANDOR SZENTIVANYI

Department of Immunology, Faculty of Medicine, University of Manitoba, Winnipeg, Manitoba, R3E0W3, Canada; and Department of Internal Medicine, Faculty of Medicine, University of South Florida, Tampa, Florida 33612, USA

ABSTRACT

Nerve growth factor (NGF) shares tyrosine kinase receptors with other neurotropic factors. These receptors belong to the TNF receptor superfamily and are present in lymphoid cells and cells of the nervous system. In general, NGF stimulates immune responses with the exception of IgE production which is inhibited. Nerve growth factor is stimulatory for mast cells and neutrophilic leukocytes and locally exerts a proinflammatory effect. However, systematically applied NGF suppresses inflammation. During inflammatory reactions, lymphocyte-derived NGF protects the nervous system and other tissues from damage.

Leptin (LEP) is produced primarily by adipocytes and structurally is related to the GLH cytokine family. LEP signals by a class I cytokine receptor of which two isoforms are known. Leptin regulates energy metabolism, reproductive function and lymphocyte development and function. It stimulates TH-1 mediated immune responses and upregulates phagocytosis and the production of proinflammatory cytokines. During the acute phase response, LEP rises rapidly in response to elevated TNF levels. LEP appears to contribute significantly to survival in sepsis by moderating glucocorticoid and IL-6 production, stimulating IL-1 receptor antagonist and potentiating the immune response.

Arginine vasopressin (AVP) exerts an immunostimulatory effect through V_1 type receptors. AVP is produced by lymphocytes. Substance P (SP) is produced in both the nervous system and in the immune system. SP is a major mediator of neurogenic inflammation and is capable of causing mast cell discharge, increased capillary permeability and smooth muscle construction. SP has a stimulatory effect on immune phenomena in general. Its effect on immunoglobulin secretion is variable. SP stimulates bone marrow cytokines and the formation of granuloma tissue. Calcitonin gene related peptide (CGRP) stimulates mast cell discharge and inflammatory response whereas it has an inhibitory effect on immune reactions. T lymphocytes synthesize CGRP. Somatostatin (SOM) is an antagonist of SP and inhibits inflammatory and immune reactions. Vasoactive intestinal peptide (VIP) is produced within the immune system and the receptors are present on monocytes and lymphocytes. In general, VIP functions as an immunosuppressive neuropeptide and is involved in the immunosuppressive properties of aqueous humour in the eye. Its effect on immunoglobulin secretion and NK cell activity is variable. VIP downregulated the inflammatory response. Pituitary adenylate cyclase (PACAP) activating peptide has a similar immunosuppressive and antiinflammatory effect.

1. NERVE GROWTH FACTOR (NGF)

NGF was discovered in murine submandibular glands as a factor stimulating the growth of sensory and sympathetic ganglia [1]. Low affinity pan-neurotropic receptor (P75) and specific high affinity receptor tyrosine kinases, designated as gp140 trkA for NGF, gp145 trkB primarily for brain derived neurotropic factor (BDNF) and gp145 trkC for neurotropin 3 are known. These receptors belong to the TNF receptor superfamily which includes the Fas and CD40 receptors on lymphoid tissue. trkA is expressed in the thymus and lymph nodes, trkB in the spleen, localized primarily to the stroma of these organs. Some expression was also found in splenocytes and thymocytes. B lymphocytes and follicular dendritic cells also express NGF receptors [2–6].

NGF is mitogenic for human B lymphoblastoid cell lines which is associated with tyrosine phosphorylation, the activation of mitogen activated protein kinase (MAPK) [7]. NGF stimulated cell proliferation in mixed lymphocyte cultures [8], increased the response of rat spleen cells to T and B cell mitogens [9] and the growth of unstimulated human B lymphocytes [10]. The production of IgM and IgG4 by B cells was stimulated by NGF [10,11]. NGF inhibited the induction of IgE by IL-4 in human peripheral B lymphocytes and IL-4 prevented the NGF induced increase in IgA and IgM production [12]. Murine T and B lymphocytes and rat peritoneal mast cells are capable of synthesizing biologically active NGF [13,14].

NGF treatment increased the number and size of mast cells in neonatal mice. It also stimulated histamine release from mast cells, both *in vivo* and *in vitro* [15–19]. NGF stimulated the development of hemopoietic colonies and the chemotactic and phagocytic activity of neutrophilic leukocytes [20,21].

The stimulatory effect of NGF on mast cells and neutrophilic leukocytes suggests a proinflammatory role. However, the suppression of inflammation has been observed *in vivo* by systemically applied NGF in several experimental models [22,23]. The reason for this paradox is not well understood. Recent studies revealed that lymphocyte-derived NGF provides protection for the nervous system and to other host tissues during inflammatory reactions. This phenomenon implies the existence of beneficial 'autoimmune' reactions [23,24].

2. LEPTIN

Leptin (LEP) is produced primarily by fat cells (adipocytes). Structurally LEP belongs to the GLH/CTK family and signals by a class I cytokine receptor (Ob-R). Two receptor isoforms are known: Ob-Ra and Ob-Rb. Leptin regulates energy metabolism, reproductive function, lymphoid development and function. Under normal physiological conditions the secretion of LEP is regulated by insulin, cortisol and sex steroids, mainly testosterone. In rats centrally administered LEP suppressed the mitogenic response of splenic lymphocytes. This was mediated through CRF-sympathetic activation. Leptin plays an important role in linking nutritional state and T cell function. In starving mice, which show immunosuppression, treatment with LEP enhanced TH1-mediated immune responses, in spite of the catabolic state of the animals. Starving animals have reduced LEP levels and show an increased sensitivity to endotoxin shock. Fasting mice respond to LPS with a blunted corticosterone and exaggerated TNF production. This could be corrected by LEP treatment [25,26].

During acute phase responses (eg. sepsis) the serum level of LEP rises rapidly. Cytokines, especially TNF-α, causes this elevation. LEP exerts an inhibitory effect on glucocorticoid and IL-6 production. Blood levels of LEP correlate positively with the survival of patients with septicaemia. LEP stimulates the production of IL-1 receptor antagonist (IL-1ra), which protects against LPS toxicity in mice. In murine glial cells LEP stimulated the production of IL-1β. In animal experiments exogenous LEP upregulated both phagocytosis and the production of proinflammatory cytokines. Leptin is also involved in wound healing and angiogenesis [27–36].

3. NEUROPEPTIDES

3.1. Arginine vasopressin (AVP)

Human peripheral blood mononuclear cells express V_1 type AVP receptors. AVP enhanced the autologous mixed lymphocyte culture, stimulated the production of prostaglandin E2 by human mononuclear phagocytes, and the production of β-endorphin by human peripheral blood mononuclear cells. AVP is antipyretic and is capable of attenuating fever after central administration [37–39]. AVP was detectable in rat thymocytes and splenic lymphocytes [40,41].

3.2. Substance P (SP)

Substance P receptors are present on thymocytes, B and T lymphocytes, macrophages, mast cells and astrocytes. SP is one of the major mediators of neurogenic inflammation and it is capable of inducing mast cell degranulation, plasma extravasation and bronchoconstriction. Uterine mast cells of rats released histamine and cytokines in response to substance P at diestrus, but not during proestrus or on day 4 of pregnancy. In the latter case preincubation of uterine mast cells with SP significantly enhanced mediator release in response to anti-IgE antibodies [42].

SP acts directly on lymphocytes, macrophages and neutrophils. In general, lymphocyte proliferation and lymphokine production are enhanced by SP, whereas its effect on immunoglobulin secretion is variable. SP increased Fcγ and ε receptors and decreased C3b on eosinophils, released TNFα from macrophages and modified macrophage function during stress [43–45]. In polymorphonuclear leukocytes SP stimulated the respiratory burst and chemotactic and phagocytic activities [46–50]. SP increased the release of PGE2 and collagenase from rheumatoid synoviocytes [51] and PGA and thromboxane B2 from astrocytes [52]. SP induced IL-3 and GM-CSF secretion by bone marrow cells. This was partially mediated by the stimulation of IL-1 and IL-6 in marrow cells [53]. SP activated platelets for cytotoxicity against *Shistosoma mansoni* larvae [54]. The SP receptor was necessary for the normal granulomatous response to *Schistosoma mansoni.*

3.3. Calcitonin gene related peptide (CGRP)

Mature lymphocytes, macrophages and bone marrow cells express receptors for CGRP. These receptors are functionally coupled to adenylate cyclase [55,56]. CGRP inhibited the proliferative response of murine thymus cells to Con A. [57]. In the skin CGRP can trigger mast cell discharge, induce slow onset intense erythema and vasodilation [47]. CGRP has an inhibitory effect on antigen presentation, lymphocyte proliferation, IL-2 production, and mRNA synthesis for TNFα, -β and IFN-γ. IFN-γ induced H_2O_2 production by human monocytes was also suppressed by CGRP [58–60].

CGRP containing nerve fibers innervated Langerhans cells in the human epidermis and exerted an inhibitory effect on antigen presentation by these cells [61]. Locally applied CGRP was found to inhibit the induction of cutaneous hypersensitivity reactions in mice by some investigators [62], but others observed increased cutaneous inflammation to allergens and irritants and an enhanced sensitization process [63]. CGRP inhibited the expression of NFκB and promoted apoptosis. T lymphocytes from rat thymus and mesenteric lymph nodes sythesized CGRP [64].

3.4. Somatostatin (SOM)

T and B lymphocytes and mast cells possess receptors for SOM. SOM inhibits the IgE dependent mediator release from human basophils and mast cells. SOM acts as an antagonist of substance P and it has beneficial effects in models of autoimmune disease and of chronic inflammation. Lymphocyte proliferation, endotoxin induced leukocytosis, IgA secretion, and IFNγ production are all inhibited by SOM. Its effect on antibody dependent cytotoxicity is variable. SOM exerts a regulatory influence on macrophages [46,47,65–67].

3.5. Vasoactive intestinal peptide (VIP) and Pituitary adenylate cyclase activating peptide (PACAP)

VIP receptors are present on monocytes and lymphocytes. Human and murine thymocytes were protected by VIP against the cytolytic effect of prednisolone [68]. VIP regulated T cell homing to gut associated lymphoid tissue. VIP stimulated macrophage chemotaxis by PKC activation and inhibited lymphocyte chemotaxis through the activation of adenylate cyclase [69]. VIP enhanced phagocytosis by mouse peritoneal macrophages [70], inhibited lymphocyte mitogenesis and had a variable effect on immunoglobulin secretion and NK cell activity [71,72]. The addition of VIP to cultures of human B lymphocytes from the colonic mucosa increased IgA production and reduced significantly IgG levels. *In vitro* VIP induced IgA production by isotype switching and inhibited the production of IL-2 and IL-4 by murine thymocytes [73–75]. Immunoreactive VIP was detected in rat thymus, spleen and lymph nodes in both lymphoid and non-lymphoid cells [76].

VIP and α-MSH contribute to the immunosuppressive properties of aqueous humour in the eye. Antigen presentation in the eye is deviated towards antibody production. Cell-mediated immune responses are suppressed. SP acted in the aqueous humour as an antagonist of VIP. Antigen or TNFα induced VIP release into the aqueous humour [77,78].

VIP and PACAP inhibited the nuclear translocation of NFκB in stimulated macrophages by inhibiting Jak/Stat phosphorylation. This antagonised the effect of IFN-γ and down-regulated the inflammatory response. The production of TGF-β1, IL-4, -6, -12, TNF-α and nitric oxide (NO) were inhibited by both peptides [79–84]. IL-6 production was enhanced by VIP/PACAP in unstimulated macrophages [Martinez et al. 1998]. VIP and PACAP inhibited antigen induced apoptosis in CD4+ (but not CD8+) T lymphocytes by downregulating Fas [79]. In man topically injected VIP inhibited the development of contact dermatitis to nickel sulphate [85].

REFERENCES

1. Levi-Montalcini R. The nerve growth factor, its mode of action on sensory and sympathetic nerve cells. Harvey Lectures 1966;60:217–259.
2. Brodie C, Gelfand EW. Functional nerve growth factor receptors on human B lymphocytes: Interaction with IL-2. J Immunol 1992;148:3492–97.
3. Pezzati P, Stanisz AM, Marshall JS. Expression of nerve growth factor receptor

immunoreactivity on follicular dendritic cells from human mucosa associated lymphoid tissues. Immunology 1992;76:485–490.
4. Hannestad J, Levanti MB, Vega JA. Distribution of neurotrophin receptors in human palatine tonsils : an immunohistochemical study. J Neuroimmunol 1995;58:131–37.
5. Lomen-Hoerth C, Shooter EM. Widespread neutrophin receptor expression in the immune system and other nonneuronal rat tissues. J Neurochem 1995;64:1780–89.
6. Cosman D. A family of ligands for the TNF receptor superfamily. Stem Cells Dayt 1994;12:440–55.
7. Franklin RA, Brodie C, Melamed I. Nerve growth factor induces activation of MAP-kinase and $p90^{rsk}$ in human B lymphocytes. J Immunol 1995;154:4965–72.
8. Manning PT, Russell JH, Simmons B, Johnson EM. Protection from Guanethidine-induced neuronal destruction by nerve growth factor: effect of NGF on immune function. Brian Res 1985;340:61–69.
9. Thorpe LW, Perez-Polo JR. The influence of nerve growth factor on the *in vitro* proliferative response of rat spleen lymphocytes. J Neurosci Res 1987;18:134–39.
10. Otten U, Ehrhard P, Peck R. Nerve growth factor induces growth and differentiation of human B lymphocytes. Proc Natl Acad Sci USA 1989;86:10059–63.
11. Kimata H, Yoshida A, Ishioka C. Nerve growth factor specifically induces human IgG4 production. Eur J Immunol 1991;21:137–41.
12. Brodie C, Oshiba A, Renz H, Bradley K, Gefland EW. Nerve growth factor and anti-CD40 provide opposite signals for the production of IgE in interleukin-4-treated lymphocytes. Eur J Immunology 1996;26:171–8.
13. Santambrogio L, Beneditti M, Chao M, Muzaffar R, Kulig K, Gabellini N, Hochwald G. Nerve growth factor production by lymphocytes. J Immunol 1994;153:4488–95.
14. Leon A, Buriani A, DalToso R. Mast cells synthesize, store and release nerve growth factor. Proc Natl Acad Sci USA 1994;91:3739–43.
15. Aloe L, Levi-Montalcini R. Mast cells increase in tissues of neonatal rats injected with the nerve growth factor. Brain Res 1977;133:358–366.
16. Bohm A, Aloe L, Levi-Montalcini R. Nerve growth factor enhances precocious differentiation and numerical increase in mast cells in cultures of rat splenocytes. Acad Naz Lineei 1986;80:1.
17. Pearce FL, Thompson HL. Some characteristics of histamine secretion from rat peritoneal mast cells stimulated with nerve growth factor. J Physiol 1986;372:379–93.
18. Marshall JS, Stead RH, McSharry C. The role of mast cell degranulation products in mast cell hyperplasia. I. Mechanism of action of nerve growth factor. J Immunol 1990;144: 1886–92.
19. Bischoff SC, Dahinden CA. Effect of nerve growth factor on the release of inflammatory mediators by mature human basophils. Blood 1992;79:2662–69.
20. Gee AP, Boyle MDP, Munger KL. Nerve growth factor: stimulation of polymorphonuclear leukocyte chemotaxis *in vitro*. Proc Natl Acad Sci USA 1983;80:7213–18.
21. Boyle MDP, Lawman MJP, Gee AP, Young M. Nerve growth factor: a chemotactic factor for polymorphonuclear leukocytes *in vivo*. J Immunol 1985;135:564–68.
22. Banks BEC, Vernon CA, Warner JA. Nerve growth factor has anti-inflammatory activity in the rat hind paw oedema test. Neurosci Lett 1984;47:41–45.
23. Amico-Roxas M, Caruso A, Leone MG, Scifo R, Vanella A, Scapagnini U. Nerve growth factor inhibits some acute experimental inflammations. Arch Int Pharmacodyn Ther 1989;299:269–285.

24. Hohlfeld R, Kerschensteiner M, Stadelmann C, Lassmann H, Wekerle H. The neuroprotective effect of inflammation: implications for the therapy of multiple sclerosis. J Neuroimmunol 2000;107:161–66.
25. Reinshagen M, Rohm H, Steinkamp M, Lieb K, Geerling I, Von Herbay A, Flamig G, Eysselein VE, Adler G. Protective role of neurotrophins in experimental inflammation of the rat gut. Gastroenterology 2000;119:368–76.
26. Matarese G. Leptin and the immune system: how nutritional status influences the immune response. Eur Cytokine Netw 2000;11:7–14.
27. Bornstein SR, Licinio J, Tauchnitz R, Engelmann L, Negrao AB, Gold P, Chrousos GP. Plasma leptin levels are increased in survivors of acute sepsis: associated loss of diurnal rhythm, in cortisol and leptin secretion. J Clin Endocrinol Metab 1998;83:280–3.
28. Giovambattista A, Chisari AN, Corro L, Gaillard RC, Spinedi E. Metabolic, neuroendocrine and immune functions in basal conditions and during the acute-phase response to endotoxic shock in undernourished rats. Neuroimmunomodulation 2000;7:92–8.
29. Okamoto S, Irie Y, Ishikawa I, Kimura K, Masayuki Saito. Central leptin suppresses splenic lymphocyte functions through activation of the corticotropin-releasing hormone-sympathetic nervous system. Brain Res 2000;855:192–97.
30. Chautard T, Spinedi E, Voirol M, Pralong FP, Gaillard RC. Role of glucocorticoids in the response of the hypothalamo-corticotrope, immune and adipose systems to repeated endotoxin administration. Neuroendocrinology 1999;69:360–69.
31. Torpy DJ, Bornstein SR, Chrousos GP. Leptin and interleukin-6 in sepsis. Horm Metab Res 1998;30:726–9.
32. Faggioni R, Fantuzzi G, Gabay C, Moser A, Dinarello CA, Feingold KR, Grunfeld C. Leptin deficiency enhances sensitivity to endotoxin-induced lethality. Am J Physiol 1999;276:R136–42.
33. Takahashi N, Waelput W, Guisez Y. Leptin is an endogenous protective protein against the toxicity exerted by tumor necrosis factor. J Exp Med 1999;189:207–12.
34. Flinck BN, Kelley KW, Dantzer R, Johnson RW. *In vivo* and *in vitro* evidence for the involvement of tumor necrosis factor-alpha in the induction of leptin by lipopolysaccharide. Endocrinology 1998;139:2278–83.
35. Loffreda S, Yang SQ, Lin HZ, Karp CL, Brengman ML, Wang DJ, Klein AS, Bulkley GB, Bao C, Noble PW, Lane MD, Diehl AM. Leptin regulates proinflammatory immune responses. FASEB J 1998;12:57–65.
36. Fantuzzi G, Faggioni R. Leptin in the regulation of immunity, inflammation, and hematopoiesis. J Leukoc Biol 2000;68:437–46.
37. Berczi I. Pituitary Function and Immunity. Boca Raton, FL:CRC Press;1986.
38. Malkinson, TJ, Bridges TE, Lederis K, Veale WL. Perfusion of the septum of the rabbit with vasopressin antiserum enhances endotoxin fever. Peptides 1987;8:385–89.
39. Torres BA, Johnson HW. Arginine vasopressin (AVP) replacement of helper cell requirement in IFN-γ production: evidence for a novel AVP receptor on mouse lymphocytes. J Immunol 1988;140:2179–83.
40. Jessop D, Chowdrey HS, Lightman SL, Larsen PJ. Vasopressin is located within lymphocytes in the rat spleen. J Neuroimmunol 1995a;56:219–23.
41. Jessop D, Murphy D, Larsen PJ. Thymic vasopressin (AVP) transgene expression in rats: a model for the study of thymic AVP hyper-expression in T cell differentiation 1995;62: 85–90.
42. Cocchiara R, Albeggiani G, Azzolina A, Bongiovanni A, Lampiasi N, DiBlasi F, Geraci D.

Effect of substance P on uterine mast cell cytokine release during the reproductive cycle. J Neuroimmunol 1995;60:107–15.
43. Lee HR, Ho WZ, Douglas SD. Substance P augments tumor necrosis factor release in human monocyte-derived macrophages. Clin Diagn Lab Immunol 1994;1:419–223.
44. Nair MPN, Schwartz SA. Immunomodulatory effects of corticosteroids on natural killer and antibody-dependent cellular cytotoxic activities of human lymphocytes. J Immunol 1984;132:2876–82.
45. Chancellor-Freeland C, Zhu GF, Kage R, Beller DI, Leeman SE, Black PH. Substance P and stress-induced changes in macrophages. Ann NY Acad Sci 1995;771:472–84.
46. Payan DG, McGillis JP, Renold FK, Mitsuhashi M, Goetzl EJ. The immunomodulating properties of neuropeptides. In: Berczi I, Kovacs K, editors. Hormones and Immunity. Lancaster, England: MTP Press;1987.
47. Foreman, JC. Neuropeptides and the pathogenesis of allergy. Allergy 1987;42:1–11.
48. Shanahan F, Lee TDG, Bienenstock J, Befus AD. Mast cell heterogeneity: Effect of anti-allergic compounds on neuropeptide-induced histamine release. Int Arch Allergy Appl Immunol 1986;80:424–26.
49. Jancso G, Obal Jr. F, Toth-Kasa I, Katona M, Husz S. The modulation of cutaneous inflammatory reactions by peptide-containing sensory nerves. Int J Tissue React 1985;7: 449–57.
50. Serra M, Bazzoni F, Della Bianca V, Greskowiak M, Rossi F. Activaiton of human neutrophils by substance P: Effect on oxidative metabolism, exocytosis, cytosolic Ca^{2+} concentration and inositol phosphate formation. J Immunol 1988;141:2118–24.
51. Lotz M, Carson DA, Vaughan JH. Substance P activation of rheumatoid synoviocytes: Neural pathway in pathogenesis of arthritis. Science 1987;235:893–95.
52. Hartung HP, Heininger K, Schafer B, Toyka KV. Substance P and astrocytes: stimulation of the cyclooxygenase pathway of arachidonic acid metabolism. Fed Am Soc Exp Bio J 1988;2:48–51.
53. Rameshar P, Ganea D, Gascon P. Induction of IL-3 and granulocyte-macrophage colony-stimulating factor by substance P in bone marrow cells is partially mediated through the release of IL-1 and IL-6;J Immunol 1994;152:4044–54.
54. Damonneville M, Monte D, Auriault C, Capron A. The neuropeptide substance-P stimulates the effector functions of platelets. Clin Exp Immunol 1990;81:346–51.
55. Mullins MW, Ciallella J, Rangnekar V, McGillis JP. Characterization of a calcitonin gene-related peptide (CGRP) receptor on mouse bone marrow cells. Regul Pept 1993;49: 65–72.
56. Abello J, Kaiserlian D, Cuber JC, Revillard JP, Chayvialle JA. Characterization of calcitonin gene-related peptide receptors and adenylate cyclase response in the murine macrophage cell line P388 D1. Neuropeptides 1991;19:43–49.
57. Bulloch K, McEwen BS, Diwa A, Baird S. Relationship between dehydroepiandrosterone and calcitonin gene-related peptide in the mouse thymus. Am J Physiol 1995;268:E168–73.
58. Nong YH, Titus RG, Ribeiro JMC, Remold HG. Peptides encoded by the calcitonin gene inhibit macrophage function. J Immunol 1989;145:45–49.
59. Wang F, Millet I, Bottomly K, Vignery A. Calcitonin gene-related peptide inhibits interleukin 2 production by murine T lymphocytes. J Biol Chem 1992;267:21052–57.
60. Wang J, Klein JR. Thymus-neuroendocrine interactions in extrathymic T cell development. Science 1994;265:1860–62.

61. Hosoi J, Murphy GF, Egan CL, Lerner EA, Grabbe S, Asahina A, Granstein RD. Regulation of Langerhans cell function by nerves containing calcitonin gene-related peptide. Nature 1993;363:159–63.
62. Asahina A, Hosoi J, Beissert S, Stratigos A, Granstein RD. Inhibition of the induction of delayed-type and contact hypersensitivity by calcitonin gene-related peptide. J Immunol 1995;154:3056–61.
63. Gutwald J, Goebeler M, Sorg C. Neuropeptides enhance irritant and allergic contact dermatitis. J Invest Dermatol 1991;96:695–98.
64. Millet I, Phillips RJ, Sherwin RS, Ghosh S, Voll RE, Flavell RA, Vignery A, Rincon M. Inhibition of NF-kappaB activity and enhancement of apoptosis by the neuropeptide calcitonin gene-related peptide. J Biol Chem 2000;275:15114–21.
65. Foris G, Gyimesi E, Komaromi I. The mechanism of antibody-dependent cellular cytotoxicity stimulation by somatostatin in rat peritoneal macrophages. Cell Immunol 1985;90:217–25.
66. Muscettola M, Grasso G. Somatostatin and vasoactive intestinal peptide reduce interferon gamma production by human peripheral blood mononuclear cells. Immunobiology 1990;180:419–30.
67. Church MK, El-Lati S, Caulfied JP. Neuropeptide-induced secretion from human skin mast cells. Int Arch Allergy Appl Immunol 1991;94:310–18.
68. Ernstrom U, Gafvelin G, Mutt V. Rescue of thymocytes from cell death by vasoactive intestinal peptide. Regul Pept 1995;57:99–104.
69. de la Fuente M, Delgado M, del Rio M, Garrido E, Leceta J, Hernanz A, Gomariz RP. Vasoactive intestinal peptide modulation of adherence and mobility in rat peritoneal lymphocytes and macrophages. Peptides 1994;15:1157–63.
70. Ichinose M, Asai M, Imai K, Sawada M. Enhancement of phagocytosis in mouse macrophages by pituitary adenylate cyclase activating polypeptide (PACAP) and related peptides. Immunopharmacology 1995;30:217–24.
71. Ottaway CA. *In vitro* alterations of receptors for vasoactive intestinal peptide changes the *in vivo* localization of mouse T cells. J Exp Med 1984;160:1054–69.
72. O'Dorisio MS. Biochemical characteristics of receptors for vasoactive intestinal pollypeptide in nervous endocrine and immune systems. Fed Proc 1987;46:192–95.
73. Kimata H, Yoshida A. Effect of growth hormone and insulin-like growth factor-I on immunoglobulin production by and growth of human B cells. J Clin Endocrinol Metab 1994;78:635–41.
74. Xin Z, Tang H, Ganea D. Vasoactive intestinal peptide inhibits interleukin (IL)-2 and IL-4 production in murine thymocytes activated via the TCR/CD3 complex. J Neuroimmunol 1994;54:59–68.
75. Boirivant M, Fais S, Annibale B, Agostini D, Delle Fave G, Pallone F. Vasoactive intestinal polypeptide modulates the *in vitro* immunoglobulin: A production by intestinal lamina propria lymphocytes. Gastroenterology 1994;106:576–82.
76. Leceta J, Martinez MC, Delgado M, Garrido E, Gomariz RP. Lymphoid cell subpopulations containing vasoactive intestinal peptide in the rat. Peptides 1994;15:791–97.
77. Taylor AW, Streinlein JW, Cousins SW. Immunoreactive vasoactive intestinal peptide contributes to the immunosuppressive activity of normal aqueous humor. J Immunol 1994;153:1080–86.
78. Ferguson TA, Fletcher S, Herndon J, Griffith TS. Neuropeptides modulate immune deviation induced via the anterior chamber of the eye. J Immunol 1995;155:1746–56.

79. Delgado M, Ganea D. Inhibition of IFN-gamma-induced janus kinase-1-STAT1 activation in macrophages by vasoactive intestinal peptide and pituitary adenylate cyclase-activating polypeptide. J Immunol 2000; 165:3051–7.
80. Sun W, Tadmori I, Yang L, Delgado M, Ganea D. Vasoactive intestinal peptide (VIP) inhibits TGF-beta1 production in murine macrophages. J Neuroimmunol 2000;107: 88–99.
81. Xin Z, Sriram S. Vasoactive intestinal peptide inhibits IL-12 and nitric oxide production in murine macrophages. J Neuroimmunol 1998; 89:206–12.
82. Kawamura N, Tamura H, Obana S, Wenner M, Ishikawa T, Nakata A, Yamamoto H. Differential effects of neuropeptides on cytokine production by mouse helper T cell subsets. Neuroimmunomodulation 1998;5:9–15.
83. Martinez C, Delgado M, Pozo D, Leceta J, Calvo JR, Ganea D, Gomariz RP. Vasoactive intestinal peptide and pituitary adenylate cyclase-activating polypeptide modulate endotoxin-induced IL-6 production by murine peritoneal macrophages. J Leukoc Biol 1998; 63:591–601.
84. Tang H, Sun L, Xin Z, Ganea D. Down-regulation of cytokine expression in murine lymphocytes by PACAP and VIP. Ann N Y Acad Sci 1996;805:768–78.
85. Lundeberg L, Mutt V, Nordlind K. Inhibitory effect of vasoactive intestinal peptide on the challenge phase of allergic contact dermatitis in humans. Acta Derm Venereol 1999;79: 178–82.

Edited by I. Berczi and A. Szentivanyi

Cytokines and Chemokines

ISTVAN BERCZI and ANDOR SZENTIVANYI

Department of Immunology, Faculty of Medicine, University of Manitoba, Winnipeg, Manitoba, R3E 0W3, Canada; and Department of Internal Medicine, Faculty of Medicine, University of South Florida, Tampa, Florida 33612, USA

ABSTRACT

Cytokines are locally acting tissue hormones that normally function as autocrine and paracrine regulators. Growth, differentiation, function, inhibition and apoptosis of cells in various tissues are all regulated by cytokines.

Type I Cytokines: Interleukin (IL)-2, -4, -7, -9 and -15 are immunoregulatory cytokines that are recognized by type I membrane bound glycoprotein receptors. These receptors are heterodiamers consisting of a cytokine specific chain and a common γ_c chain which is involved in signaling. IL-3, -5 and granulocyte-macrophage colony stimulating factor (GM-CSF) are hemopoietic cytokines with similar receptors that share a signaling β_c chain. Interleukin –6, IL-11, oncostatin M (OSM), leukemia inhibitory factor (LIF), ciliary neurotropic factor (CNTF) and cardiotrophin-1 have diverse targets that range from the hematopoietic and immune systems to the cardiovascular and central nervous systems. Receptors for these cytokinases share glycoprotein (gp) 130. Interleukin-12 promotes cellular immunity and hemopoiesis. It is a covalent dimer of a 35 and 50 kDa peptide, the latter is analogous to the soluble IL-6Rα chain. IL-21 and 23 are new members of this cytokine family.

Type II Cytokines: Type I interferons (IFNα, β, ω) and the Type II IFNγ belong to this group. Type I interferons share receptors and induce the same signals. They exert anti-proliferative and anti-viral effects and stimulate cytolysis by lymphocytes, NK cells and macrophages. Interferon-γ forms a homodimer (34kDa). The functional receptor consists of 2 chains (IFNGR-1 and IFNGR-2). The IFNγ receptor is ubiquitously expressed even by platelets.

Interleukin-10 is an inhibitory cytokine regulating the growth and differentiation of B cells, T cells, mast cells, granulocytes, dendritic cells, keratinocytes and endothelial cells. The IL-10 receptor is related to the IFN receptor. IL-19, 20 and 21 are newly discovered cytokines related to IL-10.

Proinflammatory Cytokines: 1) Tumor necrosis factor family – Tumor necrosis factor (TNF), lymphotoxins (LTα, β) belong in this family. TNF and LTα share receptors though LTβ has a specific receptor as well. These receptors, the p75 nerve growth factor receptor and a number of other related receptors share a "death domain" and are capable of inducing apoptosis. LTα and other TNF ligands bind to their receptors as trimers. TNF exists both as a soluble and membrane bound protein and LTβ is membrane bound. Numerous cells can produce TNF in response to noxious agents, infections or immune stimuli. These cytokines regulate the growth of some target cells while inhibit others, promote inflammation and exert a neuroendocrine and a metabolic effect during acute phase reactions.

2) The interleukin-1 family – IL-1α, IL-1β and the IL-1 receptor antagonist (IL-1ra) belong in this family. These cytokines coordinate the response to infection and injury. IL-1ra is the only known natural cytokine receptor antagonist. Pro-IL-1 is stored in the cytoplasm of various cells and needs to be enzymatically activated before it can be secreted. Various enzymes, including the IL-1β converting enzyme (ICE), can activate pro-IL-1. IL-1 is secreted in response to diverse stimuli, including infection, immunological and inflammatory processes and endogenous mediators of cell and tissue injury. Two receptors, IL-1RI and IL-1RII and a receptor associated protein, IL-1RacP, with putative signal transducing function, are known. The type II receptor serves as a "decoy" receptor with no signaling capacity. Immune and inflammatory cells express receptors. IL-1 is a pleiotropic cytokine that coordinates host response to infection and injury. IL-1 is a major messenger between the neuroendocrine and the immune systems. Interleukin 18 is a new member of the IL-1 family of cytokines.

3) Chemokines (CEM) – Chemokines are the largest cytokine group showing chemotactic and proinflammatory properties. They play a major role in acute and chronic inflammation. They also promote immune responses, contribute to lympho- and hematopoiesis, vascularization, cell-to-cell adhesion, leukocyte recirculation and homing. Injury, polyclonal lymphocyte stimulation, antigens and endogenous cytokines stimulate CEM production. Chemokines are classified into 4 groups on the basis of their receptor specificity: CXC(α), CC(β), C(γ) and CX3C. Chemokine receptors are seven pass transmembrane proteins and are coupled with G-proteins. Chemokines are stored intracellularly as inactive precursors and have to be enzymatically cleaved before secretion in their active form. Interleukin-16 and –17 are novel members of the chemotactic and proinflammatory cytokine family.

4) Transforming growth factor-β (TGF-β) – This superfamily includes over 30 members, all of which are multifunctional cytokines regulating growth, differentiation, inflammation and immune function. Three isotypes are present in mammals: TGF-β2 and TGF-β3 are important in embryogenesis and cellular differentiation, whereas TGF-β1 is an immunoregulator. TGF-β is produced as an inactive protein and is bound to cell surfaces or to matrix components before it is activated enzymatically. Type I, II and III receptors exist. The first two are transmembrane serine/threonine kinases that are involved in signal transduction; type III receptors (beta glycans) present TGF-β to receptors I and II. TGF-β is a powerful immunosuppressor and has a systemic anti-inflammatory effect. It promotes the induction of oral immunological tolerance, promotes tissue repair and wound healing and regulates cell proliferation and differentiation. Macrophage inhibitory protein-1 (MIC-1) is a divergent member of the TGF-β superfamily.

5) Migration inhibitory factor (MIF) – MIF is a pleiotropic cytokine that is produced within the immune and neuroendocrine systems, including the pituitary gland. MIF is secreted by the pituitary gland during infection and injury. It regulates glucocorticoid action systemically and locally and it is a major regulator of immune function. MIF promotes cytokine production, angiogenesis, and the immune/inflammatory process.

1. INTRODUCTION

Cytokines (CTK) were originally discovered within the immune system as humoral mediators between leukocytes (interleukins, IL). Now it is clear that cytokines are produced in all tissues and organs in the body. Under physiological conditions these mediators function locally regulating either the cytokine-producing cell itself (autocrine regulation) or neighboring target cells (paracrine regulation). This regulatory effect is mediated by specific cell surface receptors on target cells. Cytokines exert a wide range of actions on their targets including growth stimulation, differentiation, cytolytic activity, apoptosis and chemotaxis [1,2].

2. TYPE I CYTOKINES

Type I cytokines share a similar 4-α helical structure. The receptors for these cytokines also have common features and now have been designated as the type I cytokine receptor superfamily. Other regulatory peptides, including growth hormone (GH), prolactin (PRL), erythropoietin, thrombopoietin and leptin have a similar structure to type I cytokines and their receptors also belong to the same receptor superfamily.

Type I cytokines may be further divided as either short-chain or long-chain for α-helical bundles. Short chain CTK include IL-2, -3, -4, and -5 granulocyte macrophage colony-stimulating factor (GM-CSF), IL-7, -9, -13, and -15 monocyte-CSF (M-CSF) and stem cell factor (SCF). Long chain hormone/cytokines include GH, PRL, erythropoietin (EPO), thrombopoietin, leptin, IL6-11, leukemia inhibitory factor (LIF), oncostatin M (OSM), ciliary neurotropic factor (CNTF), cardiotrophin-1 (CT1) and G-CSF [1,2].

Interleukin 21 is produced by CD3+ T lymphocytes. It has a four helix bundle structure and is closely related to IL-2 and IL-15. The receptor for this cytokine has been demonstrated to be the IL-2αR. *In vitro* experiments indicated that IL-2αR has a role in the proliferation and maturation of natural killer (NK) cell populations from bone marrow, in the proliferation of mature B-cell populations co-stimulated with anti-CD40, and in the proliferation of T cells co-stimulated with anti-CD3 [3]

2.1. Type I Cytokine receptors (CTK-IR)

Type I Cytokine receptors are type I membrane-spanning glycoproteins with extracellular N-terminal and intracellular C-terminal regions. An exception to this is the CNTF receptor α-chain which has no cytoplasmic domain and is anchored by glycosylphosphoinositol (GPI). Four conserved cysteine residues and a tryptophan residue located two amino acids C-terminal from the second conserved cysteine, are characteristic of the extracellular domain of these receptors. In addition, a membrane proximal WSXWS (trp-ser-X-trp-ser) motif was found to be generally conserved, except for the growth hormone receptor where the motif is a YGEFS (tyr-gly-glu-phe-ser) sequence. Yet another shared feature of type I CTK receptors is the presence of fibronectin type III domains. Although the various receptor molecules belonging to this group have an extremely limited sequence identity they form similar structures based on the known structures of the GH, PRL and erythropoietin receptors. The short chain cytokines that signal to homodimeric receptors are SCF and M-CSF, yet in reality, the receptors (c-kit and CSF-1R respectively) are different from Type I CTK receptors in that they contain intrinsic tyrosine kinase domains. Most cytokines have heterodimeric receptors and in several distinct groups, at least one common receptor chain is shared [1,2,4].

2.1.1. Cytokines that signal through γ_c

The cytokines IL-2, IL-4, IL-7, IL-9 and IL-15 share a common cytokine receptor γ-chain, γ_c (CD132). IL-2 is a peptide 133 amino acids in length and is produced exclusively by activated T lymphocytes. It functions as a major T cell growth factor, however, it can also increase immu-

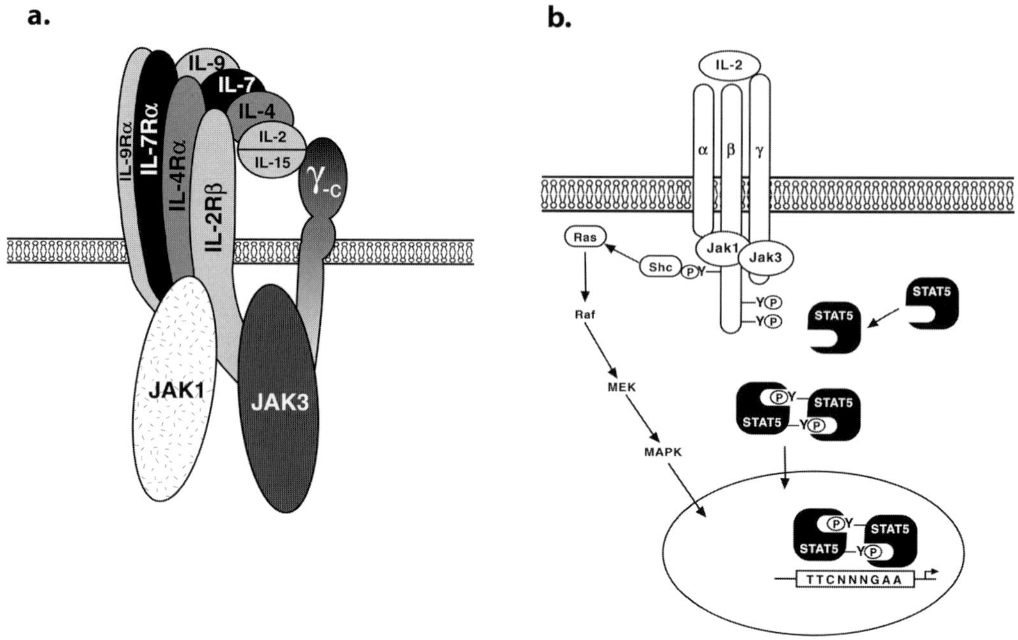

Figure 1A. Schematic Representation of Type I Cytokine Receptors. 1B.Three-Chain Receptor for IL-2. Interleukins 2, 4, 7, 9 and 15 share the γc chain IL-2Rα and IL-15Rα are not shown. The distinctive chains are associated with Jak1 and γc with Jak3. Mutations of γc cause x-linked severe combined or more moderate combined immunodeficiency diseases. Mutations of Jak3 cause an autosomal recessive form of severe combined immunodeficiency (Modified from Warren 1999).

noglobulin synthesis and J-chain transcription in B cells. It augments the cytolytic activity of NK cells and induces cytotoxicity by lymphokine-activated killer (LAK) cells. The antigen receptor determines the specificity of the T cell response whereas IL-2 determines the magnitude and the duration of the response by acting on high affinity receptors. IL-2 may act either by an autocrine or a paracrine fashion. IL-2 recognizes three different classes of receptors formed by three different chains: IL-2Rα, IL-2Rβ and a third chain initially called IL-2Rγ, but now it is recognized as the common cytokine receptor γ-chain (γ_c) [1,5,6].

IL-4 is produced principally by activated CD-4$^+$ T cells (Th2), CD-4$^+$NK1.1$^+$ natural T cells and by mast cells and basophils. Interleukin 4 is the major B cell growth factor and is vital for immunoglobulin class switch by enhancing the production of IgG-1 and IgE. IL-4 can induce the expression of class II MHC molecules and increase the expression of Fcε RII on B cells. IL-4 also functions as a T cell growth factor and acts as an antagonist of IL-2. In addition, IL-4 exerts actions on macrophages, hematopoietic precursor cells, stromal cells and fibroblasts. The IL-4 receptor on T cells and other hematopoietic cells consist of 140 kDa, IL-4Rα protein and γ_c. This is known as a type I IL-4 receptor. An alternate receptor comprising IL-4Rα and IL-13Rα1 is expressed on a number of cell types other than T cells [1,7–9].

IL-7 is produced by stromal cells, B cells, Moc/Mf, dendritic cells, keratinocytes and gut epithelial cells. It promotes the growth, survival and differentiation of thymocytes and has an influence on the growth of immature T cells, especially of γδ T cells. It is a co-factor in myelopoiesis, activates monocyte/macrophages and NK cells. Interleukin-7 transgenic mice develop acute and chronic colitis. IL-7 is vital for the growth of murine pre-B cells, but not of human pre-B cells [10–13].

IL-9 is an 126 amino acid long peptide that supports the growth of activated helper T (Th2) but not cytolytic T lymphocytes. It is produced much later than IL-2 during immune reactions. It also exerts an effect on erythro-progenitors, B cells, mast cells and fetal thymocytes. IL-9 synergizes with IL-3, and it has a major anti-apoptotic effect on thymic lymphocytes. The IL-9 receptor consists of 64 kDa IL-9Rα binding protein plus γ_c [1,14].

IL-15 has been recently discovered. Although mRNA is produced by a range of non-lymphocytic cell types in the body, the protein is seldom detected. IL-15 receptors are expressed widely and on T cells it contains IL-2Rβ, γ_c and IL-15Rα. An alternate form of IL-15R denoted as IL-15RX has been detected on mast cells [1,15,16].

The type I cytokines collectively exhibit overlapping roles related to T and B lymphocyte NK and mast cell activation and play key roles in the normal development and function of these cells. The γ_c chain is located on the X chromosome, and its mutation is responsible for X-linked severe combined immunodeficiency (XSCID) [1,5].

2.1.2. Cytokine Receptors with a Common Beta Chain (β_c)

Cytokine receptors with a common beta chain include the hematopoietic cytokines IL-3, IL-5 and GM-CSF that are synthesized by T lymphocytes. These pluripotent cytokines are vital for the proliferation and differentiation of myeloid precursor cells. IL-3 is the most pluripotent, stimulating most lineages in the bone marrow. These include stem cells, macrophages, eosinophils, mast cells, megakaryocytes and erythroid lineages. IL-3 also enhances phagocytosis and cytotoxicity. GM-CSF is mainly restricted to the granulocyte and monocyte-macrophage lineages and serves as both a growth and survival factor. It also stimulates the growth of dendritic cells and thereby enhances antigen presentation. IL-5 stimulates primarily eosinophilic lineages and the release of eosinophils from the bone marrow. It enhances the killing of *Schistosoma mansoni* by eosinophils, induces immunoglobulin production, B cell activation, but does not stimulate human B cell growth. IL-5 plays a key role in eosinophilic inflammation of the lung and in bronchial asthma [17,1,18,19,20].

Eosinophilic progenitors express receptors for IL-3, IL-5, GM-CSF and murine *pre-B* cells for IL-3 and IL-5. In these cells, the signals induced by these different receptors are indistinguishable. Each of these three cytokines has its own unique 60 to 80 kDa α receptor chain (IL-3Rα, IL-5Rα and GM-CSFRα, respectively) and shares a common 120 to 130 kDa β_c receptor chain. The α chains have the binding capacity for the cytokines, whereas the β_c chain augments binding affinity, but does not exhibit any binding in the absence of the proper α chain. The α chains have relatively short cytoplasmic domains (approximately 55 amino acids) and are unlikely to play major roles in signaling. The β_c chain has a cytoplasmic domain of 432 amino acids which is involved in signaling. One of the interesting features of these hemopoietic cytokines is that there appears to be a considerable redundancy in their function, so that knockout mice that lack GM-CSF, IL-3 or IL-5 actions still exhibit relatively normal hematopoiesis. These observations do not diminish the importance of these particular cytokines in hematopoiesis, but rather underscore the substantial redundancy for a particularly important set of functions. It is noteworthy that these mice are sensitive to some infectious diseases, which suggests the role of these hematopoietic cytokines in immune functions [1,21,22].

2.1.3. Cytokine Receptors That Share gp130

Six cytokines are known to belong to the cytokine receptor family that shares gp 130, which is sometimes referred to as the IL-6 family. These include IL-6, IL-11, OSM, LIF, CNTF and CT-1. These are multifunctional cytokines that act on diverse targets ranging beyond the hematopoietic and immune system and include the central nervous system and the cardiovascular system [1,23,24].

IL-6 was first discovered as a B cell differentiation factor, but now is known to stimulate T cell growth and differentiation, induce monocyte differentiation into macrophages, induce acute phase protein synthesis by hepatocytes, exert actions on keratinocytes, mesangial cells, hematopoietic stem cells and to promote the development of osteoclast and the neural differentiation of PC-12 cells. IL-6 binds to an 80 kDa binding protein denoted as IL-6 Rα which has an 82 amino acid long cytoplasmic domain. This IL-6Rα chain interacts with the 130 kDa signal transducing molecule, gp130, which can form a functional IL-6 receptor together with IL-6Rα. The gp130 molecule is the common receptor component for these cytokines. It contains a total of 6 fibronectin Type III modules, 4 conserved cysteine residues and the WSXWS motif located in the sequence, starting from the N terminus. The cytoplasmic domain of IL-6Rα is superfluous for signaling. A soluble form of IL-6Rα exocellular domain is sufficient for ligand binding and coordination with gp130. Thus, independent of soluble IL-6Rα and IL-6, many cell types that express gp130, but not IL-6Rα are capable of receiving signals. It was observed that IL-6 signaling requires the trimerization of gp130 [1,25,26].

IL-11 is produced by stromal cells and it influences hematopoiesis particularly in combination with IL-3. It also stimulates the proliferation of lymphoid and hematopoietic progenitor cells, megakaryocytic progenitors and megakaryocyte maturation stimulates erythroid progenitors. IL-11 can induce acute phase proteins and augment antigen specific B cell responses. IL-11 is anti-inflammatory and inhibits adipogenesis. The IL-11 receptor consists of IL-11Rα and gp130. The IL-11Rα mRNA is spliced to yield a form lacking the cytoplasmic domain (IL-11Rα2). A soluble form of IL-11Rα2 can coordinate with IL-11 to signal cells expressing gp130 [1,27,28].

Leukemia inhibitory factor suppresses the differentiation of pluripotent embryonic stem cells, inhibits adipogenesis and induces monocyte differentiation of the M-1 murine leukemia cell line. It exerts a number of actions in the central nervous system and it is identical with cholinergic neural differentiation factor. This is a factor that can induce acetylcholine synthesis while simultaneously suppressing catecholamine production. This leads to the induction of cholinergic function while suppressing noradrenergic function. LIF has a hematopoietic effect, affects bone cell metabolism, energy homeostasis and reproduction. The LIF receptor consists of the heterodiamerization of LIFRβ and gp130 [1,29,30,31].

Ciliary neurotropic factor was discovered on the basis of its ability to promote neural survival. It is produced by glial cells, and acts on glia, hepatocytes, skeletal muscle, stem cells, bone marrow stromal cells, lactosomatotroph cells, pituitary stellate cells, Sertoli cells and gonocytes. Rat brain mast cells, but not peritoneal mast cells, released histamine in response to neurotropins with an order of BDNF > CNTF > NGF. Intracerebroventricular infusion of CNTF to male rats for 6 to 7 days (2.5 μg/day) reduced body weight. Chronic infusion of higher doses (5.0 μg/day) produced sustained anorexia (2.9% body weight loss). CNTF receptors consist of LIFRβ and gp130, but also require a specific binding protein now denoted as CNTFRα. This latter chain lacks transmembrane and cytoplasmic domains, but rather it is a GPI-linked receptor molecule. The CNTFRα does not require its own cytoplasmic domain for signaling [1,32–36].

Oncostatin M was identified by its ability to inhibit the growth of human breast carcinoma, glioma and melanoma cells. OSM is a growth regulator and is a potent growth factor for Kaposi's sarcoma in patients with AIDS. OSM supports the development of the liver and hematopoiesis in the fetus, participates in bone remodeling, and supports the proliferation, differentiation and survival of a wide variety of cells. OSM downregulates inflammation and participates in the acute phase response. Human OSM binds directly to gp130 and signals through a receptor combination of gp130 and LIFRβ. However, OSM also signals by an alternative receptor consisting of specific OSMRβ and gp130. Murine OSM acts only through this second type of receptor [1,37–40].

The common chains of γ_c, β_c and gp130 in cytokine receptors make it possible to share signaling pathways, whereas the cytokine binding chain exerts unique specificity. The leptin receptor and IL-12 receptor also resemble gp130 [41,42].

Cardiotrophin-1 (CT-1) induces hypertrophy of cardiac muscle, stimulates the survival of cardiac and neural cells. CT-1 belongs to the IL-6 family and signals by the p42/p44 MAPK and Jak/STAT pathways. CT-1 signals through LIFRβ and gp130. Circulating CT-1 is produced by the heart. The level of CT-1 increases in response to heart failure, peripheral neural injury or during endotoxin shock. CT-1 protected rats against endotoxin induced injury of the lung, it is cardioprotective and reduces blood pressure by the stimulation of inducible nitric oxide synthase (iNOS). CT-1 is highly expressed in embryonic skeletal muscle; it is secreted by myotubes and promotes the survival of embryonic motoneurons. CT-1 stimulates the production of atrial natriuretic peptide in the heart.

In the mouse, the CT-1 gene is 5.4 kilobases in length and has three exons and two introns. This gene was stimulated in cardiac myocytes by epinephrine. Promoter analysis revealed a cAMP responsive element (CRE) in the gene. Oncostatin M, LIF and CT-1 induce osteoclast activation and differentiation, which is supported synergistically by glucocorticoids. Oncostatin M, but not LIF or CT-1 stimulated [43–53].

Cytokine-induced inhibitors of the Jak-STAT signal transduction pathway have been discovered. These include the cytokine-inducible SH2 protein (CIS), suppressor of cytokine signaling (SOCS) and STAT-induced STAT inhibitor (SSI). Jak-binding protein (JAB/SOCS-1/SS-1) and CIS-3/SOCS-3/SSI-3 serve as endogenous inhibitors of CT-1-mediated Jak-STAT signaling in the cardiovascular system [54].

IL-12 is primarily produced by phagocytic cells, in response to bacterial and intracellular parasites, however, it may also be produced by B cells and neutrophils. IL-12 induces the production of IFNγ by NK cells and T cells and serves as a growth factor for activated NK and T cells but not for resting cells. IL-12 also induces IL-2, IL-3, GM-CSF, IL-9, TNFα and MCSF. IL-12 is important for both the innate immune response and adaptive immunity. It also supports the proliferation and survival of hematopoietic stem cells. Structurally, it is a covalent dimer of 35 and 40 kDa peptides. The p40 chain is a functional homologue of the soluble p80 IL-6Rα chain. Thus, it appears that the cytokine-binding part of the receptor has become part of the cytokine. Mutations of IL-12 RβI and IL-12p40 have been described in man [1,55–58].

A novel sequence was discovered that appears distantly related to the p35 subunit of IL-12. This factor, which is termed p19, shows no biological activity by itself, however, it combines with the p40 subunit of IL-12 to form a novel, biologically active, composite cytokine, termed IL-23. Activated dendritic cells secrete detectable levels of IL-23. IL-23 binds to IL-12Rβ1, and it activates Stat4 in PHA blast T cells. IL-23 induces strong proliferation of mouse memory ($CD4^+$ $CD45Rb^{low}$) T cells. Similar to IL-12, human IL-23 stimulates IFNγ production and proliferation of PHA blast T cells and CD45RO (memory) T cells (59).

2.1.4. Cytokine Pleiotropy and Redundancy
Pleiotropy refers to the ability of a cytokine to exert many different types of responses often on different cell types; whereas, redundancy indicates that many different cytokines can induce similar signals. Pleiotropy is exhibited by the cytokine's host receptors containing $\gamma_{c,}$ which influences a number of lymphoid populations.

2.1.5 Soluble Receptors
For many cytokines, soluble receptors have been identified (e.g., IL-1, -2, -4, -5, -6, -7, GM-CSF, IFNs and TNF). Soluble receptors can be created by alternative splicing that truncates the protein N-terminal to the transmembrane domain. Alternatively, soluble receptors may be created by the proteolytic cleavage of the membrane receptor. At this point, there is little information available for about the role of soluble receptors. Some may have a negative regulatory role whereas soluble IL-6R α exerts an agonistic role because of its signaling capacity [1].

3. TYPE II CYTOKINES

Interferons are type II cytokines. Type I interferons include IFNα (originally known as leukocyte IFN) and IFNβ (originally fibroblast IFN) and IFNω (which is closely related to IFNα and was formerly designated as such). At least 13 functional genes have been identified for IFNα. IFNβ and IFNω are each encoded by single human and murine genes. In addition, another interferon denoted as IFNδ has been reported. Type I interferons share the same receptor and induce the same signals. Their biological effects include anti-proliferative and anti-viral activities as well as the ability to stimulate cytolytic activities in lymphocytes, NK cells and macrophages and to regulate IFNγ [1,60–63].

Interferon γ is designated as Type II interferon, and it has a distinct receptor from Type I interferons. Interferon receptors are referred to as Type II cytokine receptors in order to reflect their substantial difference between these receptors and Type I receptors. Both Type I and Type II interferons bind to such Type II cytokine receptors. Because the IL-10 receptor is similar to the IFNγ receptors, IL-10 is considered to be a Type II cytokine [1].

Interferon receptors are now designated as IFNAR-1 (previously known as IFN-αRi or IFN-Rα) and IFNAR-2 (previously known as IFN-α-β receptor, or IFN-αR2, or IFN-Rβ). IFNAR-2 has both short and long forms and a soluble form. IFNAR-1 cannot bind IFNα, and IFNAR-2 binds with low affinity. Both chains are required for high affinity binding and signal transduction. IFNAR-1 binds the Janus family tyrosine kinase (Jak) Tyk2, and IFNAR-2 binds Jak1 and signaling is mediated by STAT1. Interferon-α variants are species specific with the exception of IFNα-8. The human IFNα-8 can bind to the mouse receptor [1,64].

Interferon γ forms a homodiamer with an apparent molecular weight of 34 kDa. Each IFNγ monomer has 6 α helices, 4 of which resemble the short chain helical cytokines and there is no β-sheet structure. The subunits interact in an antiparallel fashion. IFNγ is produced by NK cells, CD-8^{+}T and Th1 subclass of CD-4^{+}T cells. The receptors for IFNγ are ubiquitously expressed, with the exception of erythrocytes. Even platelets express IFNγ receptors. The functional receptor consists of 2 chains: IFNGR-1 (IFNγR-1 or IFNγRα) which is a 90 kDa protein. The second chain is denoted as IFNGR-2 (IFNγRβ). IFNGR-1 is required for ligand binding and IFNGR-2 is involved in signaling by association with the Jak1 and Jak2 kinases. There is an additional IFN receptor family member denoted as CRF2-4 which is now known to be a component of the IL-10 receptor [1,61,62,65].

3.1. Interleukin 10

Interleukin 10 is produced by activated T cells, B cells, monocytes and keratinocytes. IL-10 is an 18 kDa protein. It inhibits the production of a number of cytokines including IL-2, IL-3, IFNγ, GM-CSF, and TNF. It falls into the category of TH-2 cytokines. While IL-10 inhibits monocyte-dependent T-cell proliferation, it appears to exert a direct stimulatory effect on thymocytes and T cells *in vitro*. Interleukin 10 is a major inhibitor of cellular immunity, and it favors the development of the antibody response by suppressing interleukin 12. IL-10 regulates the growth and differentiation of B cells, cytotoxic and helper T cells, mast cells, granulocytes, dendritic cells, keratinocytes and endothelial cells. It plays a key role in the differentiation and function of T regulatory cells. The IL-10 receptor is related to interferon receptors and contains the CRF2-4 chain [1,66].

3.2. Interleukin 19 (IL-19)

A novel homologue of IL-10, designated IL-19 has been cloned. IL-19 has 21% amino acid identity with IL-10. The exon/intron structure of the IL-19 gene is comprised of five exons and four introns within the coding region of the cDNA. There are at least two distinct IL-19 mRNA species that differ in their 5'-sequences. IL-19 mRNA can be induced in monocytes by LPS-treatment. The treatment of monocytes with IL-4 or IL-13 did not induce *de novo* expression of IL-19, but these cytokines did potentiate IL-19 gene expression in LPS-stimulated monocytes. In addition, GM-CSF directly induced IL-19 gene expression in monocytes. IL-19 does not bind or signal through the IL-10 receptor complex, suggesting existence of an IL-19 specific receptor complex [67].

3.3. Interleukin 20 (IL-20)

Interleukin 20 is a novel IL-10 homologue. The chromosomal localization of IL-20 led to the discovery of an IL-10 family cytokine cluster. IL-20 transgenic (TG) mice show neonatal lethality and suffer from skin abnormalities including aberrant epidermal differentiation. Recombinant IL-20 stimulated signal transduction through STAT3 in a keratinocyte cell line. The IL-20 receptor was identified as a heterodimer of two orphan class II cytokine receptor subunits. Both receptor subunits are expressed in skin and are dramatically upregulated in psoriatic skin. These results demonstrate a role for IL-20 in epidermal function and in psoriasis [68].

3.4. Interleukin 22 (IL-22)

IL-22 (also called IL-TIF) is produced by activated T lymphocytes, and shows 22 per cent amino acid identity with IL-10. Both the mouse and human IL-22 genes consist of 6 exons spreading over approximately 6 Kb. The IL-TIF gene is a single copy gene in humans, and is located on chromosome 12q15. It is also a single copy gene in BALB/c and DBA/2 mice, but is duplicated in other strains, such as C57Bl/6, FVB and 129 [69]. Interleukin 22 is a ligand for CRF2-4, a member of the class II cytokine receptor family. A new member of the interferon receptor family, IL-22R, functions as a second component and together with CRF2-4 enables IL-22 signaling. IL-22 is not recognized by the IL-10R. In responding cells IL-22 activates STATs 1, 3, and 5. Unlike IL-10, IL-22 does not inhibit the production of proinflammatory cytokines by monocytes, nor does it impact IL-10 function on monocytes, and it has modest inhibitory effects on IL-4 production by Th2 T cells (70)

4. PROINFLAMMATORY CYTOKINES

This group of cytokines promotes innate and subsequent immunologically regulated inflammatory responses. These cytokines play important roles in host defense; however, when present in excess or produced during the course of inappropriate autoimmune responses, they can have a self-destructive effect. The tumor necrosis factor and interleukin 1 families, chemokines, transforming growth factor β (TGF-β), and the recently discovered interleukins 16, 17 and 18 belong to this category [71–75].

4.1. The Tumor Necrosis Factor family

Tumor necrosis factor (TNF) was originally described as a factor responsible for lipopolysaccharide (LPS) – induced hemorrhagic necrosis of tumors in animals. A factor called "cachectin," which was responsible for wasting of animals during parasitic infections, later turned out to be TNF. A closely related factor, originally termed "lymphotoxin" is produced by T lymphocytes, is now termed "LTα" (also known as TNFβ). TNF and LTα utilize the same receptors and produce similar biological activities. However, LTα also forms a hetero-complex with LTβ, and this complex utilizes a different receptor (LTβ receptor) with different biologic activity of LTα alone [76].

At present, a large family of ligands and receptors have been identified that belong to the TNF family [15]. Except for LTα, the ligands of this family are Type II membrane proteins consisting of C-terminal extracellular domain and transmembrane and intracellular domains. Only LTα, TNF and FasL exist in a soluble form. Members of the TNF receptor family are Type I membrane proteins with structural homology in extracellular ligand binding domains characterized by varying numbers of repeating cysteine rich domains. Intracellularly, some of these receptors share some 80 amino acid "death domain" regions which play a role in the induction of the apoptotic signal. This is present in the TNF-RI, Fas/Apo1, DR3/WSL-1, DR4, CAR1, TRAMP and p75 NGF-R. The p75 NGF-R is a receptor for nerve growth factor and for neurotrophins that regulates neuronal differentiation as well as cell death [71,74,77].

The human TNF gene is located within the class III MHC region on chromosome 6 flanked by the genes encoding LTα and LTβ, respectively. The murine genes are similarly arranged within the murine MHC gene complex. In addition to transcriptional regulation, the rate of TNF synthesis is controlled by the level of mRNA elongation, mRNA processing and by the extent of translation. Enzymatic cleavage of the transmembrane TNF protein results in the release of soluble TNF. Human TNF is synthesized as a 233 amino acid long transmembrane protein, which is biologically active and appears to preferentially activate TNF-RII rather than TNF-RI. The soluble form of TNF is 157 amino acids long derived from the membrane precursor by proteolytic cleavage. Current evidence suggests that LTα and other members of the TNF ligand family bind to their receptors as trimers and that their site for receptor interaction is at the base of the trimers. LTα also forms a heterotrimer with LTβ because LTβ is a membrane protein and is not secreted, it serves as an anchor for LTα at the cell surface. The FasL ligand also exists in two forms, a 40-kDa transmembrane form and a 26-kDa soluble form [71,74].

Mononuclear phagocytes are major cellular producers of TNF although T cells, B cells, NK cells, endothelial cells, keratinocytes, smooth muscle cells, mast cells, neutrophils, astrocytes and glial cells are also capable of production. Bacterial products, especially LPS, phorbol esters, zymosan, ultraviolet light, virus infection, allergenic B cells and T cells, protozoa and other microorganisms up-regulate the production of TNF. Of the cytokines, TNF itself, IL-1 IFNγ, IFNα, GM-CSF, IL-2, TGFβ, substance P and platelet activating factor stimulate TNF production. Inhibitors of TNF synthesis are prostaglandins, corticosteroids, IL-4, IL-6 and TGFβ [71].

Lymphotoxin α is produced by activated B cells, T cells and possibly NK cells. Lymphotoxin β is also expressed on lymphocytes, and it appears that the level of LT-α/β heterodimer expression on the surface of lymphocytes is regulated mainly by the level of LTα synthesis. FasL is mainly produced by activated T cells, whereas B cells, macrophages, endothelial cells and many other cell types do not produce FasL. However, FasL expression was found in the central nervous system, testes and the eye [71].

Two separate receptors, TNF-RI and TNF-RII exist that bind both TNF and LTα homotrimers. No structural homologies exist in the intracellular domains of these two receptors indicating that they signal by distinct mechanisms. TNF-RI contains the death domain region in its cytosolic portion. TNF receptors are widely distributed among different tissues, and many cells express both of these receptors. Resting T lymphocytes do not express TNF receptors, but expression may be induced by IL-2. IFNγ up-regulates TNF receptor expression on many cell types. TNF receptors are downregulated by phorbol ester, IL-1, IL-6, TGFβ, retinoids and TNF itself. Most of these agents also induce the shedding of TNF-RI and TNF-RII from the cell surface. Shed TNF receptors bind TNF and neutralize its biological activity. Soluble TNF receptors are present in the plasma and urine of healthy individuals. The level of soluble receptors becomes elevated in patients with various infections, immune disorders and cancer. Soluble receptors are also able to prolong the retention of TNF in the body. The cross-linking of two or three TNF receptors is required for signaling. This is followed by the recruitment of a variety of intracellular signaling proteins to the receptor complex. TNF-RI is the main signaling chain and induces cytotoxicity, fibroblast proliferation and activation of NF-κ B. TNF-RII can also transmit some signals, but primarily it functions as a "ligand-binding" receptor [71].

The soluble LT-α homotrimer binds to TNF-RI and TNF-RII, while the LT-$\alpha_1\beta_2$ heterotrimer binds to a separate receptor termed LT-βR. FasL recognizes a receptor termed Fas, whose major function is to trigger apoptosis. Whereas FasL is predominantly expressed on lymphoid cells, Fas is expressed in a wide variety of different tissues including tumor cells. The expression of Fas in some cells is increased after treatment with IFNγ or TNF [71,74]. Tumor necrosis factor-alpha induces neuronal death by interfering with survival signals generated by the type I insulin-like growth factor receptor [78].

Signal transduction by the TNF receptor family is initiated by the recruitment of various intracellular adaptor proteins/signal transducers to the intracellular domains of the activated receptors. The number of these proteins is far greater than those involved in the transduction of signals for other cytokine receptors. This leads to a complexity of signals which matches the enormous diversity of biological actions generated by these receptors. The earliest intracellular signaling protein is TNF receptor associated factor 1 and 2 (TRAF1 and TRAF2). Four more TRAF proteins have been identified, all of which participate in signal transduction by different receptors. TRAF6 plays a key role in IL-1 signaling. TRAF proteins bind to specific sequences in the intracellular regions of receptors, and contain a RING finger motif and a Zinc finger sequence. TRAF2 appears to play a role in TNF induced activation of NFκB transcription factor by both TNFRI and TNFRII. Other proteins interact with the death domains of receptors. A 34 kDa protein, TRADD, interacts with TNF-RI; the protein FADD/MORT and RIP with the Fas receptor. RAIDD is also a protein associated with death domains. Some of these proteins have

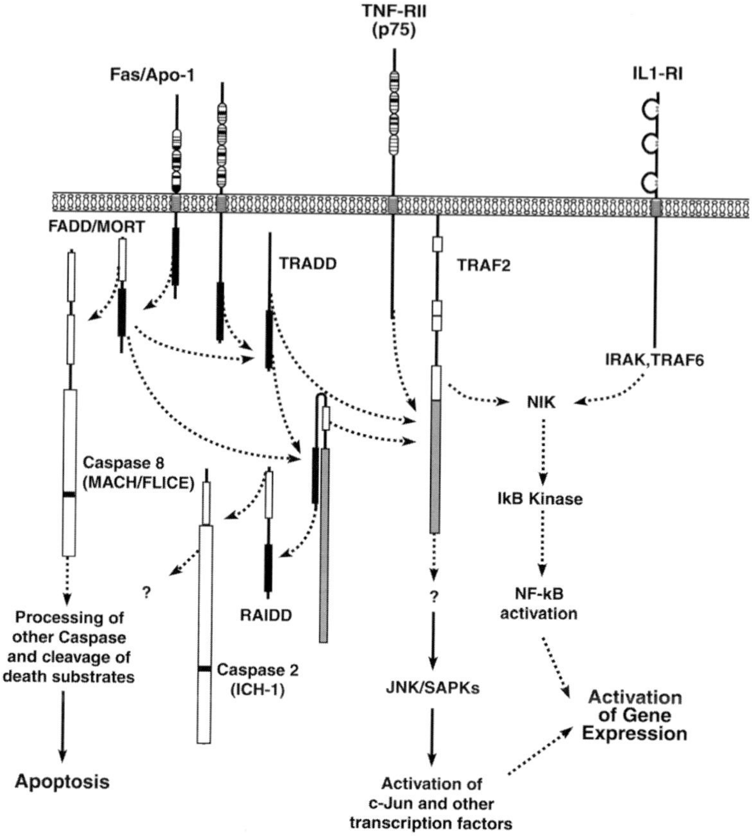

Figure 2. Signal Transduction by the TNF-Receptor Family.
The P-55 and the P-75 TNF receptors share no sequence homology, yet they signal cooperatively by the interactions of intracellular adapter proteins. The death domain-containing proteins (such as TRADD) bind to the death domain of TNF-RI, but this protein also interacts with other death domain-containing proteins (FADD/MORT1 and RIP). TRAF2 in turn, binds directly to TNF-RII and serves as the downstream activator of JNK/SAPK family of MAP kinases. TRAF2 also activates directly NIK kinase which then phosphorylates the IκB kinase leading to IκBα and IκBβ phosphorylation and resulting in NFκB activation. The recruitment of death domain proteins also leads to the activation of caspases that in some cells will trigger cell death via apoptosis. The intracellular domain of the Fas/Apo1 comprises a death domain similar to that found in the TNF-RI, and it activates similar intracellular signaling pathways to the TNF receptor but it is much more efficient in triggering apoptosis than activating gene expression. Signaling by the IL-1 receptor through the involvement of the IRAK and of TRAF6 is also shown. This leads to NIK activation and subsequent NFκB activation, which is a shared pathway with TNF receptors. The IL-1 receptor also signals by JNK/SAPK activation which is largely unknown and not shown. The pathways of p38MAP kinase activation by the TNF and IL-1 receptors are also unknown and not shown here (After Krauker et al. 1999, with minor modifications).

enzymatic activity and are capable of triggering the apoptotic enzymatic cascade whereas others are able to promote NFκB activation, which induces the synthesis protective proteins and prevents cell death [71,79–81].

The activation of cellular transcription factors such as AP-1 and especially NFκB play a major role in gene activation. Receptor activated TRAF2 activates the serine-threonine kinase termed, NFκB activating kinase (NIK). NIK activates another serine-threonine kinase called IκB-α kinase. IκB-α is the major inhibitory subunit of NFκB, and its degradation is triggered by this enzyme. The same enzyme also phosphorylates IκB-β which results in degradation. Current evidence suggests that NIK is at the point of convergence of TNF and IL-1 signaling pathways. Like TRAF2 in TNF-treated cells, TRAF6 is likely to function as the immediate off stream activator of NIK in IL-1-stimulated cells. This would explain in molecular terms why TNF and IL-1 cytokines, that are structurally unrelated and bind two completely distinct receptors, activate so many identical target chains and produce so many identical biologic actions [71,74].

Mitogen-associated protein (MAP) kinases play prominent roles in trans-cytoplasmic signaling to the nucleus by activating genes through the phosphorylation of transcription factors. Some of these kinases are associated with protection from apoptosis, whereas others can promote cell death. MAP kinases are considered to be important in TNF induced arachidonic acid production and the generation of pro-inflammatory metabolites [71,74].

4.2. Interleukin-1 family

The IL-1 family consists of three structurally related cytokines, IL-1α, IL-1β and a specific antagonist termed, IL-1 receptor antagonist (IL-1ra). These cytokines have multiple biologic activities and coordinate the response of the host to infection and injury. At this time IL-1ra is the only known naturally occurring cytokine antagonist that acts as a competitive inhibitor of IL-1 receptors [71,82].

The regulatory sequences of the genes for IL-1α, IL-1β and IL-1ra differ, which implies differential regulation. A novel LPS-inducible site has been identified in the IL-1β promoter. Glucocorticoid regulatory elements are expressed in both the IL-1α and IL-1β genes. The IL-1β promoter expresses AP-1, NF-IL-6 and CRE sites. A lipophosphoglycan-responsive promoter sequence is located downstream from AP-1 and acts as a "gene silencer". This sequence down-regulates LPS induced IL-1β mRNA. The intracellular IL-1ra (icIL-1ra) gene lacks a TATA box, whereas the secretory IL-1ra (sIL-1ra) gene has a TATA box, with consensus sequences for possible NFκB, NF-IL-1βA, AP-1 and CRE binding sites. The intracellular variant is generated by alternately splicing the 5' exons. The production of IL-1 and IL-1ra are differently regulated at both the transcriptional and post-transcriptional levels which depend on the particular cell type and the stimulus used for induction [71,83,84].

IL-1 is not secreted via the classical secretory pathway, it is found in the cytoplasm, but not in the endoplasmic reticulum or Golgi apparatus. Intracellular pro-IL-1α is stored, while the rest is either phosphorylated or myristoylated and becomes membrane bound by a lectin interaction involving mannose residues. Pro-IL-1α and membrane bound IL-1α is biologically active, but pro-IL-1β has no biological activity. Serum proteases, elastase, plasmin, chymotrypsin, cathepsin G or a specialized intracellular cysteine protease named IL-1β converting enzyme (ICE) cleave Pro-IL-1β to generate the 17-kDa mature IL-1β. Both LPS and IFN-γ upregulate the expression of ICE. ICE does not cleave pro-IL-1α, but mice lacking the ICE gene fail to secrete both IL-1β and IL-1α, presumably because IL-1β induces IL-1α. Pro-IL-1α is cleaved by calpain and other extracellular nonspecific proteases. In some cells, the release or processing of IL-1 may not occur and Pro-IL-1 may accumulate intracellularly which is then released as a consequence of cell injury or death [71,72].

IL-1ra is secreted through the endoplasmic reticulum and Golgi as three structural variants of 20 to 80 kDa that differ in the extent of glycosylation. Receptor binding is identical by the three variants. The intracellular variant of IL-1ra is released only by the process of cell disruption. Mice lacking IL-1ra spontaneously develop vasculitis and arthritis [82,85].

Although mononuclear phagocytes are the major cell sources of IL-1, all other nucleated cells are capable of producing it. Normal cells must be stimulated for production. In general, infections, immunological and inflammatory processes, endogenous mediators and cell and tissue injury would trigger IL-1 production. In macrophages, the stimuli triggering IL-1 production include bacterial cell wall products, LPS, microbial superantigens, zymosan, phorbol ester, muramyl dipeptide, antigens, mitogens, endogenous cytokines, leukotrienes, activated complement components, immune complexes, ultraviolet irradiation, celiaca particles, viruses, parasites and other microorganisms. Activated lymphocytes can also stimulate macrophages to produce IL-1 either by direct conduct or by cytokines such as TNF-α, GM-CSF or IL-1 itself. IL-1ra is also produced by many cell types, and tissue macrophages and keratinocytes are good producers. In human monocytes, the production of IL-1β and IL-1ra is reciprocally regulated. For example, IL-4, IL-10 and TGF-β increase the synthesis of IL-1ra and decrease IL-1β production in monocytes. Corticosteroids suppress the production of IL-1α and IL-1β but do not affect IL-1ra. The production of IL-1β usually precedes that of IL-1α and IL-1ra. IL-1β is an inducer of IL-1ra production [71,72,86].

Two distinct receptors, IL-1RI and IL-1RII, and a receptor accessory protein, IL-1RacP which is the putative signal transducing unit have been identified. Both the Type I and Type II receptors are members of the immunoglobulin superfamily. The Type I receptor has a long cytoplasmic tail and participates in signal transduction. Its extracellular ligand binding region contains three immunoglobulin-like domains. The Type II receptor has only a short cytoplasmic tail and is incapable of signaling. This receptor competes for IL-1 by acting as a "decoy" receptor and downregulates the level of extracellular IL-1. Type I receptors are found on mast cells, T lymphocytes, endothelial cells, hepatocytes, fibroblasts and keratinocytes. Type II receptors are predominant on B cells, monocytes and neutrophils. The extracellular domain for both Type I and Type II receptors are shed by activated neutrophils and monocytes, and these soluble receptors inhibit IL-1β function [71,83].

The signal transducing receptor consists of IL-1RI and IL-1RAcP that binds both IL-1α and IL-1β with higher affinity than that of IL-1ra. Two transcription factors are activated by the receptor, NFκB and AP-1 which translocate to the nucleus and activate many of the cytokine inducible genes. In most cells, NFκB is retained in the cytoplasm by inhibitory proteins designated as I-κB. I-κB is degraded in response to IL-1 and NFκB is released to enter the nucleus where it binds DNA and activates transcription. Although IL-1RI does not posess kinase activity, the binding of IL-1 by this receptor leads to rapid phosphorylation of serine and threonine residues in a number of proteins. After ligand binding an IL-1 receptor associated kinase activity (IRAK) homes rapidly to IL-1RI and becomes auto-phosphorylated. TRAF-6 which is a member of the TNF receptor-associated protein family also functions as a signal transducer for IL-1. Numerous secondary messengers and signaling cascades have been proposed to mediate the biological activity of IL-1. One of these is the sphingomyelinase-ceramide pathway. Other kinases including p38MAPK, JNK/SAPK (c-Jun N-terminal kinase/stress activated protein kinase or p54 MAPK), I-κB kinase and β-casein kinase are also activated by IL-1. The precise order of signaling cascades for these kinases has not been determined. Different cell types may utilize diverse signal transducing pathways for IL-1 which result in cellular specificity. Two additional transcription factors, Jun-1 and Jun-2, that lead to increased gene transcription are also activated as well as small c-Fos. The Jun and Fos proteins join to form AP-1 [71,84].

The biologic activities of IL-1 are mostly indirect due to the induction of numerous other mediators. There is a major overlap with TNF-α and IL-6 in biologic activities. Interleukin-1 is a pleiotropic mediator of host response to infections and injury/ insults, and it coordinates the activities of other cytokines during an immune/inflammatory response. IL-1 also stimulates a variety of cells within the immune system including T lymphocytes, B cells, monocytes, neutrophils and dendritic cells, and plays a major role in the recruitment of leukocytes to sites of inflammation. Finally, IL-1 is a major mediator between the neuroendocrine and immune systems. It is involved in hematopoiesis; it functions as an endogenous pyrogen, influences slow-wave sleep, and decreases appetite. Neurons and glial cells can synthesize IL-1, and synthesis is increased during development as well as in response to injuries. IL-1 stimulates the hypothalamus to produce CRF that results in the expression of the POMC gene by the pituitary gland [71,87,88].

Knockout mice defective in the production of IL-1α or IL-1β or in the expression of IL-1RI receptor are viable and show only minor deficiencies. IL-1β deficient mice are resistant to fever induction and have an impaired acute phase response to turpentine but respond normally to LPS. The normal response to LPS is based on redundancy in the activities of IL-1α, IL-1β, TNF-α, LT-α and IL-6. IL-1ra deficient mice are more susceptible than normal animals to lethal endotoxemia, arthritis and vasculitis, but are less susceptible to *L. monocytogenes*. Transgenic IL-1ra over producer mice are protected from the lethal effects of endotoxin, but are susceptible to listeriosis. IL-1ra acts as a positive regulator of IL-1 levels to balance the potentially beneficial effect of IL-1 against excessive production during infection [71,82,89].

3.3. Interleukin 18 (IL-18)

Interleukin-18 is related to the IL-1 family in terms of its structure, processing, receptor, signal transduction pathway and pro-inflammatory properties. Functionally it is also related to IL-12, as it induces the production of Th1 cytokines and participates in cell-mediated cytotoxicity. Mice that lack IL-18 are deficient in natural killer cell activity and Th1 mediated immune responses [90,91].

Interleukin-18, as IL-1β, does not possess a signal sequence allowing direct secretion through the plasma membrane. The IL-1β converting enzyme (ICE, also called caspase-1) cuts intracellular proIL-1β after the aspartic acid residue in position 116, resulting in a highly active mature IL-1 beta that is secreted into the extracellular space. IL-18 is also initially synthesized as an inactive proIL-18 lacking a signal peptide and like IL-1β, proIL-18 is cleaved by ICE to yield an active molecule. However, IL-18 is not an endogenous pyrogen following intraperitoneal injection to mice. Nevertheless, IL-18 may contribute to inflammation and fever because it is a potent inducer of TNF, chemokines and IFNγ [92,93].

Interleukin-18 exerts its biological activity by means of a signaling chain of the IL-18 receptor (IL-18R) complex. This receptor is made up of a binding chain termed IL-18Rα, a member of the IL-1R family previously identified as the IL-1R-related protein (IL-1Rrp), and a signaling chain, the IL-18Rβ, also a member of the IL-1R family. The IL-18R complex recruits IL-1R-activating kinase and TNF receptor-associated factor-6, which phosphorylates NFκB-inducing kinase with subsequent activation of NFκB [94].

The toll-like receptor (TLR) proteins represent a conserved family of innate immune recognition receptors and show extended homology with receptors for the cytokines IL-1 and IL-18. These receptors are conserved in mammals, insects, and plants, mediate cellular activation and stimulate innate immune defenses. Various bacterial and fungal products have been identified that serve as TLR ligands. Endogenous protein ligands for TLR receptors have also been identified [95].

Interleukin 18 and IL-12 exhibit a marked synergism in IFNγ induction in T cells. IL-12 upregulates expression of the IL-18 receptor on cells producing IFNγ. IL-18 is essential for the effective induction and activation of Th1 cells by IL-12. Natural killer cells are activated by IL-18 independently of IL-12. The excessive production of IL-12 and IL-18 by activated macrophages may induce multiple organ disorders including disruption of the immune system [96]. After exposure to IL-18, T lymphocytes and NK cells produce substantial amounts of IFNγ. IL-18 seems to be a stronger IFN inducer than IL-12. IL-18 stimulated Fas ligand expression, the induction of both CC and CXC chemokines, and increased the production of competent human immunodeficiency virus [93,97].

In animal models the neutralization of IL-18 reduces the severity of inflammatory disease. Endotoxin lethality can be prevented by anti-IL-18. Even in IFNγ independent models, the neutralization of IL-18 prolongs survival. Anti-IL-18 also prevents liver injury induced by toxins or activated T cells. The neutralization of IL-18 reduces the adherence of malignant hepatoma cells to vascular endothelium by preventing the upregulation of vascular endothelial adhesion-1 molecule expression. IL-18 and IL-12 act synergistically in the stimulation of NK cells for the production of IFNγ, which is prevented by the neutralization of IL-18. IL-18 enhances host defense against tumors in mice by a mechanism that is most often IFNγ-dependent. Nevertheless, it is the proinflammatory effect of IL-18 that most frequently contributes to enhanced host defenses. In models of arthritis, lung injury or inflammatory bowel disease, IL-18 plays important roles in mediating inflammation [98].

Propionibacterium acnes-primed, LPS-challenged liver injury is the prototype of IL-18-induced tissue injury. In this model IL-18 acts on natural killer cells to increase Fas ligand (FasL) that causes liver injury by induction of Fas-dependent hepatocyte apoptosis. LPS induces IL-18 secretion by Kupffer cells, which is dependent on caspase-1. Caspase-1-deficient mice are resistant to *P. acnes* and LPS-induced liver injury [99].

5. CHEMOKINES

These are chemo-attractant cytokines that comprise the most diverse and largest subset of cytokines identified to date. They are capable of inducing the directional migration of activated leukocytes as well as other somatic cell types and play a major role in acute and chronic inflammation. Chemokines also promote humoral and cell mediated immune reactions; regulate cell-to-cell adhesion, angiogenesis, leukocyte re-circulation and homing, and contribute to lymphopoiesis and hematopoiesis, vascularization, organogenesis and embryogenesis. Leukocytes and other cell types produce chemokines in response to irritation, polyclonal stimulants, antigens and endogenous cytokines. Chemokines play a central role in host defense against infectious disease and participate in the pathogenesis of diverse conditions that involve immune inflammatory processes. In contrast to other pro-inflammatory cytokines, chemokines in general do not induce other cytokines, acute phase response or fever, but are capable of inducing effector molecules, such as histamine, enzymes and defensins. High concentrations of circulating chemokines have a desensitizing effect on local inflammation [71,100–102].

Four chemokine subfamilies have been identified on the basis of their structure: CXC(α), CC(β), C(γ) and CX_3C. Chemokines are produced as inactive precursors, cleaved enzymatically intracellularly and secreted by the classical pathway. The carboxyl ends of basic chemokine molecules exhibit low affinity heparin binding properties, bind to glycosaminoglycan molecules

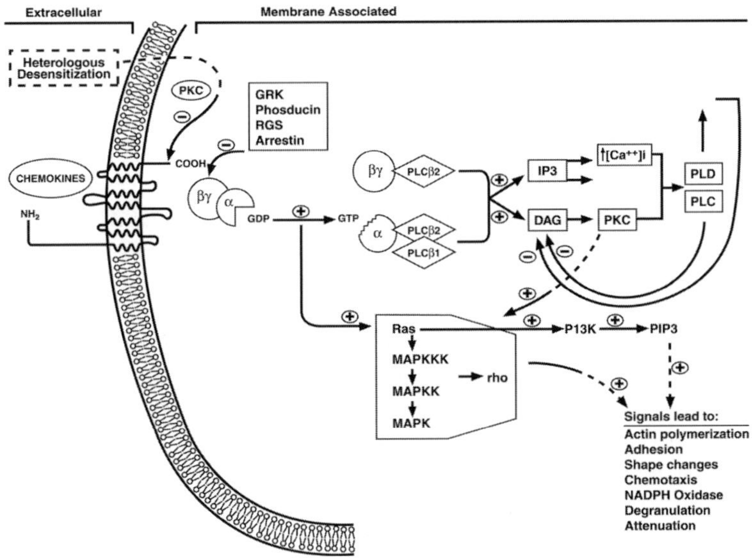

Figure 3. Signal Transduction by Chemokines.
The seven transmembrane-spanning receptor (7TMR) will activate G proteins after binding to chemokines. This results in exchange of the G protein from GDP to GTP (bound state). A number of down regulatory molecules such as protein kinase C are capable of uncoupling G proteins from GTP, which results in desensitization. The α and βγ subunits of activated G protein will then activate PLCβ1 and PLCβ2 which in turn generate IP_3 and DAG. IP_3 mobilizes calcium and DAG stimulates PKC2 phosphorylate downstream effectors including MAP kinases (After Krauker et al. 1999, with minor modifications).

and to other negatively charged sugar molecules on cell surfaces and to tissue matrix glycoproteins. This enables chemokines to induce directional chemotactic migration of cells along glycosaminoglycan coated surfaces [71].

Chemokines are recognized by a growing subfamily of high affinity seven pass transmembrane G protein-coupled receptors (7TMR). Other pro-inflammatory chemoattractants, such as platelet activating factor (PAF), leukotriene B-4 (LTB-4) and C-5a and FMLP also use seven TMR, but with lower homology to the chemokine receptors. The chemokine receptors consist of extracellular N-terminus, 7TMR domains, which form three extracellular and three intracellular loops, and an intracellular carboxyl terminus that contains a number of serine and threonine residues. These residues are phosphorylated during signaling and desensitization. Chemokine ligands usually interact with the N-terminal and a second site located on one of the extracellular loops of the receptors. The heterodimer G protein is usually coupled to one intracellular loop of the receptor. The DRY motif (Asp-Arg-Tyr) which is usually located in the second intracellular loop also contributes to signal transduction. Chemokines exhibit considerable redundancy in their receptor utilization and leukocytes express multiple receptors. Therefore, chemokines, like other cytokines, are pleiotropic in their activities. With some exceptions, activated leukocytes express more receptors than do unstimulated leukocytes [71,101,102].

Signal transduction by chemokine receptors requires an initial coupling with the G proteins which initiates a cascade of cellular signals and a biological response. G proteins are a large gene family coding for at least 16 α, 4β and multiple γ subunits. Receptor coupling typically starts with an exchange in the α subunit of the G protein from GDP to GTP which dissociates the α from the β and γ subunits. These subunits then activate PLC-β1 and PLC-β2 followed by hydrolysis of phosphatidylinositol 4,5-biphosphate (PIP_2) which yields inositol triphosphate (IP_3) and diacylglycerol (DAG). IP_3 mobilizes intracellular calcium (Ca^{++}) and DAG stimulates PKC which phosphorylates and activates a number of effector molecules such as MAP kinases. The chemokines MCP-1 and -3 are dependent on external Ca^{++} to increase intracellular Ca^{++}, whereas MCP-2 and MIP-1β are weak inducers of Ca^{++} fluxes in monocytes. Chemokines frequently induce tyrosine phosphorylation, which participates in the downstream activation of the MAP kinase cascade. G protein dependent signals lead to actin polymerization, the reconfiguration of adhesion proteins and to other cellular responses that contribute to cell migration [71,101,102].

CXC chemokines may be divided into two subgroups. Those with an ELR sequence (glu-leu-arg) are positioned at 4, 5 and 6 near the N terminus; and all of them interact with CXCR-2 with equally high affinity. Interleukin 8, GRO (α, β, γ)/MGSA, NAP-2, ENA-78 and GCP-2 belong to this group. CXC-ELR negative chemokines are PF-4, IP-10, MIG, I-TAC, NSDF-1. Lymphotactin is a type C chemokine and fractalkine (neurotactin) is a CX3C chemokine [71]

Human CC chemokines are as follows: MCP-1, -2, -3, -4, -5, MIP-1α, -1β, LARC/MIP-3A, exodus, MIP-3β/ERC, RANTES, eotaxin, I-309, PARC, SLC, TARC (TECK). The mRNA for CXC chemokines contains a 3' AU-rich sequence that causes rapid degradation with a half-life of less than one hour [71].

Numerous exogenous agents stimulate CXC chemokine production in a wide variety of cell types by stabilization of their mRNA as well as by activation of gene transcription. Interleukin 1 and TNF are major stimulators of ELR^+CXC chemokines, whereas interferons are the major inducers of non-ELR^+CXC chemokines. Interferons actively suppress the production of IL-8. Glucocorticoids are potent inhibitors of chemokines which may be due to action on glucocorticoid-responsive promoter elements, but also by the inhibition of NFκB activity [71,101–104].

5.1. Interleukin 8

IL-8 is a major contributing cytokine in a variety of acute as well as chronic inflammatory states. IL-8 can bind with equal affinity to CXCR1 and CXCR2. Mice lack a clear cut homologue for IL-8. In addition to inflammation, IL-8 regulates myeloid and B lymphocyte production. Chemokines such as IL-8 are capable of degranulating neutrophils and initiate an immune stimulating cascade that helps to convert an innate immune response to a specific adaptive immune response [71,105].

5.2. Stromal Cell-Derived Factor 1 (SDF-1)

SDF-1 was originally discovered in bone marrow stromal cells. SDF-1 is homologous to CXC and CC chemokines, and it is conserved evolutionarily so it could be considered a primordial chemokine. SDF-1 is produced constitutively. It is present in the plasma and is a potent chemoattractant of lymphocytes and monocytes; and indeed, it may play a greater role in mononuclear cell homing and development than in inflammation. SDF-1 is also chemotactic for hematopoietic progenitor cells and is capable of mobilizing them from the bone marrow to the peripheral blood. Mice with a disrupted SDF-1 show severely reduced numbers of B cell progenitors, T cell deficiencies, reduced bone marrow myelopoiesis, ventricular septal defect and they die prematurely. SDF-1 utilizes the orphan receptor (CXCR-4), which also acts as a co-receptor for T lymphocyte and neuronal cell invasion by lymphotropic HIV-1. CXCR-4 is expressed during embryogenesis and in many organs and T cells during adult life in mice [71,106–107].

5.3. CC chemokines

More than 25 human and murine CC chemokines are known. The genes are transcribed and then cleaved off to yield mature proteins that show 28 to 73 percent homology. CC chemokines are potential regulators of NFκB, NF-IL6, AP-1, AP-2 and MCP-1. Although the CC chemokines are produced by a wide variety of cell types, a significant number of them are T cell derived. Eight receptors that react to one or more of the CC chemokines are called CCR1-8. With some exceptions, CC chemokines are chemoattractant for T cells and monocytes, but not neutrophils. A subgroup of them participate in allergic reactions. Eotaxin-1, -2, MCP-4 and to a lesser extent RANTES, MCP-2, -3 and MIP-1α attract eosinophils whereas eotaxin-1, -2, RANTES and MCP-1, -2, -3 and –4 are all chemotactic for basophils and are capable of inducing degranulation and rapid histamine release. MCP-1, -2, -3, RANTES and MIP-1α are chemotactic for murine mast cells. A number of these chemokines downregulate hematopoiesis, whereas, eotaxin-1, Mu MIP-2 and IL-8 apparently enhance early hematopoietic progenitor cell growth. These chemokines appear to act on hematopoiesis in a redundant fashion [71,82].

CC chemokines play a role in immunologically induced inflammation; however, these chemokines are necessary, but not sufficient, to induce an inflammatory response. A human CC chemokine is abundantly expressed by dendritic cells and attract $CD45RA^+$ T cells specifically. Chemokines promote the release of granular proteins by neutrophils and monocytes which contribute to tumor cell cytolysis. A number of CC chemokines also enhance CTL and NK cell-mediated killing of target cells. This was dependent on the expression of β_2 integrin, which presumably promotes target-killer cell interaction. Chemokines can also induce the release of serine esterases from lymphocytes [71,108].

MCP-1 enhanced the production of IL-4 by TH-2 lymphocytes and diminished the production of IL-12 and IFN-γ by TH-1 cells. MCP-1 participates in the TH-2 based chronic secondary inflammatory response to *Schistosoma mansoni* eggs whereas MIP-1α plays a greater role in TH-1 dependent granulomatous responses. MIP-1α induced the differentiation of antigen stimulated TH-0 cells into TH-1 phenotype whereas MCP-1 drove the same cells into IL-4 producing TH-2 cells [108].

Heterologous desensitization occurs in response to physiologic doses of other chemoattractant ligands interacting with distinct G protein-coupled receptors that results in cross-phosphorylation and deactivation of chemokine receptors. For example, prior exposure for more than one hour to opiates that act on μ and δ opioid receptors can inhibit chemokine receptor dependent chemotaxis in response to IL-8 or RANTES. This is not due to the internalization of chemokine receptors; but rather, it is associated with phosphorylation by the chemokine receptors in response to opiates. Furthermore, pre-incubation of monocytes with GP-120 also suppresses their chemotactic response to a variety of chemokines and fMLP. Consequently, a prolonged exposure to heterologous ligands of G protein coupled receptors may depress the capacity of cells to mobilize in response to chemoattractants. These data illustrate that chemokines may be antagonized comparatively with chemokine receptor interactions, or desensitization may be achieved by receptor-dependent signal transduction [71,109,110].

5.4. Transforming Growth Factor Beta (TGF-β)

The transforming growth factor-β superfamily includes more than 30 members which have a broad array of biological activities. The members of this family are multifunctional cytokines that regulate the growth and differentiation of many cell types and also exert a pro-inflammatory and an immunosuppressive effect. Many normal cells produce TGF-β constitutively which may be enhanced by stimulation. Platelets, macrophages, B and T lymphocytes, fibroblasts, osteoblasts and osteoclasts, astrocytes and microglial cells produce TGF-β. Of the five isoforms identified, three are present in mammals: TGF-β1, TGF-β2, TGF-β3. TGF-β2 and TGF-β3 are important regulators of cellular differentiation and affect development and embryogenesis whereas TGF-β1 is an immunoregulator. Growth/differentiation factor-15 (GDR-15) or macrophage inhibitory cytokine-1 (MIC-1) is a novel member of this family [111,112].

The promoter regions of the genes for the three TGF-β isoforms show little homology, which suggests different regulation of their expression. The TGF-β2 and TGF-β3 promoters contain classic TATA box and CRE/ADF elements. TGF-β1 contains AP-1, NF-1 and Sp-1 binding sites and GC-rich regions, but lack a TATA consensus sequence. The AP-1 sites mediate auto-induction of TGF-β1 and lay ground for the increase of TGF-β1 transcription in pathologic conditions [71].

The biologically active forms of TGF-β are disulfide linked dimers of 25 kDa highly conserved subunits. The pro-domain is also well conserved and may assist in folding dimerization and secretion of TGF-β. The precursor protein is cleaved by a peptidase to release the mature cytokine. The dimerized pro-domain, also called latency associated protein (LAP) remains non-covalently associated with the C terminal fragment to produce a latent complex with another glycoprotein of 125 kDa called latent TGF-β binding protein (LTBP). TGF-β is bound to the cell surface and to extracellular matrix via binding to mannose-6-phosphate-insulin like growth factor type 2 receptors or to other glycosylated binding proteins [383]. It is converted to active TGF-β at these sites by unknown, but presumably highly regulated mechanisms. Latent TGF-β can be converted to active TGF-β *in vitro* by plasmin, cathepsin D, or thrombospondin. In some cells, activation requires the cascade action of other molecules including tissue Type II transglutaminase LTB, plasmin and urokinase-plasminogen activator. Latent TGF-β can also be activated by cell interaction. Homodimeric forms of mature TGF-β can interact with cell surface receptors and generate signal transduction [71,111,113].

TGF-β receptors are expressed by most cell types, including lymphoid and hematopoietic cells. Three types of receptors, Type I, II and III have been identified. Type I and II receptors are transmembrane proteins with a short cysteine-rich extracellular region, a hydrophobic transmembrane helix and cytoplasmic domain with serine-threonine kinase activity. The type II receptor binds ligand efficiently, but the type I receptor can bind ligand only in the presence of type II receptors. Homodimers of either type I or type II receptors are incapable of generating biological responses. The type II receptor is an active kinase and phosphorylates and activates the type I receptor which then propagates the signal to intracellular substrates. Many cells express type III receptors, which are also known as beta glycans and a set of isoform-specific TGF-β binding proteins. Beta glycan binds all three TGF-β isoforms with the same affinity. The function of type III receptors is to present TGF-β to the signaling receptors. The affinity of TGF-β2 for the signaling receptors is increased 20 times by its association with beta glycan. Other TGF-β binding proteins are endoglin, decorin, α2-macroglobulin, and thrombospondin. These proteins modulate the access of TGF-β to the signaling receptor complex. TGF-β receptors are internalized rapidly upon binding of the diametric ligand. So far, 2 signal transducers, Smad-2 and DPC-4 have been identified for these receptors [71,111,114,115].

TGFβ ligands bind to type II and type I serine/threonine kinase receptors and signal via Smad proteins. Receptor-regulated Smads (R-Smads) are classified into two subclasses, those activated by activin and TGFβ signaling pathways (AR-Smads), and those activated by bone morphogenetic protein (BMP) pathways (BR-Smads). The numbers of type I and type II receptors and Smad proteins are limited, which results in signal convergence at the receptor and Smad levels. Signaling by Smads is modulated by various other signaling pathways, which allows TGFβ superfamily ligands to elicit diverse effects on target cells [116].

TGF-β suppresses the immune system at the systemic level, but stimulates immune and inflammatory responses at the local level. TGF-β suppresses inflammation by downregulating E-selectin and IL-8 on endothelial cells, thereby inhibiting the adhesion and transmigration of neutrophils and lymphocytes to the inflammatory site. TGF-β suppresses hematopoiesis by inhibiting bone marrow stem cell proliferation. TGF-β can deactivate macrophages, downregulate the expression of MHC Class II and inhibit the production of TNF-α and IL-1. TGF-β promotes the induction of oral immunological tolerance during oral antigen administration. Regulatory cells designated as Th3 cells are producing TGF-β1, IL-4 and IL-10. These regulatory T cells distribute all over the body and inhibit cell-mediated immune and inflammatory reactions. At sites of inflammation and tissue injury, TGF-β induces the chemotaxis of neutrophils, monocytes, lymphocytes, mast cells and fibroblasts. It activates these cells to produce inflammatory cytokines (IL-1, TNF, and IL-6) and enhances leukocyte adhesion to endothelial cells and to the extracellular matrix [71,111,117].

TGF-β is a potent regulator of tissue repair and a promoter of wound healing. TGF-β increases the synthesis of integrin receptors thereby enhancing interaction between cells and the extracellular matrix. TGF-β3 controls mesenchymal-epithelial cell interaction and regulates cellular differentiation and proliferation. It promotes the growth of mesenchymal cells, while it is inhibitory for most other cell types including epithelial cells, endothelial cells, lymphocytes and hematopoietic progenitor cells. TGF-β1 promotes the development of dendritic cells, and it is essential for normal murine Langerhans cell development. TGF-β normally functions as a tumor suppressor in human malignancies. The role of TGF-β in growth, development and immunoregulation is well substantiated by knockout mice [71].

TGFβ1 controls liver size, by inducing apoptotic cell death in hepatocytes. This apoptosis is a receptor-mediated event, it operates through the mitochondrial/Apaf-1 caspase activation pathway that appears to act as a common execution pathway for many diverse apoptotic stimuli [118].

Macrophage inhibitory cytokine-1 (MIC-1) is a divergent member of the TGFβ superfamily. While it is synthesised in a pre-pro form, it is unique in this superfamily as it does not require its propeptide for correct folding or secretion of the mature peptide [119].

6. INTERLEUKIN 16 (IL-16)

Interleukin 16 (IL-16) was the first T cell chemoattractant described. It is secreted from a variety of immune cells that includes mast cells, CD8 cells, CD4 cells and eosinophils. It interacts with CD4+ as the receptor, which is coupled to G-proteins. IL-16 aggregates into a tetrameric form essential for direct interaction and cross linking of its receptor. It is now clear that IL-16 is an immunomodulatory cytokine that contributes to the regulatory process of CD4+ cell recruitment and activation at sites of inflammation. Therefore, CD4+ is a sentinel receptor that can switch T cells between immune and inflammatory functions. IL-16 is a chemoattractant also for eosinophils and monocytes, promotes T cell adhesion and survival, the expression of IL-2Rα, HLA-DR, IL-3, GM-CSF and IFNγ. It plays a role in asthma and in several autoimmune diseases. In atopic asthmatics IL-16 is a major source of the lymphocyte chemotactic activity early after antigen challenge. Here the major cell of origin is the epithelium. The presence of IL-16 directly correlates with the number of infiltrating CD4+ T cells. IL-16 suppresses antigen-induced lymphocyte proliferation and the transcription of HIV and simian immunodeficiency virus [120–123].

7. INTERLEUKIN 17 (IL-17)

Interleukin-17 is a homodimeric cytokine about 32 kDa in size, which is secreted by activated memory T cells or by αβTCR+CD4-CD8- thymocytes. The IL-17 receptor, a novel cytokine receptor, is ubiquitously distributed and is most abundant in spleen and kidney. The T lymphotropic *Herpesvirus Saimiri* (HVS) captures IL-17. This cytokine induces the secretion of IL-6, IL-8, PGE2, MCP-1 and G-CSF by fibroblasts, keratinocytes, epithelial and endothelial cells. IL-17 also induces ICAM-1 surface expression, the proliferation of T cells, and growth and differentiation of CD34+ human progenitors into neutrophils in the presence of irradiated fibroblasts. IL-17 synergizes with other proinflammatory cytokines, such as TNFα for GM-CSF induction, and with CD40-ligand for IL-6, IL-8, RANTES and MCP-1 secretion. In mice the injection of IL-17 induces neutrophilia, except in IL-6-knockout animals. IL-17 is involved in the rejection of kidney grafts [124]. Interleukin-17 links the activation of certain T-lymphocytes to the recruitment and activation of airway neutrophils. IL-17 acts via induced CXC chemokine release through steroid-sensitive mechanisms, which is modulated by endogenous tachykinins. These effects of IL-17 are potentiated by IL-1β and TNFα [125].

8. MIGRATION INHIBITORY FACTOR (MIF)

Macrophage migration inhibitory factor (MIF) was first identified nearly 40 years ago. It is a pleiotropic protein, participating in inflammatory and immune responses. MIF was originally recognized as a cytokine involved in delayed hypersensitivity and various macrophage functions, including phagocytosis, spreading, and tumoricidal activity. MIF functions as a proinflammatory cytokine and pituitary-derived hormone that potentiates endotoxemia. A considerable amount of MIF is stored in granules of anterior pituitary cells and is released by CRF. LPS releases pituitary MIF and raises the normal plasma level of 10 to 15 ng/ml to 50 ng/ml. MIF is also stored in peripheral monocytes and T lymphocytes. It is ubiquitously expressed in various organs, such as the brain and kidney. MIF antagonizes glucocorticoids. Glucocorticoids stimulate the release of MIF from T cells and macrophages in order to override the effects of glucocorticoids. MIF overrides, in a dose-dependent manner, glucocorticoid induced inhibition of TNF-α, IL-1β, IL-6 and IL-8 production by monocytes at the inflammatory site. MIF also overrides the anti-proliferative effect of glucocorticoids on antigen-stimulated T cells. The MIF gene contains enhancer-promoter motifs for transcription factors such as NFκB, Ck-1, nGRE, and CREB. MIF appears to have enzymatic activity and catalyzes an isomerization reaction [126–131].

Anti-MIF antibodies effectively suppress tumor growth and tumor-associated angiogenesis. MIF levels were significantly elevated on day 1 in septic shock, in contrast with trauma and control patients. Nonsurvivors of septic shock showed significantly higher MIF levels than survivors. Patients with septic adult respiratory distress syndrome (ARDS) showed higher MIF levels than those who did not develop ARDS. In acute respiratory distress syndrome high concentrations of MIF in the alveoli might contribute to pathogenesis. Regression analysis showed significant correlations between MIF and cortisol as well as between MIF and IL-6 levels and disease severity scores. In multitrauma patients MIF levels were not elevated at any time point. Therefore, MIF may serve as a marker for infection in trauma patients [132,133].

The levels of MIF in the sera in patients with ulcerative colitis (UC) were significantly higher than in normal controls, or in C-reactive protein (CRP)-positive cases with UC than in CRP-negative cases with UC, or in patients with severe colitis with UC than in mild colitis with UC. There was a positive relationship between serum MIF levels with the CRP levels and activities of colitis. Infiltrating cells at the colonic mucosa in UC expressed MIF, suggesting a role of MIF in the pathogenesis of UC [134].

REFERENCES

1. Warren JL. Type I cytokines, interferons and their receptors. In: Paul WE, editor. Fundamental Immunology 4th edition. New York: Lippincott-Raven, 1999;741–74.
2. Leonard WJ, Lin JX. Cytokine receptor signaling pathways. J Allergy Clin Immunol 2000;105:877–88.
3. Parrish-Novak J, Dillon SR, Nelson A, Hammond A, Sprecher C, Gross JA, Johnston J, Madden K, Xu W, West J, Schrader S, Burkhead S, Heipel M, Brandt C, Kuijper JL, Kramer J, Conklin D, Presnell SR, Berry J, Shiota F, Bort S, Hambly K, Mudri S, Clegg C, Moore M, Grant FJ, Lofton-Day C, Gilbert T, Rayond F, Ching A, Yao, L, Smith D, Webster P, Whitmore T, Maurer M, Kaushansky K, Holly RD, Foster D. Interleukin 21 and its receptor are involved in NK cell expansion and regulation of lymphocyte function. Nature 2000 Nov 2;408(6808):57–63.
4. Imada K, Leonard WJ. The Jak-STAT pathway. Mol Immunol 2000;37:1–11.
5. Notarangelo LD, Gilliani S, Mazza C, Mella P, Savoldi G, Rodriguez-Perez C, Mazzolari E, Fiorini M, Duse M, Plebani A, Ugazio AG, Vihinen M, Candotti F, Schumacher RF. Of genes and phenotypes: the immunological and molecular spectrum of combined immune deficiency. Defects of the gamma (c) -JAK3 signaling pathway as a model. Immunol Rev 2000;178:39–48.
6. Ellery JM, Kempshall SJ, Nicholls PJ. Activation of the interleukin 2 receptor: a possible role for tyrosine phosphatases. Cell Signal 2000;12:367–73.
7. McKenzie AN. Regulation of T helper type 2 cell immunity by interleukin-4 and interleukin 13. Pharmacol Ther 2000;88:143–51.
8. Jiang H, Harris MB, Rothman P. IL-4/IL-13 signaling beyond JAK/STAT. J Allergy Clin Immunol 2000;105:1063–70.
9. Murphy KM, Ouyang W, Farrar JD, Yang J, Ranganath S, Asnagli H, Afkarian M, Murphy TL. Signaling and transcription in T helper development. Annu Rev Immunol 2000;18: 451–94.
10. Appasamy PM. Biological and clinical implications of interleukin-7 and lymphopoiesis. Cytokines Cell Mol Ther 1999;5:25–39.
11. Akashi K, Kondo M, Weissman IL. Role of interleukin-7 in T-cell development from hematopoietic stem cells. Immunol Rev 1998 165:13–28.

12. Maeurer MJ, Lotz MT. Interleukin-7 (IL-7) knockout mice. Implications for lymphopoiesis and organ-specific immunity. Int Rev Immunol 1998;16:309–22.
13. Watanabe M, Euno Y, Yamazaki M, Hibi T. Mucosal IL-7-mediated immune responses in chronic colitis-IL-7 transgenic mouse model. Immunol Res 1999;20:251–59.
14. Demoulin JB, Renauld JC. Interleukin 9 and its receptor: an overview of structure and function. Int Rev Immunol 1998;16:345–64.
15. Fehniger TA, Caligiuri MA. Interleukin 15 : biology and relevance to human disease. Blood 2001;97:14–32.
16. Waldmann TA, Tagaya Y. The multifaceted regulation of interleukin-15 expression and the role of this cytokine in NK cell differentiation and host response to intracellular pathogens. Annu Rev Immunol 1999;17:19–49.
17. Reddy EP, Korapati A, Chaturvedi P, Rane S. IL-3 signaling and the role of Src kinases, JAKs and STATs: a covert liaison unveiled. Oncogene 2000;19:2532–47.
18. Weltman JK, Karim AS. IL-5: biology and potential therapeutic applications. Expert Opin Investig Drugs 2000;9:491–96.
19. Hamelmann E, Gelfand EW. Role of IL-5 in the development of allergen-induced airway hyperresponsiveness. Int Arch Allergy Immunol 1999;120:8–16.
20. Binder R, Kress A, Kan G, Herrmann K, Kirschfink M. Neutrophil priming by cytokines and vitamin D binding protein (Gc-globulin): impact on C5a-mediated chemotaxis, degranulation and respiratory burst. Mol Immunol 1999;36:885–92.
21. de Groot RP, Coffer PJ, Koenderman L. Regulation of proliferation, differentiation and survival by the IL-3/IL-5/GM-CSF receptor family. Cell Signal 1998;10:619–28.
22. Coffer PJ, Koenderman L, de Groot RP. The role of STATs in myeloid differentiation and leukemia. Oncogene 2000;19:2511–22.
23. Hirano T, Ishihara K, Hibi M. Roles of STAT3 in mediating the cell growth, differentiation and survival signals relayed through the IL-6 family of cytokine receptors. Oncogene 2000;19:2548–56.
24. Bravo J, Heath JK. Receptor recognition by gp130 cytokines. EMBO J 2000;19:2399–411.
25. Horn F, Henze C, Heidrich K. Interleukin-6 signal transduction and lymphocyte function. Immunobiology 2000;202:151–67.
26. Ershler WB, Keller ET. Age-associated increased interleukin-6 gene expression, late-life diseases, and frailty. Annu Rev Med 2000;51:245–70.
27. Nandurkar HH, Robb L, Begley CG. The role of IL-11 in hematopoiesis as revealed by a targeted mutation of its receptor. Stem Cells 1998;16 Suppl 2:53–65.
28. Schwertschlag US, Trepicchio WL, Dykstra KH, Keith JC, Turner KJ, Dorner AJ. Hematopoietic, immunomodulatory and epithelial effects of interleukin-11. Leukemia 1999;13:1307–15.
29. Auernhammer CJ, Melmed S. Leukemia-inhibitory factor-neuroimmune modulator of endocrine function. Endocr Rev 2000;21:313–45.
30. Smith SK, Charnock-Jones DS, Sharkey AM. The role of leukemia inhibitory factor and interleukin-6 in human reproduction. Hum Reprod 1998;13 Suppl 3:237–43; discussion 244–46.
31. Gadient RA, Patterson PH. Leukemia inhibitory factor, Interleukin 6, and other cytokines using the GP130 transducing receptor : roles in inflammation and injury. Stem Cells 1999;17:127–37.
32. Sleeman MW, Anderson KD, Lambert PD, Yancopoulos GD, Wiegand SJ. The ciliary

neurotrophic factor and its receptor, CNTFR alpha. Pharm Acta Helv 2000;74:265–72.
33. Perez Castro C, Nagashima AC, Pereda MP, Goldberg V, Chervin A, Largen P, Renner U, Stalla GK, Artz E. The gp130 cytokines interleukin-11 and ciliary neurotropic factor regulate through specific receptors the function and growth of lactosomatotropic and folliculostellate pituitary cell lines. Endocrinology 2000;141:1746–53.
34. De Miguel MP, De Boer-Brouwer M, Paniagua R, van den Hurk R, De Rooij DG, Van Dissel-Emiliani FM. Leukemia inhibitory factor and ciliary neurotropic factor promote the survival of Sertoli cells and gonocytes in coculture system. Endocrinology 1996;137: 1885–93.
35. Purcell WM, Westgate C, Atterwill CK. Rat brain mast cells: an *in vitro* paradigm for assessing the toxic effects of neurotropic therapeutics. Neurotoxicology 1996; 17:845–50.
36. Pu S, Dhillon H, Moldawer LL, Kalra PS, Kalra SP. Neuropeptide Y counteracts the anorectic and weight reducing effects of ciliary neurotropic factor. J Neuroendocrinol 2000;12:827–32.
37. Miyajima A, Kinoshita T, Tanaka M, Kamiya A, Mukouyama Y, Hara T. Role of oncostatin M in hematopoiesis and liver development. Cytokine Growth Factor Rev 2000;11:177–83.
38. Gomez-Lechon MJ. Oncostatin M : signal transduction and biological activity. Life Sci 1999;65:2019–30.
39. Grant SL, Begley CG. The oncostatin M signaling pathway: reversing the neoplastic phenotype? Mol Med Today 1999;5:406–12.
40. Loy JK, Davidson TJ, Berry KK, Macmaster JF, Danle B, Durham SK. Oncostatin M: development of a pleitropic cytokine. Toxicol Pathol 1999;27:151–55.
41. Heymann, D, Rousselle AV. gp130 Cytokine family and bone cells. Cytokine 2000;12: 1455–68.
42. Heinrich PC, Horn F, Graeve L, Dittrich E, Kerr I, Muller-Newen G, G J, Wollmer A. Interleukin-6 and related cytokines: effect on the acute phase reaction. Z Ernahrungswiss 1998;37:43–9.
43. Arce V, Garces A, de Bovis B, Filippi P, Henderson C, Pettmann B, de Lapeyriere O. Cardiotrophin-1 requires LIFRbeta to promote survival of mouse motoneurons purified by a novel technique. J Neurosci Res 1999;55:119–26.
44. Pulido EJ, Shames BD, Pennica D, O'Leary RM, Bensard DD, Cain BS, McIntyre RC Jr. Cardiotrophin-1 attenuates endotoxin-induced acute lung injury. J Surg Res 1999;84: 240–46.
45. Asai S, Saito Y, Kuwahara K, Mizuno Y, Yoshimura M, Higashikubo C, Tsuji T, Kishimoto I, Harada M, Hamanaka I, Takahashi N, Yasue H, Nakao K. The heart is a source of circulating cardiotrophin-1 in humans. Biochem Biophys Res Commum 2000;279:320–23.
46. Latchman DS. Cardiotrophin-1: a novel cytokine and its effects in the heart and other tissues. Pharmacol Ther 2000;85:29–37.
47. Richards CD, Langdon C, Deschamps P, Pennica D, Shaughnessy SG. Stimulation of osteoclast differentiation *in vitro* by mouse oncostatin M, leukaemia inhibitory factor, cardiotrophin-1 and interleukin 6: synergy with dexamethasone. Cytokine 2000;12: 613–21.
48. Funamoto M, Hishinuma S, Fujio Y, Matsuda Y, Kunisada K, Oh H, Negoro S, Tone E, Kishimoto T, Yamauchi-Takihara K. Isolation and characterization of the murine

cardiotrophin-01 gene: expression and norepinephrine-induced transcriptional activation. J Mol Cell Cardiol 2000;32:1275–84.
49. Talwar S, Squire IB, Downie PF, O'Brien RJ, Davies JE, Ng LL. Elevated circulating cardiotrophin-1 in heart failure: relationship with parameters of left ventricular systolic dysfunction. Clin Sci (Colch) 2000;99:83–88.
50. Hamanaka I, Saito Y, Nishikimi T, Magaribuchi T, Kamitani S, Kuwahara K, Ishikawa M, Miyamoto Y, Harada M, Ogawa E, Kajiyama N, Takahashi N, Izumi T, Shirakami G, Mori K, Inobe Y, Kishimoto I, Masuda I, Fukuda K, Nakao K. Effects of cardiotrophin-1 on hemodynamics and endocrine function of the heart. Am J Physiol Heart Circ Physiol 2000;279:H388–96.
51. Ito Y, Yamamota M, Li M, Mitsuma N, Tanaka F, Doyu M, Suzumura A, Mitsuma T, Sobue G. Temporal expression of mRNAs for neuropoietic cytokines, interleukin-11 (IL-11), oncostatin M (OSM), cardiotrophin-1 (CT-1) and their receptors (IL-11Ralpha and OSMRbeta) in peripheral nerve injury. Neurochem Res 2000;25:1113–18.
52. Ghosh S, Ng LL, Talwar S, Squire IB, Galinanes M. Cardiotrophin-1 protects the human myocardium from ischemic injury. Comparison with the first and second window of protection by ischemic preconditioning. Cardiovasc Res 2000;48:440–47.
53. Oppenheim RW, Wiese S, Prevette D, Armanini M, Wang S, Houenou LJ, Holtmann B, Gotz R, Pennica D, Sendtner M. Cardiotrophin-1, a muscle-derived cytokine, is required for the survival of subpopulations of developing motoneurons. J Neurosci 2001; 21:1283–91.
54. Hamanaka I, Saito Y, Yasukawa H, Kishimoto I, Kuwahara K, Miyamoto Y, Harada M, Ogawa E, Kajiyama N, Takahashi N, Izumi T, Kawakami R, Masuda I, Yoshimura A, Nakao K. Induction of JAB/SOCS-1/SSI-1 and CIS3/SOCS-3/SSI-3 is involved in gp130 resistance in cardiovascular system in rat treated with cardiotrophin-1 *in vivo*. Circ Res 2001;88:727–32.
55. Trinchieri G. Interleukin-12. A proinflammatory cytokine with immunoregulatory functions that bridge innate resistance and antigen-specific adaptive immunity. Annu Rev Immunol 1995;13:251–76.
56. Braun MC, Kelsall BL. Regulation of interleukin-12 production by G-protein-couple receptors. Microbes Infec 2001;3:99–107.
57. Lammas DA, Casanova JL, Kumararatne DS. Clinical consequences of defects in the IL-12-dependent interferon-gamma (IFN-gamma) pathway. Clin Exp Immunol 2000;121: 417–25.
58. Dorman SE, Holland SM. Interferon-gamma and interleukin-12 pathway defects and human disease. Cytokine Growth Factor Rev 2000;11:321–33.
59. Oppmann B, Lesley R, Blom B, Timans JC, Xu Y, Hunte B, Vega F, Yu N, Wang J, Singh K, Zonin F, Vaisberg E, Churakova T, Liu M, Gorman D, Wagner J, Zurawski, S, Liu Y, Abrams JS, Moore KW, Rennick D, de Waal-Malefyt R, Hannum C, Bazan JF, Kastelein RA. Novel p19 protein engages IL-12p40 to form a cytokine, IL-23, with biological activities similar as well as distinct from IL-12. Immunity 2000;13:715–25.
60. Durbin JE, Fernandez-Sesma A, Lee CK, Rao TD, Frey AB, Moran TM, Vukmanovic S, Garcia-Sastre A, Levy DE. Type I IFN modulates innate and specific antiviral immunity. J Immunol 2000;164:4220–28.
61. Nguyen KB, Cousens LP, Doughty LA, Pien GC, Durbin JE, Biron CA. Interferon alpha/beta-mediated inhibition and promotion of interferon gamma : STAT1 resolves a paradox. Nat Immunol 2000;1:70–76.

62. Edwards L. The interferons. Dermatol Clin 2001;19:139–46.
63. Cooper MA, Fehniger TA, Ponnappan A, Mehta V, Wewers MD, Caligiuri MA. Interleukin-1beta costimulates interferon-gamma production by human natural killer cells. Eur J Immunol 2001;31:792–801.
64. Usacheva A, Smith R, Minshall R, Baida G, Seng S, Croze E, Colamonici OR. The WD motif-containing protein RACK1 is required for recruitment and activation of STAT1 through the type I iterferon receptor. J Biol Chem 2001.
65. Bernabei P, Allione A, Rigamonti L, Bosticardo M, Losana G, Borghi I, Forni G, Novelli F. Regulation of interferon-γ receptor (IFN-γR) chains: a peculiar way to rule the life and death of human lymphocytes. Eur Cytokine Netw 2001;12:6–14.
66. Moore KW, de Walal Malefyt R, Coffman RL, O'Garra A. Interleukin-10 and the interleukin-10 receptor. Annu Rev Immunol 2001;19:683–765.
67. Gallagher G, Dickensheets H, Eskdale J, Izotova LS, Mirochnitchenko OV, Peat JD, Vazquez N, Pestka S, Donnelly RP, Kotenko SV. Cloning, expression and initial characterization of interleukin-19 (IL-19), a novel homologue of human interleukin-10 (IL-10). Genes Immun 2000;1:442–50.
68. Blumberg H, Conklin D, Xu WF, Grossmann A, Brender T, Carollo S, Eagan M, Foster D, Haldeman BA, Hammond A, Haugen H, Jelinek L, Kelly JD, Madden K, Maurer MF, Parrish-Novak J, Prunkard D, Sexson S, Sprecher C, Waggie K, West J, Whitmore TE, Yao L, Kuechle MK, Dale BA, Chandrasekher YA. Interleukin 20: discovery, receptor identification, and role in epidermal function. Cell 2001 12;104:9–19.
69. Dumoutier L, Van Roost E, Ameye G, Michaux L, Renauld JC. IL-TIF/IL-22: genomic organization and mapping of the human and mouse genes. Genes Immun 2000;1:488–94.
70. Xie MH, Aggarwal S, Ho WH, Foster J, Zhang Z, Stinson J, Wood WI, Goddard AD, Gurney AL. Interleukin (IL)-22, a novel human cytokine that signals through the interferon receptor-related proteins CRF2-4 and IL-22R. J Biol Chem 2000;275:31335–9.
71. Krauker T, Vilcek J, Oppenheim JJ. Proinflammatory cytokines. In: Paul WE, editor. Fundamental Immunology 4th edition. New York: Lippincott-Raven, 1999;775–811.
72. Dinarello CA. The role of the interleukin-1-receptor antagonist in blocking inflammation mediated by interleukin-1. N Engl J Med 2000a;343:732–4.
73. Frontiers
74. Locksley RM, Killeen N, Lenardo MJ. The TNF and TNF receptor superfamilies: integrating mammalian biology. Cell 2001;104:487–501.
75. Papadakis KA, Targan SR. Tumor necrosis factor: biology and therapeutic inhibitors. Gastroenterology 2000;119:1148–57.
76. Shakhov AN, Nedospasov SA. Expression profiling in knockout mice: lymphotoxin versus tumor necrosis factor in the maintenance of splenic microarchitecture. Cytokine Growth Factor Rev 2001;12:107–19.
77. Schmitz I, Kirchhoff S, Krammer PH. Regulation of death receptor-mediated apoptosis pathways. Int J Biochem Cell Biol 2000;32:123–36.
78. Ventors HD, Dantzer R, Kelley KW. Tumor necrosis factor-alpha induces neuronal death by silencing survival signals generated by the type I insulin-like growth factor receptor. Ann N Y Acad Sci 2000;917:210–20.
79. Wu H, Park YC, Ye H, Tong L. Structural studies of human TRAF2. Cold Spring Harb Quant Biol 1999;64:541–49.
80. Cao A, Tanaka M, Regnier C, Rothe M, Yamit-heiz A, Woronicz JD, Fuentes ME, Durnin MH, Dalrymple SA, Goeddel DV. NF-kappa B activation by tumor necrosis factor and

interleukin-1. Cold Spring Harb Symp Quant Biol 1999;64:473–83.
81. Wajant H, Scheurich P. Tumor necrosis factor receptor-associated factor (TRAF)2 and its role in TNF signaling. Int J Biochem Cell Biol 2001;33:19–32.
82. Arend WP, Gabay C. Physiologic role of interleukin-1 receptor antagonist. Arthritis Res 2000;2:245–48.
83. Daun JM, Fenton MJ. Interleukin-1/Toll receptor family members: receptor structure and signal transduction pathways. J Interferon Cytokine Res 2000;20:843–55.
84. O'Neill L. The Toll/interleukin-1 receptor domain: a molecular switch for inflammation and host defence. Biochem Soc Trans 2000;28:557–63.
85. Arend WP, Guthridge CJ. Biological role of interleukin 1 receptor antagonist isoforms. Ann Rheum Dis 2000;59 Suppl 1:i60–4.
86. Hacham M, Argov S, White RM, Segal S, Apte RN. Distinct patterns of IL-1 alpha and IL-1 beta organ distribution--a possible basis for organ mechanisms of innate immunity. Adv Exp Med Biol 2000;479:185–202.
87. Neveu PJ, Liege S. Mechanisms of behavioral and neuroendocrine effects of interleukin-1 in mice. Ann N Y Acad Sci 2000;917:175–85.
88. Rothwell NJ, Luheshi GN. Interleukin 1 in the brain: biology, pathology and therapeutic target. Trends Neurosci 2000;23:618–25.
89. Schiff MH. Role of interleukin 1 and interleukin 1 receptor antagonist in the mediation of rheumatoid arthritis. Ann Rheum Dis 2000;59 Suppl 1:i103–8.
90. Lebel-Binay S, Berger A, Zinzindohoue F, Cugnenc P, Thiounn N, Fridman WH, Pages F. Interleukin-18: biological properties and clinical implications. Eur Cytokine Netw 2000;11:15–26.
91. Akira S. The role of IL-18 in innate immunity. Curr Opin Immunol 2000;12:59–63.
92. Dinarello CA. Interleukin-1 beta, interleukin-18, and the interleukin-1 beta converting enzyme. Ann N Y Acad Sci 1998;856:1–11.
93. Golab J. Interleukin 18--interferon gamma inducing factor--a novel player in tumour immunotherapy? Cytokine 2000;12:332–8.
94. Dinarello CA. IL-18: A TH1-inducing, proinflammatory cytokine and new member of the IL-1 family. J Allergy Clin Immunol 1999;103(1 Pt 1):11–24.
95. Means TK, Golenbock DT, Fenton MJ. Structure and function of Toll-like receptor proteins. : Life Sci 2000;68:241–58.
96. Okamura H, Kashiwamura S, Tsutsui H, Yoshimoto T, Nakanishi K. Regulation of interferon-gamma production by IL-12 and IL-18. Curr Opin Immunol 1998;10:259–64.
97. Dinarello CA, Novick D, Puren AJ, Fantuzzi G, Shapiro L, Muhl H, Yoon DY, Reznikov LL, Kim SH, Rubinstein M. Overview of interleukin-18: more than an interferon-gamma inducing factor. J Leukoc Biol 1998;63:658–64.
98. Dinarello CA. Interleukin-18, a proinflammatory cytokine. Eur Cytokine Netw 2000;11: 483–86.
99. Tsutsui H, Matsui K, Okamura H, Nakanishi K. Pathophysiological roles of interleukin-18 in inflammatory liver diseases. Immunol Rev 2000;174:192–209.
100. Murdoch C. CXCR4: chemokine receptor extraordinaire. Immunol Rev 2000;177:175–84.
101. Mackay CR. Chemokines: immunology's high impact factors. Nat Immunol 2001;2: 95–101.
102. Yoshie O. Immune chemokines and their receptors: the key elements in the genesis, homeostasis and function of the immune system. Springer Semin Immunopathol 2000;22:

371–91.
103. Thelen M. Dancing to the tune of chemokines. Nat Immunol 2001;2:129–34.
104. Moser B, Loetscher P. Lymphocyte traffic control by chemokines. Nat Immunol 2001;2: 123–28.
105. Mukaida N. Interleukin-8: an expanding universe beyond neutrophil chemotaxis and activation. Int J Hematol 2000;72:391–98.
106. Wang JF, Park IW, Groopman JE. Stromal cell-derived factor-1alpha stimulates tyrosine phosphorylation of multiple focal adhesion proteins and induces migration of hematopoietic progenitor cells: roles of phosphoinositide-3 kinase and protein kinase C. Blood 2000;95:2505–13.
107. Tilton B, Ho L, Oberlin E, Loetscher P, Baleux F, Clark-Lewis I, Thelen M. Signal transduction by CXC chemokine receptor 4. Stromal cell-derived factor 1 stimulates prolonged protein kinase B and extracellular signal-regulated kinase 2 activation in T lymphocytes. J Exp Med 2000;192:313–24.
108. Taub DD, Ortaldo JR, Turcovski-Corrales SM, Key ML, Longo DL, Murphy WJ. Beta chemokines costimulate lymphocyte cytolysis, proliferation, and lymphokine production. J Leukoc Biol 1996; 59:81–89.
109. Grimm MC, Ben-Baruch A, Taubb DD, Howard OM, Resau JH, Wang JM, Ali H, Richardson R, Snyderman R, Oppenheim JJ. Opiates transdeactivate chemokine receptors: delta and mu opiate receptor-mediated heterologous desensitization. J Exp Med 1998;188: 317–25.
110. Wang JM, Ueda H, Howard OM, Grimm MC, Chertov O, Gong X, Gong W, Resau JH, Broder CC, Evans G, Arthur LO, Ruscetti FW, Oppenheim JJ. HIV-1 envelope gp120 inhibits the monocyte response to chemokines through CD4 signal-dependent chemokine receptor down-regulation. J Immunol 1998;161:4309–17.
111. Bottner M, Krieglstein K, Unsicker K. The transforming growth factor-betas: structure, signaling, and roles in nervous system development and functions. J Neurochem 2000;75: 2227–40.
112. Strelau J, Sullivan A, Bottner M, Lingor P, Falkenstein E, Suter-Crazzolara C, Galter D, Jaszai J, Krieglstein K, Unsicker K. Growth/differentiation factor-15/macrophage inhibitory cytokine-1 is a novel trophic factor for midbrain dopaminergic neurons *in vivo*. J Neurosci 2000;20:8597–603.
113. Dennis PA, Rifkin DB. Cellular activation of latent transforming growth factor beta requires binding to the cation-independent mannose 6-phosphate/insulin-like growth factor type II receptor. Proc Natl Acad Sci USA 1991;88:580–84.
114. Mangasser-Stephan K, Gartung C, Lahme B, Gressner AM. Expression of isoforms and splice variants of the latent transforming growth factor beta binding protein (LTBP) in cultured human liver myofibroblasts. Liver 2001;21:105–113.
115. Blobe GC, Schiemann WP, Pepin MC, Beauchemin M, Moustakas A, Lodish HF, O'Connor-McCourt MD. Functional roles for the cytoplasmic domain of the Type III TGF-{beta} receptor in regulating transforming growth factor {beta} signaling. J Biol Chem 2001;276:24627–37.
116. Miyazono K, Kusanagi K, Inoue H. Divergence and convergence of TGF-beta/BMP signaling. J Cell Physiol 2001;187:265–276.
117. Wong SF, Lai LC. The role of TGFbeta in human cancers. Pathology 2001;33:85–92
118. Cain K, Freathy C. Liver toxicity and apoptosis: role of TGF-beta(1), cytochrome c and the apoptosome. Toxicol Lett 2001;120:307–315.

119. Fairlie WD, Zhang HP, Wu WM, Pankhurst SL, Bauskin AR, Russell PK, Brown PK, Breit SN. The propeptide of the TGF-{beta} superfamily member, macrophage inhibitory cytokine-1 (MIC-1) is a multifunctional domain which can facilitate protein folding and secretion. J Biol Chem 2001;276: 16911–8.
120. Cruikshank WW, Kornfeld H, Center DM. Signaling and functional properties of interleukin-16. Int Rev Immunol 1998;16:523–40.
121. Center DM, Kornfeld H, Ryan TC, Cruikshank WW. Interleukin 16: implications for CD4 functions and HIV-1 progression. : Immunol Today 2000;21:273–80.
122. Cruikshank WW, Kornfeld H, Center DM. Interleukin-16. J Leukoc Biol 2000;67:757–66.
123. Center DM, Kornfeld H, Cruikshank WW. Interleukin-16. : Int J Biochem Cell Biol 1997;29:1231–4.
124. Fossiez F, Banchereau J, Murray R, Van Kooten C, Garrone P, Lebecque S Interleukin-17. Int Rev Immunol 1998;16:541–51.
125. Linden A, Hoshino H, Laan M. Airway neutrophils and interleukin-17. Eur Respir J 2000;15:973–7.
126. Bloom BR, Bennett B, Mechanism of a reaction in vitro associated with delayed-type hypersensitivity. Science 1966:153:8–32.
127. David J. Delayed hypersensitivity in vitro: its mediation by cell-free substances formed by lymphoid cell-antigen interactions. Proc Natl Acad Aci USA 1966;56:72–77.
128. Donnelly SC, Bucala R. Macrophage migration inhibitory factor: a regulator of glucocorticoid activity with a critical role in inflammatory disease. Mol Med Today 1997;3:502–7.
129. Weiser WY, Temple PA, Witeek-Giannotti JS, Remold HG, Clark SC, David JR. Molecular cloning of cDNA encoding a human macrophage migration inhibitory factor. Proc Natl Acad Sci USA 1989;86:7522–26.
130. Nishihira J. Macrophage migration inhibitory factor (MIF): its essential role in the immune system and cell growth. J Interferon Cytokine Res 2000;20:751–62.
131. Mitchell RA, Bucala R Tumor growth-promoting properties of macrophage migration inhibitory factor (MIF). Semin Cancer Biol 2000;10:359–66.
132. Joshi PC, Poole GV, Sachdev V, Zhou X, Jones Q. Trauma patients with positive cultures have higher levels of circulating macrophage migration inhibitory factor (MIF). Res Commun Mol Pathol Pharmacol 2000;107(1–2):13–20.
133. Beishuizen A, Thijs LG, Haanen C, Vermes I. Macrophage migration inhibitory factor and hypothalamo-pituitary-adrenal function during critical illness. J Clin Endocrinol Metab 2001;86:2811–6.
134. Murakami H, Akbar SM, Matsui H, Onji M.Macrophage migration inhibitory factor in the sera and at the colonic mucosa in patients with ulcerative colitis: clinical implications and pathogenic significance. Eur J Clin Invest 2001;31:337–43.

Edited by I. Berczi and A. Szentivanyi

Steroid Hormones

[1]ISTVAN BERCZI, [1]EVA NAGY, [2]EDWARD BARAL and [3]ANDOR SZENTIVANYI

Departments of [1]Immunology and [2]Medicine, Faculty of Medicine, The University of Manitoba, Winnipeg, Manitoba, Canada, R3E 0W3 and [3]Department of Internal Medicine, The University of South Florida, Tampa, Florida 33612, USA

ABSTRACT

The steroid hormones that play major roles in the neuroimmune regulatory network include glucocorticoids, aldosterone, estrogens, androgens and vitamin D. These hormones have cytoplasmic or nuclear receptors, which are transcription factors that directly regulate gene expression. Glucocorticoids are produced in the adrenal gland as well as the thymus and are fundamental to immune function. Physiological levels are required for the normal development and functioning of the immune system. Under pathophysiological conditions the serum level of glucocorticoids is elevated, which plays an important role in the suppression of the adoptive immune response that depends on T lymphocytes. Glucocorticoids also exert a powerful anti-inflammatory effect. Elevated glucocorticoids during acute phase reactions augment the production of natural antibodies and support the production of acute phase proteins by the liver.

Luteinizing hormone releasing hormone and gonadotropins exert a direct regulatory influence on the immune system, in addition to the regulation of sex steroid hormones. In turn immune-derived cytokines regulate the production of gonadotropins. These mechanisms insure the coordination of reproduction with health status and prevent inopportune conception.

Estrogens regulate the thymus and suppress cell-mediated immune reactions. The antibody response and natural immunity (NK cytotoxicity, phagocytosis) are augmented by estradiol. Testosterone is immunosuppressive during trauma and shock. Many of the immunological effects of testosterone are due to its conversion to estradiol by aromatase in the thymus and in other lymphoid organs. The adrenal androgen, dehydroepyandrosterone stimulates immune reactions in experimental animals and in man. It antagonizes the immunosuppressive effect of glucocorticoids and its age-related decline may contribute to the immunodeficiency that develops in elderly individuals. Progesterone is a powerful immunosupressive hormone. It plays a major role in the protection of the fetus during mammalian reproduction. Progesterone also contributes to the generation of self tolerance and protects against the excessive activation of the immune system.

1, 25-dihydroxy vitamin D3 (VD3) is a major immunoregulator with a powerful immunosuppressive potential. Deficiency of VD3 plays an important role in the development of autoimmune disease. VD3 is also required for normal immune function and for proper defense against infectious disease and cancer.

1. GLUCOCORTICOIDS (GC)

Glucocorticoid treatment exerts an immunosuppressive effect in a number of species, including man, mouse, rat, guinea pig, rabbit, chicken, lizard and frog. Serum immunoglobulin and humoral and cell-mediated immune reactions are depressed in animals chronically treated with GC. However, immunological memory cells and the cells responsible for the maintenance of graft versus host reactions are resistant to the suppressive effect of GC [1].

1.1. Glucocorticoid receptors

Leukocytes express GC receptors and expression is highest in thymic epithelial cells and bone marrow cells. The spleen has lower levels of expression and neutrophils have fewer receptors than lymphocytes and monocytes. Receptor numbers increase during lymphocyte activation and expression is regulated by plasma corticosteroid levels, hormones, cytokines and by genetic factors [2,3].

Steroid and thyroid hormones, as well as vitamins A and D and their metabolites, are recognized by intracellular (cytoplasmic or nuclear) receptors that belong to the same receptor family. These receptors are nuclear transcription factors, which directly regulate gene expression in the nucleus. Steroid and thyroid hormones, as well as vitamins A and D, are small lipophylic molecules and for this reason are able to diffuse into cells and bind intracellularly to their receptors. The GC receptor (GR) is located in the cytoplasm where it is bound to heat shock protein 90 (HSP90). Glucocorticoids dissociate the receptor from HSP90 and induce its translocation into the nucleus. The receptors for thyroid hormones and retinoic acid are located in the nucleus. The regulation of the transcriptional activator AP-1, which is composed of homo- and heterodimers of Jun and Fos proteins, is one pathway to the regulation of gene transcription by these receptors [4].

The impact of the proinflammatory cytokine interleukin-1β IL-1α was examined on GR translocation from the cytoplasm to the nucleus, on cytosolic radioligand GR binding, and on GR-mediated gene transcription. Pretreatment of L929 cells with IL-1α followed by dexamethasone (DEX) led to about 20% inhibition of DEX-induced GR-mediated gene transcription, whereas coincubation of IL-1α plus DEX for 24 hours inhibited DEX-induced GR-mediated gene activity by up to 42%. The latter effect was reversed by interleukin-1 receptor agonist (IL-1ra) [Pariante et al. 1999]. Interleukin-1 up-regulated GR mRNA expression in hypothalamic corticotropin releasing hormone (CRH) secreting cells. Macrophage migration inhibitory factor (MIF) counter-regulated GC action within the immune system. Interleukin-2 and IL-6 significantly up-regulated the number of GR binding sites and the expression of GR mRNA in peripheral blood mononuclear cells and in osteoblast-like Saos-2 cells. Cytokine-induced autocrine transcription of GR in immunocytes and/or osteoblasts could be a mechanism that restrains excess cytokine production [5].

Membrane GR in both mouse and human lymphoma cells are necessary for the initiation of GC-induced apoptosis which is related to the developmental phenomenon of thymic involution [6].

Differences have been observed in the activation of adrenal steroid receptors in the brain and in immune tissues of Sprague-Dawley, Fischer 344 and Lewis rats. F344 rats exhibited the greatest magnitude of type II receptor activation in brain tissues; Lewis rats showed the lowest magnitude in immune tissues during stress [7].

1.2. Glucocorticoid action

1.2.1. Molecular mechanisms

The immunosuppressive activity of GC is due to the transcriptional activation of IκBα, which is the cytoplasmic inhibitor of NF-κB. NF-κB activates many immunoregulatory genes in response to inflammatory stimuli. I-κBα traps activated NF-κB and renders it inactive in the cytoplasm. The inactivation of NF-κB markedly decreases cytokine secretion and is largely responsible for the anti-inflammatory activity of GC [8,9]. A third nuclear transcription factor regulated by glucocorticoids is the cAMP-responsive element binding protein (CREB) [10]. The GR and the AP-1 transcription factors interact on target gene promoters, which contain only a binding site for either one of the two transcription factors. A negative interference of both factors with each other's activity was frequently observed. However, synergism is also possible under cell-specific conditions and when AP-1 is a homodimer of c-Jun. Numerous other examples of transcription factor interactions have been described. The transcription factor families of NF-κB/Rel as well as Stat, Oct, and C/EBP are engaged in cross-talk with steroid receptors [11].

IFN-γ and GC regulate inflammatory and immune responses through Stat1 and GR transcription factors, respectively. The GR stimulated Stat1-dependent transcription in a ligand-dependent manner, while Stat1 did not influence GR-dependent transcription. This costimulatory effect was not dependent on a consensus GC response element in the Stat1-responsive promoters, but required the DNA-binding and transactivation functions of GR, and DEX-induced protein synthesis. The activation of the FcγRI promoter required both Stat1 and the Ets family transcription factor PU.1 [12].

Lipocortin 1 is a GC-induced protein, which inhibits neurogenic inflammation. In adddition, lipocortin 1 plays a role in GC-induced inhibition of HPA response to cytokines [13,14].

In asthma and in other inflammatory and immunological diseases, resistance to the anti-inflammatory effects of GC may develop. In this case there is a marked reduction in the binding of GC receptors to DNA, which is the consequence of increased binding of GR to AP-1. This cross-talk ability of glucocorticoid receptors is essential for mouse development, while the activation of target promoters carrying a glucocorticoid response element (GRE) is dispensable. The cross-talk function is responsible for almost all regulatory actions of cortisol in the immune system. Cytokine release that activates AP-1 may induce secondary steroid resistance in asthma. High concentrations of β2-adrenoreceptor agonists activate another GR binding transcription factor CREB [15,16].

Dexamethasone induced IκBα mRNA expression in rat brain. Lipopolysaccharide (LPS) but not DEX induced IL-1β mRNA expression. Pretreatment with DEX 30 minutes before LPS injection enhanced the expression of IκBα mRNA in the brain in a dose-dependent manner. Brain IκBα expression induced by LPS was attenuated by the GC antagonist RU-486. In the brain of adrenalectomized rats, LPS increased the expression of IL-1β mRNA. Therefore, in the brain GC selectively induce IκBα mRNA expression, which inhibits the production of LPS-induced proinflammatory cytokines [17].

In the marmoset, which is a primate paradigm of peripheral GC resistance, IL-1 stimulated the secretion of adrenocorticotropic hormone (ACTH) in a dose-dependent fashion. Maximal blood level was achieved at 120 minutes after injection, and lasted up to 240 minutes; ACTH levels returned to baseline at 300 minutes. Plasma cortisol levels were elevated in accordance with ACTH levels. Body temperature elevation occurred 10–15 minutes after injection in a dose-dependent fashion and lasted 3 hours. These results suggest that the pyrogenic effect of interleukin-1 is associated with integrated activation of the HPA axis [18].

1.2.2. Effect on the thymus

Glucocorticoids induce thymus involution by increasing the rate of apoptosis in thymocytes [19]. The GC sensitivity of thymocytes is influenced by thymic hormones and genes of the major histocompatibility (MHC) complex [20]. Corticosteroids suppress PGE2 and 6-keto $PGF_{1\alpha}$ production by phagocytic cells in the thymus [21].

The expression of the GC receptor antisense transgene was targeted to the thymus with the aid of the proximal *lck* promoter, which is predominantly active in immature thymocytes. There was a decrease in thymic size in embryonic and pospartum mice, which was due primarily to the decrease in the number of $CD4^+$ $CD8^+$ cells. The susceptibility to T cell receptor (TCR)-mediated apoptosis was enhanced, and thymocyte loss was also detected before the expression of TCRαβ. This suggests that GC are necessary for the transition of thymocytes from the $CD4^-$ $CD8^-$ stage to the $CD4^+$ $CD8^+$ stage and that GC support the survival of the latter cell population in the thymus [22].

The molecular interaction between immune activators (cytokines and TCRs) and the GR in cells in which GC play a key regulatory function, namely cells susceptible to tumor necrosis factor (TNF)-induced cytotoxicity, pituitary cells and thymocytes was studied. Cytokines, such as TNF-α and IL-1, increased GC-induced transcriptional activity of the GR via the DNA GREs. The priming of fibroblasts with a low dose of TNF significantly increased their sensitivity to glucocorticoid inhibition of TNF-induced apoptosis (without involving NF-κB). Priming of AtT-20 murine pituitary corticotroph cells and Cushing pituitary cells with IL-1 increased their sensitivity to GC inhibition of CRH-induced ACTH/POMC expression. In thymocytes the activation of the T cell receptor counteracts the GC-induced thymic apoptosis by downregulating the GC action on GRE-driven apoptotic genes. Thus, cytokines and immune mediators prevent their own deleterious effects not only by stimulating GC production, but also by modifying the sensitivity of the target cells for the GC counter-regulatory action [23].

The thymic epithelium is capable of synthesizing GC. This plays a significant role in the normal development of thymocytes [24,22]. Thymus-derived GC antagonize T lymphocyte antigen receptor (TCR)-mediated cell deletion, suggesting a role for endogenous thymic GC in promoting survival of thymocytes. The inhibition of thymus GC biosynthesis causes an increase in thymocyte apoptosis, which is directly proportional to the number of MHC-encoded molecules present, and therefore available for TCR recognition. The expression of CD5 on CD4+CD8+ thymocytes (an indicator of TCR-mediated activation) increased in a TCR- and MHC-dependent manner when corticosteroid production or responsiveness was decreased. These results indicate that thymus-derived GC determine where the window of thymocyte selection occurs in the TCR avidity spectrum [24].

1.2.3. Effect on phagocytic cells and antigen presentation

In the monocyte-macrophage lineage, metabolism, chemotaxis, phagocytosis, cytotoxic reactions, the capacity to present antigen and the secretion of IL-1, IL-1ra, IL-6, and the ability to respond to lymphokines are all inhibited by GC. In macrophages GC suppress the production of collagenase, elastase, plasminogen activated TNFα, superoxide and nitric oxide [20,25–27]. The treatment of human beings, or the exposure of human eosinophils to GC *in vitro* inhibited the chemotactic response. Human neutrophilic granulocyte function was also impaired by GC. Receprotods for the Fc portion of immunoglobulins (FcR), complement receptors and chemotactic receptors were rendered nonfunctional and β-adrenergic receptors were uncoupled from adenylate cyclase after GC treatment [20].

The treatment of human peripheral blood monocytes with GC induced apoptosis which required the activation of caspases. Proteolytic activation of caspase-8 and caspase-3 was observed. CD95 and CD95 ligand (CD95L) were up-regulated in a dose-dependent manner on the cell membrane and also released after treatment with GC. The GC receptor antagonist, mifepristone, diminished monocyte apoptosis as well as CD95/CD95L expression and subsequent caspase-8 and caspase-3 activation. The caspase inhibitor N-acetyl-Asp-Glu-Val-Asp-aldehyde suppressed caspase-3 activation and apoptosis, but did not downregulate caspase-8 activation and expression of CD95 and CD95L. The GC-induced monocyte apoptosis was abolished by a neutralizing CD95L monoclonal antibody. This indicates that GC-induced monocyte apoptosis is at least partially mediated by an autocrine or paracrine pathway involving the CD95/CD95L system [28].

GC selectively inhibited the expression of B7-1 by human monocytes, while leaving B7-2 unaffected. This effect occured at concentrations that are reached during therapeutic applications of GC in humans. Inhibition of B7-1 occurs at the transcriptional level. Up-regulation of the molecule can similarly be inhibited by GC [29].

The function of dendritic cells (DC) was impaired in mice after treatment with DEX [30]. In LPS-stimulated dendritic cells, GC strongly reduced the secretion of IL-12p70 and, to a lesser extent, the production of IL-6 and TNFα. GC did not influence the cell surface expression of HLA-DR or the costimulatory molecules CD40 and CD80 or the ability of DC to take up Ag. GC pretreatment of DC did not affect their ability to stimulate CD4+ helper T (Th) cell proliferation in response to superantigen. However, GC-pretreated DC had a reduced ability to promote the production of IFNγ in CD4+ Th lymphocytes. IFNγ production could be restored by exogenous IL-12. GC treatment of DC enhanced the secretion of the anti-inflammatory cytokine IL-10 and the type 2 cytokine IL-5 by T cells [31]. GC increased the endocytic activity of human dendritic cells *in vitro*. This favored the scavenging of antigen from the external milieu, decreased antigen concentration and availability, and simultaneously inhibited the capacity of DC to stimulate T cells [32]. The treatment of DC with DEX resulted in apoptosis. Monocytes, which are precursor cells of DC, are resistant to DEX-mediated apoptosis. This apoptosis of DC was independent of caspase activation as it was not inhibited by the caspase inhibitor, Z-VAD-fmk. Agonistic CD40 antibody completely inhibited DEX-induced cell death. This suggests that CD40 signaling may selectively modulate GC-mediated DC apoptosis [33].

Glucocorticoids increased the expression of HLA antigens and IFN-γ receptors by monocytes [34,35], induced high affinity IL-1 receptors on human blood lymphocytes [36], upregulated high affinity IL-6 receptors on human epithelial cells [37].

Low concentrations of GC induce the production of MIF from macrophages. MIF is capable of overriding the GC-mediated inhibition of cytokine secretion by LPS activated monocytes and antagonizes the protective effect of GC in lethal endotoxemia [38].

GC act through the type I corticosteroid receptor (GCR-I), which is also known as the mineralocorticoid receptor. Thus, the type I receptor antagonist, spironolactone and the type II receptor (GCR-II) antagonist RU38486, both partially reversed the inhibitory effect of GC on IL-1ra by LPS stimulated monocytes. In this case aldosterone was also inhibitory, and this inhibition could be blocked by spironolactone [27].

1.2.4. Effect on T lymphocytes

Glucocorticoids inhibit the helper, suppressor and killer function of T lymphocytes. The production of IL-2, IFN-γ, CSF, and macrophage activating factor (MAF) by T lymphocytes was inhibited by GC, which could be abrogated by additional treatment with IL-2. GC, at concentrations analogous to plasma levels in stress, cause a shift in type 1/type 2 cytokine production by human peripheral blood mononuclear cells (PBMCs) towards a type 2 response. This is primarily the result of downregulation of type 1 cytokines, which allows for the enhanced production of type 2 cytokines. By virtue of their downregulation of FasL expression, GC inhibit activation-induced T cell apoptosis. Glucocorticoids are also potent inducers of apoptosis, and can cause the death of CD4+CD8+ thymocytes. Thymic epithelial cells produce GC, and participate in antigen-specific thymocyte development and promote positive and negative selection [39,40–42].

Corticosterone (CON) is the principal GC in the rat. Physiological concentrations of CON accelerated the anti-T cell receptor antibody-induced growth of rat lymphocytes after 2–3 days in culture and inhibited cell growth after 5–7 days. Corticosterone had to be present within 60 minutes after the initiation of TCR activation in order to produce maximal enhancement. CON treatment increased the expression of the IL-2 receptor α chain on $TCR\alpha\beta^+$ and $CD4^+$ T cells after 48–72 hours in culture [43].

Human mature T cells are considered to be resistant to the apoptotic effect of GC, however, mature T cells show apoptosis if exposed to GC soon after mitotic stimulation. These cells are also sensitive to apoptosis induced by growth factor deprivation. $CD8^+$ cells were more sensitive than $CD4^+$ cells. Apoptosis was increased by macromolecular synthesis inhibitors (cyclohexamine and puromycin) and antagonized by treatment with IL-2, -4, and -10 [44]. Glucocorticoids have two different effects on the vulnerability of human antigen-specific T cells. Firstly, GC induce T cell apoptosis in a CD95-independent, but caspase-dependent manner. Secondly, GC protect T cells from CD95-mediated apoptosis, which is also caspase-dependent. GC-mediated inhibition of apoptosis requires gene transcription, in particular of GITR and GILZ, which affect NFκB transcription activity and the expression of the Fas/FasL system. An increase in BCL-2 expression is observed upon incubation with GC [45,46].

DEX suppressed the production of IL-2, IL-4, and IFNγ in spleen cells from BALB/c mice. These data challenge the hypothesis that GC can differentially regulate Th1-like versus Th2-like cytokine production [47].

DEX inhibited IL-12-induced IFNγ secretion and IFN regulatory factor-1 expression in both NK and T cells. DEX markedly inhibited IL-12-induced phosphorylation of Stat4 without altering its expression, which was inhibited by the GC antagonist RU486. Transfection experiments showed that DEX reduced the responsiveness to IL-12 through the inhibition of Stat4-dependent IFN regulatory factor-1 promoter activity. The blocking of IL-12-induced Stat4 phosphorylation, without altering IL-4-induced Stat6 phosphorylation, targets the suppressive action of GC on the Th1 cellular immune response and may explain the GC -induced shift toward the Th2 humoral immune response [48]. Preincubation of peripheral blood mononuclear cells with hydrocortisone sodium succinate for 1 hour, suppressed LPS induced IL-10 production in a dose-dependent manner both in normal volunteers and in cancer patients. The production of IL-10 was higher by PBMC from cancer patients than from normal individuals [49].

Granzyme B gene transcription is induced in activated lymphocytes upon antigenic stimulation. Several regulatory regions, including CBF, AP-1, and Ikaros binding sites are essential in the control of granzyme B promoter activation. DEX abolished the binding of nuclear transcription factors to the Ikaros binding site and reduced AP-1 binding activity and thereby abrogated the transcriptional activity of the human granzyme B gene promoter [50].

1.2.5. Effect on B lymphocytes

Monoclonal antibodies to IgD induce polyclonal B cell activation in mice. In Balb/c mice treated with monoclonal antibodies to IgD cortisol (200 mg/kg) increased IL-4 and IgE serum levels significantly [51].

1.2.6. Cell mediated immunity

GC inhibited NK cell-mediated cytotoxicity and antibody dependent cellular cytotoxicity (ADCC). This inhibitory effect was potentiated by PGE2 and abrogated by IFN-γ or IL-2 [20,52]. In mice, GC play a role in the development of T cell anergy during graft versus host reactions [53].

1.2.7. Effects on the antibody response

Adult male Sprague Dawley rats received adrenalectomies (ADX) and were given either basal CON replacement in their drinking water (25 μg/ml) or no CON. ADX rats with no CON replacement had reduced anti-KLH IgM and IgG responses compared with sham-operated controls. ADX rats that received basal CON replacement had partially restored anti-KLH IgM, but still had suppressed anti-KLH IgG. GC receptor type I (RU28318) and type II (RU40555) antagonists also reduced the anti-KLH IgM and IgG responses. ADX rats that received both basal CON replacement and low dose injections of CON on days 5 and 7 after KLH had anti-KLH IgG levels equal to those of sham-operated controls. GC elevation 4–7 days after immunization may play a role in stimulating the IgM to IgG2a switch, as GC receptor blockade reduced the anti-KLH IgG2a and splenic IFN-γ, but not the anti-KLH IgG1, response. Because IFNγ is an important regulator of the IgM to IgG2a switch, it is possible that the small rise in GC found 4–7 days after KLH facilitates IgG2a isotype switching [54].

Serum Ig levels were decreased in GC treated animals and patients. This was followed by an overshoot reaction after the termination of treatment. In GC treated animals the number of B lymphocytes are decreased in the spleen and lymph nodes but are not affected in the bone marrow. Early B cell activation was inhibited by GC, but preactivated B cells were not affected. The mucosal secretion of IgA and IgE was inhibited by GC. The production of secretory component (SC), which is responsible for the transportation of IgA onto mucosal surfaces, was stimulated by GC in rat hepatocytes and inhibited by estradiol [55].

Cutaneous application of GC (50 μg diflucortolone valerate in ethanol) to mice on the flank seven times on alternate days, augmented the expression of passive cutaneous anaphylactic reaction on the ear skin, which was induced by intravenous applications of monoclonal IgE anti-DNP antibody and local DNP challenge. In contrast, GC inhibited the reactions when applied on the challenged sites. GC induced a late phase but not an early phase cutaneous reaction in mast cell deficient WBB6F1 v/v mice by IgE anti-DNP antibody [56].

1.2.8. Cytokine production

Endothelial cells (EC) constitutively produce leukemia inhibitory factor (LIF), which is enhanced by IL-1. GC inhibited the basal production of LIF by EC and suppressed its IL-1-induced augmentation. GC acted mainly by decreasing the LIF mRNA level via a rapid increase in the degradation rate. The early LIF mRNA destabilizing activity of GC was translation dependent [57].

Acute stress stimulated the production of IL-18 mRNA in the glucocorticoid-producing cells of the adrenal cortex. Four hours after subcutaneous ACTH injection to adult male rats, IL-18 mRNA was increased 4-fold in the adrenal gland and pro-IL-18 was also elevated. Chronic ACTH treatment for 6 days had the opposite effect [58].

1.2.9. Lymphocyte recirculation

Glucocorticoids cause neutrophilia and decrease the number of other leukocytes in the circulation. This is attributed to the influence of GC on the vascular endothelial cells, rather than affecting lymphocytes directly [59]. Hydrocortisone treatment of mice daily for 2 weeks (0.5 or 1.0 mg/day/mouse) induced severe leukocytopenia in all organs of the immune system. In contrast, granulocytes (Gr-1$^+$Mac-1$^+$) were increased in number in the bone marrow and peripheral blood. This was due to an elevated level of myelopoiesis in the bone marrow. The granulocytes were functionally activated as estimated by the Ca^{2+} influx and superoxide production. The proportion of primordial T cells (CD3^{int}IL-2R$^+$) in the thymus and in the bone marrow was also increased. Mice that were administered hydrocortisone became susceptible to stress and developed gastric ulcers when they were exposed to restraint stress for 12 hours. [60].

1.2.1. Inflammation

Glucocorticoids have a powerful anti-inflammatory effect. This is the result of GC action on a variety of pathophysiological mechanisms that play a role in inflammation, including the inhibition of cytokine and other mediator secretion, inhibition of leukocyte priming, reduction of vascular permeability, and synergism with other anti-inflammatory mediators, such as catecholamines, β-END and α-MSH. This multifaceted effect on the inflammatory response explains the unparalleled inhibitory effect of GC on inflammatory disease [61,62]. Glucocorticoids also inhibit neurogenic inflammation, which is initiated by sensory nerves by the release of substance P and other peptide mediators [13,63].

DEX suppressed both gene expression and the production of protein by fibroblast-like rheumatoid synoviocytes (FLS). This suppression was independent of NF-κB and IκB-α [64].

Lymphocytes from patients with sepsis or septic shock were more sensitive to the antiproliferative effect of GC in the mitogen stimulated lymphocyte proliferation assay when compared with normal controls. Suppression was antagonized by additional treatment with IL-2, and to a lesser extent with the combination of IL-1, IL-6 and TNFα. It was suggested that the general hypersensitivity to GC is eliminated at the site of inflammation by locally produced cytokines [65].

GC treatment of rats induced mast cell destruction and local GC treatment led to the dramatic depletion of cutaneous mast cells in humans [66]. GC inhibited mediator release from mast cells and basophilic leukocytes [67]. DEX downregulated IgE receptors and inhibited the IgE dependent release of mediators and the synthesis and release of leukotriens LTC4 and LTD4 in murine bone marrow derived mast cells. The release of PGD2 was significantly increased [68].

GC also contributes to the inhibition of immune and inflammatory reactions by the aqueous humor of the anterior chamber of the eye, which is an immunologically privileged site. Significant concentrations of GC and little or no cortisol binding globulin were found in the aqueous humor of mice, rats and humans [69]. GC also contributes to strain specific variations in the immune response and, along with other steroid hormones, to the age-related abnormalities of immune inflammatory reactions [70].

Neonatal GC treatment of rats increases the severity ($p \leq 0.01$) and incidence ($p \leq 0.01$) of the inflammatory autoimmune disease, experimental autoimmune encephalomyelitis (EAE), in adult life. Neonatal GC treatment reduced the corticosterone response after an LPS challenge in adult rats ($p \leq 0.001$). LPS-stimulated macrophages of GC-treated rats produced less TNF-α and IL-1β in adult life than control rats ($p<0.05$). Splenocytes obtained from adult rats express increased mRNA levels of the proinflammatory cytokines IFNγ ($p<0.01$) and TNF-β ($p<0.05$) after neonatal GC treatment [71]. Antenatal treatment with GC did have immunomodulatory effects in human babies as well: T cell proliferation was decreased in infants born preterm (gestational age 26–31 weeks) and in infants born between 32 and 36 weeks of gestation. The activity of NK cells was increased in GC-treated infants born between 26 and 31 weeks. Antenatal

GC treatment had no significant effect on the capacity to produce IL-6 [72].

1.2.11. Infectious disease

The glucocorticoid response peaks 36 hours following murine cytomegalovirus (MCMV) infection, coincident with circulating levels of the cytokines IL-12, IFN-γ, TNF, and IL-6, and is dependent on IL-6 for maximal release. ADX mice were more susceptible than intact mice to MCMV-induced lethality, and the increased sensitivity was reversed by hormone replacement. Lack of endogenous GC resulted in increases in IL-12, IFN-γ, TNF, and IL-6 production as well as in mRNA expression for a wider range of cytokines, including IL-1α and IL-1β. TNF, but not IFN-γ, was required for increased lethality in the absence of endogenous hormone [73].

2. ALDOSTERONE

In the spleen and thymus of ADX rats, GR-II binding activity was significantly increased and binding could be blocked by aldosterone administration. This indicates that GR-I may play a tonic inhibitory role on GR-II expression in immune cells, which express both receptor subtypes [3]. The effect of selective GR-I and GR-II agonists on lymphocyte distribution in the peripheral blood and spleen of young adult rats has been studied. The GR-II agonist RU48362 decreased T and B cell and NK cell numbers to a very low absolute level. Aldosterone, which is a GCR-I agonist, significantly reduced the number of lymphocytes and monocytes and, unlike RU48362, also decreased the number of neutrophils. T helper cells and NK cells were decreased by aldosterone, but there was no effect on B cells and T suppressor/cytotoxic cells. Corticosterone at physiological doses behaved like a GR-II agonist in these experiments [74]. The response of spleen cells from ADX animals to concanavalin A (Con-A) was stimulated by pretreatment by low concentrations (3–30 mM) of aldosterone if RU48362 was also present. Corticosterone had a similar effect. These effects were evident in the presence of GR antagonist and after short corticosteroid exposure times. Immunosuppression was observed after prolonged preincubation with high concentrations of corticosteroids [75].

MRL/MpJ-Fas(lPr) autoimmune mice suffer from autoimmune hearing loss, which is preventable by prednisone treatment. The treatment of such mice with the mineralocorticoid aldosterone was equally effective in the prevention of the disease. Aldosterone therapy may offer advantages over prednisone for long-term management of autoimmune hearing loss and other forms of nonimmune-related deafness for which steroids are currently prescribed [76].

The induction of cytotoxic T lymphocytes and the activation of HSV-specific memory cytotoxic T lymphocytes (CTL) is inhibited by sympathectomy. Glucocorticoids, mineralocorticoids and beta-adrenergic agents play a role in mediating this sympathectomy-induced suppression of CTL activity. Both Type I and Type II corticosteroid receptors play a role in the regulation of HSV-specific immunity. These findings indicate the need to exercise caution in attributing the effects of chemical sympathectomy solely to the absence of sympathetic innervation of lymphoid tissues [77].

3. SEX HORMONES

3.1. Gonadotropins

LHRH contributes to sex-dependent changes in immune responsiveness during the estrous-menstrual cycle as well as pregnancy. Lymphocytes secrete LHRH and express receptors for it. The LHRH system in lymphocytes is steroid sensitive. LHRH may act as an immunological response modifier in the brain-pituitary-lymphoid-gonadal axis [78].

Intravenous bolus of LHRH and TRH significantly increased plasma IFNγ production in five normoprolactinemic women with irregular menstrual cycles. *In vitro* LHRH and TRH, separately and together, significantly enhanced IFNγ production by *Staphylococcus* enterotoxia A (SEA)- and ConA-activated peripheral blood mononuclear cells [79].

Follicle stimulating hormone (FSH) and luteinizing hormone (LH) modify immune reactions through the regulation of sex steroid hormone production and also by direct effect on immunocytes. LH significantly increase the proliferation of murine lymphocytes in response to mitogens and stimulate the secretion of IL-1 and IL-2. Primary and secondary antibody responses were stimulated by LH [80]. The response of human peripheral blood mononuclear cells to IL-2 was enhanced by physiological concentrations of FSH and LH. At supra-physiological concentrations, FSH significantly decreased the percentage of $CD4^+$ cells in unstimulated cultures, whereas it enhanced the percentage of $CD8^+$ cells following PHA stimulation. FSH and LH caused the release of IL-6 from human peripheral blood monocytes *in vitro* [81].

Pyrogenic cytokines, especially IL-1β, are significantly influenced by exogenous gonadal steroids and gonadotropins. In turn, cytokines influence gonadotropin secretion by the pituitary and steroidogenesis by the ovaries and testes. The inhibitory influence of IL-1β on the LH surge may prevent inopportune conception, and the abortive influences of TNFα and IFNγ may terminate pregnancy during periods of infection [82]. IL-1β, generated in the central nervous system (CNS) during inflammation, upregulates opioids and tachykinins in the hypothalamus. These neuropeptides are critically involved in relaying the cytokine signal to neuroendocrine neurons and causing the suppression of hypothalamic LHRH and pituitary LH release [83]. Immune/inflammatory stimulation by endotoxin disrupts the follicular phase of the estrous cycle of the ewe. Endotoxin interrupts the preovulatory estradiol rise and delays or blocks the subsequent LH and FSH surges. Endotoxin suppresses LH pulsatility, although the ovarian responsiveness to gonadotropin stimulation appears to be disrupted as well [84,85].

Human chorionic gonadotropin (hCG) suppresses delayed type hypersensitivity reactions and NK cell activity in mice. hCG induced suppressor cells in murine lymphocytes *in vitro*. Some attempts to modulate various immune reactions in rats by treatment with hCG led to negative results. In guinea pigs hCG inhibited the development of delayed type hypersensitivity reactions and hCG treated guinea pigs showed depressed lymphocyte responses to phytohemagglutinin (PHA) and to purified protein derivative *in vitro*. Several investigators concluded that commercial preparations of hCG contained impurities which are responsible for the suppressive effect on lymphocyte reactions *in vitro* [86].

After immune activation, IL-1β circulates in high concentrations, and blunts hCG-stimulated testosterone secretion. Cytokine treatment of male rats blunted *ex vivo* testosterone secretion in response to hCG, indicating that the mechanism is independent of circulating cytokines. The steroidogenic acute regulatory (StAR) protein of the inner mitochondrial memebrane, which mediates the first step of testosterone biosynthesis, was reduced in Leydig cells isolated from rats treated with IL-1β [87].

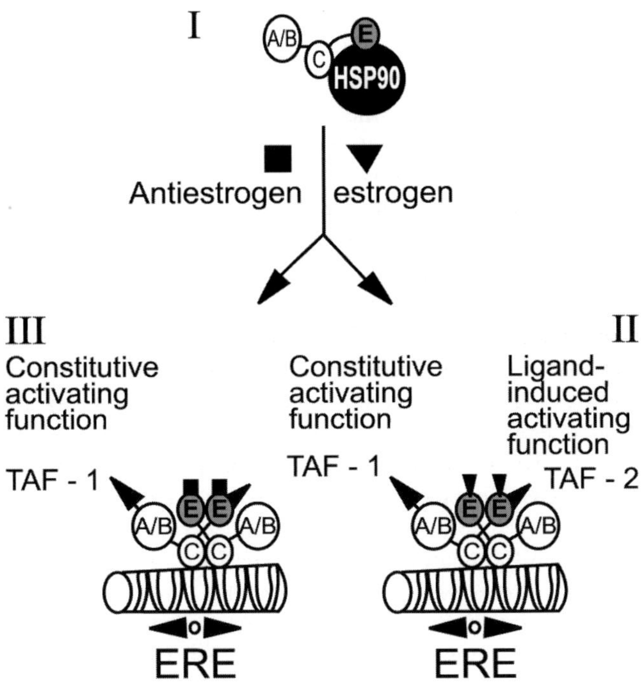

Figure 1. Nuclear signaling by estrogen and the effect of antiestrogen.
The estrogen receptor (E) is dissociated from its complex with heat shock protein 90 by estrogen, dimerized and the dimer binds to estrogen responding elements (ERE) in the nucleus. The dimeric receptor itself exerts constitutive activating function. In addition, there is evidence for ligand induced functions in gene activation. Tamoxifen functions as both an antagonist and an agonist. If a cell type requires activating factors (TAF) 1 and 2 of the estrogen receptor to be activated, tamoxifen is antagonistic. If a cell requires only TAF 1 to interact with transcription factors at the promoter, tamoxifen is agonistic (After Gallo and Kaufman 1997).

Immune cells, particularly macrophages, accumulate in the corpus luteum during luteolysis. In the human corpus luteum, macrophages were localized primarily to the vascular connective tissue and theca-lutein areas, although some were found in the granulosa-lutein cell layer. Macrophage numbers increased throughout the luteal phase and reached a maximum in the late-luteal phase ($p<0.05$). Luteal 'rescue' with HCG treatment of the donors was associated with a marked reduction in the numbers of tissue macrophages when compared with those of the late-luteal phase ($p<0.001$). One of the effects of HCG during maternal recognition of pregnancy is to prevent the normal influx of macrophages into the corpus luteum. As LH/HCG receptors are localized to the steroidogenic cells, this implies a fundamental role for steroidogenic cell products in the control of macrophage influx into the human corpus luteum [88].

3.2. Sex hormone receptors and signaling

3.2.1. Estrogen receptors

Estrogen and androgen receptors are present in the thymus and in the bursa of Fabricius. The classical estrogen receptor, now known as ERα, is detectable in lymphoid tissue. Progesterone also acts through GC receptors, in addition to its own specific receptors. At high concentrations, estrogens and androgens may also act on GC receptors [89–91]. A second estrogen receptor called ERβ has been recently described, which is more abundant in lymphoid cells. Membrane bound steroid receptors, which include the polyglycoprotein (PGP) pump have also been described. Triphenylethylene-derived antiestrogens are capable of inhibiting lymphocyte proliferation under conditions where estradiol has no effect. A membrane bound antiestrogen binding site, for which the natural ligand is unknown, seems to be involved in this phenomenon [92].

After stimulation by phorbol myristate acetate (PMA), human myeloid monocytic THP-1 cells expressed ER, together with markers of monocytic cell differentiation, such as CD68, CD54 and HLA-DR [93].

The estrogen receptor binds to estrogen response elements in target genes and recruits a coactivator complex called CBP-pl60 that mediates the stimulation of transcription. ER also activates transcription at AP-1 sites that bind the Jun/Fos transcription factors. ER increases the activity of Jun/Fos [94,95]. ER interacts directly with the transcription factors NF-IL6 and NFκB, and inhibits their ability to bind DNA, which is likely to be the molecular basis for repression of IL-6 gene expression by estrogens. Unlike estrogens, tamoxifen does not inhibit the induction of the IL-6 promoter [96].

Two forms of ER exist in the mammalian uterus: (i) ERα, which influences transcriptional process; and (ii) the non-activated estrogen receptor (naER), an alternative form of ER with no DNA binding function, localized in the plasma membrane. The steps involved in ER signaling include: (i) transport from the cytoplasm to the nucleus; (ii) interaction with the nuclear transcription machinery; (iii) involvement in post-transcriptional control mechanisms; and (iv) degradation through ubiquitination. naER moves from the plasma membrane to the nucleus. Within the nucleus it dimerizes with a DNA-binding protein, called the estrogen receptor activation factor (E-RAF). Deglycosylated naER fails to dimerize with the E-RAF. This dimerized naER is identified as nuclear estrogen receptor II (nER II). nER II phosphorylates certain subunits in RNA polymerase [97]. Membrane steroid receptors are candidates for mediating steroid effects, which do not fit the paradigm of nuclear transcription factor mechanisms. The membrane estrogen receptors are involved in the estrogen-induced release of prolactin [6].

Different residues of the ER ligand binding domain are involved in the recognition of structurally distinct estrogens and antiestrogens. An ER-specific coregulator that potentiates antiestrogen antagonist effectiveness and represses ER transcriptional activity has been isolated. Quinone reductase (QR) is a detoxifying phase II antioxidant enzyme, and its activity is up-regulated by antiestrogens, in an ER-dependent manner, in breast cancer cells. Estrogens rapidly up-regulate the expression of several genes associated with cell cytoarchitectural changes including NHE-RF, the sodium hydrogen exchanger regulatory factor, also known as EBP50 [98].

Estrogens suppress IL-6-induced osteoporosis and the growth of multiple myeloma cells by repressing IL-6 and IL-6 receptor gene expression. The IL-6-induced activation of STAT3 activity and STAT3-mediated gene expression were suppressed by 17beta-estradiol in breast cancer cells. Estradiol (E2)-mediated inhibition of STAT3 activation was reversed by tamoxifen. The inhibitory action of ER on STAT3 activity was due to direct physical interactions between STAT3 and ER, which represents a novel form of cross-talk between STAT3 and ER signaling pathways [99].

There are differences in estrogen receptors between BALB/c mice, which do not get autoimmune diseases, and two strains that do, MRL/MP-lpr/lpr and NZB/W mice. There is evidence to indicate that in one model of SLE, the normal regulation of estrogen action by progesterone may be impaired [100,101].

Normal male, female, autoimmune (lpr, Fas deficient) and testicular-feminized (Tfm, androgen receptor deficient) mice were compared for ER content in lymphoid cells and autoantibody production. ERα was expressed in all lymphocyte subsets examined. ER β was expressed at low levels in thymic CD4+/CD8- T cells in wild-type mice and at high levels in the peripheral CD4-/CD8- T cells in lpr mice. Autoantibody production was higher in female lpr mice than in male lpr mice [102].

ERα knockout (ER-/-), wild-type (ER+/+), and radiation chimeric (ERα positive or negative in either nonhematopoietic or hematopoietic elements, or both) male mice were used to study ERα involvement in the hematopoietic and immune systems. ERα knockout and ER+/+ mice showed similar hematopoietic progenitor profiles, but the ER-/- animals had fewer cells in all B lymphocyte subpopulations. Animals receiving a pharmacological dose (5 mg/kg BW) of 17 beta-estradiol with both elements ER+/+, had decreased early hematopoietic progenitors and a shift toward a mature B cell subpopulation, whereas ER-/- animals showed changes only in early hematopoietic progenitors. Hematopoietic element ER+/+ animals exhibited greater E2-induced hematopoietic progenitor and B lymphocyte alterations than those having only nonhematopoietic ERα expression [103,104].

3.2.2. The androgen receptor

Male C3H/HeN mice were sham operated or subjected to trauma and hemorrhagic shock (TH) followed by adequate fluid resuscitation. The aromatase inhibitor, 4-hydroxy-androstenedione (4-OHA) (5 mg/kg) or vehicle was administrated subcutaneously just before resuscitation. In another group, sepsis was induced by cecal ligation and puncture (CLP) 3 days after resuscitation, and survival was measured over a period of 10 days. In vehicle-treated mice a significant decrease in splenocyte proliferation, IL-2, and IFNγ release and increased release of IL-10 were observed. In animals treated with 4-OHA there was an increase in splenocyte proliferation, IL-2, and IFNγ release, and a decrease in IL-10 release. Decreased expression of the cytosolic androgen receptor (AR), but no significant difference in the cytosolic and nuclear ER-α and -β expression was observed in the vehicle-treated group after shock. AR and ER-β mRNA expression was increased, whereas ER-α expression decreased in the vehicle-treated group after TH. ER-α expression decreased and ER-β expression increased in the cells of 4-OHA treated mice. There was no difference in the cytosolic AR expression in the 4-OHA-treated group after TH. AR and ER-β mRNA expression was unaffected, whereas ER-α expression increased under such conditions. The increased mortality rate after TH and subsequent sepsis was significantly reduced by 4-OHA treatment [105].

In the bone marrow of human males and females, stromal cells, macrophages, endothelial cells, myeloblasts, myelocytes, neutrophils, and megakaryocytes express AR. AR was not detected in lymphoid or erythroid cells, or in eosinophils [106].

Four novel proteins that recognize AR zinc-finger regions both *in vivo* and *in vitro* have been identified. One is a small nuclear RING-finger protein that interacts with AR and with other transcription activators, such as Sp1. The second is a nuclear serine/threonine protein kinase called androgen-receptor-interacting nuclear protein kinase (ANPK). The AR itself does not seem to be a substrate for this kinase. The third one is named androgen-receptor-interacting protein 3 (ARIP3). This is a novel member of the PIAS (protein inhibitor of activated STAT) protein family. The fourth protein, termed ARIP4, is a nuclear ATPase that belongs to the SNF2-like family of chromatin remodelling proteins. Each protein modulates AR-dependent transactivation in co-transfection experiments, but their activating functions are not restricted to AR [107].

The X-linked androgen receptor gene was examined for mutations and polymorphisms in infertile men. Screening of over 400 patients and controls showed that defects in the AR gene lead to the production of dysfunctional receptor protein in up to 10% of males with abnormally low sperm production and male infertility. The mutations and polymorphisms uncovered were associated with subtly reduced intrinsic AR activity, and are of two main categories: polymorphic changes in length of a trinucleotide CAG tract in the N-terminal transactivation domain, and missense mutations in the C-terminal ligand-binding domain. These polymorphisms and mutations are associated with defective intermolecular protein-protein interactions with coactivator molecules [289]. X-linked androgen receptor gene mutations cause complete androgen insensitivity, impair virilization with or without infertility, and sometimes yield a wide phenotypic spectrum among siblings [108].

3.2.3. Progesterone receptors

In humans and other vertebrates, the biological activities of progesterone are mediated by two receptors, PR-A and PR-B. Both receptors arise from the same gene and function as progesterone activated transcription factors that exhibit different transcription regulatory activities. These receptors reside in the cytoplasm where they are associated with heat shock proteins (hsp) of 70 and 90 kDa molecular weight. Hormone binding leads to the dissociation of hsp70 and hsp90 from the receptors. The receptors then become phosphorylated and dimerized by progesterone, which in turn bind to DNA in the nucleus. Three dimeric species, namely A/A, A/B, and B/B may be formed and bind to progesterone response elements (PRE) and subsequently regulate transcription. Receptor dimerization is obligatory for binding to PRE, but it is not sufficient to activate transcription without the hormone. In pure heterodimers, A receptors are dominant negative inhibitors of B receptors [109–111]. Numerous proteins termed coactivators are recruited by the ligated progesterone receptor and enhance receptor-dependent transactivation. Coactivators are a diverse group of molecules with multiple structural and enzymatic functions [112].

Mice lacking both PR isoforms exhibit reproductive abnormalities. PR-B mediates a subset of the reproductive functions. Ablation of PR-A does not affect responses of the mammary gland or thymus to PS, but results in severe abnormalities in ovarian and uterine function [113].

The transcription factor RelA(p65) subunit of NFκB and PR are mutually repressed. This trans-repression occurs independent of PR isoform, reporter construct, or cell type used. PR and RelA interact *in vitro*, suggesting that the mutual repression is due to a direct interaction between these proteins. The activation of NFκB by TNFα also results in repression of PR, while PR is able to repress TNFα-induced NFκB activity. NF-κKB-regulating cytokine receptors are expressed in progesterone target tissues, like breast and endometrium and as such, this regulatory circuit has the potential to fulfill an important physiological role [114].

Rat spleen and peripheral blood lymphocytes express progesterone receptors (PR). Lymphocyte PR concentration is increased greatly during the early phase of pregnancy and is stimulated by progesterone treatment *in vitro*. Progesterone stimulated lymphocytes to release a soluble factor that inhibits cellular immune reactions, cell proliferation and the synthesis of anti-DNP antibodies by mouse hybridoma cells [115].

In human nasal polyps only mast cells tested positive for estrogen and progesterone receptors by immunocytochemistry, performed with polyclonal and monoclonal antibodies. Lymphocytes, macrophages, or other immune cells, did not label for ER/PR in human upper airways [116].

3.3. Estrogens

Estradiol suppresses bone marrow function, causes thymic involution, inhibits cell-mediated immune reactions, including the helper, suppressor and effector (delayed type hypersensitivity) functions of T lymphocytes. Natural killer cells, neutrophils and mast cell degranulation are also inhibited by E2. Phagocytosis, humoral immune reactions and certain autoimmune diseases are enhanced by E2 in laboratory animals [86,117,118].

3.3.1. Effects on phagocytes and antigen presentation

In murine splenic macrophages, 17-β-estradiol decreased LPS-induced IL-1α, IL-6, and TNFα production but not IL-10, IL-12, or macrophage inflammatory protein (MIP). This inhibition was associated with a decreased LPS-induced NFκB-binding activity [119].

Rat peritoneal macrophages treated with 10^{-2} and 10^{-3} ng/ml E2 secreted significantly greater amounts of TNFα, whereas those treated with 10^{-4} ng/ml of E2 showed a significantly reduced TNF secretion when compared to untreated macrophages [120].

The release of superoxide by rat peritoneal macrophages was not affected by E2 at concentrations between 10^{-10} and 10^{-9} M, but stimulation occurred at concentrations above or below that range. Estradiol significantly inhibited nitrite release by macrophages at most concentrations used, except 10^{-10} M, which had no effect. In the presence of anti-TNF antibody superoxide and hydrogen peroxide release was moderately reduced, but nitrite release was inhibited dramatically by E2. Male rat peritoneal macrophages were treated with 0.1 μg/ml of LPS and with E2 or progesterone. At 10^{-2} ng/ml E2 significantly ($p<0.05$; $n=6$) enhanced, while 10^{2} ng/mL or 10^{3} ng/mL of E2 and progesterone at less than 10^{-3} ng/ml or greater than 10^{-1} ng/ml significantly ($P < 0.05$; $n = 6$) inhibited TNF release. E2 (10^{-4} and 10 ng/ml) and progesterone (10^{-6}–10^{-4} ng/ml and 10^{2} ng/ml) each significantly ($p<0.05$, $n=8$) enhanced H2O2 release and significantly inhibited NO2 release ($p<0.05$; $n=6$) (E2<10^{-2} and >10 ng/mL; progesterone 10^{-7}–10^{4} ng/mL) by LPS-treated macrophages [121].

Total superoxide dismutase (SOD) and catalase (CAT) activities were markedly higher in macrophages of female than in male rats. Removal of the gonads in both males and females (comparison between castrated groups) increased this difference. Castration and testosterone administration did not significantly modify enzyme activities. Ovariectomy did not affect SOD or glutathione peroxidase activity but markedly reduced CAT activity. This latter change was fully reversed by estrogen administration, whereas progesterone had a smaller effect [122].

In rats, ovariectomy (OVX) significantly inhibited the spontaneous production of NO by alveolar macrophages (AM). In orchidectomized rats, the NO response by AMs to LPS stimulation relative to spontaneous NO production was significantly downregulated. Pharmacological doses of E2 inhibited *in vitro* LPS-stimulated NO production in AMs of both saline and LPS treated rats, and basal NO production only in LPS-treated male rats. Progesterone inhibited basal and *in vitro* LPS-stimulated NO generation by AMs of both saline- and LPS-treated male rats. In LPS-treated female rats *in vitro* LPS-stimulated NO production was not affected by estradiol treatment. In OVX LPS-treated female rats progesterone significantly inhibited NO production by *in vitro* stimulated AMs [123].

Estradiol and progesterone increased survival and prevented apoptosis induced by TNFα in both undifferentiated and macrophage-like PMA-differentiated U937 cells. The steroid receptor antagonist drugs such as raloxifene, RU486, and ICI 182,780 in PMA-differentiated cells displayed agonist activity by preventing TNFα induced apoptosis as efficiently as did the hormones [124].

In OVX rats E2 stimulated antigen presentation by uterine epithelial cells and at the same time presentation by stromal uterine cells was inhibited [125]. Estradiol, given to OVX rats, inhibited vaginal cell antigen presentation within 24 hours after a single injection. The number of MHC class II positive cells was not affected and macrophage, dendritic cells and granulocytes increased in the vagina in response to estradiol treatment. Antigen presentation by vaginal cells from ovariectomized rats was mediated through MHC class II and involved the expression of B7.1 and B7.2. In other studies, vaginal APCs interacted with thymus APCs to synergistically enhance antigen presentation under conditions in which vaginal antigen presentation was inhibited by estradiol. Estradiol tretment also inhibited the response of vaginal lymphocytes to mitogens [126].

High doses of estradiol and ethynylestradiol, but not of estrone, suppressed phagocytosis in the teleost fish, tilapia. No suppressive effect on phagocytosis was observed with progesterone, testosterone, and 11-ketotestosterone. A combination of estradiol and cortisol potentiatingly suppressed phagocytosis. Cortisol had weaker suppressive effects on the phagocytosis of leukocytes in common carp than tilapia [127].

The chemotactic response of human polymorphonuclear leukocytes to FMLP was significantly reduced by physiological concentrations of estradiol (10^{-10} M). Two estrogen receptor antagonist drugs, clomiphene and tamoxifen, eliminated this inhibitory effect [128]. Estradiol suppressed inflammatory responses in mice by the inhibition of homing and activation of inflammatory cells and their production of TNFα and IFNγ [129].

3.3.2. Effects on T lymphocytes

In rats, the effects of varying estrogenic states on T-cells are variable and transient, except for bone marrow T lymphocytes, which are significantly reduced after OVX. Estrogen replacement therapy at the low-dose reversed the effects of OVX and high-dose E2 treatment increased T-cell subsets in both the sham and OVX groups, an effect that was more pronounced at day 14 compared with day 28 [130].

3.3.3. Effects on cellular immunity

Estradiol reduced the NK cell-mediated cytotoxicity in a dose dependent manner in various mouse strains. C3H/N, DBA/1 and NZB/W mice showed a high (more than 50%) degree of suppression, whereas C57BL/6 and MRL/lpr/lpr mice exhibited low susceptibility (< 30%). The injection of mice with anti-asialo GM1 antibodies for 4 weeks, which affects NK cells, raised several-fold the frequency of IgG producing cells. It was suggested that estrogen mediated suppression of NK cells, which are inhibitory for B cells, led to the increase of Ig producing cells [131].

In mice, E2 treatment decreased the PPD-specific delayed type hypersensitivity (DTH) response and diminished lymphocyte numbers in the draining lymph nodes (DLN) and spleen compared with control mice. E2 also suppressed the purified protein derivative (PPD)-specific proliferative response of DLN and spleen cells from PPD-primed mice. E2 suppressed IL-2 and IFNγ production in response to PPD *in vitro*. In contrast, IL-4 and IL-10 gene expression by DLN cells of E2-treated mice was enhanced. Spleen APC from E2-treated mice failed to induce optimum proliferation of the PPD-primed T cells in response to PPD *in vitro*. The addition of the normal spleen APC to APC from E2-treated mice restored the proliferative response of the PPD-primed T cells in response to PPD [132]. In mice, suboptimal doses of DEX and E2 have an additive effect on suppression of delayed type hypersensitivity. Such additive effects are not observed on granulocyte-mediated inflammatory responses, immunoglobulin production and antigen specific antibody responses [133].

Murine $CD4^+$ T cells required estrogen for cytotoxic activity. This estrogen dependency was also observed in β2-microglobulin deficient mice infected with lymphocytic choriomeningitis virus. Such mice do not have functional $CD8^+$ cytotoxic T cells and must rely on $CD4^+$ CTL for anti-viral immunity. Fatal immune meningitis was observed when orchidectomized and virus infected animals were treated with estrogen [134].

Rhesus monkeys (aged 13.7+/-2.6 years) were ovariectomized and received either sham or E2 replacement implants. Nine months post-ovariectomy, natural killer cell activity was reduced and serum soluble gp130 levels were increased. There was no difference in leukocyte formation and distribution in serum TNF, TNF-R1, TNF-R2, IL-6, soluble IL-6R, and IL-1 between the groups [135].

Kidneys from female or male Fischer 344 rats were orthotopically transplanted into OVX female Lewis recipients and were treated for 16 weeks with either E2, testosterone, or vehicle. Testosterone induced increased urinary protein excretion independently of the donor gender, as well as extended glomerular sclerosis, interstitial fibrosis, and severe vascular lesions. Mononuclear cell infiltration was most pronounced in these animals and there was increased expression of intercellular adhesion molecule-1 (ICAM-1), fibronectin, laminin, and TGF-β in the grafts. Estradiol treatment resulted in an improved graft function, reduced glomerular sclerosis, diminished cellular infiltration, and reduced ICAM-1, fibronectin, laminin, and TGF-β expression. In animals treated with vehicle, grafts of male origin had good graft function and histology, whereas grafts from female donors developed severe proteinuria and glomerular, interstitial, and vascular damage. This indicates that a male kidney may benefit from the absence of testosterone, whereas the function of a female kidney deteriorates in the absence of E2 [136].

New Zealand White rabbits were immunosuppressed with cyclosporine and given cardiac allografts from Dutch Belted rabbits. The recipients received either E2 or placebo daily for 6 weeks, and then they were killed. MHC class II antigen was not detectable in allograft coronary arteries from any estradiol-treated recipients, whereas MHC II was present in the allograft coronary arteries of all the placebo-treated recipients. There was a 60% decrease in allograft coronary artery myointimal thickening (determined by morphometry) in the E2-treated compared with the placebo-treated group. T-lymphocyte infiltration was decreased by E2 only in the myointima [290]. Chronic E2 treatment of transplant recipients inhibits arteriosclerosis by attenuating both IGF-I expression and the immune response, particularly MHC class II expression [137].

3.3.4. Effects on antibody formation

Estradiol increased the production of immunoglobulin by IL-2 stimulated human tonsilar lymphocytes *in vitro*. T cell DNA synthesis was also increased significantly by E2 alone. The PHA induced DNA synthesis by T cells was enhanced by pharmacological concentrations of E2 [138]. Estradiol enhanced IgG and IgM production by human PBMCs both from men and women, without altering cell viability and proliferation. Anti-IL-10 antibody blocked the E2 effect on immunoglobulin production of PBMCs. Exogenous IL-10 (1 U/ml) further enhanced the E2-induced increase of immunoglobulin production by B cells, although this level of IL-10 alone had no effect [139]. Estradiol treatment of various mouse strains for 31 weeks, but not for 4 weeks, significantly increased the frequency of IgG and IgM producing cells [131]. Immunoglobulin production and antigen-specific antibody responses were increased in mice after exposure to E2 and suppressed by DEX. The upregulation of B lymphocytes was still evident when mice were treated with both hormones [133].

Estradiol can break B cell tolerance and induce a lupus-like phenotype in nonautoimmune mice transgenic for the heavy chain of a pathogenic anti-DNA antibody. E2 induced a rise in anti-DNA serum titers and in Ig deposition in renal glomeruli. Hybridomas generated from E2-treated mice produced high-affinity, unmutated anti-DNA antibodies, indicating that naive B cells that are normally deleted or anergized are rescued from tolerance induction. Bcl-2 expression was increased in splenic B cells after E2 treatment [140]. The estradiol-induced breakdown in B-cell tolerance can be abrogated by bromocriptine, which induces anergy in the high-affinity DNA-reactive B cells. This demonstrates that some of the effects of estrogen on naive autoreactive B cells require the presence of prolactin [141].

Estradiol benzoate significantly enhanced the humoral immune response to *Escherichia coli* and sheep erythrocytes (SRBC) in chicken. Flutomide and tamoxifen strongly inhibited humoral immune response to *E. coli* and SRBC antigen [142].

3.3.5. Effects on cytokines and inflammation

Neuroantigen-specific T cell clones were isolated from normal control subjects and patients with the demyelinating disease, multiple sclerosis. E2 enhanced the antigen-stimulated IL-10 and IFNγ secretion. The effect of E2 on the secretion of TNFα and TNFβ was biphasic, with enhancement occurring at low doses, and inhibition present at high concentrations. E2 had no effect on the secretion of IL-4 or TGFβ. These data indicate that E2 is capable of modulating both pro- and anti-inflammatory cytokine activities [143]. In women, serum estrogen levels showed a correlation with IL-4 secretion by PBMC, which fluctuated with the menstrual cycle [144].

3.3.6. Reproduction biology

In granulosa cells obtained from the ovaries of immature diethylstilbestrol (DES)-treated rats, a dose dependent inhibition was found by TNFα and IL-1β on gonadotropin induced P450 aromatase (P450arom) and 17beta-hydroxysteroid dehydrogenase (HSD) enzyme synthesis. The downregulation of the expression of these estrogen biosynthetic enzymes opposed the stimulation of follicular development by gonadotropin. The cytokine effect on P450arom expression was mainly derived from a decrease in gonadotropin-induced cAMP production, while the inhibitory mechanisms on 17HSD type 1 expression involved distal sites from cAMP generation. The protein kinase C and tyrosine kinase pathways were not involved [145].

The potent regulatory effects of IL-4 and IL-6 on both estrogenic and androgenic 17beta-HSD/KSR kinase suppressor of Ras enzyme activities in breast cancer cells depend on cell-specific gene expression. Exposure to IL-4 and IL-13 of both ER-positive (ZR-75-1, T-47D) and ER-negative (MDA-MB-231, BT-20) human breast cancer cells caused a rapid and potent induction of 3beta-HSD type 1 gene expression. Such an induction was also observed in normal human mammary and prostate epithelial cells in primary culture as well as in human HaCaT immortalized keratinocytes, ME-180 cervix cancer cells, and HT-29 colon cancer cells. The DNA-binding activity of Stat6 was activated after a 30 minute exposure to IL-4 in all the cell types where IL-4 induced 3beta-HSD expression, but not in those that failed to respond to IL-4. These data indicate that IL-4 and IL-13 may play a role in the biosynthesis of active sex steroids from the inactive adrenal steroid dehydroepiandrosterone (DHEA) in various cell types derived from peripheral target tissues [146].

Diestrus levels of 17beta-estradiol (<100 pg/ml in serum) significantly reduced the clinical manifestations of active EAE in both male and female mice. Estriol was also effective but at doses below those previously established for pregnancy. A coincidental reduction in the number and size of inflammatory foci in the CNS of estrogen (17beta-estradiol or estriol)-treated mice was observed. The recipients of encephalitogenic T cells from low-dose estrogen-treated mice developed less severe paralysis than mice receiving T cells from placebo-treated mice. A modest shift in Th1/Th2 balance suggested that low dose estrogen therapy could bias the immune reaction towards an anti-inflammatory cytokine response. Estrogen treatment at the onset of active EAE failed to reduce disease severity [147]. Estriol treatment reduced the severity of EAE in mice significantly. Estriol-treated mice had significantly higher levels of serum IgG1 antibodies to myelin basic protein (MBP). MBP-specific T-lymphocytes from estriol-treated EAE mice produced significantly more IL-10 after antigen stimulation [148].

In women, controlled ovarian hyperstimulation leads to neutrophil activation, which correlates with the degree of luteinization. Neutrophil L-selectin expression negatively correlated with serum progesterone levels [149].

3.3.7. Aging

Post-menopausal women taking estrogen/progestin combinations as hormone replacement therapy (HRT) had more B cells ($p<0.05$), higher mitogen-induced T cell proliferation ($p<0.05$), and higher levels of induced TNFα ($p<0.05$) when compared to non-terated controls. There was a trend towards a lower proportion of CD4+ T cells. These findings represent a reversal of immune alterations associated with normal aging, suggesting that preservation or improvement of immune function may be associated with the use of HRT [150].

3.3.8. Estrogens and disease

Cell-mediated immune function was suppressed in female, but not male, mice 10 days after burn injury. This was mediated, in part, by increased production of IL-6. In female mice after burn an increased circulating E2 corresponded with suppressed DTH and splenocyte-proliferative responses and with increased concentrations of IL-6. Ovariectomy restored the suppressed DTH response and decreased IL-6 concentrations. E2 treatment of both ovariectomized females and intact male mice resulted in a suppressed DTH response. *In vitro* E2 suppressed splenocyte proliferation in a macrophage-dependent manner and enhanced the production of IL-6 by macrophages [151].

FasL expression by breast tumors induces apoptosis of infiltrating Fas positive T cells. An estrogen recognizing element-like motif has been identified in the promoter region of the FasL gene. FasL RNA and protein levels increased in hormone sensitive breast cancer cells treated with E2 and the increased expresion of Fas by cancer cells augmented the apoptosis of Fas bearing T cells. Tamoxifen blocked this estrogenic effect. Tamoxifen also inhibited FasL expression in estrogen-depleted conditions [152].

Myelin basic protein is a major autoantigen from brain tissue. 17β-estradiol had a synergistic effect with myelin basic protein in the induction of degranulation of rat mast cells. This effect was more pronounced in Lewis rats, which are susceptible to the development of EAE, than in Sprague-Dawley rats, which are resistant [153]. Sex hormones do not appear to have an influence on the susceptibility to and the severity of experimental myasthenia gravis in Lewis rats [154].

In short-term unstimulated cultures, the apoptosis of PBMC from patients with SLE was accelerated compared to that of cells from normal individuals. Compared to untreated cultures, estrogen significantly reduced apoptosis of both patient and normal PBMCs if the donors had normal menstrual cycles. Estrogen did not decrease apoptosis in noncycling patients or normal cells. Apoptosis of T cells cultured alone was not affected by estrogen. Patients' estrogen-treated PBMCs secreted significantly less TNFα than untreated cells. TNFα secretion from normal cell

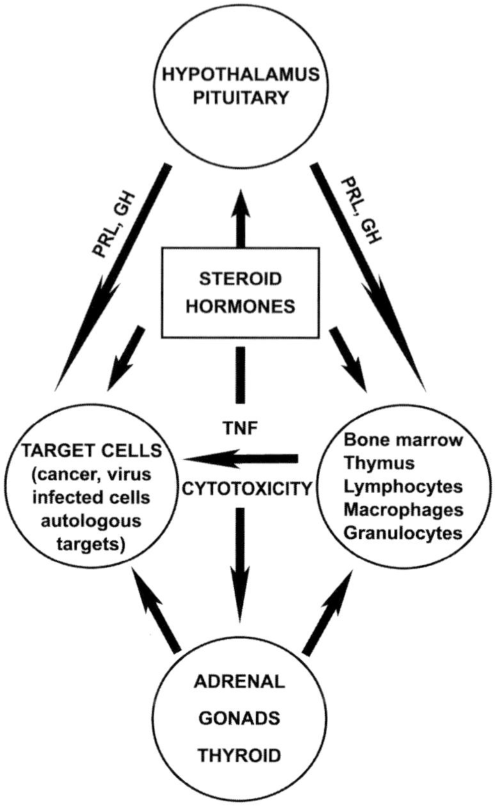

Figure 2. Immunomodulation by steroid hormones.
Some of the pathways by which steroid hormones interact within the neuroimmune regulatory network are shown. Steroid hormones affect the immune system directly and indirectly, through their regulatory influence on the neuroendocrine system. Apart from the powerful immunoregulatory effect, a very important function of steroids is the regulation of cytotoxicity and apoptosis at the target cell level. This has major physiological roles and is also significant for host defense. After Nagy et al. [118].

cultures was unchanged. Estrogen-induced decreases in apoptosis combined with decreased TNFα may allow the survival of autoimmune cells in SLE patients [155].

One year old male rats were treated with E2 (250 μg/kg s.c.) or vehicle for 30 days followed by the isolation and culture of the splenocytes in the presence of Con-A. Approximately 10^7 splenocytes were transferred into young naive syngeneic rats. The recipients of splenocytes from E2-treated animals presented evidence of inflammation, lymphocytic infiltration and presence of degranulated mast cells in their prostate. T cells stimulated by anti-CD3 antibody produced a similar degree of inflammation. These results demonstrate that hormonal imbalance in male rats produce the symptoms of autoimmune prostatitis [156].

3.3.9. Immunomodulation by antiestrogens

The non-steroidal antiestrogen tamoxifen (TX) has an antiproliferative effect on lymphocytes and inhibits the induction of humoral and cell mediated immune reactions. TX also inhibits suppressor T lymphocytes, which could lead to enhanced immunoglobulin secretion in response to mitogens. TX interferes with the stimulatory effect of E2 on phagocytosis, inhibits giant cell formation by monocytes and blocks H_2O_2 production by human neutrophils. The LPS induced production of TNFα by normal human monocytes and by rat peritoneal cells is enhanced by TX treatment [92,118,157,158].

Tamoxifen and toremifene (TO) decrease serum PRL, GH and IGF-I levels, influence the expression of hormone receptors and their binding proteins, which also affect the immune system. TX-induced immunosuppression in rats could be reversed by treatment with either GH or PRL, suggesting that these hormones are effective antagonists of TX-induced immunosuppression. TX also antagonized the stimulatory effect of PRL on lymphoid cells [92,159].

The inhibitory effect of E2 on NK cells is antagonized by TX. NK, LAK and CTL effector cells remain cytolytic after treatment with 1 μM TX or 5 μM TO, which are the concentrations commonly achievable during cancer therapy. In murine tumor-host systems both TX and TO enhanced the cell mediated rejection (e.g. by NK, LAK or CTL effectors) of ERα negative tumors, which led to the cure and long-term survival of a significant proportion (50–75%) of animals that bore lethal cancer [160–166].

Killer and suppressor cells ($CD8^+$ T lymphocytes) and NK cells express classical estrogen receptors (ERα) and TX or TO may act on these receptors as anti-estrogens. However, the antiproliferative effect of TX on lymphocytes is independent of ERα and of estrogen action. The sensitizing effect of TX and TO on ERα negative target cells for cell mediated lysis is mediated by yet another receptor on which E2 has a similar effect to that of TX and TO. This receptor is likely to be ERβ [148,157,158,161].

Anti-Fas monoclonal antibody mediated cytotoxicity was enhanced by antiestrogens in the majority of human ovarian carcinoma cells examined. This indicates that apoptotic cell death is amplified by these drugs. The perforin/granzyme pathway of immune cytolysis was also enhanced in the case of NK mediated cytotoxicity against K562 target cells. MHC-I and MHC-II expression by human ovarian carcinomas was not affected by antiestrogens [157,167,168].

MRL-lpr/lpr mice treated with TX had less severe proteinuria, reduced serum titre of anti-dsDNA autoantibodies and an increased survival rate. The TX-treated groups (200 μg per mouse, 5.5 mg/kg, was given every 2 weeks s.c.) had smaller spleens and lymph nodes. Splenocytes had a significantly lower percentage of T cells and $CD4^-CD8^-$ T cells. There was no difference in cytokine production (IL-2, -4, -5, -10 and IFN-γ) by splenocytes stimulated with Con A, or in IL-6 secretion by peritoneal exudate cells when stimulated with LPS. However, IL-2 production by lymph node cells was significantly higher in TX-treated mice. Splenocytes or purified T cells were stimulated with anti-CD3 antibody plus cross-linking IgG. The TX-treated group had higher 3H-incorporation compared to controls. The IL-2-activated proliferation of $CD4^-CD8^-$ lymph node T cells was inhibited by TX treatment in a dose-dependent manner [169].

3.4. Androgens

In general, testosterone suppresses immune reactions. The development of the bursa of Fabricius is prevented by testosterone treatment of chicken embryos. Testosterone inhibits the enhancing effect of estrogens in a number of autoimmune disease models. It has been proposed that testosterone selectively favors the differentiation of suppressor T lymphocytes in the thymus [86,170]. MHC-linked genes influence the effect of androgens on the immune system [171]. Androgens stimulate hemopoiesis in the bone marrow [172]. Human lymphocytes metabolize sex hormones and are capable of synthesizing androgens [173].

3.4.1. Effect on the thymus

In aging rats, the thymus was regenerated after castration and this regeneration was inhibited by testosterone treatment. Testosterone is converted to E2 by aromatase and E2 is a powerful inducer of thymic involution. When young adult and aging (18 months old) orchidectomized male rats were treated with testosterone and with aromatase inhibitors (1,4,6-androstatriene-3,17-dione [ATD]; or 4-hydroxyandrostenedione [4-OHA]), the effects of testosterone were blocked by ATD in both the thymus and spleen. These inhibitors restored the thymus in aging rats with normal histological appearance. ATD enlarged the thymus in young intact animals. In old rats ATD blocked the effects of testosterone in the thymus and spleen. Cytosolic estrogen receptors were not detectable in the thymus of old rats but were measurable in orchidectomized or ATD-treated old animals [174].

3.4.2. Effect on T lymphocytes

Dihydrotestosterone (DHT) treatment of murine T lymphocytes reduced the amount of IL-4, -5 and IFN-γ in response to activation by anti-CD3 antibodies, while the production of IL-2 was not affected. Androstenedione or testosterone, which are metabolic precursors of DHT, had no effect in this system. The enzyme, 5α-reductase, which converts testosterone to DHT, has been detected in macrophages. Macrophages generated DHT and the inhibition of this conversion blocked IL-4 production. Moreover, transdermal treatment of mice with DHT altered the capacity of T cells in the draining lymph nodes to produce lymphokines. Combined treatment of aged mice with DHT and DHEA restored the capacity of T cells to produce IL-2, IL-4 and IFN-γ to the levels present in younger mice [175]. Testosterone and DHT treatment of male and female neonatal rats increased the reactivity of lymphocytes in the mixed lymphocyte reaction, which could be further increased by additional gonadectomy. Serum immunoglobulin levels were increased by this hormonal treatment in males, but was suppressed in females [176].

3.4.3. Effect on phagocytic cells and antigen presentation

Testosterone inhibited nitrite release and stimulated the release of reactive oxygen intermediates from rat peritoneal macrophages *in vitro*. Anti-TNF antibody moderately reduced superoxide and hydrogen peroxide release and profoundly inhibited nitrite release when added to this system [25]. Testosterone (0.1–40 μM) showed a dose dependent inhibition of inducible NO synthesis in the RAW 264.7 murine macrophage cell line [177].

In guinea pigs, androgens (testosterone, DHT, mesterolone) impaired the clearance of IgG-coated cells by decreasing splenic macrophage FcγR expression. Antiandrogens (flutamide, nilutamide, cyproterone acetate, spironolactone, and finasteride) did not significantly alter macrophage FcγR expression, but counteracted the inhibitory effects of androgens [178].

3.4.4. Effect on cytokines

Transsexual men and women were studied before and after 4 months of cross-sex hormone administration. In men, estrogen plus antiandrogen treatment increased free cortisol levels in the urine, decreased NK cell numbers, and slightly inhibited the mitogen-induced IFNγ/IL-4 ratio, but upregulated the expression of the chemokine receptors, CCR1, CXCR3, and CCR5. In women, androgens slightly decreased free urinary cortisol levels and enhanced the mitogen-induced IFNγ/IL-4 ratio and TNFα production. The upregulation of Th1 cytokines was accompanied by down-regulation of CCR1, CXCR3, and CCR5 expression [179].

Bioactive TGF-β_1 fell to approximately 50% after castration of male mice and was normalized 1 week after testosterone treatment. Testosterone modulated the production of TGF-β by thymocytes [180].

3.4.5. Trauma and infection

The testosterone receptor antagonist, flutamide, restored depressed immune function in males after trauma and hemorrhage. Flutamide improved plasma corticosterone content, restored cAMP content, and attenuated adrenal morphological alterations. Flutamide also improved the diminished ACTH-induced corticosterone release and adrenal cAMP response [181,182]. Testosterone-treated female mice that had experienced hemorrhage showed significant depression in splenic and peritoneal macrophage IL-1 and IL-6 production, comparable with the values seen in macrophages from male mice after hemorrhage [183].

In male C3H/HeN mice DHT-treatment significantly decreased the releases of IL-1β and IL-6 by splenic and peritoneal macrophages after trauma-hemorrhage, compared with cells of sham operated mice. DHT-treated animals exhibited increased Kupffer cell IL-6 release. IL-10 release was increased in DHT-treated hemorrhaged animals compared with sham-operated animals but was decreased in estrogen-treated mice under such conditions [184]. Estrogen prevents this androgen-induced immunodepression in castrated male mice [185]. Male C3H/HeN mice were subjected to castration 2 weeks before trauma-hemorrhage and resuscitation. Twenty four hours after hemorrhage an elevated macrophage IL-1 and IL-6 release was only observed in sham-castrated mice, as opposed to basal levels of cytokine release from castrated animals. Only sham-castrated animals showed a significantly increased levels of IL-6 release from Kupffer cells. Short-term therapy of male mice with the androgen receptor blocker, flutamide at 25 mg/kg body weight, following trauma-hemorrhage has protective effects on immune functions [186].

The sexually dimorphic immune response to trauma-hemorrhage in young males and females reverses as sex hormone levels decline with age [187]. Amongst 84 patients that sustained blunt injuries a sex difference in posttraumatic cytokine release by peripheral blood mononuclear cells was not detectable. Male trauma patients developing severe sepsis (n=16) showed a significantly increased cytokine producing capacity in the early posttraumatic period (≤24 hours after admission to the emergency room) when compared with males with an uncomplicated recovery. In females, no differences were detectable between these subgroups. In males, the development of severe sepsis later was found to coincide with significantly decreased testosterone and increased E2 serum levels, whereas in females, differences in sex hormone levels were not detectable [188]. In male patients the risk of major infections following trauma is dramatically increased. This effect is most significant following injuries of moderate severity and persists in all age groups [189].

In healthy men, subcutaneous IL-6 administration, which caused acute elevations in circulating IL-6 levels of a similar magnitude to severe inflammatory and noninflammatory disease, induced prolonged suppression in testosterone levels without apparent changes in gonadotropin levels [190].

A novel mouse cDNA encodes IAP38, a plasma membrane protein of 38 kDa in splenic macrophages, B cells and T cells. IAP38 is induced by blood-stage infections of *Plasmodium chabaudi* malaria and is testosterone-sensitive in non-immune mice. When the mice have acquired testosterone-resistant immunity to *P. chabaudi*, there is a 40-fold increase in the expression of IAP38, which has then largely lost its responsiveness to infection and testosterone [191].

Female C57BL/10 mice were fed a diet containing 17α-methyl-testosterone for 3 weeks. Circulating testosterone levels were raised from 0.28 ng/ml to 2.69 ng/ml on the average. In these mice, blood stage infections with *P. chabaudi* had a lethal outcome, whereas protective immunity developed in about 80% of mice fed a control diet. Dietary testosterone reduced the capacity of peritoneal cells to generate reactive oxygen intermediates after stimulation, increased TNF secretion in response to *Salmonella typhimurium,* whereas the production of IL-10 by spleen cells was largely suppressed and no effect was found on the production of IFNγ and IL-4 [192].

3.4.6. Reproduction biology

Subcutaneous testosterone implants, which inhibit Leydig cell function through suppression of gonadotrophin secretion, reduced αβT cells and the cytotoxic $CD8^+$ T cell subsets in both scrotal and cryptorchid testes. Testicular T cell numbers were positively correlated with the number of testicular resident macrophages and were also reduced by testosterone implants. The concentration of lymphocyte-inhibiting activity in the testicular interstitial fluid was reduced by 80% in short-term (1 month) and longer-term (3 months) cryptorchidism [193]. $CD8^+$ T lymphocytes have a potentially significant role in testicular function. T cell-inhibiting activity is present in the testis interstitium, which is not substantially affected by changes in pituitary hormones or Leydig cell function, but rather, is related to local changes in immune activity [194].

In comparison with rat peritoneal macrophages, testicular macrophages displayed only a very small IL-1 response to LPS (2.8% output/cell). The addition of testosterone (10–1,000 ng/ml) did not affect IL-1 production [195].

Both TNFα and IL-1β inhibited cAMP-stimulated testosterone production, mRNA and protein levels of cholesterol side chain cleavage enzyme (P450scc) and 17 alpha-hydroxylase/C17,20 lyase (P450c17) in mouse Leydig cells. Basal testosterone production was inhibited by TNF, but not IL-1 [196].

In rats injected i.c.v. by IL-1β the steroidogenic acute regulatory protein (StAR) plays a role in the suppression of testicular function [197]. Unconjugated 3beta,7alpha-dihydroxy-5-androstene-17-one (7alpha-OH-DHEA) and its 7beta-hydroxyisomer, which are proven immunomodulators, have been detected and quantified in semen [198].

Adult spring chinook salmon (*Oncorhynchus tshawytscha*) have high plasma concentrations of testosterone during sexual maturation. Such high levels of testosterone have been shown to reduce the salmonid immune response *in vitro*. Physiological levels of testosterone induced a significant loss of leukocytes in cultures after 5 days of contact [199].

3.4.7. Autoimmune disease

Male SJL mice immunized with proteolipid protein (PLP) peptide 139–151 developed an initial episode of paralysis, but failed to relapse. EAE relapses could be induced, however, by orchidectomy [200].

Human antibodies specific for digoxigenin, estradiol, testosterone and progesterone have been isolated from a small combinatorial IgM library. The affinities of both the anti-estradiol and anti-progesterone antibodies were approximately 10^8 M^{-1}. Naive IgM genes were highly represented, showing that it is possible for completely naive antibodies to bind steroids with affinities comparable to those obtained after a secondary immune response [201].

The administration of testosterone propionate or DHT had no noticeable effect on the humoral immune response, whereas DHT suppressed growth of the bursa of Fabricius in both sexes of HC chickens [142].

3.4.8. Dehydroepyandrosterone (DHEA) and its derivatives

Recent studies revealed that the weak androgen, DHEA, and its metabolites are major immunoregulatory hormones. Cholesterol is converted to pregnenolone (PREG) in the adrenal gland, which serves as a precursor for the synthesis of DHEA. DHEA is regarded as a "mother steroid" from which other hormones of high biological activity, such as testosterone (TS), androstenediol (AED), and DHT are derived [202]. Both PREG and DHEA are metabolized by the spleen and the 7α-hydroxy derivatives of DHEA and PREG are some 160 times more active in stimulating the antibody response in mice than are the precursor hormones [203]. Androstenediol and androstenetriol (AET), which are also derived from DHEA, each have been shown to be more potent immunoregulators than DHEA itself. DHEA suppressed Con-A or LPS-driven lymphocyte proliferation, AED had little influence and AET potentiated these responses. The secretion of IL-2 and IL-3 by Con-A activated lymphocytes was depressed by DHEA and potentiated by AET. The suppression of Con A induced lymphocyte proliferation by hydrocortisone was strongly antagonized by AET; AED showed a weak anti-glucocorticoid effect and DHEA had no effect [204].

Calcitonin gene related peptide (CGRP) and DHEA both inhibited Con A induced thymocyte proliferation. The CGRP antagonist, CGRP [8–37] significantly enhanced the proliferative response over a range of Con A concentrations. In addition, the CGRP antagonist blocked the inhibitory effect of DHEA in this system. None of these agents affected thymocyte proliferation in the absence of Con A [205].

DHEA affects Th-1 type helper cells as the primary target, which are stimulated for proliferation and IL-2 secretion. DHEA antagonizes the suppression of IL-2 secretion by glucocorticoids. The production of IFN-γ is also promoted, whereas IL-4 and IL-6 production is inhibited by DHEA. The deficient production of IL-2, IL-4 and IFN-γ of lymphocytes from old mice in response to anti-CD3 antibodies was restored to the level of production by lymphocytes from young donors, when combined treatment with DHEA and DHT was given [175,202,206]. A single subcutaneous injection of DHEA to mice, abolished the high dose tolerance of delayed type hypersensitivity to sheep red blood cells. Furthermore, the treatment of spleen cells from tolerant mice *in vitro* with DHEA prevented the transfer of tolerance to naive recipients. Lymphocytes from DHEA-treated tolerant mice produced more IFN-γ and less IL-4 and IL-6 when compared to lymphocytes of non-treated tolerant animals [207].

Dehydroepiandrosterone as well as its 7alpha-hydroxylated metabolite significantly decreased the number of plaque forming cells (PFC). 7beta-hydroxy-DHEA decreased PFC at low doses, but this effect was less pronounced at higher concentrations. In addition, 7beta-hydroxy-DHEA counteracted the suppressive effect of dexamethasone on PFC. None of the natural steroids affected cell viability [208]. Both 7 alpha-hydroxy-DHEA and 7 alpha-hydroxy-epiandrosterone (EPIA) triggered a significant increase of IgG antibodies to tetanus toxoid and Bordetella pertussis by activated B + T cells from human tonsils. 7 alpha-hydroxy-pregnenolone and 7 alpha-hydroxy-DHEA increased the immune response in mice [209]. Male rats are prone to develop sperm autoantibodies after vasectomy and during recovery from treatment with exogenous testosterone [210].

Human monocytes express receptors for DHEA. In monocytes DHEA enhanced the induction of cytotoxicity, IL-1 secretion, reactive nitrogen intermediate release, and the expression of complement receptor-1 and TNFα protein [211].

In animals, DHEA increased host resistance against viral, bacterial and parasitic infections and potentiated vaccination, including mucosal immunization [70,202]. DHEA protected mice against thermal injury, whereas prednisone increased mortality [212].

Male NMRI mice were subjected to laparotomy (sham) or cecal ligation and puncture (CLP). DHEA administration improved the survival rate (87% vs. 53% after 48 hrs; $p<0.001$) of septic mice. DHEA restored the depressed DTH reaction and reduced TNFα serum concentrations [213]. In male mice DHEA improved splenocyte function after trauma and hemorrhage by directly stimulating T cells and also by preventing a rise in serum corticosterone. The stimulatory effect of DHEA on splenocyte proliferation was completely abrogated by the estrogen antagonist tamoxifen [214].

DHEA levels decrease during chronic illness and DEX also suppresses plasma levels of dehydroepiandrosterone sulfate (DHEA-S), androstenediol and testosterone in patients. ACTH treatment corrects this suppressive effect [202]. DHEA-S levels are depressed in postmenopausal women with rheumatoid arthritis [215].

DHEA declines progressively in the blood after the second decade by approximately 20% per decade and will decrease to 10–15% of maximum by age 85–90. The immune system is gradually disregulated in aging animals and man, which is referred to as *immune senescence*. The proportion of $CD5^+$ B cells is increased with age. These B cells produce natural antibodies with frequent autoreactivity. Th1 type helper activity and the production of IL-2 decline, whereas the activity of Th2 cells and the production of IL-4, -5 and -6 increase. Treatment of aged animals with DHEA sulphate led to the reversal of immunosenescence and restored the reactivity of old animals to the level of young ones. Currently DHEA is being investigated for the potentiation of immune defence in patients [70,216–218].

The activity of IFNγ-secreting cells varied as a function of serum DHEA-S levels in men and pre-menopausal women. Post-menopausal women had fewer cells actively secreting cytokines and the activity of these cells did not correlate with sex hormone levels [144]. Postmenopausal women were treated with physiologic doses of DHEA for 3 weeks in a prospective randomized double blind crossover study, separated by a 2 week washout period. There was a decrease in $CD4^+$ T cells and an increase in $CD8^+/CD56^+$ (natural killer) cells in treated individuals. T cell mitogenic and IL-6 responses were inhibited, whereas NK cell cytotoxicity was increased dramatically by DHEA [219].

3.5. Progesterone (PS)

The overall effect of progesterone on systemic immunity is suppressive. Lymphocyte proliferation in response to *Candida albicans* was reduced by approximately 50% in the presence of luteal phase levels of PS (25 ng/ml) as opposed to proliferative phase levels (0.15 ng/ml). This inhibitory effect of PS was eliminated by the removal of monocytes from the culture [220]. Progesterone (10^{-5} M) suppressed the anti-*Candida* activity of neutrophils from mice, whereas TS, E2 and estriol did not [221]. Progesterone significantly inhibited nitrite release and stimulated the release of reactive oxygen intermediates from rat macrophages with the exception of 10^{-10} and 10^{-9} M concentrations, which had no effect. Progesterone (10^{-2} ng/ml) stimulated TNF release from rat peritoneal macrophages [25]. Normal human PBMC, stimulated with LPS, produced less IL-1 upon exposure to PS, whereas IL-6 secretion was not altered. However, PS failed to inhibit IL-1 secretion by PBMC from male donors with rheumatoid arthritis [222].

Progesterone plays a fundamental role in the protection of the fetus against maternal immune reactions. Infertile women treated with the fertility drug, clomiphene citrate, were studied. It was observed that only those women conceived who reacted to clomifen with immunosuppression. Women with no immunosuppressive effect did not achieve pregnancy. Progesterone levels were significantly different in these two groups of women. Other steroid levels were not different [223]. In women suffering from primary habitual abortions, the incidence of pregnancies after immunotherapy was increased when additional PS treatment was given, suggesting a synergistic effect [224].

During pregnancy, progesterone promotes the production of IL-4 and IL-5, whereas relaxin promotes the production of IFNγ by T cells in the uterus. LIF is a Th2 cytokine and is essential for embryo implantation. It is upregulated by IL-4 and PS and down-regulated by Th1 inducers (e.g. IL-12, IFNγ and IFNα). There is a decreased production of leukemia inhibitory factor (LIF), IL-4 and IL-10 by decidual T cells in women with unexplained recurrent abortions in comparison with women with normal gestation. A decreased production of LIF, IL-4 and IL-10 was not found in peripheral-blood T cells [225].

The increased progesterone sensitivity of lymphocytes during pregnancy is due to the expression of progesterone receptors by γδ T lymphocytes in response to fetal antigens. Progesterone binding by γδ T cells results in the synthesis of progesterone-induced blocking factor (PIBF). PIBF acts on the phospholipase A2 enzyme, interferes with arachidonic acid metabolism, induces a Th2 biased immune response, and exerts an anti-abortive effect by controlling NK activity [226].

The endometrium of the ewe secretes a progesterone-induced member of the serpin superfamily of serine proteinase ovine uterine serpin (OvUS) during pregnancy, which has immunosuppressive properties. OvUS has a wide range of action and inhibits the mixed lymphocyte reaction, mitogen-stimulated lymphocyte proliferation, and T cell-dependent antibody production. OvUS functions by inhibiting protein kinase C and interleukin-2-mediated events. OvUS and similar genes present in cattle and pigs diverged from other serpins prior to the divergence of artiodactyls. Several lines of evidence suggest that, in sheep, OvUS functions to mediate the immunosuppressive effects of progesterone and to prevent the immunological rejection of the fetal allograft [227].

Progesterone given concurrently with estrogen to OVX mice for 4 days antagonized the ability of estrogen to recruit macrophages and neutrophils into the uterus. In progesterone receptor knockout mice progesterone was devoid of such action and neutrophils recruited by estrogen were found to be degranulated, partially explaining the edema that is observed with long-term treatment. B lymphocytes and the neutrophil chemotactic chemokine, MIP-2, were unchanged after estrogen and progesterone treatment [228].

A synthetic progestin (medroxyprogesterone acetate) inhibited the production of monocyte chemotactic protein-1 (MCP-1) by choriodecidual cells. This inhibition was reversed by the antiprogestin RU486. MCP-1 release from T47D breast cancer cells was stimulated by IL-1 and this production was inhibited by progesterone. DEX had no effect on MCP-1 release in this system [229].

Male DBA mice have a significantly higher percentage of glucocorticoid-induced apoptotic thymocytes (46.1+/-3.8%) than their female counterparts (31.6+/-3.1%; P=0.012). *In vitro*, both estrogen and testosterone increased thymocyte apoptosis. In contrast, PS not only inhibited spontaneous thymocyte apoptosis, but prevented glucocorticoid-induced apoptosis *in vitro*. Progesterone treatment of mice suppressed glucocorticoid-induced thymocyte apoptosis [230].

Estrogen and progesterone reduce the consequences of brain injury by enhancing antioxidant mechanisms, reducing excitotoxicity (altering glutamate receptor activity, reducing immune inflammation, providing neurotrophic support, stimulating axonal remyelinization), enhancing synaptogenesis and dendritic arborization. Estrogen seems more effective as a prophylactic treatment in females at risk for cardiac and ischemic brain injury, whereas progesterone is helpful in the post-injury treatment of both male and female subjects with acute traumatic brain damage. However, estradiol replacement therapy in elderly women who have a history of cerebrovascular disease did not protect against reoccurrence of ischemia and did not reduce mortality [231].

In male and randomly cycling female rats, a similar increase in body temperature occurred after intraperitoneal IL-1β injection. However, the febrile response was found to be significantly higher and more prolonged in females at proestrus than at diestrus. Bilaterally ovariectomized females, supplemented with sequential injections of E2 and PS, showed a significantly higher IL-1β fever than did OVX rats receiving E2 only [232]. Progesterone decreases host resistance against viral and fungal infections [233,234]. Progesterone increased, and estradiol decreased, the susceptibility of rats to intrauterine chlamydial infection [235].

Progesterone deficiency is present in female patients suffering from thyroid and ovarian autoimmune disease and from rheumatoid arthritis [236]. In three patients a severe pre-menstrual exacerbation of asthma was reported which did not respond to treatment with high dose glucocorticoids. Treatment by intravenous PS corrected the decreased airflow and led to the reduction of daily doses of prednisone [237].

4. 1,25-DIHYDROXYVITAMIN D3 (VD3)

The precursor of VD3, cholecalciferol, is present in the diet and also synthesised in the skin under the influence of ultraviolet (UV) radiation. 25-Hydroxyvitamin D3 is generated in the liver, which is processed further by 1-hydroxylase in the kidney. 1-hydroxylase is also present in monocyte/macrophages, in keratinocytes, bone marrow, placenta, glia cells and pneumocytes [238–240].

4.1. Receptor

The vitamin D receptor (VDR) is of 50 kDa molecular weight and belongs to the superfamily of steroid/thyroid hormone receptors. VDR interacts with vitamin D responsive elements (VDRE) on DNA and also with the retinoic X receptor (RXR). Positive (stimulation) and negative (inhibition) regulation are both possible through VDRE. Protein kinase C is involved in receptor mediated signaling and VD3 regulates the DNA binding subunit of NFκB. Within the immune system, monocyte/macrophages, activated T lymphocytes, B lymphocytes and bone marrow cells express VDR [238–240].

In T lymphocytes VDR levels are increased when activated and treated with VD3. The highest concentrations of VDR are found in CD8+ lymphocytes, although significant amounts are also present in CD4 lymphocytes. B lymphocytes do not contain detectable amounts of VDR. Monocyte/macrophages possess small amounts of VDR that are not affected by activation but are increased by treatment with VD3 [241].

1,25-Dihydroxyvitamin D3 down-regulates activated IL-2 expression at the level of transcription, through direct antagonism of the transactivating complex NFAT-1/AP-1 by the VDR. GMCSF transcription in Jurkat T cells is also directly repressed by VD3 and VDR. One of four NFAT/AP-1 elements in the granulocyte-macrophage-colony-stimulating-factor (GM-CSF) enhancer, when multimerized, is sufficient to mediate both the activation by NFAT-1 and AP-1 and repression in response to VD3. Although this element does not contain any recognizable VDREs, high affinity DNA binding by recombinant VDR is present. This binding is independent of VDR's heterodimeric partner, the retinoid X receptor [242].

The GM-CSF enhancer is transcriptionally repressed by VD3 through the VDR. The DNA element involved is a composite site that is recognized by both Fos-Jun and NFAT1. VDR outcompetes NFAT1 for binding to the composite site. Overexpression of NFAT1 is able to antagonize the VD3-dependent repression. VDR stabilizes the binding of a Jun-Fos heterodimer to the adjacent AP-1 portion of the element. This occurs through a direct interaction between VDR and c-Jun. *In vivo*, the overexpression of c-Jun, but not c-Fos, leads to a rescue of VD3-mediated repression [243].

The IFNγ promoter is down-regulated by VD3. This effect is enhanced by the RXR. A functional VDR DNA-binding domain is necessary for repression. Two important promoter regions are mainly involved: the first binds the heterodimer VDR-RXR. The other negative regulatory interaction was detected at the level of the promoter fragment -108 to +64 base pairs from the transcription start site. The activity of the IFNγ enhancer from -108 to -36 bp was not affected. Binding activity for VDR-RXR has been detected in the IFNγ minimal promoter, suggesting a possible mechanism of interference with transcription initiation/progression. These data indicate that direct modulation of the IFNγ promoter activity is one of the mechanisms involved in the repressive effect of VD3 on IFNγ gene expression [244].

The polymorphic and pleotropic vitamin D3 receptor plays a role in the pathogenesis of immunological disorders (e.g. primary biliary cirrhosis, inflammatory bowel disease, type 1 diabetes mellitus) and of malignancies (malignant melanoma, breast cancer). This sheds light on the importance of common nuclear receptors [245].

In Japanese patients with type 1 diabetes, a significantly higher prevalence of the F allele, the FF genotype of VDR was found (p=0.0069 and p=0.014, respectively). This suggests that vitamin D receptor initiation codon polymorphism influences the genetic susceptibility to diabetes among the Japanese [246].

4.2. *The biology of VD3*

VD3 is a potent anti-proliferative agent and pro-differentiation mediator for macrophages, lymphocytes and other cell types [247]. In monocytes/macrophages adherence, chemotaxis, phagocytosis, cytotoxicity, H_2O_2, oxygen radicals, and heat shock protein production are all stimulated by VD3. In contrast, antigen presentation, the production of IL-1, -6, -12, and TNFα are inhibited. The function of Th-1 cells and the production of IL-2, IFN-γ, and TNF are inhibited by VD3. Likewise, supressor T cell function is inhibited, whereas the effect of VD3 on cytotoxic T lymphocytes is not clear. Natural killer cell cytotoxicity is stimulated, whereas B lymphocyte proliferation and immunoglobulin secretion is inhibited [238,239,248].

4.2.1. Effects on phagocytes and antigen presentation

VD3 profoundly upregulated phagocytic activity while downregulating HLA-DR antigen and CD13 expression by human monocytes [249]. VD3 inhibited the differentiation and maturation of human monocytes into potent antigen presenting dendritic cells (DC) [250]. Viatmin D3 mediated physiologically relevant inhibition of DC maturity that is resistant to maturational stimuli and modulated antigen-specific immune responses *in vivo* [251]. VD3 prevented the differentiation of human monocytes into immature DCs when cultured with GM-CSF and IL-4. The capacity of DCs to activate alloreactive T cells was reduced, as determined by decreased proliferation and IFNγ secretion in CD4+ cells. VD3 also inhibited IL-12p75 and enhanced IL-10 secretion by DC upon activation by CD40 ligation. In addition, VD3 promoted the spontaneous apoptosis of mature DCs [252].

Macrophage activation for phagocytosis and antigen processing is the first step of antibody formation. Macrophage activation requires the participation of serum vitamin D3-binding protein (DBP; human DBP is known as Gc protein) and glycosidases of B and T lymphocytes [253].

After 3 and 7 days of infecting human macrophages with *Mycobacterium avium*, VD3 supplementation of the culture medium increased the numbers of bacteria in cells from control subjects. There was no change or even a decrease in the number of bacteria in cells from HIV-infected patients [254]

The activation of THP-1 human monocytic cells with IFNγ enhances VD3 synthesis 30-fold, blocks VD3 suppression of its synthesis and reduces VD3 induction of its degradation by 42.2%. In THP-1 cells and in normal monocytes, IFNγ inhibits VD3 induction of 24-hydroxylase mRNA levels without reducing mRNA stability. These results explain the mechanism of VD3 overproduction in granulomatosis and demonstrate the potent inhibitory effect of IFNγ on VD3 action in immune cells [255].

In hemodialysis patients, oral VD3 pulse therapy resulted in a marked decrease in the HLA-DR expression by peripheral blood monocytes 2 and 4 days after the first administration ($p<0.01$). PMA-induced superoxide production was significantly enhanced 2 days after the treatment ($p<0.01$). Phagocytosis by neutrophils and the mitogenic response of lymphocytes were not affected [256].

4.2.2. Effect on T lymphocytes

Vitamin D3 exerts its immunosuppressive activity by preferentially influencing helper T cell activity (Th1) and inhibiting the secretion of IL-2 and IFNγ and IL-12. VD3 further inhibits class II antigen expression and enhances suppressor cell activity. VD3 prevents the development of EAE, reduces the incidence of diabetes, attenuates murine lupus and prolongs graft survival in animal models. In humans, VD3 has an antiproliferative effect in psoriasis, antineoplastic effect in prostate cancer and immunomodulatory effect in scleroderma [257].

Vitamin D3 treatment of mice fed adequate or high dietary calcium resulted in a decreased number of lymphocytes in the lymph nodes and increased IL-4 and TGFβ1 mRNA levels. When calcium was omitted from the diet, VD3 supplementation increased TGFβ1 mRNA. [258].

VD3 has a significant additive effect on DEX-mediated inhibition of lymphocyte proliferation and on the inhibition of Th1 cytokine production (IFNγ). However, VD3 upregulates Th2 cytokines (IL-4, -13) and inhibits the GC-mediated suppression of these cytokines [259]. VD3 up-regulated IL-4 and its effects were blunted in IL-4 deficient mice. This indicates that IL-4 is a significant mediator of the immunological action of VD3 [260].

A marked Th1/Th2 shift occurred in VD3-treated mice after priming with the autoantigen GAD65 (p524-543), but not after immunization with ovalbumin (OVA). Cytokine profiles were measured after rechallenge with a panel of autoantigens (GAD65, heat shock protein 65, insulin B-chain) and control antigens (OVA, keyhole limpet hemocyanine, myelin proteolipid protein, tetanus toxin) and the Th1/Th2 shift in autoantigen-injected mice was confirmed, but not in control antigen-injected mice [261].

PBMCs from healthy donors were cultured with or without VD3 for up to 21 days. On days 0, 7, 14, and 21, the percentage of cytokine-producing T lymphocytes was analyzed. VD3 (10^{-8} M) had no significant effects on IFNγ- or IL-4-producing cells, and IL-12 producing cells were reduced in both the CD4+ and CD8+ populations. There was an increase in the low percentage of CD4+ and CD8+ T cells producing IL-13. IL-6-producing CD4+ and CD8+ T cells could only be detected in cultures with VD3, reaching a plateau after 14 days. Production remained stable for at least 7 weeks. About 70% of IL-6-producing CD4+ and CD8+ cells were also positive for IL-2, but more than 90% were negative for IFNγ, IL-4, or IL-13, respectively [262].

After weaning, mice lacking the 25(OH)D-1alpha-hydroxylase enzyme developed hypocalcemia, secondary hyperparathyroidism, retarded growth, and the skeletal abnormalities characteristic of rickets. Altered non-collagenous matrix protein expression and reduced numbers of osteoclasts were also observed in the skeleton. Female mice were infertile and exhibited uterine hypoplasia and absence of corpora lutea. Enlarged lymph nodes in the vicinity of the thyroid gland and a reduction in CD4- and CD8-positive peripheral T lymphocytes were observed [263].

4.3.3. Effect on cytokines

Bone-resorbing human osteoclasts synthesize and secrete high constitutive and inflammation-stimulated levels of the chemokine, IL-8. Production of IL-8 was decreased by IFNγ, VD3, and the anti-inflammatory glucocorticoid, DEX [264].

VD3 augmented M-CSF and LIF mRNA in C6 glioma cells, while LPS had minimal effects. When LPS and VD3 were used in combination, a strong synergistic effect upon the induction of M-CSF and LIF genes was observed. No TNFα transcript has been detected. VD3 had no pronounced effect on M-CSF, LIF, and TNFα transcripts in primary rat astrocytes when used as a sole stimulus. Treatment with LPS strongly enhanced the levels of these cytokines. VD3 partially reduced the LPS-induced levels of M-CSF and TNFα mRNA [265].

4.3.4. Cell mediated immunity

Exogenous VD3 or an analog, 19-nor-1,25-dihydroxyvitamin D2 markedly prolonged the survival of neonatal mouse heart allografts. Similar results were obtained with heterotopic heart transplants in rats. In mice, cyclosporine (25 mg/kg) was less effective than VD3 [266].

In mice, vitamin D deficiency suppressed the induction of contact sensitivity reaction by dinitrofluorobenzene. The ingestion of a vitamin D sufficient diet for 8 weeks corrected this deficiency [267]. Tuberculin skin reactivity developed more slowly in VD3 deficient guinea pigs immunized with purified protein derivative, but eventually the reaction achieved normal levels. The antigen-induced proliferation of lymphocytes was suppressed both by deficiency and excess dietary VD3. Nevertheless, the vitamin D status did not affect the ability of naive animals to control primary pulmonary tuberculosis or to develop protective immunity after BCG vaccination [268].

4.3.5. The antibody response

Lymph node cells of mice exposed *in situ* to VD3 produced less IL-2 and IFN-γ and significantly more IL-4, IL-5 and IL-10 than did lymphocytes from control animals. When vaccination with hepatitis B surface antigen was coupled with VD3 treatment, systemic and mucosal immune responses were induced. These systemic and mucosal antibody responses were augmented if DHEA was also added to the immunization regime as an immunomodulator [269]. In mice, VD3 was effective as an immunological adjuvant for subcutaneous vaccination when combined with *Haemophilus influenzae* type b oligosaccharide conjugated to diphtheria toxoid. Both mucosal (IgA, IgG) and the systemic antibody response was stimulated [270].

4.3.6. Tumor immunity

The combination of VD3 and adoptive immunotherapy significantly reduced metastasis in mice with established tumors, and reduced both metastasis and recurrence after surgical excision of the primary tumor [271]. The suppressive activity of CD34+ cells toward the cytolytic activity of tumor-reactive lymph node cells could be blocked by *in vitro* culture of CD34+ cells with VD3. Similarly, treatment of LLC-LN7 tumor bearing mice with VD3 alone diminished the levels of CD34+ suppressor cells within regional lymph nodes, spleens and tumors. IFNγ production by lymph node cells was increased in response to LLC-LN7 cells. The extent of tumor metastasis in VD3 treated mice was reduced [272].

The treatment of the human renal carcinoma cell line SK-RC-29 with VD3 (100 nM) in addition to TNFα (30 ng/ml) resulted in a 2.6 +/- 0.2-fold (mean +/- S.E.) increase in the cytotoxic effect of TNF. This potentiation was dose-dependent, and significant at VD3 concentrations equal to or higher than 10 nM. VD3 did not modulate the cytostatic effect of IFNα or the susceptibility of SK-RC-29 to killing by lymphokine-activated killer cells [273].

4.3.7. Aging

In healthy nonagenarians and centenarians the highest number of NK cells and the best preserved cytolytic function also presented a preserved metabolism of thyroid hormones and vitamin D and integrity of muscle mass. The NK cell number and/or cytolytic activity of healthy subjects greater than 90 years old was positively associated with serum levels of vitamin D, while T3, FT4, i-PTH hormones and lean body mass were associated only with NK cell number [274].

Postmenopausal women were treated with VD3, group one (16 subjects) received 1 μg/day for 14 days, and group two (14 participants) was treated with 2 μg/day for the same period. The placebo group comprised another 10 postmenopausal women. The lower dose of VD3 did not change any of the measured immune parameters. After the higher dose the mean values of CD3+ and CD8+ lymphocytes increased. Total lymphocytes, the CD4+ subset and the immune response did not change [275].

4.3.8. VD3 and disease

Vitamin D3 levels were significantly higher in patients with sarcoidosis versus normal controls, active sarcoidosis (n=20) versus inactive disease (n=17) and active sarcoidosis versus controls [276].

Vitamin D treatment prevents autoimmune disease in animals, including nephritis, experimental allergic encephalitis, diabetes, thyroiditis and systemic lupus erythematosus. However, if therapy is attempted by VD3 treatment, the disease will only be delayed but not cured. In man, the treatment of psoriasis with VD3 has a 100% success rate. However, treatment with the natural vitamin is toxic and for this reason new synthetic structural analogues with much reduced toxicity are employed. Combination of the treatment with cyclosporine A, FK506, or rapamycin improves the efficiency of treatment. In sarcoidosis the hypercalcemia, soft tissue granulomatosis and calcification are due to extrarenal production of VD3 [238,239,277,278].

VD3 completely prevented EAE, significantly prolonged allograft survival, and prevented collagen-induced arthritis in experimental animals [279].

Non-obese diabetic (NOD) mice treated with VD3 (5 μg/kg every 2 days) from the time of weaning were protected against diabetes induced by cyclophosphamide. Splenocytes from VD3-treated mice were less capable of transferring diabetes to young, irradiated NOD mice. The better elimination of effector cells in VD3-treated mice could be explained by a restoration of the sensitivity to cyclophosphamide-induced apoptosis in both thymocytes and splenocytes, in normally apoptosis-resistant NOD mice. Nonhypercalcemic analogs of VD3 administered to NOD mice when the autoimmune diabetes is already active can prevent clinical diabetes if cyclophosphamide treatment is also given [280,281].

Human pancreatic islets were treated with a combination of three cytokines, IL-1β+TNFα+IFNγ, in the presence or absence of VD3, and compared with untreated control cells. The addition of VD3 significantly reduced nitrite release, IL-6 production and MHC class I expression to the level of controls [282].

A large multi-center trial consistently showed a protective effect of vitamin D supplementation in infancy against the development of type I diabetes. These findings indicate that activated vitamin D prevents the development of, or arrests, an ongoing immune process that leads to the disease [283]

F1(NZBxW) mice, which are prone to developing SLE, were treated with VD3. A worsening of the histopathological changes was observed in treated animals [Vaisberg et al. 2000]. In the CNS of rats suffering from EAE the monocyte/macrophage lineage continuously expressed the inducible form of nitric oxide synthase (iNOS), whereas astrocytes only expressed iNOS during relapse (day 23). VD3 exerted a drastic inhibitory effect on iNOS expression, both at the protein and the mRNA levels. This effect was region-specific, most pronounced in the cerebellum and brainstem, but non-existent in cerebral cortex. iNOS down-regulation occurred in macrophages, activated microglia and in astrocytes [284].

In SLE patients a higher percentage of T, B, and NK cells express VDR compared to normal individuals. Incubating SLE PBMC with VD compounds significantly reduced proliferation, polyclonal and anti-dsDNA IgG production, and the percentages of $CD3^+/DR^+$ T and B ($CD19^+$) cells. NK ($CD16^+$) cells were elevated. 1,25 VD3 analogs were more potent in this respect than the natural compound. These data suggest that 1,25 VD3 compounds can offer novel approaches to the clinical management of SLE [285].

In postmenopausal patients with rheumatoid arthritis, IL-6 was found to be a critical determinant of increased bone resorption with high disease activity. Serum levels of VD3 were inversely related to T-cell activation [286].

In Sjögren's (SS) syndrome the serum levels of VD3 correlated inversely with the levels of soluble interleukin-2 receptor, status indices for global disease, total exocrine disease, surface exocrine disease, internal organ exocrine disease, and mediator-induced disease. There was no correlation with clinical/immunopathological status. The correlations found with measures of clinical and immunoinflammatory status supports the notion that vitamin D metabolism may be involved in the pathogenesis of primary SS [287].

Psoriasis patients were treated with topical calcitrol (0.005%) and a vehicle control for 8 weeks. Calcitrol treatment decreased keratinocyte proliferation, normalized keratinocyte differentiation and decreased immune activation in the psoriatic plaques [288].

The geographical areas with low supplies of vitamin D (e.g Scandinavia) correlate with regions with high incidences of multiple sclerosis, arthritis and diabetes. The active form of vitamin D has been shown to suppress the development of autoimmunity in experimental animal models [279].

REFERENCES

1. Berczi I. Pituitary Function and Immunity. Berczi I, editor, Boca Raton, Florida CRC Press, 1986.
2. Homo-Delarche F. Duval D. Glucocorticoid receptors in lymphoid tissue. In Hormones and Immunity, Berczi I, Kovacs K, editors, Lancaster, England, MTP Press, 1987;1–19.
3. Miller AH, Spencer RL, Husain A, Rhee R, McEwen BS. Stein M. Differential expression of type I adrenal steroid receptors in immune tissues is associated with tissue-specific regulation of type II receptors by aldosterone, Endocrinology 1993; 133: 2133–40.
4. Karin M. Ynag-Yen H-F, Chambard J-C, Deng T, Saatcioglu F. Various modes of gene regulation by nuclear receptors for steroid and thyroid hormones. Eur J Clin Pharmacol 1993;45:S9–15.

5. Angeli A, Masera RG, Sartori ML, Fortunati N, Racca S, Dovio A, Staurenghi A, Frairia R. Modulation by cytokines of glucocorticoid action. Ann NY Acad Sci 1999;876:210–20.
6. Watson CS, Gametchu B. Membrane estrogen and glucocorticoid receptors--implications for hormonal control of immune function and autoimmunity. Int Immunopharmacol 2001;1:1049–63.
7. Dhabhar FS, Miller AH, McEwen BS, Spencer RL. Differential activation of adrenal steroid receptors in neural and immune tissues of Sprage Dawley, Fischer 344, and Lewis rats. J. Neuroimmunol 1995;56: 77–90.
8. Scheinman RI, Cogswell PC, Lofquist AK, Baldwin Jr. AS. Role of transcriptional activation of IκBα in mediation of immunosuppression by glucocorticoids, Science 1995;270: 283–6.
9. Auphan N, DiDonato JA, Rosette C, Helmberg A, Karin M. Immunosuppression by glucocorticoids: inhibition of NF-κB activity through induction of IκB synthesis, Science 1995;270: 286–90.
10. Adcock IM, Brown CR, Gelder GM, Shirasaki H, Peters MJ, Barnes PJ. Effects of glucocorticoids on transcription factor activation in human peripheral blood mononuclear cells. Am J Physiol 1995;268: C331–8.
11. Gottlicher M, Heck S, Herrlich P.Transcriptional cross-talk, the second mode of steroid hormone receptor action. J Mol Med 1998;76:480–9.
12. Aittomaki S, Pesu M, Groner B, Janne OA, Palvimo JJ, Silvennoinen O. Cooperation among Stat1, glucocorticoid receptor, and PU.1 in transcriptional activation of the high-affinity Fc gamma receptor I in monocytes. J Immunol 2000;164:5689–97.
13. Ahluwalia A, Newbold P. Brain SD, Flower RJ. Topical glucocorticoids inhibit neurogenic inflammation: involvement of lipocortin 1. Eur J Pharmacol 1995;283: 193–8.
14. Taylor AD, Loxley HD, Flower RJ, Buckingham JC. Immunoneutralization of lipocortin 1 reverses the acute inhibitory effects of dexamethasone on the hypothalamo-pituitary-adrenocortical responses to cytokines in the rat in vitro and in vivo. Neuroendocrinology 1995;62: 19–31.
15. Barnes PJ. Adcock IM. Steroid resistance in asthma. QJ Med 1995;88:455–68.
16. Herrlich P. Cross-talk between glucocorticoid receptor and AP-1. Oncogene 2001;20: 2465–75.
17. Quan N, He L, Lai W, Shen T, Herkenham M. Induction of IkappaBalpha mRNA expression in the brain by glucocorticoids: a negative feedback mechanism for immune-to-brain signaling. : J Neurosci 2000;20:6473–7.
18. Bernardini R, Johnson EO, Kamilaris T, Chiarenza A, Cantarella G, Calogero AE, Lempereur L, Chrousos GP, Giuffrida R, Gold PW. Increased ACTH and cortisol secretion after interleukin-alpha injection in the common marmoset (Callithrix jacchus jacchus). Life Sci 2001;68:1657–65.
19. Thompson EB. Apoptosis and steroids. Mol Endocrinol 1994;8:665–673.
20. Berczi I. The influence of pituitary-adrenal axis on the immune system. In Pituitary Function and Immunity, Berczi I, editor, Boca Raton, FL, CRC Press, 1986;pp. 49–132.
21. Homo-Delarche F. Duval D, Papiernik M. Prostaglandin production by phagocytic cells of the mouse thymic reticulum in culture and its modulation by indomethacin and corticosteroids. J Immunol 1985;135: 506–512.
22. King LB, Vacchio MS, Dixon K, Hunziker R, Margulies DH, Ashwell JD. A targeted glucocorticoid receptor antisense transgene increases thymocyte apoptosis and alters thymocyte development. Immunity 1995;3: 647–56.
23. Arzt E, Kovalovsky D, Igaz LM, Costas M, Plazas P, Refojo D, Paez-Pereda M, Reul

JM, Stalla G, Holsboer F.Functional cross-talk among cytokines, T-cell receptor, and glucocorticoid receptor transcriptional activity and action. Ann N Y Acad Sci 2000;917: 672–7.
24. Vacchio MS, Lee JY, Ashwell JD. Thymus-derived glucocorticoids set the thresholds for thymocyte selection by inhibiting TCR-mediated thymocyte activation. J Immunol 1999;163:1327–33.
25. Chao TC, Van-Alten PJ, Walter RJ. Steroid and sex hormones and macrophage function: modulation of reactive oxygen intermediates and nitrite release. Amer. J. Reprod Immunol 1994;32: 43–52.
26. Munck A, Naray-Fejes-Toth A, Guyre PM. Mechanisms of glucocorticoid actions on the immune system. In Hormones and Immunity. Berczi I, Kovacs K, editors, Lancaster, England, MTP Press, 1987;pp. 20–37.
27. Sauer J. Castren M, Hopfner U, Holsboer F, Stalla GK, Arzt E., Inhibition of lipopolysaccharide-induced monocyte interleukin-1 receptor antagonist synthesis by cortisol: involvement of the mineralocorticoid receptor. J Clin Endocrinol Metab 1996;81: 73–9.
28. Schmidt M, Lugering N, Lugering A, Pauels HG, Schulze-Osthoff K, Domschke W, Kucharzik T. Role of the CD95/CD95 ligand system in glucocorticoid-induced monocyte apoptosis. J Immunol 2001;166:1344–51.
29. Girndt M, Sester U, Kaul H, Hunger F, Kohler H.Glucocorticoids inhibit activation-dependent expression of costimulatory molecule B7-1 in human monocytes. Transplantation 1998;66:370–5.
30. Moser M, De Smedt T, Sornasse T, Tielemans F, Chentoufi AA, Muraille E, Van Mechelen M. Urbain J, Leo O. Glucocorticoids down-regulate dendritic cell function in vitro and in vivo. Eur J Immunol 1995;25: 2818–24.
31. Vieira PL, Kalinski P, Wierenga EA, Kapsenberg ML, de Jong EC. Glucocorticoids inhibit bioactive IL-12p70 production by in vitro-generated human dendritic cells without affecting their T cell stimulatory potential. J Immunol 1998;161:5245–51.
32. Piemonti L, Monti P, Allavena P, Leone BE, Caputo A, Di Carlo V. Glucocorticoids increase the endocytic activity of human dendritic cells. Int Immunol 1999;11:1519–26.
33. Kim KD, Choe YK, Choe IS, Lim JS.Inhibition of glucocorticoid-mediated, caspase-independent dendritic cell death by CD40 activation. J Leukoc Biol 2001;69:426–34.
34. Gerrard TL, Cupps TR, Jurgensen CH, Fauci AS. Increased expression of HLA-DR antigens in hydrocortisone-treated monocytes. Cell Immunol 1984;84:311–316.
35. Strickland RW, Wahl LM, Finbloom DS. Corticosteroids enhance the binding of recombinant interferon-gamma to cultured human monocytes. J Immunol 1986;137: 1577–1580.
36. Akahoshi T, Oppenheim JJ, Matsushima K. Induction of high-affinity interleukin 1 receptor on human peripheral blood lymphocytes by glucocorticoid hormones. J Exp Med 1988;167: 924–936.
37. Snyers L, De Wit L, Content J. Glucocorticoid up-regulation of high-affinity interleukin 6 receptors on human epithelial cells. Proc Natl Acad Sci 1990;87: 2838–2842.
38. Calandra T, Bernhagen J, Metz CN, Spiegel LA, Bacher M, Donnelly T, Cerami A, Bucala R. MIF as a glucocorticoid-induced modulator of cytokine production, Nature 1995:377: 68.
39. Schleimer RP, Jaques A, Shin HS, Lichtenstein LM, Plaut M. Inhibition of T cell-mediated cytotoxicity by anti-inflammatory steroids. J Immunol 1984;132: 266–271.
40. Jiayi D, Shikun Y, Renbao X. The inhibitory effect of hydrocortisone on interferon production by rat spleen cells. J Steroid Biochem 1989;33: 1139–41.

41. Agarwal SK, Marshall GD Jr. Dexamethasone promotes type 2 cytokine production primarily through inhibition of type 1 cytokines. J Interferon Cytokine Res 2001;21: 147–55.
42. Ashwell JD, Lu FW, Vacchio MS.Glucocorticoids in T cell development and function. Annu Rev Immunol 2000;18:309–45.
43. Wiegers GJ, Labeur MS, Stec LEM, Klinkert EEF, Holsboer F, Reul JHHM. Glucocorticoids accelerate anti-T cell receptor-induced T cell growth. J Immunol 1995;155: 1893–1902.
44. Brunetti M, Martelli N, Colasante A, Piantelli M, Musiani P, Aiello FB. Spontaneous and glucocorticoid-induced apoptosis in human mature T lymphocytes, Blood 1995;86: 4199–205.
45. Riccardi C, Cifone MG, Migliorati G. Glucocorticoid hormone-induced modulation of gene expression and regulation of T-cell death: role of GITR and GILZ, two dexamethasone-induced genes. Cell Death Differ. 1999;6:1144–5.
46. Zipp F, Wendling U, Beyer M, Grieger U, Waiczies S, Wagenknecht B, Haas J,Weller M. Dual effect of glucocorticoids on apoptosis of human autoreactive and foreign antigen-specific T cells. J Neuroimmunol 2000;110:214–22.
47. Moynihan JA, Callahan TA, Kelley SP, Campbell LM. Adrenal hormone modulation of type 1 and type 2 cytokine production by spleen cells: dexamethasone and dehydroepiandrosterone suppress interleukin-2,interleukin-4, and interferon-gamma production in vitro. Cell Immunol 1998;184:58–64.
48. Franchimont D, Galon J, Gadina M, Visconti R, Zhou Y, Aringer M, Frucht DM, Chrousos GP, O'Shea JJ. Inhibition of Th1 immune response by glucocorticoids: dexamethasone selectively inhibits IL-12-induced Stat4 phosphorylation in T lymphocytes. J Immunol 2000;164:1768–74.
49. Kohka H, Iwagaki H, Yoshino T, Kobashi K, Saito S, Isozaki H, Takakura N, Tanaka N. Hydrocortisone sodium succinate suppressed production of interleukin-10 by human peripheral blood mononuclear cells: clinical significance. Acta Med Okayama 1999;53: 55–9.
50. Wargnier A, Lafaurie C, Legros-Maida S, Bourge JF, Sigaux F, Sasportes M, Paul P.Down-regulation of human granzyme B expression by glucocorticoids. Dexamethasone inhibits binding to the Ikaros and AP-1 regulatory elements of the granzyme B promoter. : J Biol Chem 1998;273:35326–31.
51. Padgett DA, Sheridan JF, Loria RM, Steroid hormone regulation of a polyclonal T_H2 immune response. Ann NY Acad Sci 1995;774: 323–5.
52. Gatti G, Cavallo R, Sartori ML, Marinone C, Angeli A. Cortisol at physiological concentrations and prostaglandin E_2 are additive inhibitors of human natural killer cell activity. Immunopharmacology 1986;11: 119–128.
53. Desbarats J, You-Ten KE, Lapp WS, Levels of $p56^{lck}$ and $p59^{fyn}$ are reduced by a glucocorticoid-dependent mechanism in graft-versus-host reaction-induced T cell anergy. Cell Immunol 1995;163: 10–18.
54. Fleshner M, Deak T, Nguyen KT, Watkins LR, Maier SF.Endogenous glucocorticoids play a positive regulatory role in the anti-keyhole limpet hemocyanin in vivo antibody response. J Immunol 2001;166:3813–9.
55. Wira CR, Rossoll RM. Glucocorticoid regulation of mucosal immunity: Effect of dexamethasone on IgA and secretory component (SC), In Advances in Psychoneuroimmunology, Berczi I, Szelenyi J, editors, New York, Plenum Press, 1994;pp. 203–212.
56. Katayama I, Igawa K, Minatohara K, Nishioka K. Topical glucocorticoid augments IgE-

mediated passive cutaneous anaphylaxis in Balb/C mice and mast cell deficient WBB6F1 v/v mice. Clin Exp Allergy 1997;27:1477–83.
57. Grosset C, Taupin JL, Lemercier C, Moreau JF, Reiffers J, Ripoche J.Leukaemia inhibitory factor expression is inhibited by glucocorticoids through post-transcriptional mechanisms. Cytokine 1999;11:29–36.
58. Conti B, Sugama S, Kim Y, Tinti C, Kim H, Baker H, Volpe B, Attardi B, Joh T. Modulation of IL-18 production in the adrenal cortex following acute ACTH or chronic corticosterone treatment. Neuroimmunomodulation 2000;8:1–7.
59. Chung H-T, Samlowski WE, Daynes RA. Modification of the murine immune system by glucocorticosteroids: Alterations of the tissue localization properties of circulating lymphocytes. Cell Immunol 1986;101: 571–585.
60. Maruyama S, Minagawa M, Shimizu T, Oya H, Yamamoto S, Musha N, Abo W, Weerasinghe A, Hatakeyama K, Abo T. Administration of glucocorticoids markedly increases the numbers of granulocytes and extrathymic T cells in the bone marrow. Cell Immunol 1999;194:28–35.
61. Schleimer RP. An overview of glucocorticoid anti-inflammatory actions. Eur J Clin Pharmacol 1993;45:S3–7.
62. Sternberg EM. Neuroendocrine factors in susceptibility to inflammatory disease: focus on the hypothalamic-pituitary-adrenal axis. Horm Res 1995;43:159–61.
63. Karlais K, Crofford L, Wilder RL, Chrousos GP. Glucocorticoid and/or glucocorticoid antagonist effects in inflammatory disease-susceptible Lewis rats and inflammatory disease-resistant Fischer rats, Endocrinology 1995;136:3107–12.
64. Han CW, Choi JH, Kim JM, Kim WY, Lee KY, Oh GT. Glucocorticoid-mediated repression of inflammatory cytokine production in fibroblast-like rheumatoid synoviocytes is independent of nuclear factor-kappaB activation induced by tumour necrosis factor alpha. Rheumatology (Oxford) 2001;40:267–73.
65. Molijn GJ, Spek JJ, van Uffelen JCJ, de Jong FH, Brinkmann AO, Bruining HA, Lamberts SWJ, Koper JW. Differential adaptation of glucocorticoid sensitivity of peripheral blood mononuclear leukocytes in patients with sepsis or septic shock. J Clin Endocrinol Metab 1995;80:1799–1803.
66. Lavker RM, Schechter NM. Cutaneous mast cell depletion results from topical corticosteroid usage. J Immunol 1985;135: 2368–2373.
67. Bergstrand H, Bjornsson A, Lundquist B, Nilsson N, Brattsand R. Inhibitory effect of glucocorticosteroids on anti-IgE-induced histamine release from human basophilic leukocytes: Evidence for a dual mechanism of action. Allergy 1983;39: 217–230.
68. Robin JL, Seldin DC, Austen KF, Lewis RA. Regulation of mediator release from mouse bone marrow-derived mast cells by glucocorticoids, J Immunol 125: 2719–2726 (1985).
69. Knisely TL, Hosoi J, Nazareno R, Granstein RD. The presence of biologically significant concentrations of glucocorticoids but little or no cortisol binding globulin within aqueous humor: relevance to immune privilege in the anterior chamber of the eye. Invest Ophthalmol Vis Sci 1994;35:3711–23.
70. Daynes RA, Araneo BA, Hennebold J, Enioutina E, Mu HH. Steroids as regulators of the mammalian immune response. J Invest Dermatol 1995;105: 4S–19S.
71. Bakker JM, Kavelaars A, Kamphuis PJ, Cobelens PM, van Vugt HH, van Bel F, Heijnen CJ. Neonatal dexamethasone treatment increases susceptibility to experimental autoimmune disease in adult rats. J Immunol 2000;165:5932–7.
72. Kavelaars A, van der Pompe G, Bakker JM, van Hasselt PM, Cats B, Visser GH, Heijnen

CJ. Altered immune function in human newborns after prenatal administration of betamethasone: enhanced natural killer cell activity and decreased T cell proliferation in cord blood. Pediatr Res 1999;45:306–12.
73. Ruzek MC, Pearce BD, Miller AH, Biron CA. Endogenous glucocorticoids protect against cytokine-mediated lethality during viral infection. J Immunol 1999;162:3527–33.
74. Miller AH, Spencer RL, Hassett J, Kim C, Rhee R, Ciurea D, Dhabhar F, McEwen B, Stein M. Effects of selective type I and II adrenal steroid agonists on immune cell distribution, Endocrinology 1994;135:1934–44.
75. Wiegers GJ, Reul JM, Holsboer F, de Kloet ER. Enhancement of rat splenic lymphocyte mitogenesis after short term preexposure to corticosteroids in vitro. Endocrinology 1994;135: 2351–7.
76. Trune DR, Kempton JB, Kessi M. Aldosterone (mineralocorticoid) equivalent to prednisolone (glucocorticoid) in reversing hearing loss in MRL/MpJ-Fas1pr autoimmune mice. Laryngoscope 2000;110:1902–6.
77. Leo NA, Bonneau RH. Mechanisms underlying chemical sympathectomy-induced suppression of herpes simplex virus-specific cytotoxic T lymphocyte activation and function. J Neuroimmunol 2000;110:45–56.
78. Marchetti B, Gallo F, Farinella Z, Tirolo C, Testa N, Romeo C, Morale MC.Luteinizing hormone-releasing hormone is a primary signaling molecule in the neuroimmune network. Ann N Y Acad Sci 1998;840:205–48.
79. Grasso G, Massai L, De Leo V, Muscettola M. The effect of LHRH and TRH on human interferon-gamma production in vivo and in vitro. Life Sci 1998;62:2005–14.
80. Rouabhia M, Chakir J, Deschaux P. Interaction between the immune and endocrine systems: Immunodulatory effects of luteinizing hormone. Prog Neuroendocrinimmunol 1994;4: 86–91.
81. Komorowski J, Stepien H. FSH and LH induce interleukin-6 (IL-6) release from human peripheral blood monocytes cultures in vitro. A dose-response study. Hormone Metab Res 1994;26: 438–439.
82. Cannon JG. Adaptive interactions between cytokines and the hypothalamic-pituitary-gonadal axis. Ann N Y Acad Sci 1998;856:234–42.
83. Kalra PS, Edwards TG, Xu B, Jain M, Kalra SP.The anti-gonadotropic effects of cytokines: the role of neuropeptides. Domest Anim Endocrinol 1998;15:321–32.
84. Battaglia DF, Beaver AB, Harris TG, Tanhehco E, Viguie C, Karsch FJ. Endotoxin disrupts the estradiol-induced luteinizing hormone surge: interferencewith estradiol signal reading, not surge release. Endocrinology 1999;140:2471–9.
85. Battaglia DF, Krasa HB, Padmanabhan V, Viguie C, Karsch FJ. Endocrine alterations that underlie endotoxin-induced disruption of the follicular phase in ewes. Biol Reprod 2000;62:45–53.
86. Berczi I. Gonadotropins and sex hormones. In Pituitary Function and Immunity, Berczi I, editor, Boca Raton, FL, CRC Press, 1986;pp. 133–159.
87. Ogilvie KM, Held Hales K, Roberts ME, Hales DB, Rivier C.The inhibitory effect of intracerebroventricularly injected interleukin 1beta on testosterone secretion in the rat: role of steroidogenic acute regulatory protein. Biol Reprod 1999;60:527–33.
88. Duncan WC, Rodger FE, Illingworth PJ.The human corpus luteum: reduction in macrophages during simulated maternal recognition of pregnancy. Hum Reprod 1998;13: 2435–42.
89. Stimson WH. Sex steroids, steroid receptors and immunity. In Hormones and Immunity

Berczi I, Kovacs K, editors, Lancaster, England, MTP Press, 1987;pp. 43–53.
90. Myers MJ, Heim MC, Hirsch KS, Queener SF, Petersen BH. Translocatable estrogen receptors in rat splenocytes Life Sci 1986;39: 313–320.
91. Kovacs WJ, Olsen NJ. Androgen receptors in human thymocytes. J Immunol 1987;139: 490–493.
92. Baral E, Nagy E, Berczi I. The effect of tamoxifen on the immune response. In: Tamoxifen: Beyond the Antiestrogen, J.A. Kellen (ed), Birkhauser, Boston, MA, 1996;pp. 137–178.
93. Cutolo M, Villaggio B, Bisso A, Sulli A, Coviello D, Dayer JM.Presence of estrogen receptors in human myeloid monocytic cells (THP-1 cell line). Eur Cytokine Netw 2001;12:368.
94. Kushner PJ, Agard DA, Greene GL, Scanlan TS, Shiau AK, Uht RM, Webb P.Estrogen receptor pathways to AP-1. J Steroid Biochem Mol Biol 2000;74:311–7.
95. Kato SS. Estrogen receptor-mediated cross-talk with growth factor signaling pathways. Breast Cancer 2001;8:3–9.
96. Ray P, Ghosh SK, Zhang DH, Ray A. Repression of interleukin-6 gene expression by 17 beta-estradiol: inhibition of the DNA-binding activity of the transcription factors NF-IL6 and NF-kappa B by the estrogen receptor. FEBS Lett 1997;409:79–85.
97. Govind AP, Thampan RV. Proteins interacting with the mammalian estrogen receptor: proposal for an integrated model for estrogen receptor mediated regulation of transcription. J Cell Biochem 2001;80:571–9.
98. Katzenellenbogen BS, Choi I, Delage-Mourroux R, Ediger TR, Martini PG, Montano M, Sun J, Weis K, Katzenellenbogen JA.Molecular mechanisms of estrogen action: selective ligands and receptor pharmacology. J Steroid Biochem Mol Biol 2000;74:279–85.
99. Yamamoto T, Matsuda T, Junicho A, Kishi H, Saatcioglu F, Muraguchi A.Cross-talk between signal transducer and activator of transcription 3 and estrogen receptor signaling. FEBS Lett 2000;486:143–8.
100. Greenstein B, Roa R, Dhaher Y, Nunn E, Greenstein A, Khamashta M, Hughes GR.Estrogen and progesterone receptors in murine models of systemic lupus erythematosus. Int Immunopharmacol 2001;1:1025–35.
101. Greenstein BD. Lupus: why women? J Womens Health Gend Based Med 2001;10:233–9.
102. Tornwall J, Carey AB, Fox RI, Fox HS. Estrogen in autoimmunity: expression of estrogen receptors in thymic and autoimmune T cells. J Gend Specif Med 1999;2:33–40.
103. Staples JE, Gasiewicz TA, Fiore NC, Lubahn DB, Korach KS, Silverstone AE.Estrogen receptor alpha is necessary in thymic development and estradiol-induced thymic alterations. J Immunol 1999;163:4168–74.
104. Thurmond TS, Murante FG, Staples JE, Silverstone AE, Korach KS, Gasiewicz TA. Role of estrogen receptor alpha in hematopoietic stem cell development and B lymphocyte maturation in the male mouse. Endocrinology 2000;141:2309–18.
105. Schneider CP, Nickel EA, Samy TS, Schwacha MG, Cioffi WG, Bland KI, Chaudry IH. The aromatase inhibitor, 4-hydroxyandrostenedione, restores immune responses following trauma-hemorrhage in males and decreases mortality from subsequent sepsis. Shock 2000;14:347–53.
106. Mantalaris A, Panoskaltsis N, Sakai Y, Bourne P, Chang C, Messing EM, Wu JH.
107. Localization of androgen receptor expression in human bone marrow. J Pathol 2001;193: 361–6.
108. Janne OA, Moilanen AM, Poukka II, Roulcau N, Karvonen U, Kotaja N, Hakli M, Palvimo JJ. Androgen-receptor-interacting nuclear proteins. Biochem Soc Trans 2000;28:401–5.

109. Gottlieb B, Pinsky L, Beitel LK, Trifiro M.Androgen insensitivity. Am J Med Genet 1999 ;89:210–7.
110. DeMarzo AM, Beck CA, Onate SA. Edwards DP. Dimerization of mammalian progesterone receptors occurs in the absence of DNA and is related to the release of the 90-k-Da heat shock protein. Proc Natl Acad Sci USA 1991;88: 72–6.
111. Mohamed MK, Tung L, Takimoto GS, Horwitz KB. The leucine zippers of c-fos and c-jun for progesterone receptor dimerization: A-dominance in the A/B heterodimer. J Steroid Biochem Mol Biol 1994;51:241–50.
112. Conneely OM. Perspective: female steroid hormone action. Endocrinology 2001;142: 2194–9.
113. Rowan BG, O'Malley BW. Progesterone receptor coactivators. Steroids 2000;65:545–9.
114. Conneely OM, Mulac-Jericevic B, Lydon JP, De Mayo FJ. Reproductive functions of the progesterone receptor isoforms: lessons from knock-out mice. Mol Cell Endocrinol 2001;179:97–103.
115. Kalkhoven E, Wissink S, van der Saag PT, van der Burg B. Negative interaction between the RelA(p65) subunit of NF-kappaB and the progesterone receptor. J Biol Chem 1996;271:6217–24.
116. Borel IM, Freire SM, Rivera E, Canellada A, Binaghi RA, Margni RA.Modulation of the immune response by progesterone-induced lymphocyte factors. Scand J Immunol 1999;49:244–50.
117. Zhao XJ, McKerr G, Dong Z, Higgins CA, Carson J, Yang ZQ, Hannigan BM. Expression of oestrogen and progesterone receptors by mast cells alone, but not lymphocytes, macrophages or other immune cells in human upper airways. Thorax 2001;56:205–11.
118. Nelson JL Steinberg AD. Sex steroids, autoimmunity and autoimmune disease. In Hormones and Immunity, Berczi I, Kovacs K, editors, Lancaster, England, MTP Press, 1987; pp. 93–119.
119. Nagy E, Baral E. Berczi I. Immune System. In: "Handbook of Experimental Pharmacology", Estrogens and anti-estrogens. Section IV. Physiology and Pathophysiology of estrogens. M. Oettel, ed. in chief, Springer-Verlag 1999;135/I:343–351 (A).
120. Deshpande R, Khalili H, Pergolizzi RG, Michael SD, Chang MD. Estradiol down-regulates LPS-induced cytokine production and NFkB activation in murine macrophages. Am J Reprod Immunol 1997;38:46–54.
121. Chao TC Van Alten PJ, Greager JA, Walter RJ. Steroid sex hormones regulate the release of tumor necrosis factor by macrophages. Cell Immunol 1995;160:43–9.
122. Chao TC, Chao HH, Chen MF, Greager JA, Walter RJ. Female sex hormones modulate the function of LPS-treated macrophages. Am J Reprod Immunol 2000;44:310–8.
123. Azevedo RB, Lacava ZG, Miyasaka CK, Chaves SB, Curi R. Regulation of antioxidant enzyme activities in male and female rat macrophages by sex steroids. Braz J Med Biol Res 2001;34:683–7.
124. Robert R, Spitzer JA. Effects of female hormones (17beta-estradiol and progesterone) on nitric oxide production by alveolar macrophages in rats. Nitric Oxide 1997;1:453–62.
125. Vegeto E, Pollio G, Pellicciari C, Maggi A. Estrogen and progesterone induction of survival of monoblastoid cells undergoing TNF-alpha-induced apoptosis. FASEB J 1999 May;13(8):793–803.
126. Wira CR, Rossoll RM. Antigen-presenting cells in the female reproductive tract: influence of the estrous cycle on antigen presentation by uterine epithelial and stromal cells. Endocrinology 1995;136: 4526–34.

127. Wira CR, Rossoll RM, Kaushic C.Antigen-presenting cells in the female reproductive tract: influence of estradiol on antigen presentation by vaginal cells. Endocrinology 2000;141:2877–85.
128. Law WY, Chen WH, Song YL, Dufour S, Chang CF. Differential in vitro suppressive effects of steroids on leukocyte phagocytosis in two teleosts, tilapia and common carp. Gen Comp Endocrinol 2001;121:163–72.
129. Ito I, Hayashi T, Yamada K, Kuzuya M, Naito M, Iguchi A. Physiological concentration of estradiol inhibits polymorphonuclear leukocyte chemotaxis via a receptor mediated system. Life Sci 1995;56:2247–53.
130. Salem ML, Hossain MS, Nomoto K. Mediation of the immunomodulatory effect of beta-estradiol on inflammatory responses by inhibition of recruitment and activation of inflammatory cells and their gene expression of TNF-alpha and IFN-gamma. Int Arch Allergy Immunol 2000;121:235–45.
131. Safadi FF, Dissanayake IR, Goodman GG, Jago RA, Baker AE, Bowman AR, Sass DA, Popoff SN, Epstein S. Influence of estrogen deficiency and replacement on T-cell populations in rat lymphoid tissues and organs. Endocrine 2000;12:81–8.
132. Nilsson N, Carlsten H. Estrogen induces suppression of natural killer cell cytotoxicity and augmentation of polyclonal B cell activation. Cell Immunol 1994;158:131–9.
133. Salem ML, Matsuzaki G, Kishihara K, Madkour GA, Nomoto K. Beta-estradiol suppresses T cell-mediated delayed-type hypersensitivity through suppression of antigen-presenting cell function and Th1 induction. Int Arch Allergy Immunol 2000;121:161–9.
134. Carlsten H, Verdrengh M, Taube M. Additive effects of suboptimal doses of estrogen and cortisone on the suppression of T lymphocyte dependent inflammatory responses in mice. Inflamm Res 1996;45:26–30.
135. Muller D, Chen M, Vikingsson A, Hildeman D, Pederson K. Oestrogen influences $CD4^+$ T-lymphocyte activity in vivo and in vitro in β2-microglobulin-deficient mice. Immunology 1995;86:162–7.
136. Keller ET, Zhang J, Yao Z, Qi Y. The impact of chronic estrogen deprivation on immunologic parameters in the ovariectomized rhesus monkey (Macaca mulatta) model of menopause. J Reprod Immunol 2001;50:41–55.
137. Muller V, Szabo A, Viklicky O, Gaul I, Portl S, Philipp T, Heemann UW. Sex hormones and gender-related differences: their influence on chronic renal allograft rejection. Kidney Int 1999;55:2011–20.
138. Saito S, Foegh ML, Lou H, Aras R, Ramwell PW.Oestrogen and transplant vascular disease. Clin Exp Pharmacol Physiol 1999;26:137–43.
139. Evagelatou M, Farrant J. Effect of 17β-estradiol on immunoglobulin secretion by human tonsillar lymphocytes in vitro. J Steroid Biochem Molec Biol 1994;48:171–7.
140. Kanda N, Tamaki K. Estrogen enhances immunoglobulin production by human PBMCs. J Allergy Clin Immunol 1999;103(2 Pt 1):282–8.
141. Bynoe MS, Grimaldi CM, Diamond B. Estrogen up-regulates Bcl-2 and blocks tolerance induction of naive B cells. Proc Natl Acad Sci U S A 2000;97:2703–8.
142. Peeva E, Grimaldi C, Spatz L, Diamond B. Bromocriptine restores tolerance in estrogen-treated mice. J Clin Invest 2000;106:1373–9.
143. Leitner G, Landsman T, Blum O, Zaltsmann N, Heller ED. Effects of gonadal steroids and their antagonists on the humoral immune response of immune-selected broiler chicks. Poult Sci 1996;75:1373–82.
144. Gilmore W, Weiner LP, Correale J. Effect of estradiol on cytokine secretion by proteolipid

protein-specific T cell clones isolated from multiple sclerosis patients and normal control subjects. J Immunol 1997;158:446–51.
145. Verthelyi D, Klinman DM. Sex hormone levels correlate with the activity of cytokine-secreting cells in vivo. Immunology 2000;100:384–90.
146. Ghersevich S, Isomaa V, Vihko P. Cytokine regulation of the expression of estrogenic biosynthetic enzymes in cultured rat granulosa cells. Mol Cell Endocrinol 2001;172: 21–30.
147. Simard J, Gingras S. Crucial role of cytokines in sex steroid formation in normal and tumoral tissues. Mol Cell Endocrinol 2001;171:25–40.
148. Bebo BF Jr, Fyfe-Johnson A, Adlard K, Beam AG, Vandenbark AA, Offner H. Low-dose estrogen therapy ameliorates experimental autoimmune encephalomyelitis in two different inbred mouse strains. J Immunol 2001;166:2080–9.
149. Kim S, Liva SM, Dalal MA, Verity MA, Voskuhl RR. Estriol ameliorates autoimmune demyelinating disease: implications for multiple sclerosis. Neurology 1999;52:1230–8.
150. Orvieto R, Ben-Rafael Z, Abir R, Bar-Hava I, Fisch B, Molad Y. Controlled ovarian hyperstimulation: a state of neutrophil activation. Am J Reprod Immunol 1999;42:288–91.
151. Porter VR, Greendale GA, Schocken M, Zhu X, Effros RB. Immune effects of hormone replacement therapy in post-menopausal women. Exp Gerontol 2001;36:311–26.
152. Gregory MS, Duffner LA, Faunce DE, Kovacs EJ. Estrogen mediates the sex difference in post-burn immunosuppression. J Endocrinol 2000;164:129–38.
153. Mor G, Kohen F, Garcia-Velasco J, Nilsen J, Brown W, Song J, Naftolin F.Regulation of fas ligand expression in breast cancer cells by estrogen: functional differences between estradiol and tamoxifen. J Steroid Biochem Mol Biol 2000;73:185–94.
154. Theoharides TC, Dimitriadou V, Letourneau R, Rozniecki JJ, Vliagoftis H, Boucher W. Synergistic action of estradiol and myelin basic protein on mast cell secretion and brain myelin changes resembling early stages of demyelination, Neuroscience 1993;57: 861–71.
155. Leker RR, Karni A, Brenner T, Weidenfeld J, Abramsky O. Effects of sex hormones on experimental autoimmune myasthenia gravis. Eur J Neurol 2000;7:203–6.
156. Evans MJ, MacLaughlin S, Marvin RD, Abdou NI. Estrogen decreases in vitro apoptosis of peripheral blood mononuclear cells from women with normal menstrual cycles and decreases TNF-alpha production in SLE but not in normal cultures. Clin Immunol Immunopathol 1997;82:258–62.
157. Seethalakshmi L, Bala RS, Malhotra RK, Austin-Ritchie T, Miller-Graziano C, Menon M, Luber-Narod J. 17 beta-estradiol induced prostatitis in the rat is an autoimmune disease. J Urol 1996;156:1838–42.
158. Nagy E, Berczi I, Baral E. Combination immunotherapy of cancer. In Neuroimmune Biology: New foundation of Biology. I. Berczi, R. Gorczynski (eds), Elsevier,2001. pp. 417–432.
159. Berczi I, Chow DA, Baral E, Nagy E. Neuroimmunoregulation and cancer. Int. J. Oncol. 1998;13:1049–1051.
160. Nagy E, Berczi I. Immunomodulation by tamoxifen and pergolide. Immunopharmacology 1986; 12:145–153.
161. Baral E, Nagy E, Berczi I. Modulation of natural killer cell mediated cytotoxicity by tamoxifen and estradiol. Cancer 1995;75:591–599.
162. Baral E, Nagy E, Berczi I. Modulation of lymphokine-activated killer cell-mediated

cytotoxicity by estradiol and tamoxifen. Int J Cancer 1996;66:214–218.
163. Baral E, Nagy E, Kangas L, Berczi I. Anti-estrogens enhance the therapeutic effect of lymphokine activated killer cells on the P815 murine mastocytoma. Int J Cancer 1996; 67: 580–585.
164. Nagy E, Baral E, Kangas L, Berczi I. Anti-estrogens potentiate the immunotherapy of the P815 murine mastocytoma by cytotoxic T lymphocytes. Anticancer Res 1997;17: 1083–1088.
165. Baral E, Nagy E, Kangas L, Berczi I. Immunotherapy of the SL2–5 murine lymphoma with natural killer cells and tamoxifen or toremifene. Anticancer Res. 1997;17:77–84.
166. Baral E, Nagy E, Kangas L, Berczi I. Combination therapy of the H2712 murine mammary carcinoma with cytotoxic T lymphocytes and anti-estrogens. Anticancer Res. 1997;17: 3647–3652.
167. Baral E, Nagy E, Kangas L, Berczi I. Combination immunotherapy of the P815 murine mastocytoma with killer cells, IL-2 and anti-estrogens. Anticancer Res. 1997;17:3653–3658.
168. Baral E, Nagy E, Krepart GE, Lotocki RJ, Unruh HW, Berczi I. Antiestrogens sensitize human ovarian and lung carcinomas for lysis by autologous killer cells. Anticancer Res 2000;20:2027–31.
169. Haeryfar SMM, Nagy E, Baral E, Krepart GE, Lotocki R J, Berczi I. Antiestrogens affect both pathways of killer cell-mediated oncolysis. Anticancer Res 2000;20:1849–53.
170. Wu WM, Suen JL, Lin BF, Chiang BL. Tamoxifen alleviates disease severity and decreases double negative T cells in autoimmune MRL-lpr/lpr mice. Immunology 2000;100:110–8.
171. Ravche ES, Steinberg AD. Sex hormones and autoimmunity. In Pituitary Function and Immunity, Berczi I, editor, Boca Raton, FL, CRC Press, 1986;pp. 283–301.
172. Ahmed SA, Talal N. Christadoss P. Genetic regulation of testosterone-induced immune suppression. Cell Immunol 1987;104: 91–98.
173. Fried W, Morley C. Effects of androgenic steroids on erythropoiesis. Steroids 1985;46: 799–826.
174. Zhou Z, Shackleton CH, Pahwa S, White PC, Speiser PW. Prominent sex steroid metabolism in human lymphocytes. Mol Cell Endocrinol 1998;138:61–9.
175. Greenstein BD, de Bridges EF, Fitzpatrick FTA. Aromatase inhibitors regenerate the thymus in aging male rats. Int J Immunopharm 1992;14:541–53.
176. Araneo BA, Dowell T, Diegel M, Daynes RA. Dihydrotesterone exerts a depressive influence on the production of interleukin-4 (IL-4), IL-5, and γ-interferon, but not IL-2 by activated murine T cells. Blood 1991;78: 688–99.
177. Konstadoulakis MM, Syrigos KN, Baxevanis CN, Syrigou EI, Papamichail M, Peveretos P, Anapliotou M, Golematis BC. Effect of testosterone administration, pre- and postnatally, on the immune system of rats. Horm Metab Res 1995;27: 275–8.
178. Friedl R, Brunner M, Moeslinger T, Spieckermann PG. Testosterone inhibits expression of inducible nitric oxide synthase in murine macrophages. Life Sci 2000;68:417–29.
179. Gomez F, Ruiz P, Lopez R, Rivera C, Romero S, Bernal JA. Effects of androgen treatment on expression of macrophage Fcgamma receptors. Clin Diagn Lab Immunol 2000;7: 682–6.
180. Giltay EJ, Fonk JC, von Blomberg BM, Drexhage HA, Schalkwijk C, Gooren LJ. In vivo effects of sex steroids on lymphocyte responsiveness and immunoglobulin levels in humans. J Clin Endocrinol Metab 2000;85:1648–57.
181. Olsen NJ, Zhou P, Ong H, Kovacs WJ. Testosterone induces expression of transforming

growth factor-beta$_1$ in the murine thymus. J Steroid Biochem Molec Biol 1993;45: 327–32.

182. Ba ZF, Wang P, Koo DJ, Zhou M, Cioffi WG, Bland KI, Chaudry IH. Testosterone receptor blockade after trauma and hemorrhage attenuates depressed adrenal function. Am J Physiol Regul Integr Comp Physiol 2000;279:R1841–8.
183. Messingham KA, Shirazi M, Duffner LA, Emanuele MA, Kovacs EJ. Testosterone receptor blockade restores cellular immunity in male mice after burn injury. J Endocrinol 2001;169:299–308.
184. Angele MK, Ayala A, Monfils BA, Cioffi WG, Bland KI, Chaudry IH.Testosterone and/or low estradiol: normally required but harmful immunologically for males after trauma-hemorrhage. J Trauma 1998;44:78–85.
185. Angele MK, Knoferl MW, Schwacha MG, Ayala A, Cioffi WG, Bland KI, Chaudry IH. Sex steroids regulate pro- and anti-inflammatory cytokine release by macrophages after trauma-hemorrhage. Am J Physiol 1999;277(1 Pt 1):C35–42.
186. Angele MK, Schwacha MG, Ayala A, Chaudry IH. Effect of gender and sex hormones on immune responses following shock. Shock 2000;14:81–90.
187. Wichmann MW, Ayala A, Chaudry IH. Male sex steroids are responsible for depressing macrophage immune function after trauma-hemorrhage. Am J Physiol 1997;273(4 Pt 1): C1335–40.
188. Kahlke V, Angele MK, Schwacha MG, Ayala A, Cioffi WG, Bland KI, Chaudry IH.Reversal of sexual dimorphism in splenic T lymphocyte responses after trauma-hemorrhage with aging. Am J Physiol Cell Physiol 2000;278:C509–16.
189. Majetschak M, Christensen B, Obertacke U, Waydhas C, Schindler AE, Nast-Kolb D. Schade FU. Sex differences in posttraumatic cytokine release of endotoxin-stimulated whole blood: relationship to the development of severe sepsis. J Trauma 2000;48:832–9; discussion 839–40
190. Offner PJ, Moore EE, Biffl WL. Male gender is a risk factor for major infections after surgery. Arch Surg 1999;134:935–8; discussion 938–40.
191. Tsigos C, Papanicolaou DA, Kyrou I, Raptis SA, Chrousos GP. Dose-dependent effects of recombinant human interleukin-6 on the pituitary-testicular axis. J Interferon Cytokine Res 1999;19:1271–6.
192. Krucken J, Schmitt-Wrede HP, Markmann-Mulisch U, Wunderlich F. Novel gene expressed in spleen cells mediating acquired testosterone-resistant immunity to Plasmodium chabaudi malaria. Biochem Biophys Res Commun 1997;230:167–70.
193. Mossmann H, Benten WP, Galanos C, Freudenberg M, Kuhn-Velten WN, Reinauer H, Wunderlich F. Dietary testosterone suppresses protective responsiveness to Plasmodium chabaudi malaria. Life Sci 1997;60:839–48.
194. Hedger MP, Meinhardt A. Local regulation of T cell numbers and lymphocyte-inhibiting activity in the interstitial tissue of the adult rat testis. J Reprod Immunol 2000;48:69–80.
195. Hedger MP, Wang J, Lan HY, Atkins RC, Wreford NG. Immunoregulatory activity in adult rat testicular interstitial fluid: relationship with intratesticular CD8+ lymphocytes following treatment with ethane dimethane sulfonate and testosterone implants. Biol Reprod 1998;58:935–42.
196. Hayes R, Chalmers SA, Nikolic-Paterson DJ, Atkins RC, Hedger MP.Secretion of bioactive interleukin 1 by rat testicular macrophages in vitro. J Androl 1996;17:41–9.
197. Xiong Y, Hales DB. Differential effects of tumor necrosis factor-alpha and interleukin-1 on 3 beta-hydroxysteroid dehydrogenase/delta 5-->delta 4 isomerase expression in mouse

Leydig cells. Endocrine 1997;7:295–301.
198. Ogilvie KM, Held Hales K, Roberts ME, Hales DB, Rivier C. The inhibitory effect of intracerebroventricularly injected interleukin 1beta on testosterone secretion in the rat: role of steroidogenic acute regulatory protein. Biol Reprod 1999;60:527–33.
199. Hampl R, Hill M, Sterzl I, Starka L. Immunomodulatory 7-hydroxylated metabolites of dehydroepiandrosterone are present in human semen. J Steroid Biochem Mol Biol 2000;75:273–6.
200. Slater CH, Schreck CB. Physiological levels of testosterone kill salmonid leukocytes in vitro. Gen Comp Endocrinol 1997;106:113–9.
201. Bebo BF Jr, Zelinka-Vincent E, Adamus G, Amundson D, Vandenbark AA, Offner H.Gonadal hormones influence the immune response to PLP 139–151 and the clinical course of relapsing experimental autoimmune encephalomyelitis. J Neuroimmunol 1998;84:122–30.
202. Dorsam H, Rohrbach P, Kurschner T, Kipriyanov S, Renner S, Braunagel M, Welschof M, Little M. Antibodies to steroids from a small human naive IgM library. FEBS Lett 1997;414:7–13.
203. Regelson W, Loria R, Kalimi M. Dehydroepiandrosterone (DHEA) - the "mother steroid." I. Immunologic action. Ann NY Acad Sci 1994;719:553–63.
204. Morfin R, Courchay G. Pregnenolone and dehydroepiandrosterone as precursors of native 7-hydroxylated metabolites which increase the immune response in mice. J Steroid Biochem Molec Biol 1994;50: 91–100.
205. Padgett DA, Loria RM. In vitro potentiation of lymphocyte activation by dehydroepiandrosterone, androstenediol, and androstenetriol. J Immunol 1994;153: 1544–1552.
206. Bulloch K, McEwen BS, Diwa A, Baird S. Relationship between dehydroepiandrosterone and calcitonin gene-related peptide in the mouse thymus. Am J Physiol 1995;268: E168–73.
207. Suzuki T, Suzuki N, Daynes RA, Engleman EG. Dehydroepiandrosterone enhances IL2 production and cytotoxic effector function of human T cells. Clin Immunol Immunopathol 1991;61: 202–11.
208. Kim HR, Ryu SY, Kim HS, Choi BM, Lee EJ, Kim HM, Chung HT. Administration of dehydroepiandrosterone reverses the immune suppression induced by high dose antigen in mice. Immunol Invest 1995;24:583–93.
209. Sterzl I, Hampl R, Sterzl J, Votruba J, Starka L. 7Beta-OH-DHEA counteracts dexamethasone induced suppression of primary immune response in murine spleenocytes. J Steroid Biochem Mol Biol 1999;71:133–7.
210. Morfin R, Lafaye P, Cotillon AC, Nato F, Chmielewski V, Pompon D. 7 alpha-hydroxy-dehydroepiandrosterone and immune response. Ann N Y Acad Sci 2000;917:971–82.
211. Verajankorva E, Martikainen M, Saraste A, Sundstrom J, Pollanen P. Sperm antibodies in rat models of male hormonal contraception and vasectomy. Reprod Fertil Dev 999;11: 49–57.
212. McLachlan JA, Serkin CD, Bakouche O. Dehydroepiandrosterone modulation of lipopolysaccharide-stimulated monocyte cytotoxicity. J Immunol 1996;156: 328–35.
213. Gianotti L, Alexander JW, Fukushima R, Pyles T. Steroid therapy can modulate gut barrier function, host defense, and survival in thermally injured mice. J Surg Res 1996;62: 53–8.
214. Oberbeck R, Dahlweid M, Koch R, van Griensven M, Emmendorfer A, Tscherne H, Pape HC. Dehydroepiandrosterone decreases mortality rate and improves cellular immune function during polymicrobial sepsis. Crit Care Med 2001;29:380–4.

215. Catania RA, Angele MK, Ayala A, Cioffi WG, Bland KI, Chaudry IH. Dehydroepiandrosterone restores immune function following trauma-haemorrhage by a direct effect on T lymphocytes. Cytokine 1999;11:443–50.
216. Hall GM, Perry LA, Spector TD. Depressed levels of dehydroepiandrosterone sulphate in postmenopausal women with rheumatoid arthritis but no relation with axial bone density. Ann Rheum Dis 1993;52: 211–14.
217. Weksler ME. Immune senescence and adrenal steroids: immune dysregulation and the action of dehydroepiandrosterone (DHEA) in old animals, Eur J Clin Pharmacol 1993;45: S21–3.
218. Araneo B, Dowell T, Woods ML, Daynes R, Judd M, Evans T. DHEAS as an effective vaccine adjuvant in elderly humans. Ann NY Acad Sci 1995;774: 232–48.
219. Danenberg HD, Ben-Yehuda A, Zakay-Rones Z, Friedman G. Dehydroepiandrosterone (DHEA) treatment reverses the impaired immune response of old mice to influenza vaccination and protects from influenza infection. Vaccine. 1995;13:1445–8.
220. Casson PR, Andersen RN, Herrod HG, Stentz FB, Straughn AB, Abraham GE, Buster JE. Oral dehydroepiandrosterone in physiologic doses modulates immune function in postmenopausal women. Am J Obstet Gynecol 1993;169:1563–9.
221. Kalo-Klein A, Witkin SS. Regulation of the immune response to Candida albicans by monocytes and progesterone. Am J Obstet Gynecol 1991;164:1351–4.
222. Nohmi T, Abe S, Dobashi K, Tansho S, Yamaguchi H. Suppression of anti-Candida activity of murine neutrophils by progesterone in vitro: a possible mechanism in pregnant women's vulnerability to vaginal Candidiasis. Microbiol Immunol 1995;39:405–9.
223. Li ZG, Danis VA, Brooks PM. Effect of gonadal steroids on the production of IL-1 and IL-6 by blood mononuclear cells in vitro. Clin Exp Rheumatol 1993;11: 157–62.
224. Scarpellini F, Scarpellini L, Dino N, Benvenuto P. Progesterone immunosuppressive levels and luteal steroid profiles in the cycles induced with clomiphene citrate. Clin Exp Obstet Gynecol 1983;20:182–8.
225. Check JH, Tarquini P, Gandy P, Lauer C. A randomized study comparing the efficacy of reducing the spontaneous abortion rate following lymphocyte immunotherapy and progesterone treatment versus progesterone alone in primary habitual aborters. Gynecol Obstet Invest 1995;39: 257–61.
226. Piccinni MP, Scaletti C, Maggi E, Romagnani S. Role of hormone-controlled Th1- and Th2-type cytokines in successful pregnancy. J Neuroimmunol 2000;109:30–3.
227. Szekeres-Bartho J, Barakonyi A, Par G, Polgar B, Palkovics T, Szereday L. Progesterone as an immunomodulatory molecule. Int Immunopharmacol 2001;1:1037–48.
228. Peltier MR, Hansen PJ. Immunoregulatory activity, biochemistry, and phylogeny of ovine uterine serpin. Am J Reprod Immunol 2001;45:266–72.
229. Tibbetts TA, Conneely OM, O'Malley BW.Progesterone via its receptor antagonizes the pro-inflammatory activity of estrogen in the mouse uterus. Biol Reprod 1999;60:1158–65.
230. Kelly RW, Carr GG, Riley SC. The inhibition of synthesis of a beta-chemokine, monocyte chemotactic protein-1 (MCP-1) by progesterone. Biochem Biophys Res Commun 1997;239:557–61.
231. McMurray RW, Wilson JG, Bigler L, Xiang L, Lagoo A. Progesterone inhibits glucocorticoid-induced murine thymocyte apoptosis. Int J Immunopharmacol 2000;22: 955–65.
232. Stein DG. Brain damage, sex hormones and recovery: a new role for progesterone and estrogen? Trends Neurosci 2001;24:386–91.

233. Mouihate A, Chen X, Pittman QJ.Interleukin-1beta fever in rats: gender difference and estrous cycle influence. Am J Physiol 1998;275:R1450–4.
234. Fleming MW, Gamble HR. Consequences of dose-dependent immunosuppression by progesterone on parasitic worm burdens in lambs. Am J Vet Res 1992;54: 1299–302.
235. Parr MB, Kepple L, McDermott MR, Drew MD, Bozzola JJ, Parr EL. A mouse model for studies of mucosal immunity to vaginal infection by herpes simplex virus type 2. Lab Invest 1994;70: 369–80.
236. Kaushic C, Zhou F, Murdin AD, Wira CR. Effects of estradiol and progesterone on susceptibility and early immune responses to Chlamydia trachomatis infection in the female reproductive tract. Infect Immun 2000;68:4207–16.
237. Valentino R, Savastano S, Tommaselli AP, Riccio A, Mariniello P, Pronesti G, DeDivitiis PM, Lombardi G. Hormonal pattern in women affected by rheumatoid arthritis. J Endocrinol Invest 1993;16: 619–24.
238. Benyon HLC, Garbett ND, Barnes PJ. Severe premenstrual exacerbations of asthma: effect of intramuscular progesterone. Lancet 1988;2:370–2.
239. Manolagas SC, Yu X-P, Girasole G, Bellido T. Vitamin D and the hematolymphopoietic tissue: a 1994 update. Sem Nephrol 1994;14:129–43.
240. Bouillon B, Garmyn M, Verstuyf A, Segaert S, Casteels K, Mathieu C. Paracrine role for calcitriol in the immune system and skin creates new therapeutic possibilities for vitamin D analogs. Eur J Endocrinol 1995;133: 7–16.
241. Jones G, Strugnell SA, DeLuca HF. Current understanding of the molecular actions of vitamin D. Physiol Rev 1998;78:1193–231.
242. Veldman CM, Cantorna MT, DeLuca HF. Expression of 1,25-dihydroxyvitamin D(3) receptor in the immune system. Arch Biochem Biophys 2000;374:334–8.
243. Towers TL, Freedman LP. Granulocyte-macrophage colony-stimulating factor gene transcription is directly repressed by the vitamin D3 receptor. Implications for allosteric influences on nuclear receptor structure and function by a DNA element. J Biol Chem 1998;273:10338–48.
244. Towers TL, Staeva TP, Freedman LP. A two-hit mechanism for vitamin D3-mediated transcriptional repression of the granulocyte-macrophage colony-stimulating factor gene: vitamin D receptor competes for DNA binding with NFAT1 and stabilizes c-Jun. Mol Cell Biol 1999;19:4191–9.
245. Cippitelli M, Santoni A. Vitamin D3: a transcriptional modulator of the interferon-gamma gene. Eur J Immunol 1998;28:3017–30.
246. Csaszar A, Abel T. Receptor polymorphisms and diseases. Eur J Pharmacol 2001;414: 9–22.
247. Ban Y, Taniyama M, Yanagawa T, Yamada S, Maruyama T, Kasuga A, Ban Y. Vitamin D receptor initiation codon polymorphism influences genetic susceptibility to type 1 diabetes mellitus in the Japanese population. BMC Med Genet 2001;2:7.
248. Holick MF. Noncalcemic actions of 1,25-dihydroxyvitamin D3 and clinical applications. Bone 17: 107S–111S.
249. Saini A, Cohen DJ, Ooi BS. Regulation of macrophage growth responses to colony-stimulating factor-1 by 1,25-dihydroxyvitamin D3. J Am Soc Nephrol 1995;5:2091–3.
250. Tokuda N, Levy RB. 1,25-dihydroxyvitamin D3 stimulates phagocytosis but suppresses HLA-DR and CD13 antigen expression in human mononuclear phagocytes. Proc Soc Exp Biol Med 1996;211:244–50.
251. Piemonti L, Monti P, Sironi M, Fraticelli P, Leone BE, Dal Cin E, Allavena P, Di Carlo V.

Vitamin D3 affects differentiation, maturation, and function of human monocyte-derived dendritic cells. J Immunol 2000;164:4443–51.
252. Griffin MD, Lutz W, Phan VA, Bachman LA, McKean DJ, Kumar R. Dendritic cell modulation by 1alpha,25 dihydroxyvitamin D3 and its analogs: a vitamin D receptor-dependent pathway that promotes a persistent state of immaturity in vitro and in vivo. Proc Natl Acad Sci U S A 2001;98:6800–5.
253. Penna G, Adorini L. 1-alpha,25-dihydroxyvitamin D3 inhibits differentiation, maturation, activation, and survival of dendritic cells leading to impaired alloreactive T cell activation. J Immunol 2000;164:2405–11.
254. Yamamoto N, Naraparaju VR. Structurally well-defined macrophage activating factor derived from vitamin D3-binding protein has a potent adjuvant activity for immunization. Immunol Cell Biol 1998;76:237–44.
255. Haug CJ, Muller F, Aukrust P, Froland SS.Different effect of 1,25-dihydroxyvitamin D3 on replication of Mycobacterium avium in monocyte-derived macrophages from human immunodeficiency virus-infected subjects and healthy controls. Immunol Lett 1998;63: 107–12.
256. Dusso AS, Kamimura S, Gallieni M, Zhong M, Negrea L, Shapiro S, Slatopolsky E. gamma-Interferon-induced resistance to 1,25-(OH)2 D3 in human monocytes and macrophages: a mechanism for the hypercalcemia of various granulomatoses. J Clin Endocrinol Metab 1997;82:2222–32.
257. Tokuda N, Kano M, Meiri H, Nomoto K, Naito S. Calcitriol therapy modulates the cellular immune responses in hemodialysis patients. Am J Nephrol 2000;20:129–37.
258. Lemire J. 1,25-Dihydroxyvitamin D3--a hormone with immunomodulatory properties. Z Rheumatol 2000;59 Suppl 1:24–7.
259. Cantorna MT, Humpal-Winter J, DeLuca HF. Dietary calcium is a major factor in 1,25-dihydroxycholecalciferol suppression of experimental autoimmune encephalomyelitis in mice. J Nutr 1999;129:1966–71.
260. Jirapongsananuruk O, Melamed I, Leung DY. Additive immunosuppressive effects of 1,25-dihydroxyvitamin D3 and corticosteroids on TH1, but not TH2, responses. J Allergy Clin Immunol 2000;106:981–5.
261. Cantorna MT, Humpal-Winter J, DeLuca HF. In vivo upregulation of interleukin-4 is one mechanism underlying the immunoregulatory effects of 1,25-dihydroxyvitamin D(3). Arch Biochem Biophys 2000;377:135–8.
262. Overbergh L, Decallonne B, Waer M, Rutgeerts O, Valckx D, Casteels KM, Laureys J, Bouillon R, Mathieu C. 1-alpha,25-dihydroxyvitamin D3 induces an autoantigen-specific T-helper 1/T-helper 2 immune shift in NOD mice immunized with GAD65 (p524–543). Diabetes 2000;49:1301–7.
263. Willheim M, Thien R, Schrattbauer K, Bajna E, Holub M, Gruber R, Baier K, Pietschmann P, Reinisch W, Scheiner O, Peterlik M. Regulatory effects of 1-alpha,25-dihydroxyvitamin D3 on the cytokine production of human peripheral blood lymphocytes. J Clin Endocrinol Metab 1999;84:3739–44.
264. Panda DK, Miao D, Tremblay ML, Sirois J, Farookhi R, Hendy GN, Goltzman D. Targeted ablation of the 25-hydroxyvitamin D 1-alpha -hydroxylase enzyme: Evidence for skeletal, reproductive, and immune dysfunction. Proc Natl Acad Sci U S A 2001;98:7498–503.
265. Rothe L, Collin-Osdoby P, Chen Y, Sunyer T, Chaudhary L, Tsay A, Goldring S, Avioli L, Osdoby P. Human osteoclasts and osteoclast-like cells synthesize and release high basal and inflammatory stimulated levels of the potent chemokine interleukin-8. Endocrinology

1998;139:4353–63.
266. Furman I, Baudet C, Brachet P. Differential expression of M-CSF, LIF, and TNF-alpha genes in normal and malignant rat glial cells: regulation by lipopolysaccharide and vitamin D. J Neurosci Res 1996;46:360–6.
267. Hullett DA, Cantorna MT, Redaelli C, Humpal-Winter J, Hayes CE, Sollinger HW, Deluca HF. Prolongation of allograft survival by 1,25-dihydroxyvitamin D3. Transplantation 1998;66:824–8.
268. Yang S, Smith C, Prahl JM, Luo X, DeLuca HF. Vitamin D deficiency suppresses cell-mediated immunity in vivo. Arch Biochem Biophys 1993;303:98–106.
269. Hernandez-Frontera E, McMurray DN. Dietary vitamin D affects cell-mediated hypersensitivity but not resistance to experimental pulmonary tuberculosis in guinea pigs. Infect Immun 1993;61:2116–21.
270. Daynes RA, Enioutina EY, Butler S, Mu H-H, McGee ZA,Araneo BA. Induction of common mucosal immunity by hormonally immunomodulated peripheral immunization. Infect Immun 1996;64: 1100–9.
271. Enioutina EY, Visic D, McGee ZA, Daynes RA. The induction of systemic and mucosal immune responses following the subcutaneous immunization of mature adult mice: characterization of the antibodies in mucosal secretions of animals immunized with antigen formulations containing a vitamin D3 adjuvant. Vaccine 1999;17:3050–64.
272. Wiers KM, Lathers DM, Wright MA, Young MR. Vitamin D3 treatment to diminish the levels of immune suppressive CD34+ cells increases the effectiveness of adoptive immunotherapy. J Immunother 2000;23:115–24.
273. Wiers K, Wright MA, Vellody K, Young MR. Failure of tumor-reactive lymph node cells to kill tumor in the presence of immune-suppressive CD34+ cells can be overcome with vitamin D3 treatment to diminish CD34+ cell levels. Clin Exp Metastasis 1998;16: 275–82.
274. Yacobi R, Koren R, Liberman UA, Rotem C, Wasserman L, Ravid A. 1,25-dihydroxyvitamin D3 increases the sensitivity of human renal carcinoma cells to tumor necrosis factor alpha but not to interferon alpha or lymphokine-activated killer cells. J Endocrinol 1996;149: 27–33.
275. Mariani E, Ravaglia G, Forti P, Meneghetti A, Tarozzi A, Maioli F, Boschi F, Pratelli L, Pizzoferrato A, Piras F, Facchini A.Vitamin D, thyroid hormones and muscle mass influence natural killer (NK) innate immunity in healthy nonagenarians and centenarians. Clin Exp Immunol 1999;117:206.
276. Zofkova I, Kancheva RL. The effect of 1,25(OH)2 vitamin D3 on CD4+/CD8+ subsets of T lymphocytes in postmenopausal women. Life Sci 1997;61:147–52.
277. Bansal AS, Bruce J, Hogan PG, Allen RK. An assessment of peripheral immunity in patients with sarcoidosis using measurements of serum vitamin D3, cytokines and soluble CD23. Clin Exp Immunol 1997;110:92–7.
278. Casteels K, Bouillon R, Waer M, Mathieu C. Immunomodulatory effects of 1,25-dihydroxyvitamin D3. Curr Opin Nephrol Hypertens 1995;4:313–8.
279. Branisteanu DD, Waer M, Sobis H, Marcelis S, Vandeputte M, Bouillon R. Prevention of murine experimental allergic encephalomyelitis: cooperative effects of cyclosporine and 1α,25-$(OH)_2D_3$. J Neuroimmunol 1995:61: 151–60.
280. Cantorna MT. Vitamin D and autoimmunity: is vitamin D status an environmental factor affecting autoimmune disease prevalence? Proc Soc Exp Biol Med 2000;223:230–3.
281. Casteels K, Waer M, Bouillon R, Depovere J, Valckx D, Laureys J, Mathieu C. 1,25-

Dihydroxyvitamin D3 restores sensitivity to cyclophosphamide-induced apoptosis in non-obese diabetic (NOD) mice and protects against diabetes. Clin Exp Immunol 1998;112: 181–7.

282. Casteels KM, Mathieu C, Waer M, Valckx D, Overbergh L, Laureys JM, Bouillon R.Prevention of type I diabetes in nonobese diabetic mice by late intervention with nonhypercalcemic analogs of 1,25-dihydroxyvitamin D3 in combination with a short induction course of cyclosporin A. Endocrinology 1998;139:95–102.
283. Riachy R, Vandewalle B, Belaich S, Kerr-Conte J, Gmyr V, Zerimech F, d'Herbomez M, Lefebvre J, Pattou F. Beneficial effect of 1,25 dihydroxyvitamin D3 on cytokine-treated human pancreatic islets. J Endocrinol 2001;169:161–8.
284. The EURODIAB Substudy 2 Study Group. Vitamin D supplement in early childhood and risk for Type I (insulin-dependent) diabetes mellitus. Diabetologia 1999;42:51–4.
285. Garcion E, Nataf S, Berod A, Darcy F, Brachet P. 1,25-Dihydroxyvitamin D3 inhibits the expression of inducible nitric oxide synthase in rat central nervous system during experimental allergic encephalomyelitis. Brain Res Mol Brain Res 1997;45:255–67.
286. Linker-Israeli M, Elstner E, Klinenberg JR, Wallace DJ, Koeffler HP. Vitamin D(3) and its synthetic analogs inhibit the spontaneous in vitro immunoglobulin production by SLE-derived PBMC. Clin Immunol 2001;99:82–93.
287. Oelzner P, Franke S, Muller A, Hein G, Stein G. Relationship between soluble markers of immune activation and bone turnover in post-menopausal women with rheumatoid arthritis. Rheumatology (Oxford) 1999;38:841–7.
288. Bang B, Asmussen K, Sorensen OH, Oxholm P.Reduced 25-hydroxyvitamin D levels in primary Sjogren's syndrome. Correlations to disease manifestations. Scand J Rheumatol 1999;28:180–3.
289. Lu I, Gilleaudeau P, McLane JA, Heftler N, Kamber M, Gottlieb S, Krueger JG, Gottlieb AB. Modulation of epidermal differentiation, tissue inflammation, and T-lymphocyte infilitration in psoriatic plaques by topical calcitriol. J Cutan Pathol 1996;23:419–30.
290. Yong EL, Lim LS, Wang Q, Mifsud A, Lim J, Ong YC, Sim KS. Androgen receptor polymorphisms and mutations in male infertility. J Endocrinol Invest 2000;23:573–7.
291. Lou H, Kodama T, Zhao YJ, Maurice P, Wang YN, Katz N, Foegh ML. Inhibition of transplant coronary arteriosclerosis in rabbits by chronic estradiol treatment is associated with abolition of MHC class II antigen expression. Circulation 1996;94:3355–61.

The Immune-Neuroendocrine Circuitry
Edited by I. Berczi and A. Szentivanyi

Regulatory Enzymes

ISTVAN BERCZI and EDRIS SABBADINI

Department of Immunology, Faculty of Medicine, The University of Manitoba, Winnepeg, Manitoba, Canada R3E 0W3

ABSTRACT

Enzymes of the serine protease family induce a wide range of biological effects by activating several hormones and growth factors and by participating in enzyme cascades for inflammation, blood coagulation, complement activation, and other reactions. The action of these proteases is regulated by the endocrine system and by several inhibitors, also under endocrine control. Many inhibitors of these enzymes belong to a single family of proteins called serine protease inhibitors or serpins. This review will deal first with the components of the contact system and with the mechanisms of activation of its primary enzymes and then with the functions of other serine proteases and their inhibitors. Finally, a brief discussion about the roles of serine proteases in physiological reactions and in disease states will be presented.

1. THE CONTACT SYSTEM

The main components of the contact system are three zymogens: Hagemann factor (HF), also referred to as coagulation factor XII (FXII), plasma pre-kallikrein (PK), and coagulation factor XI (FXI). These zymogens can be activated by partial proteolysis into enzymatically active serine proteases. In addition, a fourth primary component, high molecular weight kininogen (HMWK), acts as a co-factor in the activation of these enzymes and as a substrate for the active plasma kallikrein (K). The activation of the primary enzymes leads to activation of the intrinsic coagulation cascade, to the triggering of the complement cascade via the classical pathway, fibrinolysis via plasminogen activation, and inflammatory reactions via the activation of plasma kallikrein with the subsequent production of vasodilatatory kinins. Effects on blood pressure may also be induced by the activation of prorenin and by the hypotensive effects of kinins (Figure 1).

HF is a β-globulin of 74–80 kDa molecular weight [1,2] contained in plasma. The molecule can be cleaved at various points by limited proteolysis giving rise to enzymatically active factors, the main ones being αHFa of 80 kDa and βHFa of 28 kDa. The initial activation of the contact system is probably due to the binding of HF to negatively charged surfaces where the inactive molecules generate small amounts of enzymatically active αHFa which activates more HF molecules. The reaction is further amplified by plasma kallikrein (K) which is activated by the two active HF products and, in turn, activates HF. Of the two main activation products of HF, αHFa retains large numbers of positively charged amino acids in its heavy chain and remains attached to the negatively charged surface where the reaction was initiated. On this surface,

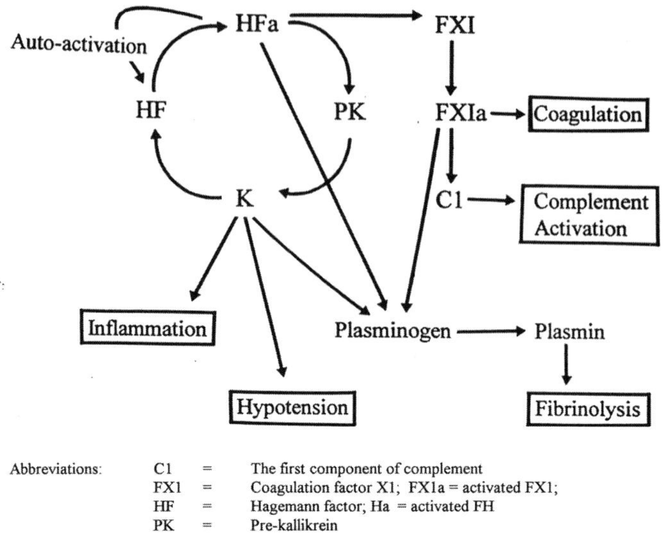

Figure 1. Schematic representation of the contact activation cascade.

αHFa can activate FXI and trigger the intrinsic blood coagulation cascade. On the other hand, βHFa is soluble and returns to the fluid phase. It does not participate in blood coagulation but can induce the activation of plasma PK and of C1.

Kallikrein circulates in blood as the zymogen PK, a gamma globulin of 88 kDa [2,3]. In the circulation, PK is normally bound to its co-factor high molecular weight kininogen (HMWK). Since this is rich with positively charged amino acids, it will bind to negatively charged surfaces and bring PK in contact with HF, which also binds to such cell surfaces. PK is then activated through limited proteolysis by the products of HF activation. The active K molecule has a molecular weight of 36 kD. Its main substrate is HMWK from which the nonapeptide bradykinin is generated. Besides this action on kininogen, K interacts with HF which it activates, resulting in the amplification of the contact activation cascade. It also acts on FXI which it activates into FXIa and with plasminogen which it activates into plasmin, the main enzyme of the fibrinolytic system. The last action proceeds slowly and may have only minor physiological significance.

FXI is a gamma globulin of 80 kDa, [4] that in circulation is bound to HMWK and can be activated through limited proteolysis by αHFa on a negative surface. Amino acid sequencing of FXI shows significant homologies with PK, indicating that the two proteins may have a common evolutionary origin. The main function of FXI is the proteolytic activation of factor IX which is an essential component of the intrinsic pathway of blood coagulation. Deficiencies in FXI produce a hemophilia like syndrome [5]. Activated FXI also activates HF and plasminogen.

There are two major kininogens in humans and most other mammalian species, HMWK of 110 kDa and low molecular weight kininogen (LMWK) of 60 kD, [3,6] the latter being contained in plasma in concentrations 4–5 times higher than HMWK. The two molecules originate from a single gene which gives rise to two messenger RNA molecules of different lengths. HMWK is the main substrate of plasma K. Plasma K cleaves HMWK at a fast rate but cleaves LMWK slowly. On the other hand, tissue kallikrein cleaves both kininogens with equal efficiency. From the action of plasma K on kininogens, the nonapeptide bradykinin is obtained, while the action of tissue kallikrein results in the production of the decapeptide kallidin (lys-bradykinin). Kininogens are multifunctional molecules. Besides being the source of kinins, they can inhibit the function of several peptidases. Moreover, HMWK circulates in association with PK and FXI, which it prevents from being activated while in plasma. Because of its ability to bind to negatively charged surfaces, HMWK brings these factors in contact with HF whenever a tissue injury exposes such negatively charged surfaces. Thus, the contact system is activated only when a trauma or infection produces sufficient tissue disruption and exposes negatively charged surfaces.

The kinins bradykinin and lys-bradykinin have similar action and act via the same receptors expressed on various cells but particularly on smooth muscle cells. There are three well characterized kinin receptors BK_1, BK_2, and BK_3, of which the BK_2 receptor is the most abundant. The BK_2 receptor is homologous to the H_2 histamine receptor and to the neurotensin and tachykinin receptors. The effects of the two kinins consist of vasodilation, hypotension and bronchial smooth muscle contraction. Pain is also elicited either directly or via the stimulation of prostaglandin production. All these actions result in strong inflammatory effects. Moreover, the hypotensive effects, effects on electrolytes and modulatory effects on the renin-angiotensin system are thought to be important in counteracting the hypertensive effects of the renin-angiotensin system and, therefore, in the regulation of blood pressure. Kinins have a very short half life in blood and tissues. They are rapidly degraded by enzymes, collectively referred to as kininases, which limit their potentially harmful effects. Kininase I is a carboxypeptidase that accounts for most of the bradykinin destroying activity of human plasma. Kininase II was shown to be identical to the angiotensin converting enzyme (ACE).

Several inhibitors of the contact system components are present in the circulation [2]. The main ones are: C1 inhibitor, which is a potent inhibitor of C1 but also inhibits HFa quite strongly; K, and FXIa, the α_2-plasmin inhibitor or α_2-antiplasmin, which strongly inhibits plasmin activity and more weakly HFa activity; α_2-macroglobulin, which inhibits HFa and K; antithrombin III, an inhibitor of HFa and K; and α_1-antitrypsin inhibitor, an inhibitor of FXIa.

2. TISSUE KALLIKREINS

Tissue or glandular kallikreins are a group of serine proteases that differ from plasma K in terms of enzymatic action and genes of origin. They form a family of proteins controlled by homologous genes that appear to originate from a common ancestor gene, probably via a series of gene duplications and differentiations. The best studied of these enzymes is true glandular kallikrein, usually referred to as K1, found in all mammalian species studied so far. The other members of the family, called K2, K3, etc., differ from species to species. There are three well characterized tissue K in humans, hK1, hK2, and hK3, and a fourth gene has been described. In mice 24 genes have been described and 20 in rats [7,8]. Though highly homologous, the different members of the kallikrein family act on different substrates and have different tissue distributions. Their actions are believed to be mainly local, in the sense that they are believed to take place either in the tissue itself or in the local circulation immediately after release. Most members of the tissue K family have been shown to be dependent on several hormones [2,9]. Thus, androgens increase the production of many K family enzymes in salivary glands and in the prostate, and probably in other tissues. Prolactin and growth hormone have been suggested to have similar function of promoting K family enzyme synthesis. Estrogens increase the production of K1 in the pituitary gland and in the brain.

In humans, as in most other species, K1 is particularly abundant in the pancreas, in the kidney and salivary glands [10]. It is also found in various body fluids such as saliva, urine, plasma and bile. Kidney K1 is secreted in urine. It is believed to have a significant role in the maintenance of normal blood pressure by contributing to the regulation of blood flow in the kidneys, by helping in the control of electrolyte and water transport, and by interacting (in a manner not completely clarified) with the renin-angiotensin system. In salivary glands, K1 is contained in the granules of convoluted tubules. It can be secreted in saliva and, in smaller quantities, in plasma. It probably contributes to regulation of blood flow in the salivary glands and, at least in the rat, appears to have a major role in the maintenance of the normal functions of the mucosa associated lymphoid tissue of the gastrointestinal tract. K1 was also found to be produced in several other tissues. These include the stomach, intestine, heart, blood vessels, brain, and endocrine organs such as the pituitary, ovary, testis and the adrenals. Pituitary K1 was shown to be estrogen dependent, blocked by bromocriptine, and to be involved in the processing of prolactin and proopiomelanocortin (POMC).

K1 was shown to have several effects on the immune system; therefore, several authors have described mitogenic and co-mitogenic effects of kallikrein and other serine proteases. Such mitogenic effects were observed in thymocytes [11], T cells and B cells [12]. Although bradykinin may also have mitogenic effects [13], the involvement of this kinin in kallikrein-induced mitogenesis is not well investigated. Moreover, several proteases, including kallikrein, were shown to be involved in immunoglobulin isotype control. The addition of kallikrein and other serine proteases to cultures of B cells stimulated with LPS and IL-4, depending on concentration, enhanced the production of IgE, IgG1, or IgG3 [14]. When injected subcutaneously, K1 suppressed cell mediated and humoral immune responses [15]. On the other hand, if given orally K1 favored the induction of tolerance to antigens administered orally either at the same time or shortly after K1.

Human K3, also known as prostate specific antigen (PSA), is found in high concentrations in the prostate gland, from which small amounts are released into the plasma [16,17]. During the development of prostatic neoplasias, marked increases in plasma concentrations of PSA are observed. These increases in plasma concentration are thought to be mainly due to the tissue disruption induced by the invasive neoplasia. As a result, PSA concentration can be used as a diagnostic tool for the early detection and for the post-surgical follow up of prostatic cancer. The normal function of human K3 appears to be the induction of the liquefaction of spermatic fluid, thanks to its action on the protein semenogelin, the major component of the sperm coagulum. Human K2 is also found in significant amounts in the prostate. Its function is not clear, although it has been suggested that it may synergize with K3 in sperm liquefaction [16,17].

In species other than humans, only some members of the K family have clear functions. In mice, the two kallikreins γ-NGF and β-NGF form a complex with the nerve growth factor (NGF) molecule and are probably involved in its activation. Similarly, the so called epidermal growth factor binding proteins (EGF-BPs) are kallikrein family members that form complexes with and activate the EGF molecule. Mouse renin and rat tonin are also members of the K family and have hypertensive effects.

3. OTHER SERINE PROTEASES

Other serine proteases have a variety of functions. Trypsin and chymotrypsin are serine proteases with well known digestive functions. These are not discussed in any detail here. Several serine proteases are involved in the destruction and reshaping of the extracellular matrix.

Enzymes such as elastase, collagenase, cathepsin G can solubilize elastin and collagen in various connective tissues. Cathepsin G, elastase released by neutrophils, and chymase released by mast cells, appear to be involved in several inflammatory diseases, such as arthritis and gingivitis, in emphysema and in the invasion of tissues by malignant tumors.

In the immune system, activated cytotoxic T cells and natural killer (NK) cells contain granules rich with the serine protease Granzyme B. This enzyme acts in concert with enzyme, perforin, another enzyme also contained in the granules of these cells, to induce apoptosis of target cells [18]. Mast cells contain chymase, a serine protease with neuropeptide degrading functions, capable of generating angiotensin II from angiotensin I. This would constitute an alternative pathway to the one provided by the angiotensin converting enzyme (ACE) in the renin/angiotensin system. Some not well identified serine proteases have been shown to modify or suppress immune responses. Ishizaka described a kallikrein-like factor called glycosylation-enhancing factor, which induced CD4+ T cells to produce IgE-potentiating factor and to favor the production of IgE memory B cells [19]. Serine proteases from Schistosoma mansoni schistosomula were reported to enhance IgE production [20].

4. SERPINS

Serpins are a family of more than 50 proteins [21] sharing similar amino acid sequences and are found in vertebrates, invertebrates, viruses and plants. There are two groups of serpins, those which inhibit serine proteases and those which do not. Proteins such as ovalbumin, angiotensinogen and thyroxine binding protein, have strong homologies with the rest of the family but are devoid of protease inhibiting activity. On the other hand, most serpins are strong inhibitors of serine protease activity. They bind in a stoichiometric (1:1) fashion to the active site of their specific target enzyme(s) and exert their inhibitory function in this way. This wide phylogenetic distribution and the variety of functions supported by these proteins suggest that they have a significant role in evolutionary fitness.

The functions of several serpins in the regulation of inflammatory processes, in the coagulation cascade, in complement activation and in the fibrinolysis process have been described above. Protease nexin I (PNI), thrombin inhibitor produced by glia and found in cerebrospinal fluid, was suggested to be contained in the amyloid plaques of patients with Alzheimer disease. Aprotinin, a serpin from bovine pancreas, has been extensively studied [22]. It has been used therapeutically in situations, such as major surgical procedures and pancreatitis, in which the extensive activation of the fibrinolytic system may lead to complications. Several serpins have been obtained from hematophagous invertebrates which use them to block coagulation of blood in the parasitized species.

REFERENCES

1. Cochrane CG, Griffin JH. The biochemistry and pathophysiology of the contact system of plasma. Adv Immunol 1982;33:241–304.
2. Kozin F, Cochrane CG. The contact activation system of plasma. Biochemistry and pathophysiology. In: Inflammation. Basic principles and clinical correlates. Gallin JI, Goldstein IM, and Snyderman R, Eds. 1992;103–121.
3. Bhoola KD, Figueroa CD, Worthy K. Bioregulation of kinins: kallikreins kininogens, and

kininases. Pharmacological Reviews 1992; 44:1–80.
4. Bouma BN, Griffin JH. Human blood coagulation factor XI. Purification, properties and mechanisms of activation by activated factor XII. J Biol Chem 1986;252:6432–6437.
5. Rimon A, Schiffman S, Feinstein DI, et al Factor XI activity and factor XI antigen in homozygous and heterozygous factor XI deficiency. Blood 1976;48:165–174.
6. Nakanishi S. Substance P precursor and kininogen: their structures, gene organizations, and regulation. Physiological Reviews 1987;67:1117–1142.
7. Wines DR, Brady JM, Pritchett DB, Roberts JL, MacDonald RJ. Organization and expression of the rat kallikrein gene family. J Biol Chem 1989;264:7653–7662.
8. Fuller PJ, and Funder JW. The cellular physiology of glandular kallikrein. Kidney 1986;29: 953–964.
9. Nagy E, Berczi I, Sabbadini E. Endocrine control of the immunosuppressive activity of the submandibular gland. Brain Behav Immun 1992;6:418–428.
10. Orstavik TB. The kallikrein-kini system in exocrine organs. J Histochem Cytochem 1980;28:881–889.
11. Naughton MA, Geczy C, Bender V, Hoffman H, Hamilton E. Esteropeptidase and thymotropic activity of a protein isolated from the mouse submaxillary gland. Biochem Biophys Acta 1972;263:106–114.
12. Hu ZQ, Murakami K, Ikigai H, Shimamura T. Enhancement of lymphocyte proliferation by mouse glandular kallikrein. Immunol Lett 1992;32:85–90.
13. Perris AD, Whitfield JF. The mitogenic action of bradykinin on thymic lymphocytes and its dependence on calcium. Proc Soc Exp Biol Med 1969;130:1198–1201.
14. Matsushita S, Katz DH. Biphasic effect of kallikrein on IgE and IgG1 syntheses by LPS/ IL4 stimulated B cells. Cell Immunol 1993;146:210–214.
15. Nagy E, Berczi I, Sabbadini E. Immunoregulatory effects of glandular kallikrein from salivary submandibular glands of rats. Neuroimmunomodulation 1997;4:101–112.
16. Carbini LA, Scicli AG, Carretero OA. The molecular biology of the kallikrein-kinin system: III. The human kallikrein gene family and kallikrein substrate. J Hypertension 1993;11:893–898.
17. Evans BA, Yun ZX, Close JA et al. Structure and chromosomal localization of the human renal kallikrein gene. Biochemistry 1988;27:3124–3129.
18. Berke G. Lymphocyte-triggered internal target disintegration. Immunol Today 1991;12: 396–399.
19. Ishizaka K. Twenty years with IgE: from the identification of IgE to regulatory factors for the IgE response. J Immunol 1985;135:i–x.
20. Verwaerde C, Auriault C, Neyrinck JL, Capron A. Properties of serine proteases of Schistosoma mansoni. Shistosomula involved in the regulation of IgE synthesis. Scand J Immunol 1988;27:17–24.
21. Salzet M, Vieau D, Stefano GB. Serpins: an evolutionarily conserved survival strategy. Immunol Today 1999;20:541–544.
22. Fritz H, Wunderer G. Biochemistry and applications of aprotinin, the kallikrein inhibitor from bovine organs. Drug Res 1983;33:479–494.

PART II: NEUROIMMUNE FUNCTION AND THE NEUROIMMUNE REGULATORY NETWORK

IV. NEUROIMMUNE FUNCTION

A. PHYSIOLOGY

The Immune-Neuroendocrine Circuitry
Edited by I. Berczi and A. Szentivanyi

Immunocompetence

ISTVAN BERCZI and ANDOR SZENTIVANYI

Department of Immunology, Faculty of Medicine, The University of Manitoba, Winnipeg, Manitoba RE3 0W3, Canada; and Department of Internal Medicine, Faculty of Medicine, The University of South Florida, Tampa, Florida 33612, USA

ABSTRACT

The evidence indicating that immune reactions, which are based on lymphocyte proliferation, are regulated by mechanisms that are involved in the growth control of all cells in higher animals is reviewed. Growth and lactogenic hormones (GLH) are required for the development and function of the immune system. It is suggested that GLH deliver the first signal to cells, including lymphocytes, that prepares them for proliferation, differentiation and function. This signal is designated as the *competence signal.* It is required for lymphocyte growth and is obligatory for the maintenance of immunocompetence. The second group of signals that control cell growth, are delivered by cell-to-cell and cell-to-matrix signaling and are designated as *stromal or adherence signals.* Adhesion molecules, tissue bound hormones, cytokines and matrix components mediate these signals. Within the immune system antigen presentation represents such a signal for which cell-to-cell interaction is obligatory. Adhesion molecules are fundamental to the organization of multi-cellular organisms and the signals delivered by them serve as the basis of species, organ and tissue specific recognition. This recognition system has been perfected during evolution from self-recognition to individually specific antigen recognition. This system also plays a role in the elimination of degenerated and neoplastic cells. Cell-to-cell signaling has a dominant power over other signals to commit the cell to proliferation. The cell cycle is then completed after the delivery of *cytokine signals.* Cytokines are tissue hormones, which are usually, but not always, secreted by the same cells that deliver the second signal. The nature and combination of these three groups of signals will determine the fate of each cell, which may be survival, proliferation, differentiation and function or alternately apoptosis. Hormones and neurotransmitters that alter signal delivery, modulate further this basic pattern of animal cell growth. It is reasonable to conclude on the basis of current evidence that GLH maintain immunocompetence, which enables the immune system to respond to specific antigenic and mitogenic stimuli.

1. INTRODUCTION

It has been known for over a hundred years now, since the pioneering experiments of Cushing and Ashner [1,2], that pituitary growth hormone (GH) is capable of stimulating the proportional growth of higher animals and man. Excess production of growth hormone will lead to gigantism or to acromegaly, whereas the deficiency of GH secretion will result in pituitary dwarfism. The

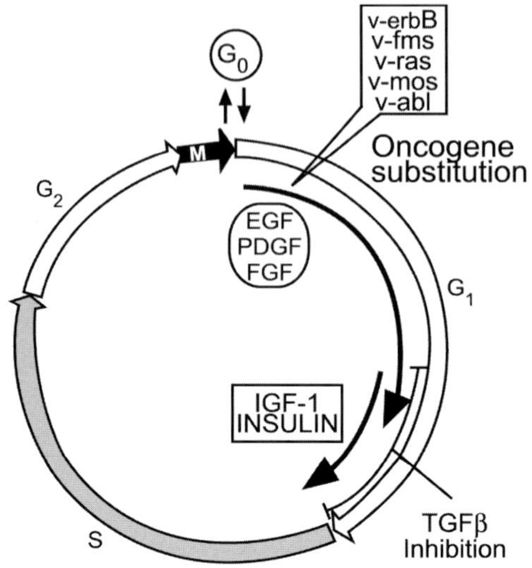

Figure. 1 Growth factors and the cell cycle.

A schematic representation is given of the coordinated action of two complementing growth factors to induce DNA synthesis. In BALB-MK cells several oncogenes can substitute for the competence factor requirement. The ability of TGFβ to inhibit the onset of DNA synthesis, even when added in late G_1 phase, is also shown. (After Aaronson 1991 [11]).

serum level of GH is highest in infancy and early childhood and gradually declines with ageing. Elderly people may show GH deficiency. For a long time body growth was solely attributed to the secretion of GH and its decline with ageing was regarded as a sign for lack of demand for this hormone during adult life. However, recent observations indicate that all tissues bear receptors for GH and/or for prolactin (PRL), which may be regarded as a modified growth hormone. Under proper experimental conditions the biological activity of GH on adult tissues and cells is easily demonstrated [3,4].

2. GROWTH AND LACTOGENIC HORMONES AND CELL GROWTH

Growth hormone stimulates the production of insulin-like growth factor I (IGF-I) in the liver and in other tissues and organs, including the immune system. IGF-I is a peptide hormone with structural relation to insulin and with crossreactivity at the receptor level. Insulin and IGF-I do not only bind to each other's receptor, but also mutually regulate receptor levels, which is proportional to their ability to occupy that particular receptor. Prolactin also has the capacity to stimulate IGF-I in the liver and in other tissues, which has been named by Nicoll and coworkers as synlactin. With the discovery of IGF many investigators assumed that growth hormone action is really mediated by IGF. This view is still held by a significant proportion of investigators [4–10].

Studies on the growth of cultured mammalian cell lines led to the discovery that serum derived growth factors were necessary. In each case at least two growth promoting hormones were required for the completion of the mitotic cell cycle. The first hormone (e.g. platelet derived growth factor, epidermal growth factor, or fibroblast growth factor) rendered the cells *competent*

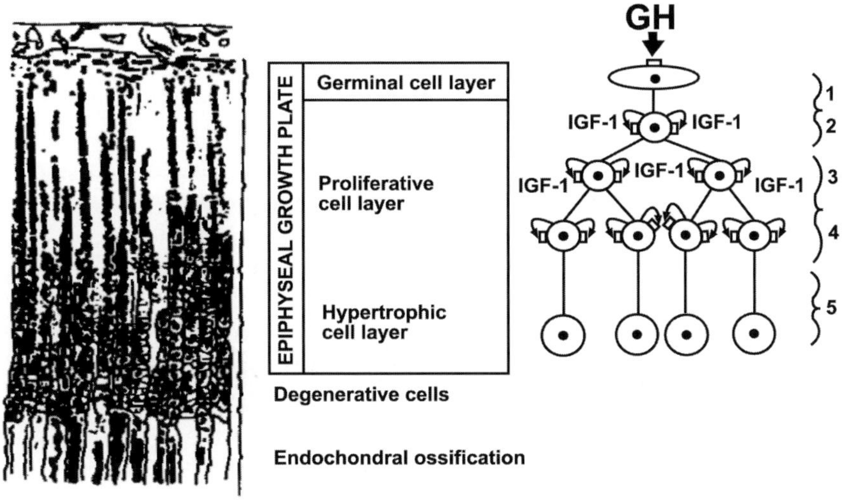

Figure 2. A hypothetical model for the stimulatory effect of GH on longitudinal bone growth.
GH stimulates the germinal cell layer of the epyphiseal growth plate to produce insulin like growth factor-1 (IGF-1), which in turn acts in a autocrine/paracrine manner to stimulate cell growth in the proliferative layer. After proliferation the cells will undergo hypertrophy, degeneration and endochondral ossification takes place (After Casauneva 1992 [12]).

for proliferation, but cell division would not take place, unless the cells were stimulated also with a second hormone (e.g. IGF or insulin), which act in the late G1 phase. After the second stimulus the cells will *progress* in the cell cycle and mitosis takes place. Exposure to the competence hormone is mandatory for the cell in order to acquire the ability to respond to the progression growth factor in this model [11]. These observations led to a two signal hypothesis known as the *competence-progression model of cell proliferation* (Figure 1).

A similar two signal hypothesis has been proposed for GH action on the growth of long bones (Figure 2). According to this model the biological effect of GH is necessary for target cell activation. One important aspect of GH action is the stimulation of IGF-I in the responding cells. In turn, the locally induced IGF-I will act as an autocrine and paracrine growth factor, completing the mitogenic cycle in GH targets [12]. By now this model of GH action is supported by a compelling body of experimental evidence. It is also apparent that there are other hormones/ cytokines that are capable of inducing IGF-I in certain target cells. This implies that such hormones/cytokines also activate their targets similarly to that of GH, in a manner analogous to the competence-progression model of cell proliferation. Such hormones are IL-3, GM-CSF and TGFβ1 in the bone marrow and estradiol for the mammary gland [13–16]. Thus the tropic effect of these hormones on their specific target organs is also based on two signals, analogous to the competence-progression model of cell proliferation.

The PRL-dependent Nb2 rat T cell lymphoma cell line obeys the rules of the competence-progression model for cell proliferation. It responds to serum factors (IGF-I) only after being primed by PRL. Most importantly, the magnitude of the mitogenic response is directly proportional to the concentration of PRL used for priming (Figure 3) [17]. In vivo observations in hypophysectomysed (Hypox) rats revealed that organ weight, DNA and RNA synthesis and cell proliferation in the bone marrow, thymus and spleen (Figures 4 and 5) and immune reactivity

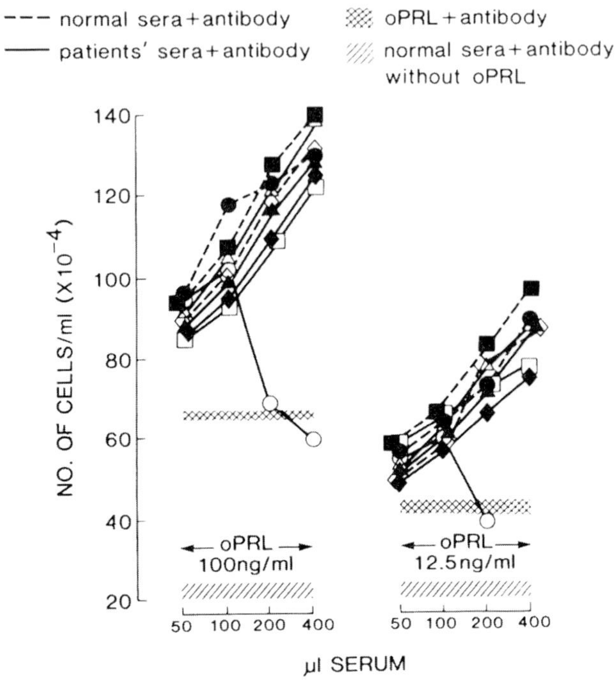

Figure 3. Enhancement of oPRL bioactivity in increasing concentrations of rheumatoid and normal human sera. The Nb2 rat lymphoma cells were cultured in Fischer's medium plus 10% horse serum in duplicate. This medium did not support Nb2 growth unless prolactin (ovine in this case) was added. The growth response to two concentrations of oPRL (12.5 and 100 ng/ml) was dose dependent. The addition of human serum (50–400 μl/2 ml culture), after the neutralization of both PRL and GH by specific monoclonal antibodies, increased cell proliferation further, again in a dose dependent manner. This was true for samples from both normal and rheumatoid individuals, although one rheumatoid serum was inhibitory if 100 μl or more was added per culture. None of the antibody treated human samples tested stimulated Nb2 growth in the absence of oPRL (After Berczi et al. 1987 [76]).

require the presence of pituitary GH or PRL. These experiments show that lymphocyte proliferation in primary and secondary lymphoid tissue and immunocompetence are dependent on pituitary GLH. *Redundancy* was also demonstrated as either PRL or GH was sufficient to restore the immune system. Placental lactogen was also capable of restoration [18].

Like GH, PRL is also a pleiotropic hormone with multiple targets. On the basis that GH is able to cause the proportional growth of the entire organism one may conclude that the entire body is a target for GH. PRL seems to affect most tissues except to stimulate body growth, though a limited growth promotion is present. Pituitary GLH induces IGF-I within the immune system as observed by several investigators. Both GH and PRL were capable of inducing IGF-I within the thymus and spleen of Hypox rats (Figure 6). This furnishes additional evidence for the ability of these hormones as growth promoters of the immune system according to the competence-progression model of cell proliferation [4,19–21].

There is evidence to indicate that GLH are essential for the maintenance of vital bodily functions [22,23]. In mammals the fetus is exposed to high levels of placental GLH as they are present in the amniotic fluid in high concentrations [24]. During embryonic life placental GLH

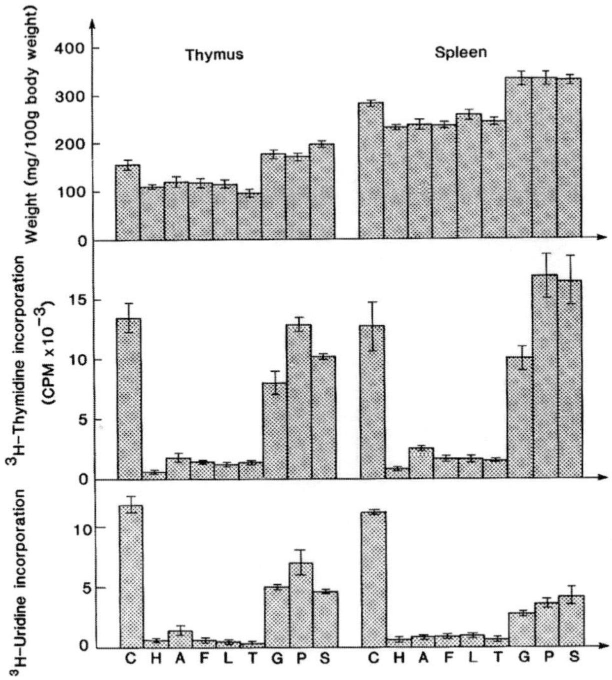

Figure 4. The effect of hypophysectomy and additional treatment with pituitary hormones on relative weights, DNA and RNA synthesis of thymus and spleen of rats.

Female Fischer rats (175–200 g body weight) were hypophysectomized and selected groups of 5 animals were treated as follows: C = control; H = hypophysectomy; A = H + ACTH; F = H + FSH; L = H + LH; T = H + TSH; G = H + GH; P = H + PRL; S = H + all the above listed hormones. ACTH, FSH and LH were given at 20 µg, GH and PRL at 40 µg, and TSH at 0.66 IU/rat/day from days 12 to 19 after hypophysectomy. DNA synthesis was determined by ^{3}H-thymidine and RNA synthesis by ^{3}H-uridine incorporation, respectively.

play a fundamental role in the growth and development of the fetus and of the immune system and pituitary hormones are not required as fetuses with the congenital lack of the pituitary gland develop normally *in utero* [25]. During postnatal life the pituitary gland assumes the role of growth control in the organism [26]. Bone marrow and thymus function and the maintenance of immunocompetence during postnatal life depend entirely on pituitary GLH [4]. On this basis it has been proposed that *GH and PRL are the hormones of immunocompetence* [27,28].

3. GROWTH SIGNALING IN THE IMMUNE SYSTEM

Talmage made theoretical calculations and concluded that the mammalian genome could not possibly accommodate the enormous multitude of immunological specificity displayed by antibodies [29]. On this basis Burnet proposed the clonal selection theory [30]. This theory postulated that minor groups of lymphocytes (clones) are generated through somatic mutation. After the self-reactive clones had been eliminated in the thymus, the remaining clones will recognize

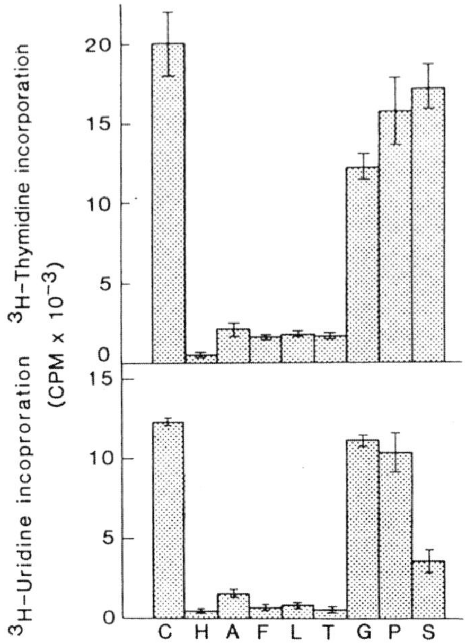

Figure 5. Nucleic acid synthesis in the bone marrow is dependent on growth hormone and/or prolactin.
The treatment schedule of the various groups is explained in the legends of Figure 4. The animals were killed on day 20 and bone marrow cells from the femurs were tested for nucleic acid synthesis in vitro (After Nagy and Berczi, 1989 [20]) .

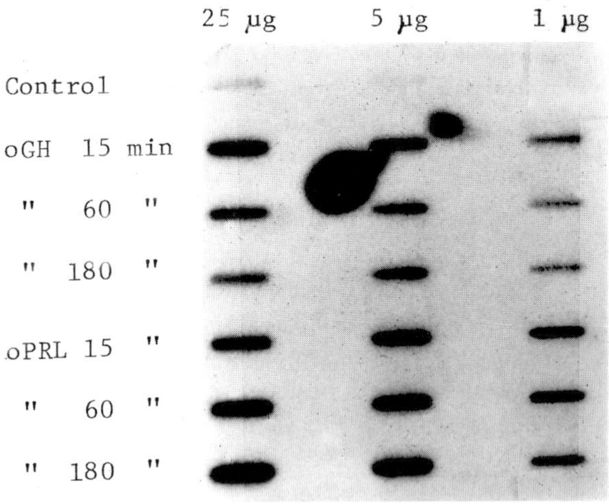

Figure 6. Insulin like growth factor-I mRNA expression in the thymus of hypophysectomized rats after treatment with ovine growth hormone or prolactin.
Groups of 3 Fischer rats were hypophysectomized (day 0) and on day 10 treated with oPRL (NIADDK-oPRL-17) or with oGH (NIADDK-oGH-12) i.p. at 100 µg/100 g body weight. The animals were killed at 15, 60 and 180 minutes after hormone injection. Total RNA was prepared from pooled thymuses of 3 animals, blotted onto nitrocellulose paper, and slot blot analysis was performed using 25, 5 and 1 µg of RNA per slot. A rat cDNA probe encoding prepro-IGF-I from amino acids minus 3 to 105 was used.

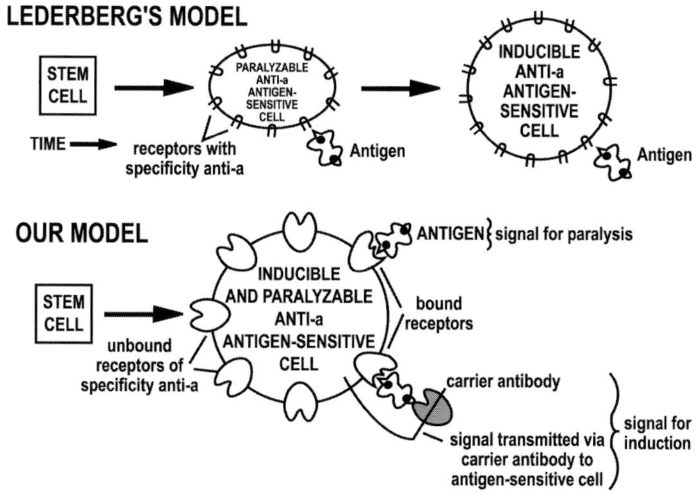

Figure 7. One and two signal theories of the immune response.
Lederberg's one signal model and the two signal theory of Bretcher and Cohn is shown. (After Bretcher and Cohn, 1970 [34]).

and respond to specific antigenic determinants (epitopes) of a foreign antigen by proliferation (clonal expansion). The cells of the expanded clone will then differentiate into plasma cells and produce specific antibodies [30]. An overwhelming body of experimental evidence, including the molecular structure of antigen receptors of lymphocytes, supports the validity of this hypothesis [31]. It is now of general consensus that lymphocyte growth is a prerequisite of the adaptive immune response. If lymphocytes are unable to grow, immune reactions are not possible.

Since the seminal work of Landsteiner [32] it has been recognized that small molecules called *haptens* are not immunogenic, but become immunogenic if coupled covalently to macromolecules, called *carriers*, that express multiple epitopes. On the basis of this long recognized fact, in 1970 Bretcher and Cohn [34] proposed the two signal model for the initiation of the immune response (Figure 7). These authors argued that the first signal is delivered by the antigen and the second signal by a carrier-specific antibody, although they acknowledge that it could come from T lymphocytes that stimulate antibody production by B cells. Meanwhile, Claman and co-workers discovered that the collaboration of thymus derived (T) and bone marrow derived (B) lymphocytes is necessary for an antibody response to occur [33]. This led to the "bridging" hypothesis of Mitchison, whereby the contact interaction of T and B lymphocytes was necessary for an antibody response to occur (Figure 8) [35,36].

Today it is clear that the adherence interaction of an antigen presenting cell (APC) with antigen specific T lymphocytes is required for the initiation of an immune response. Initially the APC interacts with helper T cells and later on the helper T cells with B cells (antibody response) or immature antigen sensitive T cells (cell-mediated immunity). This adherence interaction stimulates the secretion of cytokines from helper T lymphocytes, which function as growth factors. The antigen signal is modified by additional "co-stimulatory" and inhibitory signals, which are also delivered by adhesion molecules [37,38].

The *cytokine signal* completes the mitogenic stimulus and enables the immature cell to proliferate (clonal expansion). Multiple cytokines are available to deliver this signal, which varies

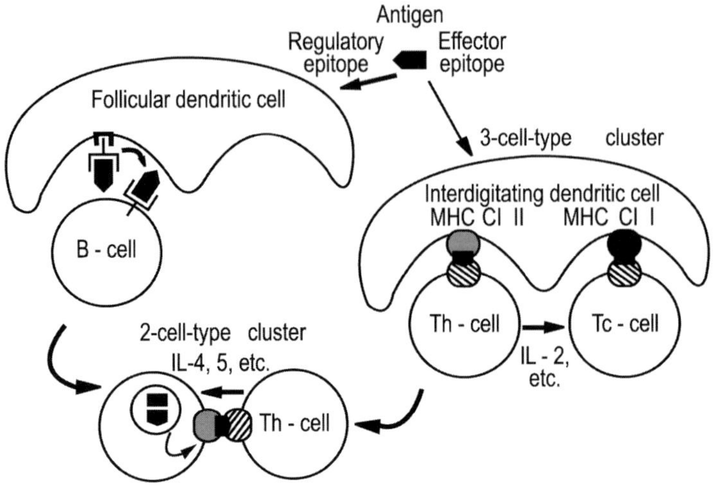

Figure 8. The bridge hypothesis in its later form.
T-B and T-T lymphocyte cooperation is shown. The epitopes are derived from the antigen by processing and presented by surface MHC molecules of B lymphocytes or intedigitating dendritic cells to T lymphocytes that respond to the specific epitope. Some cytokines that are also required to initiate the cell cycle are indicated. (Mitchison 1989 [35]).

according to the type and stage of the immune response [31,37,38].

4. THE THREE-SIGNAL HYPOTHESIS

If one carefully examines the theories for cell growth and for the initiation of an immune response, in fact they define three groups of signals: (i) the competence signal, (ii) stromal or adherence signals, and (iii) cytokine signals (Figure 9). The model of Bretcher and Cohn [34], which was constructed on the basis of *in vivo* experiments ignores the first signal, which is delivered to lymphocytes and antigen presenting cells by GLH. This signal was always available in the animal and immunocompetent lymphocytes and could produce additional GLH as required for the paracrine-autocrine support of rapid lymphocyte proliferation. The situation was similar during *in vitro* experiments on cell collaboration. From the very beginning it was well recognized that fetal calf serum, which again contains GLH, was necessary in order to initiate immune reactions. Once the lymphocytes got activated they could produce their own GLH. But at that time nobody thought that this should be the case, as hormones were considered to fall outside of immunoregulators.

The antigen signal is in actual fact an adherence signal as T cells must recognize antigen in the context of self-MHC cell surface molecules. The TCR/CD3 complex, MHC-I and -II, and the CD4/CD8 accessory molecules and immunoglobulin (Ig) itself are all members of the immunoglobulin family of adhesion molecules [37].

The competence progression model for cell proliferation ignores the necessity for cell-to-cell contact in the regulation of proliferation. This is known in the old literature as "contact inhibition". Rapidly growing cells in culture lose their sensitivity to contact inhibition, and often generate tumors if injected into animals. Some of the genes controlled by adhesion molecules are

Figure 9. Growth control in higher animals.
This hypothesis proposes that 3 groups of signals are required for the initiation of cell proliferation in normal animal cells. The first signal provides the cell with competence to respond to additional (adherence and cytokine) signals. The second signal is delivered by cell-to-cell and/or cell-to-matrix interaction (adherence signal). This signal is dominant and determines what the cell is going to do. The third signal is delivered by cytokines. This is required for the completion of the cell cycle. This basic signaling process is modulated further by hormones, neurotransmitters and neuropeptides, as discussed later. (Berczi 1993 [28]).

classified as "tumor suppressor" genes that are capable of inhibiting cell proliferation [39–41]. Hence, the competence progression model does identify signal number 1 and signal number 3 in our model for the growth of animal cells.

4.1. The significance of GLH in signalling.

Mice with targeted gene disruption (knockout) that lack either PRL or its receptor (PRLR), or GHR or IGF-I are immunocompetent. It was proposed, therefore, that these hormones are not obligate immunoregulators, but rather, affect immune reactions as anabolic and stress modulating agents [42–44]. In actual fact the data obtained in knockout mice is a powerful confirmation of our original observations that *GLH show redundancy* in the maintenance of immunocompetence. Current evidence indicates the IGF-I may be substituted for in the immune system by IGF-II or insulin. Clearly, immune function, as many other functions in the body, are maintained by multiple hormones and cytokines that show overlap and redundancy [4,45].

Functional overlap and redundancy is the rule for type I cytokines (and for other cytokines as well) in the immune system. The receptor for type I cytokines consists of a ligand specific chain and of a shared signal transducing chain. For instance in the first group, where IL-2, -4, -7, -9 and -15 belong, there is a common gamma chain (γ_c). Signal transduction is possible only if the two chains are crosslinked by the specific cytokine. Knockout experiments in this system showed that the elimination of specific cytokines or their specific receptor chains produced minimal if any abnormalities. However, knocking out the shared γ_c chain resulted in severe combined immunodeficiency [46,47]. These observations collectively indicate that type I cytokines as a group are indispensable for normal immune function.

Prolactin and growth hormone do not share receptor chains with any of the above cytokines. Human GH and other primate GH are known to act on the PRL receptors and to exert lactogenic activity in many species, which indicates functional overlap within GLH hormones. The major signal transduction pathway, which involves the Janus kinase (JAK) and signal transducers and activators of transcription (STAT) nuclear regulatory factors, is shared between cytokines and growth and lactogenic hormones. STAT knockout mice show severe developmental and immune deficiencies [48,49]. This emphasizes the significance of this signal transduction pathway in development and immune function.

Clevenger and co-workers produced data indicating that PRL is capable of nuclear signalling [50]. Similarly, it has been proposed that IGF binding proteins are capable of nuclear signalling [51]. The possibility of nuclear signalling by peptide hormones is not yet generally accepted. However, if correct, such a mechanism for PRL and IGF-IBP would mean that these molecules are capable of bypassing the extensive regulatory mechanisms of signal transduction by membrane bound receptors. One may visualize that for a hormone, which acts as a survival factor for cells and indeed for the entire organism, nuclear signalling would be of major advantage or perhaps, even a necessity.

Taken together, current evidence indicates that GLH, as a group, are indispensable for normal development of higher animals, including the development of the immune system. The experiments performed on knockout mice are fully compatible with this conclusion. Therefore, in order to prove or disprove the relevance of PRL to immunity, the entire GLH system should be disabled. However, we predict on the basis of our own observations that the disabling of GLH genes in the same animal would have lethal consequences [23].

Lymphocytes are capable of producing both PRL and GH. There is evidence to suggest that lymphocyte-derived GLH is capable of paracrine and autocrine growth stimulation [52,53]. It is suggested that during immune reactions, such locally generated GLH is required to support the rapid proliferation of lymphocytes in the interest of promotion of an effective immune response. This situation has parallels with the development of the mammalian embryo, which gradually grows independent of the pituitary hormones and relies mostly if not exclusively on placental lactogenic hormones. Moreover, placental GLH actually have a major influence on maternal metabolism, which serves the interest of the fetus [54].

Naïve lymphocytes are in a quiescent state and do not synthesize nor do they respond to cytokines [55]. It is suggested that these cells are dependent on pituitary GLH and on IGF-I for survival in lymphoid organs and rapidly undergo apoptosis in Hypox animals. On the other hand memory cells are metabolically active and remain functional in Hypox animals. Memory T lymphocytes are glucocorticoid resistant and produce cytokines readily after stimulation with the specific antigen. Memory cells are capable of recruiting naïve antigen sensitive T cells [55]. The lymphocyte derived PRL gene was shown to have placental promoting sequences [56]. If such a gene becomes activated in memory cells they would be able to synthesize PRL for autocrine stimulation in a pituitary independent manner and to assure survival even in hypophysectomized animals. The glucocorticoid resistance of memory cells would also be explained as PRL is an antagonis of glucocorticoids. This possibility remains to be established.

4.2. The importance of cell-to-cell signalling

Self recognition has been thought for a long time to be the exclusive feature of the immune system. However, self recognition is easily demonstrable in the most primitive multi-cellular animals, sponges. Sponges also control cell proliferation and differentiation. They are capable of rejection of grafts from another species of sponges, and show self defence against infection, which is mediated by phagocytic cells. One may easily disintegrate sponges by passing them through a screen. When brought together, the cells are capable of re-aggregation and regeneration to form a functional sponge unit. These facts demonstrate the ability of these seemingly loosely aggregated cells to behave in a highly coordinated fashion. Sponge cells will grow and differentiate into functional cells according to their topographical localization [57]. Similar observations were made in higher animals. For instance, if cells from different anlages of the amphibian embryo are mixed, they will sort out in a pattern that resembles the initial organization of embryonic tissue. Such aggregation experiments may also be performed with embryonic cells from birds or mammals. Cell adhesion molecules present in embryonic tissue mediate re-aggregation and cell motility and play a key role in morphogenesis [58,59].

Embryonic morphogenesis is governed by cell-to-cell contact and by diffusible mediators. Adhesion molecules are non-diffusible and for this reason are capable of signalling single cells very specifically. It is very clear from embryonic development and from antigen-induced lymphocyte proliferation that adherence signals are dominant over growth factor signalling. This is an absolute requirement for morphogenesis, which is based on the positional relationship of cells/tissues to each other. In general, adherence signals determine, according to the local tissue/organ requirements, whether or not the cell is going to divide, differentiate and take up a function, or simply be on standby (survive), or perhaps be committed to the pathway of programmed cell death (apoptosis) [28,60]. At the cellular level, this means that only certain cells will divide at any given time while others go into differentiation and take up the appropriate function according to their location in the body, or be on standby (stem cells, as well as differentiated cells) or may even be eliminated. Thus, the general growth stimulus is modified according to local needs so that the morphological and functional integrity of the organism is maintained at all times. GH is well recognized as a hormone capable of stimulating the proportional growth of all tissues and organs. This dominance of local regulatory mechanisms, which also include tissue bound hormones and cytokines, over the systemic (GLH) signals assures the development of a fully functional animal or human being.

Injured nerve cells in the CNS can be reinduced to grow axons and establish functional connections if exposed to non-neural elements of the peripheral nervous system [61]. This illustrates very well that even in adult tissues that lost their capacity to grow stromal adherence signals are capable of inducing growth and regeneration.

Plants show a remarkable morphological and functional differentiation. Some proteins extracted from plants and collectively named lectins [62] have the capacity to activate animal cells, especially lymphoid cells for proliferation and function, including immunoglobulin secretion, cytotoxicity, helper or suppressor activity [38]. Therefore, plant lectins function as regulatory molecules on animal cells and probably fulfil similar functions in the plants as well. Animal tissues also contain lectin-like molecules [62].

Adherence signals mediate positional regulation, which is species-, organ- or tissue-specific, and in the case of MHC and antigen receptor molecules, individually specific. The antigen receptor also shows *epitope specificity*. The antigen receptors of B and T lymphocytes, MHC antigens, CD4 and CD8, some receptors of natural killer cells and numerous other cell bound molecules belong to the immunoglobulin family of adhesion molecules. Both activating and inhibitory receptors are found in this group that carry ITAM (immunorecptor tyrosine-based activation motif) and ITIM inhibitory motifs respectively (Figures 10, 11). Receptors containing ITAM motifs promote tyrosine phosphorylation and are stimulatory. Receptors with ITIM dephosphorylate tyrosine residues and are inhibitory. During lymphocyte activation stimulatory and inhibitory receptors co-aggregate (capping) and the result will depend on the ratio of enzymatic phosphorylation/dephosphorylation leading to stimulation/inhibition, respectively [63,64]. This "higher order" of receptor complexity may revolutionize our understanding of cell signaling. Surface receptors capable of initiating apoptosis, such as the Fas/Fas ligand, are also involved in regulation [38].

The restrictive power of cell-to-cell signaling is also fundamental to the immune response. Clearly, an antigen specific lymphocyte clone must not proliferate unless it is triggered by the

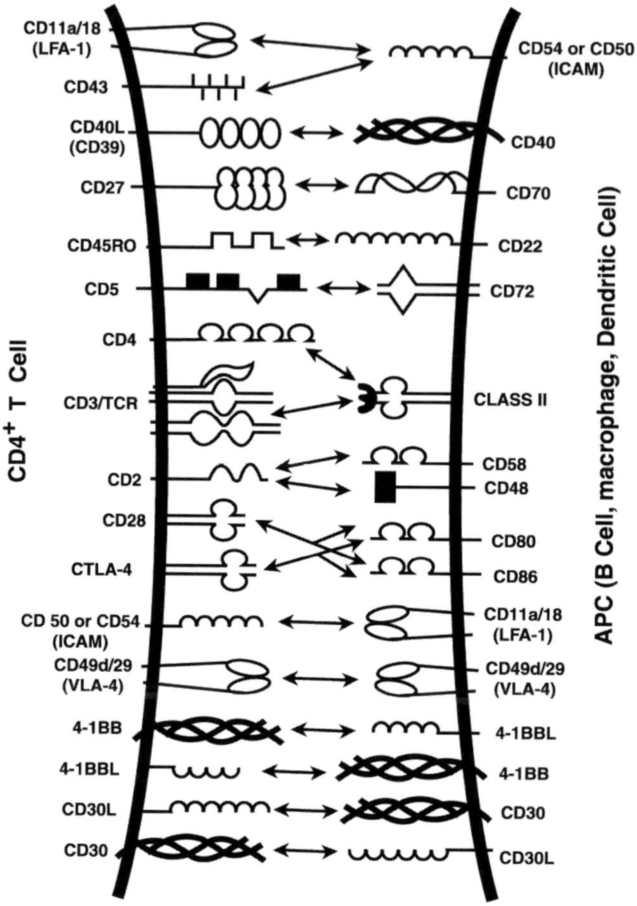

Figure 10. Selected examples of accessory molecules between CD4+ T cells and antigen presenting cells. The schematic structures shown are representative but not a genuine reflection of the individual molecular structures (After Bluestone et al. 1999 [64]).

specific antigenic epitope in the context of self-MHC molecules. Without this restriction antigen-specific immune reactions would not be possible. Further, MHC recognition by suppressor T lymphocytes and killer inhibitory receptors in natural killer cells serves as safeguards against the killing of normal non-infected and non-cancerous cells [38].

4.3. Cytokine signalling

Bone marrow cells change their hemopoietic growth factor requirement as they go through the various differentiation pathways [65–67]. Cell-to-cell interaction between stromal cells and B cell precursors and IL-7 play an essential role in B lymphocyte development in the bone marrow. The thymic microenvironment is essential for normal T cell development. Here MHC-T cell receptor interactions and a special cytokine and neuroendocrine milieu regulate T cell maturation [38,68–71]. During adaptive immune responses the cytokine requirement for the growth

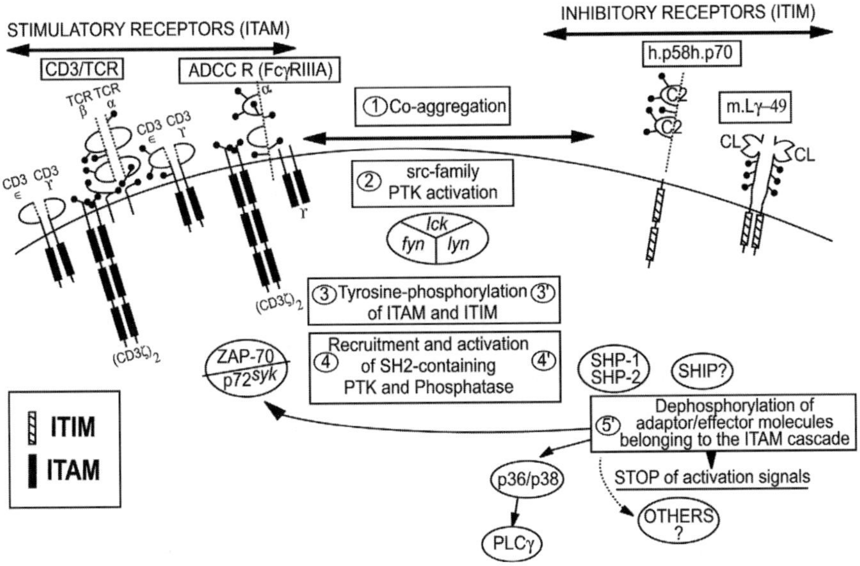

Figure 11. Capping of stimulatory and inhibitory receptors prior to T and NK cell activation.
The black cylinders represent immunoreceptor tyrosine based activation motifs (ITAM) and the striped cylinders stand for – inhibitory motifs (ITIM). Upon co-aggregation (capping) of stimulatory and inhibitory receptors the activation of *src*- related protein kinases (PTK) induce tyrosine phosphorylation of ITAM and ITIM. Phosphorylated ITAM recruits tandem SH2-PTK, such as ZAP-70 or $p72^{syk}$, which become activated. Following inhibitory receptor engagement the phosphorylation of ITIM leads to the recruitment of tandem SH2-phosphatases, such as SHP-1 and SHP-2. The co-aggregation of activating and inhibitory receptors by capping facilitates the interaction of enzymes that stimulate cell activation by phosphorylation and inhibit activation by dephosphorylation (After Renard et al. 1997 [65]).

and differentiation of mature T and B lymphocytes is also subject to change according to the type and/or phase of the immune reaction elicited [31,37,38]. Therefore, cytokines do not only complete the signalling process for cell proliferation, but also appear to play a role in the determination of the type of effector cells that are generated during the proliferative response.

Cytokine signals are usually, but not always, delivered by locally generated mediators. These signals antagonize apoptosis and complete the positional signalling cycle for growth, differentiation, functional activation, etc. *Insulin-like growth factors* are both systemic and locally generated and *insulin* is a systemic hormone. As systemic mediators, these hormones deliver third signals to all cells in the body. The cells affected may be in an inactive quiescent state incapable of synthesis and circulating IGF and INS enable such quiescent cells to metabolize and survive. It was suggested that small naïve lymphocytes survive this way. IGF-I is also produced readily in tissues after GH/PRL stimulation as a cytokine and is able to complete the growth cycle in any tissue or cell. Both GH and PRL stimulate IGF-I in a variety of target cells (including T, B cells and macrophages) and thus generate signal number 1 and number 3 for cell proliferation. Moreover, GH itself can also exert an insulin-like action. However, given the heterogeneity of GLH and of their receptors, these hormones may be capable of delivering a number of signals other than the competence signal [3,4,45,72,73]. A detailed discussion of these questions is beyond the scope of this chapter.

4.4. Signal regulation and signal modulation

Thyroid hormones, glucocorticoids, sex hormones and vitamin D regulate immune reactions via controlling nuclear transcription processes. These steroid hormones may mediate activation, inhibition and apoptosis within the immune system. This regulation is fundamental for normal immune function and for the adaptability of immune system to reproduction and for protection against stressful insults and of various other pathological processes. These hormones may be designated as *nuclear signal regulators* [71,73,74].

Catecholamines regulate cyclic AMP, GMP and calcium influx. A number of other mediators, including opioid peptides, act through G-proptein linked adenylate cyclase receptors as well. Because Ca^{2+} is needed for cellular activation and phosphorylation is a fundamental mechanism of signal transduction by membrane bound receptors, these mediators are designated as *signal modulators* in immune activation and in cellular activation in general [73,75].

5. CONCLUSIONS

Lymphocyte growth obeys the general rules of growth regulation of the entire organism. In primitive multi-cellular animals, like the sponge, cell-to-cell recognition and communication is the basis of organization, and little is known about the role of soluble mediators. However, in embryos of higher animals soluble mediators are clearly important, which are necessary for cell growth and differentiation, which must fulfil the functional requirements posed by the position of a given cell in the body. During evolution and during embryonic development growth control is gradually taken over by the pituitary gland. In mammals this happens abruptly at parturition, when pituitary GH assumes the regulation of body growth and PRL joins in as a similar regulator, but with a very limited ability to promote body growth.

The adaptive immune response is based on the recognition of MHC antigens presenting either self-related or foreign epitopes to T lymphocytes. Both the MHC molecules and antigen receptors are adhesion molecules, that belong to the immunoglobulin family. Therefore, the fundamental mechanism of self recognition has been modified to recognize non-self in the context of self. According to recent consensus, suppressor T lymphocytes actively recognize self antigens and exert a local suppression in the various organs and tissues against the development of effector (autoimmune) cells. When the self-MHC presents a foreign epitope, no suppressor cells would be present against such modified self antigen and consequently effector T lymphocytes can freely develop against them [38]. Therefore, the immune system is an essential part of growth control in higher organisms. It exercises quality control, allowing the growth of normal cells and destroying cells which display an altered phenotype (e.g. infected, cancer and degenerated cells). This way the homeostasis of the immune system, which is based on the equilibrium of effector and suppressor mechanisms is closely linked to the homeostasis and growth control of the entire organism.

The growth promoting effect of GH and PRL is no different on lymphocytes from the effect of other cells in the body. This control is based on cell activation, which in practical terms must mean to shift the signalling balance by adhesion molecules towards activation. This is followed up by the production of IGF-I, which completes the mitogenic signal. Quiescent cells (e.g. naïve lymphocytes) will not be activated, but use serum-derived IGF-I and insulin for survival. Immature lymphocytes and leukocytes also depend on pituitary GH and PRL for survival. Once differentiation is initiated, new adhesion molecules are induced (e.g. the antigen receptor) and new cytokines will take over the stimulation of cell growth. The production of autocrine PRL and GH by activated cells boosts further cell proliferation. Some cells that show resistance to hypophysectomy, such as memory cells, are likely to have their GLH genes under the control of the placental promoter. Cytokines, such as interleukin-2, IL-3 and GM-CSF may actually be able to boost the competence signal in lymphocytes and in the bone marrow, respectively. Thus during the evolution of an immune response the immune system may gradually escape pituitary control by the production of competence hormones locally, which are capable of maintaining immune function in an autonomous fashion.

In spite of this autonomy of memory lymphocytes, the immune system requires the continuous production of naïve lymphocytes to fight novel pathogens that might occur in the environment. A stable output of leukocytes of the bone marrow is also required for the maintenance of immune function and health. These remain pituitary dependent. Interestingly, during acute phase responses the GH/PRL-IGF-I axis is suppressed, which results in a profound suppression of the thymus. However, the bone marrow is actually activated. Under these conditions IL-6 and insulin are elevated. Because bone marrow function is glucocorticoid resistant, these systemic growth promoting mediators are free to act and, combined with locally produced growth factors, could explain the intensive bone marrow activation. This remains to be validated.

Steroid, thyroid hormones and vitamin D regulate nuclear signalling proteins and for this reason are required for normal immune function. These hormones are the ultimate *regulators of signaling* in all cells in the body, including the immune system. Other hormones, cytokines and neuropeptides (e.g. catecholamines, chemokines and opioid peptides) *modulate signalling* by the regulation of Ca^{2+} and of the rate of phosphorylation.

It may be concluded on the basis of available evidence that immunocompetence depends on pituitary and lymphocyte-derived GLH. Without these hormones the immune system looses cellularity and the ability to respond to antigenic and mitogenic stimuli. Moreover, the entire organism ceases to function if no GLH is available and death will ensue within a short period of time. This hypothesis is supported by observations in animals and man. Therefore, it is postulated that GLH provide *vitality* not only to the immune system, but also to the entire organism.

REFERENCES

1. Cushing SJ. The function of the pituitary body. Amer J Med Sci 1910;139:473–484.
2. Aschner B. Uber die Funktion der Hypophyse. Pflueger's Arch Ges Physiol 1912;146: 1–146.
3. Matera L, Rapaport R., editors. Growth and lactogenic hormones. In: Berczi I, Szentivanyi A, series editors. Neuroimmune Biology, Volume 2. Amsterdam:Elsevier, 2002.
4. Berczi I, Szentivanyi A. Growth and Lactogenic Hormones, Insulin-Like Growth Factor and Insulin. In: Berczi I, Szentivanyi A, series editors. The immune-neuroendocrine circuitry. History and progress. Neuroimmune Biology, Volume 3. Amsterdam:Elsevier, 2003;129–153.
5. Phillips LS, Vassilopoulou-Sellin R. Somatomedins. N Engl J Med 1980;302:371–380 and 438–446.
6. Blundell TL, Humbel RE. Hormone families: pancreatic hormones and homologous growth factors. Nature (London) 1980;287:781–787.
7. Rosenfeld R, Thorrson AV, Hintz RL. Increased somatomedin receptor sites in newborn circulating mononuclear cells. J Clin Endocrinol Metab 1979;48:456–461.

8. Rosenfeld R, Hintz RL, Dollar LA. Insulin-induced loss of insulin-like growth factor-I receptors on IM-9 lymphocytes. Diabetes 1982;31:375–381.
9. Mick CCW, Nicoll CS. Prolactin directly stimulates the liver in vitro to secrete a factor (synlactin) which acts synergistically with the hormone. Endocrinology 1985;116:2049–2053.
10. Nicoll CW, Herbert NJ, Russell SM. Lactogenic hormones stimulate the liver to secrete a factor that acts synergistically with prolactin to promote growth of the pigeon crop-sac mucosal epithelium in vivo. Endocrinology 1985;116:1449–1453.
11. Aaronson SA. Growth factors and cancer. Science 1991;254:1146–1153.
12. Casauneva FF. Physiology of growth hormone secretion and action. Endocr Metab Clin N Amer 1992;21:483–517.
13. Kelley KW, Arkins S, Minshall C, Liu Q, Dantzer R. Growth hormone, growth factors and hematopoiesis. Horm Res 1996;45(1–2):38–45.
14. Kelley KW, Meier WA, Minshall C, Schacher DH, Liu Q, VanHoy R, Burgess W, Dantzer R. Insulin-like growth factor-I inhibits apoptosis in hematopoietic progenitor cells. Implications in thymic aging. Ann N Y Acad Sci 1998 May 1;840:518–24.
15. Fournier B, Gutzwiller S, Dittmar T, Matthias G, Steenbergh P, Matthias P. Estrogen receptor (ER)-alpha, but not ER-beta, mediates regulation of the insulin-like growth factor I gene by antiestrogens. J Biol Chem 2001;276:35444–9.
16. Kveiborg M, Flyvbjerg A, Eriksen EF, Kassem M. Transforming growth factor-beta1 stimulates the production of insulin-like growth factor-I and insulin-like growth factor-binding protein-3 in human bone marrow stromal osteoblast progenitors. J Endocrinol 2001;169:549–61.
17. Berczi I, Nagy E. Neurohormonal control of cytokines during injury. In: Rothwell NJ, Berkenbosch F, editors. Brain Control of the Response to Injury. Cambridge University Press 1994;32–107.
18. Berczi I, Nagy E, Matusik RJ, Friesen HG. Pituitary hormones regulate c-myc and DNA synthesis in lymphoid tissue. J Immunol 1991;146:2201–2206.
19. Kelly PA, Djiane J, Postel-Vinay MC, Edery M. The prolactin/growth hormone receptor family. Endocr Rev 1991;12:235–251.
20. Nagy E, Berczi I. Pituitary dependence of bone marrow function. Br J Haematol 1989;71: 457–462.
21. Nagy E, Berczi I, Sabbadini E. Endocrine control of the immunosuppressive activity of the submandibular gland. Brain Behav Immun 1992;6:418–428.
22. Sinha YN, VanderLaan WP. Effect on growth of prolactin deficiency induced in infant mice. Endocrinology 1982;110:1871–1878.
23. Nagy E, Berczi I. Hypophysectomized rats depend on residual prolactin for survival. Endocrinology 1991;146:2776–2784.
24. Hill DJ. What is the role of growth hormone and related peptides in implantation and the development of the embryo and fetus. Horm Res 1992;38:28–34.
25. Potter EL, Craig JM. Pathology of the fetus and the infant. 3rd edition. Chicago:Year Book, 1975.
26. Glasscock GF, Gelber SE, Lamson G, McGee-Tekula R, Rosenfeld RG. Pituitary control of growth in the neonatal rat: Effects of neonatal hypophysectomy on somatic and organ growth, serum insulin-like growth factors (IGF)-I and -II levels, and expression of IGF binding proteins. Endocrinology 1990;127:1792–1803.
27. Berczi I, Nagy E. The effect of prolactin and growth hormone on hemolymphopoietic tissue

and immune function. In: Berczi I, Kovacs K editors. Hormones and Immunity. Lancaster, UK:MTP Press, 1987; 145–171.
28. Berczi I. The role of the growth and lactogenic hormone family in immune function. Neuroimmunomodulation 1994;1:201–216.
29. Talmage DW. Allergy and immunology. Ann Rev Med 1957;8:239–256.
30. Burnet FM. The clonal selection theory of acquired immunity. Cambridge University Press, 1959.
31. Paul WE. The immune system, An introduction. In: Paul WE editor. Fundamental Immunology. 4th edition. Philadelphia:Raven Press, 1999; 1–18.
32. Landsteiner K. The Specificity of Serological Reactions. Harvard University Press, 1945.
33. Claman HN, Chaperon EA, Triplett RF. Thymus-marrow cell combinations. Synergism in antibody production. Proc Soc Exp Biol Med 1966;122:1167–71.
34. Bretscher P, Cohn M. A theory of self-nonself discrimination. Science 1970;169:1042–1049.
35. Mitchison NA. T cells in transplantation immunity. Immunol Lett 1989;21:15–19.
36. Mitchison NA. Unique features of the immune system: Their logical ordering and likely evolution. In: Burger MM, Sordat B, Zinkernagel RM editors. Cell to Cell Interaction. Basel: Karger, 1990; 201–214.
37. Berczi I, Szentivanyi A. Antigen presentation. In: Berczi I, Szentivanyi A. Series editors. The immune-neuroendocrine circuitry. History and progress. Neuroimmune Biology, Volume 3. Amsterdam:Elsevier, 2003;301–313.
38. Berczi I, Szentivanyi A. Immune reactions. In: Berczi I, Szentivanyi A. Series Editors. The immune-neuroendocrine circuitry. History and progress. Neuroimmune Biology, Volume 3. Amsterdam:Elsevier, 2003;315–377.
39. Klein G. The approaching era of the tumor suppressor genes. Science 1987;238:1539–1545.
40. Klein G. Tumor Suppressor Genes. New York:Dekker, 1990; 273.
41. Weinberg RA. Tumor suppressor genes. Science 1981;254:1138–1146.
42. Horseman ND, Zhao W. Montecino-Rodrigez E, Tanaka M, Nakashima K, Engle SJ, Smith F, Markoff E, Dorshkind K. Defective mammopoiesis, but normal hematopoiesis, in mice with a targeted disruption of the prolactin gene. EMBO J 1997;16:6926–35.
43. Bole-Feysot C, Goffin V, Edery M, Binart N and Kelly PA. Prolactin (PRL) and its receptor: Actions, signal transduction pathways and phenotypes observed in PRL-receptor knockout mice. Endocrine Rev 1998; 19:225–68.
44. Dorshkind K, Horseman ND, The roles of prolactin, growth hormone, insulin-like growth factor-I, and thyroid hormones in lymphocyte development and function. Insights from genetic models of hormone and hormone receptor deficiency. Endocrine Rev 21:292–312, 2000.
45. Berczi, I. Neuroimmune Biology – An introduction. In: Berczi I, Gorczynski R, editors. New Foundation of Biology; Neuroimmune Biology Series, Vol. 1. Amsterdam:Elsevier, 2001; 3–45.
46. Warren JL. Type I cytokines, interferons and their receptors. In: Paul WE, editor. Fundamental Immunology 4th edition. New York: Lippincott-Raven, 1999;741–74.
47. Notarangelo LD, Gilliani S, Mazza C, Mella P, Savoldi G, Rodriguez-Perez C, Mazzolari E, Fiorini M, Duse M, Plebani A, Ugazio AG, Vihinen M, Candotti F, Schumacher RF. Of genes and phenotypes: the immunological and molecular spectrum of combined immune deficiency. Defects of the gamma (c) -JAK3 signaling pathway as a model. Immunol Rev

2000;178:39–48.
48. Freeman ME, Kanyicska B, Lerant A, Nagy G. Prolactin: structure, function, and regulation of secretion. Physiol Rev 2000;80:1523–631.
49. Ivashkiv LB. Jak-STAT signaling pathways in cells of the immune system. Rev Immunogenet 2000;2:220–30.
50. Clevenger CV, Sillman AL, Prystovsky MB. Interleukin 2 driven nuclear translocation of prolactin in cloned lymphocytes-T. Endocrinology 1990;127:3151–3159.
51. Weinzimer SA, Cohen P. Biological significance of insulin-like growth factor binding proteins. In: Matera L, Rapaport R. editors. Growth and lactogenic hormones. Neuroimmune Biology Volume 2, Berczi I, Szentivanyi A, series editors, Amsterdam: Elsevier, 2002.
52. Hartmann DP, Holaday JW, Bernton DW. Inhibition of lymphocyte proliferation by antibodies to prolactin. FASEB J 3:2194–2202, 1989.
53. Sabharwal P, Glaser R, Lafuse W, Varma S, Liu Q, Arkins S, Kooijman R, Kutz L,Kelley KW, Malarkey WB. Prolactin synthesized and secreted by human peripheral blood mononuclear cells: an autocrine growth factor for lymphoproliferation. Proc Natl Acad Sci U S A 1992;89:7713–6.
54. Handwerger S, Freemark M. The roles of placental growth hormone and placental lactogen in the regulation of human fetal growth and development. J Pediatr Endocrinol Metab 2000;13:343–56.
55. Tough DF, Sun S, Zhang X, Sprent J. Stimulation of memory T cells by cytokines. Vaccine 2000 Feb 25;18(16):1642–8.
56. DiMattia GE, Gellersen B, Duckworth ML, Friesen HG. Human prolactin gene expression: The use of an alternative noncoding exon in decidua and the IM-9-P3 lymphoblastoid cell line. J Biol Chem 1990;265:16412–421.
57. Barnes RD. Invertebrate Zoology. Third Edition. Philadelphia:WB Saunders, 1974.
58. Hood L, Huang HV, Dreyer WJ. The area code hypothesis. The immune system provides clues to understanding the genetic and molecular basis of cell recognition during development. J Systamol Struct 1987;7:531–559.
59. Thiery JP. Cell adhesion molecules as morphoregulators. A short survey. In: Burger MM, Sordat B, Zinkernagel RM, editors. Cell to Cell Interaction. Basel:Karger, 1990;242–246
60. Arends MJ, Wyllie AH. Apoptosis: Mechanisms and roles in pathology. Int Rev Exp Pathol 1991;32:223–254.
61. Aguayo AJ, Villegas-Perez MP, Vidal-Sanz M, Carter D, Bray GM. Influences of non-neuronal tissues on the regeneration of injured nerve cells in the central nervous system of adult mammals. In Burger MM, Sordat B, Zinkernagel RM, editors. Cell to Cell Interaction. Basel:Karger, 1990; 40–61.
62. Sharon N. Lectin-carbohydrate complexes of plants and animals: Anatomic view. Trends Biochem Sci 1993;18:221–226.
63. Berczi I, Szentivanyi A. Adhesion molecules. In: Berczi I, Szentivanyi A, series editors. The immune-neuroendocrine circuitry. History and progress. Neuroimmune Biology, Vol. 3. Amsterdam:Elsevier, 2003;99–115.
64. Bluestone JA, Khattri R, Seventer GA, van. Accessory molecules. In: Paul WE, editor. Fundamental Immunology. 4th edition. Philadelphia:Raven Press, 1999;449–478.
65. Renard V, Cambiaggi A, Vély F, Bléry M, Olcese L, Oliviero S, Bouchet M, Vivier E. Transduction of cytotoxic signals in natural killer cells: a general model of fine tuning between activatory and inhibitory pathways in lymphocytes. Immun Rev, 155:205–221.

66. Sachs L. The molecular control of blood cell development. Science 1987;238:1374–1379.
67. Clark SC, Kamen R. The human hematopoietic colony stimulating factors. Science 1987;236: 1229–1237.
68. Metcalf D. The granulocyte-macrophage colony-stimulating factors. Science 1985;229: 16–21.
69. Melchers F. Cell to cell cooperations in B-cell development and B-cell responses. In: Burger MM, Sordat B, Zinkernagel RM, editors. Cell to Cell Interaction. Basel:Karger, 1990; 162–169.
70. Owen JJT. Lymphocyte interaction in the thymus. In: Burger MM, Sordat B, Zinkernagel RM, editors. Cell to Cell Interaction. Basel:Karger, 1990;152–161.
71. Berczi I, Nagy E, Baral E, Szentivanyi A. Steroid hormones. In: Berczi I, Szentivanyi A. Series editors. The immune-neuroendocrine circuitry. History and progress. Neuroimmune Biology, Vol. 3. Amsterdam:Elsevier, 2003;221–270.
72. Lewis UJ. Growth hormone - What is it and what does it do. Trends Endocrinol Metab 1992;3:117–121.
73. Berczi I. The immunology of prolactin. Sem Reprod Endocrinol 1992;10:196–219.
74. Berczi I. Pituitary hormones and immune function. Acta Paediatr 1997;(Suppl) 423: 70–75.
75. Szentivanyi A, Berczi I, Nyanteh H. Neurotransmission by conventional neurotransmitters.
76. Berczi I, Cosby H, Hunter T, Baragar F, McNeilly AS, Friesen HG. Decreased bioactivity of circulating prolactin in patients with rheumatoid arthritis. Brit J Rheum 1987;26:433–436.

Edited by I. Berczi and A. Szentivanyi

Antigen Presentation

ISTVAN BERCZI and ANDOR SZENTIVANYI

Department of Immunology, Faculty of Medicine, The University of Manitoba, Winnipeg, Manitoba R3E OW3, Canada; and Department of Internal Medicine, Faculty of Medicine, The University of South Florida, Tampa, Florida 33612, USA

ABSTRACT

Antigen presentation takes place by cell-to-cell interaction whereby a complex signaling process via cell surface adhesion molecules initiates the adaptive (antigen specific) immune responses. Without exception the macromolecular antigen undergoes proteolytic breakdown and peptide fractions (~9–24 residues) are presented by antigen presenting cells (APC) in association with surface MHC antigens. External antigens are engulfed by phagocytic mononuclear cells and are digested in endosomes and the peptide fragments are joining with MHC-II molecules in endocytic vesicles (endocytic pathway of processing) prior to expression on the cell surface. The APC of this pathway are resposible for the induction of the antibody response and delayed type hypersensitivity reactions. Endogenous antigens are processed in the cytoplasm, in enzyme-containing proteosomes (cytosolic pathway) and the peptides generated are associated with MHC-I in the endoplsmic reticulum, which in turn is expressed on the surface of all nucleated cells in the body. Cytotoxic T lymphocytes recognize MHC-I-peptide complexes on the cell surface and destroy infected and cancer cells. This antigen presenting system requires the digestion of the antigen after phagocytosis, which protects against extracellular and intracellular pathogens, it exposes hidden antigenic determinents, decreases the impact of mutations by employing short peptides and allows for self-non-self discrimination. Non-classical MHC antigens are also involved in antigen presentation and specialized surface molecules (CD1) mediate carbohydrate and lipid presentation. Heat shock proteins normally serve to eliminate dead cells from tissues but also form complexes with antigen which are taken up by APC through a specific receptor (CD91). Such complexes are taken up by macrophages and dendritic cells, digested and are presented by MHC-I. The dominant role of antigen presentation in the adaptive immune system illustrates the fundamental regulatory function of adherence signals in multi-cellular organisms.

1. INTRODUCTION

Adaptive immune reactions are mediated by lymphocyte clones that have specific receptors for the determinants (epitopes) of the antigen. Initially, the antigen was assumed to be the sole signal that instructed immunocytes to make antibodies by some sort of recognizing and copying mechanism [1]. Later it was realized that antibody specificity is genetically coded and that the antigen has to be digested and presented by specialized antigen presenting cells (APC) to T lymphocytes. Phagocytic mononuclear cells (monocyte-macrophages), B lymphocytes and some specialized cells (e.g., dendritic cells, Langerhans cells, Kuppfer cells) are "professional" APC that present antigen via major histocompatibility (MHC)-I and MHC-II. In the central nervous system the macrophage-related microglia and to a lesser extent, astrocytes, are involved in antigen presentation as well as in inflammatory responses. However, all nucleated cells are capable of antigen presentation by MHC-I, which is constitutively expressed by such cells. In addition, IFNγ is capable of inducing the expression of MHC-II in somatic cells, which enables them to present antigens by this pathway. Major histocompatibility antigens bind peptides during their intracellular biosynthesis and carry them to the surface of APC. In turn the digested (processed) antigenic peptide-MHC complexes are recognized by T cells as "altered self" [2,3,4,5,6].

The conventional view is that the T cell receptor recognizes the MHC antigen as "self" and the peptide as "non-self" in this process. This self-nonself recognition has evolved to provide protection against autoimmune reactions while foreign invaders are attacked. The requirement for T cells to recognize antigens in the context of MHC assures that infected and cancerous cells are specifically eliminated by killer cells. However, soluble antigen, which is not processed and not associated with MHC antigens, is not capable of triggering killer T lymphocytes. This mechanism prevents the exhaustion of T cells by viremia or by other antigen that is present in the circulation [2].

2. THE SIGNIFICANCE OF PROCESSING

The requirement for processing has several advantages: (i) It involves the phagocytosis and digestion of the antigen, which leads to inactivation of microbes, viruses and other pathogens; (ii) Digestion and presentation allow for the presentation of epitopes from the antigens of intracellular pathogens to T cells. This leads to the selective elimination of infected or cancerous cells. (iii) Epitopes that are hidden in the native molecule, viruses etc. may be recognized after processing and presentation; (iv) It reduces the chance of avoiding recognition by mutation of the antigen; (v) It allows for self-nonself discrimination [2].

3. ANTIGEN RECOGNITION BY T CELLS

The current consensus is that T lymphocytes recognize self-MHC antigens via their receptors (TCR). During the development of T cells, those clones that are triggered for an immune response by self-MHC are killed (selected out) in the thymus and also in the periphery. The cells that remain "recognize", but are not activated by self-MHC. The various clones will, however, be activated if the MHC molecule contains a peptide fragment for which the T cell clone shows *specificity.* The TCR of these cells recognizes the peptide antigen in the "context" of self-MHC as "altered self". This phenomenon is known as the *MHC-restriction* of T-cell activation. If the MHC component is missing, the T cell will not be activated. MHC-II presents antigens to CD4+ and MHC-I to CD8+ T cells. CD4 and CD8 are accessory recognition molecules on the surface of these T cells, which contribute to MHC recognition by the T lymphocyte. These rules apply to αβTCR [7].

MHC-II is involved in the presentation of foreign antigen-derived peptides. This is called the *endocytic pathway* as the antigens are ingested by endocytosis, digested and the resulting peptides presented by APC to T cells. MHC-I presents peptides by the *cytosolic pathway,* which are derived from the cytosol of the cell itself. While the endocytic pathway is present only in specialized APC, the cytosolic pathway is present in all cells of the body, as MHC-I is expressed by all somatic cells. This latter pathway serves the purpose of *immunological surveillance* against intracellular pathogens (e.g., viruses, some bacteria, protozoa and fungi) and cancer. CD8+ cytotoxic T lymphocytes patrol the entire body by penetrating the tissues via the high endothelial cells in capillaries and detect and kill abnormal cells expressing altered MHC-I antigens [2,3,4].

4. PEPTIDE BINDING BY MHC

In the MHC-I molecule the α1 and α2 domains combine to form a single peptide binding site supported by a β-pleated sheet floor containing eight strands bound by two α helices, one from α1 and the other from α2. β2-microglobulin makes contact with the immunoglobulin-like α3 domain and also with the β sheet floor of the α1α2 peptide-binding region. These structures show a series of pockets studding the peptide-binding groove. Occasionally the pockets extend deep between the floor and helical walls of the binding domain [8,9,10,11]. These pockets have now been designated from A to F. MHC polymorphism is primarily related to the diversification of their peptide-binding specificities. Different alleles are meant to bind distinct sets of peptides. Hydrogen bonds, hydrophobic interactions, and generic sequence independent interactions play major roles in peptide binding. MHC molecule polymorphism directly influences recognition by TCR. Distinct alleles of MHC molecules bind and present distinct peptides to T lymphocytes [9,10,12,13].

The interaction of peptides with MHC-I is very stable and the peptides bound are of fixed length of mostly 9 (8–10) amino acids. Predominant amino acids are found in comparable positions, which are called motif or anchor amino acids, that promote binding to a particular allele of MHC-I. Most peptides presented by MHC-I are derived from cytoplasmic or nuclear proteins. Physiologically stable class I molecules are actually trimers of the MHC heavy chain, β2-microglobulin and the peptide. These components synergize in forming a long-lived, properly-folded complex [15,16,17].

MHC-II bound peptides are longer and more heterogeneous in size, ranging from 12 to more than 24 residues. Anchor residues play a role in binding. These peptides are derived from proteins that have access to the endocytic pathway of antigen processing. Intracellular proteins access the class II pathway inefficiently compared with the class-I pathway, although exceptions have been reported. For class-II molecules the peptide is not needed for the maintenance of chain association at physiologic temperature and pH. MHC-II molecules bind peptides mostly via main chain atoms and not by the ends of the peptide [18,19,20,21].

5. SYNTHESIS OF MHC-I AND MHC-II

Class-I MHC and β2-microglobulin (β2M) are type I proteins and enter the endoplasmic reticulum (ER) via the signal recognition and transport apparatus. The MHC chain associates rapidly with an ER resident protein now termed *calnexin* [22]. This protein is Ca^{2+} sensitive and is associated with the *signal sequence dependent translocation apparatus* and interacts with various membrane and secreted proteins in ER. Calnexin's binding to the MHC-I chain involves both carbohydrate and protein structure recognition [23,24]. In the case of human MHC-I, this chaperone dissociates from the molecule upon binding β2M and another chaperone, *calreticulin,* will combine with the complex [25]. At this stage the MHC-β2M complex also associates with an MHC-encoded molecule called *tapasin* [26]. Once the MHC-β2M complex is associated with the antigenic peptide, calreticulin and tapasin dissociate from the complex, which exits the ER and moves through the Golgi network on to the cell surface by the default secretory pathway.

MHC class II α and β chains associate soon after synthesis in the presence of the *invariant chain (Ii),* which is a nonpolymorphic type II membrane glycoprotein [27,28]. Ii is a family of proteins. Ii forms noncovalently associated trimes in ER, and Class II αβ dimers are quickly associate with such trimers [29,30]. Non-class II associated Ii binds to calnexin, which dissociates upon the formation of class II-Ii oligomers [31,32]. The αβ Ii complex then moves out of the ER, through the Golgi complex , where N- and O- (for Ii) linkage to glycans takes place. Ii is removed from class II after egress from the trans-Golgi apparatus to the cell surface by sequential COOH terminal proteolytic cleavage [33].

Ii contains a segment encoded by exon 3, which is termed *class-II associated invariant chain-derived peptide* (CLIP). CLIP occupies the peptide binding groove of class II molecules in a manner that is indistinguishable from that of antigenic peptides. The CLIP region is essential for the transportation of class II from the ER [2].

6. PEPTIDE GENERATION AND MHC ASSOCIATION

The first possible site of peptide binding by MHC-I is in the ER lumen. There are genes in the MHC complex coding for *transporters associated with antigen processing* (TAP-1, -2) [34,35,36]. Embryonic stem cells lack TAP transcripts. These transporters show size and chemical selectivity. Transportation is optimal with peptides approximately 12 residues in length. TAP are translocated into the ER lumen during transportation. *Tapasin* provides linkage between the nascent MHC-β2M complex, the available peptide binding site and TAP. This is important for the efficient loading of the peptide onto MHC-I.

The antigenic peptide is generated in proteosomes, which is a large assembly of 16–20 components and is capable of degradation of proteins in the cytosol as part of housekeeping and of regulated functions of the cell [37,38]. Two proteosome subunits, LMP-1 and LMP-7, are encoded next to TAP-1 and TAP-2 in the MHC complex. The LMP proteins and the related MECL-1 are interferon-γ (IFN-γ) inducible [39,40,41,42]. The antigenic peptides presented by MHC-II are the result of endocytic proteolysis. The Ii (CLIP) chain occupies the peptide binding site of MHC-II and thus interferes with peptide binding in the ER lumen, preventing association with endogenous peptides [43,44]. Newly synthesized MHC-II molecules localize to unique endocytic vesicles distinct from typical endosomes where peptide loading takes place before expression at the cell surface [45,46,47,48].

Progressive proteolysis of the Ii chain by cathepsins within the endocytic pathway results in the *in situ* generation of MHC-II molecules with CLIP in their binding sites [49]. CLIP must be removed prior to binding the antigenic peptide. CLIP removal is mediated by a *class-II-like heterodimer*, called *DM*, whose component chains are encoded within the MHC complex [50,51,52]. There is evidence to indicate that initially MHC-II captures large protein fragments just after unfolding in the endosomes, which are then digested while the fragment bound in the MHC-II groove is protected from degradation [53]. In most cases class-II antigen presentation is mediated by newly synthesized αβ dimers, Ii and DM. However, there is also evidence for the internalization and recycling of mature class-II molecules through early endocytic compartments [54,55,56].

HLA-DM catalyzes the exchange and selection of ligands for MHC class II molecules in mature endosomal/lysosomal compartments. DM edits peptides in early endosomes, and by this influences presentation by recycling class II molecules. Maximal class II-restricted presentation of an albumin-derived peptide by recycling class II molecules was observed in cells lacking HLA-DM. DM editing of this epitope was observed in early endocytic compartments as revealed by inhibitors of early to late endosomal transport. Editing was tempered by coexpression of HLA-DO [57].

Peptide occupancy controls the lifetime of cell surface MHC-II molecules. B lymphocytes use their surface immunoglobulin to bind and internalize antigen [58,59]. Macrophages and dendritic cells have receptors that recognize common features of pathogen antigens, such as mannose residues [60]. Non-B cells can also use antibodies to capture antigen via FcγRIII that triggers rapid endocytosis in macrophages [61]. There are also examples of presentation of "endogenous" antigens by MHC-II molecules [62].

7. CYTOKINES AND ANTIGEN PRESENTATION

It is generally held that different subsets of T cells and cytokines are involved in the cellular and humoral immune response. Th1 cells, which evoke cell-mediated immunity and phagocyte-dependent inflammation, produce interferon (IFN)-gamma, interleukin (IL)-2 and tumor necrosis factor (TNF)-beta. Th2 cells, which evoke strong antibody responses (including those of the IgE class) and eosinophil accumulation, but inhibit several functions of phagocytic cells, produce IL-4, IL-5, IL-6, IL-9, IL-10, and IL-13 [63].

IL-15 has been implicated in the differentiation of Langerhans cells from monocytes [64]. IL-13 is also an important regulatory factor involved in both early innate and late adaptive responses [65]. IL-10-treated DCs and, to a lesser extent, hydrocortisone (HC)-treated DCs showed a decreased expression of MHC-II molecules, the costimulatory molecule CD86, the DC-specific marker CD83 and IL-12 secretion is markedly reduced. As a result, T-cell proliferation was reduced after stimulation with either IL-10- or HC-treated DCs. However, IL-10 inhibited the production of both TH1 and TH2 cytokines, whereas HC inhibited the production of IFNγ, but increased the release of IL-4 and there was no change in IL-5. Both effects were long-lasting [66]. The expression of MHC-I is enhanced by IFNγ. In general, IFNγ slows the maturation rates of class I complexes and causes a prolonged retention of molecules in the ER, because it regulates the expression of ER-residing proteins during maturation. Consequently, IFNγ induces more rigorous ER quality control [67].

Chemokines also participate in immune activation. Thus, TH1 reactions depend upon IFNγ-induced CXC chemokines: IFN- inducible protein (IP)-10, IFN-inducible T cell-alpha chemoattractant (iTAC) and monokine induced by IFNγ (MiG), which bind to chemokine receptor CXCR3. TH2 lymphocytes express the chemokine receptors CCR4 and CCR8 for which thymus- and activation-regulated chemokine (TARC), macrophage-derived chemokine (MDC) and I-309 function as ligands, respectively [68].

8. ANTIGEN PRESENTATION BY "NONCLASSICAL" MOLECULES

It is now emerging that in addition to the "classical" MHC-I and MHC–II, "non-classical" MHC molecules, such as Qa-2, also present antigen in both innate and adaptive immune responses, as well as participate in embryonic development [69]. In the eye B cells are necessary for *tolerogenic antigen presentation*. Anterior chamber associated B cells do not directly suppress the expression of delayed-type hypersensitivity. Instead, antigen-specific regulatory T cells are induced by B cells that require the identity of histocompatibility antigens at the TL/Qa-1 region [70].

Similarly, CD1 molecules, a family of cell surface-associated glycoproteins, are now recognized as having a role in antigen presentation. These glycoproteins are distinct from, yet have some similarities to classical major histocompatibility complex class I and class II molecules. They can present nonprotein antigens to certain subpopulations of T cells [71]. CD1d constitutes a carbohydrate antigen processing system enabling T cells to recognize processed fragments of complex glycolipids [72]. CD1d molecules control the function of natural T cells, which are an early source of cytokines that stimulate type 1 or type 2 differentiation of helper T lymphocytes. CD1d is postulated to sense alterations in cellular lipid content by virtue of its affinity for such ligands. The presentation of altered-self glycolipid, presumably after infectious assault of antigen-presenting cells, activates natural T cells, which promptly release pro-inflammatory and anti-inflammatory cytokines and jump-start the immune system [73].

Glycolipid-specific, CD1a-, b- and c-dependent cytotoxic T cells have recently been shown to be involved in the host immunity against tuberculosis. These CD1 molecules present mycobacterial glycolipids from different intracellular sites in the infected cell. CD1d-dependent natural killer T cells promptly produce cytokines and perform regulatory activities [74].

Human histocompatibility leukocyte antigen (HLA)-G is an antigen-presenting molecule and *down-modulates* CD8(+) and CD4(+) T-cell responsiveness. HLA-G modulates innate immunity by interacting with immunoglobulin-like receptors and by regulating HLA-E expression and its interaction with CD94/NKG2 receptors [75].

9. BIOLOGICAL SIGNIFICANCE OF ANTIGENIC PEPTIDE PRESENTATION

It is generally accepted that MHC presenting self-peptides are involved in the selection of the T cell repertoire in the thymus. In the periphery T cell regulatory and effector responses are dependent on MHC-peptide presentation. Some pathogens mimic self-antigens and thus escape immune attack. Such mimicry may also lead to the activation of autoimmune reactions. Continuous mutation of the antigens is another way by which pathogens escape immune attack. These mutations lead to epitope substitution and give advantage to the pathogen (e.g. in the case of HIV). Some mutations may lead to the loss of MHC binding of the peptide. Many pathogens have the ability to produce molecules that modulate the immune response [76]. Viruses have developed many ways to escape immune surveillance and to downregulate the immune response in their hosts. All the viruses that induce generalized immunosuppression appear to do so by hindering antigen presentation to T cells and/or hematopoiesis [77]. The diversity of MHC and of lymphocyte antigen receptors provides protection against the mutational evasion of antigen presentation.

Intracellular pathogens produce proteins from which peptides are derived that enter the ER and are presented by MHC-I molecules. This presentation stimulates CD8+ T cells that differentiate into cytotoxic T lymphocytes capable of killing the infected target cells. CD8+ cells also produce cytokines, such as TNF-β and IFN-γ. Antibody production is not stimulated by MHC-I-peptide complexes. Cytotoxic T cells of the CD4+ subset also exist and such cells contribute to host defence. However, their major function is related to the termination of immune reactions by inducing apoptosis of excess lymphocytes via the Fas-FasL pathway. Cytokines provide protection against pathogens (e.g. by inducing NO) but pathogens also use cytokines in order to sway the immune response to their advantage [2].

Extracellular debris that may contain foreign material is taken up endocytically by MHC-II bearing cells, processed and presented to CD4+ effector/regulatory cells, including those helping antibody production by B cells. The antibody response is especially useful for the neutralization of exotoxins. Antigen specific B cells are capable of recognizing the *conformation* of the antigen and form antibodies to conformational epitopes. Such antibodies are capable of recognizing proteins in their native folded state, which is of major significance from the point of view of fighting pathogenic microorganism.

Pre-existing antibodies facilitate the uptake of antigen by macrophages and dendritic cells via Fcγ receptors that promote endocytosis. Passively acquired maternal antibodies can compete for antigen and prevent antigen presentation by B lymphocytes [2]. The high-affinity IgG receptor (CD64 or FcγRI) is constitutively expressed exclusively on professional APC. When antigen is targeted specifically to FcγRI, presentation is markedly enhanced. FcγRI-targeted antigens converge upon a class I processing pathway. Such targeting can lead to Ag-specific activation of cytotoxic T cells [78].

B lymphocytes play an essential role as APC also for T cell expansion in lymph nodes and for systemic T cell responses to low concentrations of antigen [79]. B-cell interaction with antigens that are immobilized on the surface of a target cell leads to binding (the formation of a synapse) and the acquisition of membrane-integral antigens from the target. B-cell antigen receptor accumulates at the synapse, segregated from the CD45 co-receptor, which is excluded. B cells concentrate antigen by this mechanism and thereby potentiate antigen processing and presentation to T cells with high efficacy [80]. The indication is that B cell-deficient mice have a diminished IL-2 production and do not develop normal frequencies of *memory cells* after immunization. The transfer of B cells to such mice restored memory cell development. Antigen presentation was not essential for this function, since B cells activated *in vitro* with irrelevant antigen also restored the frequencies of memory cells [81].

The stimulation of antibody production by B-cells was observed in the presence of antigen-pulsed macrophages. Similar levels of stimulation were detected following depletion of Thy 1.2(+) cells from spleen cell preparations. Stimulation was inhibited by antigen specific IgG antibodies. This B cell-macrophage interaction was mediated by MHC-II [82]. A new B7 family member, B7-DC, whose expression is highly restricted to DCs, was identified. B7-DC binds PD-1, a receptor for B7-H1/PD-L1. B7-DC costimulates T cell proliferation more efficiently than does B7.1 and induces IFNγ but not IL-4 or IL-10 production by isolated naive T cells. These properties of B7-DC may account for the ability of DC to initiate potent T helper cell type 1 responses [83].

Necrotic but not apoptotic cells release heat shock protein (HSP) gp96, calreticulin, hsp90 and hsp70. HSP stimulates cytokine secretion by macrophages and induces the expression of antigen-presenting and co-stimulatory molecules on the DC. HSP gp96 and hsp70 act differentially, and each induces some but not all molecules. HSP interacts with APC through the highly conserved NFκB pathway. Normally HSP are intracellular, abundant and soluble. Their appearance in the *extra-cellular milieu* and the consequent activation of APC represents an effective immune mechanism for the handling of cell death. HSP are conserved from bacteria to mammals. Therefore, the ability of HSP to activate APC provides a unified mechanism for response to internal stimuli as well as to microbes [84].

Complexes of the heat shock protein, gp96, and antigenic peptides are taken up by antigen-presenting cells and presented by MHC class I molecules. The uptake of gp96 had been postulated to occur through a receptor, identified recently as CD91. Complexes of peptides with heat shock proteins hsp90, calreticulin, and hsp70 are also taken up by macrophages and dendritic cells and presented by MHC class I molecules. All heat shock proteins utilize the CD91 receptor. Processing of gp96-chaperoned peptides requires proteosomes and the transporters associated with the *classical endogenous antigen presentation pathway* [85].

The question how CD4+ T cells provide help to CD8+ T cell responses is not resolved. The most compelling model, which is based on *in vivo* observations, is that a single APC, most likely a dendritic cell, stimulates both CD4+ and CD8+ T cells by displaying antigenic epitopes associated by MHC-II and MHC-I respectively. The proximity allows for the interaction of CD4+ and CD8+ cells via cytokines, which leads to the initiation of cell mediated immunity [86].

There are indications that immune responses do not show exquisite specificity as it is generally assumed. Although singular TCR can discriminate as exquisitely among antigens as the most specific antibodies but also exhibit "degeneracy": i.e., it can react with many disparate antigens (peptide-MHC complexes). An explanation for this duality (specificity and degeneracy) can be found in: (i) The powerful amplifying signal transduction cascades that allow a T cell to respond to the stable engagement of very few TCR molecules, initially perhaps only one or two out of around 100,000 per cell, and (ii) The inverse relationship between TCR affinity for epitopes and epitope density. B cells also exhibit degeneracy, as well as specificity [87].

10. CONCLUSIONS

The recent observations discussed here emphasizes further the importance of antigen presentation as a fundamental signal in immune function. It is now indicated that this signal is not only capable of initiating a response, but also determines the type of the response elicited. The concept of *"tolerogenic" antigen presentation* is novel and exciting, as are the indications that there are specialized APC molecules that would efficiently present *carbohydrates* or *lipids*. Another important finding is that there are APC molecules with capabilities of activating both the *innate* and the *adaptive immune system* and that natural immune T cells instantaneously produce cytokines in response to such APC that serve to *"jump start" the specific immune system*. Finally, the *heat shock protein system* serves as an ultimate example to illustrate what the immune system is about. The HSP system serves to eliminate dead cells from tissues under physiological or pathophysiological conditions and it fights microorganisms that may invade the host.

Clearly our advanced understanding of antigen presentation reconfirms the dominant nature of this signal for immune function. The original extrapolations from antigen presentation that adherence signals are dominant in cell biology is well illustrated by these recent developments. The general validity of these signals as antigen and cell specific adherence signals that initiate immune reactions have been substantiated by more examples than ever before. The response to these signals, which seem now qualitatively different, may be immunity, tolerance, anergy and even lymphocyte death (apoptosis). Apparently, the signaling is dependent on the physiological (homeostatic) and pathological (allostatic) requirements in higher animals and in man [88,89]. Indeed, it is now emerging that APC adherence signals have major roles not only in governing the protection of the host, but also in the physiological and pathophysiological regulation of bodily functions.

REFERENCES

1. Silverstein AM. The history of immunology. In: Paul WE, editor. Fundamental Immunology, 4th edition, Philadelphia:Lippincott-Raven, 1999;19–35.
2. Germain RN. Antigen processing and presentation, 4th edition. In: Fundamental Immunology, Paul WE, editor. Philadelphia: Lippincott-Raven, 1999;287–340.
3. Cresswell P. Assembly, transport and function of MHC class II molecules. Ann Rev Immunol 1994;12:259–293.
4. York IA, Rock KL. Antigen processing and presentation by the class 1 major histocompatibility complex. Ann Rev Immunol 1996;14:369–396.
5. Aloisi F. Immune function of microglia. Glia 2001;36:165–179.
6. Gebicke-Haerter PJ. Microglia in neurodegeneration: molecular aspects. Microsc Res Tech 2001;54:47–58.
7. Benoist C, Mathis D. T-lymphocyte differentiation and biology. In: Paul WE, editor. Fundamental Immunology, 4th edition. Philadelphia: Lippincott-Raven, 1999;367–409.
8. Garrett TP, Saper MA, Bjorkman PJ, Strominger JL, Wiley DC. Specificity pockets for the side chains of peptide antigens in HLA-Aw68. Nature 1989; 42:692–696.
9. Fremont DH, Matsumura M, Stura EA, Peterson PA, Wilson IA. Crystal structures of two vital peptides in complex with murine MHC class I H-2K^b. Science 1992;257:919–927.
10. Matsumura M, Fremont DH, Peterson PA, Wilson IA. Emerging principles for the recognition of peptide antigens by MHC class I molecules. Science 1992;257:927–934.
11. Madden DR, Garboczi DN, Wiley DC. The antigen identity of peptide-MHC complexes. A comparison of the conformation of five viral peptides presented by HLA-A2. Cell 1993;75:693–708.
12. Fischer-Lindahl K, Hermel E, Loveland BE, Wang C-R. Maternally transmitted antigen of mice. A model transplantation antigen. Ann Rev Immunol 1991;9:351–372.
13. Beckman EM, Porcelli SA, Morita CT, Behar SM, Furlong ST, Brenner MB. Recognition of a lipid antigen by CD1-restricted $\alpha\beta^-$ T cells. Nature 1994;372:691–694.
14. Sieling PA, Chatterjee D, Porcelli, SA, et al. CD-1 restricted T cell recognition of microbial lipoglycan antigens. Science 1995;269:227–230.
15. Hunt DF, Henderson RA, Shabanowitz J, et al. Characterization of peptides bound to the class I MHC molecule HLA-A2.1 by mass spectrometry. Science 1992;255:1261–1263.
16. Boehncke WH, Takeshita T, Pendleton CD, et al. The importance of dominant negative effects of amino acid side chain substitution in peptide-MHC molecule interactions and T cell recognition. J Immunol 1993;150:331–341.
17. Ruppert J, Sidney J, Celis E, Kubo RT, Grey HM, Sette A. Prominent role of secondary anchor residues in peptide binding to HLA-A2.1 molecules. Cell 1993;74:929–937.
18. Falk K, Rotzschke O, Stevanovic S, Jung G, Rammensee HG. Pool sequencing of natural HLA-DR, DQ, and DP ligands reveals detailed peptide motifs, constraints of processing, and general rules. Immunogenetics 1994;39:230–242.
19. Nelson CA, Vidavsky I, Viner NJ, Gross ML, Unanue ER. Amino-terminal trimming of peptides for presentation on major histocompatibility comples class II molecules. Proc Natl Acad Sci USA 1997;94:628–633.
20. Hammer J, Bono E, Gallazzi F, Belunis C, Nagy Z, Sinigaglia F. Precise prediction of major histocompatibility complex class II-peptide interaction based on peptide side chain scanning. J Exp Med 1994;180:2353–2358.
21. Alexander J, Sidney J, Southwood S, et al. Development of high potency universal DR-

restricted helper epitopes by modification of high affinity DR-blocking peptides. Immunity 1994;1:751–761.
22. Hammond C, Helenius A. A chaperone with a sweet tooth. Curr Biol 1993;3:884–886.
23. Zhang Q, Tector M, Salter RD. Calnexin recognizes carbohydrate and protein determinants of class I major histocompatibility complex molecules. J Biol Chem 1995;270:3944–3948.
24. Ware FE, Vassilakos A, Peterson PA, Jackson MR, Lehrman MA, Williams DB. The molecular chaperone, calnexin binds GlcMan9GlcNAc2 oligosaccharide as an initial step in recognizing unfolded glycoproteins. J Biol Chem 1995;270:4697–4704.
25. Sadasivan B, Lehner PJ, Ortmann B, Spies T, Cresswell P. Roles for calreticulin and a novel glycoprotein, tapasin, in the interaction of MHC class I molecules with TAP. Immunity 1996;5:103–114.
26. Ortmann B, Copeman J, Lehner PJ, et al. A critical role of tapasin in the assembly and function of MHC class I-TAP complexes. Science 1997;277:1306–1309.
27. Singer PA, Lauer W, Dembic Z, et al. Structure of the murine Ia associated invariant (Ii) chain as deduced from a CDNA clone. EMBO J 1984;3:873–877.
28. Strubin M, Mach B, Long EO. The complete sequence of the mRNA for the HLA-DR-associated invariant chain reveals a polypeptide with an unusual transmembrane polarity. EMBO J 1984;3:869–872.
29. Marks MS, Bloom JS, Cresswell P. Invariant chain trimers are sequestered in the rough endoplasmic reticulum in the absence of association with HLA class II antigens. J Cell Biol 1990;111:839–855.
30. Roche PA, Marks MS, Cresswell P. Formation of a nine subunit complex by HLA class II glycoproteins and the invariant chain. Nature 1991;354:392–394.
31. Anderson KS, Cresswell P. A role for calnexin (IP90) in the assembly of class II MHC molecules. EMBO J 1994;13;675–682.
32. Romagnoli P, Germain RN. Inhibition of invariant chain (Ii)-calnexin interaction results in enhanced degradation of Ii but does not prevent the assembly of αβ complexes. J Exp Med 1995;182:2927–2036.
33. Blum JS, Cresswell P. Role of intracellular proteases in the processing and transport of class II HLA antigen. Proc Nat Acad Sci USA 1988;85:3975–3979.
34. Spies T, DeMars R. Restored expression of major histocompatibility class I molecules by gene transfer of a putative peptide transporter. Nature 1991;351:323–324.
35. Powis SJ, Townsend AR, Deverson EV, Bastin J, Butcher GW, Howard JC. Restoration of antigen presentation to the mutant cell line RMA-S by an MHC-linked transporter. Nature 1991;354:528–531.
36. Spies T, Cerundolo V, Colonna M, Cresswell P, Townsend A, DeMars R. Presentation of viral antigen by MHC class I molecules is dependent on a putative peptide transporter heterodimer. Nature 1992;355:644–646.
37. Peters JM. Proteosomes: Protein degradation machines of the cell. Trends Biochem Sci 1994;19:377–382.
38. Goldberg AL, Rock KL. Proteosomes, proteolysis and antigen presentation. Nature 1992;357:375–379.
39. Belich MP, Glynne RJ, Senger G, Sheer D, Trowsdale J. Proteosome components with reciprocal expression to that of the MHC-encoded LMP proteins. Curr Biol 1994;4: 769–776.
40. Fruh K, Gossen M, Wang K, Bujard H, Peterson PA, Yang Y. Displacement of housekeeping

proteosome subunits by MHC-encoded LMPs: A newly discovered mechanism for modulating the multicatalytic proteinase complex. EMBO J 1994;13:3236–3244.
41. Groettrup M, Kraft R, Kostka S, Standers S, Stohwasser R, Kloetzel PM. A third interferon-gamma-induced subunit exchange in 20S proteosome. Eur J Immunol 1996;26:863–869.
42. Nandi D, Jiang H, Monaco JJ. Identification of MECL-1 (LMP-10) as the third interferon-gamma-inducible proteosome subunit. J Immunol 1996;156:2361–22364.
43. Ghosh P, Amaya M, Mellins E, Wiley DC. The structure of an intermediate in class II MHC maturation: CLIP bound to HLA-DR3. Nature 1995;378:457–462.
44. Morkowski S, Goldrath AW, Eastman S, et al. T cell recognition of major histocompatibility complex class II complexes with invariant chain processing intermediates . J Exp Med 1995;182:1403–1413.
45. Amigorena S, Drake JR, Webster P, Mellman I. Transient accumulation of new class II MHC molecules in a novel endocytic compartment in B lymphocytes. Nature 1994;369: 113–120.
46. Qiu Y, Xu X, Wandinger-Ness A, Dalke DP, Pierce SK. Separation of subcellular compartments containing distinct functional forms of MHC class II. J Cell Biol 1994;125: 595–605.
47. Tulp A, Verwoerd D, Dobberstein B, Ploegh HL, Pieters J. Isolation and characterization of the intracellular MHC class II compartment. Nature 1994;369:120–126.
48. West MA, Lucocq JM, Watts C. Antigen processing and class II MHC peptide-loading in human B-lymphoblastoid cells. Nature 1994;369:147–151.
49. Avva RR, Cresswell P. In vivo and in vitro formation and dissociation of HLA-DR complexes with invariant chain-derived peptides. Immunity 1994;1:763–774.
50. Morris P, Shaman J, Attaya M, et al. An essential role for HLA-DM in antigen presentation by class II major histocompatibility molecules. Nature 1994; 368:551–554.
51. Fling SP, Arp B, Pious D. HLA-DMA and –DMB genes are both required for MHC class II/peptide complex formation in antigen-presenting cells. Nature 1994;368:554–558.
52. Denzin LK, Robbins NF, Carboy-Newcomb C, Cresswell P. Assembly and intracellular transport of HLA-DM and correction of the class II antigen-processing defect in T2 cells. Immunity 1994;1:595–606.
53. Deng H, Apple R, Clare-Saltzer M, et al. Determinant capture as a possible mechanism of protection afforded by major histocompatibility complex class II molecules in autoimmune disease. J Exp Med 1993;178:1675–1680.
54. Roche PA, Teleski CL, Stang E, Bakke O, Long EO. Cell surface HLA-DR invariant chain complexes are targeted to endosomes by rapid internalization. Proc Nat Acad Sci USA 1993;90:8581–8585.
55. Reid P, Watts C. Cycling of cell surface MHC glycoproteins trough primaquine-sensitive intracellular compartments. Nature 1990;346:655–657.
56. Weber DA, Buck LB, Delohery TM, Agostino N, Pernis B. Class II MHC molecules are spontaneously internalized in acidic endosomes by activated B cells. J Mol Cell Immunol 1990;4:255–266.
57. Pathak SS, Lich JD, Blum JS. Cutting edge: editing of recycling class II: peptide complexes by HLA-DM. J Immunol 2001;167:632–635.
58. Lanzavecchia A. Antigen specific interaction between T and B cells. Nature 1985;314: 537–539.
59. Rock KL, Benacerraf B, Abbas AK. Antigen presentation by hapten-specific B lymphocytes. 1. Role of surface immunoglobulin receptors. J Exp Med 1984;160:1102–

1113.
60. Jiang W, Swiggard WJ, Heufler C, et al. The receptor DEC-205 expressed by dendritic cells and thymic epithelial cells is involved in antigen processing. Nature 1995;375:151–155.
61. Amigorena S, Salamero J, Davoust J, Fridman WH, Bonnerot C. Tyrosine containing motif that transduces cell activation signals also determines internalization and antigen presentation via type-III receptors for IgG. Nature 1992;358:337–341.
62. Long EO. Antigen processing for presentation to CD4+ T cells. New Biol 1992;4:274–282.
63. Romagnani S. T-cell subsets (Th1 versus Th2). Ann Allergy Asthma Immunol 2000;85: 9–18.
64. Mohamadzadeh M, Berard F, Essert G, Chalouni C, Pulendran B, Davoust J, Bridges G, Palucka AK, Banchereau J. Interleukin 15 skews monocyte differentiation into dendritic cells with features of Langerhans cells. J Exp Med 2001;194:1013–20.
65. Brubaker JO, Montaner LJ. Role of interleukin-13 in innate and adaptive immunity. Cell Mol Biol 2001;47:637–51.
66. Bellinghausen I, Brand U, Steinbrink K, Enk AH, Knop J, Saloga J. Inhibition of human allergic T-cell responses by IL-10-treated dendritic cells: differences from hydrocortisone-treated dendritic cells. J Allergy Clin Immunol 2001;108:242–249.
67. Fromm SV, Ehrlich R. IFN-gamma affects both the stability and the intracellular transport of class I MHC complexes. J Interferon Cytokine Res 2001;21:199–208.
68. Kaplan AP. Chemokines, chemokine receptors and allergy. Int Arch Allergy Immunol 2001;124:423–31.
69. He X, Tabaczewski P, Ho J, Stroynowski I, Garcia KC. Promiscuous antigen presentation by the nonclassical MHC Ib Qa-2 is enabled by a shallow, hydrophobic groove and self-stabilized peptide conformation. Structure 2001;9:1213–1224.
70. D'Orazio TJ, Mayhew E, Niederkorn JY. Ocular immune privilege promoted by the presentation of peptide on tolerogenic B cells in the spleen. II. Evidence for presentation by Qa-1. J Immunol 2001;166:26–32.
71. Rhind SM. CD1--the pathology perspective. Vet Pathol 2001;38:611–9.
72. Prigozy TI, Naidenko O, Qasba P, Elewaut D, Brossay L, Khurana A, Natori T, Koezuka Y, Kulkarni A, Kronenberg M. Glycolipid antigen processing for presentation by CD1d molecules. Science 2001;291:664–7.
73. Joyce S. CD1d and natural T cells: how their properties jump-start the immune system. Cell Mol Life Sci 2001;58:442–469.
74. Schaible UE, Kaufmann SH. CD1 and CD1-restricted T cells in infections with intracellular bacteria. Trends Microbiol 2000;8:419–25.
75. Le Bouteiller P, Solier C. Is antigen presentation the primary function of HLA-G? Microbes Infect 2001;3:323–332.
76. Spriggs MK. One step ahead of the game: Viral immunomodulatory molecules. Annu Rev Immunol 1996;14:11101–130.
77. Naniche D, Oldstone MB. Generalized immunosuppression: how viruses undermine the immune response. Cell Mol Life Sci 2000;57:1399–407.
78. Guyre CA, Barreda ME, Swink SL, Fanger MW. Colocalization of Fc gamma RI-targeted antigen with class I MHC: implications for antigen processing. J Immunol 2001;166: 2469–78.
79. Rivera A, Chen CC, Ron N, Dougherty JP, Ron Y. Role of B cells as antigen-presenting cells in vivo revisited: antigen-specific B cells are essential for T cell expansion in lymph

nodes and for systemic T cell responses to low antigen concentrations. Int Immunol 2001;13:1583–1593.
80. Batista FD, Iber D, Neuberger MS. B cells acquire antigen from target cells after synapse formation. Nature 2001;411:489–94.
81. Linton PJ, Harbertson J, Bradley LM. A critical role for B cells in the development of memory CD4 cells. J Immunol 2000;165:5558–5565.
82. Weaver DJ Jr, Voss EW Jr. A novel cellular interaction involving antigen-pulsed macrophage and antigen-specific B-lymphocytes. Mol Immunol 2000;7:311–320.
83. Tseng SY, Otsuji M, Gorski K, Huang X, Slansky JE, Pai SI, Shalabi A, Shin T, Pardoll DM, Tsuchiya H. B7-DC, a new dendritic cell molecule with potent costimulatory properties for T cells. J Exp Med 2001;193:839–846.
84. Basu S, Binder RJ, Suto R, Anderson KM, Srivastava PK. Necrotic but not apoptotic cell death releases heat shock proteins, which deliver a partial maturation signal to dendritic cells and activate the NF-kappa B pathway. Int Immunol 2000;12:1539–1546.
85. Basu S, Binder RJ, Ramalingam T, Srivastava PK. CD91 is a common receptor for heat shock proteins gp96, hsp90, hsp70, and calreticulin. Immunity 2001;14:303–13.
86. Mitchison NA. Linked help in the cytotoxic T-cell response revealed by adoptive transfer. Transplant Proc 1983;15:2121–2124.
87. Eisen HN. Specificity and degeneracy in antigen recognition: yin and yang in the immune system. Annu Rev Immunol 2001;19:1–21.
88. Berczi I. The role of the growth and lactogenic hormone family in immune function. Neuroimmunomodulation 1994;1:201–216.
89. Berczi I. Neuroimmune Biology - An introduction. In: Berczi I, Gorczyski RM, editors, Berczi I, Szentivanyi A, series editors, Neuroimmune Biology, Vol.1, New Foundation of Biology, Amsterdam: Elsevier Science B.V. 2001;3–45.

Immune Reactions

ISTVAN BERCZI and ANDOR SZENTIVANYI

Department of Immunology, Faculty of Medicine, The University of Manitoba, Winnipeg, Manitoba R3E OW3, Canada; and Department of Internal Medicine, Faculty of Medicine, The University of South Florida, Tampa, Florida 33612, USA

ABSTRACT

Under physiological conditions the immune system provides continuous defense against infectious agents and cancer, and is part of the homeostatic neuroimmune regulatory circuit that coordinates the normal function of the entire organism. The immune system is a local and mobile defense and regulatory system that has enormous capacity to deliver defense and regulatory molecules to sites that are in need. Every organ and tissue possesses stromal lymphoid elements that intervene locally to control autoimmune reactions, inflammation, and in general, participate in the physiological processes, as it is becoming obvious from recent investigations. Adaptive cell mediated and humoral immunity and immunological memory are reactions exerted by T and B lymphocytes in concert with members of the leukocyte series. Innate host defense relies on non-immune mechanisms and on specialized immune cells, such as natural killer cells, γδT lymphocytes and $CD5^+$ B cells. The alternately activated complement system and of T and B lymphocytes are also part of the natural immune system. The natural immune system relies on germ-line coded receptors that recognizes evolutionarily highly preserved homologous epitopes (homotopes) on microbes and also on self components. Finally, immunocytes are present in various tissues and organs and participate in the normal physiological regulation in addition to providing defense.

Cell-to-cell interactions by the antigen receptor and of MHC molecules and by other adhesion molecules are fundamental to immune activation as well as to stromal regulation. Cytokines and complement fine-tune this regulatory system. In addition, the immune system interacts with the neural elements and mediators, parenchymal and stromal cells as part of the local gerulatory circuits that govern the organ/tissue under physiological circumstances.

1. INTRODUCTION

According to current consensus immune reactions are classified into *adaptive reactions* and *innate or natural immune responses*. Adaptive responses involve the specific recognition of antigenic epitopes by T lymphocytes (cell mediated immunity), or by B cells (antibody response). It is the antigenic signal that triggers the proliferation of epitope-specific lymphocyte clones. This results in cell proliferation and differentiation into effector T lymphocytes or into antibody forming plasma cells. In contrast, the natural immune system recognizes highly conserved crossreactive (homologous) epitopes, or *homotopes,* with the aid of germ-line gene coded receptors that are constant. Therefore, in contrast with the adaptive system, which uses diversified receptors, the natural immune system is hard wired in terms of antigen recognition [1].

The third category of immune activity, which is just emerging, may be defined as *regulatory responses.* The immune system has an elaborate internal regulatory network itself in which regulatory cells, immunooglobulins, cytokines and chemokines play an important roles. In addition, cells of the immune system are present in all tissues and organs as *stromal lymphoid elements* and exert local protection as well as play important roles in functional regulation both under physiological and pathophysiological conditions.

2. CELL-MEDIATED IMMUNITY

In early experiments it was established that this type of immunity could not be transferred from immune to normal animals by antibodies, but rather, the transfer of lymphocytes was required. Hence the designation of *cell-mediated immunity*. We may distinguish two basic forms of CMI. The first form is *delayed-type hypersensitivity reactions* (DTH). This reaction is termed delayed, because it will appear 24–48 hours after the deposition of the antigen in the skin. This is in contrast with immediate type reactions mediated by the release of inflammatory mediators by mast cells, which appears minutes after antigenic challane. The second form is immune reactions mediated CD8+ *cytotoxic T lymphocytes* (CTL).

2.1. Delayed type hypersensitivity

DTH in essence is the recruitment of effector αβ-T cells out of vessels into sites of antigen deposition in peripheral tissues. These T cells then are activated by APC and produce pro-inflammatory cytokines. Recent studies show that three other lymphocyte subsets are involved: Contact sensitivity (CS)-inducing NK T cells, CS-initiating B-1 cells, and CS-assisting γδ-T cells. αβ-T cells cannot be recruited into the tissues without the local action of IgM antibodies produced by B-1 cells within 1 day post-immunization. The IgM complexes with the challenge antigen within 1–2 hours in the elicitation phase to locally activate complement, which leads to vascular activation required for T cell recruitment. CS-inducing NK T cells activate the B-1 cells by producing IL-4 1 hour after immunization. γδ-T cells assist the local inflammatory function of the recruited CS-effector αβ-Tcells cells [2]. $CD8^+$ T cells are now known to mediate DTH responses in allergic contact dermatitis, drug eruptions, asthma, and autoimmune diseases. This capability of $CD8^+$ cytotoxic T cells was previously poorly recognized, but there is now considerable evidence that these diseases may be mediated by $CD8^+$ DTH. The difference between $CD8^+$ T cells and $CD4^+$ T cells mediating DTH relates to the molecular mechanisms by which antigens are presented to the T cells [3].

Antibodies against CD44 epitopes arrest Langerhans cells in the epidermis, prevent the binding of activated lymphocytes and dendritic cells to the T cell zones of lymph nodes, and severely inhibit their capacity to induce a DTH reaction to a skin hapten *in vivo* [4]. Cutaneous T cell-attracting chemokine (CTACK) and CCR4 can both support homing of T cells to skin, and that either one or the other is required for lymphocyte recruitment in cutaneous delayed type hypersensitivity [5]

The roles of IL-1α and IL-1β in contact hypersensitivity to 2,4,6-trinitrochlorobenzene was analyzed in gene-targeted mice. Ear swelling was suppressed in IL-1α (-/-) but not in IL-1β (-/-) mice. Only IL-1α but not IL-1β, producing epidermal antigen-presenting cells could prime allergen-specific T cells [6].

Macrophage migration inhibitory factor (MIF) is an important mediator of several diseases, including gram-negative septic shock and delayed-type hypersensitivity reactions, and modulates host macrophage and T and B cell responses. MIF production is induced rather than suppressed by glucocorticoids, and it overrides the immunosuppressive effects of glucocorticoids [7]. MHC-II-deficient skin grafts were placed on allogeneic severe combined immunodeficient (SCID) recipients adoptively transferred with naive $CD4^+$ T cells. The recipients were treated with neutralizing anti-MIF monoclonal antibody or isotype control IgG. In this model graft rejection can only occur indirectly as the graft cells lack MHC-II for recognition by the recipient $CD4^+$ T cells. The blockade of MIF inhibited indirect $CD4^+$ cell-mediated skin graft destruction and alloreactive DTH but did not prevent T cell priming or IFNγ release by primed T cells [8]. The introduction of adenoviral interleukin-10 (ad-vIL-10) gene into one joint of an animal with polyarthritis suppressed disease in both treated and untreated joints (the "contralateral effect"). Similarly, the delivery of the vIL-10 gene suppressed DTH reactions in injected and contralateral paws. T lymphocytes from immunized mice injected with ad-vIL-10 were unable to transfer the DTH response, but were not inhibitory. Peritoneal exudate cells from i.p. injected ad-vIL-10 mice inhibited DTH reactions in recipient mice, but only when the donor mice had been sensitized to the antigen used to initiate the DTH response. Dendritic cells (DC) behaved similarly. Bone-marrow-derived DCs cultured ex vivo with antigen and ad-vIL-10 or recombinant mouse IL-10 also suppressed DTH reactions after adoptive transfer [9].

DTH leads to pathologic responses, such as granulomatous inflammation, calcification, caseation, necrosis, and cavity formation. Granulomas usually form after a DTH response to a persisting nondegradable product. DTH may contribute to both the clearance of the etiologic agent and the healing of tissue damage [10] .

Acute stress administered immediately before antigen exposure significantly enhances skin DTH. Chronic stress significantly suppresses skin DTH. Adrenalectomy (ADX) eliminates the stress-induced enhancement of DTH. Low doses of corticosterone and/or epinephrine significantly enhances skin DTH in ADX and stressed animals. Acute administration of high dose corticosterone or chronic administration of moderate doses suppressed skin DTH [11]. The time course of allergic contact dermatitis (ACD) was examined in normal, ADX and capsaicin (CAP)-treated rats. ADX and CAP-treated rats developed contact dermatitis more rapidly with a faster recovery, than did normal controls. This indicates a biphasic action of innervation and of the HPA axis on ACD, i.e. inhibition in the early phase but augmentation in the late phase of ACD [12].

In male mice with combined ethanol and burn injuries 17 beta-estradiol treatment restored delayed-type hypersensitivity and splenocyte-proliferative responses, reduced macrophage interleukin-6 (IL-6), and increased survival after bacterial challenge ($P<0.01$). The beneficial effects of estrogen treatment were mediated through modulation of macrophage IL-6 production [13].

Pre-treatment of Langerhans cells (LC) with calcitonin gene-related peptide (CGRP) inhibits the elicitation of delayed type hypersensitivity (DTH) in mice. Upregulation of B7-2 expression on LC is suppressed by CGRP, which might be, in part, responsible for the inhibitory effect of CGRP in the functional assay. The functional effect of CGRP appears to be at least partially mediated through the autocrine regulation of IL-10. Treatment of mice with alpha-MSH produced inhibitory effects in contact hypersensitivity, which appeared to be mediated also by IL 10 [14].

2.2. Cytotoxic T lymphocytes

2.2.1. CTL activation and regulation

Cytotoxic T lymphocytes react to altered self MHC antigens. This alteration may be due to the presentation of an antigenic peptide or may be due to histocompatibility differences, because the target cell comes from an allogeneic donor (e.g. transplantation immunity). After recognition of the target in an "MHC-restricted" way, the CTL will adhere to the target via antigen receptors and additional adhesion molecules: CD8-MHC, LFA1 recognizing ICAM-1 and –2, and CD2 recognizing LFA3 on the target cell [15]. These adhesion molecules contribute significantly to the signaling process [16]. Next, the CTL delivers to the target cell a signal that initiates programmed cell death, or *apoptosis* (APO) in sensitive targets. This signal may be delivered through the Fas receptor for which the T cell expresses a cell surface ligand (FasL). This signal initiates in the target a complex enzymatic cascade that involves *caspases,* which in turn leads to the activation of endonuclease that produces brakes in the DNA of the target cell. This is followed by the enzymatic cross-linking (insolubilization) of cytoplasmic proteins. Eventually the apoptotic cell will fall apart by nuclear condensation and blebbing of the cytoplasm, which is followed by phagocytosis. With the exception of IL-1, apoptotic cells do not release inflammatory mediators and pathological signs do not occur, even when cell death is extensive (e.g. involution of organs). Other cell types besides killer cells (e.g. the brain, cancer cells) can express FasL and are able to kill target cells *in vitro* [17].

Single amino acid substitutions at TCR contacts may transform a natural peptide antigen in CTL activation with partial agonist, antagonist, or no activity. The variants that associated more stably to HLA-A2/molecules did not activate any CTL function, behaving as null ligands. T cell stimulations performed with the combination of null ligands and the natural epitope produced significantly higher specific CTL reactivation than reactivation of CTLs induced by the wild-type epitope alone. In addition, these particular variants activated memory CTL responses in the presence of natural epitopes that did not induce T cell responses on their own. Null ligands increased ZAP-70 tyrosine kinase activation induced by the natural epitope [18]. Human CTLs triggered by a partially agonistic peptide ligand die through a death receptor-independent apoptotic program. This APO exhibits some features of T cell death due to lymphokine deprivation and is prevented by exogenous IL-2. The choice between APO or cytotoxicity is determined by the agonistic potency of MHC-peptide ligand [19] .

Human CD8+ lymphocytes express killer inhibitory receptors (KIR) and ILT2/LIR1 inhibitory receptors. Initially ILT2/LIR1 is expressed in effector T cells and KIR gene transcription occurs in a minor proportion of expanded clones, which survives activation-induced cell death to become long term memory T cells [20]. CD30 signals inhibit cytotoxic NK and T cell effector function. Using gene microarrays, 750 gene products were induced and 90 gene products were suppressed >2-fold by CD30 signals. CD30 signals by both TNFR-associated factor 2-dependent and -independent pathways. CD30 can down-modulate lymphocyte effector function and proliferation while directing the cells to lymph nodes and increasing their susceptibility to certain apoptotic signals [21].

In most peripheral sites in the body the interruption of CD40-CD40L interactions can lead to the induction of T cell tolerance whereas in mucosal tissues this interruption can lead to abrogation of T cell tolerance. Moreover, CD40 activation *in vivo* can convert specific T cell tolerance following peptide vaccination into efficient T cell priming [22]. CD154 (CD40 ligand, gp39) interaction with CD40 is critically important for the generation of cell-mediated and humoral immunity. The production of IL-2, IFNγ and TNF was significantly suppressed in splenic cells from CD154-deficient mice, but CD86 expression and IL-12 production was normal. The treatment of CD154-/- mice with the agonistic anti-CD40 mAb, FGK45, generated activated APCs yet failed to restore either the CTL or the alloantibody responses. Immunization with B7-transfected P815 tumor cells also failed to generate expansion of the CTL effector population in CD154-/- mice [23].

Although only a fraction of CD3+ cells express the CD85/LIR-1/ILT2 inhibitory receptor on their surface, it is present in the cytoplasm of all T lymphocytes, and that it is detectable on the surface of all T cell clones. The CD85/LIR-1/ILT2 protein is tyrosine-phosphorylated in all T clones analyzed. The HP-F1 anti-ILT2 monoclonal antibody (mAb) added to the cytolytic assays enhances target cell lysis by Ag-specific CTL. In contrast, cross-linking by HP-F1 mAb triggers inhibitory signals that reduce cytotoxicity. CD85/LIR-1/ILT2 also affects responses to recall Ags and, in low responders, it sharply increases T cell proliferation [24].

Intestinal intraepithelial $CD8^+$ lymphocytes (IEL) of mice express the OX40 surface receptor and OX40 ligand following CD3 stimulation. Intracellular IFNγ synthesis was increased and cell-mediated cytotoxicity was markedly enhanced. IL 10 secretion was suppressed by CD3 activated cells. This could be reversed by the blockage of OX40L by a monoclonal antibody [25].

The chemokine RANTES, when aggregated on the cell surface, led to cell activation, including up-regulation of cell surface markers, and secretion of IFNγ and MIP1β. Specific inhibitors of RANTES-induced activation, such as soluble glycosaminoglycans, MIP1α and MIP1β, acted by preventing the binding of RANTES on the cell surface [26].

IL-4 increases FasL expression on T cells results in a shift of the mechanism of CTL killing from a dominant perforin-mediated cytolytic pathway to a dominant FasL-mediated cytolytic pathway [27]. IFNα up-regulates FasL mRNA and protein synthesis in mitogen-activated PBMC of hepatitis C virus infected individuals and in healthy subjects. Stimulation with IFNα increases perforin mRNA levels in PBMC. In NK cytotoxicity assays, the enhancement of cytotoxicity by IFNα is mainly due to perforin, while FasL plays only a minor role. In CTL cytotoxicity experiments neither the FasL nor the perforin pathway is further enhanced by IFNα [28].

In lymphocyte cultures from leprosy patients the presence of IL-12 during the first 24 h of antigen stimulation amplified the hsp65 cytotoxic response whenever IFNγ and TNFα were present. These three cytokines abrogated the inhibitory effect of IL-10 on hsp65 specific CTL in cells from paucibacillary patients (PB) but not that of IL-4 in PB and normal controls. Both IL-12, or anti IL-4 enhanced the cytotoxic activity in cells from multibacillary patients (MB). Anti IL-4 upregulated IFNγ but did not modify TNFα. Cells from MB patients taking thalidomide (MB-T) did bind neither IFNγ nor TNFα even when antigen or anti-IL-4 was added. CD56 effector cells developed during hsp65 stimulation of PB and N lymphocytes by the addition of IL-12 plus IFNγ and TNFα, while for MB and MB-T cells anti-IL-4 was also required [29].

In a mouse model of the CTL response using the OVA-transfected tumor cell line EL4 (EG.7), cultures of splenocytes obtained from EG.7-primed mice secrete high levels of MIF following Ag stimulation *in vitro*. Cultures treated with neutralizing anti-MIF mAb showed a significant increase in the CTL response directed against EG.7 cells. This effect was accompanied by elevated expression of IFNγ. The EG.7 tumors from anti-MIF-treated animals showed a prominent increase in both $CD4^+$ and $CD8^+$ T cells as well as apoptotic tumor cells. This increased CTL activity was associated with enhanced expression of the common γ_c-chain of the IL-2R that mediates $CD8^+$ T cell survival [30].

The neutralization of TNFα in mice blocked the lymphocytopenic features characteristic of acute graft-vs-host disease (GVHD) and induced a lupus-like chronic GVHD phenotype (lymphoproliferation and autoantibody production). These effects resulted from complete inhibition of detectable anti-host CTL activity, indicating that TNFα plays a critical role in the induction of CTL. The *in vivo* blockade of TNFα preferentially inhibited the production of IFNγ and blocked IFNγ-dependent up-regulation of Fas [31]. Murine heart or kidney allografts given to recipients with disrupted IFNγ genes showed infiltrate but little or no induction of donor MHC and developed extensive necrosis. The necrosis was immunologically mediated. The administration of recombinant IFNγ to deficient hosts reduced but did not fully prevent the effects of gene disruption [32]

Vasoactive intestinal peptide (VIP) and pituitary adenylate cyclase-activating polypeptide (PACAP) did not affect perforin/granzyme-mediated cytotoxicity, perforin gene expression, or granzyme B enzymatic activity, but drastically inhibited FasL/Fas-mediated cytotoxicity against allogeneic or syngeneic Fas-bearing targets. VIP/PACAP inhibit CTL generation, but not the activity of competent CTLs [33].

The roles of corticosterone and norepinephrine in sympathectomy-mediated modulation of both the primary and memory cellular immune responses to herpes simplex virus type 1 (HSV-1) infection was investigated. Both of these immunomodulators play a role in sympathectomy-induced suppression of the generation of HSV-specific primary CTL and the activation of HSV-specific memory CTL. A role for both Type I and Type II corticosteroid receptors was established in the regulation of HSV-specific immunity.

2.2.2. *Target cell lysis by CTL*

Killer T lymphocytes also contain granules that are released into the intercellular cleft, which is formed after the CTL is bound to the target cell. This is known as a membrane attack pathway (MAP) of cytotoxicity. The granules release membrane active peptides called *cytolysisn* or *perforin,* which dissolve into the membrane of the target cell and polymerize with the aid of Ca^{2+} to form microchannels (pores), which destroys membrane function and causes cell death [34,35,36]. Granule associated enzymes called *granzymes* are also released. These are serine proteinases designated by capital letters from A to H. Granzymes are synthesized as inactive precursors and must be activated by proteolytic cleavage [37,38,39,40]. *Dipeptidyl peptidase* I (cathepsin C) belonging to the cystein protease family is also present in granules and activates granzymes and confers susceptibility to cytolysis by forming esters of hydrophobic amino acids and dipeptides (e.g. leucine-leucine methyl ester) [41,42]. *Proteogycans* are also found in granules and are known to form an insoluble complex visualized as the granule core. A variety of normal *lysosomal enzymes* are also detected in cytotoxic granules [43]. *Granulysin* (NK-lysin/519), which is a small protein in CTL granules, has lytic activity on tumor cells and on certain bacteria [44]. *Calreticulin,* a 46 kD Ca^{2+} binding protein and molecular chaperone is also present in and released from granules [45]. Its function is uncertain at this time. Pore forming agents cause necrotic death in target cells. However, most cells killed by CTL or NK cells acquire an apoptotic phenotype [46,47,48]. Perforin is necessary for the rapid killing of Fas^- target cells by CTL or NK cells [49]. It permealizes the membrane even to macromolecules, such as granzymes [50]. The expression of granzymes A and B are also required for cytotoxicity [51,52]. Granzymes are capable of causing DNA damage of target cells, leading to APO, which is due to the activation of caspases [53,54].

Transgenically expressed human perforin gene locus DNAs demonstrated that IL-2R signals target two IL-2-dependent enhancers approximately 15 and 1 kilobase upstream of the promoter. In transient transfections, both enhancers required two identically spaced Stat-like elements for their activation. This was abolished by the expression of a dominant negative regulator, Stat5, which is constitutively active, bypassing the requirement for IL-2R signals [55].

Plasma of patients with activated CTLs and NK cells contained elevated levels of extracellular granzyme A (GrA). Part of the extracellular GrA circulated in the mature conformation and was bound to proteoglycans that protected it against inactivation by protease inhibitors such as antithrombin III and alpha-2-macroglobulin. Another part of GrA circulated as a complex with antithrombin III. Some plasma samples with high levels of extracellular GrA contained active GrA [56].

Perforin is not only an important effector molecule of cytotoxic T lymphocytes (CTL) but also involved in the down-regulation of peripheral T cells. Mice deficient for both perforin and FasL spontaneously develop infiltrations by highly activated $CD8^+$ T cells in kidney and liver and die between 5 and 12 weeks of age. In perforin-deficient mice *Staphylococcal enterotoxin B* (SEB) results in dramatically increased selective expansion and prolonged persistence of $CD8^+$, but not $CD4^+$, SEB-reactive T cells. Secondary immunization of TCR transgenic perforin-deficient mice with the lymphocytic choriomeningitis virus glycoprotein-derived epitope peptide leads to an increased proliferation of transgenic $CD8^+$ T cells. This is not explained by failure to deplete professional antigen-presenting cells [57].

Perforin-deficient donor T cells induced initially the features of acute GVHD in F1 recipients, e.g., engraftment of both donor CD4+ and CD8+ T cells, upregulation of Fas and FasL, production of antihost CTL, and secretion of both Th1 and Th2 cytokines. The donor cells failed to eliminate host B cells, and, by 4 weeks of disease, cytokine production in pfp $\rightarrow$ F1 mice had polarized to a Th2 response. Such mice eventually developed features of chronic GVHD with increased numbers of B cells, persistence of donor CD4 T cells, autoantibody production, and lupus-like renal disease. These results indicate that perforin plays an important immunoregulatory role in the prevention of humoral autoimmunity through the elimination of both autoreactive B cells and antigen-specific T cells [58].

The adherence of the CTL to its target may last from minutes to hours until target cell damage become detectable. Granule secretion is mediated by membrane fusion between the granule membrane and plasma membrane, termed *exocytosis*. This process is dependent on Ca^{2+} and protein kinase C. Some basic aspects of exocytosis are still poorly understood. Granzyme A secretion is not blocked by inhibitors of RNA and protein synthesis but IFNγ is inhibitory [59,60]. Nucleated cells have a cytoplasmic system of protection against membrane injury [61]. CTLs are resistant to perforin-mediated lysis, relative to other lymphoid cells [62,63,64].

Although $CD8^+$ T cells are the professional killer cells, $CD4^+$ T cells are capable of using both the FasL and perforin pathway, especially in animals that are devoid of functional CD8+cells. Such animals respond to viral infections by $CD4^+$ CTL [65]. CD4+ T cells may exert cytotoxicity also on antigen presenting cells [66,67]. Thus, some suppressor cells may actually be cytotoxic cells that lyse antigen-presenting targets. IFNγ activated CTL respond to 3–5 peptide-MHC complexes on the target cell surface [68]. In addition of the TCR, CD44 mediates the lysis of some CTL. The over-expression of the combination of B7.1 and ICAM on target cells may also trigger TCR independent lysis [69,70].

Memory CTL [mCTL] do develop after immune reactions, apparently from naïve T cells. The mechanism of mCTL development is under investigation. There is considerable debate on the role of antigen in the long-term maintenance of mCTL *in vivo* [71,72].

2.2.3. CTL and disease

CTL provides protection against virus infections that persist in the infected cell. Cytokine production by CTL, but not cytotoxicity, appears to play a major role in protection against such viral infections [73]. Lymphocyte cytotoxicity does not play a role in the control of cytopathic viruses

[74]. CTL provides protection against small inocula of intracellular bacteria, such as *Listeria* [75]. Lymphocyte cytotoxicity is not required for graft rejection [76]. Lymphocyte cytotoxicity plays a major role in graft versus host disease with the involvement of both the MAP and FAS-FasL pathways [77,78] Perforin and FasL deficient mice show a dramatic and lethal autoimmune disease [77]. CTL are involved in the induction of autoimmune diabetes [79,80]. In general, CTL are effective against tumors though exceptions do occur [81]. Some tumors express FasL and are capable of inducing death in Fas bearing lymphocytes recognizing tumor antigens [82].

Chronic graft-versus-host disease (GVHD), which was induced by the transfer of DBA/2 spleen cells into (C57BL/6 x DBA/2)F1 mice, could be prevented by IL-18. IL-18 significantly decreased immunological parameters indicative of chronic GVHD, such as elevated serum IgG antinuclear Abs, IgG1, and IgE levels, and host B cell numbers and their activation. Importantly, IL-18 partially but significantly restored the production of IL-2 and IFNγ production by Con A-stimulated spleen cells [83].

Mycobacterium tuberculosis (MTB)-reactive $CD4^+$ and $CD8^+$ T lymphocytes were analyzed from healthy tuberculin skin test-positive persons. After stimulation *in vitro* with MTB, both $CD4^+$ and $CD8^+$ T cells up-regulated mRNA expression for granzyme A and B, granulysin, perforin, and FasL. Increased perforin and granulysin protein expression was detected both in $CD4^+$ and $CD8^+$ T cells. Both T cell subsets lysed MTB-infected monocytes, which was perforin and FasL dependent. The intracellular growth of MTB was also inhibited by both type of effector cells, which was independent of perforin and Fas-FasL [84].

After culture with influenza virus, $CD45RO^+$ $CD8^+$ T cells from human peripheral blood increased in frequency and exhibited significant major MHC-I-restricted CTL activity, whereas $CD45RA^+$ $CD8^+$ T cells did not. The virus-stimulated T cells contained significantly higher levels of IFNγ-producing than did $CD45RA^+$ $CD8^+$ T cells. Recombinant human IL-12 further enhanced CTL activity and IFNγ production [85].

3. HUMORAL IMMUNITY

Early experiments over a century ago by von Behring indicated that protection could be provided against infectious disease, such as diphtheria in children, with immune serum produced in animals. This form of immunity has been named humoral as it was transferable with bodily fluid. Later it became apparent that a special subset of globulins, called *antibodies* mediate this form of immunity [86]. Today it is clear that bone marrow-derived (B) lymphocytes are responsible for the production of antibodies and thus these cells are responsible for the maintenance of humoral immunity.

3.1. B Lymphocytes

B lymphocytes are generated in the bone marrow and, after proper selection, B cells exit to the periphery. Such B cells go through a brief *immature phase* and are very sensitive to inactivation by antigen. This contributes to the establishment of *self tolerance*. The immature phase lasts less than a week, after which B cells become mature long-lived recirculating cells, competent to make antibody responses after receiving the proper signals from antigen and helper cells. Antigen contact triggers transmembrane signaling by the *B cell antigen receptor* (BCR). This leads to early events in B cell activation, such as increased expression of MHC-II molecules, progress from the resting (G_0) to the G_1 phase of the cell cycle, and, in cases of adequate antigenic stimulus, to proliferation. BCR molecules are rapidly internalized after antigen binding and the ingested antigen is degraded in *endosomes* and *lysosomes*. Peptides derived from protein antigens bind to the groove of MHC-II molecules and this complex is exposed on the cell surface and serves as a specific stimulus for *type 2 helper T lymphocytes* (TH2). TH2 form a tight and long-lasting interaction with the B cell and synthesize B-cell growth and differentiation factors. These cytokines can be secreted into the narrow space between the interacting cells. The activated T cells also express cell surface molecules, the most significant being the CD40 ligand (CD40L). CD40L interacts with B cell receptors and provides important activation signals. After such activation, B cells proliferate and differentiate into antibody forming plasma cells. In lymphoid organs the foci of proliferating B lymphocytes in response to antigenic stimulation are known as *germinal centers*. In germinal centers B cells activate the hypermutation of their immunoglobulin genes followed by stringent selection, which results in the secretion of high affinity antibody molecules. Memory B lymphocytes are also produced in germinal centers [87].

B lymphocytes have extraordinarily diverse functions within the immune system. B cells display a variety of characteristics, other than antibody production. They play a role in the development of lymphoid architecture, regulate T-cell subsets and dendritic cell (DC) function through cytokine production, and activate T cells. Receptor editing in B cells aids immunity to infection and, possibly, prevention of autoimmunity. Transgenic animal models revealed that B cells are necessary for the development of many autoimmune diseases although autoantibodies are not required in some cases. Intrinsic abnormalities of B cells have been described in autoimmune conditions. Negative signalling by CD5 and other molecules, such as CD22, in maintaining tolerance has also been documented [88].

Germinal centre B cells represent a stage of differentiation that can be defined by a gene expression that is not shared by other highly proliferative B-cell populations such as mitogenically activated peripheral blood B cells and lymphoma cells. This gene expression programme does not require ongoing interactions with other germinal centre cell types [89].

IgG Fc receptors, namely FcγRII (CD32) and FcγRIII (CD16) can regulate murine T- and B-lineage development at stages before the expression of clonal antigen receptors. The physiologically relevant ligand is the IgG-antigen complex, but it is also known that alternative, non-Ig ligands exist for Fc receptors [90].

The Pax5 gene encoding the transcription factor BSAP is required for progression of B-lymphopoiesis beyond the pro-B cell stage. Pax5-deficient pro-B cells have a broad lymphomyeloid developmental potential. Pax5 appears to mediate B-lineage commitment by repressing the transcription of non-B-lymphoid genes and by simultaneously activating the expression of B-lineage-specific genes [91]. The Rel/NFκB family of transcription factors have been implicated in the regulation of a wide variety of genes, in particular those encoding proteins crucial to immune function. Rel, RelA and NFκB1 operate as differentiation stage-specific regulators of B-cell development, survival, division and immunoglobulin expression [92]. The E2A transcription factors are required for the development of committed B-lineage cells and regulate the expression of essential B-lineage genes at multiple stages of differentiation and activation. The E2A gene products function in the regulation of (i) transcription factors required for B-lineage determination; (ii) essential proteins involved in pro-B and pre-B-cell development; (iii) accessibility and recombination of the IgH and IgL chain loci, and (iv) isotype switching in activated, mature B lymphocytes [93].

Plasma cells are cellular factories devoted entirely to the manufacture and export immunoglobulin (Ig) and play a critical role in adaptive immunity. Little information has been available on the regulation and requirements of plasma cell development [94].

Antibody response to some antigens can occur in the absence of T cell help. These antigens are called T-independent (TI) antigens and categorized as type 1 or type 2. TI-2 antigens have highly repetitive structures, polysaccharides being the prime examples. These antigens are able to trigger vigorous and prolonged signaling via the BCR. TH-1 antigens contain epitopes that are potent polyclonal activators of B lymphocytes. For both type of TI antigens cytokines are a requirement to trigger an antibody response. The cytokine source may be a macrophage [87].

3.2. B cell activation

Antigen activates B lymphocytes by cross-linking cell surface immunoglobulin receptors (oligomerizaton). This leads to the phosphorylation of ITAM tyrosines of the BCR associated signaling chains, Ig-α and Ig-β, by Src family tyrosine kinases, which provides binding sites to recruit Syk. Syk SH2 domains bind to the YxxL sequences of ITAMs. Syk becomes activated by the phosphorylation of tyrosine in it's kinase domain. Several other Src family kinases are activated, such as Lyn and Fyn, with the subsequent activation of a third type kinase, Btk [95,96]. Src family kinases are also negatively regulated by the phosphorylation of a tyrosine near the C-terminus by the tyrosine kinase, Csk. This negative regulation is reversed by the removal of the phosphate by CD45 [97]. Btk plays an important role in BCR-induced phosphoinositidine hydrolysisis. Btk is deficient in X-liked agammaglobulinemia in which there is greatly impaired B-cell development and very poor antibody production [98,99].

The major signaling pathways activated by BCR are: the phospholypase-Cγ2 – pospatidyl inositol (PIP_2) – diacylgycerol (DAG) – protein kinase C – inositol triphosphate (IP_3) pathway that releases Ca^{2+} from internal stores into the cytoplasm. The Ras pathway is also activated, which leads to the activation of the Erk1/Erk2 MAP kinases. The activation of phosphatidyl inositol 3 (PI3) – kinase leads to the activation of PIP_3, which may serve as a ligand for various proteins containing PH domains, such as the Btk kinase and the serine/threonine protein kinase B (PkB). Additional signaling pathways that are activated by BCR are: Vav, Cbl and HS1. Vav and Cb1 are protooncogenes and HS1 may be a nuclear transcription factor. The exact role of these pathways in B cell activation remains to be elucidated [87].

Antigen-complement conjugates with C3 can bind to both the BCR and to the CR2/CD19/TAPA-1 receptors simultaneously. This induces the phosphorylation of the cytoplasmic tyrosine of CD 19 by Syk, which in turn serves as binding sites for the signaling components PI 3 and Vav. These in turn enhance the B cell response to antigenic stimulation [100,101,102].

The FcγRIIb receptor on B cells is co-ligated by BCR when the predominant form of antigen consists of soluble immune complexes. This has an inhibitory effect on B cell activation *in vitro* and promotes the apoptosis of B cells [103,104]. This inhibitory effect is attributed to the immunoreceptor tyrosine-based inhibitory motif (ITIM) present in the cytoplasmic domain of FcγRIIb, which activates the phosphatases SHP-1 and SHP-2, and the inositol-5 phophatases SHIP [105,106,107,108]. There is evidence to indicate that FcγRIIb exerts its suppressive effect on B cell activation by inducing dephosphorylation of molecules carrying activation signals. SHIP was identified as the major inhibitory enzyme. Three other receptors, CD22, CD45 and CD5 also regulate signal transduction in B lymphocytes. The extracellular interactions that modulate these receptors are unknown at the present time [87].

B cells may be activated *in vitro* by anti-IgM or anti-IgD to enter the mitotic cycle. A number of cytokines will support the proliferation of human B cells. For murine B cells the best co-stimulant is IL-4 [109,110]. Among the preexisting transcription factors that are regulated by B cell signaling events are NF-AT, NF-κB, SRF/TCF, and CREB [111–115].

The activation of B cells increases the expression of IL-2 receptors and the ability to proliferate in response to IL-2. B7-2 expression is also increased, which is recognized by CD28 on helper T cells and provides important costimulatory signals to the T cell as part of antigen presentation [116,117].

Antigen structure, localization, dose, and time during which antigen is available are all important factors in regulating an immune response [118]. Dendritic cells (DC) have important roles in B-cell activation and regulation of antibody synthesis. Antibody responses induced by antigen-pulsed DC *in vivo* are skewed towards IgG. Naive B cells incubated with antigen-pulsed DC *in vitro* secrete Ig M and IgG when cultured with an antigen-specific CD4+ T-cell line, whereas if B cells are incubated with antigen only, IgM is produced [119].

The B cell antigen receptor triggers adaptive immune responses and controls a series of antigen-independent checkpoints during B cell development [120].

Inhibitory receptors are important in immunoregulation. Their mode of signal transduction is by immunoreceptor tyrosine-based inhibitory motif (ITIM), which recruits and activates phosphatases that mediate the receptors' function. The protein tyrosine phosphatases SHP-1 (Src-homology-2-domain-containing protein tyrosine phosphatase 1) and SHP-2, function to dephosphorylate key intermediaries in antigen receptor signaling pathways [121]. Antigen-induced activation of the B cell antigen receptor co-aggregates with the inhibitory FcγRIIB. The inositol 5-phosphatase SHIP and its effector p62i(Dok), a RasGAP adapter protein are important for FcγRIIB-mediated inhibition as well as for feedback regulation of signals generated through the BCR [122].

In B cells CD40 engagement produces upregulation of Fas expression and marked sensitivity to Fas-induced cell death. Antigen receptor engagement, or IL-4R engagement produce Fas resistance. Surface immunoglobulin (sIg) relies on PKC and IL-4R on STAT6 signaling. sIg signaling for inducible Fas resistance requires NFκB and depends on new macromolecular synthesis. Proximate mediators for Fas resistance include Bcl-xL and FLIP (but not Btk), and a gene that encodes Fas apoptosis inhibitory molecule (FAIM). FAIM-S and faim-L are alternatively spliced forms. The former is broadly expressed, whereas faim-L expression is tissue specific [123].

The *Complement Receptor Type 2* (Cr2-145, CR2, CD21) facilitates the internalization of immune complexes by B cells to enhance antigen presentation. CD21, in association with CD19/CD81, also serve as accessory molecules to the B-cell antigen receptor, permitting maximal B-cell activation at lower antigen concentrations. Furthermore, CD21 traps immune complexes on the surface of follicular dendritic cells and displays them to activated B cells in germinal centres [124]. Complement fragments interact with their receptors, CD21/CD35, to enhance humoral immune responses by: (i) focusing antigens for optimal presentation; (ii) enhancing co-receptors for B-cell antigen receptor signaling; (iii) regulating B-cell responses, by CD19 aggregation. CD21/CD35-deficient mice express a BCR with very low affinity [125].

Cells in general and B lymphocytes in particular maintain a balance between cell proliferation and cell death. When activated, cells control their own death through activation of an internally coded suicide program that initiates APO. This allows the elimination of cells that have been produced in excess, that have developed improperly, or that have sustained genetic damage. Apoptosis is the most common physiological form of cell death and occurs during embryonic development, tissue remodelling immune regulation, cell activation and tumor regression [126].

Two enzymes which act on plasma membrane phospholipids, phosphatidylinositol 3-kinase (PI3K) and phospholipase Cγ (PLCγ), have been implicated as critical mediators of B-cell activation and differentiation signals. These pathways are regulated by B-lymphocyte-specific signal transduction proteins, such as CD 19 and B-cell linker protein. Since PI3K and PLCγ(have common downstream targets such as the NF-AT and NFκB transcription factors, it is becoming clear that these two pathways are interconnected at several levels [127].

Linker or adaptor proteins are important in the signaling pathways activated by the B cell antigen-receptor. These proteins direct the subcellular localization of enzymatic complexes, amplify signaling pathways and integrate the functions of distinct signaling complexes. Linker proteins upregulate the BCR signalling and facilitate the influences of co-receptors that augment or dampen the BCR signaling pathway [128]. Adapter molecules orchestrate the assembly of signaling complexes. Adapter proteins play critical roles in both positive and negative regulation of antigen-receptor signaling, influencing lymphocyte development and function [129].

T cell dependent B lymphocyte activation involves the interaction of numerous receptor-ligand pairs on the surface of the two cell types [130]. Antigen-specific immunity develops through a series of intercellular information exchanges organized around cognate T cell receptor-peptide/MHC interactions. Complex molecular events take place in a serial synapsis model of adaptive immunity [131].

CD19 activates a novel Src-family kinase amplification loop that regulates basal signal transduction thresholds in resting B cells as well as in B cells activated by BCR ligation. The amplification of Lyn activity by CD19 leads to phosphorylation of CD19 and of downstream substrates, including CD22. Phosphorylated CD19 recruits other effector molecules including Vav, Grb2, phosphoinositide 3-kinase, phospholipase Cγ2, and c-Abl. The CD19/Lyn complex formation also regulates the phosphorylation of FcγRIIB, which inhibits B cell signal transduction through the recruitment of the SHPI and SHIP phosphatases [132].

CD27 is a TNF receptor family member whose expression is limited to cells of the lymphoid lineage. It is constitutively expressed on T lymphocytes, and is a costimulatory molecule for a regulatory subset. CD27 is induced on B lymphocytes after antigenic challenge, and is a marker of memory cells. CD70 is the CD27 ligand, a TNF related trans-membrane protein induced upon activation on T and B cells. CD27/CD70 interaction plays a key role in T dependent B cell responses and is responsible for plasma cell differentiation. Thus, B lymphocyte responses are controlled by different T cell subsets expressing CD154 (CD40 ligand), CD27, or CD70 (CD27 ligand) [133].

The critical role of CD40 in T cell-dependent humoral immune responses was demonstrated by patients with the hyper-IgM syndrome, as well as by gene targeting in mice. CD40 is also expressed on monocytes, dendritic cells, endothelial cells, and epithelial cells. The CD40-ligand (CD40L/CD154), a member of the TNF family, is also expressed more widely than activated $CD4^+$ T cells only. Therefore it is now thought that CD40-CD40-L interactions play a more general role in immune regulation [134]. Antigen receptor (BCR) transduces either pro-apoptotic or anti-apoptotic signals of mature B cells. B cells also undergo apoptosis by signaling through CD95. These apoptotic signals are blocked CD40, BCR, CD21 and IL-4 receptor signalling [135].

Suppressor of cytokine signaling-1 (SOCS-1) can inhibit the activation of IL-4 signaling through the inhibition of JAKs. The Fes tyrosine kinase is activated by IL-4 and regulates IL-4-induced proliferation through the phosphorylation of *insulin receptor substrate* (IRS) molecules. IRS is essential for IL-4-induced proliferation through their ability to recruit phosphoinositol-3 kinase to the activated IL-4 receptor kinase. IL-4 can also activate a number of phosphatases including SH2-containing inositol phosphatase (SHIP), SHP-1, and SHP-2. B-cell lymphoma gene-6 (BCL-6) appears to regulate a subset of IL-4-induced genes [136].

Activin A, a member of the TGFβ superfamily, causes apoptotic death of tumor myeloma cells and inhibits B cell spread and accumulation [137].

3.3. The antibody response

3.3.1. T-cell independent antigens

The initial IgM response to many infectious agents is likely to be based on T-cell independent antibody responses. Many bacterial cell wall antigens fall into the TI-1 and TI-2 categories of antigens. The initial IgM responses to some viruses, such as hepatitis, vesicular stomatitis and polyoma, were shown to be T cell independent [138,139,140]. TI responses are slow to appear after birth, which may be accelerated by adjuvants [141,142]. At this point there is no absolute empirical criteria to separate TI-1 and TI-2 antigens. Nevertheless this categorization is useful and almost certainly reflects important mechanistic differences in the antibody response to these two types of antigens [87].

3.3.2. TI-1 antigens

These antigens are *polyclonal B cell activators*, directly mitogenic for B cells irrespective of antigen specificity and are capable to induce the differentiation of antibody-secreting cells. Many of these antigens, such as LPS, peptidoglycan, lipoprotein, porin, etc. are components of bacterial cell walls. These responses are likely to represent an evolutionary adaptation of immune host defence against bacteria, as they are also present in lower animals [143–146]. LPS is capable of activating macrophages and murine B cells via specific receptors. Macrophages respond with the production of IL-1, IL-6, IL-8, TNFα, IFNα and macrophage inflammatory protein 1 and 2 and by the release of prostaglandins [147–150].

3.3.3. TI-2 antigens

Type 2 T cell independent antigens are *typically repeating polymers*, such as polysaccharides. These antigens lack polyclonal B cell activating capacity. Bacterial cell wall components and viral antigens have highly repetitive epitopes [151]. A key feature of TI-2 antigens may be the formation of persistent and efficient BCR signaling complexes (immunons) at the cell surface. Many TI-2 antigens are able to fix complement, either by the alternate pathway, or by binding natural antibody and may also activate the CD2/CD19/CD81 complex in addition to BCR. This enhances B cell activation. These antigens typically show the least requirement for additional signals from T cells or macrophages [87]. However, the addition of IL-2, IL-3, IL-5, IFNγ and GM-CSF promote the response, although none of them are essential. The stimulation of B cells with anti-Ig-dextran induces the expression of IL-5 receptors on B cells [152]. *In vitro* IFNγ stimulates Ig-class switching to IgG3, which is the predominant class seen in mouse responses to TI-2 antigens. The source of the cytokines is not well established. In viral infections NK cells may play a role [143]. TI-1 and TI-2 antigens are synergistic in B cell activation [143].

3.3.4. T-cell dependent antigens

The antibody response to most protein antigens depends on helper T cells. Naïve, long lived B cells circulate through the blood and lymph nodes until they contact their specific antigen, which typically happens in T cell areas of lymph nodes and spleen. Antigen is transported to these locations by macrophages and/or dendritic cells. Antigen contact stops B cell migration and causes them to remain in the T cell zone [153,154]. Specific B cells bind and internalize antigen, digest it and present it to helper T cells. Naïve, antigen-specific helper T cells are also activated in this location, probably by interdigitating antigen presenting dendritic cells. These activated helper T cells seek for B cells presenting the antigen, form stable conjugates with them and deliver soluble and membrane-bound helper signals for B cell proliferation and differentiation into antibody forming cells. The induction of CD40 ligand on T cells and of CD40 on B cells is required for this pathway. The related OX40 (on T) and OX40L (on B) may promote the non-germinal center pathway [155]. In the germinal center a rapid proliferation and rapid somatic mutation of the B cell's immunoglobulin genes take place. After mutation the B cells undergo stringent selection, with only those cells expressing high affinity membrane Ig surviving. Class switching to other Ig heavy chain isotypes is also frequent in such B cells. Germinal Center B cells differentiate into antibody forming plasma cells and live only for a few days, or some become quiescent memory cells. Plasma cells in the gut or bone marrow may live for more than 20 days [87].

3.3.5. The T-cell dependent antibody response

IgM response to the treatment of resting B cells with anti-IG requires IL-4, IL-5 and a third, late-acting cytokine, such as IL-2 or IL-6 [156]. IL-4 promotes proliferation, IL-2 and IL-6 promote differentiation. IL-6 induces the expression of cell-cycle inhibitor protein, $p18^{INK4c}$, which induces growth arrest as a prelude for terminal differentiation [157]. IL-2 up-regulates Blimp-1, which is a key factor in differentiation of B cells [158,159]. The role of IL-5 is less clear. It may also promote differentiation [160]. Many protein antigens do not stimulate sufficient BCR signaling to support an antibody response in the presence of IL-4, -5, -2, -6. For such a response a very potent engagement of BCR is required, such as by dextran-anti-Ig conjugates, for instance. Nonetheless, soluble proteins are capable of significant B cell signaling, as revealed by increased tyrosine phosphorylation, phosphoinositide breakdown, Ca elevation and *c-myc* induction in antigen specific B cells [161–164].

The antibody response to sheep red blood cells (SRBC) required T cells, which could be substituted for by IL-2 and IFNγ [165]. In many cases IFNγ suppresses the antibody response, while in others, like above, it has an enhancing effect. IFNγ often antagonizes the effects of IL-4 on B cells [166].

One surprising feature of these two models is that the antibody response to SRBC requires TH1 cytokines, whereas in the anti-Ig system TH2 cytokines are operative. The anti-Ig system also functions when IL-2 is substituted with IL-6, and with the cytokines produced by a third subset of helper T cells (TH0), namely IL-2, IL-4 and IL-5, along with other cytokines. The antibody response to anti-Ig may proceed with a TH2 or a TH0 helper cell, whereas the SRBC response requires TH1 helper cells [87].

The antibody response to soluble proteins requires contact and activation of B cells by helper T cells [167]. The major activating signal delivered by TH cells is mediated by the CD40 ligand that interacts with CD40 on B lymphocytes [168]. TCR stimulation induces CD40L within 1 to 2 hours on T cells [166]. Oligomeric, cross-linked or immobilized CD40L induce proliferation of purified B cells, which is enhanced by cytokines, notably IL-4. CD40L does not induce terminal differentiation to the plasma cell stage. Cytokines are also required, the best combination being IL-4 plus IL-5 [169]. CD40 signaling stimulates the expression of Fas by B cells and renders them susceptible to killing by FasL. BCR signaling does not induce Fas and protects B cells from such killing [170]. The blocking of CD40L by antibodies results in a greatly diminished thymus dependent (TD) antibody response [169]. The CD40L gene is on the X-chromosome and humans with a defective gene suffer from *X-linked hyper-IgM syndrome.* Such patients have low levels of IgG, IgA and IgE, but elevated titers of IgM. Mice lacking CD40 or CD40L show similar phenotypes, although IgM is usually not elevated [170].

CD40 and CD40L belong to the TNF receptor-ligand family, where Fas-FasL and CD30-CD30L also belong [172]. Signaling is achieved by receptor cross-linking, which involves TNF-receptor associated factors (TRAF2 and –3). TRAF-2 appears to activate NF-κB and TRAF-3 mediates isotype switching [173]. CD40 may also activate protein kinases, such as Lyn and Btk [169]. Two type of MAP kinases, the c-Jun N-terminal kinases (JNK-1 and JNK-2) and the p38 MAP kinases, and to a lesser extent the Erk1/Erk2 MAP kinases, are also activated [174,175].

A large number of adhesion molecules on the surface of B cells and helper T cells appear to participate in the interaction of these two cells. The molecules that have been shown to play a role on B cells include: MHC-II (recognized by TCR and CD4); ICAM-1 (recognizing LFA-1 on T cells); B7 (CD28); LFA-1 (ICAM-1); LFA-3 (CD2) [176]. These interacting molecules are involved in crucial information transfer. Thus B7-CD28 is needed for the activation of TH2 cells. Anergy results if this signal is missing [117]. Also, MHC-II can provide a strong proliferation signal to B cells [177,178].

3.3.6. Regulation of Ig isotype expression

In mice IL-4 promotes isotype switching to IgG1 and to IgE. IL-4 is compulsory for IgE but not for IgG1 switching [179]. Two signals regulate the immunoglobulin switch to IgE. The first signal that initiates the heavy chain switch to Cε is delivered by the cytokines IL-4 or IL-13. CD40 on B cells and CD40L expressed on activated T cells deliver the second signal. These signals activate deletional switch recombination [180]. IL-5 regulates isotype expression of IgA under some circumstances. In this case TGF-β also plays a role. IFNγ promotes class switching to IgG2a and interferes with switching by IL-4 [181,87].

3.3.7. Cytokine production by B cells

Activated B cells produce IL-1, TNFα, lymphotoxin-"α and GM-CSF [182].

3.3.8. Thymus-dependent antibody response in vivo

Normal T cells exposed to antigen in vivo acquire the ability to secrete large amounts of cytokines, when re-stimulated several days later. Various cytokines may drive these cells to differentiate into TH0, TH1 or TH2 type cells [183,184]. These cells are able to help B cells. Experiments with blocking antibodies indicated that TH0 and TH1-induced responses were inhibited by blocking IL-2 and to a lesser extent, IL-6. TH1 cells did not make IL5. Blocking of IL-5 inhibited TH0-induced responses by 50%. TH-1-like cells did make low amounts of IL-4 and blocking decreased the response by about 2/3rd. Anti-IL-4 did not inhibit the TH-2 response, whereas anti-IL-5 had a strong inhibitory effect [185,186].

Almost all TD responses are dependent on CD40-CD40L interaction *in vivo*. This interaction can induce B7-2, which in turn leads to greater cytokine production by TH2 [169]. Direct activation signals are also delivered to T cells by CD40L [187].

In vivo IL-4 and IFN-γ clearly play a role in the antibody response. However, the ablation of the genes producing IL-2 and IL-4 in mice did not lead to significant alterations of the antibody response [189]. IL-4-deficient mice were impaired in antibody production to mucosal immunization. The interpretation of these results is that in IL-4 knockout mice TH1 cells are good helpers for antibody production as long as they do not produce high amounts of IFN-γ, which inhibits the antibody response. The TH1 cytokines that support the humoral immune response under these circumstances remain to be elucidated [189]. Even when both TH1 and TH2 activity is inhibited by IL-2 and IL-4 double knockout in mice, the antibody response is normal [188]. The TD-IgG response to protein and viral antigens is decreased 5–10-fold in IL-6 deficient mice

[190]. Similar results were obtained in IL-5Rα chain mutant mice [191].

3.3.9. *B-cell inactivation*

If self-antigen comes into contact with immature B cells in the bone marrow immediately after the expression of BCR, it leads to mutational arrest, to the re-exposure of *rag-1* and *rag-2* and resumption of rearrangement of the light chain loci. This could lead to the deletion of the initially successfully rearranged light chain gene and to the creation of a new rearranged gene, which may contribute to a new BCR that is not self-reactive. This process is known as *receptor editing* [192]. Such immature B cells may now mature and leave the bone marrow. Such B cells are short lived and are very sensitive to antigen-induced apoptosis. Strong activation by antigen leads to death, whereas weak antigenic stimulus causes clonal anergy [193,194]. In Ig transgenic mice expressing self-reactive membrane Ig on almost all of their B cells both clonal deletion and anergy are found. Soluble antigens induce anergy, whereas membrane-bound antigens cause deletion [195]. Similarly, deletion of B cells may be induced by antigen in the absence of T cell help [196].

Anergic B cells may be re-activated if the antigen appears in a more immunogenic form than the anergizing self antigen. Thus, soluble antigen may maintain the anergy, whereas antigen primed helper T cells would induce B cell activation [197,161]. Deficient BCR signaling also leads to FasL induced apoptosis of B cells. BCR stimulation induces resistance to such killing [170].

4. IMMUNOLOGICAL MEMORY

Memory cells are present both in the B and T cell compartments. Memory T cells are heterogeneous, both CD4+ and CD8+ cells are involved. In contrast with other lymphocytes that are programmed to die after antigen-induced proliferation ('replicative senescence') [198], memory cells have the capacity of self-renewal, much like stem cells do [199]. Although it is recognized that the survival and death of T cells is strictly regulated [200], there is no adequate explanation at present for the long-term survival and self-renewal capacity of memory cells of either the T or the B lineage.

The CD4+ population of memory T cells may be divided further into T helper type 1 (TH1) and TH2. TH1 cells induce cell-mediated immunity, whereas TH2 cells promote the antibody response. T cell receptor signaling, co-stimulatory molecules (e.g. CD28-B7) and cytokines are involved in the induction of CD4+ memory cells. Once CD4+ memory cells are generated they do not require antigen or stimulation by MHC-II determinants to survive or to expand. CD4+T cells are fundamental to the maintenance of immunological memory [201–204]. Resting CD4+ T cells from spleens of mice immunized 6 weeks earlier were restimulated *in vitro* with the specific antigen and produced IL-2 and IL-3 in high amounts, but only low levels of IL-4 and IFNγ (memory effector TH cells). In the case of humoral immunity the evolution of CD4 naive T cells into memory effector cells was characterized in a shift from IL-2 to IL-4 production. There was an inverse correlation between IL-2 and IL-4 production, although both cytokines were produced by both cell types [205].

CD8+ memory T cells are cytotoxic and provide defense by killing virus infected and cancer cells. MHC-I molecules but not antigens are required for the survival of CD8+ long term memory cells, which constitutively express cytotoxic effector function [206]. Memory T cells turn over more quickly than are naive T cells that do not respond to cytokines. In mice the turnover of CD8+ memory cells can be considerably stimulated by polyinosinic polycytidylic acid, lipopolysaccharide (LPS), immunostimulatory DNA (CpG DNA) and by type I interferons (e.g. IFNα or IFNβ). However, IFN does not induce CD8+ T cell proliferation *in vitro.* Memory, but not naive-type T cells are stimulated by infection-induced cytokines [207,208].

The mechasnism(s) for the long-term survival and self-renewal capacity of memory B and T lymphocytes is not understood. It was hypothesised that memory B lymphocytes are stimulated by idiotypes, which are unique determinants of antigen receptors [209]. For memory T cells MHC presenting self peptides has been suggested to fulfill a stimulatory role [210]. The role of cytokines in T cell longevity is also recognized and IL-15 was suggested to be necessary for CD8+ memory cells [211]. The fact that T lymphocyte apoptosis is inhibited by IFN and IFN has been suggested to play a major role. These cytokines are able to maintain T cells without an antigenic stimulus [212]. Cho and co-workers [213] demonstrated in recombinase deficient (RAG-1 -/-) mice that are lymphopenic, that naïve T lymphocytes undergo *"homeostasis-stimulated" proliferation*, which is MHC restricted, and develop into memory cells in the absence of antigenic stimulation. These cells acquire the phenotypic and functional characteristics of antigen-induced memory CD8+ T cells and lyse target cells directly and respond to lower doses of antigen than naïve cells and secrete IFNγ faster upon restimulation. Interleukin-2 or co-stimulation by CD28 is not required and effector cells are not formed during this homeostatic differentiation. These findings indicate that memory T cells may be generated and maintained under the influence of physiological immunoregulatory mechanisms, in the complete absence of immune stimulation by antigen, adhesion molecules and cytokines.

Immature thymocytes of rodents are killed by glucocorticoids, whereas mature thymocytes are saved. The helper, suppressor and killer functions of T lymphocytes and the production of interleukins by them are all inhibited by glucocorticoids. In contrast, the function of memory cells and of cells mediating the graft-versus-host reaction is not inhibited by glucocorticoids [214,215]. We have established earlier that PRL antagonizes the immunosuppressive effects of the ACTH-adrenal axis [216–218]. Taking all the evidence in consideration, one may suggest that naive T cells are maintained in the absence of antigenic stimulation by pituitary GLH, whereas memory T cells are autonomous and survive and resist glucocorticoids, possibly via autocrine growth stimulation.

5. NATURAL IMMUNITY

Innate resistance may be divided into host defense mediated by non-immune mechanisms and to natural host defense by the immune system. Non-immune mechanisms include: behavioral defense (e.g. avoidance, aversion, cough, vomiting, diarrhea, etc.); defense mechanisms of the skin (for instance the reaction to UV irradiation); non-immune mucosal protection (the secretion of mucus, antimicrobial proteins, enzymes, low pH in the stomach, etc.); secreted defense molecules by the salivary, lacrimal and sweat glands, the liver, mammary gland and male exocrine glands; and non-immune systemic host defense mechanisms, which include oxygen radicals, nitric oxide, heat shock proteins, non-immune interferons, inductive enzyme synthesis, properdin, chemokines, prostaglandins and leukotrienes, blood clotting and species-related natural resistance [219–224, 405].

The natural immune system is comprised of highly specialized cells, such as NK cells, $\gamma\delta$T lymphocytes and B cells secreting natural antibodies (Nab). However, antigen specific T and B lymphocytes may also participate when activated in a non-antigen specific manner. Neutrophilic,

Table I Natural immunity.

Non-immune defense	Immune defense
Behavioral, neural: avoidance, cough, vomiting, diarrhea. Skin: pigmentation, vitamin-D, sweat-gland and keratinocyte-derived mediators, etc. Mucosal, exocrine: mucus, enzymes, defensins and other antimicrobial substances, bile acids, HCl in stomach, cytokines, hormones, neuropeptides, etc. Systemic: nitric oxide, oxygen radicals, heat shock proteins, non-immune interferons, enzymes, properdin, prostaglandins, leukotrienes, chemokines, blood clotting, species-related resistance.	Natural killer cells Tγδ cells CD5+ B lymphocytes producing natural antibody (NAb). T lymphocytes after activation by superantigens, chemokines, cytokines, etc. B lymphocytes activated by LPS, superantigens, cytokines, etc. Mast cells triggered by physical, chemical factors and by neuropeptides. Phagocytes reacting to bacterial substances, NAb, opsonins, etc. Complement activated by natural antibodies and by the alternate pathway. Acute phase proteins: C-reactive protein, endotoxin binding protein, mannose binding protein.

Adapted from reference [406].

Table II Examples of homotopes, their receptors and their functions.

Homotope	Source	Homotope Receptor	Immune activation
Lipid-A of LPS	Gram neg. bacteria	LBP	B, MF, C', PLT EDT, CNS, ENG
Phosphoryl choline, galactose cationic proteins	Microbes, degraded cells	CRP	B, T, MF, C' PMN, NK, PLT
Diverse carbohydrates	Microbes	MBP	MF, PMN, C'
Diverse epitopes; charged molecules, carbohydrates, superantigens	Microbes, cancer cells, self components	Natural antibodies	B, T, MF, C' PMN, NK
Superantigens ("toxins")	Microbes, self cmponents?	TcR, BcR (sIg)	T, B Inflammatory cells
Lipid, glycolipid, peptides	Microbes, self?	CD1	T
Phosphate-containing nonpeptides	Microbes, self?	γδ TcR	T γδ

Adapted from reference [406].

eosinophilic and basophilic leukocytes and mast cells participate in the natural immune system as well [223,225–227] (Table I) [405]. As it is obvious from this table the entire immune system may be activated by natural immune mechanisms.

Natural immune reactions are initiated by the recognition of evolutionarily highly preserved *homologous epitopes, or homotopes*. Homotopes are recognized by serum proteins and germline encoded cellular receptors that belong to several molecular families (Table II) [405]. Homotopes are found in association with microorganisms, cancer cells, degenerated cells and with other cell components. Some serum proteins, such as lipopolysaccharide endotoxin binding

Table III Immune activation by LPS, CRP and Nab

Cells/systems activated	Lipopolysaccharide (LBP)	C-reactive protein	Natural antibody
1. Complement	+	+	+
2. Monocyte/macrophage,Kupfer cells in liver etc.	+	+	+
3. T cells	+ (T)	+ (CTL)	+
4. B-cells	+	+	+
5. Mast cells discharge	+ (via C3a, C5a)	+ (via C3a, C5a)	+ (via C3a, C5a)
6. Polymorphonuclear cells	+ (via C3a, C5a)	+	+ (via C3a, C5a)
7. NK cells	—	+	+
8. Platelets	+	+	+?
9. Endothelium	+		
10. Peritoneum	+		
11. Adipocyte	+		
12. CNS (astrocytes, glia cells)	+		
13. Vagus nerve	+		
14. Endocrine glands ("Resident Mf" follicular stellate cells)	+		

NOTE: + action and/or stimulation – no effect.
Adapted from reference [406].

protein (LBP), C-reactive protein (CRP) [228], mannose-binding protein (MBP) [229] or Nab [230,231] can serve as recognition molecules of homotopes. After combination with the appropriate homotope these soluble receptors acquire the capacity to activate various cells within the immune system and elsewhere. Specific cell surface receptors, such as CD1 [232], or the antigen receptor of T [233,234], and B [235] lymphocytes recognize homotopes. As a rule, these homotope receptors are polyspecific and allow the recognition of a vast variety of closely related molecular structures on pathogens and of abnormal cell components with the aid of relatively few germ-line encoded molecules. The *microbial homotopes* involved in natural recognition are stable because they are essential for the survival or for the pathogenicity of the microorganism. An example is the lipid-A moiety of bacterial endotoxin, which is present in all Gram-negative bacteria, regardless of pathogenicity [222,223,225,227].

5.1. Natural immunity and host defense

The first line of host defense against infectious disease, tissue injury and against a variety of noxious agents is natural immunity. Also, there is evidence to indicate that natural immune mechanisms are part of normal physiological regulation of various bodily functions. For example, LPS is capable of activating the entire immune system after complexing with LBP, which in turn will be recognized by the CD14 cell surface molecule on macrophages and on some other cells (Table III) [405]. Alternate LPS receptors, such as the Toll-like receptor 4, also exist. LPS is capable of acting on numerous other targets in the body that include platelets, endothelium, peritoneum, adipocytes, the central nervous system, the vagus and possibly other peripheral nerves, and endocrine glands [223,227,236–242]. It seems most remarkable that LPS is capable of activating multiple targets in concert, all of which play essential roles in neuroimmune host defense. C-reactive protein is capable of complete immune activation, similarly to NAb. Therefore, CRP may be considered at least as efficient as NAb in host defense. Clearly CRP, LBP and

Table IV Neuroendocrine regulation of natural immunity.

Hormone/ mediator	NK	CD5+B/NAb	Phacogytosis	APP
GLH hormones				
GH	↑		↑	
PRL	↑		↑	
IGF-I	↑		↑	
INS			↑	
GLU				↑
T3	↑	↑		↓
HPA axis				
ACTH			↓	
GC	↓	↑	0↓	↑
α-MSH		↑		↓
β -END	↑↓		↑	
CAT				↑
Steroid Hormones				
E2	↓	↑	↑	
DHEA	↑	↑		
PS			↓	
VD3	↑			
Other Mediators				
SP				↑
ANP				↑
MEL				↑

NOTE: ↑ = increase, ↓ = decrease, ↑↓ = variable effect, 0 = no effect, ? = predicted change.
Adapted from reference [406].

MBP are capable of directing immune reactions against well-defined targets, just as antibodies are. Once target identification has been achieved, the immune effector mechanisms (e.g. inflammation, phagocytosis, cytotoxicity) are shared with adaptive immune mechanisms. This fact also indicates that natural immune host defence can be as effective as adaptive immunity [223]. B-1/ CD5$^+$ B cells constitute a distinctive peripheral B cell population, enriched for natural autoreactive specificities that are encoded by particular germline V(H)-V(L) combinations [243].

Most cells in the body are capable of producing IFN-α and -β after viral infections and secreting chemotactic and proinflammatory cytokines (chemokines) after injury. *Sensory nerve fibers* are capable of triggering inflammation in response to irritation, which does not involve tissue injury. This phenomenon is known as *neurogenic inflammation,* which does not only function as a potent defense reaction, but in some cases it may be regarded as a pre-emptive response (Table IV) [228,244,245].

Mucosal secretions defend the host outside the body, on mucosal surfaces. Natural immune defense is fundamental to mucosal protection, although adaptive immune reactions also play a role. For instance, the *submandibular gland* in rodents produces non-immune resistance factors, such as enzymes, antimicrobial proteins and mucus, CRP and Nab that belong mainly to the IgA class. In this case the immune reactions take place on mucosal surfaces. For this reason self-nonself discrimination by the immune system is of limited value. Rather, neuroendocrine mechanisms play a major role in the regulation of mucosal immune and inflammatory reactions.

Thus, it is clear that in rodents the submandibular gland plays a key role in the regulation of mucosal immunity and in the coordination of mucosal defense with systemic immune reactions [222,246].

Current evidence indicates that natural immune mechanisms are in the frontline of host defense at all times. The natural immune system is capable of detecting instantaneously pathogenic agents, infected and injured cells by polyspecific cellular receptors and by serum recognition proteins. The Nervous System enables this defense system even to react to irritants (Table IV). It is clear that the entire immune system may be activated by natural immune pathways. The polyspecificity of receptors and soluble recognition molecules mediating natural immunity is a distinct advantage as they are capable of providing a wide spectrum of host defense with minimal usage of the genome. The prerequisite for adaptive immunity is prior exposure to the pathogenic agent. This is not necessary for natural immunity because of its innate characteristics.

Bacteria, such as mycobacteria in Freund's adjuvant, killed Gram negative bacteria, corynebacterium, etc., have been used as immunological adjuvants for decades. Some bacterial products, such as LPS and its "detoxified" forms, are being applied as therapeutic agents to boost host defense under various situations. There is evidence to indicate, that sub-lethal doses of endotoxin, or its detoxified forms, increase resistance against infectious agents and injury, protect against X-irradiation, enhance the antibody response, exert an antitumor effect, stimulate bone marrow function, activate the complement system and blood clotting, promote regeneration and healing in the liver, and possibly even in the central nervous system [221,223]. LBP was shown to provide defense against Gram-negative bacterial infection in mice [247]. Mice transgenic for human CRP showed increased resistance against infection with staphylococci [248]. Monoclonal antibody-staphylococcal enterotoxin A fusion protein has been tested for superantigen-based immunotherapy of human pancreatic and colorectal cancer [249].

6. REGULATORY IMMUNE FUNCTION

6.1. Clonal selection

Talmage proposed first on the basis of careful assumptions and calculations that antibody molecules are heterogeneous with specifities to a large number of antigens [250]. Burnet developed the theory of somatic mutation followed by clonal selection as a mechanism for the immune system to generate the diversity that was required for host defence against microorganisms [251]. These theories are now extensively supported by facts and clonal selection may be regarded today as a fundamental mechanism by which self tolerance is established, while the reactivity against foreign antigens is maintained. The thymus is the central organ for T cell maturation and T cells govern the adaptive immune system. Therefore, the thymic selection of T lymphocytes is essential to normal immune function.

In the thymus T cell precursors express their TCR for the first time and the immature T cells are presented immediately by self antigen in the context of MHC-I or MHC-II. The cells that react to TCR stimulation with full activation will not be able to undergo mitosis, but rather, will undergo apoptosis. This is known as *negative selection,* which is widely regarded as a preventive mechanism against autoimmune disease. Some other cells will recognize MHC plus self peptide, but only to the extent that they will survive and make no attempt to respond to this signal by proliferation. These cells undergo *positive selection* and will mature and be released from the thymus to the periphery as self tolerant T lymphocytes. The third category of cells that do not recognize the MHC signal at all will die of neglect [252]. Once the T cells are in the periphery, they can still undergo APO, if presented by antigen without the necessary accessory and cytokine signals. Therefore, negative selection is going on also in the periphery until T cell reactivity is adjusted to the desirable level for normal immune function. There is evidence to indicate that B cells also undergo positive and negative selection in the bone marrow and that antigen induced negative selection of B cells also happens in the periphery.

The differentiation events in the thymus that lead to the development and export of mature T cells are both dynamic and complex. Many aspects remain poorly defined, such as the precise relationship between stromal cells and thymic selection [252]. Thymus cellularity and thymic hormone secretion peak during the first year of life and then decline gradually until the age of 50–60 years. This process is known as "thymic involution". A rapid involution occurs in young patients following injuries, chemotherapy and other forms of stress. This involution is dependent on factors intrinsic to the thymic tissue, such as the production of cytokines and chemoattractants that promote recruitment, growth, and differentiation of bone marrow-derived T cell progenitors. The systemic levels of endocrine hormones and mediators released by intrathymic nerves also play an important role [254].

In both humans and mice major MHC molecules show a highly restricted thymic expression. Transgenic mice with MHC expression targeted to specific thymic sites show altered TCR repertoire selection. While HLA-DQ is poorly expressed by most peripheral APC, the thymus stains strongly for HLA-DQ as well as for HLA-DR. HLA-DM is abundant in the medulla, but is weakly expressed by cortical cells. Class II expression in Hassall's corpuscles (HC) is encircled by HLA-DO+ epithelial cells and are negative for HLA-DR and HLA-DP, but often positive for HLA-DQ and HLA-DM. Apoptotic thymocytes are restricted to the cortex and the corticomedullary junction, but a minority of apoptotic cells are visible in the medulla, these being found in the HLA-DQ positive HC. The apoptotic thymocytes in HC can be CD4+ or CD8+ positive, or CD4+CD8+ double-positive. Thus HC within human thymic medulla are noteworthy both for their unusual hierarchy of HLA class II expression and because they are the only medullary site of thymocyte apoptosis. It was proposed that HC are sites at which mature thymocytes receive activation/tolerization signals from peptides reprocessed from apoptotic cells. The differential HLA transcriptional control at this site may indicate that specific T-cell subpopulations are affected [255]. In the human thymus MHC-II (HLA-DR) antigens are expressed by stromal cells: epithelial cells, dendritic-like cells and macrophages as well as HLA-DR IP lymphocytes in all thymic regions [256]. Human T-cells and thymocytes, unlike mouse T-cells, express MHC-II molecules on their surfaces and can present antigen in vitro. Thus T-T interaction may play a role in thymic selection [257].

In vivo experiments established that positive selection of CD4+ thymocytes is a peptide-specific process, which is based on the intrathymic TCR recognition of self-peptide:self-MHC molecular complexes. TCR reactivity to non-self MHC molecules or alloreactivity does not result from intrathymic T cell selection, but rather is an intrinsic property of germline-encoded TCR domains. In secondary lymphoid organs, a repeated TCR-MHC low affinity interaction is required to maintain the mature peripheral T cell pool and therefore the mature TCR repertoire. This TCR-MHC interaction-induced protection from apoptosis is remarkably reminiscent of the intrathymic positive selection phenomenon. The role of self-MHC recognition in TCR repertoire development and survival may account for the influence of MHC genotype on susceptibility to specific autoimmune diseases [258].

Negative selection caused by an endogenously expressed self-determinant can effectively occur in the thymic cortex in vivo. Both cortical epithelial cells and cortical dendritic cells were able to support negative selection of CD4(+)CD8(+) thymocytes, albeit with a distinct efficiency [259].

The intracellular class II heterodimer, H2-O, is associated with H2-M and co-localized in MHC-II+ vesicles, which suggest a related function in peptide exchange. In mice with disrupted H2-Oa gene the array of endogenous peptides bound by class II was altered and a defect in antigen presentation through H2-A to T cells was seen on the 129/Sv/ C57BL/6 mixed strain background but not in 129/Sv pure strain mice. H2-O-null mice showed enhanced selection of CD4+ single positive thymocytes. H2-O should be regarded as a modulator of peptide editing [260].

Experiments with TCR transgenic mice revealed that:(1) thymic selection produces a repertoire with marked overrepresentation of a subset of sequences; (2) MHC class I- and class II-restricted TCRs can be distinguished by single-residue changes in CDR3α, and (3) homeostatic expansion and survival in the periphery can markedly remold the postselection TCR repertoire [261]. Other experiments in transgenic mice showed that partial negative but no positive selection events can be induced by thymic B-lymphocytes in vivo [262].

The thymus provides an optimal cellular and humoral microenvironment for the development of immunocompetent T lymphocytes. Yolk sac derived pre-T stem cells enter the thymus using a *homing receptor*, which is aided by *thymotaxin* secreted by the cells of the *reticulo-epithelial* (RE) network in the thymus. The thymic RE cells are derived from a common, endodermal epithelial precursor cell. The subcapsular, endocrine RE cell layer (giant or nurse cells) is filled with PAS positive granules and also express A2B5/TE4 cell surface antigens and MHC-I (HLA A, B, C) molecules. These subcapsular nurse cells also produce thymosins (3 and 4). The thymic nurse cells (TNCs) display a neuroendocrine cell specific immunophenotype (IP): Thy-1+, A2B5+, TT+, TE4+,UJ13/A+, UJ127.11+, UJ167.11+, UJ181.4+, and the common leukocyte antigen (CLA+). Medullar RE cells display MHC-II molecules (HLA-DP, HLA-DQ, HLA- DR). These cells also contain TGFβ type II receptors and are involved in the positive selection of T cells. Four functional subtypes of medullary RE cells are distinguished: undifferentiated, squamous, villous and cystic. All subtypes are connected with desmosomes. The thymic hormones, thymulin, thymosinα-"1 and thymopoietin are produced by RE cells. Thymic RE cells also produce numerous cytokines, including IL-1, IL-6, G-CSF, M-CSF, and GM-CSF. These cells also produce pituitary hormones and neuropeptides, which include: GH, PRL, ACTH, TSH, T3, somatostatin, oxytocin, FSH, LH, arginine vasopressin, GHRH, CRH, NGF, VIP, pro-encephalin, and beta-endorphin. RE cells express receptors for all these mediators [263].

Steroidogenic organelles were isolated from murine thymic tissue and incubated with radioactive precursor steroids in vitro. The thymus possessed all enzymes and cofactors required for glucocorticoid production. However, an intact thymic architecture was necessary for glucocorticoid production, since 11β-hydroxylase was not detected in irradiated thymi or in a thymic epithelial cell line [264]. Thymic epithelial cells produce glucocorticoids, and it has been proposed that these locally produced glucocorticoids participate in antigen-specific thymocyte development by inhibiting activation-induced gene transcription and thus increasing the TCR signaling thresholds required to promote positive and negative selection [265,266].

Thymic stromal cells are capable of synthesizing hormones and neuropeptides, which belong to the specific microenvironment in which T cells differentiate. These mediators contribute to lymphocyte signaling, but also are processed and presented to maturing T cells in order to induce tolerance towards them. The thymic presentation of some neuroendocrine self-antigens does not seem to be restricted by MHC alleles [267].

Both MHC-peptide complexes and accessory molecules must be provided by thymic epithelium for efficient positive selection of T cells. Thymic and non-thymic stromal cells expressing

MHC-II molecules present the same limited peptide array. This suggests that the unique ability of thymic epithelium to mediate positive selection of CD4+8- cells is not solely due to presentation of a specialised peptide repertoire, but is dependent upon provision of specialised accessory interactions [268].

The adhesion molecule, ICAM-1, provides a selective boost to thymocyte signaling for differentiation without substantially affecting induction of death among CD4+CD8+ cells, a property that is lost as thymocytes mature further. In contrast, B7 engagement enhances both cell activation and death. Thus ICAM-1 permits the efficient initiation of positive selection, whereas B7-CD28 dependent co-signaling helps to enforce negative selection [269].

Self-tolerance of CD8(+) T lymphocytes does not develop in transgenic mice expressing MHC-I molecules in the thymus exclusively on cortical epithelial cells. In spite of the absence of MHC-I expression on professional APC, normal numbers of CD8(+) cells were observed in the periphery. Upon specific activation, transgenic $CD8^+$ T cells efficiently lysed syngeneic MHC class I^+ targets, indicating that thymic cortical epithelium (in contrast to medullary epithelium and antigen-presenting cells of hematopoietic origin) is incapable of tolerance induction. Thus, compartmentalization of the antigen-presenting cells is involved in thymic positive and negative selection [270].

Repertoire selection in the thymus is initiated following "priming" of the thymic microenvironment by plasma proteins, thereby ensuring the onset of positive selection to be delayed until the entry of extrathymic proteins to which self tolerance must be established [271]. Studies in transgenic mice revealed that diverse, low-abundance self-peptides play an important role in thymic positive selection and do not support the notion that dominant, high-abundance peptides may be critical for shaping the TCR repertoire [272].

How hematopoietic grafts exert their influence over the regenerating immune system is not completely understood. Hematopoietic stem cell (HSC) engraftment results in mixed T cell chimerism. HSC grafts induce tolerance to donor-matched neonatal heart grafts. T cells in chimeric mice rejected grafts only in the context of the donor MHC type. These findings conflict with the conventionally held view that radio-resistant host elements primarily dictate positive selection [273].

Although self MHC-II molecules do have a modest effect on long-term survival of individual $CD4^+$ T cells, interactions with specific TCR ligands are not required for peripheral naive $CD4^+$ T cell maintenance. The selective engagement of TCRs by self-peptide:MHC complexes does promote proliferation of $CD4^+$ T cells under severe lymphopenic conditions. This growth is associated with an activation marker phenotype that is different from that induced by antigenic stimulation. Most importantly, the ability of naive T cells to divide in response to *homeostatic stimuli* does not appear to be stringently dependent on TCR-self peptide:MHC interactions. Therefore, the factors regulating survival and *homeostatic expansion* of naive T cells in the periphery are not identical. A novel form of T cell proliferation exists that can occur independently of TCR signaling. This reflects another mechanism regulating homeostatic T cell expansion [274].

Apoptosis of thymocytes associated with thymic nurse cells (TNCs) is well-documented. TNCs selectively bind and internalize immature αβTCRlo $CD4^+$ $CD8^+$ thymocytes in vitro. A subset of the internalized population matures to the αβTCRhi CD69hi stage of development while the fraction that remains within the cytoplasm dies through the process of apoptosis. The TNC line tsTNC-1 maintains the ability to internalize αβTCRlo $CD4^+$ $CD8^+$ cells in vitro. TsTNC-1 cells express both FasL and TNFα. Apoptosis of triple positive cells was reduced approximately 50–68% in co-cultures of tsTNC-1 cells and thymocytes in the presence of either anti-TNFα or Fas-Fc [275].

In the mouse thymus CD4+ CD25+ regulatory T cells appear during natural selection of CD4+ T cells. The proportion of CD4+ CD25+ cells in the population of CD4+ thymocytes remains constant during maturation [276]. The development of regulatory CD4+ 25+ T cells is dependent on MHC class II-positive thymic cortical epithelium. A subset of CD4+ 25+ T cells is subject to negative selection on hematopoietic antigen-presenting cells [277].

Interactions with a single self-peptide can induce thymocytes that bear an autoreactive T cell receptor (TCR) to undergo selection to become CD4+CD25+ regulatory T cells. Selection requires a TCR with high affinity for a self-peptide. Thymocytes that bear TCRs with low affinity do not undergo selection into this pathway. This specificity for self-peptides directs the selection of CD4+CD25+ regulatory thymocytes by a process that is distinct from positive selection and deletion [278].

NK1.1+ T cells develop and function through interactions with CD1complexes. If the invariant chain (Ii) processing enzyme, cathepsin S is missing, NK1.1+ T cell selection in the thymus and function are impaired [279]. The differentiation of NK1.1+ αβT cells in the thymus is influenced strongly by productive α-chain rearrangements. This is especially true for developing thymocytes, which operates through positive selection on self-determinants. It applies to both classical and nonclassical αβT cells and therefore, represents a general phenomenon in intrathymic αβT lymphocyte development [280].

Epithelial T cells of the murine skin, called dendritic epidermal T cells (DETC), express exclusively an invariant γδTCR. Fetal thymic precursors of DETC immigrate to the skin before birth, and in adult mice γδ T cells are not found in other lymphoid or epithelial tissues. DETC precursors migrate to the gut as well as to the skin during fetal periods, but preferentially survive and expand in the skin after birth [281].

Mutation of T cell receptor signaling (Tec) kinases causes graded defects in thymocyte selection, leading to a switch from negative to positive selection. The reduction in both positive and negative selection and decreased CD4/CD8 ratios correlates with decreased biochemical parameters of TCR signaling [282].

During the maturation of thymocytes an increase occurs in activation thresholds, which correlates with TCR down-regulation. TCR efficiency is down-regulated during T cell maturation and will increase again with the differentiation of T cells into memory cells. It is an unexpected finding that the ability of the TCR to signal fully is in inverse correlation by its level of expression [283].

E2F1, one member of the E2F transcriptional factor family, plays an important role in the induction of apoptosis. E2F1 is required for the apoptosis of autoimmune immature T cells during thymic negative selection in vivo. TCR-mediated apoptosis coincides with the E2F1-dependent increase of p19-ARF mRNA and p53 protein levels. E2F1 is not required for the induction of apoptosis by glucocorticoids or DNA damage. Thus E2F1 has a specific role in a physiological apoptosis pathway that is uncoupled from a normal proliferative event [284]. Mice lacking the transcription factor, Schnurri-2 (Shn-2), had severely defective positive selection of CD4+ and CD8+ cells [285].

Mice lacking the orphan nuclear hormone receptor RORγ lose thymic expression of the anti-apoptotic factor Bcl-xL. RORγ thus regulates the survival of CD4+ 8+ thymocytes. RORγ was also required for development of lymph nodes and Peyer's patches, but not of splenic follicles. In the absence of RORγ, there was loss of a population of CD3-CD4+ CD45+ cells that normally express it [286].

The 2C murine T cell line bearing TCR alloreactive to H2Ld was reactive not only with H-

2Ld and H-2Kb, but also with nonclassical MHC class I products to enable positive selection of 2C+ T cells in the fetal and newborn thymus and to support their maintenance in the intestine [287]. TCR transgenic CD4+CD8+ thymocytes from MHC-deficient mice were stimulated using various presenting cells and ligands. The production of CD4+ T cells required continuous TCR signaling for full differentiation. Dendritic cells could initiate positive selection and could also support the TCR signaling necessary for late maturation. Unique epithelial ligands were not necessary for positive selection [288].

The CD45 protein tyrosine phosphatase regulates antigen receptor signaling in T and B lymphocytes. In the absence of CD45, TCR coupling to downstream signaling cascades is profoundly reduced. Moreover, in CD45-null mice, the maturation of CD4+CD8+ thymocytes into $CD4^+$ or $CD8^+$ thymocytes is severely impaired. The modulation of the strength of TCR-mediated signals can alter the outcome of thymic selection. This demonstrates that CD45 directly participates in the shaping of the T cell repertoire [289].

The homozygous Fas mutation (lpr/lpr) induced a marked reduction in $CD4^+$ $CD8^+$ double-positive thymocytes in mice with nonselecting background and the thymus showed severe cortical atrophy. This mutation also inhibited the activation of $CD4^+$ $CD8^+$ thymocytes in the process of positive selection, as indicated by the lower levels of CD5 and CD69 expressions on double positive thymocytes in lpr/lpr mice than in those of lpr/+ mice. A significant skewing from $CD4^+$ to $CD8^+$ single-positive thymocytes was found in lpr/lpr mice with nonselecting background compared with lpr/+ mice. Taken together, these results indicate that Fas is involved in thymic positive selection, as a co-stimulatory molecule but is not involved in the deletion of neglected thymocytes [290].

The NF-κB transcriptional pathway plays an important role in the negative and positive selection of immature thymocytes. This pathway can regulate the efficiency of TCR signaling [289]. TNF regulates mouse thymocyte production, inducing both apoptosis and proliferation of the most immature $CD3^-$ $CD4^-$ $CD8^-$ triple negative subset within a broad range of dosages (10^1–10^5 pg/ml) in the presence of IL-7. TNF receptor knockout mice showed thymic hypertrophy with a 60% increase in total thymocytes, with no effect on the CD4/CD8 subsets [292].

A cellular ligand for herpes virus entry mediator and lymphotoxin receptor (LIGHT), a member of the TNF superfamily, can delete immature T cells in vivo and thus play a critical role in negative thymocyte selection [293].

Transforming growth factor β1 was found to control the life and death decisions of T lymphocytes. Both thymic and peripheral T cell apoptosis was increased in mice lacking TGFβ1. TGFβ1 was present in the mitochondria of normal T cells, and the absence of TGFβ1 resulted in disruption of mitochondrial membrane potential, which marks the point of no return in a cell condemned to die. This TGFβ1-dependent regulation of viability is independent of the TGFβ1 membrane receptor-Smad3 signaling pathway, but associated with a mitochondrial antiapoptotic protein, Bcl-XL [294]. A novel thymocyte-specific protein, named TARPP, has been described. It is a cytosolic protein expressed at high levels in immature thymocytes and its expression is switched off as a consequence of TCR engagement during positive selection [295].

During age-related involution an increased density of sympathetic noradrenergic (NA) innervation and concentration of norepinephrine (NE) have been observed in the thymus. Two-month and 18-month old BALB/c mice were treated with the non-selective β-adrenoceptor antagonist, nadolol. In old, but not in young mice, the frequency of the immature CD4-8- population was increased, and the intermediate $CD4^+8^+$ population was reduced. A corresponding increase in the frequency of mature $CD4^-8^+$, but not $CD4^+8^-$ cells was observed [296].

6.2. Suppression

That immune reactions may be actively suppressed by specialized regulatory cells was first proposed by Gershon [297]. After an enthusiastic study of the subject, which produced a voluminous literature, doubts were raised about the existence of true suppressor cells. It was felt that the phenomenon of suppression could be explained by other means, like cytotoxicity, for instance. However, the field has been rejuvenated with the discovery of true regulatory cells. These regulatory T cells are required to control immune responses to certain self-antigens. Regulatory T cells also have the potential to control inflammation. In spite of these developments, it is not well understood how the adaptive immune system achieves self-non-self discrimination [298].

There is now abundant evidence indicating that the normal T cell repertoire contains overtly autoreactive T cells whose pathogenic potential is held in check by the activity of a distinct subset of peripheral T cells, the so-called *regulatory or suppressor T cells (TR)* [299]. Regulatory CD4(+) T cells are generated in the thymus and can be distinguished from effector cells by the expression of unique membrane antigens [300]. CD4+ regulatory T cells that specialize in the suppression of immune responses play a key role in the control of immune pathology in animals and in man. These cells can inhibit harmful immunopathological responses directed against self or foreign antigens [301–303].

$CD4^+CD25^+$ regulatory T cells are potent suppressors of $CD4^+CD25^-$ T cell activation *in vitro* and inhibit organ-specific *autoimmune diseases*. $CD4^+CD25^+$ T cells also suppress both the proliferation and IFNγ production by $CD8^+$ T cells induced either by polyclonal or antigen-specific stimuli. $CD4^+CD25^+$ T cells inhibit the activation of $CD8^+$ responders by inhibiting both IL-2 production and up-regulation of IL-2Rα-chain (CD25) expression. Antigen specific suppression is mediated via a T-T interaction in the complete absence of APC [304]. $CD4^+CD25^+$ T cells regulate the activity of autoreactive T cells. If these cells are depleted a wide-spectrum of organ-specific autoimmune disease will develop. Following their activation via their TCR these suppressor cells downregulate the responses of $CD4^+CD25^-$ effectors by a T-T interaction [305].

Specialized regulatory T (TR) cells exist as part of the normal immune repertoire, preventing the development of pathogenic responses to both self- and intestinal antigens. The TR cells that control *intestinal inflammation* express the same phenotype ($CD25^+CD45RB^{low}CD4^+$) as those that control autoimmunity. The immune-suppressive function of these cells *in vivo* is dependent on signaling via the negative regulator of T cell activation, *cytotoxic T lymphocyte-associated antigen 4* (CTLA-4), as well as on secretion of the immune-suppressive cytokine TGFβ [306].

Suppressor T cells inhibit the proliferative response of naive $CD4^+$ T cells. Cell contact was required for efficient suppression, and known inhibitory cytokines such as IL-4, IL-10, and TGFβ were not involved. Suppressor T cells inhibited IL-2 production by naive $CD4^+$ T cells, and the addition of exogenous IL-2 diminished the suppressed activity while having little activity on tolerant T cells. Suppression was abolished by the elimination of $CD25^+$ T cells in the tolerant $CD4^+$ T cell subset. Suppressor T cells apparently control only a small population of nonanergic cells in the periphery and may be viewed as a *homeostatic mechanism* [307,308].

In mice, allergen-specific *TGFβ-producing $CD4^+$ T cells* induced in the respiratory tract controlled *cutaneous inflammatory responses* by Th1 or Th2 cells either directly by TGFβ or indirectly through *IL-10* induction [309]. $CD4^+CD25^+$ T cells do produce high levels of TGFβ1 and IL-10 compared with $CD4^+CD25^-$ T cells. When stimulated, TGFβ1 (but not other cytokines), is further enhanced by costimulation of CTLA-4. $CD4^+CD25^+$ T cells suppress the proliferation of $CD4^+CD25^-$ T cells. This suppression is abolished by the presence of anti-TGFβ. $CD4^+CD25^+$ T cells also suppress B cell immunoglobulin production which depends on TGFβ as well.

Stimulated $CD4^+CD25^+$ T cells, but not $CD4^+CD25^-$ T cells, express high and persistent levels of TGFβ1 on the cell surface. This, plus the fact that no evidence was found for a soluble factor that mediates suppression, strongly suggests that $CD4^+CD25^+$ T cells exert immunosuppression by a cell-cell interaction involving *cell surface TGFβ1* [310]. Murine $CD4^+$ T cells were tolerized to alloantigen via *ex vivo* CD40 ligand/CD40 or D28/CTLA- 4/B7 blockade. This resulted in *tolerance to alloantigen in vivo*. $CD4^+$ $CD25^+$ T cells were potent regulators of alloresponses. The depletion of $CD4^+$ $CD25^+$ T cells from the $CD4^+$ responder population completely abrogated tolerance induction. This could be reverted by adding back $CD4^+$ $CD25^+$ T cells to $CD4^+$ $CD25^-$ cultures [311].

$CD4^+CD25^+CD45RO^+$ T cells (mean 6% of $CD4^+$ T cells) are present in the blood of adult healthy volunteers. These $CD4^+$ $CD25^+$ T cells do not constitute conventional memory cells, but rather, they are *regulatory cells* exhibiting properties identical to their rodent counterparts. For example, CTLA-4 (CD152), which is essential for the *in vivo* suppressive activity of $CD4^+$ $CD25^+$ T cells, was constitutively expressed, and remained strongly up-regulated after stimulation. The regulatory cells did not proliferate in response to stimulation via their T cell receptor for antigen. This anergic state was partially reversed by IL-2 and IL-15. After stimulation with allogeneic (but not syngeneic) mature dendritic cells, or by plate-bound anti-CD3 plus anti-CD28 the $CD4^+$ $CD25^+$ T cells released *IL-10*, and in co-culture experiments suppressed the activation and proliferation of $CD4^+$ and $CD8^+$ T cells. This *suppression* proved to be IL-10 independent, yet *contact dependent* as in the mouse [312]. Human $CD25^+CD4^+$ T regulatory cells isolated from peripheral blood failed to proliferate and displayed reduced expression of CD40 ligand (CD40L) in response to T cell receptor-mediated polyclonal activation, but strongly up-regulated CTLA-4. Human $CD25^+CD4^+$ T cells also did not proliferate in response to allogeneic APC, but they produced IL-10, TGFβ, low levels of IFNγ and no IL-4 or IL-2. $CD25^+CD4^+$ T cells strongly inhibited the proliferative responses of both naive and memory $CD4^+$ T cells to alloantigens, but neither IL-10, TGFβ, nor CTLA-4 seemed to be directly required for their suppressive effects. $CD25^+CD4^+$ T cells *could be expanded in vitro* in the presence of *IL-2* and allogeneic feeder cells and maintained their suppressive capacities [313]. The human suppressor T cell clone III-1-C5 suppressed the mixed lymphocyte reaction (MLR). It did not suppress mitogen (PHA, Con A, PWM) response nor proliferation induced by exogeneous rIL-2. Suppression was achieved by *direct cell-to-cell contact* with responder cells and not by soluble factors [314]. Human *lamina propria lymphocytes from the jejunum* first respond to lactoglobulin *in vitro* by proliferation and IFNγ production, then developed antigen nonspecific suppression [315]. *Allergic contact dermatitis* is characterized by the development of T-cell-based effector and regulatory mechanisms [316].

Mature CD4-CD8- *double negative* αβ *T cells* (DNTC) in the periphery of TCR transgenic mice are resistant to clonal deletion in cognate antigen-expressing (Ag+) animals. Despite appearing functionally anergic, when challenged with Ag *in vitro*, Ag-experienced DNTC exhibit a lowered threshold of activation, cytolytic activity, and rapidly secrete IFNγ. These memory-like DNTC also possess potent immunoregulatory properties, competing effectively for bystander-produced IL-2 and suppressing autoreactive CD8+ T cell proliferation via a *Fas/FasL-dependent cytolysis*. DNTC recovered from Ag+ mice possessed markers and attributes characteristic of naïve CD8+ T cells that have undergone *homeostasis-induced proliferation*. This suggested that they might be derived from a similar peripheral expansion process. Naive DNTC adoptively transferred into Ag-bearing hosts rapidly acquire markers and functional attributes of DNTC that have continually developed in the presence of Ag. Thus, the peripheral selection and maintenance of such autoreactive cells may serve to negatively regulate potential autoimmune

T cell responses [317].

Low avidity engagement of TCR by low densities of agonist peptides induced *anergy* in T cell clones. *Memory CD4⁺ T cells* could be rendered anergic by presentation of low densities of agonist peptide-MHC complexes *in vivo*. Anergy was the most likely mechanism because IL-2 restored reactivity in specific T cells. Deletion, ignorance, or immunosuppression were ruled out. CTLA-4 played a critical role in the induction of anergy. There was an increased surface expression of CTLA-4 on anergized T cells, and the injection of anti-CTLA-4 blocking antibody restored anergy *in vivo* [318].

Fractionated irradiation directed to the lymphoid tissues and depletive anti-T cell Abs given to C57BL/6 and BALB/c mice results in a marked change in the residual host T cells, such that $NK1.1^+$ or $DX5^+$asialo-$GM1^+$ T cells become the predominant T cell subset in lymphoid tissues. These cells function as *"natural suppressor" T cells* and protect hosts from graft-vs-host disease after the infusion of allogeneic bone marrow and peripheral blood cells that ordinarily kill hosts conditioned with sublethal or lethal total body irradiation. Protected hosts become stable mixed chimeras. Protection afforded by $NK1.1^+$ and $DX5^+$asialo-$GM1^+$ T cells derived from either donors or hosts conditioned with lymphoid irradiation is dependent on their secretion of high levels of IL-4 [319]. Natural suppressor cells in the bone marrow of mice and/or humans suppress immune responses and inhibit the mixed leukocyte reaction, graft-versus-host disease, and systemic autoimmunity. These suppressor T cells are the *$CD4^-CD8^-$ TCR-$\alpha\beta^+$ phenotype*. These cells are members of the $NK1.1^+$ T-cell family that secrete high levels of IFNγ and IL-4 after initial activation. The suppressive activity of the bone marrow $NK1.1^+$ T cells is dependent on their rapid secretion of high levels of IL-4. This unique cytokine secretion is not observed in conventional $NK1.1^-$ T cells and can down-regulate the function of the latter cells [320]. Cord blood (CB) natural killer (NK) cells induced a dose-dependent inhibition of autologous T lymphocyte proliferation in mixed lymphocyte cultures (MLC). Irradiated CB NK cells were also effective. Adult blood NK cells inhibited the alloreactive T cells in the MLC using adult T lymphocytes as responders [321].

Anterior chamber (AC) tolerance is a thymus-dependent phenomenon initiated by the formation of *regulatory $\alpha\beta$ cell receptor-positive $CD4^-$ $CD8^-$ thymocytes* (THYR). $F4/80^+$ $CD1^+$ peripheral blood mononuclear cells from mice whose ACs were injected with trinitrophenol-bovine serum albumin induce the appearance of $NK1.1^+$ THYR cells in naive recipients within 24 h of i.v. infusion. These NK cells induce (or generate) *suppressor-effector T cells* in the spleens of adoptive recipients. The suppressor-effector spleen cells, but not the NK cells themselves, directly inhibit the expression of delayed-type hypersensitivity in sensitized recipients. Peripheral blood mononuclear cells from anterior chamber-injected mice do not induce immune deviation in thymectomized recipients. Therefore, tolerance induction in the AC is a centrally induced dominant tolerance mediated by *CD-1-dependent NK T cells of recent thymic origin*. This is a novel tolerance induction pathway by which blood-borne antigen-presenting cells generated by antigen injection into an immunologically privileged site transport antigen to the thymus and induce the formation and export of THYR cells [322]. *Ocular immune privilege* is based on the participation of ocular APC and splenic B and T cells. Splenic B cells, incubated with anterior chamber APC *in vitro*, were capable of inducing *ocular tolerance* when transferred to naive mice. The induction of tolerance required the normal expression of ss(2)-microglobulin on both the B cell and APC, but not on the T suppressor cells. Identical histocompatibility antigens were required between the B cells and regulatory T cells at the TL/Qa region for successful interaction [323].

Mucosal tolerance was studied in autoimmune diabetes using the non-obese diabetic mouse model. Treatment of pre-diabetic mice with insulin, by aerosol or intranasal delivery, reduced the incidence of diabetes. This was associated with induction of CD8(αα) γδ T cells, a small numbers of which prevented the adoptive transfer of diabetes. *Regulatory γδ T cells* secrete IL-10 in pancreatic lymph nodes, which could account for the anti-diabetic and bystander suppressor effect of naso-respiratory insulin. In contrast, intranasal (pro)insulin peptide, or oral insulin which is degraded in the gut, induces *CD4 regulatory cells* [324].

Myeloid cells that are double positive for the CD11b and Gr-1 markers can induce apoptotic death of $CD8^+$ T cells. The appearance of a $CD11b^+$ /$Gr\text{-}1^+$ *inhibitory macrophages* (iMacs) could be attributed to high levels of GM-CSF *in vivo*. Deletion of iMacs *in vitro* or *in vivo* reversed the depression of $CD8^+$ T-cell function. iMacs were positive for CD31, ER-MP20 (Ly-6C), and ER-MP58, markers characteristic of granulocyte/monocyte precursors. IL-4 increased iMac inhibitory activity and these cells could be differentiated into a non-adherent population of mature and activated *dendritic cells* when cultured with IL-4 and GM-CSF. Therefore, a common $CD31^+$ /$CD11b^+$ /$Gr\text{-}1^+$ progenitor can give rise to cells capable of either activating or inhibiting the function of $CD8^+$ T lymphocytes, which depends on the *cytokine milieu* that prevails during antigen-presenting cell maturation [325].

Dendritic cells that interact with antigen-specific T suppressor cells become *tolerogenic*, inducing T helper *anergy*. Immature dendritic cells stimulated by T suppressor cells differentiate into mature dendritic cells with a distinct phenotype [326]. *Anergic T cells function as suppressor cells* by inhibiting antigen presentation by DC via a cell contact-dependent mechanism. The response to co-immobilized anti-CD3 and anti-CD28 antibodies was not inhibited. Suppression was not reversed by addition of neutralizing anti-IL-4, anti-IL-10, and anti- TGFβ antibodies [327].

Anergic cells generated in mixed cultures of mouse lymphocytes by blocking CD28/B7 costimulatory pathway suppressed *islet cell allograft rejection* after adoptive transfer. The anergic state was reversed by the addition of exogenous IL-2 [328]. Beads coated with anti-CD3 and anti-CD28 antibodies as *surrogate APCs* were used to study the interaction of $CD4^+CD25^+$ and $CD4^+CD25^-$ T cells in vitro. $CD4^+CD25^+$ T cell-induced suppression was not abrogated by blocking the B7-CTLA-4 pathway. $CD4^+CD25^-$ T cells were separated from the $CD4^+CD25^+$ suppressor cells after 24 h of coactivation by the Ab-coated beads. $CD4^+CD25^-$ T cells were unable to proliferate or to produce IL-2 upon restimulation. The induction of this anergic phenotype of T cells correlated with the up-regulation of a *novel gene related to anergy* in lymphocytes, that acts as a negative regulator of IL-2 transcription. This constitutes a novel mechanism of anergy induction in the presence of costimulation [329].

E2F nuclear factors are important regulators of proliferation, differentiation, and apoptosis. E2F2(-/-) mice develop late-onset autoimmune disease, characterized by widespread inflammatory infiltrates, glomerular immunocomplex deposition, and anti-nuclear antibodies. E2F2-deficient T lymphocytes show enhanced TCR-stimulated proliferation and a lower activation threshold. This leads to the accumulation of autoreactive effector/memory T lymphocytes. Rather than functioning as a transcriptional activator, E2F2 appears to function as a transcriptional repressor acting particularly on E2F1 [330].

Interleukin-10 elicits tolerance in T cells by selective inhibition of tyrosine phosphorylation in the CD28 co-stimulatory pathway. IL-10 only inhibits T cells stimulated by low numbers of triggered TCR, which depend on CD28 co-stimulation [331].

The repeated injection of bacterial *superantigens* (SAg), such as staphylococcus enterotoxin (SE) A or B, induces in mice a state of unresponsiveness characterized by the lack of secretion of Th1 lymphokines, such as IL-2 and IFNγ following subsequent SAg challenge. The production of both IFNγ and IL-2 following SEB stimulation was greatly impaired in co-cultures containing CD4(+) T cells, but not CD8(+) T cells, isolated from unresponsive mice. The production of both IFNγ and IL-2 responses to SEB was dramatically reduced in animals adoptively transferred with unresponsive spleen cells. This suppression was abrogated in recipients injected with neutralizing anti-IL-10 antibodies. In animals made unresponsive to SEB, the reactive CD4(+) T cells expressed high levels of CTLA-4, a molecule that plays an essential role in the suppressive function of regulatory T cells [332]. Small dose (2 mg/g) of *staphylococcal enterotoxin B* (SEB) injected into mice induced an early emerging unresponsiveness in intrahepatic-lymphocyte populations (IHLs). IHLs prepared at 3 to 24 hours after SEB injection failed to proliferate in response not only to SEB but also to SEA, exhibiting ligand-nonspecific unresponsiveness. Spleen cells (SPCs) and mesenteric lymph-node cells showed transient proliferation to the same stimulus. Mac-1^{hi}Ia^{+} adherent cells, largely *Kupffer cells* activated by SEB, nonspecifically suppressed the proliferation of Th1 cells via *nitric oxide* production [333].

Copolymer 1 (Cop 1, Copaxone, glatiramer acetate) is a potent inducer of *Th2 regulatory cells* in both mice and humans. Highly reactive Cop 1-specific T cell lines were established from both brains and spinal cords of Cop 1-treated mice. These cells secreted IL-4, IL-5, IL-6, IL-10, and TGFβ and cross-reacted with myelin basic protein. Adoptively transferred labelled Cop 1-specific suppressor cells were found in brain sections 7 and 10 days after their injection to the periphery and exerted a therapeutic effect. The inflammatory cytokine IFNγ was decreased and the anti-inflammatory cytokine *IL-10 was increased* in response to the autoantigen, MBP [334].

TGFβ can induce certain CD4^{+} T cells in the naive (CD45RA^{+}RO^{-}) fraction in human peripheral blood to develop powerful, contact-dependent suppressive activity that is not antagonized by anti-TGFβ or anti-IL-10 monoclonal antibodies. On naive CD4^{+} T cells TGFβ up-regulated CD25 and CTLA-4 expression, increased their transition to the activated phenotype, and decreased activation-induced apoptosis. Suppressive activity was present in the CD25^{+} fraction. These CD4^{+}CD25^{+} regulatory cells prevented both CD8^{+} T cell proliferation in response to alloantigens and the development of cytotoxic effector cells. These regulatory cells exerted their suppressive activities in remarkably low numbers and maintained these effects even after they are expanded. Once activated, their suppressive properties were not antigen specific. Although <1% of naive CD4+ T cells expressed CD25, the depletion of this subset before priming with TGFβ markedly decreased the generation of suppressive activity [335]. *Human self-MHC-reactive (autoreactive) T cell clones* exhibited helper function for B cell Ig production when cultured with non-T cells alone, and they exhibit suppressor function when cultured with PWM- or rCD40 ligand-activated non-T cells. This suppression was mediated by TGFβ but not by IL-10 or IFNγ. The self-MHC-reactive clones also inhibited the proliferation of primary CD4+ T cells and of tetanus toxoid-specific T cell clones, but in this case the inhibition was mediated by both *IL-10 and TGFβ*. The addition of CTLA-4/Fc or blocking F(ab')2 anti-CTLA-4 monoclonal antibody, plus optimally stimulated non-T cells, to cultures of self-MHC-reactive clones inhibited the induction of TGFβ but not IL-10 or IFNγ production. The activated self-MHC-reactive T cells have the cytokine phenotype of Th3 or T regulatory cell 1[336]. A 4- kDa component with *TGFβ2 activity* recovered from uterine luminal secretions of gilts suppressed the lytic responses of porcine NK cells [337].

A novel function of *IL-6* is the negative regulation of CD4+ Th1 cell differentiation. IL-6-directed CD4+ Th2 differentiation is mediated by IL-4, but the inhibition of Th1 differentiation is independent of IL-4. IL-6 upregulates *suppressor of cytokine signaling 1* (SOCS1) expression in activated CD4+ T cells, thereby interfering with STAT1 phosphorylation induced by IFNγ.

This prevents the autoregulation of IFNγ (gene expression by IFNγ during CD4+ T cell activation, thereby preventing Th1 differentiation [338].

Vasoactive intestinal peptide (VIP) and *pituitary adenylate cyclase-activating polypeptide* (PACAP) protect CD4+ T cells against Ag-induced apoptosis by down-regulating the expression of FasL. VIP/PACAP *inhibit the FasL/Fas-mediated cytotoxicity* of T1 effectors and do not affect T2 cytotoxicity. Similar effects were observed *in vivo* [339]. PACAP greatly decreased arthritis frequency and severity in mice, improved clinical symptoms, ameliorated joint damage, and blocked both the inflammatory and autoimmune mediators [340].

1,25-dihydroxyvitamin D3 (VD3), exerts immunosuppressive activity. The hormone preferentially targets helper T cell activity (Th1) by inhibiting the secretion of both IL-2 and IFNγ by Th1 and by suppressing the secretion pro-Th1 cytokine IL-12 by antigen-presenting cells. The active metabolite further inhibits class II antigen expression and enhances suppressor cell activity. In animal models VD3 prevents the development of experimental autoimmune encephalomyelitis, reduces the incidence of diabetes, and attenuates murine lupus. The hormone also prolongs graft survival. In humans, non-classical use of VD3 has led to an anti-proliferative effect in psoriasis, antineoplastic effect in prostate cancer, and immunomodulatory effect in scleroderma [341].

In mice β-*estradiol* (E2) treatment decreased the PPD-specific DTH response, which coincided with a decrease of leukocytes in the draining lymph nodes (DLN) and spleen. E2 also suppressed *in vitro* the PPD-specific proliferative response of DLN and spleen cells from PPD-primed mice. E2 treatment suppressed IL-2 and IFNγ production in response to PPD. In contrast, *IL-4 and IL-10* gene expression was enhanced. Spleen APC from E2-treated mice failed to induce optimum proliferation of the PPD-primed T cells in response to PPD in vitro. The impaired APC function by E2 was not due to induction of suppressor cell activity [342].

The hypothesis was proposed on the basis of current evidence that autoimmune diseases result from deficits in autoreactive regulatory T cell generation and/or function. Natural tolerance is achieved through evolutionarily selected developmental genetic programs that: (i) ensure auto(multi)reactivity at the outset of the system; (ii) establish a cellular composition of thymic stroma that grows and activates regulatory (autoreactive) T cells in early development; (iii) initiate lymphocyte differentiation and population dynamics, that results in peripheral "education" of regulatory tissue-specific cells, while allowing for "unregulated" clonal responses to non-self. In the present model, the suppressor/non-suppressor discrimination is "supraclonal" and "dominant", related to other "systemic" properties such as the regulation of total lymphocyte numbers, the "open-endedness" of repertoires, and their differences in health and disease. Dominant tolerance models also solve the paradox that pathogenic autoreactivity is rare, in spite of the extensive V-region degeneracy of lymphocyte recognition. The time is ripe to move from component (clonal) analysis to system's biology [343]. Viral infections can be responsible for the autoimmune reactions. Chronic inflammation associated with viral persistence initiates a multi-step process leading to autoimmunity [344].

6.3. Apoptosis and cytotoxicity

The discovery of Fas/FasL defects in lpr and gld lupus mice was the first example of spontaneous mutations of apoptosis-promoting genes being associated with systemic autoimmunity. Today the emerging picture is that multiple genetic contributions can independently lead to systemic autoimmunity in mice, which reinforces the view that human lupus may be similarly composed of diverse genotypes [345].

The Fas receptor (APO-1 or CD95) is a key initiator of programmed cell death in a variety of cell types. $CD4^+$ T cells are unique in their ability to commit "suicide" by stimulating their own Fas receptors via secreted or membrane-bound FasL. T cells also trigger apoptosis in B cells, macrophages, and other cell types through FasL. These interactions negatively regulate the immune system and also contribute to immunopathology by causing Fas-mediated damage in hepatitis and in other organ-specific autoimmune diseases. This dual role of Fas in the immune response complicates the understanding of its role in disease states. The disruption of this pathway in mice results in systemic autoimmunity and a predisposition toward lymphoid malignancies. Similarly, Fas-deficient patients exhibit an autoimmune lymphoproliferative syndrome [346].

A requirement for T-cell receptor (TCR)-mediated signals in maintaining peripheral CD4+ T-cell longevity has been demonstrated. Naive CD4+ T-cells also require engagement of their CD4 co-receptors by peripheral, class II MHC-bearing cells for their survival. Thus, interactions between CD4 and MHC-II molecules contribute to the regulation of *homeostasis* in the peripheral immune system [347].

Antigen presenting murine dendritic cells were relatively insensitive to Fas-mediated killing by the agonist monoclonal antibody, Jo-2, as well as by Fas ligand expressed on the CD4+ T cell line, A.E7. Apoptosis could be induced with C2-ceramide, suggesting that signals proximal to the generation of ceramide might mediate resistance to Fas. The expression of the IL-1-converting enzyme-inhibitory protein, which is an intracellular inhibitor of apoptosis, was significantly higher in Fas-resistant DCs than in Fas-sensitive macrophages [348].

In mice deficient in $CD8^+$ T cells, the down-regulation of $CD4^+$ V $\beta 8^+$ T cells below baseline is not observed. Following SEB administration, CD8+ T cells emerge that preferentially kill subpopulations of activated $CD4^+$ V $\beta 8^+$, but not $CD4^+$ V $\beta 8^+$ T cells in vitro. This TCR V β8-specific cytotoxicity is dependent on β2-microglobulin and is inhibited by antisera specific for Qa-1 but not by antibody to MHC-Ia. These data suggest the idea that the specificity of immune regulation may involve $CD8^+$ T cell recognition of TCR V β determinants and Qa-1 molecules expressed on CD4+ T cells [349]. TCR peptide-primed CD8+ CTL recognized class I-restricted, natural V β/Jβ-TCR epitopes. Such anti-TCR CTL exerted V β-specific immunoregulation of the T cell system suppressing their functional reactivity (partial anergy) without deleting them [350].

Fas ligand (FasL), expressed on activated T cells, plays a central role in regulating the immune response by inducing apoptosis in activated lymphocytes through binding to its receptor, Fas. Interferon-γ-inducing factor (IGIF) [351] selectively enhances the FasL-mediated cytotoxicity of cloned murine Th1 cells, but not Th0 or Th2 cells. Anti- IFNγ (antibody (Ab) did not block the IGIF-induced cytotoxicity of Th1 cells, nor did IFNα IFNγ, or TNFα augment the cytotoxic activity. This indicates that the enhanced cytotoxicity of Th1 cells was mediated by IGIF. IL-12 was also found to enhance the FasL-mediated cytotoxicity of Th1 cells. Thus IGIF (now IL-18) might play a potential role in immunoregulation or in inflammation by augmenting the functional activity of FasL on Th1 cells [352].

NK1.1^+ cells may be subdivided by their different expression pattern of Ly-6C antigen. Both NK1.1^+ Ly-6C^+ and NK1.1^+Ly-6C^- cells responded to the stimulation with IL-2 plus IL-12 and showed strong cytotoxicity against YAC-1 cells. However, after cytokine activation only NK1.1^+ Ly-6C^+ cells, but not NK1.1^+ Ly-6C^- cells, produced high levels of IFNγ. NK1.1^+ Ly-6C^+ cells were separated further to NK1.1^+ $CD3^-$Ly-6C^+ NK cells and NK1.1^+ $CD3^+$ Ly-6C^+ NKT cells. Although, both cell types showed the same level of cytotoxicity, it was clearly demonstrated that NK1.1^+ CD3+Ly-6C^+ NKT cells were major immunoregulatory cells that produced IFNγ in response to IL-12 alone or IL-2 plus IL-12 [353].

Murine lung microvascular endothelial cells (EC) constitutively express low levels of MHC-I and CD80 molecules. IFNγ treatment of EC resulted in a marked up-regulation of MHC-I, but not CD80 expression. Resting T cells were stimulated to divide by quiescent peptide-prepulsed EC, while peptide-pulsed, cytokine-activated EC lost the ability to induce T cell division. Antigen presentation by cytokine-activated EC induced $CD8^+$ T cell hyporesponsiveness. The immunogenicity of activated EC could be restored by adding nonsaturating concentrations of anti-H2-K^k antibody. Activated peptide-pulsed EC were killed more efficiently than were resting cells by fully differentiated effector $CD8^+$ T cells [354].

The treatment of Kuppfer cells with IFNγ increased FasL mRNA levels, which was followed by an overexpression of the soluble Fas. Supernatants from 24-hr but not from 48-hr IFNγ-treated Kupffer cells killed lymphocytes by a Fas-dependent mechanism. Cyclosporin A inhibited CD95 ligand expression in Kupffer cells and lymphocyte killing. In liver parenchymal cells, IFNγ increased mRNA levels of the transmembrane Fas and the sensitivity of these cells toward Fas-mediated apoptosis. This indicates that Kupffer cells regulate programmed cell death of T lymphocytes and hepatocytes during immune/inflammatory reactions [355].

The apoptosis of microglia and CNS macrophages in experimental autoimmune encephalitis (EAE) of rats is not mediated through the Fas pathway, and Bcl-2 expression does not protect these cells from apoptosis. The high level of microglial apoptosis in EAE may be an important homeostatic mechanism for controlling the number of microglia in the CNS following microglial activation and proliferation. However, the expression of FasL by macrophages and microglia contributes to immunoregulation by interactions with Fas+ oligodendrocytes and T cells and may also be involved in the pathogenesis of EAE [356].

Cortical cells of the adrenals express MHC-II antigens, Fas and FasL. This allows for physiological and pathological interactions with T-cells, which has relevance to the immune response and to immune and/or adrenal pathology during disease [357].

Lung inflammation must be regulated tightly to preserve gas exchange. In primed C57BL/6 mice intratracheal challenge with the T cell-dependent Ag sheep erythrocytes, the apoptosis of alveolar and interstitial lymphocytes was demonstrated. The apoptotic lymphocytes were exclusively $CD4^-$, $CD8^-$, $B22^-$, but many were $CD3^+$ and Thy 1^+. Pulmonary lymphocyte apoptosis depended on the Fas-Fas-L pathway. These results demonstrate that *in situ* lymphocyte apoptosis contributes to the termination of immune responses in nonlymphoid organs and may be important in inducing tolerance to repeated antigen exposure [358].

Defective lymphocyte apoptosis caused by mutations of the Fas gene can result in a severe autoimmune lymphoproliferative syndrome (ALPS) in humans. In 9 unrelated children ALPS manifested by moderate to massive splenomegaly and lymphadenopathy, hypergammaglobulinemia, autoimmunity, B-cell lymphocytosis, and the expansion of an unusual population of $CD4^-$ $CD8^-$ T cells that express the αβ TCR. Lymphocytes from all patients showed defective apoptosis in vitro. Eight patients had heterozygous mutations of the Fas gene and one ALPS patient lacked a Fas gene mutation. Healthy relatives with Fas mutations were identified in 7 of 8 ALPS kindreds. These relatives showed *in vitro* abnormalities of Fas-mediated lymphocyte apoptosis, but clinical features of ALPS were not present in the majority of these individuals [359].

In patients infected with schistosomes cell death induced by combinations of specific antigen and cytokines corresponds with specific clinical presentations. Soluble egg antigen increased the level of apoptosis only in T cells from patients with the intestinal form. Stimulation with soluble lung worm antigen did not induce apoptosis in T cells from patients with the different clinical forms of *Schistosoma mansoni* induced disease [360].

In rats with lateral hypothalamic lesions spleen lymphocyte apoptosis significantly increased after 6 hrs, spleen weights and the number of splenocytes decreased significantly at 24 hrs [361]. Serotonin protected human NK cells from oxidatively induced inhibition inflicted by autologous monocytes *in vitro*. Serotonin maintained the activation of NK cells induced by IL-2 despite the presence of inhibitory monocytes. Serotonin scavenged reactive oxygen species derived from the H_2O_2-myeloperoxidase (-MPO) system. Thus, serotonin may serve to protect NK cells from oxidative damage at inflammatory sites [362].

6.4. Stromal regulatory mechanisms

For a long time the stroma of tissues and organs was regarded as structural elements that hold together the functional "parenchymal" cells in the various organs and tissues. With the discovery of proinflammatory and chemotactic cytokines (chemokines), which are produced widely by stromal cells as well as cells of the immune system, stromal cells have emerged as important local regulatory cells. Now it is certain that each organ and tissue has a resident or "stromal" immune system, which does not only provide local defence against invaders, but also participates in the physiology and pathophysiology of the particular organ/tissue. Phagocytic and antigen presenting cells, T lymphocytes and in certain organs and tissues (e.g. mucosa, mamary and salivary glands) B lymphocytes, reside in tissues. T lymphocytes continuously recirculate between the blood and the extravascular compartment. It is also emerging that the various organs and tissues have their own endocrine system. This, coupled with the fact that innervation is also present, allows for the local interaction of the nervous, endocrine and immune system in the regulation organ/tissue function. One should take notice that in lymphoid organs (e.g. bone marrow, thymus, spleen, lymph nodes and gut and mucosa associated lymphoid tissue) the lymphocytes and other leukocytes are regarded as "parenchymal" as they represent the dominant function of these organs. In all other tissues and organs the lymphocytes, macrophages are designated as "stromal" cells. As already alluded to specialized antigen presenting cells (e.g. Kupffer cells, Langerhans cells, dendritic cells, microglia) are resident in certain organs and tissues as are tissue mast cells, which serve as effector cells for both the immune and endocrine system as well as fulfil physiological roles. Below a brief overview is given of stromal regulatory mechanisms.

6.4.1. Stromal regulatory cells

Chemokine expression (mRNA/ protein) was examined in cultured fibroblasts isolated from a variety of human tissues and different pathologic states: normal bone marrow vs. myelometaplastic spleen, normal lung vs. metastasis stroma, and normal breast vs. radiation fibrosis and tumor stroma. In all fibroblasts, transcripts for IL-8, SDF-1, monocyte chemotactic protein (MCP)-1 and eotaxin were detected. The production of IL-8 was abundant in most of the fibroblasts studied, but fibroblasts from lung and pathologic breast tissue produced significantly less. Eotaxin production was low in most fibroblasts except in those isolated from myelometaplastic tissue where it was highly produced. MCP-4, RANTES and macrophage inflammatory protein-1α were found to be expressed only in fibroblasts from select tissues. Most of the fibroblasts expressed the CD40 antigen [363].

Pre-adipocytes exhibit functional features of macrophages, such as phagocytosis and anti-microbial activity. This suggests that pre-adipose cells could play a role in the inflammatory process or immune response. Obese (ob/ob) mice exhibited strong decrease in anti-microbial activity of both macrophages and stroma-vascular fraction (SVF) of the adipose tissue. This defect is corrected in SVF, at least in part, by an enhancement of phagocytosis that does not seem to be due to an increase in macrophages. Leptin treatment of SVF and macrophages from obese mice did not restore their immune defects. Thioglycollate induced inflammation led to an increase in macrophage activity in both the lean and obese strains. Stimulation was also observed in SVF from lean but not from obese mice [364].

Keratocytes in the corneal stroma of the mouse can function as antigen-presenting cells and, thus, may play a role in immune-mediated stromal inflammation such as herpetic stromal keratitis [365].

Thymic neural-crest-derived stromal cells are involved in a neural differentiation pathway and are able to produce neurotrophic factors and neurokines, such as epidermal growth factor, neurotransmitters and neurotropic cytokines, such as IL-6 and ciliaty neurotrophic factor. EGF acts on epithelial stromal cells and may also affect T-cell differentiation. The neurotrophin, brain-derived neurotrophic factor, and its specific receptors may have a role in sustaining thymocyte precursor survival and supporting the thymocyte differentiation process [366]. The MHC-II$^+$ thymic epithelium and fibroblasts, which are essential requirements for development of CD4$^-$8$^-$ precursors, show surface expression of extracellular matrix (ECM) components, such as fibronectin and heparan sulfate. Both cell types bind IL-7. Enzymatic disruption of specific ECM components indicated that IL-7 presentation by both cell types is dependent upon heparan sulfate. The disruption of IL-7 presentation specifically on fibroblasts does not affect their ability to contribute to T cell development [367].

Murine gut lamina propria IgM$^+$ B cells seem to be pre-committed to switching to IgA$^+$ *in vitro* as well as *in vivo*. Culturing lamina propria IgM$^+$ B cells together with lamina propria stromal cells enhanced preferential switching and differentiation of B cells to IgA+ plasma cells. Thus IgA$^+$ cells in the gut lamina propria are generated *in situ* from B220+IgM+ lymphocytes [368] .

Mast cell development is dependent on stem cell factor, on T-cells for T-dependent tissue mast cells, and on local *in situ* factors, which generate diversity [369].

Defects in stromal cell function have been demonstrated in a number of aplastic anaemia patients. This defect could be corrected by bone marrow transplantation. Restoration of stromal cell function correlated with engraftment of donor macrophages and T cells, but not fibroblasts and endothelial cells [370]. Complete hematopoietic reconstitution of lethally irradiated rats can be achieved by vascularized bone marrow transplantation (VBMT) in an orthotopic hind-limb graft. The rapid cell maturation and repopulation may be due to the presence of stromal cells transplanted. Self renewal of radioresistant host cells was seen after VBMT but not after bone marrow cell transplantation [371].

The alloimmunogenicity of the normal cornea resides within its epithelial and stromal layers. Immune privilege arises from the endothelium. In normal mice, CD95L-expressing endothelium can inhibit the stroma from reacting to MHC alloantigens. In presensitized mice the endothelium can protect itself only from immune rejection [372].

The Ikaros (Ik) gene encodes several isoforms and is a critical regulator of hematolymphoid differentiation. The expression of Ikaros 6 (Ik6) in human placental blood CD34(+) progenitors have a reduced capacity to produce lymphoid B cells in 2 independent assays: (1) in vitro re-initiation of human hematopoiesis during co-culture with the MS-5 murine stromal cell line and (2) xenotransplantation in nonobese diabetic-severe combined immunodeficient mice. These results suggest that Ikaros plays an important role in stem cell commitment in humans [373].

6.4.2. The role of adhesion molecules

Vascular cellular adhesion molecule (VCAM)-1 mediates adhesive interactions between hematopoietic progenitor cells and stromal cells in the bone marrow (BM) and between leukocytes and endothelial as well as dendritic cells. VCAM-1-deficient mice die embryonically. Interferon-induced Cre-loxP-mediated deletion of the VCAM-1 gene after birth efficiently ablates expression of VCAM-1 in most tissues including bone marrow, lymphoid organs, lung, but not in brain. VCAM-1 deficiency leads to a reduction of immature B cells in the BM and to an increase of these cells in peripheral blood but not in lymphoid organs. Mature recirculating B cells are reduced in the BM. The number of mature B cells that appears in the BM after intravenous injection is decreased. The humoral immune response to a T cell-dependent antigen is impaired. VCAM-1 serves an important role for B cell localization and in the T cell-dependent humoral immune response [374].

LFA-1, but not VLA-4, mediated blood T cell transendothelial migration (TEM) to RANTES, MIP1α and SDF-1, and across TNFα and IFNγ-stimulated endothelial cells (EC). Chemokine stimulation in combination with TNFα activation of EC induced TEM, which was partially mediated by VLA-4. SDF-1 increased a β1-integrin activation marker on T cells and enhanced VLA-4-mediated adhesion [375].

The inhibition of initial PI3-K activation by inactive H-Ras, or sustained activation of an inhibitory extracellular signal receptor-activated kinase pathway by active H-Ras prevail to abolish LFA-1 regulation and transendothelial migration induced by SDF1α in leukocytes. This indicates a complex and bimodal involvement of H-Ras [376].

6.4.3. Chemokines

The chemokine receptor CXCR4 is functionally expressed on a multitude of tissues and cell types, including different leukocyte subsets, hematopoietic progenitor cells and non-hematopoietic cells such as endothelial and epithelial cells. CXCR4 is important for lymphocyte trafficking and for recruitment at sites of inflammation. Its ligand, *stromal cell-derived factor-1* (SDF-1) play an important roles in hematopoiesis and developmental processes such as organogenesis, vascularization and embryogenesis. Chemokines have both a more widespread cellular expression pattern and a wider biological role than first envisaged [377].

Primitive subsets of human CXCR4$^+$ and CXCR4$^-$ cells were identified. The response to SDF1α *in vitro* indicated that CXCR4$^+$ and CXCR4$^-$ cells represent discrete populations of primitive blood cells. The transplantation of these sub-fractions into immune-deficient mice revealed that both possess comparable engraftment capacity. Myeloid, lymphoid, and primitive CD34$^+$ CXCR4$^+$ cells were present in chimeric animals transplanted with either subset, indicating that CXCR4$^+$ and CXCR4$^-$ stem cells have equivalent proliferative and differentiative abilities [378].

CXCR4 expression by primary mouse astrocytes is suppressed by exposure to TNFα, which causes a pronounced down-regulation of CXCR4 mRNA and protein in dose- and time-dependent manner. As a consequence, the ability of SDF1α to induce activation of MAP kinases, Erk1/2 was impaired [379].

The CXCR4 receptor is expressed on the syncytiotrophoblast surface. Leukocyte-mediated HIV-1Lai (an X4 strain) infection of syncytiotrophoblast was reduced when SDF1α was added to the cocultures [380].

Neutralizing Abs to CXCR4 reduced lung eosinophilia (bronchoalveolar lavage fluid and interstitium) by half in a mouse model of allergic airway disease (AAD). This was accompanied by a significant decrease in airway hyper-responsiveness. SDF1α neutralization resulted in similar reduction in both lung allergic inflammation and airway hyper-responsiveness. Retroviral delivery of a CXCR4 cDNA to leukocytes resulted in greater inflammation when transduced mice were subjected to a mouse model of AAD.

Fourteen children with precursor-B acute lymphoblastic leukemia (pre-B ALL) were studied. Their immunophenotypes belonged to the first three stages of B cell differentiation. CXCR-4 was highly expressed at all stages, yet the responsiveness to SDF1α varied with patients and seemed to be less significant for pre-B3 (and pre-B1) than for pre-B2. Fibronectin enhanced the SDF1α induced migration of pre-B ALL without influencing CXCR-4 expression [381].

SDF1α is a potent chemoattractant for hematopoietic progenitor cells. SDF1α rapidly (1 minute) and transiently upregulated the adhesion of $CD34^{hi}$ bone marrow (BM) cells and hematopoietic cell lines to both CS-1/fibronectin and VCAM-1, and to BM stromal cells. This upregulation was VLA-4-dependent and was inhibited by pertussis toxin and cytochalasin D, indicating the involvement of G proteins [382].

SDF1α is abundantly expressed in the thymus. In the absence of extracellular matrix, chemotactic activity of SDF1α for cortical (CD4/8 double-positive) thymocytes was almost same as that for medullary ($CD4^+$ or $CD8^+$) thymocytes. Laminin of fibronectin enhanced the chemotactic response of cortical thymocytes to SDF1α over 10-fold, compared to medullary thymocytes. This suggests that SDF1α with fibronectin and laminin regulate the distribution of cortical thymocytes [383].

SDF1α induced a progressive increase of apoptosis when added to the Jurkat $CD4^+/CXCR4^+$ human T cell line. This was observed in serum-free or serum-containing cultures, and peaked at SDF1α concentrations of 10–100 ng/ml, required 3 days to take place, and was completely blocked by the z-VAD-fmk tripeptide caspase inhibitor. SDF1α increased the expression of surface Fas and intracellular FasL significantly [384].

In adult mice *peritoneal mesothelial cells produced SDF-1.* Peritoneal B1a lymphocytes expressed a functional SDF-1 receptor. SDF-1 stimulated migration, proliferation of a minority of peritoneal B1a lymphocytes, and prevented apoptosis in a large fraction of such cells *in vitro*. B1a cells migrating in response to SDF-1 were largely of the $CD5^{high}CD43^{high}B220^-CD1d^-$ phenotype. In mice the neutralization of SDF-1 for 3 weeks significantly decreased the number of peritoneal B1 cells. SDF-1 also affected peritoneal B2 cells [385].

Interleukin-10 produced by B1a cells and SDF-1 secreted by mesothelial cells act in synergy on peritoneal B lymphocytes. The autocrine production of IL-10 was required for all effects of SDF-1 on B cells, including increased proliferation, survival, and chemotaxis. Adding IL-10 to peritoneal B lymphocytes increased the effects of SDF-1. IL-10 abolished the migration of peritoneal B lymphocytes in response to B-lymphocyte chemoattractant (BLC). The constitutive production of IL-10 by B1a lymphocytes, combined with the opposite effects of this cytokine on the responses to SDF-1 and BLC, may account for the selective accumulation of B1 lymphocytes in body cavities [386].

B lymphocytes, NK cells and, to a lesser extent, T lymphocytes inactivate SDF-1 by cathepsin G, which is associated with the membrane of lymphocytes and rapidly cleaves off five N-terminal residues. Cathepsin G is likely to play a significant role in down-regulating SDF-1 activity [387].

Examination by complementary DNA microarray revealed the genes, which are associated with DNA repair, detoxification, apoptosis, cell morphology, cell adhesion, and signal transduction in $CD4^+$ T cells are up-regulated upon SDF-1 exposure. SDF-1 promoted $CD4^+$ T cell survival through the PI3K- and MAPK-cascades without cell cycle progression. Bcl-2 was inactivated. The genes known to be associated with cell survival were up-regulated upon SDF-1 exposure. These data suggest that SDF-1 promotes cell survival and primes T cells for response to immunologic challenges [388].

In human PBL the Rho-p160ROCK (Rho coiled-coil kinase) pathway mediates signaling induced by the chemokine, SDF1α. This led to myosin light chain phosphorylation. Inhibition of the Rho- cascade at different steps blocked lymphocyte migration and the chemotaxis induced by SDF-1alpha [389]. In Wiskott-Aldrich syndrome (WAS) T lymphocytes display abnormal chemotaxis in response to the T-cell chemoattractant SDF-1. The Wiskott-Aldrich syndrome protein (WASP) and the Rho family GTPase Cdc42, control stimulus-induced actin cytoskeleton rearrangements that are involved in cell motility. SDF-1 activates Cdc42 and WASP is an effector of Cdc42. Cdc42-WASP interaction is critical for SDF-1-induced chemotaxis of T cells [390].

Jurkat T cells lacking the ZAP-70 tyrosine kinase exhibit reduced migration in response to the CXCR4 ligand CXCL12 when compared with wild-type Jurkat T cells. Expression of wild-type, but not inactive, ZAP-70 resulted in enhanced migration of ZAP-70-deficient Jurkat T cells. The tyrosine residue at position 292 in the interdomain B region of ZAP-70 exerts a negative regulatory effect migration [391]. SDF1α treatment of Jurkat T cells increased the activity of NO synthase, which catalyzes the generation of NO. The treatment of Jurkat cells or activated PBLs with several NO donors significantly enhanced the SDF1α-induced migration. Various inhibitors of NO synthase markedly abrogated the chemotactic response. Inhibitors of the transcription factor NF-kappa B, which is linked to NO signaling pathways, also significantly blocked the SDF1α-induced chemotactaxis. These compounds did not have a significant effect on SDF1α-induced mitogen-activated protein kinase activity. AKT, which has been shown to mediate NO production, was also phosphorylated upon SDF1α stimulation [392].

Memory but not naive T cells from human tonsils are CXCR5$^+$ and migrate in response to the B cell-attracting chemokine 1 (BCA-1), which is selectively expressed by reticular cells and blood vessels within B cell follicles. Tonsillar CXCR5$^+$ T cells do not respond to other chemokines present in secondary lymphoid tissues. Peripheral blood CXCR5$^+$ T cells also belong to the CD4$^+$ memory T cell subset but, in contrast to tonsillar cells, are in a resting state and migrate weakly to chemokines. Thus CXCR5$^+$ T cells are a distinct memory T cell subset with B cell helper function, designated as follicular B helper T cells (TFH) [393].

Endometriotic lesions secrete chemokines that recruit immune cells into the peritoneal cavity. The chemokine, RANTES (regulated on activation, normal T cell expressed and secreted) is synthesized by endometriotic stromal cells and circulates in peritoneal fluid, coincidently with endometriosis. The RANTES receptors CCR-1 and CCR-5 are present in the human monocyte cell line, U937, and in normal peritoneal macrophages. Over a range of 0–1000 pg/ml recombinant human RANTES had a direct, linear effect on monocyte migration. Peritoneal fluid induced dose-dependent effects on monocyte migration that were correlated with concentrations of immunoreactive RANTES and with the severity of endometriosis [394]. *Eotaxin* is a C-C-chemokine of 8.4 kDa, whose major biological activity is the chemoattraction of eosinophils. In the *endometriosis syndrome* eotaxin protein localized mainly in epithelial cells, with only very faint immunostaining in the surrounding stromal cells. Prominent eotaxin accumulation was present in the luminal epithelium of secretory endometrium. Peritoneal fluid concentrations of eotaxin were high in women with moderate or severe endometriosis and significantly different from values obtained in women with minimal or mild endometriosis or no disease [395].

6.4.4. *Cytokines and growth factors*

Human B cells inhibit the formation of osteoclasts (OC) and shorten the life span of mature OCs by secreting TGFβ, which induces apoptosis in these cells. The antiosteoclastogenic effects of B cells are abolished by addition of anti-TGFβ antibody, and may be mimicked by treatment of B cell-deprived PBSC cultures with recombinant TGFβ [396].

In normal human prostate, IL-15 and IL-15Rα expression were restricted to smooth muscle and stromal cells. In benign hyperplastic prostate (BPH) additional epithelial cells frequently exhibited discrete IL-15R and often intense, membranous IL-15 reactivity. IL-15 mRNA synthesis of prostatic stromal cell (PSC) was significantly up-regulated by IFNγ. IL-15 strongly stimulated the growth of BPH-T-lymphocytes and weakly that of carcinoma cell lines, but not of stromal cells. Thus, chronic inflammation might be triggered by this paracrine loop, involving T-cell derived IFNγ that stimulates prostatic IL-15 production [397].

Pre-pro-B cell growth-stimulating factor (PPBSF) is a heterodimer of IL-7 and a variant beta-chain of hepatocyte growth factor (HGF). HGF is a pleiotropic cytokine homologous to plasminogen that regulates cell growth, motility, and morphogenesis. In the presence of low m.w. heparin sulfate-derived oligosaccharides, rHGFβ combines with rIL-7 to form a biologically active heterodimer having the properties of PPBSF. Thus PPBSF is a hybrid cytokine consisting of the bioactive components of two unrelated cytokines. It is postulated that the HGFβ-chain in PPBSF enables IL-7 to participate in cognate interactions at the stromal cell surface and to transduce signals effectively at low levels of IL-7R [398].

Hepatocyte growth factor acts on the c-met receptor. HGF is a pleiotropic factor that is mainly produced by mesenchymal cells and acts on cells of epithelial origin. Biologically active HGF is constitutively produced by fibroblast-like stromal cells from human lymphoid tissues. HGF secretion was increased by direct contact with activated T cells, which was abrogated if the T cells were removed. The enhancement of HGF production was due to additive effects of T-cell membrane-associated IL-1 and CD40 ligand. B lymphocytes activated with CD40L/anti-mu or phorbol 12-myristate 13-acetate expressed the c-met receptor and co-culture of activated B cells with stromal cells from spleen led to enhanced production of immunoglobulins. This could be partially inhibited by anti-HGF neutralizing antibodies. The substitution of stromal cells with recombinant HGF did not produce enhancement of immunoglobulin secretion but enhanced integrin-mediated adhesion of activated B cells to vascular cell adhesion molecule (VCAM-1) and fibronectin [399].

6.4.5. *Stromal elements*

Laminin-8/9 and laminin-10/11 are the most abundant laminin isoforms synthesized by human bone marrow stromal cells. Other isoforms, if present, certainly play a minor role. In addition to its adhesion-mediating properties, laminin-10/11 preparations had a mitogenic activity for human hematopoietic progenitor cells [400].

6.4.6. *Steroid hormones*

The local intracrine formation of sex steroids from the inactive precursors, dehydroepiandrosterone (DHEA), its sulfate (DHEA-S) and 4-androstenedione (4-DIONE) plays an important role in the regulation of growth and function of peripheral target tissues. Normal tissues and solid tumors contain stromal/immune cells that secrete a wide spectrum of cytokines. These cytokines in turn regulate the host tissue and local immune reactions. The regulation by IL-4 and IL-6 of both estrogenic and androgenic 17β-HSD/KSR activities in breast cancer cells depend on the cell-specific gene expression of various types of 17β-HSD/KSR enzymes. In both ER^+ (ZR-75-

1, T-47D) and ER⁻ (MDA-MB-231, BT-20) human breast cancer cells, exposure to IL-4 and IL-13 caused a rapid and potent induction of 3β-HSD type 1 gene expression. Similar induction was observed in normal human mammary and prostate epithelial cells in primary culture as well as in human HaCaT immortalized keratinocytes, ME-180 cervix cancer cells, and HT-29 colon cancer cells. The DNA-binding activity of Stat6 was activated after a 30 min exposure to IL-4 in all the cell types where IL-4 induced 3β-HSD expression, but not in those that failed to respond to IL-4. Therefore, IL-4 and IL-13 may play a role in the biosynthesis of active sex steroids from the inactive adrenal steroid DHEA, not only in breast cells but also in various other cell types derived from peripheral target tissues [401].

NFκB regulates proinflammatory genes and may be involved in inflammation associated with reproductive events e.g. menstruation and implantation. The NFκB pathway intermediates were determined in human endometrium and first trimester decidua. IκBα and TANK-binding kinase 1 (TBK1) mRNA were increased in the perimenstrual phase of the menstrual cycle. When progesterone concentrations increased in early pregnancy; IκB kinase gamma (IKKγ, a scaffolding protein) IKKα and NFκB inducing kinase (NIK) mRNA levels increased in decidua whilst IKKβ and MEKK1 mRNA levels declined. IKKα and NIK proteins were detected in glandular epithelium and endothelium of endometrium. In decidua, both proteins were present in epithelium and decidualized stromal cells [402]. In human endometrial stromal cells (ESC) a significant increase in IL-15 mRNA levels was present after treatment with progesterone (PS) alone or with estradiol (E2) plus PS. PS induced IL-15 mRNA expression in ESC in a dose-dependent manner. E2 alone did not increase IL-15 mRNA. IL-15 protein secretion was stimulated by PS and further enhanced by combined treatment with E2 and PS, whereas E2 alone was ineffective. This indicates that IL-15 is deeply involved in the hormonal control of the human endometrium by PS and E2 [403].

Immature B cells from normal mouse bone marrow were not responsive to the suppressive effect of androgens unless they were co-cultured with marrow stromal cells or with supernatants from androgen-treated stromal cells. Bone marrow stromal cells produced TGFβ in response to dihydrotestosterone (DHT), and neutralization of TGFβ in the DHT-treated stromal cells reversed the suppressive effects. In chimeric mice androgens only affected B cell development if androgen-sensitive stromal cells were present [404].

Treatment of immature hens with diethylstilbestrol for 3 or 6 days increased the population of both IgGγ mRNA-expressing cells and IgG-containing cells in the oviducal stroma [405].

REFERENCES

1. Berczi I. Neuroendocrine response to endotoxin. Ann. N.Y. Acad. Sci. 1998;851:411–415.
2. Askenase PW. Yes T cells, but three different T cells (alphabeta, gammadelta and NK T cells), and also B-1 cells mediate contact sensitivity. Clin Exp Immunol 2001;125: 345–50.
3. Kalish RS, Askenase PW. Molecular mechanisms of CD8+ T cell-mediated delayed hypersensitivity: implications for allergies, asthma, and autoimmunity. J Allergy Clin Immunol 1999;103(2 Pt 1):192–9.
4. Weiss JM, Renkl AC, Sleeman J, Dittmar H, Termeer CC, Taxis S, Howells N, Schopf E, Ponta H, Herrlich P, Simon JC. CD44 variant isoforms are essential for the function of epidermal Langerhans cells and dendritic cells. Cell Adhes Commun 1998;6(2–3):

157–60.
5. Reiss Y, Proudfoot AE, Power CA, Campbell JJ, Butcher EC. CC chemokine receptor (CCR)4 and the CCR10 ligand cutaneous T cell-attracting chemokine (CTACK) in lymphocyte trafficking to inflamed skin. J Exp Med 2001;194:1541–7.
6. Nakae S, Naruse-Nakajima C, Sudo K, Horai R, Asano M, Iwakura Y. IL-1alpha, but not IL-1beta, is required for contact-allergen-specific T cell activation during the sensitization phase in contact hypersensitivity. Int Immunol 2001;13:1471–8.
7. Bernhagen J, Calandra T, Bucala R. Regulation of the immune response by macrophage migration inhibitory factor: biological and structural features. J Mol Med 1998;76:151–61.
8. Hou G, Valujskikh A, Bayer J, Stavitsky AB, Metz C, Heeger PS. In vivo blockade of macrophage migration inhibitory factor prevents skin graft destruction after indirect allorecognition. Transplantation 2001;72:1890–7.
9. Whalen JD, Thomson AW, Lu L, Robbins PD, Evans CH. Viral IL-10 gene transfer inhibits DTH responses to soluble antigens: evidence for involvement of genetically modified dendritic cells and macrophages. Mol Ther 2001;4:543–50.
10. Kobayashi K, Kaneda K, Kasama T. Immunopathogenesis of delayed-type hypersensitivity. Microsc Res Tech 2001;53:241–5.
11. Dhabhar FS. Acute stress enhances while chronic stress suppresses skin immunity. The role of stress hormones and leukocyte trafficking. Ann N Y Acad Sci 2000;917:876–93.
12. Katafuchi T, Kondo T, Hori T. Comparable effects of adrenalectomy and C fiber depletion on delayed-type hypersensitivity in rats. Neuroimmunomodulation 2001;9:157–62.
13. Messingham KA, Heinrich SA, Kovacs EJ. Estrogen restores cellular immunity in injured male mice via suppression of interleukin-6 production. J Leukoc Biol 2001;70:887–95.
14. Torii H, Tamaki K, Granstein RD. The effect of neuropeptides/hormones on Langerhans cells. J Dermatol Sci 1998;20:21–8.
15. Berg NN, Ostergaard HL. Characterization of intercellular adhesion molecule-1 (ICAM-1)-augmented degranulation by cytotoxic T cells. ICAM-1 and anti CD3 must be co-localized for optimal adhesion and stimulation. J Immunol 1995;155:1694–1702.
16. Griffiths GM, Namikawa R, Mueller C. et al. Grnazyme A and perforin as markers for rejection in cardiac transplantation. Eur J Immunol 1991;21:687–692.
17. Henkart PA. Cytotoxic T lymphocytes. In: Paul WE, editor. Fundamental Immunology, 4th edition. Lippincott-Raven, 1999;1021–1049.
18. Micheletti F, Canella A, Vertuani S, Marastoni M, Tosi L, Volinia S, Traniello S, Gavioli R. Supra-agonist peptides enhance the reactivation of memory CTL responses. J Immunol 2000;165:4264–71.
19. Wei CH, Yagita H, Masucci MG, Levitsky V. Different programs of activation-induced cell death are triggered in mature activated CTL by immunogenic and partially agonistic peptide ligands. J Immunol 2001;166:989–95.
20. Young NT, Uhrberg M, Phillips JH, Lanier LL, Parham P. Differential expression of leukocyte receptor complex-encoded Ig-like receptors correlates with the transition from effector to memory CTL. J Immunol 2001;166:3933–41.
21. Muta H, Boise LH, Fang L, Podack ER. CD30 signals integrate expression of cytotoxic effector molecules, lymphocyte trafficking signals, and signals for proliferation and apoptosis. J Immunol 2000;165:5105–11.
22. Diehl L, Den Boer AT, van der Voort EI, Melief CJ, Offringa R, Toes RE. The role of CD40 in peripheral T cell tolerance and immunity. J Mol Med 2000;78:363–71.

23. Shepherd DM, Kerkvliet NI. Disruption of CD154:CD40 blocks generation of allograft immunity without affecting APC activation. J Immunol 1999;163:2470–7.
24. Saverino D, Fabbi M, Ghiotto F, Merlo A, Bruno S, Zarcone D, Tenca C, Tiso M, Santoro G, Anastasi G, Cosman D, Grossi CE, Ciccone E. The CD85/LIR-1/ILT2 inhibitory receptor is expressed by all human T lymphocytes and down-regulates their functions. J Immunol 2000;165:3742–55.
25. Wang HC, Klein JR. Multiple levels of activation of murine CD8(+) intraepithelial lymphocytes defined by OX40 (CD134) expression: effects on cell-mediated cytotoxicity, IFN-gamma, and IL-10 regulation. J Immunol 2001;167:6717–23.
26. Appay V, Dunbar PR, Cerundolo V, McMichael A, Czaplewski L, Rowland-Jones S. RANTES activates antigen-specific cytotoxic T lymphocytes in a mitogen-like manner through cell surface aggregation. Int Immunol 2000;12:1173–82.
27. Aung S, Graham BS. IL-4 diminishes perforin-mediated and increases Fas ligand-mediated cytotoxicity in vivo. J Immunol 2000;164:3487–93.
28. Kaser A, Enrich B, Ludwiczek O, Vogel W, Tilg H. Interferon-alpha (IFN-alpha) enhances cytotoxicity in healthy volunteers and chronic hepatitis C infection mainly by the perforin pathway. Clin Exp Immunol 1999;118:71–7.
29. Aleman M, De La Barrera S, Fink S, Finiasz M, Farina MH, Pizzariello G, Sasiain MD. Interleukin-12 amplifies the M. leprae hsp65-cytotoxic response in the presence of tumour necrosis factor-alpha and interferon-gamma generating CD56+ effector cells: interleukin-4 downregulates this effect. Scand J Immunol 2000;51:262–70.
30. Abe R, Peng T, Sailors J, Bucala R, Metz CN. Regulation of the CTL response by macrophage migration inhibitory factor. J Immunol 2001;166:747–53.
31. Via CS, Shustov A, Rus V, Lang T, Nguyen P, Finkelman FD. In vivo neutralization of TNF-alpha promotes humoral autoimmunity by preventing the induction of CTL. J Immunol 2001;167:6821–6.
32. Halloran PF, Miller LW, Urmson J, Ramassar V, Zhu LF, Kneteman NM, Solez K, Afrouzian M. IFN-gamma alters the pathology of graft rejection: protection from early necrosis. J Immunol 2001;166:7072–81.
33. Delgado M, Ganea D. Vasoactive intestinal peptide and pituitary adenylate cyclase-activating polypeptide inhibit T cell-mediated cytotoxicity by inhibiting Fas ligand expression. J Immunol 2000;165:114–23.
34. Dourmashkin RR, Dete P, Simone CB, Henkart PA. Electron microscopic demostration of lesions on target cell membranes associated with antibody-mediated cytotoxicity. Clin Exp Immunol 1980;43:554–560.
35. Podack ER, Dennert G. Assembly of two types of tubules with putative cytolytic function by cloned natural killer cells. Nature 1983;302:442–445.
36. Dennert D, Podack ER. Cytolysis by H2 specific T killer cells. Assembly of tubular complexes on target cell membranes. J Exp Med 1963;157:1483–1495.
37. Smyth MJ, O'Connor MD, Trapani JA. Granzymes: a variety of serine protease specificities encoded by genetically distinct subfamilies. J Leukoc Biol 1996;60:555–562.
38. Jenne DE, Masson D, Zimmer M, Haefliger JA, Li WH, Tschopp J. Isolation and complete structure of the lymphocyte serine protease, granzyme G, a novel member of the granzyme multigen family in murine cytotoxic T lymphocytes. Evolutionary origin of lymphocyte proteases. Biochemistry 1989;28:7953–7961.
39. Lutzelschwab C, Pejler G, Aveskogh M, Hellman L. Secretory granule proteases in rat mast cells. Cloning of 10 different serine proteases and a carboxypeptidse A from various

rat must cell populations. J Exp Med 1997;184:13–29.
40. Henkart PA. CTL effector functions. Semin Immunol. 1997;9:85–6.
41. Kummer JA, Kamp AM, Citarella F, Horrevoets AJG, Hack CE. Expression of human recombinant granzyme A zymogen and its activation by the cysteine proteinase cathepsine C, J Biol Chem 1966;271:9281–9286.
42. Theile DL, Lipsky PE. The action of leucyl-leucine methyl ester on cytotoxic lymphocytes requires uptake by a novel depeptide specific facilitataed transport system and dipeptidyl peptidase-I mediated conversion to membranolytic products. J Exp Med 1990; 172:183–194.
43. Masson D, Peters PJ, Geuze HJ, Borst J, Tschopp J. Interaction of chondroitin sulfate with perforin and granzymes of cytotoxic T cells is dependent on pH. Biochemistry 1990;29;11229–11235.
44. Pena SV, Hanson DA, Carr BA, Goralski TJ, Krensky AM. Processing, subcellular localization and function of 519 (granulolysin), a human late T cell activation molecule whith homology to small lytic granule proteins. J Immun 1997;158:2680–2688.
45. Dupuis M, Schaerer E, Krause KH, Tschopp J. The calcium binding protein, calreticulin is a major constituent of lytic granules of cytotoxic T lymphocytes. J Exp Med 1993;177: 1–7.
46. Sanderson CJ, Glaueret AM. The mechanism of T cell mediated cytotoxicity. V. Morphological studies by electron microscopy. Proc Roy Soc London 1977;198:315–323.
47. Russel JH. Internal disintegration model of cytotoxic lymphocyte-induced target damage. Immunol Rev 1983;72:97–118.
48. Duke RC, Persechini PM, Chang S, Liu CC, Cohen JJ, Young JD. Purified perforin induces target cell lysis but not DNA fragmentation. J Exp Med 1989;170:1451–1456.
49. Kagi D, Ledermann B, Burki K, et al. Cytotoxicity mediated by T cells and natural killer cells is greatly impaired in perforin-deficient mice. Nature 1994;369:31–37.
50. Shi L, Mai S, Israels S, Browne K, Greenberg AH. Granzyme B (GraB) autonomously crosses the cell membrane and perforin initiates apoptosis and GraB nuclear localization. J Exp Med 1997;185:855–866.
51. Shiver JW, Su L, Henkart PA. Cytotoxicity with target DNA breakdown by rat basophilic leukemia cells expressing both cytolysin and granzyme A. Cell 1992;71:315–322.
52. Nakajima H, Park HL, Henkart PA. Synergistic roles of granzymes A and B in modulating target cell death by RBL mast cell tumors expressing cytolysin and perforin. J Exp Med 1995;181:1037–1046.
53. Shi L, Kam CM, Powers JC, Aebersold R, Greenberg AH. Purification of three cytotoxic granule serine proteases that induce apoptosis through distinct substances and target cell interactions. J Exp Med 1992;176:1521–1529.
54. Darmon AJ, Nicholson DW, Bleackley RC. Activation of the apoptotic protease CPP32 by cytotoxic T cell-derived granzyme B. Nature 1995;377,446–448.
55. Zhang J, Scordi I, Smyth MJ, Lichtenheld MG. Interleukin 2 receptor signaling regulates the perforin gene through signal transducer and activator of transcription (Stat)5 activation of two enhancers. J Exp Med 1999;190:1297–308.
56. Spaeny-Dekking EH, Kamp AM, Froelich CJ, Hack CE. Extracellular granzyme A, complexed to proteoglycans, is protected against inactivation by protease inhibitors. Blood 2000;95:1465–72.
57. Kagi D, Odermatt B, Mak TW. Homeostatic regulation of CD8+ T cells by perforin. Eur J

Immunol 1999;29:3262–72.
58. Shustov A, Luzina I, Nguyen P, Papadimitriou JC, Handwerger B, Elkon KB, Via CS. Role of perforin in controlling B-cell hyperactivity and humoral autoimmunity. J Clin Invest 2000;106:R39–47.
59. Schneider SW, Sritharan KC, Greibel JP, Oberleithner H, Jena HP. Surface dynamics of living acinar cells imaged by atomic force microscopy: identification of plasma membrane structures involved in exocytosis. Proc Natl Acad Sci USA 1997;94:316–321.
60. Tkayama H, Trenn G, Sitkovsky MV. A novel cytotoxic T lymphocyte activation assay. Optimized conditions for antigen receptors triggered granule enzyme secretion. J Immunol Methods 1987;104:183–190.
61. Miyake K, McNeil PL. Vesicle accumulation and exocytosis at sites of plasma membrane disruption. J Cell Biol 1995;131:1737–1745.
62. Verret CR, Firmenich AA, Kranz DM, Eisen HN. Resistance of cytotoxic T lymphocytes to the lytic effect of their toxic granules. J Exp Med 1987;166:1536–1547.
63. Shinkai Y, Ishizaka H, Hattori M, Okumura K. Resistance of mouse cytotoxic cells to pore-forming protein-mediated cytolysis. Eur J Immunol 1988;18:29–33.
64. Preschini PM, Young JD, Almers W. Membrane channel formation by the lymphocyte pore forming protein: comparison between susceptible and resistant target cells. J Cell Biol 1990;110:2109–2116.
65. Muller D, Koller BH, Whitton JL, LaPan KE, Brigman KK, Frelinger JA. LCMV-specific class II-restricted cytotoxic T cells in beta 2-microblobulin deficient mice. Science 1992; 255:1576–1578.
66. Hahn S. Gehri R, Erb P. Mechanism and biological significance of CD4-mediated cytotoxicity. Immunol Rev 1995;146:57–79.
67. Yagita H, Hanabuchi S. Asano Y, Tamura T, Nariuchi H, Okumura K. Fas-mediated cytotoxicity – a new immunoregulatory and pathogenic function of Th1 CD4-T cells. Immunol Rev 1995;146:221–239.
68. Brower RC, England R, Takeshita T, et al. Minimal reqirements for peptide medited activation of CD8+ CTL. Mol Immunol 1994;31:1285–1293.
69. GagliardiMC, De Petrillo G, Salemi S, et al. Presentation of peptides by cultured monocytes or activated T cells allows specific priming of human cytotoxic T lymphocytes *in vitro*. Int Immunol 1995;7:1741–1752.
70. Nishio M, Spielma J, Lee RK, Nelson DL, Podack ER. CD80 (B 7.1) and CD54 (intercellular adhesion molecule-1) induce target cell susceptibility to promiscuous cytotoxic T cell lysis. J Immunol 1996;157:4347–4353.
71. Zinkernagel RM, Bachmann MF, Kundig TM, Oehen S, Pirchet H, Hengartner H. On immunological memory. Annu Rev Immunol 1996;14:333–367.
72. Ahmed R, Gray D. Immunological memory and protective immunity: understanding their relation. Science 1996;271:54–60.
73. Kagi D, Lederman B, Burki K, Zinkernagel RM, Hengartner H. Molecular mechanisms of monocyte-mediated cytotoxicity and their role in immunological protection and pathogenesis *in vivo*. Annu Rev Immunol 1996;14:207–232.
74. Kagi D, Seiler P, Pavlovic J, et al. The roles of perforin- and Fas-dependent cytotoxicity in protection against cytophatic and noncytopathic viruses. Eur J Immunol 1995;25: 3256–3262.
75. Kagi D. Lederman B, Burki K, Hengartner H, Zinkernagel RM. CD8+ T cell-mediated protection against an intracellular bacterium by perforin-dependent cytotoxicity. Eur J

Immunol 1994;24:3068–3072.

76. Selvaggi G. Ricordi C. Podack ER, Inveradi L. The role of the perforin and Fas pathways of cytotoxicity in skin graft rejection. Transplantation 1996;62:1912–1915.
77. Braun MY, Lowin B, French L, Acha-Orbea H, Tscopp J, Cytotoxic T cells deficient in both Fas ligand and perforin show residual cytotoxic activity yet loose their capacity to induce lethal acute grft-versus-host disease. J Exp Med 1996;183:657–661.
78. Baker MB, Altman NH, Podack ER, Levy RB. The role of cell-mediated cytotxicity in acute GVHD after MHC-matched allogeneic bone marrow transplantation in mice. J Exp Med 1966;183:2645–2656.
79. van Herrath MG, Oldstone MB. Interferon-gamma is essential for destruction of beta cells and development of insulin-dependent diabetes mellitus. J Exp Med 1997;185:531–539.
80. Chernovsky AV, Wang Y, Wong FS, Visintin I, Janeway CA, Matis LA. The role of Fas in autoimmune disease. Cell 1997;89:17–24.
81. van den Broek M, Kagi D, Ossendrop F, et al. Decreased tumor surveillance in perforin-deficient mice. J Exp Med 1996;184:1781–1790.
82. Walker PR. Saas P. Dietrich PY. Role of Fas Ligand (CD95L) in immune escape. The tumor cell strikes back. J Immunol 1997;158:4521–4524.
83. Kurimoto I, Streilein JW. Tumor necrosis factor-alpha impairs contact hypersensitivity induction after ultraviolet B radiation via TNF-receptor 2 (p75). Exp Dermatol. 1999;8: 495–500.
84. Canaday DH, Wilkinson RJ, Li Q, Harding CV, Silver RF, Boom WH. CD4(+) and CD8(+) T cells kill intracellular Mycobacterium tuberculosis by a perforin and Fas/Fas ligand-independent mechanism. J Immunol 2001;167:2734–42.
85. Mbawuike IN, Fujihashi K, DiFabio S, Kawabata S, McGhee JR, Couch RB, Kiyono H. Human interleukin-12 enhances interferon-gamma-producing influenza-specific memory CD8+ cytotoxic T lymphocytes. J Infect Dis 1999;180:1477–86.
86. Silverstein AM. The history of immunology. In Fundamental Immunology, 4th edition Paul WE, Editor, Phyladelphya, Raven Press, 1999, pp.19–35.
87. DeFranco A. B-lymphocyte activation. In: Paul WE, editor. Fundamental Immunology, 4th edition. Lippincott-Raven, 1999;225–261.
88. Porakishvili N, Mageed R, Jamin C, Pers JO, Kulikova N, Renaudineau Y, Lydyard PM, Youinou P. Recent progress in the understanding of B-cell functions in autoimmunity. Scand J Immunol 2001;54:30–8
89. Ma C, Staudt LM. Molecular definition of the germinal centre stage of B-cell differentiation. Philos Trans R Soc Lond B Biol Sci 2001;35683–9.
90. Lynch RG. Regulatory roles for FcgammaRIII (CD16) and FcgammaRII (CD32) in the development of T- and B-lineage lymphoid cells. J Leukoc Biol 2000;67:279–84.
91. Nutt SL, Eberhard D, Horcher M, Rolink AG, Busslinger M. Pax5 determines the identity of B cells from the beginning to the end of B-lymphopoiesis. Int Rev Immunol 2001;20: 65–82.
92. Gugasyan R, Grumont R, Grossmann M, Nakamura Y, Pohl T, Nesic D, Gerondakis S. Rel/NF-kappaB transcription factors: key mediators of B-cell activation. Immunol Rev 2000 ;176:134–40.
93. Kee BL, Quong MW, Murre C. E2A proteins: essential regulators at multiple stages of B-cell development. Immunol Rev 2000;175:138–49.
94. Calame KL. Plasma cells: finding new light at the end of B cell development. Nat Immunol 2001;2:1103–8.

95. Sefton BM, Taddie JA. Role of tyrosine kinase in lymphocyte activation. Curr Opin Immunol 1944;6:372–379.
96. Pawson T. Protein modules and signalling network. Nature 1995;373:573–580.
97. Neel BG. Role of phosphatases in lymphocyte activation. Curr Opin Immunol 1997;9: 405–420.
98. Tarakhovsky A. Xid and Xid-like immunodeficiencies from a signalling point of view. Curr Opin Immunol 1997;9:319–323.
99. Takata M, Kurosaki T. A role of Burton's tyrosine kinase in B cell antigen recptor mediated activation of phopholipase C-gsmms 2. J Exp Med 1996;184: 31–40.
100. Matsumoto AK, Kopicky-Burd J, Carter RH, Tuveson DA, Tedder TF, Fearon DT. Intersection of the complement and the immune system: a signal transduction complex of the B lymphocyte-containing complement receptor type 2 and CD19. J Exp Med 1991;173:55–64.
101. Carter RH, Spycher MO, Ng YC, Hoffman F, Fearon DT. Synergistic interaction between complement receptor type 2 and membrane IgM on B lymphocytes. J Immunol 1988;141: 457–463.
102. Carter RH, Fearon DT. CD19: Lowering the threshold for antigen receptor stimulation of B lymphocytes. Science 1992;256: 105–107.
103. Phillips NE, Parker DC. Fc-dependent inhibiton of mouse B cell activation by whole anti-μ antibodies. J Immunol 1983;130:602–606.
104. Ashman RF, Packham D, Stunz LL. Fc receptor off signal in the B cell involves apoptosis. J Immunol 1996;157:5–11.
105. Chacko GW, Tridandapani S, Damen JE, Liu L, Krystal G, Coggeshall KM. Negative signalling in B lymphocytes induces tyrosine phosphorylation of the 145-kDa inositol 5-phosphatase SHIP. J Immunol 1996;157:2234–2238.
106. D'Ambrosio D, Hippen KH, Minskoff SA et al. Recruitment and activation of PTPIC in negative regulation of antigen receptor signalling by FcγRIIB1. Science 1995;268: 293–297.
107. Ono M, Bolland S, Tempst P, Ravetch JV. Role of the inositol phosphate SHIP in negative regulation of the immune system by receptor Fc(gamma)RIIB. Nature 1996;383:263–266.
108. Ono M, Odaka H, Bolland S, Yanagi S, KurosakiT, Ravetch JV. Deletion of SHIP or SHIP-1 reveals two distinct pathways for inhibitory signalling. Cell 1997;90:293–301.
109. Mongini PKA, Blessinger CA, Dalton JP. Affinity requirements for induction of sequential phases of human B cell activation by membrane IgM-cross-linking ligands. J Immunol 1991;146:1791–1800.
110. Anderson JS, Teutsch M, Dong Z, Wortis HH. An essential role for Bruton's tyrosine kinase in the regulation of B-cell apoptosis. Proc Acad Sci USA 1996;93:10966–10971.
111. Rao A, Luo C, Hogan PG. Transcription factors of the NFAT family. Regulation and function. Ann Rev Immunol 1997;15:707–747.
112. Rooney JW, Dubois PM, Sibley CH. Crosslinking of surface IgM activates NF-κB in B lymphocytes. Eur J Immunol 1991;21:2993–3008.
113. Liu J, Chiles TC, Sen R, Rothstein TL. Inducible nuclear expression of NF-κB in primary B cells stimulated through the surface Ig receptor. J Immunol 1991;146:1685–1691.
114. Treisman R. Regulation of transcription by MAP kinase cascades. Curr Opin Cell Biol 1996;8:205–215
115. Montminy, M. Trnascriptional regulation by cyclic AMP. Ann Rev Biochem 1997;66:

807–822.
116. Suzuki T, Cooper M. Comparison of the expression of IL-2 receptors by human T and B cells: Induction by the polyclonal mitogens, phorbol myristate acetate and anti-μ antibody. J Immunol 1985;325:537–540.
117. Liu Y, Linsley PS. Costimulation of T cell growth. Curr Opin Immunol 1992;4:265–270.
118. Zinkernagel RM, Hengartner H. Regulation of the immune response by antigen. Science 2001;293:251–3.
119. MacPherson G, Kushnir N, Wykes M. Dendritic cells, B cells and the regulation of antibody synthesis. Immunol Rev 1999 Dec;172:325–34.
120. Meffre E, Casellas R, Nussenzweig MC. Antibody regulation of B cell development. Nat Immunol 2000;1:379–85.
121. Tamir I, Dal Porto JM, Cambier JC. Cytoplasmic protein tyrosine phosphatases SHP-1 and SHP-2: regulators of B cell signal transduction. Curr Opin Immunol 2000;12:307–15.
122. Brauweiler AM, Tamir I, Cambier JC. Bilevel control of B-cell activation by the inositol 5-phosphatase SHIP. Immunol Rev 2000;176:69–74.
123. Rothstein TL, Zhong X, Schram BR, Negm RS, Donohoe TJ, Cabral DS, Foote LC, Schneider TJ. Receptor-specific regulation of B-cell susceptibility to Fas-mediated apoptosis and a novel Fas apoptosis inhibitory molecule. Immunol Rev 2000;176:116–33.
124. Zabel MD, Weis JH. Cell-specific regulation of the CD21 gene. Int Immunopharmacol 2001;1:483–93.
125. Chen Z, Koralov SB, Kelsoe G. Regulation of humoral immune responses by CD21/CD35. Immunol Rev 2000;176:194–204.
126. Bras A, Ruiz-Vela A, Garcia-Domingo D, Martinez C. Apoptosis as a scaffold for building up the B cell repertoire. Ann N Y Acad Sci 2000;926:13–29.
127. Marshall AJ, Niiro H, Yun TJ, Clark EA. Regulation of B-cell activation and differentiation by the phosphatidylinositol 3-kinase and phospholipase Cgamma pathway. Immunol Rev 2000;176:30–46.
128. Kelly ME, Chan AC. Regulation of B cell function by linker proteins. Curr Opin Immunol 2000;12:267–75
129. Myung PS, Boerthe NJ, Koretzky GA. Adapter proteins in lymphocyte antigen-receptor signaling. Curr Opin Immunol 2000;12:256–66.
130. Bishop GA, Hostager BS. B lymphocyte activation by contact-mediated interactions with T lymphocytes. Curr Opin Immunol 2001;13:278–85.
131. McHeyzer-Williams MG, McHeyzer-Williams LJ, Fanelli Panus J, Bikah G, Pogue-Caley RR, Driver DJ, Eisenbraun MD. Antigen-specific immunity. Th cell-dependent B cell responses. Immunol Res 2000;22:223–36.
132. Fujimoto M, Poe JC, Hasegawa M, Tedder TF. CD19 regulates intrinsic B lymphocyte signal transduction and activation through a novel mechanism of processive amplification. Immunol Res 2000;22:281–98.
133. Jacquot S. CD27/CD70 interactions regulate T dependent B cell differentiation. Immunol Res 2000;21:23–30.
134. van Kooten C, Banchereau J. CD40-CD40 ligand. J Leukoc Biol 2000;67:2–17.
135. Tsubata T. Apotosis of mature B cells. Int Rev Immunol 1999;18:347–65.
136. Jiang H, Harris MB, Rothman P. IL-4/IL-13 signaling beyond JAK/STAT. J Allergy Clin Immunol 2000 Jun;105(6 Pt 1):1063–70.
137. Zipori D, Barda-Saad M. Role of activin A in negative regulation of normal and tumor B

lymphocytes. J Leukoc Biol 2001;69:867–73.
138. Bachmann M, Hengartner H, Zinkernagel R. T helper cell-independent neutralizing B cell response against vesicular stomatitis virus: Role of antigen patterns in B cell induction? Eur J Immunol 1995;25:3445–3451.
139. Milich D, McLachlan A. The nucleocapsid of hepatitis B virus is both a T-cell-independent and a T cell-dependent antigen. Science 1986;234:1398–1401.
140. Szomolanyi-Tsuda E, Welsh RM. T cell-independent antibody-mediated clearance of polyoma virus in T cell deficient mice. J Exp Med 1996;183:403–411.
141. Mond JJ, Lees A, Snapper C. T cell-independent antigens type 2. Ann Rev Immunol 1995;13:655–692.
142. Ahokhai VI, Lukacs LJ, Jonas LC, Calandra GB. Clinical experience with Ped-vaxHIB, a conjugate vaccine of Haemophylus influenzae type b polysaccharide-Neisseria meningitidis outer protein. Vaccine 1991;9 (suppl):38–41.
143. Snapper CM, Mond JJ. A model for induction of T cell-independent immunity in response to polysaccharide antigen. J Immunol 1996;157:2229–2233.
144. Medzhitov R, Janeway CA Jr. Innate immunity: The values of a non-clonal system of recognition. Cell 1997; 91:295–298.
145. Ip YT, Levine M. Molecular genetics of Drosophila immunity. Curr Opin Genet Dev 1994;4:672–677.
146. Muta T, Iwanga S. Clotting and immune defense in Limulidae. Prog Mol Sub-cell Biol 1996;15:154–189.
147. Morrison DC, Ryan JL. Bacterial endotoxin and host immune responsees. Adv Immunol 1979;28:293–450.
148. Morrison DC, Ryan JL. Endotoxin and disease mechanisms. Ann Rev Med 1987;38: 417–432.
149. Ulevitch RJ, Tobias PS. Receptor-dependent mechanisms of cell stimulation by bacterial endotoxin. Ann Rev Immunol 1995:13:437–457.
150. Jelinek DF, Lipsky PE. Regulation of human B lymphocyte activation, proliferation, and differentation. Adv Immunol 1987;40:1–59.
151. Bachmann MF, Zinkernagel RM. Neutralizing antiviral B cell responses. Ann Rev Immunol 1997;15:235–270.
152. Allison KC, Strober W, Harriman GR. Induction of IL-5 receptors on normal B-cells by cross-linking surface Ig with anti-Ig-dextran. J Immunol 1991;146:4197–4203.
153. Kelsoe G, Zheng B. Sites of B-cell activation in vivo. Curr Opin Immunol 1993;5: 418422.
154. Liu Y-J. Sites a B lymphocyte selection, activation and tolerance in spleen. J Exp Med 1997:186:625–629.
155. Stuber E, Strober W. The T cell-B cell interaction via OX40-OX40L is necessary for the T cell-dependent humoral immune response. J exp Med 1996:183:979–989.
156. Nakanishi K, Howard M, Muraguchi A, et al. Soluble factors involved in B cell differentiation: Identification of two distinct T cell-replacing factors (TRF). J Immunol 1983;130:2219–2224.
157. Morse L, Chen D, Franklin D, Xiong Y, Chen-Kiang S. Induction of cell cycle arrest and B cell terminal differentiation by CDK inhibitor p18(INK4c) and IL-6. Immunity 1997;6: 47–56.
158. Turner CA Jr. Mack DH, Davis MM. Blimp-1, a novel zinc finger-containing protein that can drive the maturation of B lymphocytes into immunoglobulin secreting cells. Cell

1994;77:297–306
159. Schliephake DE, Scimpl A. Blimp-1 overcomes the block in IgM secretion in lipopolysaccharide/anti-mu F(ab')2-co-stimulated B lymphocytes. Eur J Immunol 1996;26:268–271.
160. Phillips C, Klaus GGB. Soluble anti-μ monoclonal antibodies prime resting B cells to secrete immunoglobulins in response to interleukin-4 and –5. Eur J Immunol 1992;22: 1541–1545.
161. Cooke MP, Health AW, Shokat KM et al. Immunoglobulin signal transduction guides the specificity of B cell interactions and is blocked in tolerant self-reactive B cells. J Exp Med 1994;179:425–428.
162. Grupp SA, Snow EC, Harmony JAK. The phosphatidylinositol response is an early event in the physiologically relevant activation of antigen-specific B lymphocytes. Cell Immunol 1987;109:181–191.
163. Snow EC, Fetherston JD, Zimmer S. Induction of the c-myc protooncogen after antigen binding to hapten-specific B cells. J Exp Med 1986;164:944–949.
164. Myers CD, Kriz MK, Sulllivan TJ, Vitetta ES. Antigen-induced changes in phospholipid metabolism in antigen binding B lymphocytes. J Immunol 1987:138:1705–1711.
165. Leibson HJ, Gefter M, Zlotnik A, Marrack P, Kappler JW. Role of interferon in antibody producing responses. Nature 1984;309:799–801.
166. Vitetta ES, Fernandez-Botran R, Myers CD, Sanders VM. Cellular interactions in the humoral immune response. Adv Immunol 1989;45:1–105.
167. Noelle RJ, Snow EC. Cognate interaction between helper T cells and B cells. Immunol Today 1990;11:361–368.
168. Banchereau J, Bazan F, Blanchard D, et al. The CD40 antigen and its ligand. Ann Rev. Immunol. 1994;12:881–922.
169. Clark LB, Foy TM, Noelle RJ. CD40 and its ligand. Adv Immunol 1996;63:43–78.
170. Rothstein TL, Wang JK, Panka DJ, et al. Protection against Fas-dependent Th1-mdiated apoptosis by antigen receptor engagement in B cells. Nature 1995;374:163–165.
171. Xu J, Foy TM Laman JD, et al. Mice deficient for the CD 40 ligand. Immunity 1994;1: 163165.
172. Armitage RJ. Tumor necrosis factor receptor superfamily members and their ligands. Curr Opin Immunol 1994;6:407–413.
173. Xu Y, Cheng G, Baltimore D. Targeted disruption of TRAF3 leads to postnatal lethality and defective T dependent immune responses. Immunity 1996;5:407–415.
174. Kyriakis JM, Avruch J. Sounding the alarm. Protein kinase cascades activated by stress and inflammation. J Biol Chem 1996;271:243113–24316.
175. Sakata N, Patel HR, Terada N, Aruffo A, Johnson GL, Gelfand EW. Elective activation of c-jun kinase mitogen-activated protein kinase by CD40 on human B cells. J Biol Chem 1995;270:30823–30828.
176. Clark EA, Ledbetter JA. How B and T cells talk to each other. Nature 1994;267:425–428.
177. Baluyut AR, Subbarao B. The synergistic effect of anti-Ig and monoclonal anti-Ia anibodies in induction of murine B lymphocyte activation. J Mol Cell Immunol 1988;1:45.
178. Cambier JC, Lehmann KR. Ia mediated signal transduction leads to proliferation of primed B cells. J Exp Med 1989;170:877–886.
179. Paul WE, Sendeer RA. Lymphocyte response and cytokines. Cell 1994;76:241–251.
180. Bacharier LB, Geha RS. Molecular mechanisms of IgE regulation. J Allergy Clin Immunol 2000;105(2 Pt 2):S547–58.

181. Finkelman FD, Holmes J, Katona IM, et al. Lymphokine control of in vivo immunoglobulin isotype selection. Ann Rev Immunol 1990;8:303–333.
182. Pistoia V. Production of cytokines by human B cells in health and disease. Immunon Today 1997;18:343–350.
183. Swain SL, Bradley LM, Croft M, et al. Helper T-cell subsets. Phenotype, function and the role of lymphokines in regulating their development. Immunol Rev 1991;123:115–144.
184. Seder RA, Paul WE, Davis MM, Fazekas de St Groth B. The presence of interleukin 4 during in vitro priming determines the lymphokine producing potential of $CD4^+$ T cells from T cell receptor trasgenic mice. J Exp Med 1992:176:1091–1098.
185. Croft M, Swain SL. B cell response to T helper cell subsets. II. Both the stage of T cell differentiation and the cytokines secreted determine the extent and nature of helper activity. J Immunol 1991;147:3679–3689.
186. Croft M, Swain SL. Analysis of CD4+ T cells that provide contact-dependent bystander help to B cells. J Immunol 1992;149:3157–3165.
187. Gray D, Bergthorsdottir S, van Essen D, Wykes M, Poudrier J, Siepmann K. Observations on memory B –cell development. Semin Immunol 1997;9:249–254.
188. Bachmann MF, Schorle H, Kuhn R, et al. Antiviral response in mice deficient for both interleukin-2 and interleukin-4. J Virol 1995;69:4842–4846.
189. Snapper CM, Rosas F, Moormman MA, et al. IFN-gamma is a potent inducer of Ig secretion by sort-purified murine B cells activated through the mIg, but not the CD40, signaling pathway. Int Immunol 1996;8:877–885.
190. Kopf M, Le Gros G, Coyle AJ, Kosco-Vilbois M, Brombacher F. Immune responses of IL-4, IL-5, IL-6 deficient mice. Immunol Rev 1995;148:45–69.
191. Yoshida T, Ikuta K, Sugaya H, et al. Defective B-1 cell development and impaired immunity against Angilostrongilus cantonensis in IL-%R alpha-deficient mice. Immunity 1996;4:483–494.
192. Radic MZ, Zouali M. Receptor editing, immune diversification, and self-tolerance. Immunity 1996;5:505–511.
193. Klinman NR. The "clonal selection hypothesis" and current concepts of B cell tolerance. Immunity 1996;5:189–195.
194. Nossal GJV. Cellular mechanisms of immunologic tolerance. Ann Rev Immunol 1983;1: 33–62.
195. Hartley SB, Crosbie J, Brink R, Kantor AB, Basten A, Goodnow CC. Elimination from peripheral lymphoid tissues of self-reactive B lymphocytes recognizing membrane-bound antigens. Nature 1991;353:7765–769.
196. Cyster JG, Goodnow CC. Antigen-induced exclusion from follicles and anergy are separate and complementary processes that influence peripheral B cell fate. Immunity 1995;3:691–701.
197. Healy JI, Dolmetsch RE, Timmermann LA, et al. Different nuclear signals are activated by the B cell receptor during positive versus negative signaling. Immmunity 1997;6: 419–428.
198. Effros RB. Long-term immunological memory against viruses. Mech Ageing Dev 2000;121(1–3):161–171.
199. Fearon DT, Manders P, Wagner SD. Arrested differentiation, the self-renewing memory lymphocyte, and vaccination. Science 2001;293:248–250.
200. Sprent J, Tough DF. T cell death and memory. Science 2001;293:245–248.
201. Harris NL, Ronchese F. The role of B7 costimulation in T-cell immunity. Immunol Cell

Biol 1999;77:304–11.
202. Farber DL. T cell memory: heterogeneity and mechanisms. Clin Immunol 2000;95: 173–81.
203. Bevan MJ, Goldrath AW. T-cell memory: You must remember this ... Curr Biol 2000;10: R338–40.
204. Swain SL. CD4 T-cell memory can persist in the absence of class II. Philos Trans R Soc Lond B Biol Sci 2000;355:407–11.
205. Bradley LM, Duncan DD, Yoshimoto K, Swain SL. Memory effectors: a potent, IL-4-secreting helper T cell population that develops in vivo after restimulation with antigen. J Immunol. 1993;150:3119–30.
206. Kung JT, Recent developments in CD8+ T-lymphocyte memory research. J Formos Med Assoc 1999;98:736–9.
207. Sprent J, Zhang X, Sun S, Tough D. T-cell turnover in vivo and the role of cytokines. Immunol Lett 1999;65:21–5.
208. Tough DF, Sun S, Zhang X, Sprent J. Stimulation of memory T cells by cytokines. Vaccine 2000;18:1642–8.
209. Nayak R, Mitra-Kaushik S, Shaila MS. Perpetuation of immunological memory: a relay hypothesis. Immunology 2001;102:387–395.
210. Ashton-Rickardta PG, Opferman JT. Memory T lymphocytes. Cell Mol Life Sci 1999;56: 69–77.
211. Sprent J, Surh CD. Generation and maintenance of memory T cells. Curr Opin Immunol 2001;13:248–254.
212. Akbar AN, Lord JM, Salmon M. IFN-alpha and IFN-beta: a link between immune memory and chronic inflammation. Immunol Today 2000;21:337–342.
213. Cho BK, Rao VP, Ge Q, Eisen HN, Chen J. Homeostasis-stimulated proliferation drives naive T cells to differentiate directly into memory T cells. J Exp Med 2000;192:549–556.
214. Berczi I. The influence of pituitary-adrenal axis on the immune system. In: Berczi I, ed. Pituitary function and immunity. Boca Raton, FL: CRC Press, 1986, 49–132.
215. Claman HN. Corticosteroids-immunologic and anti-inflammatory effects. In: Berczi I, Kovacs K, editors. Hormones and immunity. Lancaster, UK: MTP Press, 1987;38–42.
216. Berczi I, Nagy E, Kovacs K, Horvath E. Regulation of humoral immunity in rats by pituitary hormones. Acta Endocrinol. 98:506–513, 1981.
217. Berczi I, Nagy E, Asa SL, Kovacs K. Pituitary hormones and contact sensitivity in rats. Allergy 38:325–330, 1983.
218. Berczi I, Nagy E, Asa SL, Kovacs L. The influence of pituitary hormones on adjuvant arthritis. Arthritis Rheum. 27:682–688, 1984.
219. Roitt I. 1997. Innate immunity. *In* Essential Immunology, Roitt I. Ed. Ninth edition. Blackwell Scientific. Pp 3–21.
220. Bertok L. 1983. Stimulation of nonspecific resistance by radio-detoxified endotoxin. *In* Beneficial effects of endotoxins. A. Nowotny, Ed. Plenum Press, New York. Pp 213–226.
221. Berczi I. Neuroendocrine defence in endotoxin shock. Acta Microbiol Hung 1993;40: 265–302.
222. Sabbadini ER, Berczi I. The submandibular gland: a key organ in the neuro-immunoregulatory network? Neuroimmunomodulation, 1985;2:184–202.
223. Berczi I. Neurohormonal host defence in endotoxin shock Ann NY Acad Sci 1998: 840: 787–802.
224. Berczi I, Nagy E. Neurohormonal control of cytokines during injury. *In* Brain control of the

response to injury, Rothwell, N.J.& F. Berkenbosch, Ed.: 32–107. Cambridge University Press, 1994.
225. Medzhitov R, Janeway CA. Innate immunity: impact on the adaptive immune response. Curr Opin Immunol 1997;9**:**4–9.
226. Boismenu R, Havran WL. An innate view of T cells. Curr Opin Immunol **9:**57–63.
227. Berczi I, Chow DA, Sabbadini ER. Neuroimmunoregulation and natural immunity. Domestic Animal Endocrin. 1997;15:273–281.
228. Berczi I. Nagy E. Hormones as immune modulating agents. *In* Handbook of Immune Modulating Agents, Krezina, T. Ed. Marcel Dekker, New York. 1997, pp 75–120.
229. Turner MW. Mannose-binding lectin: the pluripotent molecule of the innate immune system. Immunol. Today 1996;7**:**532–539.
230. Chow DA, Wang H, Zhang Z. Natural antibody surveillance of neoplastic and activated cells. Biotecnologia Aplicada 1999;16**:**18–19.
231. Wang H, Chow DA, Protein kinase C expression links natural antibody binding with surveillance of activated and preneoplastic cells. Scand J Immunol 1999;49**:**381–390.
232. Hughes AL, Yeager M, Elshof AE, Chorney MJ. A new taxonomy of mammalian MHC class I molecules. Immunol. Today 1999;20**:**22–26.
233. Herman A, Kappler JW, Marrack P, Pullen AM. Superantigens - mechanism of T-cell stimulation and role in immune responses. Annu Rev Immunol 1991;9**:**745–772.
234. Chien Y, Jores R, Crowley MP. Recognition by γδ T cells. Annu Rev Immunol 1996;14**:** 511–532.
235. Silverman GJ. B-cell superantigens. Immunol Today 1997;18:379–386.
236. Whiteside MB, Quan N, Herkenharn M. Induction of pituitary cytokine transcripts by peripheral lipopolisaccharide. J Neuroendocrinol 1999;11:115–120.
237. Goehler LE, Gaykema RP, Nguyen KT, Lee JE, Tilders FJ, Maier SF, Watkins LR. Interleukin-1-beta in immune cells of the abdominal vagus nerve: a link between the immune and the nervous system? J Neurosci 1999;19:2799–806.
238. Faggioni R, Fantuzzi G, Gabay C, Moser A, Dinarello CA, Feingold KR, Gunfeld C. Leptin deficiency enhances sensitivity to endotoxin induced lethality. Amer J Physiol 1999; 276:R136–42.
239. Haupt W, Riese J, Denzel C, Zowe M, Gusinde J, Siassi M, Hohenberger W. Culture of human peritoneum – a new method to measure the local cytokine response and the effect of immunomodulators. Infection 1998;26:345–8.
240. Lahn M, Kalataradi H. Mittelstadt P, Pflum E, Wollner M, Cady C, Mukasa A. Wella AT, Ikle D, Harbeck R, O'Brien R, Born W. Early preferential stimulation of gamma delta T cells by TNF-alpha. J Immunol 1998;160:5221–30.
241. Leclercq G, Plum J. Stimulation of TCR V 3 cells by gram-negative bacteria. J Immunol 1995;154: 5313–5319.
242. Rivest S, Nadeau S, Lacroix S, Laflamme N. Proinflammatory signal transduction pathways in the CNS during systemic immune reactions. In New foundation of biology, Berczi I, Gorczynski R. Editors. Neuroimmune Biology, Volume 1, Berczi I, Szentivanyi A. Series editors, Amsterdam, Elsevier, 2001, pp 163–173.
243. Hardy RR, Hayakawa K. B cell development pathways. Annu Rev Immunol 2001;19: 595–621.
244. Locati M, Murphy PM. Chemokines and chemokine receptors: biology and clinical relevance in inflammation and AIDS. Annu Rev Med 1999;50:425–40.
245. Jancso N. Neurogenic inflammation. Acta Physiol. Acad. Sci. Hung. 1964:23:3.

246. Ruedl C, Wolf H. Features of oral immunization. Int Arch Allergy Immunol 1995;108: 334–9.
247. Lamping M, Dettmer R, Schroder NW, Pfeil D, Hallatschek W, Burger R, Schumann RR. LPS binding protein protects mice from septic shock caused by LPS or gram-negative bacteria. J Clin Invest 1998;101:2065–71.
248. Szalai AJ, Briles DE, Volantakis JE. 1996. Role of complement in C-reactive protein-mediated protection of mice from *Streptococcus pneumoniae*. Infect Immun **64:**4850–3.
249. Talmage DW. Allergy and immunology. Ann Rev Med 1957;8:239–256.
250. Burnet FM: The Clonal Selection Theory of Acquired Immunity. Cambridge University Press, 1959.
251. von Boehmer H, Kisielow P. Self-nonself discrimination by T cells. Science. 1990;248: 1369–73.
252. Chidgey AP, Boyd RL. Thymic stromal cells and positive selection. APMIS 2001;109: 481–92.
253. Berthiaume F, Aparicio CL, Eungdamrong J, Yarmush ML. Age- and disease-related decline in immune function: an opportunity for “thymus-boosting” therapies. Tissue Eng 1999;5:499–514.
254. Douek DC, Altmann DM. T-cell apoptosis and differential human leucocyte antigen class II expression in human thymus. Immunology 2000;99:249–56.
255. Marinova T, Altankova I, Dimitrova D, Pomakov Y. Presence of HLA-DR immunopositive cells in human fetal thymus. Arch Physiol Biochem 2001;109:74–9.
256. Choi EY, Bae Y, Lee DS, Park SH. T-T interaction during thymic selection. Cell Mol Biol 2001;47:135–43.
257. Viret C, Janeway CA Jr. MHC and T cell development. Rev Immunogenet 1999;1:91–104.
258. Viret C, Sant’Angelo DB, He X, Ramaswamy H, Janeway CA Jr. A role for accessibility to self-peptide-self-MHC complexes in intrathymic negative selection. J Immunol 2001;166: 4429–37.
259. Perraudeau M, Taylor PR, Stauss HJ, Lindstedt R, Bygrave AE, Pappin DJ, Ellmerich S, Whitten A, Rahman D, Canas B, Walport MJ, Botto M, Altmann DM. Altered major histocompatibility complex class II peptide loading in H2-O-deficient mice. Eur J Immunol 2000;30:2871–80.
260. Correia-Neves M, Waltzinger C, Mathis D, Benoist C. The shaping of the T cell repertoire. Immunity 2001;14:21–32.
261. Kleindienst P, Chretien I, Winkler T, Brocker T. Functional comparison of thymic B cells and dendritic cells in vivo. Blood 2000;95:2610–6.
262. Bodey B, Bodey B Jr, Siegel SE, Kaiser HE. The role of the reticulo-epithelial (RE) cell network in the immuno-neuroendocrine regulation of intrathymic lymphopoiesis. Anticancer Res 2000;20(3A):1871–88.
263. Lechner O, Wiegers GJ, Oliveira-Dos-Santos AJ, Dietrich H, Recheis H, Waterman M, Boyd R, Wick G. Glucocorticoid production in the murine thymus. Eur J Immunol 2000 Feb;30(2):337–46.
264. Ashwell JD, Lu FW, Vacchio MS. Glucocorticoids in T cell development and function. Annu Rev Immunol 2000;18:309–45.
265. Chmielewski V, Drupt F, Morfin R. Dexamethasone-induced apoptosis of mouse thymocytes: prevention by native 7alpha-hydroxysteroids. Immunol Cell Biol 2000;78: 238–46.

266. Geenen V, Kecha O, Brilot F, Hansenne I, Renard C, Martens H. Thymic T-cell tolerance of neuroendocrine functions: physiology and pathophysiology. Cell Mol Biol 2001;47: 179–88.
267. Hare KJ, Jenkinson EJ, Anderson G. Specialisation of thymic epithelial cells for positive selection of CD4+8+thymocytes. Cell Mol Biol 2001;47:119–27.
268. Lucas B, Germain RN. Opening a window on thymic positive selection: developmental changes in the influence of cosignaling by integrins and CD28 on selection events induced by TCR engagement. J Immunol 2000;165:1889–95.
269. Capone M, Romagnoli P, Beermann F, MacDonald HR, van Meerwijk JP. Dissociation of thymic positive and negative selection in transgenic mice expressing major histocompatibility complex class I molecules exclusively on thymic cortical epithelial cells. Blood 2001;97:1336–4.
270. Fairchild PJ, Waldmann H. Extrathymic signals regulate the onset of T cell repertoire selection. Eur J Immunol 2000;30:1948–56.
271. Chmielowski B, Muranski P, Kisielow P, Ignatowicz L. On the role of high- and low-abundance class II MHC-peptide complexes in the thymic positive selection of CD4(+) T cells. Int Immunol 2000;12:67–72.
272. Shizuru JA, Weissman IL, Kernoff R, Masek M, Scheffold YC. Purified hematopoietic stem cell grafts induce tolerance to alloantigens and can mediate positive and negative T cell selection. Proc Natl Acad Sci U S A 2000;97:9555–6.
273. Clarke SR, Rudensky AY. Survival and homeostatic proliferation of naive peripheral CD4+ T cells in the absence of self peptide:MHC complexes. J Immunol 2000;165:2458–64.
274. Pezzano M, Samms M, Guyden JC. TNF and Fas-induced apoptosis during negative selection in thymic nurse cells. Ethn Dis 2001;11:154–6.
275. Pacholczyk R, Kraj P, Ignatowicz L. Peptide specificity of thymic selection of CD4+CD25+ T cells. J Immunol. 2002;168:613–20.
276. Bensinger SJ, Bandeira A, Jordan MS, Caton AJ, Laufer TM. Major histocompatibility complex class II-positive cortical epithelium mediates the selection of CD4(+)25(+) immunoregulatory T cells. J Exp Med 2001;194:427–38.
277. Jordan MS, Boesteanu A, Reed AJ, Petrone AL, Holenbeck AE, Lerman MA, Naji A, Caton AJ. Thymic selection of CD4+CD25+ regulatory T cells induced by an agonist self-peptide. Nat Immunol 2001;2:301–6.
278. Riese RJ, Shi GP, Villadangos J, Stetson D, Driessen C, Lennon-Dumenil AM, Chu CL, Naumov Y, Behar SM, Ploegh H, Locksley R, Chapman HA. Regulation of CD1 Function and NK1.1(+) T Cell Selection and Maturation by Cathepsin S. Immunity 2001;15:909–19.
279. Viret C, Lantz O, He X, Bendelac A, Janeway CA Jr. A NK1.1+ thymocyte-derived TCR beta-chain transgene promotes positive selection of thymic NK1.1+ alpha beta T cells. J Immunol 2000;165:3004–14.
280. Minagawa M, Ito A, Shimura H, Tomiyama K, Ito M, Kawai K. Homogeneous epithelial gamma delta T cell repertoire of the skin is shaped through peripheral selection. J Dermatol Sci 2001;25:150–5.
281. Schaeffer EM, Broussard C, Debnath J, Anderson S, McVicar DW, Schwartzberg PL. Tec family kinases modulate thresholds for thymocyte development and selection. J Exp Med 2000;192:987–1000.
282. Dautigny N, Lucas B. Developmental regulation of TCR efficiency. Eur J Immunol 2000;30:2472–8.

283. Zhu JW, DeRyckere D, Li FX, Wan YY, DeGregori J. A role for E2F1 in the induction of ARF, p53, and apoptosis during thymic negative selection. Cell Growth Differ 1999;10: 829–38.
284. Takagi T, Harada J, Ishii S. Murine Schnurri-2 is required for positive selection of thymocytes. Nat Immunol 2001;2:1048–53.
285. Sun Z, Unutmaz D, Zou YR, Sunshine MJ, Pierani A, Brenner-Morton S, Mebius RE, Littman DR. Requirement for RORgamma in thymocyte survival and lymphoid organ development. Science 2000;288:2369–73.
286. Maurice MM, Gould DS, Carroll J, Vugmeyster Y, Ploegh HL. Positive selection of an MHC class-I restricted TCR in the absence of classical MHC class I molecules. Proc Natl Acad Sci U S A 2001;98:7437–42.
287. Yasutomo K, Lucas B, Germain RN. TCR signaling for initiation and completion of thymocyte positive selection has distinct requirements for ligand quality and presenting cell type. J Immunol 2000;165:3015–22.
288. Trop S, Charron J, Arguin C, Lesage S, Hugo P. Thymic selection generates T cells expressing self-reactive TCRs in the absence of CD45. J Immunol 2000;165:3073–9.
289. Kurasawa K, Hashimoto Y, Kasai M, Iwamoto I. The fas antigen is involved in thymic T-cell development as a costimulatory molecule, but not in the deletion of neglected thymocytes. J Allergy Clin Immunol 2000;106:S19–31.
290. Mora AL, Stanley S, Armistead W, Chan AC, Boothby M. Inefficient ZAP-70 phosphorylation and decreased thymic selection in vivo result from inhibition of NF-kappaB/Rel. J Immunol 2001;167:5628–35.
291. Baseta JG, Stutman O. TNF regulates thymocyte production by apoptosis and proliferation of the triple negative (CD3-CD4-CD8-) subset. J Immunol 2000;165:5621–30.
292. Wang J, Chun T, Lo JC, Wu Q, Wang Y, Foster A, Roca K, Chen M, Tamada K, Chen L, Wang CR, Fu YX. The critical role of LIGHT, a TNF family member, in T cell development. J Immunol 2001;167:5099–105.
293. Chen W, Jin W, Tian H, Sicurello P, Frank M, Orenstein JM, Wahl SM. Requirement for transforming growth factor beta1 in controlling T cell apoptosis. J Exp Med 2001;194: 439–53.
294. Kisielow J, Nairn AC, Karjalainen K. TARPP, a novel protein that accompanies TCR gene rearrangement and thymocyte education. Eur J Immunol 2001;31:1141–9.
295. Madden KS, Felten DL. Beta-adrenoceptor blockade alters thymocyte differentiation in aged mice. Cell Mol Biol 2001;47:189–96.
296. Gershon RK, Cohen P, Hencin R, Liebhaber SA. Suppressor T cells. J Immunol. 1972;108:586–90.
297. Mason D. Some quantitative aspects of T-cell repertoire selection: the requirement for regulatory T cells. Immunol Rev 2001;182:80–8.
298. Seddon B. The physiological role of regulatory T cells in the prevention of autoimmunity: generation, specificity and mode of action. Arch Immunol Ther Exp (Warsz) 2000;48: 339–45.
299. Shevach EM. Suppressor T cells: Rebirth, function and homeostasis. Curr Biol 2000;10: R572–5.
300. Read S, Powrie F. CD4(+) regulatory T cells. Curr Opin Immunol 2001;13:644–9.
301. Cortesini R, LeMaoult J, Ciubotariu R, Cortesini NS. CD8+CD28- T suppressor cells and the induction of antigen-specific, antigen-presenting cell-mediated suppression of Th reactivity. Immunol Rev. 2001;182:201–6.

302. Maloy KJ, Powrie F. Regulatory T cells in the control of immune pathology. Nat Immunol 2001;2:816–22
303. Piccirillo CA, Shevach EM. Cutting edge: control of CD8+ T cell activation by CD4+CD25+ immunoregulatory cells. J Immunol 2001;167:1137–40.
304. McHugh RS, Shevach EM, Thornton AM. Control of organ-specific autoimmunity by immunoregulatory CD4(+)CD25(+) T cells. Microbes Infect 2001;3:919–27.
305. Read S, Malmstrom V, Powrie F. Cytotoxic T lymphocyte-associated antigen 4 plays an essential role in the function of CD25(+)CD4(+) regulatory cells that control intestinal inflammation. J Exp Med 2000;192:295–302.
306. Maeda H, Fujimoto S, Greene MI. Suppressor T cells regulate the nonanergic cell population that remains after peripheral tolerance is induced to the Mls-1 antigen in T cell receptor Vbeta 8.1 transgenic mice. Proc Natl Acad Sci U S A 2000;97:13257–62.
307. Shevach EM. Regulatory T cells in autoimmmunity. Annu Rev Immunol 2000;18:423–49.
308. Terui T, Sano K, Shirota H, Kunikata N, Ozawa M, Okada M, Honda M, Tamura G, Tagami H. TGF-beta-producing CD4+ mediastinal lymph node cells obtained from mice tracheally tolerized to ovalbumin (OVA) suppress both Th1- and Th2-induced cutaneous inflammatory responses to OVA by different mechanisms. J Immunol 2001;167:3661–7.
309. Nakamura K, Kitani A, Strober W. Cell contact-dependent immunosuppression by D4(+)CD25(+) regulatory T cells is mediated by cell surface-bound transforming growth factor beta. J Exp Med 2001;194:629–44.
310. Taylor PA, Noelle RJ, Blazar BR.CD4(+)CD25(+) immune regulatory cells are required for induction of tolerance to alloantigen via costimulatory blockade. J Exp Med 2001;193: 1311–8.
311. Dieckmann D, Plottner H, Berchtold S, Berger T, Schuler G. Ex vivo isolation and characterization of CD4(+)CD25(+) T cells with regulatory properties from human blood. J Exp Med 2001;193:1303–10.
312. Levings MK, Sangregorio R, Roncarolo MG. Human cd25(+)cd4(+) T regulatory cells suppress naive and memory T cell proliferation and can be expanded in vitro without loss of function. J Exp Med 2001;193:1295–302.
313. Yahata H, Fukada Y, Hayamizu K, Okimoto T, Ishikawa T, Asahara T, Ono E, Dohi K. Mechanism of suppression of cloned human suppressor T cells. Transpl Int 1994;7 Suppl 1:S590–5.
314. Ebert EC, Roberts AI. Lamina propria lymphocytes produce interferon-gamma and develop suppressor activity in response to lactoglobulin. Dig Dis Sci 2001;46:661–7.
315. Cavani A, Albanesi C, Traidl C, Sebastiani S, Girolomoni G. Effector and regulatory T cells in allergic contact dermatitis. Trends Immunol 2001;22:118–20.
316. Priatel JJ, Utting O, Teh HS. TCR/self-antigen interactions drive double-negative T cell peripheral expansion and differentiation into suppressor cells. J Immunol 2001;167:6188–94.
317. Mirshahidi S, Huang CT, Sadegh-Nasseri S. Anergy in peripheral memory CD4(+) T cells induced by low avidity engagement of T cell receptor. J Exp Med 2001;194:719–31.
318. Lan F, Zeng D, Higuchi M, Huie P, Higgins JP, Strober S. Predominance of NK1.1+TCR alpha beta+ or DX5+TCR alpha beta+ T cells in mice conditioned with fractionated lymphoid irradiation protects against graft-versus-host disease: "natural suppressor" cells. J Immunol 2001;167:2087–96.
319. Strober S. Natural killer 1.1(+) T cells and "natural suppressor" T cells in the bone marrow.

J Allergy Clin Immunol 2000;106(1 Pt 2):S113–4.
320. El Marsafy S, Dosquet C, Coudert MC, Bensussan A, Carosella E, Gluckman E. Study of cord blood natural killer cell suppressor activity. Eur J Haematol 2001;66:215–20.
321. Wang Y, Goldschneider I, O'Rourke J, Cone RE. Blood mononuclear cells induce regulatory NK T thymocytes in anterior chamber-associated immune deviation. J Leukoc Biol 2001;69:741–6.
322. D'Orazio TJ, Mayhew E, Niederkorn JY. Ocular immune privilege promoted by the presentation of peptide on tolerogenic B cells in the spleen. II. Evidence for presentation by Qa-1. J Immunol 2001;166:26–32.
323. Hanninen A, Harrison LC. Gamma delta T cells as mediators of mucosal tolerance: the autoimmune diabetes model. Immunol Rev 2000;173:109–19.
324. Bronte V, Apolloni E, Cabrelle A, Ronca R, Serafini P, Zamboni P, Restifo NP, Zanovello P. Identification of a CD11b(+)/Gr-1(+)/CD31(+) myeloid progenitor capable of activating or suppressing CD8(+) T cells. Blood 2000;96:3838–46.
325. Suciu-Foca Cortesini N, Piazza F, Ho E, Ciubotariu R, LeMaoult J, Dalla-Favera R, Cortesini R. Distinct mRNA microarray profiles of tolerogenic dendritic cells. Hum Immunol 2001;62:1065–72.
326. Vendetti S, Chai JG, Dyson J, Simpson E, Lombardi G, Lechler R. Anergic T cells inhibit the antigen-presenting function of dendritic cells. J Immunol 2000;165:1175–81.
327. Luo Z, Gotoh M, Grochowiecki T, Tanaka T, Kimura F, Kawashima H, Yagita H, Okumura K, Miyasaka M. Anergic T cells generated in vitro suppress rejection response to islet allografts. Transplantation 2000;69:2144–8.
328. Ermann J, Szanya V, Ford GS, Paragas V, Fathman CG, Lejon K. CD4(+)CD25(+) T cells facilitate the induction of T cell anergy. J Immunol 2001;167:4271–5.
329. Murga M, Fernandez-Capetillo O, Field SJ, Moreno B, R -Borlado L, Fujiwara Y, Balomenos D, Vicario A, Carrera AC, Orkin SH, Greenberg ME, Zubiaga AM. Mutation of E2F2 in Mice Causes Enhanced T Lymphocyte Proliferation, Leading to the Development of Autoimmunity. Immunity 2001;15:959–70.
330. Akdis CA, Blaser K. Mechanisms of interleukin-10-mediated immune suppression. Immunology 2001;103:131–6.
331. Noel C, Florquin S, Goldman M, Braun MY. Chronic exposure to superantigen induces regulatory CD4(+) T cells with IL-10-mediated suppressive activity. Int Immunol 2001;13: 431–9.
332. Terabe M, Shimizu M, Mabuchi A, Matui S, Morikawa H, Kaneda K, Kakiuchi T, Yokomuro K. Unresponsiveness of intrahepatic lymphocytes to bacterial superantigen: rapid development of suppressive Mac-1(high) cells in the mouse liver. Hepatology 2000;32:507–13.
333. Aharoni R, Teitelbaum D, Leitner O, Meshorer A, Sela M, Arnon R. Specific Th2 cells accumulate in the central nervous system of mice protected against experimental autoimmune encephalomyelitis by copolymer 1. Proc Natl Acad Sci USA 2000;97): 11472–7.
334. Yamagiwa S, Gray JD, Hashimoto S, Horwitz DA. A role for TGF-beta in the generation and expansion of CD4+CD25+ regulatory T cells from human peripheral blood. J Immunol 2001;166:7282–9.
335. Kitani A, Chua K, Nakamura K, Strober W. Activated self-MHC-reactive T cells have the cytokine phenotype of Th3/T regulatory cell 1 T cells. J Immunol 2000;165:691–702.
336. Segerson EC, Beetham PK. A porcine intrauterine 4-mda component with transforming

growth factor-beta activity suppresses natural killer cell responses. J Anim Sci 2000;78: 1900–6.
337. Diehl S, Anguita J, Hoffmeyer A, Zapton T, Ihle JN, Fikrig E, Rincon M. Inhibition of Th1 differentiation by IL-6 is mediated by SOCS1. Immunity 2000;13:805–15.
338. Delgado M, Ganea D. Vasoactive intestinal peptide and pituitary adenylate cyclase-activating polypeptide inhibit T cell-mediated cytotoxicity by inhibiting Fas ligand expression. J Immunol 2000;165:114–23.
339. Abad C, Martinez C, Leceta J, Gomariz RP, Delgado M. Pituitary adenylate cyclase-activating polypeptide inhibits collagen-induced arthritis: an experimental immunomodulatory therapy. J Immunol 2001;167:3182–9.
340. Lemire J. 1,25-Dihydroxyvitamin D3--a hormone with immunomodulatory properties. Z Rheumatol 2000;59 Suppl 1:24–7.
341. Salem ML, Matsuzaki G, Kishihara K, Madkour GA, Nomoto K. Beta-estradiol suppresses T cell-mediated delayed-type hypersensitivity through suppression of antigen-presenting cell function and Th1 induction. Int Arch Allergy Immunol 2000;121:161–9.
342. Coutinho A. Germ-line selection ensures embryonic autoreactivity and a positive discrimination of self mediated by supraclonal mechanisms. Semin Immunol 2000;12: 205–13; discussion 257–344.
343. Paroli M, Schiaffella E, Di Rosa F, Barnaba V. Persisting viruses and autoimmunity. J Neuroimmunol 2000;107:201–4.
344. Theofilopoulos AN, Kono DH. The genes of systemic autoimmunity. Proc Assoc Am Physicians 1999;111:228–40.
345. Siegel RM, Fleisher TA. The role of Fas and related death receptors in autoimmune and other disease states. J Allergy Clin Immunol 1999;103(5 Pt 1):729–38.
346. Shen X, Konig R. Post-thymic selection of peripheral CD4+ T-lymphocytes on class II major histocompatibility antigen-bearing cells. Cell Mol Biol 2001;47:87–96.
347. Ashany D, Savir A, Bhardwaj N, Elkon KB. Dendritic cells are resistant to apoptosis through the Fas (CD95/APO-1) pathway. J Immunol 1999;163:5303–11.
348. Jiang H, Ware R, Stall A, Flaherty L, Chess L, Pernis B. Murine CD8+ T cells that specifically delete autologous CD4+ T cells expressing V beta 8 TCR: a role of the Qa-1 molecule. Immunity 1995;2:185–94.
349. Kuhrober A, Schirmbeck R, Reimann J. Vaccination with T cell receptor peptides primes anti-receptor cytotoxic T lymphocytes (CTL) and anergizes T cells specifically recognized by these CTL. Eur J Immunol 1994;24:1172–80.
350. Okamura H, Tsutsi H, Komatsu T, Yutsudo M, Hakura A, Tanimoto T, Torigoe K, Okura T, Nukada Y, Hattori K, et al. Cloning of a new cytokine that induces IFN-gamma production by T cells. Nature. 1995;378:88–91.
351. Dao T, Ohashi K, Kayano T, Kurimoto M, Okamura H. Interferon-gamma-inducing factor, a novel cytokine, enhances Fas ligand-mediated cytotoxicity of murine T helper 1 cells. Cell Immunol 1996;173:230–5.
352. Sato N, Yahata T, Santa K, Ohta A, Ohmi Y, Habu S, Nishimura T. Functional characterization of NK1.1 + Ly-6C+ cells. Immunol Lett 1996;54:5–9.
353. Marelli-Berg FM, Scott D, Bartok I, Peek E, Dyson J, Lechler RI. Activated murine endothelial cells have reduced immunogenicity for CD8+ T cells: a mechanism of immunoregulation? J Immunol 2000;165:4182–9.
354. Muschen M, Warskulat U, Peters-Regehr T, Bode JG, Kubitz R, Haussinger D.Involvement of CD95 (Apo-1/Fas) ligand expressed by rat Kupffer cells in hepatic immunoregulation.

Gastroenterology 1999;116:666–77.
355. White CA, McCombe PA, Pender MP. Microglia are more susceptible than macrophages to apoptosis in the central nervous system in experimental autoimmune encephalomyelitis through a mechanism not involving Fas (CD95). Int Immunol 1998;10:935–41.
356. Marx C, Wolkersdorfer GW, Bornstein SR. A new view on immune-adrenal interactions: role for Fas and Fas ligand? Neuroimmunomodulation 1998;5:5–8.
357. Milik AM, Buechner-Maxwell VA, Sonstein J, Kim S, Seitzman GD, Beals TF, Curtis JL. Lung lymphocyte elimination by apoptosis in the murine response to intratracheal particulate antigen. J Clin Invest 1997;99:1082–91.
358. Sneller MC, Wang J, Dale JK, Strober W, Middelton LA, Choi Y, Fleisher TA, Lim MS, Jaffe ES, Puck JM, Lenardo MJ, Straus SE. Clincial, immunologic, and genetic features of an autoimmune lymphoproliferative syndrome associated with abnormal lymphocyte apoptosis. Blood 1997;89:1341–8.
359. Carneiro-Santos P, Martins-Filho O, Alves-Oliveira LF, Silveira AM, Coura-Filho P, Viana IR, Wilson RA, Correa-Oliveira R. Apoptosis: a mechanism of immunoregulation during human Schistosomiasis mansoni. Parasite Immunol 2000;22:267–77.
360. Tsuboi H, Miyazawa H, Wenner M, Iimori H, Kawamura N. Lesions in lateral hypothalamic areas increase splenocyte apoptosis. Neuroimmunomodulation 2001;9:1–5.
361. Betten A, Dahlgren C, Hermodsson S, Hellstrand K. Serotonin protects NK cells against oxidatively induced functional inhibition and apoptosis. J Leukoc Biol 2001;70:65–72.
362. Brouty-Boye D, Pottin-Clemenceau C, Doucet C, Jasmin C, Azzarone B. Chemokines and CD40 expression in human fibroblasts. Eur J Immunol 2000;30:914–9.
363. Cousin B, Andre M, Casteilla L, Penicaud L. Altered macrophage-like functions of preadipocytes in inflammation and genetic obesity. J Cell Physiol 2001;186:380–6.
364. Seo SK, Gebhardt BM, Lim HY, Kang SW, Higaki S, Varnell ED, Hill JM, Kaufman HE, Kwon BS. Murine keratocytes function as antigen-presenting cells. Eur J Immunol 2001;31:3318–28.
365. Maroder M, Bellavia D, Vacca A, Felli MP, Screpanti I. The thymus at the crossroad of neuroimmune interactions. Ann N Y Acad Sci 2000;917:741–7.
366. Banwell CM, Partington KM, Jenkinson EJ, Anderson G. Studies on the role of IL-7 presentation by mesenchymal fibroblasts during early thymocyte development. Eur J Immunol 2000;30:2125–9.
367. Fagarasan S, Kinoshita K, Muramatsu M, Ikuta K, Honjo T. In situ class switching and differentiation to IgA-producing cells in the gut lamina propria. Nature 2001;413(6856): 639–43.
368. Austen KF, Boyce JA. Mast cell lineage development and phenotypic regulation. Leuk Res 2001;25:511–8.
369. Scopes J, Ismail M, Marks KJ, Rutherford TR, Draycott GS, Pocock C, Gordon-Smith EC, Gibson FM. Correction of stromal cell defect after bone marrow transplantation in aplastic anaemia. Br J Haematol 2001;115:642–52.
370. Janczewska S, Ziolkowska A, Interewicz B, Majewski T, Olszewski WL, Lukomska B. Vascularized bone marrow transplanted in orthotopic hind-limb stimulates hematopoietic recovery in total-body-irradiated rats. Transpl Int 2000;13 Suppl 1:S541–6.
371. Hori J, Joyce NC, Streilein JW. Immune privilege and immunogenicity reside among different layers of the mouse cornea. Invest Ophthalmol Vis Sci 2000;41:3032–42.
372. Tonnelle C, Bardin F, Maroc C, Imbert AM, Campa F, Dalloul A, Schmitt C, Chabannon C. Forced expression of the Ikaros 6 isoform in human placental blood CD34(+) cells impairs

their ability to differentiate toward the B-lymphoid lineage. Blood 2001;98:2673–80.

373. Leuker CE, Labow M, Muller W, Wagner N. Neonatally induced inactivation of the vascular cell adhesion molecule 1 gene impairs B cell localization and T cell-dependent humoral immune response. J Exp Med 2001;193:755–68.

374. Ding Z, Xiong K, Issekutz TB. Chemokines stimulate human T lymphocyte transendothelial migration to utilize VLA-4 in addition to LFA-1. J Leukoc Biol 2001;69:458–66.

375. Weber KS, Ostermann G, Zernecke A, Schroder A, Klickstein LB, Weber C. Dual role of H-Ras in regulation of lymphocyte function antigen-1 activity by stromal cell-derived factor-1alpha: implications for leukocyte transmigration. Mol Biol Cell 2001;12:3074–86.

376. Murdoch C. CXCR4: chemokine receptor extraordinaire. Immunol Rev 2000;177:175–84.

377. Rosu-Myles M, Gallacher L, Murdoch B, Hess DA, Keeney M, Kelvin D, Dale L, Ferguson SS, Wu D, Fellows F, Bhatia M. The human hematopoietic stem cell compartment is heterogeneous for CXCR4 expression. Proc Natl Acad Sci U S A 2000;97:14626–31.

378. Han Y, Wang J, He T, Ransohoff RM. TNF-alpha down-regulates CXCR4 expression in primary murine astrocytes. Brain Res 2001;888:1–10.

379. Douglas GC, Thirkill TL. Chemokine receptor expression by human syncytiotrophoblast--a review. Placenta 2001;22 Suppl A:S24–8.

380. Sbaa-Ketata E, Vasse M, Lenormand B, Schneider P, Soria C, Vannier JP. Fibronectin increases the migration induced by stromal cell-derived factor-1 alpha (SDF-1 alpha) in pre-B acute lymphoblastic leukemia cells. Eur Cytokine Netw 2001;12:223–30.

381. Hidalgo A, Sanz-Rodriguez F, Rodriguez-Fernandez JL, Albella B, Blaya C, Wright N, Cabanas C, Prosper F, Gutierrez-Ramos JC, Teixido J. Chemokine stromal cell-derived factor-1alpha modulates VLA-4 integrin-dependent adhesion to fibronectin and VCAM-1 on bone marrow hematopoietic progenitor cells. Exp Hematol 2001;29:345–55.

382. Yanagawa Y, Iwabuchi K, Onoe K. Enhancement of stromal cell-derived factor-1alpha-induced chemotaxis for CD4/8 double-positive thymocytes by fibronectin and laminin in mice. Immunology 2001;104:43–9.

383. Colamussi ML, Secchiero P, Gonelli A, Marchisio M, Zauli G, Capitani S. Stromal derived factor-1 alpha (SDF-1 alpha) induces CD4+ T cell apoptosis via the functional up-regulation of the Fas (CD95)/Fas ligand (CD95L) pathway. J Leukoc Biol 2001;69:263–70.

384. Foussat A, Balabanian K, Amara A, Bouchet-Delbos L, Durand-Gasselin I, Baleux F, Couderc J, Galanaud P, Emilie D. Production of stromal cell-derived factor 1 by mesothelial cells and effects of this chemokine on peritoneal B lymphocytes. Eur J Immunol 2001;31:350–9.

385. Balabanian K, Foussat A, Bouchet-Delbos L, Couderc J, Krzysiek R, Amara A, Baleux F, Portier A, Galanaud P, Emilie D. Interleukin-10 modulates the sensitivity of peritoneal B lymphocytes to chemokines with opposite effects on stromal cell-derived factor-1 and B-lymphocyte chemoattractant. Blood 2002;99:427–36.

386. Delgado MB, Clark-Lewis I, Loetscher P, Langen H, Thelen M, Baggiolini M, Wolf M. Rapid inactivation of stromal cell-derived factor-1 by cathepsin G associated with lymphocytes. Eur J Immunol 2001;31:699–707.

387. Suzuki Y, Rahman M, Mitsuya H. Diverse transcriptional response of CD4(+) T cells to stromal cell-derived factor (SDF)-1: cell survival promotion and priming effects of SDF-1 on CD4(+) T cells. J Immunol 2001;167:3064–73.

388. Vicente-Manzanares M, Cabrero JR, Rey M, Perez-Martinez M, Ursa A, Itoh K, Sanchez-Madrid F. A role for the Rho-p160 Rho coiled-coil kinase axis in the chemokine stromal cell-derived factor-1alpha-induced lymphocyte actomyosin and microtubular organization and chemotaxis. J Immunol 2002;168:400–10.
389. Haddad E, Zugaza JL, Louache F, Debili N, Crouin C, Schwarz K, Fischer A, Vainchenker W, Bertoglio J. The interaction between Cdc42 and WASP is required for SDF-1-induced T-lymphocyte chemotaxis. Blood 2001;97:33–8.
390. Ottoson NC, Pribila JT, Chan AS, Shimizu Y. Cutting edge: T cell migration regulated by CXCR4 chemokine receptor signalling to ZAP-70 tyrosine kinase. J Immunol 2001;167: 1857–61.
391. Cherla RP, Ganju RK. Stromal cell-derived factor 1 alpha-induced chemotaxis in T cells is mediated by nitric oxide signaling pathways. J Immunol 2001;166:3067–74.
392. Schaerli P, Willimann K, Lang AB, Lipp M, Loetscher P, Moser B. CXC chemokine receptor 5 expression defines follicular homing T cells with B cell helper function. J Exp Med 2000;192:1553–62.
393. Hornung D, Bentzien F, Wallwiener D, Kiesel L, Taylor RN. Chemokine bioactivity of RANTES in endometriotic and normal endometrial stromal cells and peritoneal fluid. Mol Hum Reprod 2001;7:163–8.
394. Hornung D, Dohrn K, Sotlar K, Greb RR, Wallwiener D, Kiesel L, Taylor RN. Localization in tissues and secretion of eotaxin by cells from normal endometrium and endometriosis. J Clin Endocrinol Metab 2000;85:2604–8.
395. Weitzmann MN, Cenci S, Haug J, Brown C, DiPersio J, Pacifici R. B lymphocytes inhibit human osteoclastogenesis by secretion of TGFbeta. J Cell Biochem 2000;78:318–24.
396. Handisurya A, Steiner GE, Stix U, Ecker RC, Pfaffeneder-Mantai S, Langer D, Kramer G, Memaran-Dadgar N, Marberger M. Differential expression of interleukin-15, a pro-inflammatory cytokine and T-cell growth factor, and its receptor in human prostate. Prostate 2001;49:251–62.
397. Lai L, Goldschneider I. Cutting edge: Identification of a hybrid cytokine consisting of IL-7 and the beta-chain of the hepatocyte growth factor/scatter factor. J Immunol 2001;167: 3550–4.
398. Skibinski G, Skibinska A, James K. The role of hepatocyte growth factor and its receptor c-met in interactions between lymphocytes and stromal cells in secondary human lymphoid organs. Immunology 2001;102:506–14.
399. Siler U, Seiffert M, Puch S, Richards A, Torok-Storb B, Muller CA, Sorokin L, Klein G. Characterization and functional analysis of laminin isoforms in human bone marrow. Blood 2000;96:4194–203.
400. Simard J, Gingras S. Crucial role of cytokines in sex steroid formation in normal and tumoral tissues. Mol Cell Endocrinol 2001;171(1–2):25–40.
401. King AE, Critchley HO, Kelly RW. The NF-kappaB pathway in human endometrium and first trimester decidua. Mol Hum Reprod 2001;7:175–83.
402. Okada H, Nakajima T, Sanezumi M, Ikuta A, Yasuda K, Kanzaki H. Progesterone enhances interleukin-15 production in human endometrial stromal cells in vitro. J Clin Endocrinol Metab 2000;85:4765–70.
403. Olsen NJ, Gu X, Kovacs WJ. Bone marrow stromal cells mediate androgenic suppression of B lymphocyte development. J Clin Invest 2001;108:1697–704.
404. Zheng WM, Izaki J, Furusawa S, Yoshimura Y. Localization of immunoglobulin G gamma-chain mRNA-expressing cells in the oviduct of laying and diethylstilbestrol-

treated immature hens. Gen Comp Endocrinol 2000;120:345–52.
405. Berczi I, Bertok L, Chow DA. Natural immunity and neuroimmune host defence. Ann NY Acad Sci;2001;917:248–257.

The Immune-Neuroendocrine Circuitry
Edited by I. Berczi and A. Szentivanyi
Published by Elsevier Science B.V. 2003

The Hypothalamic-Pituitary-Adrenal (HPA) Axis: A Major Mediator of the Adaptive Responses to Stress

K. EDDIE GABRY[1] , GEORGE CHROUSOS[2] and PHILIP W. GOLD[1]

[1]*Clinical Neuroendocrinology Branch, National Institute of Mental Health,* [2]*Pediatric and Reproductive Endocrinology Branch, National Institute of Child Health and Human Development, U.S. National Institutes of Health Intramural Research Program, Bethesda, Maryland 20892*

ABSTRACT

Glucocorticoid hormones exert a multitude of functions that affect virtually every cell in the body. The physiological significance of glucocorticoids is most remarkable at times of stress, when the hypothalamic-pituitary-adrenal axis (HPA) is fully stimulated. The same hypothalamic hormone that stimulates the HPA axis, corticotropin-releasing hormone (CRH) also mediates behavioral, autonomic and neuroendocrine responses to stress in rodents and primates. Hyperfunction of CRH neuron, therefore, appears to underlie a variety of psychiatric, gastrointestinal, cardiovascular, metabolic and reproductive illnesses attributable to stress. Nonpeptide CRH type-1 receptor antagonists are novel tools to moderate CRH activity in the brain after systemic administration. They are currently under intensive investigation for their potential therapeutic effects against such adaptive illnesses ascribed to stress. Here we outline the physiology of the HPA axis, its interconnection with the neurocircuitry of stress and its implication in human disease.

1. THE HYPOTHALAMIC-PITUITARY-ADRENAL UNIT

1.1. An overview

The hypothalamic-pituitary-adrenal (HPA) axis refers to the brain control, via the anterior pituitary, of glucocorticoid secretion from the adrenal cortex (Figure 1). Products of the HPA axis play several roles in the body, which are most remarkable at times of stress. The principal components of the HPA axis are the hypothalamic hormone, corticotropin-releasing hormone (CRH), adrenocorticotrophic hormone (ACTH) from the anterior pituary, and glucocorticoid hormones from the adrenal cortex, mainly cortisol in primates and corticosterone in rodents.

Glucocorticoids (GCs) are steroid hormones derived from cholesterol and secreted by the zona fasciculata of the adrenal cortex. They are essential for the balance of immune function and inflammatory reactions, cardiovascular and intravscular fluid integrity, membrane stability and the maintenance of neuroendocrine homeostasis. The discovery of the physiologic roles of GCs

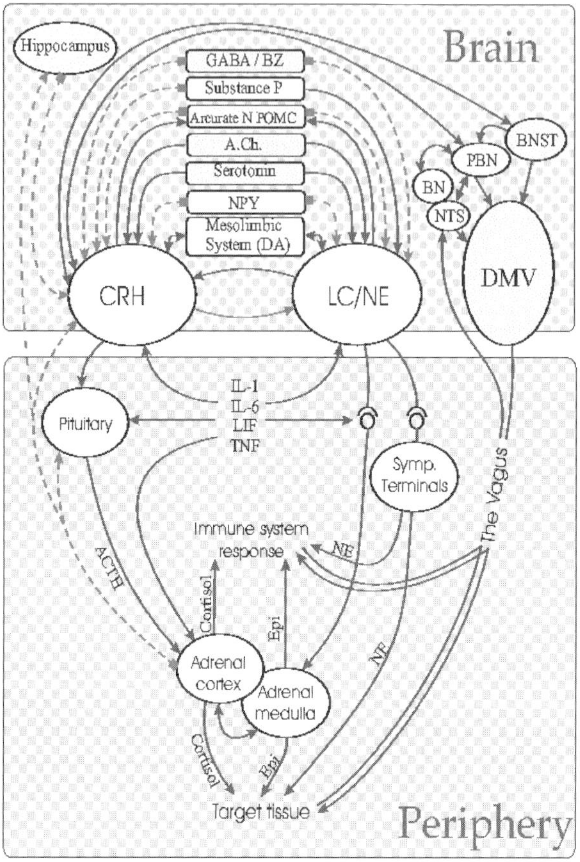

Figure 1. An illustration of the central and peripheral components of the stress system, their functional interrelations, and their relations to other central systems involved in the stress response. PVN, paraventricular nucleus of the hypothalamus; CRH, corticotropin releasing hormone; LC/NE, locus ceruleus/norepinephrine-sympathetic system; A.Ch., acetylcholine; epi, epinephrine; NE, norepinephrine; POMC, proopiomelanocortin; AVP, arginine vasopressin; GABA, gamma amino-butyric acid; BZD, benzodiazepine; ACTH, corticotropin; NPY, neuropeptide Y; SP, substance P; BNST, the bed nucleus of the stria terminalis; BA, Barrington nucleus, PBN, parabrachial nucleus, NTS, nucleus of tractus solitarius, DMV, dorsal motor nucleus of the vagus; Activation is represented by solid lines, and inhibition by dashed lines.

is a landmark in the development of endocrinology. This was demonstrated by Thomas Addison (1849), who reported an association between the clinical picture of "anaemia, general languor and debility, remarkable feebleness of the heart's action, irritability of the stomach, and a peculiar change of colour of the skin" and "adrenal disease at autopsy" [1]. To date, these features constitute the cardinal signs of what has been known as "Addison's disease" since then.

The anterior pituitary control of GCs secretion from the adrenal cortex was later on suggested by Smith (1930) [2], Ingle and Kendall (1937) [3]. The latter further showed a role of stress in stimulating the pituitary-adrenal axis and that adrenocortical extracts inhibit the adrenocorticotropic effect of the pituitary. This notion established the presence of a negative feedback through which the end-product hormones, glucocorticoids (GCs), suppress further release of the pituitary trophic hormone, ACTH. The hypothesis that a hypothalamic factor is secreted into the hypothalamic-hypophyseal portal blood to stimulate ACTH release from the anterior pituitary was suggested by Harris (1948) [4]. This extended the notion of the pituitary-adrenal axis to include the hypothalamus; the hypothalamic-pituitary-adrenal axis (HPA). The hypothalamic factor that stimulates the HPA axis was not chemically characterized until 1981 by Wylie Vale and his colleagues [5], and thence referred to as corticotropin-releasing hormone (CRH).

1.2. Components of the HPA axis

CRH, released into the hypophysial portal system, is the principal regulator of anterior pituitary ACTH secretion (Figure 1). It is a 41 amino acid peptide that is remarkably conserved in the evolutionary chain. In addition to its pituitary-adrenal stimulating effects, CRH plays widespread roles in the nervous system, which are largely related to stress. Hypothalamic arginine vasopressin (AVP) is a potent synergistic factor with CRH in stimulating ACTH secretion. However, AVP has little ACTH secretagogue activity by itself. AVP is colocalized with CRH in around half of the paraventricular nucleus (PVN) neurons. CRH and AVP neurons may also have a reciprocal positive interaction at the level of the hypothalamus, with each neuropeptide stimulating the secretion of the other [6]. CRH acts on the anterior pituitary corticotroph cells to stimulate ACTH.

The precursor gene that encodes ACTH also encodes the opioid peptide β-endorphin, as well as α, β, and γ-melanocyte stimulating hormone (MSH), and thus called proopiomelanocortin (POMC) [7]. In the corticotroph cells, the cleavage of POMC yields equimolar amounts of a carboxy-terminal glycopeptide termed 16K, ACTH 1-39, and β-lipotropin (β-LPH). Part of β-LPH molecules is further processed in these cells to β-endorphin 1–31. In the intermediate lobe, however, all of the β-LPH is converted to β-endorphin and ACTH is converted to α-MSH and corticotropin-like intermediate lobe peptide (CLIP) [7]. Although the major site of expression of the POMC gene is the pituitary gland, it is also expressed in several brain regions (e.g. the arcuate nucleus) and peripheral tissues (e.g. gut). In the pituitary, it is expressed in anterior lobe corticotropes and in the intermediate lobe. The precursor POMC undergoes a series of posttranslational processing steps to produce its tissue-specific final products [7].

ACTH is the key regulator of glucocorticoid and adrenal androgen secretion by the *zonae fasciculata* and *reticularis*, respectively, while it also participates in the control of aldosterone secretion by *the zona glomerulosa*. However, there is evidence that other hormones and/or cytokines, either originating from the adrenal medulla or coming from the systemic circulation, and/or neuronal information from the autonomic nerves of the adrenal cortex, may participate in the regulation of cortisol secretion [8].

The anterior pituitary corticotrophs release ACTH in a pulsatile fashion, thus resulting in direct increases in plasma concentrations of ACTH and subsequently cortisol. A diurnal rhythm of HPA activity has additionally been characterized [9,10]. The secretory episodes of ACTH, followed by cortisol, increase in amplitude after a few hours of sleep, reach a summit prior to awakening and gradually decline throughout the day to reach a nadir in the evening [11]. Other secretory episodes of ACTH have been observed with eating, perhaps depending on the protein content of diet [12,13].

ACTH acts on the adrenal cortex to rapidly increase cortisol synthesis and secretion because intra-adrenal cortisol storage is minimal [14]. ACTH binding to its cell-surface receptor on adrenal cells, the melanocortin-2 receptor, is associated with adenylate cyclase activation and elevation of intracellular cAMP concentration [15]. A cAMP-dependent protein kinase (pro-

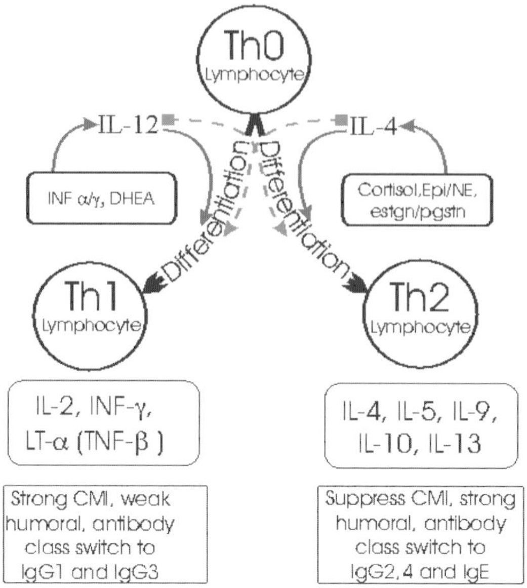

Figure 2. Immune responses and cytokine production can be classed as type 1 (TH1) or type 2 (TH2). These patterns are characterized by different patterns of cytokine production, as shown. Cytokines that stimulate TH2 responses include IL-3, IL-5, IL-4, and IL-10; cytokines that stimulate TH1 responses include IL-2, IFN-γ, and TNF-β. CMI = Cell mediated immunity. Solid arrows represent stimulation and broken lines represent inhibition.

tein kinase A) is subsequently activated to phosphorylate a number of proteins [16]. ACTH stimulates increases in the expression of proto-oncogenes such as *fos-B, jun B,* c-*fos*, and c-*jun* perhaps through its cAMP-dependent protein kinase pathway as well [17]. Chronically, ACTH increases the synthesis of most of the enzymes involved in steroidogenesis and the synthesis nucleic acids as well adrenocortical growth. Protein synthesis is maximal 24 to 36 h after ACTH stimulation [18].

In addition to stimulating the adrenal cortex, large doses of ACTH increase glucose and amino acid transport into muscle cells [19], increase hepatic protein synthesis [20], hyperpigmentation [19] and increase cAMP and stimulate lipolysis in adipocytes [21]. These extra-pituitary effects of ACTH are seen only when ACTH levels are markedly elevated as in patients with primary adrenal insufficiency.

1.3. Actions of glucocorticoid hormones

One of the most important adaptive measures the organism mounts in response to threatened homeostasis is activation of the HPA axis. This was first recognized by Selye (1936), thus extending the stress hormones to include GCs, in addition to catecholamines earlier recognized by Cannon (1914) to be released in response to "fight or flight." Via their pleiotrophic action that affect virtually every cell in the body, GCs exert a multiplicity of functions that include, but are not limited to, the following:

1.3.1. Metabolic effects

Together with catecholamines, GCs are potent stimulators of lipolysis [19]. In adrenalectomized animals, plasma free fatty acid levels are reduced and return to normal within 2 h after glucocorticoid administration [22,23]. The glycerol released during lipolysis is used as a substrate for glucose production, while released fatty acids provide an energy source for the process. GCs stimulate glycogen synthesis and inhibit glycogen breakdown by activation of glycogen synthase [24, 25], and inactivation of the glycogen-mobilizing enzyme, glycogen phosphorylase [25]. In humans, chronic glucocorticoid excess results in redistribution of body fat, a feature characteristic of Cushing's syndrome.

Glucocorticoids increase blood glucose concentrations by stimulating its production in the liver. This is largely due to activation of hepatic gluconeogenic enzymes, such as glucose-6-phosphatase and phosphoenolpyruvate carboxykinase [26]. In addition to mobilizing substrate for hepatic gluconeogenesis, glucocorticoids inhibit glucose uptake and utilization by peripheral tissues [27], possibly due to inhibition of glucose transport into the cells [28,29]. Reduction of glucose transporters mRNA in adipocytes by glucocorticoids has been reported [30]. The presence of GCs is also essential for the gluconeogenic effects of other hormones, e.g. glucagon and epinephrine, to be manifested [26,31]. The catabolic effect of GCs on muscle protein is the basis for the profound myopathy that can result from glucocorticoid excess. Muscles are major sources of amino acid substrate required for gluconeogenesis.

In the bone, GCs inhibit osteoblast and facilitate osteoclast function, thus causing decreased bone mass [32–34]. Glucocorticoids enhance the ability of osteoclasts to bind to bone surfaces as a consequence of altered expression of *N*-acetylglucosamine and *N*-acetylgalactosamine on the cell surface [35]. GCs increase urinary Ca^{++} loss by decreasing its reabsorption. Intestinal Ca^{++} absorption is additionally reduced by GCs [34,36–38], which may be a factor of GCs induced stimulation of parathyroid hormone (PTH) [34,39]. GCs also suppress the synthesis of IGF-I, which promotes osteoblast proliferation and differentiation [40].

1.3.2. Central effects

Glucocorticoids are lipid soluble hormones that easily penetrate the blood brain barrier. They affect the mood, cognitive functions and sleep in humans [41]. They may produce arousal, possible via a positive feedback effect on the CRH neuron. Around half of Cushing's patients exhibit mood disturbance, particularly in the form of depression [40,42–45], Patients with adrenal insufficiency also may suffer from psychiatric disturbances, mainly depression, apathy, and lethargy. This is in agreement with our findings of increased HPA activity in melancholic depression and its reduction in atypical depression. The duration of rapid eye movement sleep is decreased in patients with Cushing's syndrome [46] and in normal subjects given pharmacologic doses of glucocorticoid [47] or ACTH. [48].

1.3.3. Effects on immune processes and inflammation

Figure 2 summarizes the recently characterized type 1/type 2 (or Th1/Th2) immunity, which provides a unifying hypothesis to the regulation of immune system accounted to by both cell mediated and humoral immunity described several decades earlier. In 1986, Mosmann and colleagues [49] described two subpopulations of $CD4^+$ T lymphocytes, namely T helper type 1 (Th1) and T helper type 2 (Th2) lymphocytes, based on their differential patterns of cytokine production. Th1 cells produce IL-2, IFN-γ and lymphotoxin α (LT-α). Conversely, Th2 cells were found to produce IL-4, Il-5, IL-9, IL-10 and IL-13. However, the distinction between Th1 and Th2 is based on the secretion of IFN-γ by the former and IL-4 by the latter. CD4+ cells that secrete neither

IFN-γ nor IL-4 are likely to be naïve Th0 lymphocytes that have not undergone polarization into Th1 or Th2 phenotypes. Both Th1 and Th2 help B cells to produce antibody, but only Th1 cells were able to elicit delayed type hypersensitivity reaction in mice and in humans [50].
Il-12 is the principal cytokine that stimulates the polarization of Th0 cells into the Th1 phenotype [51]. On the other hand, the polarization of Th0 cells into the Th2 phenotype is facilitated by IL-4 [52]. Type 1 and type 2 immunity are reciprocal in nature. IFN-γ secreted by Th1 cells directly suppresses IL-4 secretion and inhibits the differentiation of naive Th0 cells into Th2 cells [53]. Conversely, IL-4 and IL-10 inhibit the secretion of IL-12 and IFN-γ, blocking the ability of Th0 cells to polarize into Th1 cells [54]. Moreover, activation of type 2 immunity is results active suppression of phagocytosis, characteristic of type-1 immunity.

Type 1 immunity is characterized by intense phagocytic activity and moderate levels of antibody production. Conversely, type 2 is characterized by high antibody titers and inhibition of phagocytosis. Generally, type 1 immunity is utilized in the fight against microorganisms and type 2 immune response come into play afterwards to help the resolution of the fight. Premature utilization of type 2 immunity is hazardous as it does not allow the body to mount type 1 responses, which are essential to overcome the pathogens. This phenomenon is called Th2 shift, and is commonly seen during severe stress, immunosuppression and overwhelming infection. In such cases, administration of antimicrobial chemotherapy and exogenous cytokines that augment type 1 immunity restores systemic balance and allows successful immune responses to clear the infection [50].

Glucocorticoids can potently shift the immune response from type 1 to that of type 2 (Th2 shift). They suppress the production of TNF-α, IFN-γ and IL-2 *in vitro* and *in vivo* in both humans and experimental animals [55]. Glucocorticoids also suppress the production of IL-12 by antigen presenting cells APCs [56,57]. Because IL-12 is extremely potent in enhancing IFN-γ and inhibiting IL-4 synthesis by T cells, the glucocorticoid-induced inhibition of IL-12 production may account for their ability to cause Th2 shift of the immune response. Moreover, glucocorticoids may directly stimulate IL-4 and IL-10 production from lymphocytes and from APCs [58–60]. Markedly high concentrations of glucocorticoids, on the other hands, induce lympholysis [40] and inhibit all cytokine secretion, including type 2 cytokines [40]. This may explain their efficacy in the treatment of such disease states attributed to type 2 hyper-responses as bronchial asthma [50].

The association between susceptibility and/or progression of inflammatory arthritis and HPA responsiveness to stress has been well-characterized in inbred strains of Lewis (LEW/N) and Fischer (F344/N) rats. Inbred LEW/N rats are susceptible to autoimmune inflammatory disease with a hyporesponsive HPA axis, while F344/N rats are relatively resistant to inflammation and exhibit a hyper-responsive HPA axis. Furthermore, the ability to manipulate the HPA axis and measure effects due to autoimmune disease expression in these inbred rat strains, provides evidence for a causal relationship between blunted HPA axis responses and susceptibility to inflammatory disease. Interruption of the HPA axis at any point and by many means, including pharmacologic blockade of the glucocorticoid receptor with RU 486, adrenalectomy or hypophysectomy renders inflammatory/autoimmune resistant rats susceptible to these diseases. Conversely, restitution of the HPA axis by fetal hypothalamic tissue transplantation reverses the susceptibility of the LEW/N rats to autoimmune diseases [61,62].

1.3.4. Endocrine effects:

GCs are strong modulators of reproductive functions (Figure 3). Suppression of gonadal function caused by chronic HPA activation has been demonstrated in highly trained runners of both

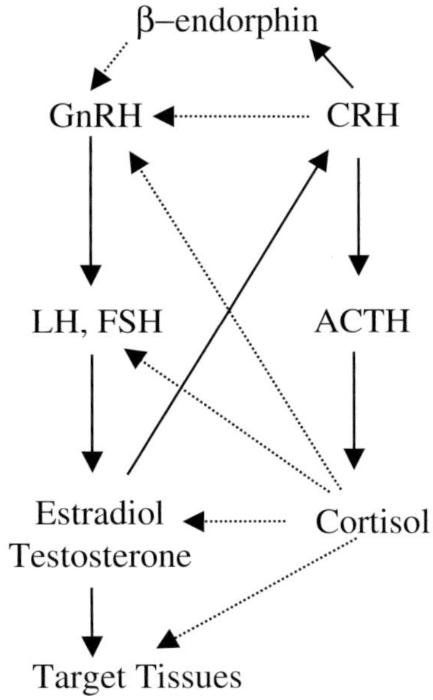

Figure 3. Interactions between the HPA axis and the reproductive system. GnRH, gonadotropin releasing hormone; LH, luteinzing hormone; FSH, follicle stimulating hormone. *(Adapted from ref.* 66: Chrousos GP, Gold PW. The concepts of stress and stress system disorders. Overview of physical and behavioral homeostasis JAMA1992;267(9):1244–52).

sexes and ballet dancers as well as in individuals suffering from anorexia nervosa or starvation. These subjects have increased evening plasma cortisol and ACTH levels, increased urinary free cortisol excretion, and blunted ACTH responses to exogenous CRH; males have low luteinizing hormone (LH) and testosterone levels, and females have hypogonadotrophic hypogonadism and amenorrhea. Characteristically, obligate athletes go through withdrawal symptoms and signs if, for any reason, they have to discontinue their exercise routine. This syndrome is possibly the result of withdrawal from the daily exercise-induced secretion of POMC-derived peptides and mesocorticolimbic system dopamine [63–65]. Glucocorticoids exert inhibitory effects at the levels of the GnRH neuron, the pituitary gonadotroph, influencing primarily the secretion of LH, and the gonads themselves, and render target tissues of sex steroids resistant to these hormones [66].

The growth axis is also inhibited at many levels during stress (Figure 4). Prolonged activation of the HPA axis leads to suppression of growth hormone secretion and inhibition of insulin-like growth factor-1 (IGF-1) and other growth factor effects on their target tissues by glucocorticoids, the latter possibly through inhibition of the c-jun/c-fos heterodimer by the ligand-bound glucocorticoid receptor [66]. However, acute elevations of growth hormone concentration in plasma may occur at the onset of the stress response or after acute administration of glucocorticoids, presumably through stimulation of the GH gene by glucocorticoids through GC response elements in its promoter region.

In several stress system-related mood disorders with a hyperactive HPA axis, such as chronic anxiety or melancholic depression, GH and/or IGF-1 levels are significantly decreased. Patients with panic disorder, compared to normal control subjects, have blunted GH responses to intravenously administered clonidine, and children with anxiety disorders might not attain their final

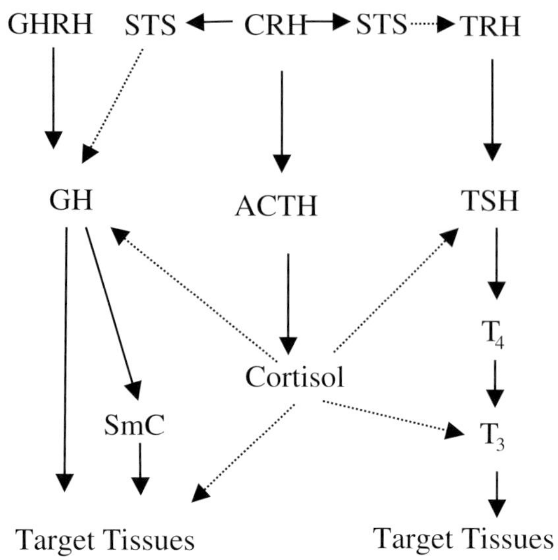

Figure 4. Interactions of the HPA axis with the growth and thyroid axes. GHRH, growth hormone releasing hormone; STS, somatostatin; TRH, thyrotropin releasing hormone; GH, growth hormone; TSH, thyroid stimulating hormone; T_4, thyroxine; T_3, triiodothyronine; SmC, somatomedin C. *(Adapted from ref.* 66: Chrousos GP, Gold PW. The concepts of stress and stress system disorders. Overview of physical and behavioral homeostasis JAMA1992;267(9):1244–52).

height potential. In melancholic depression, not only is GH secretion decreased, but also the dexamethasone-induced GH increase is blunted [67,68].

Glucocorticoids not only have profound inhibitory effects on GH and gonadal steroid production, but also antagonize the actions of these hormones on fat tissue catabolism (lipolysis), and muscle and bone anabolism. Thus, chronic activation of the stress system would be expected to increase visceral adiposity, decrease lean body (bone and muscle) mass and suppress osteoblastic activity (Figure 5). Interestingly, the phenotype of central obesity and decreased lean body mass is shared by patients with Cushing's syndrome and some patients with the combined diagnosis of melancholic depression or chronic anxiety disorder and the metabolic syndrome X (visceral obesity, insulin resistance, dyslipidemia, hypertension) or "pseudo-Cushing syndrome" [69,70].

Increased gluconeogenesis is a characteristic feature of the stress response, and GCs induce insulin resistance. Activation of the HPA axis may contribute to the poor control of diabetic patients during periods of emotional stress, or concurrently with inflammatory and other diseases. Indeed, mild, chronic activation of the HPA axis has recently been demonstrated in diabetic patients under moderate or poor glycemic control. Visceral adiposity directly causes further insulin resistance and deterioration of glycemic control in patients with diabetes mellitus. Thus, chronic activation of the stress system in this disorder participates in a vicious cycle of increasing hyperglycemia, hypercholesterolemia, and insulin needs [68].

In addition to osteoporosis and metabolic syndrome X, patients with melancholic depression develop varying degrees of innate and T helper 1-directed immunosuppression and hence certain infectious and neoplastic diseases. It is, thus, not surprising that these patients when not treated have a reduced life expectancy, which is curtailed by 15–20 years, after excluding suicides, the prevalence of which is additionally increased in this group [68,71].

1.3.5. Other effects

The presence of GCs is important for surfactant production in the lung by type II pneumocytes [72]. Glucocorticoids induce enzymes involved in phospholipid biosynthesis, and regulate transcription of the genes for the surfactant proteins (SP-A, SP-B, and SP-C) [73–76]. Mice deficient in glucocorticoid receptor genes manifest with lung atelectasis and die shortly after birth from respiratory failure [77].

Glucocorticoids suppress fibroblast DNA, RNA, and protein synthesis [78], the synthesis of the extracellular matrix components, collagen and hyaluronidate [79–82]. Chronic glucocorticoid excess thus results in impaired wound healing and friable connective tissues [83].

1.4. Mechanisms of action of glucocorticoids

Being steroids in nature, glucocorticoids exert their effects mainly through intracellular receptors. The intracytoplasmic GC receptor is a single-chain polypeptide with a molecular weight of about 94 kilodalton (kD) that specifically binds glucocorticoids with high affinity in a saturable fashion [84]. The nonactivated glucocorticoid receptor resides in the cytosol in the form of a heterooligomer with heat shock proteins and immunophilins. Upon ligand binding, the glucocorticoid receptors dissociate from the rest of the heterooligomer, and translocate into the nucleus, where they interact as homodimers with specific glucocorticoid responsive elements (GRE's) within the DNA to activate appropriate hormone-responsive genes. The activated receptors also inhibit, by protein-protein interactions, several transcription factors, such as c-jun/c-fos and NF-κB, which are positive regulators of the transcription of several genes involved in the function and growth of nonimmune and immune cells [85]. They also change the stability of mRNAs and, hence, the translation rates of several glucocorticoid-responsive proteins. Furthermore, glucocorticoids influence the secretion rates of specific proteins and alter the electrical potential of neuronal cells, through mechanisms that remain to be elucidated [86].

Glucocorticoids increase the transcriptional activity of many genes and suppress the transcription of others, such as the genes that encode POMC, prolactin, and glycoprotein hormone alpha-subunit [87]. GC may exert inhibitory effects by direct protein-protein interaction [88–90], interference with binding of another transcription factor to its own response element near the negative GRE [91,92], or direct action by binding of the receptor to negative response elements that prevent transcriptional activation [93]. Non-transcriptional effects of GCs may account for their rapid effects, particularly at times of stress (vide infra).

1.5. Corticosteroid-Binding Globulin (CBG)

CBG is a glycosylated 383-amino-acid alpha2 -globulin that binds free cortisol. The liver is the main source of CBG [94–96]. The intracellular concentration of a hormone is likely to be proportional to the concentration of free hormone in plasma, not to the plasma protein-bound hormone concentration. Because only free hormone can penetrate into tissues, protein binding may interfere with the diffusion of a hormone into tissues and effector cells [97,98].

A protein with CBG-like steroid-binding characteristics and immunoreactivity is present in glucocorticoid-responsive tissues, including the brain, pituitary, uterus, and lymphocytes [99], suggesting possible paracrine or autocrine actions. Specific, high-affinity receptors for CBG are present on target cell membranes [99–101]. Only non-ligand-bound CBG binds to these receptors, but glucocorticoids can then bind to the CBG-receptor complex [100] and activate adenylyl cyclase [102], thus enabling GCs to exert nontranscriptional effects. Cells in several parts of the central nervous system contain glucocorticoid receptors, but some neuronal responses seem too rapid to be mediated by transcriptional activation of target genes by glucocorticoid-receptor complexes. These responses include changes in electrical activity, such as hyperpolarization of the cell membrane [103] or suppression of spontaneous electrical activity [104,105]. They occur so rapidly (i.e., within 2 min of exposure to the hormone) that a direct membrane effect is likely. Hyperpolarization is blocked by antiglucocorticoids, suggesting involvement of a receptor-like molecule [98,104].

1.6. Glucocorticoid feedback

Glucocorticoids play a key regulatory role on the basal control of HPA axis activity and on the termination of the stress response, by acting on extrahypothalamic regulatory centers, such as the hippocampus and frontal cortex, the hypothalamus and the pituitary gland [68]. The inhibitory glucocorticoid feedback on the ACTH secretory response acts to limit the duration of the total tissue exposure to glucocorticoids, thus minimizing the catabolic, lipogenic, antireproductive, and immunosuppressive effects of these hormones. Interestingly, a dual receptor system exists for glucocorticoids in the central nervous system, including the glucocorticoid receptor type I, or mineralocorticoid receptor, which responds positively to low levels of glucocorticoids, and the classic glucocorticoid receptor (type II), which responds to both basal and stress levels. The latter participates in the negative feedback control of the HPA axis via activation of an afferent gabaergic pathway to the PVN [106,107] (Figure 1).

The feedback inhibition exerted by GCs on ACTH secretion appears to consist of two phases, at least in the rat [108]. Fast feedback occurs in seconds to minutes and is proportional to the rate of increase in glucocorticoid concentration rather than its absolute level. Delayed feedback occurs in hours to days and is proportional to the glucocorticoid dose, potency and duration of administration [108].

Glucocorticoid feedback occurs at the pituitary and hypothalamic levels and perhaps at higher centers [98]. In the anterior pituitary, glucocorticoids inhibit both ACTH secretion and POMC gene transcription, resulting in reduced POMC mRNA levels and reduced POMC synthesis [109]. To a lesser extent, glucocorticoids also decrease CRH and AVP mRNA and peptide levels in the hypothalamic paraventricular nuclei [110–112]. In addition, glucocorticoids block the stimulatory effect of CRH on POMC gene transcription [113] and ACTH release [114]. GCs also inhibit anterior pituitary CRH receptor expression [115,116]. GCs also exert positive feedback effects at the level of CRH neurons in the amygdala and LCINE system. This is believed to sustain and/or potentiate the arousal producing actions of CRH, particularly at times of stress (vide infra).

2. STRESS

Living organisms possess a diversity of control systems that operate in harmony to secure the most optimal *milieu* for growth, development and procreation, a phenomenon known as *homeostasis.* The regulatory centers of most of these control systems are located in the brain, where they are interconnected to one another as well as to each system's afferent and efferent limbs (Figure 1). The location of these operating units in the brain allows the higher brain centers to override the ordinary programs of a given control system when interpreted by the brain as necessary [55].

It is usually within the capacity of a given control system to respond to changes sensed by its afferent limbs without perturbing the homeostasis of other systems. However, beyond a certain threshold, stimuli interpreted by the brain as threatening to the organism's homeostasis, regardless of their modality, elicit an immediate stereotypic response conventionally referred to as the stress response. Subjects undergoing stress manifest enhanced cognition, affective immobility, vigilance, autonomic arousal and a global catabolic state. During stress, there is enhanced cardiac output and respiration, and the blood flow is redirected to provide the highest perfusion to the brain and muscular system. The brain focuses on the perceived threat and acts accordingly [117]. Moreover, endocrine programs of pleasure, growth and reproduction are shut down for the sake of saving energy for the aroused brain, heart and muscles [66].

2.1. Modalities of stress

Several physical states that potentially endanger the organism's survival mobilize the stress response. These include severe hypotension, exercise, trauma, burns or illness, major surgery, hypoglycemia, fever and cold exposure [118–122].

Psychological stress is perhaps the most common form of stress encountered in humans. Humans enjoy the highest level of cortical contribution to information processing. Because the well-developed cortical systems are responsible for sophisticated social skills and at the same time constitute the substrates for the psyche, humans are aware of subtle social cues that can potentially be threatening. Therefore, psychological stimuli recognized by the human brain as alarming mobilize the stress system. Likewise, because of their well-developed psyche and rich learning experiences, humans are particularly social, and the optimum conditions for their functioning usually involve significant others. Potential loss of such privileged relationships commonly precipitates a state of stress as well [7].

Emotions are processed in the limbic system and its widespread projections in the brain. These include the sensory thalamic nuclei, the basolateral and central amygdala nuclei, the locus ceruleus, the hippocampus, hypothalamus and their associated cortical areas as the medial prefrontal cortex. Interestingly, these same nuclei make up the stress neurocircuitry (Figure 1), which may explain why emotional threats are particularly effective in mobilizing the stress system in humans [6]. On the other hand, activation of the stress system by non-psychological stimuli elicits fear, anxiety and, in extreme cases, a depression-like behavioral pattern.

2.2. Factors determining the intensity of the stress response

It is the perception of a threat that elicits the stress circuits in the brain. Human subjects vary widely in regards to their subjective perception of threats. We believe the balance between two different forces, namely stress and resilience determines the subject's susceptibility to stress-related illness. Genetic and environmental influences, particularly at critical periods of development, contribute to such psychological make up characteristic of every individual [123].

Several factors that determine the stress responses of individuals are inherited, as quantitative genetics of human complex behaviors indicate. It has been estimated that half to two-thirds of the variance in personality traits is due to genetic influences [6]. Thus, genetic polymorphisms and/or clinically significant alterations in the expression of genes are involved in the regulation of the stress system. A significant amount of the variance of the stress responses of individuals, however, is also environmental and this includes both early life and later life events with long-term or permanent effects.

2.3. Pathological effects of stress

Stress has been repeatedly implicated in the predisposition, precipitation and/or progression of several disease states, particularly mental illness. For example, it has been estimated that more than fifty percent of depressed patients underwent a traumatic or very stressful experience just prior to manifesting the illness. A common denominator of these conditions is dysfunction of the HPA axis, which has been the strongest biological correlate of mood and forms of anxiety disorders (e.g. post-traumatic stress disorder) recognized so far [66].

2.4. Early life adversity

The intrauterine period, infancy, childhood, and adolescence are periods of increased brain plasticity; thus, abnormal activation of the stress system during these critical periods may have profound effects on its function throughout the life.

In these formative stages of development, exposure of primates to psychosocial trauma sets into motion a cascade of mental changes that frequently result in long-term predisposition to anxiety, depression, or other stress-associated somatic illnesses [124]. In spite of lacking an explanation for this phenomenon, several studies have shown an association between the patterns of activity of the hypothalamic-pituitary-adrenal axis on one hand and depression and anxiety disorders on the other hand [125].

In nonhuman primate models, separation of mother-reared infants from their mothers is a potent stressor that causes severe behavioral disruptions. Immediately after separation, a protest stage occurs, which is manifested by anxiety and agitation, with increased locomotion, vocalization and distress calling, probably reflecting the infant's attempts to return to its mother. This phase lasts for 24–48 hours and is followed by a stage of despair, as marked by significant decreases in locomotion, vocalization and food intake. Repetitive exposure to such social separation results in progressive exaggeration of the infants' behavioral and neuroendocrine responses to stress [126]. We have most recently observed an element of sensitization of CRH neuronal activity in association with this model (Gabry et al., unpublished).

In most species, parental care is essential for the offspring's adequate growth and development [7]. In primates, positive peer interactions are necessary from childhood through adulthood [127,128]. Macaque infants reared apart from their mothers and raised only with peers typically exhibit biological and behavioral changes similar to those frequently seen in depression (hypercortisolemia, withdrawal, anxiety and fearfulness), particularly upon exposure to psychosocial stress. Interestingly, disruption of the social bonds between these peer-reared infants (peer separation) results in behavioral responses that are more extreme than the mother-reared infants' responses to maternal separation. Long-term biological and behavioral features associated with these animals' responses to challenge seem strikingly similar to those reported for patients with positive history of trauma and abuse [128,129]. In this model, we have identified significantly higher concentrations of CRH in the CSF of maternally deprived infants than in their mother-reared counterparts [128].

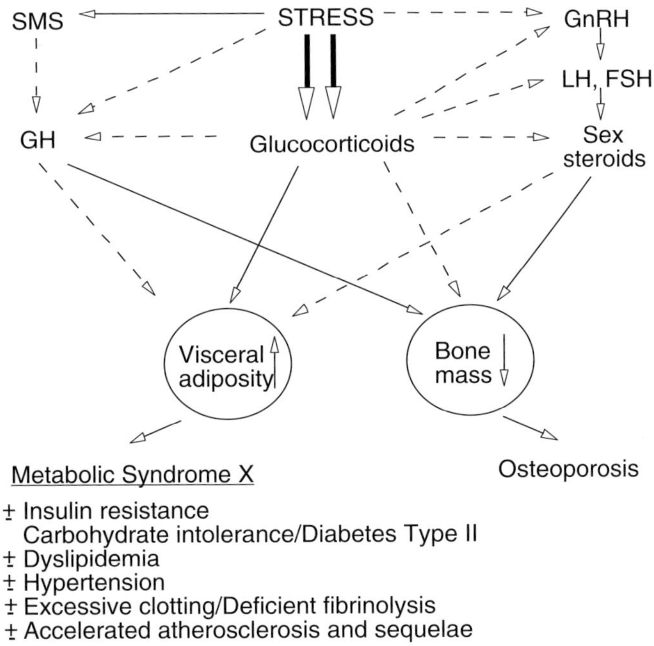

Figure 5. Detrimental effects of chronic stress on adipose tissue metabolism and bone mass; SMS, somatostatin; GH, growth hormone. Stimulation is represented by solid lines and inhibition by dashed lines *(Adapted from ref.* 6, Chrousos GP. Stressors, stress, and neuroendocrine integration of the adaptive response. The 1997 Hans Selye Memorial Lecture. Ann N Y Acad Sci 1998;851:311–35).

2.5. Physiological correlates of stress

The brain's ability to mobilize the stress response is paralleled by activation of corticotropin-releasing hormone (CRH) in several nuclei including the hypothalamus, amygdala and locus ceruleus, as well as stimulation of the locus ceruleus norepinephrine (LC/NE) system in the brain stem (Figure 1). These systems perpetuate one another, interact with several other transmitter systems in the brain and directly activate the hypothalamic-pituitary-adrenal (HPA) axis and the three components of the autonomic nervous system, namely the sympatho-adrenal, the cranio-sacral parasympathetic and the enteric nervous systems. In contrast with the global activation of the HPA and sympatho-adrenal limbs of the stress response, the activation of the parasympathetic and enteric systems appear to be selective and to affect the target tissues in a different way than it is accustomed to under non-stressful conditions. The brain also differentially activates a subset of vagal and sacral parasympathetic efferents that mediate the gut responses to stress.

During acute stress, the amplitude and synchronization of the PVN, CRH and AVP pulsations in the hypophysial portal system markedly increase [66]. Also, with strong physical stress, especially that associated with hypotension or a decrease of blood volume, recruitment of AVP of magnocellular neuron origin secreted into both the hypophyseal portal system via collateral neuraxons and into the systemic circulation takes place. Depending on the type of stress, other factors such as angiotensin II, as well as various cytokines and lipid mediators of inflammation are secreted and act on hypothalamic, pituitary, and/or adrenal components of the HPA axis to potentiate its activity [71].

3. NEUROENDOCRINOLOGY OF STRESS AND HPA ACTIVATION

3.1. An overview

As mentioned, the stress response is global and affects all body systems. Within seconds, several processes are elicited. Characteristic processes of the stress response include mobilization of stored energy with inhibition of subsequent energy storage and gluconeogenesis, sharpened focused attention on the perceived threat, increased cerebral perfusion rates and cerebral glucose utilization, enhanced cardiovascular output and respiration, enhanced delivery of energy substrates to the muscles, inhibition of reproductive physiology and behavior, modulation of immune function and decreased feeding and appetite. In case of fluid loss such as hemorrhage, there is additionally water retention through both renal and vascular mechanisms [131,131].

The brain circuits that initiate and maintain the stress response are illustrated in Figure 1. They involve the high control centers of the HPA axis and autonomic nervous systems. The hypothalamus controls the secretion of proopiomelanocortin (POMC) products that include corticotropin (ACTH) and beta-endorphin (β-end) from the anterior pituitary corticotrophs. ACTH stimulates the secretion by the adrenal cortex of glucocorticoid hormones, mainly cortisol in humans and corticosterone in rats [6]. The latters have a permissive role on the adrenal medullary secretion of epinephrine. Moreover, the other POMC product, β-end may stimulate the adrenal medulla to secrete epinephrine as well [132]. The locus ceruleus-norepinephrine (LC/NE) system controls the stress-induced stimulation of the sympathoadrenal (SA) system. Barrington nucleus (BA), the nucleus of tractus solitarius (NTS) and the dorsal motor vagal (DMV) nucleus appear to control the differential activation of vagal and sacral parasympathetic efferents that mediate the gut responses to stress [7,132]. The amygdala, acting in concert with the hippocampus and the anterior cingulate and prefrontal cortices, mediate the focused attention on the perceived threat, the affective inflexibility and fear-related behaviors [6].

The master of the HPA axis, CRH, is not confined to the hypothalamus, but is rather widely distributed in the brain, particularly in areas known to be involved in the stress response. Of its two subtypes of receptors, CRH type-1 receptor (CRH-R1) is the most abundant in the brain and the pituitary, at least in rodents, and its distribution well matches that of CRH-containing nerve endings. CRH actions in the brain are widely spread and its type-1 receptor activation appears to be central for many of the organism's responses to stress to be manifested. Thus, CRH has been considered by numerous research laboratories to be the master hormone of the stress response.

The orchestrated interplay of several neurotransmitter systems in the brain underlies the characteristic phenomenology of behavioral, endocrine, visceral, autonomic and immune responses to stress. These transmitters include corticotropin-releasing hormone, arginine vasopressin (AVP), opioid peptides, dopamine (DA) and norepinephrine (NE). Outside the brain, there is enhanced pituitary secretion of prolactin (PRL) and pancreatic secretion of glucagon. Moreover, there may be a global reduction of neuropeptide Y (NPY) in the brain and decreased hypothalamic release of gonadotropin-releasing hormone (GnRH) followed by reduced secretion of pituitary gonadotropins. Renin-angiotensin system is markedly activated in case of hemorrhagic stress [6].

3.2. Corticotropin-releasing hormone (CRH)

3.2.1. Central CRH

As mentioned, it had been deduced in the nineteen fifties that the pituitary-adrenal axis is driven by a hypothalamic factor referred to as corticotropin-releasing factor (CRF) [133,134]. In 1981, Vale's group was successful in isolating and characterizing the sequence of CRF from the hypothalami of 500,000 sheep. Subsequently, CRH was characterized in other species [135–137]. CRH is a 41-amino-acid single chain peptide whose sequence homology among species indicates its conserved nature in evolution. Rat and human CRHs are identical. Analyses of ovine [138] and rat [139] CRH cDNAs and human [140] and rat [141] genomic DNAs indicate that CRH is synthesized as a larger precursor [191 amino acids in humans], from which it is cleaved at flanking basic amino acid pairs. The single human CRH gene is located on chromosome 8 [142].

CRH is synthesized by neurons of the parvicellular division of the hypothalamic paraventricular nucleus (PVN) [143–145]. Their axons project to the median eminence, where CRH is secreted into the hypophyseal portal blood. These neurons, particularly after adrenalectomy, may also contain other ACTH secretagogues, such as AVP [146,147] and cholecystokinin [148]; opioid peptides, such as met-enkephalin [149] and dynorphin [1–8,150]; and the prolactin secretagogue PHI-27, a 27-amino-acid peptide (P) having NH2 -terminal histidine (H) and COOH-terminal isoleucine (I) amide residues (PHI-27) [149]. In addition, CRH is present in a subpopulation of oxytocin-containing neurons in the magnocellular division of the paraventricular nucleus that projects to the posterior pituitary [151].

CRH cell bodies are present in the central nucleus of the amygdala, the substantia innominata and in the bed nucleus of the stria terminalis. CRH neurons in the central nucleus of the amygdala project to the parvocellular regions of the PVN and the parabrachial nucleus of the brain stem [152]. These projections may account for CRH neuroendocrine, autonomic and behavioral effects. CRH neurons originating in the bed nucleus of the stria terminalis (BNST) send terminals to the parabrachial nuclei and dorsal vagal complex in the brain stem to coordinate autonomic activity. CRH fibers also interconnect the amygdala with the bed nucleus of the stria terminalis and the hypothalamus [153].

CRH neurons in the cerebral cortex and the cerebellum may be important for the behavioral actions of the peptide. The cerebellum is very rich in CRH-R1 whose exact physiological relevance remains to be characterized [154]. Because CRH-R1 dysfunction is implicated in the pathophysiology of depression and the fact that the cerebellum plays important roles in automaticity and possibly affective mobility, we hypothesize that cerebellar CRH-R1 may be dysfunctional in affective disorders. In the cerebral cortex, CRH interneurons are contained in the second and third layers and project to layers I and IV. In addition, scattered CRH cells are present in the deeper layers of the neocortex. The highest density of CRH-containing neurons is found in the prefrontal, insular and cingulate areas. The distribution of CRH in these areas corresponds and may explain its effects on cognitive processing [155].

Several other groups of CRH neurons occur throughout the brain stem and in the spinal cord. In the midbrain, CRH cells are found in the periaqueductal gray, Edinger-Westphal nucleus and the dorsal raphe nucleus. In the pons, CRH cell bodies in the parabrachial nucleus project to the medial preoptic nucleus of the hypothalamus [152]. The locus ceruleus (LC/NE) also contains CRH neurons that may mediate the cross talk between the two systems. In the medulla, CRH neurons are known to exist in the nucleus of the solitary tract and the dorsal vagal complex, in addition to some scattered groups in the reticular formation and the spinal trigeminal nucleus. CRH cell bodies are found in laminae V, VI, VII and X of the spinal cord as well as the intermediolateral column of the thoracic and lumbar regions. Spinal CRH neurons may play an important role in modulating sensory input via ascending projections to the thalamus and the brainstem reticular formation [6]

CRH binding sites are found in the anterior pituitary, throughout the brain and in various peripheral sites. CRH receptors belong to the seven-transmembrane G protein-coupled family of receptors, where CRH binding stimulates the intracellular accumulation of cyclic adenosine monophosphate (cAMP). Two distinct CRH receptor subtypes designated CRH-R1 and CRH-R2 have been characterized. Distinct genes that are differentially expressed encode the two receptor subtypes [156]. CRH acts on the anterior pituitary corticotrope cells by binding to its cell-surface receptor, CRH-R1 [157–159] and activating adenylate cyclase, thus increasing intracellular cAMP levels and activating cAMP-dependent protein kinase A [160], increasing the influx of extracellular calcium through L-type Ca2+ channels [161,162] and increasing the production of lipoxygenase metabolites of arachidonic acid [163]. This results in rapid release of ACTH and other POMC-related peptides [164] and subsequently in increased POMC synthesis [165–167].

Under normal conditions, there is a balance between CRH peptide availability and CRH receptor density. Stress or adrenalectomy results in hypersecretion of CRH and a consequent down-regulation of receptors in the anterior pituitary. Likewise, chronic administration of corticosterone causes dose-dependent decreases in anterior pituitary CRH receptor number. In contrast, lesions of the paraventricular nucleus that result in dramatic reductions in hypothalamic CRH secretion increase the density of pituitary CRH receptors [168]. However, the effects of CRH and products of pituitary-adrenal activation on brain CRH receptors that mediate the behavioral and autonomic aspects of the peptide may be different, i.e. upregulation rather than downregulation.

CRH-R1 is widely distributed in rat brain, mainly in neocortex and cerebellum. CRH-R1 is the most abundant subtype found in the anterior pituitary. CRH-R2 receptors are expressed mainly in the peripheral vasculature and the heart, as well as some subcortical structures in brain such as the septum, amygdala, and hypothalamus in rodents. In nonhuman primates, CRH-R2 is found in both central and peripheral locations. The CRH-R2 receptor is also found in the human amygdala. In the rhesus monkey, both CRH-R1 and CRH-R2 receptors are found in the pituitary, neocortex, amygdala, and hippocampus. CRH-R1, but not CRH-R2, receptors are present in locus ceruleus, cerebellar cortex, thalamus, and striatum. CRH-R2, but not CRH-R1, receptors are found in the choroid plexus and bed nucleus of the stria terminalis [169–172]. Most of the anxiogenic, dysphoric and pituitary-stimulatory actions of CRH are thought to result from CRH-R1 activation. CRH-R2 on the other hand appears to have antithetical effects to those of CR-R1.

The CRH family of peptides includes urocortin, sauvagine and urotensin [173–175]. Neither CRH nor any of the aforementioned peptides is specific for CRH-R1 or CRH-R2. Most recently, CRH-related peptides that are specific to CRH-R2 have been described independently by two research groups. These are stresscopin (or urocortin-3) and stresscopin-related peptide (or urocortin-2) [176–178]. The availability of such specific ligands will significantly advance our knowledge of the physiological roles of CRH-R2 in the brain and the periphery, whether under basal conditions or during stress.

CRH causes a specific, rapid, dose-dependent increase in plasma ACTH [179–181]. Human CRH is equipotent to ovine CRH, but because of its shorter circulating half-life the duration of plasma ACTH elevation is shorter and less total ACTH is secreted [179,182]. The specific binding of human CRH to the CRH-binding protein found in human plasma [183,184] may facilitate its removal from the circulating blood [185]. The ACTH released by CRH stimulates secretion of cortisol and other adrenal steroids, such as aldosterone and Dehydroepiandrosterone (DHEA) [181,186].

Central administration of small doses of CRH produces a profound increase in locomotor activity. High doses of CRH, however, produce a dramatic decrease in locomotion [187]. Intracerebral CRH injections produces additional behavioral effects, including increases in sniffing, grooming, and rearing in a familiar environment and assumption of a freeze posture in an unfamiliar environment, decreased feeding and sexual behavior and increased conflict behavior. The behavioral effects of CRH are independent of its actions on the pituitary, since they are not seen after pretreatment with doses of dexamethasone that adequately block pituitary-adrenal activation. Most of the aforementioned effects of CRH can be blocked by administration of CRH peptide antagonists, strongly supporting a specific CRH receptor-mediated event in these behaviors. Furthermore, the CRH receptor antagonist by itself attenuates many of the behavioral consequences of stress, which underscores the role of endogenous peptide in mediating many of the stress-related behaviors [188].

In nonhuman primates, exogenous administration of CRH intracerebroventricularly (icv) stimulates the pituitary-adrenal axis [189], causes anorexia [190] and enhances behavioral responses to stress [191]. Recently, we have reported that acute stress leads to increases in the endogenous concentrations of CRH in the CSF and that treatment with a CRH receptor antagonist attenuates the stress-induced CRH elevation, anxiety and sympathoadrenal activation in adult rhesus macaques [188].

CRH acts centrally to activate the sympathetic nervous system, with subsequent stimulation of epinephrine secretion from the adrenal medulla and noradrenergic outflow to the heart, kidney and vascular beds [192]. Other consequences of central administration of CRH include increases in mean arterial pressure and heart rate. In contrast, CRH acts in the brain to inhibit cardiac parasympathetic nervous activity. Peripheral administration of CRH, on the other hand, causes vasodilation and hypotension in a variety of species, including humans, possibly through activation of CRH-R2 [106,193]. The physiological role of CRH in regulating the autonomic nervous system is supported by data demonstrating central effects of CRH receptor antagonists attenuating stress-induced adrenal epinephrine secretion [118,188,194,195].

CRH stimulates the electrical activity of neurons in various brain regions that contain CRH and CRH receptors, including the locus ceruleus, hippocampus, cerebral cortex and hypothalamus, as well as in lumbar spinal cord motor neurons. In contrast, CRH has inhibitory actions in the lateral septum, thalamus and hypothalamic PVN. Centrally administered CRH increases the spontaneous discharge rate of the LC/NE in both anesthetized and unanesthetized rats.

CRH causes a generalized increase in electroencephalographic (EEG) activity associated with increased vigilance and decreased sleep time. At CRH doses below those affecting locomotor activity or pituitary-adrenal function, rats remain awake, vigilant, and display decreases in slow-wave sleep. Higher doses of CRH, however, induce seizure activity that is indistinguishable from seizures produced by electrical kindling of the amygdala, which further confirms the role of CRH in brain activation [200–202].

CRH has also been implicated in the pathophysiology of depressive [203], anxiety [204] and substance abuse disorders [205]. For example, patients afflicted with melancholic depression [206], post-traumatic stress disorder [207] and alcohol withdrawal [208] commonly exhibit features of autonomic and behavioral hyperarousal, that are thought to result from hyperactivity of CRH system hyperactivity [71].

2.2.2. *Peripheral CRH*

CRH and its receptors are located in several tissues outside the brain and the pituitary gland. Peripheral CRH receptors may mediate complementary actions to those of central CRH or exert an autoregulatory function on other CRH systems. Generally, CRH-R2 is more widely distributed in the peripheral tissues than in the CNS. The sparseness of CRH in the systemic circulation does not support an endocrine role of peripheral CRH, rather it suggests that neurocrine, autocrine or paracrine mechanisms underlie its physiological functions. CRH-R2 is likely to be the predominant receptor subtype for peripheral CRH, except for the placenta.

CRH has been localized in the adrenal medulla of a variety of species and has been reported to increase following stimulation of the splanchnic nerve and hemorrhagic stress. CRH-like immunoreactivity and CRH mRNA have been detected in lymphocytes, where they may play a role in regulating immune function. Other tissues in which CRH has been localized include the testis (Leydig cells and advanced germ cells), pancreas, stomach and small intestine [209]. While CRH is not detected in the circulation under normal circumstances, very high levels have been measured in the plasma of pregnant women, probably originating from the placenta [210].

In preclinical rodent models, stress is associated with inhibition of gastric secretion and forward motility, and enhancement of large bowel transit. Other responses include mucin depletion, diminution of mucosal blood flow, mast cell degranulation, oxidative injury, and increased susceptibility to inflammation and stress ulceration [211]. Interestingly, CRH administration reproduces many of the aforementioned responses and CRH-R1 antagonism suppress stress ulceration, mucin depletion and colon transit. These recent findings indicate that endogenous CRH-R1 activity may be responsible for the majority of gastrointestinal responses to stress [132].

CRH binding sites have also been found in the anterior pituitary, throughout the brain and in various peripheral sites such as the adrenal medulla, prostate, gut, spleen, liver, kidney, and testes. Peripheral CRH may exert local regulatory roles via autocrine or paracrine mechanisms. For example, testicular CRH inhibits human chorionic gonadotropin-stimulated androgen production by the testis [212].

The effects of CRH on the colon are manifested after its central or peripheral administration. Likewise, the effects of stress on colonic transit and are antagonized by either central or peripheral administration of CRH-R1 antagonists. These findings suggest that peripheral and central CRH-R1 are interconnected and boost the activity of one another.

Pregnancy in the third trimester is another condition characterized by hypercortisolism and is the only known physiologic state in humans, in which CRH circulates in plasma at levels high enough to cause activation of the HPA axis. Although circulating CRH, which is of placental origin, is bound with high affinity to CRH-binding protein, it appears that the circulating free fraction is sufficient to explain the observed escalating hypercortisolism when the concentration of CRH-binding protein starts to gradually decrease in plasma after the 35th week of pregnancy. Placental CRH may have a role in precipitating labor. Recently, administration of CRH-R1 antagonists in pregnant sheep has been shown to delay the onset of parturition [168]. A CRH binding protein has not been found in non-human species [183,213,214]. Despite binding to this protein, the circulating half-life of human CRH is only 4 min [179]. Ovine CRH is used for most clinical studies. Although it does not bind to the binding protein [183] it has a half-life of 55 min and, consequently, a prolonged duration of action [215,216].

3.3. Locus ceruleus – norepinephrine (LC/NE)

The locus ceruleus (LC) and other noradrenergic (NE) cell groups of the medulla and pons are collectively known as the LC/NE system (Figure 1). The LC/NE is believed to contribute the majority of NE in the brain because plasma NE cannot cross the blood brain barrier. Brain NE serves globally as an emergency or alarm system that leads to decreases in neurovegetative functions, such as eating and sleeping, and contributes to accompanying increases in autonomic and neuroendocrine responses to stress, including HPA axis activation. NE also activates the amygdala, the principal brain locus for fear-related behaviors. In addition, NE release during stress inhibits the medial prefrontal cortex and, by so doing, could interfere with two of its key functions, namely the shifting of mood from one state to the other based on internal and external cues and the generation of novel, complex behaviors [71,217]. Moreover, by enhancing the long-term storage of aversively charged emotional memories in sites such as the hippocampus and striatum, NE may contribute to subsequent vulnerability to stress in certain subjects by facilitating the recall of traumatic experiences [218,219].

Reciprocal reverberatory neural connections exist between the PVN/CRH and brainstem noradrenergic neurons of the central stress system, with CRH and norepinephrine stimulating each other, the latter primarily through β-1-noradrenergic receptors. Autoregulatory ultra-short negative feedback loops are also present in both the PVN CRH and brainstem nonadrenergic neurons, with collateral fibers inhibiting CRH and catecholamine secretion, respectively, possibly via presynaptic CRH and alpha 2-noradrenergic receptors. Both the CRH and catecholaminergic neurons receive stimulatory innervation from the serotonergic and cholinergic systems and inhibitory input from the gamma-aminobutyric acid (GABA)/benzodiazepine (BZD) and the opioid peptide neuronal systems of the brain [217].

3.4. Arginine vasopressin (AVP)

Arginine-vasopressin (AVP) producing parvocellular neurons of the paraventricular nuclei (PVN) of the hypothalamus play an important role in response to most stress modalities. In addition to the role of magnocellular AVP neurons in conserving body fluids and controlling plasma osmolality, PVN AVP serves as another hypothalamic secretagogue of ACTH. AVP synergizes with CRH during stress to stimulate the secretion of abundant quantities of ACTH [220].

AVP released from axon terminals in the median eminence enters the portal blood system and is carried to the anterior lobe of the pituitary gland, where it acts as a secretagogue for adrenocorticotropic hormone [ACTH] by binding to V1b receptors. A subset of PVN- parvocellular neurons synthesizes and secretes both CRH and AVP, while another subset secretes either CRH or AVP only. The relative proportion of the subset that secretes both neuropeptides increases significantly during stress. The terminals of the parvocellular PVN CRH and -AVP neurons project to different sites, including the noradrenergic neurons of the brainstem and the hypophysial portal system in the median eminence [221].

PVN CRH and -AVP neurons also send projections to and activate POMC-containing neurons in the arcuate nucleus of the hypothalamus, which in turn reciprocally project to the PVN CRH and AVP neurons, innervate LC/NE neurons of the central stress system in the brainstem, and terminate on pain control neurons in the hind brain and spinal cord. During hypovolemic stress, AVP becomes the principal ACTH secretagogue. Moreover, during chronic or prolonged stress, there may be a shift in control of ACTH from CRH to AVP [220–222].

In addition to its endocrine and metabolic roles, AVP has been implicated in some of the behavioral responses to stress as well as in depressive illness. Intracerebroventricular injections of picogram doses of AVP induce long-lasting facilitation of learned passive avoidance behavior in rats. In contrast, central antagonism of AVP attenuates conditioned avoidance behaviors. Additionally, AVP may play a role in the control of male territorial displays and aggression [223].

3.5. Neuropeptide Y (NPY)

NPY in the brain is emerging as an important neuromodulatory agent that affects behavior, anterior pituitary hormone secretion, autonomic control, and other neurotransmitter systems. Several studies have recently shown alterations of central NPY in stress and mental illness [224]. NPY has also been suggested to play an important role in circadian rhythms, via its pathway from the thalamic ventrolateral geniculate to the hypothalamic suprachiasmatic nucleus. Most remarkably, mammals injected centrally with NPY eat excessively, probably by modulating the feeding centers in the hypothalamus. Indeed, the magnitude of NPY-induced feeding is higher than that induced by any other known pharmacological agent [225] .

Earlier studies suggested that NPY increases hypothalamic CRH, and that intra-PVN NPY administration increases plasma adrenocorticotropic hormone (ACTH) and corticosterone in the rat [226]. However, several other lines of evidence have recently suggested an anxiolytic role for NPY in the brain [227,228]. As an example, a strong negative correlation has been seen between anxiety scores and NPY levels in the CSF of depressed patients. Anxiety is a core component of the depressive syndrome, and the two psychopathological states may share a common biological basis because both respond to classic antidepressant treatment. Local microinjections of NPY into the central nucleus of the amygdala reproduced the anxiolytic-like effect of its icv injections but did not affect food intake. Thus, the actions of NPY on appetite and anxiety can be both anatomically and functionally separated, and seem to be independent of each other. The anxiolytic-like actions of NPY are likely to be mediated by Y1 receptors in the amygdala [224,225,229,230]. NPY availability is also reduced in depression and suicide. We have recently observed progressive decrements of CSF NPY in infant primates exposed to repeated separation stress [128].

3.6. Other neuroendocrine systems

Both the mesocortical and mesolimbic components of the dopaminergic system are innervated and activated by the LC/NE and PVN/CRH systems and by glucocorticoids during stress. The mesocortical system, which includes dopaminergic neurons of the ventral tegmentum that send projections to the prefrontal cortex, is thought to be involved in anticipatory phenomena and cognitive functions, and to exert a suppressive effect upon the stress system. The mesolimbic system, which consists of dopaminergic neurons also of the ventral tegmentum that innervate the nucleus accumbens, is believed to play a principal role in motivational, reinforcement and reward phenomena. Euphoria or dysphoria are presumably mediated by the mesocorticolimbic system, which is also the target of several substances of abuse, such as cocaine [71].

The amygdala nuclei are activated during stress primarily by ascending catecholaminergic neurons originating in the brainstem, or by sensing emotional stressors possibly recognized in cortical association areas [68]. Activation of the amygdala is important for the retrieval and emotional analysis of relevant information related to any given stressor [218]. In response to emotional stressors, the amygdala can directly stimulate both central components of the stress system, and also influence the activity of the mesocorticolimbic dopaminergic system, possibly in a lateralized fashion. Interestingly, there are CRH peptidergic neurons in the central nucleus of the amygdala, which respond positively to glucocorticoids and whose activation leads to anxiety, fear and stimulation of the stress system. The hippocampus exerts an important, mostly inhibitory influence on the activity of the amygdala, as well as on the PVN CRH and LC/NE systems [231–233].

PVN CRH/AVP-producing neurons and LC/NE neurons reciprocally innnervate and are innervated by opioid peptide (POMC-producing) neurons of the arcuate nucleus of the hypothalamus [66]. Activation of the stress system stimulates hypothalamic POMC-peptide secretion, which reciprocally inhibits the activity of both central components of the stress system and in addition, through projections to the hindbrain and spinal cord, produces analgesia. POMC-peptides also stimulate the mesocorticolimbic system and hence may produce euphoria and dependence [234].

Substance P (SP) inhibits PVN CRH neurons, while it activates the LC/NE system [6]. Presumably, substance P is elevated centrally when there is peripheral activation of somatic afferent fibers and may, therefore, contribute to changes in the stress system activity in chronic inflammatory or painful states [235].

Activation of the PVN CRH/AVP and LC/NE systems elevates the core temperature of the organism [6]. Both noradrenaline and CRH given intracerebroventricularly produce this effect, possibly through prostanoid-mediated actions on the septal and hypothalamic temperature regulating centers. CRH has also been shown to mediate the pyrogenic effects of the inflammatory cytokines tumor necrosis factor-α (TNF-α), interleukin-1 (IL-1) and interleukin-6 (IL-6), all of which can be stimulated by lipopolysaccharide (LPS), which is a bacterial pyrogen [236].

Prolonged physical and psychological stress, however, is known to reduce core temperature, at least in rodents. This is thought to result from rapid energy loss that exceeds the organism's ability of generating temperature from catabolic processes. Reduction of the body core temperature may provide a protective mechanism to restrain overshooting of the stress response, by markedly slowing the metabolic functions of different systems. However, such stress-induced hypothermia may result in activation of other transmitter systems such as thyrotropin-releasing hormone (TRH), which is known to induce gastric ulceration, a common consequence of severe and prolonged stress in different species.

As mentioned, there is usually marked suppression of food intake during stress. This is possibly mediated by an effect on the appetite/satiety centers in the brain. CRH may be the endogenous mediator of the stress-induced inhibition of food intake [237]. CRH administration causes profound anorexia in macaques [190]. Suppression of NPY secretion is likely to be involved as well. NPY inhibits the LC/NE system and activates the parasympathetic system, actions that help with digestion and storage of nutrients [229,238].

4. CONCLUSION

Activation of the HPA axis is essential for the organism's adaptation to extreme stimuli. The end products of the HPA axis, glucocorticoid hormones affect virtually every cell in the body, enhancing the metabolism and facilitating a multiplicity of coping programs. Glucocorticoid secretion is under the control of ACTH secreted from the anterior pituitary corticotropes as a split product of a big molecule, POMC. Other split products of the POMC include opioid peptides and melanocortins. ACTH stimulates both the release and synthesis of GCs from the adrenal cortex. ACTH-secreting cells are in turn under the control of the hypothalamic hormones, CRH.

The characterization of the biochemical nature of CRH allowed the subsequent testing of its effects on the brain and the periphery. Interestingly, CRH was found to mediate almost the entire cascade of the stress response, not only HPA activation. CRH, therefore seems to fill our gap of knowledge of how characteristic behavioral, endocrine and autonomic programs are simultaneously expressed to enhance the organism's ability to overcome threatening conditions.

Because CRH is a peptide hormone, only peptide receptor antagonists and specific anti-CRH antibodies were the means to inhibit its actions. Such antagonists are big molecules that cannot cross biological membranes and can antagonize central CRH only if directly injected into the brain or the CSF.

One of the most remarkable advances of science is the development of selective small-molecule nonpeptide receptor antagonists that can virtually cross the blood brain barrier [239–241]. Nonpeptide CRH-R1 antagonists have been recently introduced and tested in animal models of stress [118,188,248]. Recent findings with these novel drugs have shown their promise as anti-stress agents. We and others are currently advancing CRH-R1 antagonists for clinical studies and hopefully the treatment of a variety of conditions associated with hyperactivity of the stress/HPA system such as melancholic depression [123], anxiety [188], peptic ulcer [242], gastroesophageal reflux disease, irritable bowel syndrome [118], premature labor, impotence, anovulation [132,168], and perhaps hypercholesterolemia, insulin resistance [118] and coronary heart disease [123,243,244].

ACKNOWLEDGEMENT

We thank Michael Maurer and Joel A. Bronstein for their help in preparing this manuscript.

REFERENCES

1. Addison T. On anaemia: disease of the suprarenal capsule. Lond Med Gaz 1849:517–518.
2. Smith PE. Hypophysectomy and a replacement therapy in the rat. Am J Anat 1930;45: 205–273.
3. Ingle D, Kendall E. Atrophy of the adrenal cortex of the rat produced by the administration of large amounts of cortin. Science 1937;86:245.
4. Harris GW. Neural control of the pituitary gland. Physiol Rev 1948;28:139–179.
5. Vale W, Spiess J, Rivier C, Rivier J. Characterization of a 41-residue ovine hypothalamic peptide that stimulates secretion of corticotropin and beta-endorphin. Science 1981;213(4514):1394–7.
6. Chrousos GP. Stressors, stress, and neuroendocrine integration of the adaptive response. The 1997 Hans Selye Memorial Lecture. Ann N Y Acad Sci 1998;851:311–35.
7. Gabry KE, Gold, PW, Chrousos GP. Hypothalamic pituitary adrenal axis: Introduction to physiology and pathophysiology. In Rubin R, Frank L, Pfaff D, editors. Hormones, Brain and Behavior, Academic Press, 2002;81:841–866.
8. Johnson EO, Kamilaris TC, Chrousos GP, Gold PW. Mechanisms of stress: a dynamic overview of hormonal and behavioral homeostasis. Neurosci Biobehav Rev 1992;16(2):

115–30.
9. Krieger DT, Allen W, Rizzo F, Krieger HP. Characterization of the normal temporal pattern of plasma corticosteroid levels. J Clin Endocrinol Metab 1971;32(2):266–84.
10. Weitzman ED, Fukushima D, Nogeire C, Roffwarg H, Gallagher TF, Hellman L. Twenty-four hour pattern of the episodic secretion of cortisol in normal subjects. J Clin Endocrinol Metab 1971;33(1):14–22.
11. Veldhuis JD, Iranmanesh A, Johnson ML, Lizarralde G. Amplitude, but not frequency, modulation of adrenocorticotropin secretory bursts gives rise to the nyctohemeral rhythm of the corticotropic axis in man. J Clin Endocrinol Metab 1990;71(2):452–63.
12. Quigley ME, Yen SS. A mid-day surge in cortisol levels. J Clin Endocrinol Metab 1979;49(6):945–7.
13. Slag MF, Ahmad M, Gannon MC, Nuttall FQ. Meal stimulation of cortisol secretion: a protein induced effect. Metabolism 1981;30(11):1104–8.
14. Dickerman Z, Grant DR, Faiman C, Winter JS. Intraadrenal steroid concentrations in man: zonal differences and developmental changes. J Clin Endocrinol Metab 1984;59(6): 1031–6.
15. Mountjoy KG, Robbins LS, Mortrud MT, Cone RD. The cloning of a family of genes that encode the melanocortin receptors. Science 1992;257(5074):1248–51.
16. Simpson ER, Waterman MR. Regulation by ACTH of steroid hormone biosynthesis in the adrenal cortex. Can J Biochem Cell Biol 1983;61(7):692–707.
17. Miyamoto N, Seo H, Kanda K, Hidaka H, Matsui N. A 3',5'-cyclic adenosine monophosphate-dependent pathway is responsible for a rapid increase in c-fos messenger ribonucleic acid by adrenocorticotropin. Endocrinology 1992;130(6):3231–6.
18. Simpson ER, Waterman MR. Regulation of the synthesis of steroidogenic enzymes in adrenal cortical cells by ACTH. Annu Rev Physiol 1988;50:427–40.
19. Suzuki I, Cone RD, Im S, Nordlund J, Abdel-Malek ZA. Binding of melanotropic hormones to the melanocortin receptor MC1R on human melanocytes stimulates proliferation and melanogenesis. Endocrinology 1996;137(5):1627–33.
20. Lando D, Secchi J, Roche J, Raynaud JP. Adrenocorticotropin analogs and glucocorticoids in the hypophysectomized rat. I. Effects on liver polyribosomes and rough endoplasmic reticulum. Endocrinology 1980;107(6):2055–62.
21. Grunfeld C, Hagman J, Sabin EA, al. e. Characterization of adrenocorticotropin receptors that appear when 3T3-L1 cells differentiate into adipocytes. Endocrinology 1985;116: 113–117.
22. Goodman HM, Knobil E. Some endocrine factors in regulation of fatty acid mobilization during fasting. Am J Physiol 1961;201:1–3.
23. Fain JH. Effects of dexamethasone and growth hormone on fatty acid mobilization and glucose utilization in adrenalectomized rats. Endocrinology 1962;71:633–635.
24. Hornbrook KR, Burch HB, Lowry OH. The effects of adrenalectomy and hydrocortisone on rat liver metabolites and glycogen synthetase activity. Mol Pharmacol 1966;2(2): 106–16.
25. Stalmans W, Laloux M. Glucocorticoids and hepatic glycogen metabolism. Monogr Endocrinol 1979;12:517–33.
26. Exton JH. Regulation of gluconeogenesis by glucocorticoids. Monogr Endocrinol 1979;12: 535–46.
27. Munck A. Studies on the mode of action of glucocorticoids in rats: II. The effects in vivo and in vitro on net glucose uptake by isolated adipose tissue. Biochim Biophys Acta

1962;57:318–326.
28. Olefsky JM. Effect of dexamethasone on insulin binding, glucose transport, and glucose oxidation of isolated rat adipocytes. J Clin Invest 1975;56:1499–1508.
29. Livingston JN, Lockwood DH. Effect of glucocorticoids on the glucose transport system of isolated fat cells. Biol Chem 1975;250:8353–8360.
30. Garvey WT, Huecksteadt TP, Lima FB, Birnbaum MJ. Expression of a glucose transporter gene cloned from brain in cellular models of insulin resistance: dexamethasone decreases transporter mRNA in primary cultured adipocytes. Mol Endocrinol 1989;3(7):1132–41.
31. Friedmann N, Exton JH, Park CR. Interaction of adrenal steroids and glucagon on gluconeogenesis in perfused rat liver. Biochem Biophys Res Commun 1967;29(1):113–9.
32. Frost HM, Villaneuva AR. Human osteoblastic activity. III. The effect of cortisone on lamellar osteoblastic activity. Henry Ford Hosp Med Bull 1961;9:97–99.
33. Jowsey J, Riggs BL. Bone formation in hypercortisolism. Acta Endocrinol 1970;63: 21–28.
34. Hahn TJ, Halstead LR, Teitelbaum SL, Hahn BH. Altered mineral metabolism in glucocorticoid-induced osteopenia. Effect of 25-hydroxyvitamin D administration. J Clin Invest 1979;64(2):655–65.
35. Bar-Shavit Z, Kahn AJ, Pegg LE, al. e. Glucocorticoids modulate macrophage surface oligosaccharides and their bone binding activity. J Clin Invest 1984;73:1277–1283.
36. Wajchenberg BL, Pereira VG, Kieffer J, al. e. Effect of dexamethasone on calcium metabolism and 47 Ca kinetics in normal subjects. Acta Endocrinol 1969 1969; 61: 173–192.
37. Lukert BP, Adams JS. Calcium and phosphorus homeostasis in man: effects of corticosteroids. Arch Intern Med 1976`;136:1249–1253.
38. Klein RG, Arnaud SB, Gallagher JC, al. e. Intestinal calcium absorption in exogenous hypercortisolism: role of 25-hydroxyvitamin D and corticosteroid dose. J Clin Invest 1977;60:253–259.
39. Fucik RF, Kukreja SC, Hargis GK. Effect of glucocorticoids on the function of the parathyroid glands in man. j Clin Endocrinol Metab 1975;20:152–155.
40. Delany AM, Canalis E. Transcriptional repression of insulin-like growth factor I by glucocorticoids in rat bone cells. Endocrinology 1995;136(11):4776–81.
41. McEwen BS. Influences of adrenocortical hormones on pituitary and brain function. In: Baxter JD, Rousseau GG, eds. Glucocorticoid Hormone Action. New York: Springer-Verlag 1979:467–492.
42. Jeffcoate WJ, Silverstone JT, Edwards CRW, al. e. Psychiatric manifestations of Cushing's syndrome: response to lowering of plasma cortisol. Q J Med 1979;48:465–472.
43. Cohen SI. Cushing's syndrome: a psychiatric study of 29 patients. Br J Psychiatry 1980;136:120–124.
44. Starkman MN, Schteingart DE, Schork MA. Cushing's syndrome after treatment: changes in cortisol and ACTH levels, and amelioration of the depressive syndrome. Psychiatry Res 1986;19:177–188.
45. Loosen PT, Chambliss B, DeBold CR, al. e. Psychiatric phenomenology in Cushing's disease. Pharmacopsychiatry 1992;25:192–198.
46. Krieger DT. The central nervous system and Cushing's disease. Mt Sinai J Med 1972;39: 416–442.
47. Grillin JC, Jacobs LS, Fram DH, al. e. Acute effect of glucocorticoid on human sleep. Nature 1972;237:398–399.

48. Grillin JC, Jacobs LS, Snyder F, al. e. Effects of ACTH on the sleep of normal subjects and patients with Addison's disease. Neuroendocrinology 1974;15:21–31.
49. Mosmann TR, Cherwinski H, Bond MW, Giedlin MA, Coffman RL. Two types of murine helper T cell clone. I. Definition according to profiles of lymphokine activities and secreted proteins. J Immunol 1986;136(7):2348–57.
50. Spellberg B, Edwards JE, Jr. Type 1/Type 2 immunity in infectious diseases. Clin Infect Dis 2001;32(1):76–102.
51. Seder RA, Gazzinelli R, Sher A, Paul WE. Interleukin 12 acts directly on CD4+ T cells to enhance priming for interferon gamma production and diminishes interleukin 4 inhibition of such priming. Proc Natl Acad Sci U S A 1993;90(21):10188–92.
52. Webster EL, Barrientos RM, Contoreggi C, Isaac MG, Ligier S, Gabry KE, Chrousos GP, McCarthy EF, Rice KC, Gold PW, Sternberg EM. Corticotropin releasing hormone (CRH) antagonist attenuates adjuvant induced arthritis: role of CRH in peripheral inflammation. J Rheumatol 2002;29:1252–1261.
53. Gajewski TF, Goldwasser E, Fitch FW. Anti-proliferative effect of IFN-gamma in immune regulation. II. IFN- gamma inhibits the proliferation of murine bone marrow cells stimulated with IL-3, IL-4, or granulocyte-macrophage colony-stimulating factor. J Immunol 1988;141(8):2635–42.
54. Ohmori Y, Hamilton TA. IL-4-induced STAT6 suppresses IFN-gamma-stimulated STAT1-dependent transcription in mouse macrophages. J Immunol 1997;159(11):5474–82.
55. Habib KE, Gold PW, Chrousos GP. Neuroendocrinology of stress. Endocrinology and Metabolism Clinics of North America 2001;30: 695–728.
56. Arnold J. Ein Beitrag zu der feineren Structurund dem Chemismus der Nebennieren. Arch Pathol Anat Physiol Klin Med 1866;35:64–107.
57. Kohn A. Das chromaffine Gewebe. Erg Anat Entwicklungsgesch 1902;12:253–348.
58. Roberts KD, Brandy L, Lieberman S. The occurrence and metabolism of 20alpha-hydroxycholesterol in bovine adrenal preparations. Biochemistry 1969;8:1259–1270.
59. Dixon R, Furutachi T, Lieberman S. The isolation of crystalline 22R-hydroxycholesterol and 20alpha, 22R-dihydroxycholesterol from bovine adrenals. Biochem Biophys Res Commun 1970;40:161–166.
60. Hume T, Boyd GS. Cholesterol metabolism and steroid hormone production. Biochem Soc Trans 1978;6:893–898.
61. Sternberg EM, Glowa JR, Smith MA, Calogero AE, Listwak SJ, Aksentijevich S, et al. Corticotropin releasing hormone related behavioral and neuroendocrine responses to stress in Lewis and Fischer rats. Brain Res 1992;570(1–2):54–60.
62. Webster EL, Elenkov IJ, Chrousos GP. The role of corticotropin-releasing hormone in neuroendocrine-immune interactions. Mol Psychiatry 1997;2(5):368–72.
63. Kling MA, Rubinow DR, Doran AR, Roy A, Davis CL, Calabrese JR, et al. Cerebrospinal fluid immunoreactive somatostatin concentrations in patients with Cushing's disease and major depression: relationship to indices of corticotropin-releasing hormone and cortisol secretion. Neuroendocrinology 1993;57(1):79–88.
64. Laue L, Gold PW, Richmond A, Chrousos GP. The hypothalamic-pituitary-adrenal axis in anorexia nervosa and bulimia nervosa: pathophysiologic implications. Adv Pediatr 1991;38:287–316.
65. Kaye WH, Gwirtsman HE, George DT, Ebert MH, Jimerson DC, Tomai TP, et al. Elevated cerebrospinal fluid levels of immunoreactive corticotropin- releasing hormone in anorexia nervosa: relation to state of nutrition, adrenal function, and intensity of depression. J Clin

Endocrinol Metab 1987;64(2):203–8.
66. Chrousos GP, Gold PW. The concepts of stress and stress system disorders. Overview of physical and behavioral homeostasis JAMA1992;267(9):1244–52.
67. Bremner JD, Krystal JH, Southwick SM, Charney DS. Noradrenergic mechanisms in stress and anxiety: II. Clinical studies. Synapse 1996;23(1):39–51.
68. Chrousos GP. The role of stress and the hypothalamic-pituitary-adrenal axis in the pathogenesis of the metabolic syndrome: neuro-endocrine and target tissue-related causes. Int J Obes Relat Metab Disord 2000;24 Suppl 2:S50–5.
69. Bornstein SR, Webster EL, Torpy DJ, Richman SJ, Mitsiades N, Igel M, et al. Chronic effects of a nonpeptide corticotropin-releasing hormone type I receptor antagonist on pituitary-adrenal function, body weight, and metabolic regulation. Endocrinology 1998;139(4):1546–55.
70. Gardner JD, Rothwell NJ, Luheshi GN. Leptin affects food intake via CRF-receptor-mediated pathways. Nat Neurosci 1998;1(2):103.
71. Gold PW, Gabry KE, Yasuda MR, Chrousos GP. Divergent endocrine abnormalities in melancholic and atypical depression: clinical and pathophysiologic implications. Endocrinol Metab Clin North Am 2002;31(1):37–62.
72. Ballard PL. Glucocorticoid regulation of lung maturation. Mead Johnson Symp Perinat Dev Med 1987;30:22–7.
73. Odom MJ, Snyder JM, Boggaram V, Mendelson CR. Glucocorticoid regulation of the major surfactant associated protein (SP-A) and its messenger ribonucleic acid and of morphological development of human fetal lung in vitro. Endocrinology 1988;123(4): 1712–20.
74. McCormick SM, Mendelson CR. Human SP-A1 and SP-A2 genes are differentially regulated during development and by cAMP and glucocorticoids. Am J Physiol 1994;266(4 Pt 1):L367–74.
75. Iannuzzi DM, Ertsey R, Ballard PL. Biphasic glucocorticoid regulation of pulmonary SP-A: characterization of inhibitory process. Am J Physiol 1993;264(3 Pt 1):L236–44.
76. Ballard PL, Ertsey R, Gonzales LW, Gonzales J. Transcriptional regulation of human pulmonary surfactant proteins SP-B and SP-C by glucocorticoids. Am J Respir Cell Mol Biol 1996;14(6):599–607.
77. Cole TJ, Blendy JA, Monaghan AP, Krieglstein K, Schmid W, Aguzzi A, et al. Targeted disruption of the glucocorticoid receptor gene blocks adrenergic chromaffin cell development and severely retards lung maturation. Genes Dev 1995;9(13):1608–21.
78. Pratt WB, Aronow L. The effect of glucocorticoids on protein and nucleic acid synthesis in mouse fibroblasts growing in vitro. J Biol Chem 1966;241(22):5244–50.
79. Ehrlich HP, Tarver H, Hunt TK. Effects of vitamin A and glucocorticoids upon inflammation and collagen synthesis. Ann Surg 1973;177:222–227.
80. Ponec M, Hasper I, Vianden GDNE, al. e. Effects of glucocorticosteroids on primary human skin fibroblasts: II. Effects on total protein and collagen synthesis by confluent cell cultures. 1977;259:125–134.
81. Sterling KM, Jr., Harris MJ, Mitchell JJ, DiPetrillo TA, Delaney GL, Cutroneo KR. Dexamethasone decreases the amounts of type I procollagen mRNAs in vivo and in fibroblast cell cultures. J Biol Chem 1983;258(12):7644–7.
82. Shull S, Cutroneo KR. Glucocorticoids coordinately regulate procollagens type I and type III synthesis. J Biol Chem 1983;258:3364–3369.
83. Leibovich SJ, Ross R. The role of the macrophage in wound repair. A study with

hydrocortisone and antimacrophage serum. Am J Pathol 1975;78(1):71–100.
84. Orth DN, Kovacs WJ. The adrenal cortex. In: Wilson JD, Foster DW, Kronenberg HM, Larsen PR, editors. Williams Textbook of Endocrinology. 9th ed. Philadelphia: W.B. Saunders; 1998. p. 517–664.
85. van der Saag PT, Caldenhoven E, van de Stolpe A. Molecular mechanisms of steroid action: a novel type of cross-talk between glucocorticoids and NF-kappa B transcription factors. Eur Respir J Suppl 1996;22:146s–53s.
86. Gower WR, Jr. Mechanism of glucocorticoid action. J Fla Med Assoc 1993;80(10):697–700.
87. Adler S, Waterman ML, He X, al. e. Steroid receptor-mediated inhibition of rat prolactin gene expression does not require the receptor DNA-binding domain. Cell 1988;52:685–695.
88. Jonat C, Rahmsdorf HJ, Park KK, al. e. Antitumor promotion and antiinflammation: down-modulation of AP-1 (Fos/Jun) activity by glucocorticoid hormone. Cell 1990;62: 1189–1204.
89. Yang-Yen H-F, Chambard JC, Sun YL, al. e. Transcriptional interference between c-Jun and the glucocorticoid receptor: mutual inhibition of DNA binding due to direct protein-protein interaction. Cell 1990;62:1205–1215.
90. Schule R, Rangarajan P, Kliewer S, al. e. Functional antagonism between oncoprotein c-Jun and the glucocorticoid receptor. Cell 1990;62:1217–1226.
91. Drouin J, Charron J, Gagner JP, al. e. Pro-opiomelanocortin gene: a model for negative regulation of transcription by glucocorticoids. J Cell Biochem 1987;35:293–304.
92. Akerblom IE, Slater EP, Beato M, al. e. Negative regulation by glucocorticoids through interference with a cAMP responsive enhancer. Science 1988;241:350–353.
93. Sajai DD, Helms S, Carlstedt-Duke J, al. e. Hormone-mediated repression: a negative glucocorticoid response element from the bovine prolactin gene. Genes Dev 1988(2): 1144–1154.
94. Hammond GL, Smith CL, Groping IS, al. e. Primary structure of human corticosteroid binding globulin, deduced from hepatic and pulmonary cDNAs, exhibits homology with serine protease inhibitors. Proc Natl Acad Sci USA 1987;84:5153–5157.
95. Smith CL, Hammond GL. Rat corticosteroid binding globulin: primary structure and messenger ribonucleic acid levels in the liver under different physiological conditions. Mol Endocrinol 1989;3:420–426.
96. Misao R, Hori M, Ichigo S, al. e. Corticosteroid-binding globulin mRNA levels in human uterine endometrium. Steroids 1994;59:603–607.
97. Mendel CM. The free hormone hypothesis: a physiologically based mathematical model. Endocr Rev 1989;10:232–274.
98. Orth D. The Adrenal Cortex. In: William's Textbook of Endocrinology; 1998.
99. Rosner W. The functions of corticosteroid-binding globulin and sex hormone-binding globulin: recent advances. Endocr Rev 1990;11:80–91.
100. Maitra US, Khan MS, Rosner W. Corticosteroid-binding globulin receptor of the rat hepatic membrane: solubilization, partial characterization, and the effect of steroids on binding. Endocrinology 1993;133:1817–1822.
101. Singer CJ, Khan MS, Rosner W. Characteristics of the binding of corticosteroid-binding globulin to rat cell membranes. Endocrinology 1998;122:89–96.
102. Nakhla AM, Khan MS, Rosner W. Characteristics of the binding of corticosteroid-binding globulin to rat cell membranes. Biochem Biophys Res Commun 1988;153:1012–1018.

103. Hua SY, Chen YZ. Membrane receptor-mediated electrophysiological effects of glucocorticoid on mammalian neurons. Endocrinology 1989;124:687–691.
104. Ruf K, Steiner FA. Steroid-sensitive single neurons in rat hypothalamus and midbrain: identification by microelectrophoresis. Science 1967;156:667–668.
105. Mandelbrod I, Feldman S, Werman R. Inhibition of firing is the primary effect of microelectrophoresis of cortisol to units in the rat tuberal hypothalamus. Brain Res 1974;80:303–315.
106. Korte SM, Bouws GA, Bohus B. Central actions of corticotropin-releasing hormone (CRH) on behavioral, neuroendocrine, and cardiovascular regulation: brain corticoid receptor involvement. Horm Behav 1993;27(2):167–83.
107. Michelson D, Chrousos GP, Gold PW. Type I glucocorticoid receptor blockade does not affect baseline or ovine corticotropin-releasing-hormone-stimulated adrenocorticotropic hormone and cortisol secretion. Neuroimmunomodulation 1994;1(5):274–7.
108. Keller-Wood ME, Dallman MF. Corticosteroid inhibition of ACTH secretion. Endocr Rev 1984;5:1–24.
109. Lundblad JR, Roberts JL. Regulation of proopiomelanocortin gene expression in pituitary. Endocr Rev 1988;9:135–158.
110. Davis LG, Arentzen R, Reid JM, al. e. Glucocorticoid sensitivity of vasopressin mRNA levels in the paraventricular nucleus of the rat. Proc Natl Acad Sci USA 1986;83:1145–1149.
111. Itoi K, Mouri T, Takahashi K, al. e. Suppression by glucocorticoid of the immunoreactivity of corticotropin-releasing factor and vasopressin in the paraventricular nucleus of rat hypothalamus. Neurosci Lett 1987;73:231–236.
112. Beyer HS, Matta SG, Sharp BM. Regulation of the messenger ribonucleic acid for corticotropin-releasing factor in the paraventricular nucleus and other brain sites of the rat. Endocrinology 1988;123:2117–2123.
113. Eberwine JH, Jonassen JA, Evigner MJ, al. e. Complex transcriptional regulation by glucocorticoids and corticotropin-releasing hormone of proopiomelanocortin gene expression in rat pituitary cultures. DNA Cell Biol 1987;6:483–492.
114. Oki Y, Peatman TW, Qu Z-C, al. e. Effects of intracellular Ca2+ depletion and glucocorticoid on stimulated adrenocorticotropin release by rat anterior pituitary cells in a microperifusion system. Endocrinology 1991;128:1589–1596.
115. Holmes MC, Catt KJ, Aguilera G. Involvement of vasopressin in the down-regulation of pituitary corticotropin-releasing factor receptors after adrenalectomy. Endocrinology 1987;121:2093–2098.
116. Luo X, Kiss A, Rabadan-Diehl C, al. e. Regulation of hypothalamic and pituitary corticotropin-releasing hormone receptor messenger ribonucleic acid by adrenalectomy and glucocorticoids. Endocrinology 1995;136:3877–3883.
117. Habib KE, Higley JD, Rice K, Cai J, S. L, Chrousos GP, et al. CRH type-1 receptor antagonism attenuates behavioral, endocrine and autonomic responses to stress in primates. Neuropsychopharmacology. 2000;23(S2).S119.
118. Gabry KE, Chrousos GP, Rice KC, Mostafa RM, Sternberg E, Negrao AB, Webster EL, McCann SM, Gold PW. Marked suppression of gastric ulcerogenesis and intestinal responses to stress by a novel class of drugs. Mol Psychiatry 2002;7:474–483.
119. Khalil MY, Gabry KE. Stress ulcer and depression-associated gastric damage in humans: endoscopic evidence showing suppression by anxiolytic and antidepressant treatment. Mol Psychiatry 2002;7(5):433.

120. Hassan A, Khalil MY, Yasuda MR, Rice KC, Gold PW, Habib KE. Chronic stress precipitates oxidative brain-gut damage that is suppressed by a corticotropin-releasing hormone type-1 receptor antagonist. Endo Soc Abstr 2002;84:2489.
121. Habib KE, Rice KC, Gold PW. Suppression of stress-induced oxidative neurotoxicity and gut damage by a novel class of drugs. Am College Neuropsychopharmacol (ACNP) 2001;Abstr 40:p.366.
122. Luger A, Deuster PA, Kyle SB, al. e. Acute hypothalamic-pituitary-adrenal responses to the stress of treadmill exercise: physiologic adaptations to physical training. N Engl J Med 1987;316:1309–1315.
123. Gold PW, Chrousos GP. Organization of the stress system and its dysregulation in melancholic and atypical depression: High vs low CRH/NE states. Mol Psychiatry 2002;7(3):254–275.
124. Maughan B, McCarthy G. Childhood adversities and psychosocial disorders. Br Med Bull 1997;53(1):156–69.
125. Heim C, Newport DJ, Heit S, Graham YP, Wilcox M, Bonsall R, et al. Pituitary-adrenal and autonomic responses to stress in women after sexual and physical abuse in childhood. Jama 2000;284(5):592–7.
126. Suomi SJ, Collins ML, Harlow HF, Ruppenthal GC. Effects of maternal and peer separations on young monkeys. J Child Psychol Psychiatry 1976;17(2):101–12.
127. Habib KE, Cai J, Cizza G, Pushkas J, Listwak S, Negrao AB, Habib LG, Schulkin J, Chrousos GP Suomi SJ, Higley JD, Gold PW. Sensitization of cerebrospinal fluid corticotropin-releasing hormone (CRH) with frequent exposure to psychosocial stress in infant primates. Endo Soc Abstr 2000;82:1102.
128. Habib KE, Higley JD, Chrousos GP, Gold PW. Plasticity of brain NPY, in paradox with that of CRH, during repeated exposure of nonhuman primate infants to separation stress. Endo Soc 2001;Abstr 83:p. 515.
129. Higley JD, Suomi SJ, Linnoila M. CSF monoamine metabolite concentrations vary according to age, rearing, and sex, and are influenced by the stressor of social separation in rhesus monkeys. Psychopharmacology 1991;103(4):551–6.
130. Tsigos C, Chrousos GP. Physiology of the hypothalamic-pituitary-adrenal axis in health and dysregulation in psychiatric and autoimmune disorders. Endocrinol Metab Clin North Am 1994;23(3):451–66.
131. Johnson EO, Kamilaris TC, Carter CS, Calogero AE, Gold PW, Chrousos GP. The biobehavioral consequences of psychogenic stress in a small, social primate (Callithrix jacchus jacchus). Biol Psychiatry 1996;40(5):317–37.
132. Gabry KE, Khalil MY, Ayala A, Chrousos GP. Stress in medicine. In: Martini L, editor. The Encyclopedia of Endocrinology and Endocrine Diseases, in press (2003).
133. Guillemin R, Rosenberg B. Humoral hypothalamic control of anterior pituitary: study with combined tissue cultures. Endocrinology 1955;57:599–607.
134. Saffran M, Schally AV. Release of corticotrophin by anterior pituitary tissue in vitro. Can J Biochem Physiol 1955;33:408–415.
135. Rivier J, Spiess J, Vale W. Characterization of rat hypothalamic corticotropin-releasing factor. Proc Natl Acad Sci USA 1983;80:4851–4855.
136. Esch F, Ling N, Bohlen P, al. e. Isolation and characterization of the bovine hypothalamic corticotropin-releasing factor. Biochem Biophys Res Commun 1984;122:899–905.
137. Patthy M, Horvath J, Mason Garcia M, al. e. Isolation and amino acid sequence of corticotropin-releasing factor from pig hypothalami. Proc Natl Acad Sci USA 1985;82:

8762–8766.
138. Furutachi Y, Morimoto Y, Shibahara S, al. e. Cloning and sequence analysis of cDNA for ovine corticotropin releasing factor precursor. Nature 1983;301:537–540.
139. Jingami H, Mizuno N, Takahashi H. Cloning and sequence analysis of cDNA for rat corticotropin-releasing factor precursor. FEBS Lett 1985;191:63–66.
140. Shibahara S, Morimoto Y, Furutachi Y, al. e. Isolation and sequence analysis of the human corticotropin-releasing factor precursor gene. EMBO J 1983;2:775–779.
141. Thompson RC, Seasholtz AF, Herbert E. Rat corticotropin-releasing hormone gene: sequence and tissue-specific expression. Mol Endocrinol 1987;1:363–370.
142. Arbiser JL, Morton CC, Bruns GAP, al. e. Human corticotropin releasing hormone gene is located on the long arm of chromosome 8. Cytogenet Cell Genet 1988;47:113–116.
143. Olschowka JA, O'Donohue TL, Mueller GP, al. e. Hypothalamic and extrahypothalamic distribution of CRF-like immunoreactive neurons in the rat brain. Neuroendocrinology 1982;35:305–308.
144. Swanson LW, Sawchenko PE, Rivier J, al. e. Organization of ovine corticotropin-releasing factor immunoreactive cells and fibers in the rat brain: an immunohistochemical study. Neuroendocrinology 1983;36:165–168.
145. Bloom FE, Battenberg ELF, Rivier J, al. e. Corticotropin releasing factor (CRF): immunoreactive neurons and fibers in rat hypothalamus. Regul Pept 1982;4:43–48.
146. Kiss JZ, Mezey E, Skirtboll L. Corticotropin-releasing factor-immunoreactive neurons of the paraventricular nucleus become vasopressin positive after adrenalectomy. Proc Natl Acad Sci USA 1984;81:1854–1858.
147. Sawchenko PE, Swanson LW, Vale WW. Co-expression of corticotropin-releasing factor and vasopressin immunoreactivity in parvocellular neurosecretory neurons of the adrenalectomized rat. Proc Natl Acad Sci USA 1984;81:1883–1887.
148. Mezey E, Reisine TD, Skirboll L, Beinfeld M, Kiss JZ. Role of cholecystokinin in corticotropin release: coexistence with vasopressin and corticotropin-releasing factor in cells of the rat hypothalamic paraventricular nucleus. Proc Natl Acad Sci U S A 1986;83(10):3510–2.
149. Hokfelt T, Fahrenkrug J, Tatemoto K, al. e. The PHI (PHI-27)/corticotropin-releasing factor/enkephalin immunoreactive hypothalamic neuron: possible morphological basis for integrated control of prolactin, corticotropin, and growth hormone secretion. Proc Natl Acad Sci U S A 1983;80:895–898.
150. Roth KA, Weber E, Barchas JD. Immunoreactive dynorphin-(1–8) and corticotropin-releasing factor in subpopulation of hypothalamic neurons. Science 1983;219:189–191.
151. Sawchenko PE, Swanson LW, Vale WW. Corticotropin-releasing factors: coexpression within distinct subsets of oxytocin-, vasopressin-, and neurotensin-immunoreactive neurons in the hypothalamus of the male rat. J Neurosci 1984;4:1118–1129.
152. Owens MJ, Nemeroff CB. Physiology and pharmacology of corticotropin-releasing factor. Pharmacol Rev 1991;43(4):425–73.
153. Grigoriadis DE, Heroux JA, De Souza EB. Characterization and regulation of corticotropin-releasing factor receptors in the central nervous, endocrine and immune systems. Ciba Found Symp 1993;172:85–101.
154. Palkovits M, Brownstein MJ, Vale W. Distribution of corticotropin-releasing factor in rat brain. Fed Proc 1985;44(1 Pt 2):215–9.
155. Swanson LW, Sawchenko PE, Rivier J, Vale WW. Organization of ovine corticotropin-releasing factor immunoreactive cells and fibers in the rat brain: an immunohistochemical

study. Neuroendocrinology 1983;36(3):165–86.
156. Turnbull AV, Rivier C. Corticotropin-releasing factor (CRF) and endocrine responses to stress: CRF receptors, binding protein, and related peptides. Proc Soc Exp Biol Med 1997;215(1):1–10.
157. Chen R, Lewis KA, Perrin MH, Vale WW. Expression cloning of a human corticotropin-releasing-factor receptor. Proc Natl Acad Sci U S A 1993;90(19):8967–71.
158. Wynn PC, Harwood JP, Catt KJ, al. e. Regulation of corticotropin-releasing factor (CRF) receptors in the rat pituitary gland: effect of adrenalectomy on CRF receptors and corticotropin responses. Endocrinology 1985;116:1653–1659.
159. Millan MA, Samra AB, Wynn PC, Catt KJ, Aguilera G. Receptors and actions of corticotropin-releasing hormone in the primate pituitary gland. J Clin Endocrinol Metab 1987;64(5):1036–41.
160. Giguere V, Labrie F, Cote J, Coy DH, Sueiras-Diaz J, Schally AV. Stimulation of cyclic AMP accumulation and corticotropin release by synthetic ovine corticotropin-releasing factor in rat anterior pituitary cells: site of glucocorticoid action. Proc Natl Acad Sci U S A 1982;79(11):3466–9.
161. Won JG, Orth DN. Roles of intracellular and extracellular calcium in the kinetic profile of adrenocorticotropin secretion by perifused rat anterior pituitary cells. I. Corticotropin-releasing factor stimulation. Endocrinology 1990;126(2):849–57.
162. Oki Y, Peatman TW, Qu ZC, Orth DN. Effects of intracellular Ca2+ depletion and glucocorticoid on stimulated adrenocorticotropin release by rat anterior pituitary cells in a microperifusion system. Endocrinology 1991;128(3):1589–96.
163. Won JGS, Orth DN. Role of lipoxygenase metabolites of arachidonic acid in the regulation of adrenocorticotropin secretion by perifused rat anterior pituitary cells. Endocrinology 1994;135:1496–1503.
164. Watanabe T, Orth DN. Detailed kinetic analysis of adrenocorticotropin secretion by dispersed rat anterior pituitary cells in a microperifusion system: effects of ovine corticotropin-releasing factor and arginine vasopressin. Endocrinology 1987;121(3): 1133–45.
165. Lundblad JR, Roberts JL. Regulation of proopiomelanocortin gene expression in pituitary. Endocr Rev 1988;9(1):135–58.
166. Bruhn TO, Sutton RE, Rivier CL, Vale WW. Corticotropin-releasing factor regulates proopiomelanocortin messenger ribonucleic acid levels in vivo. Neuroendocrinology 1984;39(2):170–5.
167. Loeffler JP, Kley N, Pittius CW, Hollt V. Corticotropin-releasing factor and forskolin increase proopiomelanocortin messenger RNA levels in rat anterior and intermediate cells in vitro. Neurosci Lett 1985;62(3):383–7.
168. Chrousos GP, Torpy DJ, Gold PW. Interactions between the hypothalamic-pituitary-adrenal axis and the female reproductive system: clinical implications. Ann Intern Med 1998;129(3):229–40.
169. Wynn PC, Hauger RL, Holmes MC, Millan MA, Catt KJ, Aguilera G. Brain and pituitary receptors for corticotropin releasing factor: localization and differential regulation after adrenalectomy. Peptides 1984;5(6):1077–84.
170. Smets G, Vauquelin G, Moons L, Smitz J, Kloppel G. Receptors for corticotropin-releasing hormone in human pituitary: binding characteristics and autoradiographic localization to immunocytochemically defined proopiomelanocortin cells. J Clin Endocrinol Metab 1991;73(2):268–74.

171. Wong ML, Licinio J, Pasternak KI, Gold PW. Localization of corticotropin-releasing hormone (CRH) receptor mRNA in adult rat brain by in situ hybridization histochemistry. Endocrinology 1994;135(5):2275–8.
172. Chalmers DT, Lovenberg TW, De Souza EB. Localization of novel corticotropin-releasing factor receptor (CRF2) mRNA expression to specific subcortical nuclei in rat brain: comparison with CRF1 receptor mRNA expression. J Neurosci 1995;15(10):6340–50.
173. Spiess J, Rivier J, Rivier C, Vale W. Primary structure of corticotropin-releasing factor from ovine hypothalamus. Proc Natl Acad Sci U S A 1981;78(10):6517–21.
174. Rivier C, Rivier J, Lederis K, Vale W. In vitro and in vivo ACTH-releasing activity of ovine CRF, sauvagine and urotensin I. Regul Pept 1983;5(2):139–43.
175. Lederis K. Non-mammalian corticotropin release-stimulating peptides. Ann N Y Acad Sci 1987;512:129–38.
176. Lewis K, Li C, Perrin MH, Blount A, Kunitake K, Donaldson C, et al. Identification of urocortin III, an additional member of the corticotropin-releasing factor (CRF) family with high affinity for the CRF2 receptor. Proc Natl Acad Sci U S A 2001;98(13):7570–5.
177. Reyes TM, Lewis K, Perrin MH, Kunitake KS, Vaughan J, Arias CA, et al. Urocortin II: A member of the corticotropin-releasing factor (CRF) neuropeptide family that is selectively bound by type 2 CRF receptors. Proc Natl Acad Sci U S A 2001;98(5):2843–2848.
178. Hsu SY, Hsueh AJ. Human stresscopin and stresscopin-related peptide are selective ligands for the type 2 corticotropin-releasing hormone receptor. Nat Med 2001;7(5):605–11.
179. Schurmeyer TH, Avgerinos PC, Gold PW, Gallucci WT, Tomai TP, Cutler GB, Jr., et al. Human corticotropin-releasing factor in man: pharmacokinetic properties and dose-response of plasma adrenocorticotropin and cortisol secretion. J Clin Endocrinol Metab 1984;59(6):1103–8.
180. Orth DN, Jackson RV, DeCherney GS, DeBold CR, Alexander AN, Island DP, et al. Effect of synthetic ovine corticotropin-releasing factor. Dose response of plasma adrenocorticotropin and cortisol. J Clin Invest 1983;71(3):587–95.
181. Conaglen JV, Donald RA, Espiner EA, Livesey JH, Nicholls MG. The effect of ovine corticotropin-releasing factor on catecholamine, vasopressin, and aldosterone secretion in normal man. J Clin Endocrinol Metab 1984;58(3):463–6.
182. Trainer PJ, Faria M, Newell-Price J, Browne P, Kopelman P, Coy DH, et al. A comparison of the effects of human and ovine corticotropin-releasing hormone on the pituitary-adrenal axis. J Clin Endocrinol Metab 1995;80(2):412–7.
183. Orth DN, Mount CD. Specific high-affinity binding protein for human corticotropin-releasing hormone in normal human plasma. Biochem Biophys Res Commun 1987;143(2): 411–7.
184. Potter E, Behan DP, Fischer WH, Linton EA, Lowry PJ, Vale WW. Cloning and characterization of the cDNAs for human and rat corticotropin releasing factor-binding proteins. Nature 1991;349(6308):423–6.
185. Woods RJ, Grossman AB, Saphier P, al. e. Association of human corticotropin-releasing hormone to its binding protein in blood may trigger clearance of the complex. J Clin Endocrinol Metab 1994;78:73–76.
186. Pavlov EP, Harman SM, Chrousos GP, Loriaux DL, Blackman MR. Responses of plasma adrenocorticotropin, cortisol, and dehydroepiandrosterone to ovine corticotropin-releasing hormone in healthy aging men. J Clin Endocrinol Metab 1986;62(4):767–72.
187. Eaves M, Thatcher-Britton K, Rivier J, Vale W, Koob GF. Effects of corticotropin releasing factor on locomotor activity in hypophysectomized rats. Peptides 1985;6(5):923–6.

188. Habib KE, Weld KP, Rice KC, Pushkas J, Champoux M, Listwak S, et al. Oral administration of a corticotropin-releasing hormone receptor antagonist significantly attenuates behavioral, neuroendocrine, and autonomic responses to stress in primates. Proc Natl Acad Sci U S A 2000;97(11):6079–84.
189. Insel TR, Aloi JA, Goldstein D, Wood JH, Jimerson DC. Plasma cortisol and catecholamine responses to intracerebroventricular administration of CRF to rhesus monkeys. Life Sci 1984;34(19):1873–8.
190. Glowa JR, Gold PW. Corticotropin releasing hormone produces profound anorexigenic effects in the rhesus monkey. Neuropeptides 1991;18(1):55–61.
191. Kalin NH. Behavioral effects of ovine corticotropin-releasing factor administered to rhesus monkeys. Fed Proc 1985;44(1 Pt 2):249–53.
192. Brown MR, Fisher LA, Spiess J, Rivier C, Rivier J, Vale W. Corticotropin-releasing factor: actions on the sympathetic nervous system and metabolism. Endocrinology 1982;111(3): 928–31.
193. Kregel KC, Overton JM, Seals DR, Tipton CM, Fisher LA. Cardiovascular responses to exercise in the rat: role of corticotropin- releasing factor. J Appl Physiol 1990;68(2): 561–7.
194. Habib KE, Rice KC, Chrousos GP, Gold PW. CRH-1 receptor antagonism exerts antidepressant- and anxiolytic-like effects in attenuating stress ulcerogenesis. FASEB J 2000;14(4): A388.
195. Habib KE, Negro PJ, Sciullo D, Dib A, Gold PW. Protective role of the sympathoadrenal system against stress-induced gastric mucosal injury in the rat. Gastroenterology 1999;116(4): A180.
196. Ehlers CL, Henriksen SJ, Wang M, Rivier J, Vale W, Bloom FE. Corticotropin releasing factor produces increases in brain excitability and convulsive seizures in rats. Brain Res 1983;278(1–2):332–6.
197. Siggins GR, Gruol D, Aldenhoff J, Pittman Q. Electrophysiological actions of corticotropin-releasing factor in the central nervous system. Fed Proc 1985;44(1 Pt 2):237–42.
198. Valentino RJ, Aston-Jones G. Activation of locus coeruleus neurons in the rat by a benzazocine derivative (UM 1046) that mimics opiate withdrawal. Neuropharmacology 1983;22(12A):1363–8.
199. Bishop GA. Neuromodulatory effects of corticotropin releasing factor on cerebellar Purkinje cells: an in vivo study in the cat. Neuroscience 1990;39(1):251–7.
200. Ehlers CL. EEG stability following corticotropin releasing factor in rats. Psychoneuroendocrinology 1986;11(1):121–5.
201. Holsboer F, von Bardeleben U, Steiger A. Effects of intravenous corticotropin-releasing hormone upon sleep- related growth hormone surge and sleep EEG in man. Neuroendocrinology 1988;48(1):32–8.
202. Page ME, Berridge CW, Foote SL, Valentino RJ. Corticotropin-releasing factor in the locus coeruleus mediates EEG activation associated with hypotensive stress. Neurosci Lett 1993;164(1–2):81–4.
203. Gold PW, Chrousos G, Kellner C, Post R, Roy A, Augerinos P, et al. Psychiatric implications of basic and clinical studies with corticotropin-releasing factor. Am J Psychiatry 1984;141(5):619–27.
204. Charney DS, Bremner JD. The neurobiology of anxiety disorders. In: Charney DS, Nestler EJ, Bunney BS, editors. The Neurobiology of Mental Illness. New York: Oxford University Press; 1999. p. 494–517.

205. Rodriguez de Fonseca F, Carrera MRA, Navarro M, Koob GF, Weiss F. Activation of corticotropin-releasing factor in the limbic system during cannabinoid withdrawal [see comments]. Science 1997;276(5321):2050–4.
206. Gold PW, Goodwin FK, Chrousos GP. Clinical and biochemical manifestations of depression. Relation to the neurobiology of stress (1) [see comments]. N Engl J Med 1988;319(6):348–53.
207. Bremner JD, Licinio J, Darnell A, Krystal JH, Owens MJ, Southwick SM, et al. Elevated CSF corticotropin-releasing factor concentrations in posttraumatic stress disorder. Am J Psychiatry 1997;154(5):624–9.
208. Adinoff B, Anton R, Linnoila M, Guidotti A, Nemeroff CB, Bissette G. Cerebrospinal fluid concentrations of corticotropin-releasing hormone (CRH) and diazepam-binding inhibitor (DBI) during alcohol withdrawal and abstinence. Neuropsychopharmacology 1996;15(3):288–95.
209. Aguilera G, Millan MA, Hauger RL, Catt KJ. Corticotropin-releasing factor receptors: distribution and regulation in brain, pituitary, and peripheral tissues. Ann N Y Acad Sci 1987;512:48–66.
210. Margioris AN, Grino M, Protos P, Gold PW, Chrousos GP. Corticotropin-releasing hormone and oxytocin stimulate the release of placental proopiomelanocortin peptides. J Clin Endocrinol Metab 1988;66(5):922–6.
211. Pothoulakis C, Castagliuolo I, Leeman SE. Neuroimmune mechanisms of intestinal responses to stress. Role of corticotropin-releasing factor and neurotensin. Ann N Y Acad Sci 1998;840:635–48.
212. Ulisse S, Fabbri A, Dufau ML. Corticotropin-releasing factor receptors and actions in rat Leydig cells. J Biol Chem 1989;264(4):2156–63.
213. Ellis MJ, Livesey JH, Donald RA. Horse plasma corticotrophin-releasing hormone (CRH): characterisation and lack of a late gestational rise or a plasma CRH-binding protein. J Endocrinol 1994;143(3):455–60.
214. Smith R, Chan EC, Bowman ME, Harewood WJ, Phippard AF. Corticotropin-releasing hormone in baboon pregnancy. J Clin Endocrinol Metab 1993;76(4):1063–8.
215. Nicholson WE, DeCherney GS, Jackson RV, DeBold CR, Uderman H, Alexander AN, et al. Plasma distribution, disappearance half-time, metabolic clearance rate, and degradation of synthetic ovine corticotropin-releasing factor in man. J Clin Endocrinol Metab 1983;57(6): 1263–9.
216. DeBold CR, DeCherney GS, Jackson RV, Sheldon WR, Alexander AN, Island DP, et al. Effect of synthetic ovine corticotropin-releasing factor: prolonged duration of action and biphasic response of plasma adrenocorticotropin and cortisol. J Clin Endocrinol Metab 1983;57(2):294–8.
217. Chrousos GP. Regulation and dysregulation of the hypothalamic-pituitary-adrenal axis. The corticotropin-releasing hormone perspective. Endocrinol Metab Clin North Am 1992;21(4):833–58.
218. Introini-Collison IB, Dalmaz C, McGaugh JL. Amygdala beta-noradrenergic influences on memory storage involve cholinergic activation. Neurobiol Learn Mem 1996;65(1):57–64.
219. Christianson SA, Mjorndal T. Adrenaline, emotional arousal and memory. Scand J Psychol 1985;26(3):237–48.
220. Koob GF, Lebrun C, Martinez JL, Jr., Dantzer R, Le Moal M, Bloom FE. Arginine vasopressin, stress, and memory. Ann N Y Acad Sci 1985;444:194–202.
221. Hauger RL, Irwin MR, Lorang M, Aguilera G, Brown MR. High intracerebral levels

of CRH result in CRH receptor downregulation in the amygdala and neuroimmune desensitization. Brain Res 1993;616(1–2):283–92.
222. Makino S, Smith MA, Gold PW. Increased expression of corticotropin-releasing hormone and vasopressin messenger ribonucleic acid (mRNA) in the hypothalamic paraventricular nucleus during repeated stress: association with reduction in glucocorticoid receptor mRNA levels. Endocrinology 1995;136(8):3299–309.
223. Gilligan PJ, Robertson DW, Zaczek R. Corticotropin releasing factor (CRF) receptor modulators: progress and opportunities for new therapeutic agents. J Med Chem 2000;43(9):1641–60.
224. Wettstein JG, Earley B, Junien JL. Central nervous system pharmacology of neuropeptide Y. Pharmacol Ther 1995;65(3):397–414.
225. Heilig M. Antisense inhibition of neuropeptide Y (NPY)-Y1 receptor expression blocks the anxiolytic-like action of NPY in amygdala and paradoxically increases feeding. Regul Pept 1995;59(2):201–5.
226. Haas DA, George SR. Neuropeptide Y administration acutely increases hypothalamic corticotropin-releasing factor immunoreactivity: lack of effect in other rat brain regions. Life Sci 1987;41(25):2725–31.
227. Yamada K, Shibasaki T, Tsumori C, Imaki T, Hotta M, Wakabayashi I, et al. Neuropeptide Y reverses corticotropin-releasing hormone- and psychological stress-caused shortening of sodium pentobarbital-induced sleep in rats. Brain Res 1996;725(2):272–5.
228. Westrin A, Ekman R, Traskman-Bendz L. Alterations of corticotropin releasing hormone (CRH) and neuropeptide Y (NPY) plasma levels in mood disorder patients with a recent suicide attempt. Eur Neuropsychopharmacol 1999;9(3):205–11.
229. Munglani R, Hudspith MJ, Hunt SP. The therapeutic potential of neuropeptide Y. Analgesic, anxiolytic and antihypertensive. Drugs 1996;52(3):371–89.
230. Heilig M, Soderpalm B, Engel JA, Widerlov E. Centrally administered neuropeptide Y (NPY) produces anxiolytic-like effects in animal anxiety models. Psychopharmacology 1989;98(4):524–9.
231. McGaugh JL, Cahill L, Roozendaal B. Involvement of the amygdala in memory storage: interaction with other brain systems. Proc Natl Acad Sci U S A 1996;93(24):13508–14.
232. Van Bockstaele EJ, Colago EE, Valentino RJ. Amygdaloid corticotropin-releasing factor targets locus coeruleus dendrites: substrate for the co-ordination of emotional and cognitive limbs of the stress response. J Neuroendocrinol 1998;10(10):743–57.
233. Liang KC, Lee EH. Intra-amygdala injections of corticotropin releasing factor facilitate inhibitory avoidance learning and reduce exploratory behavior in rats. Psychopharmacology 1988;96(2):232–6.
234. Millan MJ. Stress and endogenous opioid peptides: a review. Mod Probl Pharmacopsychiatry 1981;17:49–67.
235. Helke CJ, Charlton CG, Wiley RG. Studies on the cellular localization of spinal cord substance P receptors. Neuroscience 1986;19(2):523–33.
236. Elenkov IJ, Chrousos GP. Stress Hormones, Th1/Th2 patterns, Pro/Anti-inflammatory Cytokines and Susceptibility to Disease. Trends Endocrinol Metab 1999;10(9):359–368.
237. Uehara Y, Shimizu H, Ohtani K, Sato N, Mori M. Hypothalamic corticotropin-releasing hormone is a mediator of the anorexigenic effect of leptin. Diabetes 1998;47(6):890–3.
238. Kaye WH, Berrettini W, Gwirtsman H, George DT. Altered cerebrospinal fluid neuropeptide Y and peptide YY immunoreactivity in anorexia and bulimia nervosa. Arch Gen Psychiatry 1990;47(6):548–56.Ref # 239

239. Greiner E, Atkinson AJ, Ayala A, Chrousos GP, Contoreggi C, Eckelman WC, Gold PW, Habib KE, Jacobson AE, Webster EL, Rice KC. Synthesis of double 13C-labeled antalarmin isotopes for pharmacokinetics studies. J Labeled Comp Radiopharm 2002;45: 637-645.
240. Hsin LW, Tian X, Webster EL, Coop A, Caldwell TA, Jacobson A, Chrousos GP, Gold PW, Habib KE, Ayala A, Eckelman W, Contoreggi C, Rice KC. CRH-R1 receptor binding and lipophilicity of pyrrolopyrimidines: Potential nonpeptide corticotropin-releasing hormone type 1 receptor antagonists. Bioorganic & Medicinal Chemistry 2002;10:175-193.
241. Tian X, Hsin L-W, Webster EL, Contoreggi C, Chrousos GP, Gold PW, Habib KE, Ayala A, Eckelman W, Jacobson A, Rice KC. The development of a potential single photon emission computed tomography (SPECT) imaging agent for the corticotropin-releasing hormone receptor type 1. Bioorganic & Medicinal Chemistry Letters 2001;11:331-333.
242. Habib KE, Rice KC, Chrousos GP, Gold PW. CRH-1 receptor antagonism exerts antidepressant- and anxiolytic-like effects in attenuating stress ulcerogenesis. FASEB J 2000;14(4): A388.
243. Gabry KE, Quon M, Sullivan G, Yasuda M, Charney D, Gold PW. Significant hyperglycemia, hyperinsulinemia and insulin resistance in patients with depression. Am College Neuropsychopharmacol (ACNP) 2002;Abstr. 41, pending.
244. Negro PJ, Michelson D, Elliott EA, Avery S, Palladino-Negro P, Sciullo DA, Negrao AB, Habib KE, Prolo P, Ayala A, Cizza G, Fitzpatrick L, Collins M, Reynolds JC, Chrousos GP, Gold PW. Body composition and regional fat content in patients with major depression. Biol Psychiat 2000; 47 (8): 77S-78S.

The Immune-Neuroendocrine Circuitry
Edited by I. Berczi and A. Szentivanyi

Immunoregulation by Innervation

DWIGHT M. NANCE and BRIAN J. MACNEIL

Department of Pathology, University of Manitoba, Winnipeg, Manitoba R3E 0W3, Canada

ABSTRACT

All immune organs are innervated and functional neurotransmitter receptors are present on immune cells. These fundamental observations create the foundation of the research field of neuroimmunology which examines the homeostatic interactions between the nervous and immune systems in the process of host resistance. To this end, a considerable volume of literature is available which describes the effects of neurotransmitters on various *in vitro* measures of immune cells. However, the complex interactions created by neural-immune interactions at sites of antigen challenge and within immune organs cannot be accurately modeled in tissue culture systems. The objective of this review is to focus attention on the need to emphasize *in vivo* model systems which maintain the local microenvironment created when antigen is encountered in skin or mucosal membranes. In addition, we discuss the interactions of nerves and immune cells in lymphoid organs during the generation of the effector immune response.

1. INTRODUCTION

The organs and cells of the immune system constitute the primary defense of an organism against microorganisms, such as bacteria, viruses and fungi, as well as maintaining surveillance for damaged and cancerous cells. Since the cellular components of the immune system are capable of a high degree of autonomy, many of the cellular interactions and functions performed by the immune system have been demonstrated and analyzed by means of tissue cultures and *in vitro* techniques. It is not surprising, therefore, that an appreciation and understanding of the immune system as a regulated homeostatic component of complex organisms has lagged behind the enormous expansion of detailed knowledge regarding the function and activity of what has been predominately viewed as a relatively autonomous system. Despite early evidence for dynamic and powerful interactions between the nervous and immune system, such as the pioneering work of Selye on the adaptive responses of animals to stress [1,2], the contemporary research field of "Neuroimmunology" has a relatively short, albeit impressive, history.

Selye demonstrated a powerful modulatory role for the brain-pituitary-adrenal axis in the regulation of immune organs and their cellular constituents, and major physiological responses to immune activation, such as fever, were well known. However, major developments regarding our understanding of the complex interactions between the nervous and immune systems has occurred primarily within the past 25 years. Earlier developments in the field of neuroendocrinology have helped to lay the foundations as well as shape our view of the immune system as an integrated and regulated physiological system. Foremost in shaping our current view of brain-immune interactions was the experimental and theoretical work of Besedovsky [3,4,5,6]. He was the first to explicitly state that there was bi-directional communication between the brain and the immune system, and he provided important experimental data that demonstrated that signals generated by cells of the immune system acted upon the brain, which in turn exerted regulatory control over the immune system via the neuroendocrine and autonomic nervous systems.

This conceptual view of a regulatory feedback network between the immune system and the central nervous system continues to drive much of the contemporary research in the field of neuroimmunology. Since a similar regulatory feedback system for the regulation of the endocrine system had been firmly established by the time Besedovsky proposed this general model, the feasibility of this conceptual framework had a substantial amount of *prima facia* support, which was supported further by experimental evidence. In relatively simple, but elegant experiments, Besedovsky demonstrated a systematic relationship between the magnitude of an immune response and the neurotransmitter content of an immune organ. Importantly, experimental alterations in the innervation of an immune organ produced reliable changes in immune responses. Consistent with the concept of bi-directional communication between the brain and immune system, immune challenges were shown to activate both the neuroendocrine and autonomic nervous systems. Since the demonstration of these fundamental physiological processes, many of the details regarding the functional and structural organization of this proposed neural-immune regulatory system have been clarified.

2. SYMPATHETIC AND SENSORY INNERVATION

The work of Felten and colleagues [7,8,9] provided early and definitive descriptions of the sympathetic innervation of all primary and secondary immune organs. Additional studies have shown that with regard to the thymus and spleen, the innervation of these immune organs in rodents is primarily, if not exclusively, sympathetic in origin and there is no credible neuroanatomical evidence for any afferent innervation of these organs [10,11]. Since most of the cellular elements of the immune system predominantly reside in immune organs and tissue compartments at risk for microbial infection, it is likely that the direct neural regulation of immune cells occurs within these specialized regions. It may be highly relevant that no part of the body escapes innervation by the sympathetic nervous system, which cannot be stated for the parasympathetic nervous system and sensory afferents. Because of this, the sympathetic nervous system can influence the activity of immune cells throughout the entire organism, including typical sites for invasion of infectious microbes, such as skin, mucosa and the alimentary tract.

This review will focus on the efferent regulation of immune function via sympathetic innervation; however, since sensory innervation also plays a regulatory role in communicating immune signals to central integrative regions in the hypothalamus and brain stem, as well as exerting "efferent" control over immune processes via the release of neuroactive peptides from afferent nerve terminals, we will also consider the role of sensory innervation in immunoregulation. Since the skin and alimentary tract constitute primary sites of entry for microbes and associated inflammation and infection, these are critical sites where sensory, as well as sympathetic, nerve fibers can regulate localized immune cell interactions and therefore shape the nature and magnitude of subsequent immune responses. In this regard, we will consider the role of sensory fibers in the skin in modulating local inflammatory processes and in regulating subsequent immunological responses to localized antigenic challenges.

3. ORGANIZATION OF THE IMMUNE SYSTEM: INNATE AND ADAPTATIVE IMMUNITY

3.1. Innate immunity

Innate immunity is evolutionary ancient and well developed in invertebrates as well as vertebrates and it represents the first line of defense against microbes [12]. The innate immune system reacts quickly to microbes, but can express a limited number of responses to diverse organisms. This innate immune system is composed of antimicrobial chemicals (defensins) on epithelial surfaces (skin and mucosa), complement, phagocytes (macrophages and neutrophils) and natural killer cells. Granulocytes, which include neutrophils, eosinophils and basophils (mast cells), are inflammatory cells which serve as effector cells of both innate and specific immunity.

3.2. Adaptive immunity

Adaptive immunity is only found in vertebrates. This system is capable of mounting highly specific responses to the immense diversity of microbes and pathogens encountered [12]. Cellular elements of the adaptive immune system consist of T and B lymphocytes composed of cutaneous and mucosal immune systems and their associated lymph nodes. Major products of the adaptive immune system are antigen-specific antibodies produced by B cells and antigen-specific T cells which can directly kill microbes and infected cells and/or facilitate the antimicrobial capabilities of the innate immune system. The adaptive immune system can, therefore, be divided into humoral immunity, mediated by antibodies secreted by B cells, and cell-mediated immunity, which is mediated by T lymphocytes. Based upon functional differences, T-helper cells are designated as Th1 and Th2. Th1 and Th2 cells initiate adaptive immune responses in the direction of cellular or humoral immune responses, respectively. Discrimination between these divergent functional categories of T cells is based entirely upon the specific cytokines produced by these cell types. Interleukin-2 (IL-2), interferon-gamma (IFN-γ), and lymphotoxin (TNF-β) are cytokines produced by Th1 cells and generate cellular immune responses. Interleukin-4, -5, -6, -9, -10 and -13 are produced by Th2 cells and promote humoral immune responses. These cytokines and their cells of origin are mutually inhibitory such that one or the other of these adaptive responses will dominate during a specific immune response. For example, Th1 responses predominate in response to viral infections whereas Th2 responses are effective against parasites. The bias of adaptive immunity towards a humoral or cell-mediated response is directly linked with the actions of signal molecules (cytokines) generated by innate immune responses.

3.3. Cooperative actions of innate and adaptive immunity

Interleukin-12 (IL-12), which is produced by macrophages and dendritic cells in response to pathogens, directs the adaptive immune system towards Th1, or cellular, immune responses [12]. In addition to initiation of immune responses, innate and specific processes cooperate at the effector stage of host resistance. Adaptive immunity supplements the antimicrobial functions of the innate immune system and directs the actions of the innate immune system toward specific targets. Importantly, the adaptive immune system has 'memory', such that reexposure to a specific pathogen at a later date is generally dealt with more rapidly and effectively than during the initial exposure. This is the goal of immunization by vaccination, and the immunological memory for specific microbes may be maintained for the life of an organism. Given the cellular complexity of immune interactions, it is not surprising that numerous diseases are associated with, or even produced by, dysregulation of components of the innate and adaptive immune systems.

3.4. Analysis of neural-immune interactions

Even with this highly simplified overview of the immune system, it should be obvious that there are many levels within this complex interaction among widely distributed and specialized immune cells and their regulatory molecules (cytokines and growth factors) at which the endocrine and nervous systems could potentially exert regulatory control. As will be reviewed, the actions of the sympathetic nervous system on innate immunity can be inhibitory while at the same time facilitate the actions of the adaptive or specific immune system. In this context, experimental analysis of this regulatory system is further complicated by the fact that the same actions of hormones and neural innervation depends upon the anatomical region examined, the specific immune cells mediating the response, the nature of the microbe or pathogen, and no less important, the selection of dependent measures and methods of measurement (*in vitro* vs. *in vivo*). Some of the same "models" developed for the analysis of the immune system have served a similar function for analyzing the neural and endocrine regulation of immunity. For example, endotoxin has provided an experimental model for bacterial invasion, double-stranded RNA has served as a viral challenge, and immunization or antigen challenges have provided a model system for the analysis of the adaptive immune system. We will first review the regulatory role of innervation on innate immunity which will be followed by a consideration of the role of innervation on adaptive immune responses.

4. NEURAL REGULATION OF INNATE IMMUNITY

As noted above, the innate immune system constitutes the frontline defense against bacterial and viral invasion with phagocytes (monocytes, macrophages and neutrophils) representing the soldiers in this microbial warfare. For example, macrophages respond quickly to the presence of bacteria and these monocytes express specific receptors (CD14) for lipopolysaccharide (LPS), a cell-wall constituent of gram-negative bacteria. Injections of LPS, or endotoxin, produce a cascade of inflammatory cytokines, beginning within minutes with tumor necrosis factor-alpha (TNF-α) followed by interleukin-1 beta (IL-1β) and then interleukin 6 (IL-6). Other important regulatory and effector molecules are generated by macrophages at later intervals, such as interleukin-12 (IL-12), interferon-alpha (IFN-α) and nitric oxide (NO). These macrophage-dependent cytokines are both effector and signal molecules. For example, TFN-α acts upon cellular death receptors to kill infected and damaged cells, and when produced in excessive amounts results in septic shock associated with septicemia. In conjunction with IL-1β, TNF-α produces a pyrogenic response (fever) which aids further in bacterial killing. These inflammatory cytokines also produce localized effects on vascular endothelial cells including the production of adhesion molecules which rapidly recruit additional immune cells to sites of bacterial invasion. Concurrent changes in vascular permeability aids in the immigration of these cells from the blood stream. IL-6 mediates the acute phase response to infection and stimulates the production and release of various proteins from the liver. IFN-α on the other hand, is produced by macrophages in response to viral infection and represents a first-line of defense against viruses and helps to establish an "anti-viral" state. Importantly, these same cytokines set the stage for subsequent engagement and actions by the specific immune system. Macrophages, and their highly specialized associates, dendritic cells, also serve as antigen presenting cells and therefore provide a critical first step in the full engagement of the antigen-specific immune system. The ability of dendritic cells to fully activate antigen-dependent T cells requires maturation of the dendritic cell plus the production of costimulatory factors by these same cells (B7 molecules) and inflammatory cytokines, such as TNF-α, drive this process. Subsequent production of IL-12 by macrophages can also directly determine whether the specific immune system is driven in the direction of cellular versus humoral immunity. Thus, modulation of the early actions of the innate immune system will have significant impact on the magnitude and quality of the specific immune response. The long-time use and utility of 'adjuvants' in order to potentiate specific immune responses to foreign antigens is based upon this fundamental immunological process, i.e. recruitment of innate immune processes to optimize the specific immune responses. Given this pivotal role of macrophages in the regulation of the immune system, our research group has primarily focused on the neural regulation of macrophages and their primary regulatory mechanisms, the production of inflammatory cytokines.

4.1. Sympathetic nerves, norepinephrine, and the regulation of macrophages

Norepinephrine (NE) is the primary transmitter released from sympathetic nerve terminals. *In vitro* experiments with macrophages harvested from spleen and lymph nodes have consistently shown that via β-adrenergic receptors, NE can dramatically inhibit the production and secretion of TNF-α in response to LPS [13,14]. Less consistent results have been observed for IL-1 production, but norepinephrine is generally regarded as inhibitory for this cytokine [15]. Both inhibitory and facilitatory effects of norepinephrine and β-adrenergic agonists on IL-6 production have been noted, and the direction of the IL-6 response to NE may depend upon the concurrent presence or absence of LPS [16]. These same *in vitro* studies have also shown that activation of α-adrenergic receptors with specific agonists exerts a stimulatory effect on TNF-α production by macrophages in response to LPS. However, as will be summarized below, *in vivo* activation of the sympathetic nervous system by either stress or central inflammatory stimuli typically inhibits splenic macrophage function, indicating a dominance of β-adrenergic mechanisms influencing splenic macrophages.

In the absence of LPS, inflammatory cytokine production by macrophages are typically minimal or nondetectable. However, our dose-response studies have shown that i.v. administration of a low dose of LPS produced intermediate and easily detectable levels of inflammatory cytokines in the spleen and plasma [17]. High doses of LPS produced maximal levels of cytokine mRNA and even higher doses failed to produce any further increases. The use of the appropriate dose of LPS has therefore permitted us to detect experimentally induced decreases [18] as well as increases in the production of splenic cytokines [19]. Double-labeling immunocytochemistry has verified that splenic macrophages are the primary source of splenic cytokine production during the first few hours after an endotoxin challenge.

4.1.1. Central inflammatory stimuli

We have utilized two model systems to test the effects of the sympathetic nervous system on macrophage function. First, we and others have shown that intracranial injections of cytokines and related inflammatory mediators (prostaglandins) activate the sympathetic nervous system. This has been demonstrated by increased turnover rate for NE in the spleen [20] and increased splenic nerve electrical activity [21,22]. Some 10 years ago we reported that intracranial injections of IL-1β two hours prior to harvesting splenic macrophages, produced a suppression in the *in vitro* production of IL-1β in response to LPS [23]. Significantly, cutting the sympathetic nerve supply to the spleen prior to the central injection of IL-1β completely abrogated the suppression in IL-1β production by splenic macrophage in response to LPS. More recently we have shown that central injections of prostaglandin (PGE_2) produced a rapid increase in splenic sympathetic nerve activity [22] and produced a dramatic suppression in the *in vivo* production of TNF-α mRNA and protein in the spleen of animals given an i.v. injection of LPS [18]. As shown by us earlier with central injections of IL-1β cutting the splenic nerve prior to central injections of PGE_2 attenuated the suppressive effects of central injections of PGE_2 on splenic TNF-α production [24].

4.1.2. Effects of stress on splenic macrophages

Activation of the sympathetic nervous system is induced by a variety of stressful stimuli (footshock, restraint and cold water immersion, for example). We have utilized a brief (15 minute) period of intermittent footshock to examine further the sympathetic regulation of splenic macrophage function in rats [17]. We examined both splenic and plasma levels of inflammatory cytokines in response to i.v. injection of LPS administered immediately before or following the 15 minute stress period. In comparison to non-stressed controls, both splenic and plasma levels of TNF-α were dramatically suppressed in the stressed animals [17]. In subsequent studies, we found that the effects of stress on splenic TNF-α production were still present in animals that had been adrenalectomized, indicating that the inhibitory effects of stress were independent of immunosuppressive adrenal hormones. Finally, we found that cutting the splenic nerve in adrenalectomized animals abrogated the immunosuppressive effects of stress on TNF-α production induced by systemic LPS [17]. Concurrent measurements of IL-1β mRNA and protein in these experiments indicated that changes in IL-1β production followed the same pattern as found for TNF-α but the changes were much less dramatic. Interestingly, the effects of stress on splenic IL-6 production were entirely dependent upon whether the animals were stressed before receiving an i.v. injection of LPS, or immediately following the injection. If LPS was injected immediately prior to the 15 minute stress test, IL-6 production was suppressed the same as TNF-α and the suppression was abrogated by cutting the splenic nerve; however, if the rats were injected with LPS at the end of the stress test, the stressed animals showed a significant *increase* in splenic and plasma levels of IL-6, relative to non-stressed controls. This effects was entirely specific to IL-6 production, as the stress induced suppression of both TNF-α and IL-1β was observed irrespective of whether the LPS was administered immediately before or after the 15 minutes of intermittent footshock. Our results regarding IL-6 production are consistent with other reports that both stress and central inflammatory stimuli can increase the production of IL-6 in the complete absence of a systemic immune challenge [25,26].

4.1.3. Sympathetic nerves and cutaneous inflammation

The vasoconstrictor function of sympathetic postganglionic nerve (SPGN) fibers in the skin is well known regarding the regulation of local blood flow via NE release which acts on α_2-

adrenergic receptors on vascular endothelial cells. However, the role of SPGN fibers in regulating cutaneous neurogenic inflammation has received little attention despite these fibers being clearly involved in synovial inflammation. Sympathectomy has been shown to reduce the plasma extravasation (PE) resulting from intradermal injection of substance P, bradykinin or histamine and decrease the cellular and protein accumulation in cold injury-induced blisters [27,28]. In contrast, no effect of sympathectomy on PE was observed in response to nerve stimulation, skin painting with mustard oil, or injection with carrageenan [29]. However, data from cutaneous blood flow studies in the cat are able to demonstrate a vasodilator mechanism which is mediated by stimulation of the sympathetic preganglionic nerves which synapse with sympathetic postganglionic neurons [30]. This effect was present after blocking vasoconstrictor mechanisms with α-adrenergic antagonists and persisted after muscarinic receptor blockade. It would appear, therefore, that the sympathetic innervation of skin can potentially enhance inflammation through a vasodilator mechanism although the neurotransmitter is unknown. Therefore, sympathetic innervation may add an additional factor to the local regulation of the inflammatory responses to cutaneous antigen challenge.

4.2. Sensory nerves, neuropeptides, and the regulation of cutaneous inflammation

The process of neurogenic inflammation is well described in the skin. Stimulation of small, unmyelinated sensory nerves, either by topical application of capsaicin or electrical stimulation of the nerves directly, results in local release of neurotransmitters such as substance P and calcitonin gene-related peptide (CGRP). These peptides are potent vasodilators and can act on postcapillary venules to increase PE [31,32]. These responses are reproduced by intradermal injection of these peptides [33,31]. A number of intermediary agents have been proposed including histamine, nitric oxide, prostaglandins, and others although no definitive agent has been identified aside from the sensory nerve fibers themselves. Injection of sensory neuropeptides including substance P and CGRP also induces the expression of adhesion molecules and the infiltration of leukocytes [34,35]. Clearly, a distinct role for sensory innervation in regulating cutaneous inflammation can be proposed in that these factors can exert potent effects on the local vasculature and leukocyte homing mechanisms as well as the function of resident and migrating immune cells. A primary effect which is proinflammatory in nature appears to be the main contribution of sensory nerves to cutaneous aspects of host resistance.

4.2.1. Effects on macrophages and inflammatory cytokine production

Studies which have examined the *in vitro* effects of sensory neuropeptides on monocyte/macrophage function have been available for some time and have been reviewed elsewhere [36]. Of primary interest, substance P has been reported to induce the production of IL-1β, TNF-α, IL-6 and, more recently, IL-12 from monocytes and macrophages [37–40]. Neuropeptide-immunopositive nerve fibers (including substance P and CGRP) form an extensive network in the skin and are closely associated with blood vessels and immune cells [41–44]. This provides important anatomical data which supports the role of sensory neuropeptides in the regulation of cutaneous inflammatory and immune responses. Further, skin sections treated with substance P release IL-1β and TNF-α into the culture supernatant [45,46] and depletion of substance P nerve fibers or blockade of substance P receptors reduces the cytokine response to such stimuli as LPS [47]. In contrast to the proinflammatory profile of substance P, CGRP has been reported to suppress the *in vitro* LPS-induced macrophage production of TNF-α, IL-1β and IL-12 as well as the antigen presentation function of macrophages including the expression of the costimulatory molecule B-7.2 [48–50]. Further, somatostatin reduces cutaneous edema, cellular infiltration, and TNF-α production following local injection of carrageenan [51]. Unfortunately these studies did not identify the cell type which was responsible for the cytokine production. It would appear that there are no available studies which have manipulated sensory factors and assayed *in vivo* outcome measures which were subsequently confirmed to be specific functions of macrophage. Until these systematic studies are completed, the precise impact of sensory innervation on *in vivo* macrophage function (e.g. proinflammatory cytokine production) will remain uncertain. Given the importance of macrophage in bridging the innate and adaptive immune systems following antigen challenge, these studies should be well worth completing.

Sensory and sympathetic nerves play several potential roles in determining the immediate response to injury or infection, including vasodilation, adhesion molecule expression and leukocyte infiltration. The function of both resident and infiltrating immune cells will be influenced by the local environment including the tissue levels of neurotransmitters. This influence on responding cells may persist in the immune effector response to the extent that ensuing immune responses may be determined, at least in part, by the local factors where antigen is initially encountered. It is well known that adjuvants, including bacterial components, generally enhance immune responses. Several clinical examples of altered cutaneous responses after nerve injury are available which demonstrate that neural factors are a relevant mechanism regarding cutaneous inflammation and immunity [52–54]. Specifically, the selective loss of histamine-induced flare responses below the neurological level of spinal cord injury (SCI) in humans is a strong indicator of the relevance of neural factors in regulating cutaneous responses [55]. Our lab has pursued the investigation of cutaneous inflammatory and immune responses in an animal model of SCI. Using a complete SCI at the T1-T2 level to remove descending sympathetic and sensory outflow, we have demonstrated that the production of inflammatory cytokine mRNA following intradermal injection of turpentine is dramatically reduced in skin below, but not above, the level of the SCI. Preliminary data show that the cytokine mRNA response to intradermal injection of heat killed *Staphylococcus Aureus* bacteria is similarly effected by SCI. These studies indicate that descending neural factors regulate cutaneous inflammatory responses. Further experiments will be needed to determine whether the SCI effect is due to removal of descending sympathetic input or loss of descending modulation of sensory factors, or a combination of both, as well as the precise cell populations effected by the SCI.

5. NEURAL REGULATION OF ADAPTIVE IMMUNITY

5.1. Sympathetic nerves, norepinephrine, and the regulation T cells and B cells

Immunization with a soluble protein antigen has been reported to increase efferent sympathetic activity in immune organs [56]. Antigen administration has also been shown to increase the density of nerve fibers in immune organs [57]. With regard to cell-mediated immunity, the relatively limited literature suggest that NE is predominately inhibitory [58,59]. Although the regulation of humoral immune responses by NE and the sympathetic nervous system have been extensively examined, there is little consensus as to whether the sympathetic nervous system inhibits or facilitates antibody production. However, as reviewed earlier, there is general agreement that the sympathetic nervous system and NE inhibit cytokines that regulate both innate (TNF-α) and cell-mediated (IL-12) immune responses [60]. Thus, NE, as well as glucocorticoids, tend to shift adaptive immunity towards Th2 humoral responses, perhaps due in part to the mutually inhibitory interactions between Th1 and Th2-type immune responses [61]. Nonetheless, sympathetic nerve fibers have the potential of being a more important mediator of humoral immune responses than the adrenal gland [3,62–64].

Evidence that NE and the sympathetic nervous system facilitates humoral immune responses includes an early report by Kasahara *et al.* [61], and subsequently verified by Hall *et al.*, [65], in which mice treated with the NE depleting drug, 6-OHDA, showed an impaired splenic plaque forming cell response (PFC) to immunization with sheep red blood cells (SRBC) [61]. We have shown that although cutting the splenic nerve does not alter the PFC response to SRBC, the ability of IL-2 injections to potentiate the splenic PFC response to SRBC was dependent upon intact splenic innervation. Additional pharmacological studies showed that the immunoenhancing effects of IL-2 was mediated via β-adrenergic receptors [66]. *In vitro* studies have shown that NE increased LPS-induced antibody production by B cells, and this effect was independent of T cells and macrophages [67]. Also, NE has been shown to enhance LPS-induced B cell proliferation [68]. Recently it was shown in dopamine β-hydroxylase knockout mice, which are unable to synthesize NE, that these animals are more susceptible to bacterial infection and have impaired T cell function, relative to littermate controls [69]. The most compelling evidence that NE facilitates humoral immune responses comes from the work of Sanders and colleagues. They demonstrated that NE acts upon both T cells and B cells and found that when NE was added at the time of *in vitro* immunization, the PFC response to SRBC was increased [70]. This effect was blocked by a β-adrenergic receptor antagonist and reproduced by a β_2-adrenergic receptor agonist [71]. Significantly, Sanders found that only Th1 cells had β_2-adrenergic receptors ain [72,73]. The binding studies were supported further by the demonstration that a β_2-adrenergic agonist decreased IL-2 and IFN-γ expression in Th1 clones but did not affect cytokine production in Th2 cells. Since the T cells were exposed to NE prior to reconstitution with B cells, this verified the effects of the adrenergic agonist was on T cells. Importantly, these results have been verified *in vivo*. SCID (severe combined immunodeficiency) mice were reconstituted with antigen-specific T and B cells. Since the T cells were Th2 clones, only the B cells had adrenergic receptors [74]. Mice that were given a chemical sympathectomy with 6-OHDA prior to reconstitution demonstrated a decreased primary IgM response and decreased primary and secondary IgG responses. This decreased antibody response could also be demonstrated in animals treated with a β-adrenergic receptor blocker and partially restored with a β_2-adrenergic receptor agonist. They also found that B7-2, a T cell co-stimulator molecule on B cells, was increased in B cells treated with NE. These results clearly show that NE and the sympathetic nervous system can regulate both B cell and T cell function.

There are also a number of studies which indicate that NE and the sympathetic nervous system can inhibit antibody production and was first proposed by Besedovsky in 1979 [3]. He reported that surgical denervation of the spleen or 6-OHDA treatment increased the splenic PFC response to SRBCs *in vitro*. In contrast, we have reported that cutting the splenic nerve did not alter the splenic PFC response to SRBC, but rather blocked the inhibitory effects of stress on the PFC response [75]. Also, Besedovsky noted that the splenic content of NE decreased during the peak in the PFC response, and importantly, the levels of splenic NE were inversely correlated with the magnitude of the antibody response [6]. Interestingly, adrenalectomy (ADX) was found to have no effect by itself on the PFC response, but ADX augmented the facilitatory effects of chemical sympathectomy on the PFC response [6]. Our lab has also provided similar evidence for the inhibition of the humoral immunity by the sympathetic nervous system. Green-Johnson *et al.* [76] showed that epilepsy prone mice, which have a significantly higher splenic NE content than epilepsy resistant control mice, also showed decreased *in vivo* IgG antibody and reduced PFC response to SRBC, relative to controls. Consistent with this, administration of a β_2-adrenergic agonist to normal mice during the late stages of immunization decreased anti-SRBC antibody titers relative to vehicle injected controls. Interestingly, when the T cells were removed from the spleen of the epilepsy prone mice and examined *in vitro*, they responded the same as T cells from the control mice. These results indicate that it was the differential *in vivo* environment that generated the altered immune response of the epilepsy prone mice and reinforce the fact that *in vitro* immune responses do not accurately reflect the natural microenvironment for these cells. Although the exact mechanism for the inhibitory action of NE on the antibody response requires further analysis, NE has been shown to decrease MHC-II expression, a requisite molecule for antigen presentation to T cells [77,78].

Why have some labs found that NE facilitates humoral immunity while others have found that NE is inhibitory? Possible explanations include the timing of NE exposure. It is known that NE will increase intracellular cAMP via β-adrenergic receptors. However, elevated cAMP levels at the right time will increase antibody production and at the wrong time it may be inhibitory [79]. Perhaps related to this is the cell cycle in that NE has been shown to induce more B cell precursors to differentiate, but failed to produce this effect on IgG producing cells. Another possible explanation for these contradictory results involves the use of 6-OHDA to ablate the entire sympathetic nervous system versus the use of selective denervation of specific immune organs. Use of systemic 6-OHDA not only reduces NE content in immune organs, it also reduces NE at the sites of immunization, such as skin, footpads and peritoneal cavity. Thus, reduced NE levels at the site of antigen challenge may impact on the innate immune response, phagocytosis of antigen, cytokine production, antigen presentation, etc. and thereby modify the subsequent specific immune response generated in secondary immune organs which also have reduced levels of NE. At the very least, this complicates interpretation.

5.2. Sensory nerves, neuropeptides, and the regulation of T cells and B cells

Sensory neuropeptides can regulate numerous aspects of *in vitro* immune function including cytokine production and immunoglobulin synthesis [42,80,81]. Several studies have assessed the *in vivo* actions of sensory neuropeptides on the function of the specific immune system. Neonatal capsaicin treatment led to an 80% decrease in the antibody response to subcutaneous antigen [82]. The response was restored by localized substance P delivery coincident with antigen injection. This effect is consistent with the concept that the inflammatory component of neurogenic inflammation is important for recruiting immunocompetent cells to the site of antigen challenge. However, an opposite but consistent effect is seen regarding the delayed-type hypersensitivity/contact hypersensitivity reactions in that these responses are enhanced in capsaicin pretreated animals [83–85]. Enhanced DTH responses are obtained regardless of whether the capsaicin treatment occurs before or after the sensitization stage [83]. Increased DTH responses are seen despite the anticipated reduction in the initial irritant effect (edema, cellular infiltration) of the test reagents (DNFB, oxazolone). Interestingly, the potentiated DTH responses could not be adoptively transferred to non-capsaicin treated recipients; however, adoptive transfer into capsaicin pretreated recipients resulted in enhanced DTH responses [84]. These data indicated that it was the local environment that determined the magnitude of the DTH response, rather than characteristics of the transferred immune cell population. Thus, in an apparently paradoxical manner, capsaicin pretreatment results in enhanced DTH responses in the face of diminished neurogenic inflammation.

While capsaicin treatment may be best associated with depletion of substance P, it similarly reduces CGRP levels [86]. CGRP has been reported to decrease the production of several cytokines including IL-1β, TNF-α, IFN-γ and IL-12 while enhancing the synthesis of IL-10 [48,87,50]. In addition, CGRP-containing sensory nerve fibers are closely associated with Langerhans cells in skin and impair the antigen presenting function of these cells *in vitro* [42]. Consequently, local release of CGRP would be expected to inhibit DTH responses whereas blocking CGRP receptor binding should enhance DTH reactions. Data from several studies are consistent with this hypothesis. Localized injection of CGRP followed by hapten sensitization inhibits the ensuing DTH response due, at least in part, to increased production of IL-10 [88,89]. In fact, anti-IL-10 antibodies were able to abrogate the inhibitory effect of CGRP on antigen presentation [50]. These studies provide useful data with which to interpret the effects of capsaicin on the DTH response. Since CGRP will also be depleted following capsaicin treatment, it would appear that the dominant stimulatory effect of capsaicin treatment on DTH responses arises due to the loss of CGRP and therefore results in enhanced DTH reactions despite the potential for reduced initial inflammation during the DTH process. In keeping with this, lymphocytes from the draining lymph nodes obtained 2–4 hours after initial exposure to DNFB and subsequently maintained in culture exhibit increased production of IL-1β and TNF-α in capsaicin pretreated animals [85]. Thus, the removal of the inhibitory actions of CGRP by capsaicin permits enhancement of the DTH response. The elevated IL-1β and TNF-α cytokine levels would be expected to contribute to the increased DTH responses.

A potential dichotomy between substance P and CGRP actions can be proposed based on the differential effects of these sensory neuropeptides on cytokine production. As stated above, CGRP inhibits Th1-type cytokines and Th1-dominant *in vivo* immune responses (DTH) while substance P has been reported to enhance IL-4 production [87]. These data provide a mechanism with which to interpret the experimental finding stated above in which capsaicin diminished, while substance P enhanced, the humoral immune response to cutaneous antigen challenge [82]. Further, in a Th2-type model of allergic airway inflammation, substance P is expressed in the airway following allergen challenge in sensitized rats and the NK1 and NK2 receptors are expressed in the lungs of these animals [90]. *In vivo* treatment with NK1 or NK2 receptor antagonists reduced the late airway response to allergen but not the initial response indicating involvement of these receptors in immune-mediated allergic processes. It is interesting to speculate that the extent to which a given cutaneous antigen elicits substance P or CGRP release from sensory nerve terminals may influence the Th1/Th2 character of the ensuing immune response. Whether differential release of substance P and CGRP occurs after various cutaneous stimuli has not been explored.

6. ARTHRITIS: A POTENTIAL DISEASE MODEL OF NEURAL-IMMUNE INTERACTIONS?

For some time it has been believed that the nervous system has a role in the pathogenesis of rheumatoid arthritis due to the classical feature of symmetrical, bilateral presentation of inflamed joints. Support for neural involvement in the disease was provided by observations of improved disease status in limbs which were subject to neurological injury such as paralysis following stroke or polio [91,92]. Specifically, sympathetic efferent nerve fibers were implicated based upon the ability of propranolol and regional sympathetic blockade with guanethidine to ameliorate symptoms of arthritis in patients with active disease [93,94]. An additional neural factor, small, unmyelinated sensory afferent fibers were also proposed to contribute to the arthritic inflammatory process since it was known that these fibers are able to mediate the process of neurogenic inflammation [95,96]. Early experiments in animals confirmed a role for both sympathetic efferent and unmyelinated afferent nerve fibers in determining the severity of joint inflammation and erosion.

Studies of innervation and arthritis have most commonly utilized the adjuvant-induced arthritis (AA) model whereby Freund's complete adjuvant is injected into the skin at the base of the tail with arthritic changes in the hindlimb joints readily observed after 21 days. Using this model, systemic blockade of the sympathetic nervous system (sympathectomy) consistently reduces the severity of AA [97,98]. Likewise, depletion of small unmyelinated sensory fibers with the neurotoxin, capsaicin, also reduced the severity of arthritis although not to the same extent as sympathectomy [97,99]. It was also noted that the density of mast cells in joints was directly proportional to the extent of sympathetic and sensory innervation of the joint [100]. Mast cell density was reduced after depletion of either sensory or sympathetic nerves and mast cell density was increased in spontaneously hypertensive rats which also demonstrate an increased severity of AA compared to normotensive controls [97,100]. Further studies indicated that the sympathetic effect was mediated by epinephrine from the adrenal medulla acting on β_2-adrenergic receptors on the sympathetic postganglionic nerve (SPGN) terminals which innervated the joints [98,101].

The precise mechanism by which the innervation of joints alters the arthritic process is not clear but several possibilities can be stated. The first is a direct effect of neural transmitters released in the joint tissue and/or synovial fluid. Inflammation in the rat knee joint, as modeled by bradykinin (BK)-induced PE, is readily altered by neural factors and is a form of neurogenic inflammation. Surprisingly, this aspect of inflammation appears to be mediated primarily through the actions of various neural agents on the terminals of the SPGN fibers as opposed to the neural activity of the SPGN fibers themselves. For example, bradykinin mediates an increase in knee joint PE through release of such proinflammatory agents such as PGE_2, ATP and adenosine from SPGN terminals [102–104]. Depletion of the SPGN terminals with 6-OHDA reduced the action of bradykinin on PE whereas acute decentralization of the SPGN does not block the BK effect [105]. These data indicate that the local effect of BK on the SPGN terminal is responsible for the changes in PE rather than the outflow of centrally mediated sympathetic nerve activity. Further, the addition of exogenous NE inhibits the BK response indicating that BK does not generate an increase in PE via SPGN release of NE [103]. These studies indicate the role of neural-immune interactions in joints; they do not apply to the outcome of AA (joint erosion) as an inverse relationship was observed between PE and arthritis severity [106]. That is, an increase in the initial PE resulted in reduced joint damage while reduced PE lead to greater arthritis severity. The fact that sympathectomy reduces both PE and joint damage suggests that separate processes mediate these two outcome measures. However, these studies do serve to illustrate the complex network of factors which can contribute to any specific outcome measure and underline the importance of *in vivo* studies to ensure that all such factors are intact when studying interactions between physiological systems.

Aside from direct actions on the inflammatory process in the joint, a second route by which sympathetic innervation may impact on arthritis is via the immune system. It has been shown that AA is mediated by a T cell dependent immune response which is characterized by a Th1-type cytokine profile and that inhibiting Th1-type responses or enhancing Th2-type responses attenuates the arthritis severity [107–110]. Importantly, humans with rheumatoid arthritis also display a Th-1 dominated immune profile in the synovial membrane and fluid [111–114]. The Th1 nature of the arthritic response has been observed in other models of arthritis in rats as well as mice [109,115–118]. As stated above, the Th1-type T cells selectively express β_2-adrenergic receptors and the function of the Th1 cells is reduced by the binding of ligand to these receptors [72]. Given these data it can be hypothesized that sympathectomy would exacerbate arthritis since the inhibitory action of the sympathetic neurotransmitter, NE, would be removed from the Th1, proarthritic T cells. In fact, systemic sympathectomy does just the opposite; it reduces the progression of arthritis. However, the work of Lorton and colleagues has shed light on important aspects of the arthritis process. These investigators applied the sympathetic neurotoxin, 6-OHDA, locally to the lymph nodes which drained the adjuvant injection site. Doing so resulted in depletion of the sympathetic innervation of the lymph nodes, spleen and other abdominal organs as well as markedly increasing arthritis severity [119]. In a separate study, Lorton *et al.* compared local versus systemic 6-hydroxydopamine treatment and confirmed that local application of the neurotoxin led to more severe arthritis whereas systemic treatment attenuated the disease [120]. These studies are consistent with our hypothesis that the actions of the sympathetic innervation of immune organs is mediated, at least in part, through the effects of NE on Th1-type T cells and that removal of this effect would increase arthritis severity while enhanced release of NE in lymph organs would reduce the disease. In addition, NE release from SPGN terminals should reduce macrophage production of TNF-α which will remove another pro-Th1 factor from the initiating process of arthritis. The experimental observation that systemic sympathectomy reduces, rather than enhances, arthritis does not preclude these mechanisms due to the widespread effects of systemic sympathectomy on the entire organism including the innervation of immune organs and joints.

What these studies highlight is the need for selective manipulations on the various factors which can contribute to the arthritic process. Important data is likely to be gained from studies which compare the effects of selectively blocking the sympathetic input to the joint tissues versus the draining lymph nodes. We would add that the sympathetic innervation of the skin site where the adjuvant is injected should also be considered. Interestingly, Lorton *et al.* also tested the effect of local application of the sensory neurotoxin, capsaicin, to the draining lymph nodes prior to induction of arthritis [121]. This treatment was successful in reducing substance P levels in the lymph nodes without effecting that of the spleen or sciatic nerve and resulted in diminished arthritis severity. Whether specific neurokinin receptor antagonists can reproduce this effect will be necessary to determine if the primary effect of capsaicin treatment in this model was restricted to substance P or included related peptides such as CGRP. As such, these data do not follow the predicted outcome of capsaicin treatment on Th1-type immune responses as seen in enhanced DTH responses. Unfortunately, studies of this nature using localized manipulations of neural factors are very limited in number. Until the role of innervation is selectively examined at each relevant site (skin, joints, lungs, gastrointestinal tract, immune organs), the interpretation of data arising from systemic alterations of innervation will be complicated and will likely overlook important aspects of the immune response to a particular antigen and the route of antigen presentation.

REFERENCES

1. Selye H. Studies on adaptation. Endocrinology 1937;21:169.
2. Selye H. Stress and disease. Science 1955;122:625–631.
3. Besedovsky HO, Del Rey AE, Sorkin E, Da Prada M, Keller HH. Immunoregulation mediated by the sympathetic nervous system. Cell Immunol 1979;48:346–355.

4. Besedovsky HO, Sorkin E, Felix D, Haas H. Hypothalamic changes during the immune response. Eur J Immunol 1977;7:323–325.
5. Besedovsky HO, Del Rey AE, Sorkin E, Da Prada M, Burri R, Honegger C. The immune response evokes changes in brain noradrenergic neurons. Science 1983;221:564–566.
6. Del Rey AE, Besedovsky HO, Sorkin E, Da Prada M, Bondiolotti GP. Sympathetic immunoregulation: difference between high- and low-responder animals. Am J Physiol 1982;242:R30–R33.
7. Williams JM, Felten DL. Sympathetic innervation of murine thymus and spleen: a comparative histofluorescence study. Anat Rec 1981;199:531–542.
8. Williams JM, Peterson RG, Shea PA, Schmedtje JF, Bauer DC, Felten DL. Sympathetic innervation of murine thymus and spleen: Evidence for a functional link between the nervous and immune systems. Brain Res Bull 1981;6:83–94.
9. Felten DL, Felten SY, Bellinger DL, Carlson SL, Ackerman KD, Madden KS, Olschowki JA, Livnat S. Noradrenergic sympathetic neural interactions with the immune system: structure and function. Immunol Rev 1987;100:225–260.
10. Nance DM, Burns J. Innervation of the spleen: Evidence for an absence of afferent innervation. Brain Behav Immun 1989;3:281–290.
11. Nance DM, Hopkins DA, Bieger D. Re-investigation of the innervation of the thymus gland in mice and rats. Brain Behav Immun 1987;1:134–147.
12. Abbas AK, Lichtman AH, Pober JS. Cellular and molecular immunology, 3rd Edition. Philadelphia : Saunders; 1997.
13. Ignatowski TA, Gallant S, Spengler RN. Temporal regulation by adrenergic receptor stimulation of macrophage (M phi)-derived tumor necrosis factor production post-LPS challenge. J Neuroimmunol 1996;65:107–117.
14. Spengler RN, Chensue SW, Giacherio DA, Blenk N, Kunkel SL. Endogenous norepinephrine regulates tumor necrosis factor-α production from macrophages *in vitro*. J Immunol 1994;152:3024–3031.
15. Yoshimura T, Kurita C, Nagao T, Usami E, Nakao T, Watanabe S, Kobayashi J, Yamazaki F, Tanaka H, Inagaki N, Nagai H. Inhibition of tumor necrosis factor-α and interleukin-1β production by B-adrenoceptor agonists from lipopolysaccharide-stimulated human peripheral blood mononuclear cells. Pharmacology 1997;54:144–152.
16. Straub RH, Herrmann M, Frauenholz T, Berkmiller G, Lang B, Scholmerich J, Falk W. Neuroimmune control of interleukin-6 secretion in the murine spleen. Differential β-adrenergic effects of electrically released endogenous norepineprine under various endotoxin conditions. J Neuroimmunol 1996;71:37–43.
17. Meltzer JC. Neural and endocrine regulation of *in vivo* splenic immune function in the rat. 2000, Ph.D. Thesis, University of Manitoba, Winnipeg, Canada.
18. Pan L, Pylypas S, Sanders V, Greenberg AH, Nance DM. Central injections of prostaglandin activate the hypothalamus and suppress splenic cytokine mRNA production. Soc Neurosci 1997;23:593–624.
19. Vriend CAY, MacNeil BJ, Meltzer JC, Sanders V, Greenberg AH, Nance DM. Central gp120 alters the *in vivo* production of splenic cytokines following LPS. Psychoneuroimmunology Research Society Annual Meeting, June 1997, Boulder, CO
20. Vriend CY, Zuo L, Dyck DG, Nance DM, Greenberg AH. Central administration of interleukin-1β increases norepinephrine turnover in the spleen. Brain Res Bull 1993;31: 39–42.
21. MacNeil BJ, Jansen AH, Greenberg AH, Nance DM. Activation and selectivity of splenic

sympathetic nerve electrical activity response to bacterial endotoxin. Am J Physiol 1996;270:R264–R270.
22. MacNeil BJ, Jansen AH, Janz L, Greenberg AH, Nance DM. Peripheral endotoxin increases splenic sympathetic nerve activity via central prostaglandin synthesis. Am J Physiol 1997;273:R609–R614.
23. Brown R, Li Z, Vriend CY, Nirula R, Janz L, Falk J, Nance DM, Dyck DG, Greenberg AH. Suppression of splenic macrophage interleukin-1 secretion following intracerebroventricular injection of interleukin-1β: Evidence for pituitary-adrenal and sympathetic control. Cell Immunol 1991;132:84–93.
24. Nance DM, MacNeil BJ. Immunoregulation by the sympathetic nervous system. In: Berczi A and Gorczynski RM (eds) Neuroimmune Biology Vol. 1: New Foundation of Biology 2001;121–139.
25. Ando T, Rivier J, Yanaihara H, Arimura A. Peripheral corticotropin-releasing factor mediates the elevation of plasma IL-6 by immobilization stress in rats. Am J Physiol 1998;275:R1461–R1467.
26. Finck BN, Dantzer R, Kelley K, Woods JA, Johnson RW. Central lipopolysaccharide elevates plasma IL-6 concentration by an α-adrenoceptor-mediated mechanism. Am J Physiol 1997;272:R1880–R1887.
27. Gonzalez R, Coderre TJ, Sherbourne CD, Levine JD. Postnatal development of neurogenic inflammation in the rat. Neurosci Lett 1991;127:25–27.
28. Helme RD, Andrews PV. The effect of nerve lesions on the inflammatory response to injury. J Neurosci Res 1985;13:453–459.
29. Donnerer J, Amann R, Lembeck F. Neurogenic and non-neurogenic inflammation in the rat paw following chemical sympathectomy. Neurosci 1991;45:761–765.
30. Kawarai M, Koss MC. Neurogenic cutaneous vasodilation in the cat forepaw. J Auton Nerv Syst 1992;37:39–46.
31. Brain SD, Williams TJ, Tippins JR, Morris HR, MacIntyre I. Calcitonin gene-related peptide is a potent vasodilator. Nature 1985;313:54–56.
32. Lembeck F, Holzer P. Substance P as neurogenic mediator of antidromic vasodilation and neurogenic plasma extravasation. Naunyn-Schmiedeberg's Arch Pharmacol 1979;310: 175–183.
33. Gamse R, Posch M, Saria A, Jancso G. Several mediators appear to interact in neurogenic inflammation. Acta Physiol Hung 1987;69:343–354.
34. Smith CH, Barker JN, Morris RW, MacDonald DM, Lee TH. Neuropeptides induce rapid expression of endothelial cell adhesion molecules and elicit granulocytic infiltration in human skin. J Immunol 1993;151:3274–3282.
35. Iwamoto I, Tomoe S, Tomioka H, Yoshida S. Leukotriene B4 mediates substance P-induced granulocytic infiltration into mouse skin. Comparison with antigen-induced granulocyte infiltration. J Immunol 1993;151:2116–2123.
36. Eglezos A, Andrews PV, Boyd RL, Helme RD. Modulation of the immune response by tachykinins. Immunol Cell Biol 1991;69:285–294.
37. Kimball ES, Persico FJ, Vaught JL. Substance P, neurokinin A and neurokinin B induce generation of IL-1-like activity in P388D1 cells. J Immunol 1988;141:3564–3569.
38. Laurenzi MA, Persson MAA, Dalsgaard C-J, Hægerstrand A. The neuropeptide substance P stimulates production of interleukin 1 in human blood monocytes: Activated cells are preferentially influenced by the neuropeptide. Scand J Immunol 1990;31:529–533.
39. Lotz M, Vaughan JH, Carson DA. Effect of neuropeptides on production of inflammatory

cytokines by human monocytes. Science 1988;241:1218–1221.
40. Kincy-Cain T, Bost KL. Substance P-induced IL-12 production by murine macrophages. J Immunol 1997;158:2334–2339.
41. Hartschuh W, Weihe E, Reinecke M. Peptidergic (neurotensin, VIP, substance P) nerve fibres in the skin. Immunohistochemical evidence of an involvement of neuropeptides in nociception, pruritus and inflammation. Br J Dermatol 1983;109 Suppl 25:14–17.
42. Hosoi J, Murphy GF, Egan CL, Lerner EA, Grabbe S, Asahina A, Granstein RD. Regulation of Langerhans cell function by nerves containing calcitonin gene-related peptide. Nature 1993;363:159–163.
43. Bienenstock J, Tomioka M, Matsuda H, Stead RH, Quinonez G, Simon GT, Coughlin MD, Denburg JA. The role of mast cells in inflammatory processes: evidence for nerve/mast cell interactions. Int Arch Allergy Appl Immunol 1987;82:238–243.
44. Egan CL, Viglione-Schneck MJ, Walsh LJ, Green B, Trojanowski JQ, Whitaker-Menezes D, Murphy GF. Characterization of unmyelinated axons uniting epidermal and dermal immune cells in primate and murine skin. J Cutan Pathol 1998;25:20–29.
45. Okabe T, Hide M, Koro O, Yamamoto S. Substance P induces tumor necrosis factor-alpha release from human skin via mitogen-activated protein kinase. Eur J Pharmacol 2000;398: 309–315.
46. Branchet-Gumila MC, Boisnic S, Le Charpentier Y, Nonotte I, Montastier C, Breton L. Neurogenic modifications induced by substance P in an organ culture of human skin. Skin Pharmacol Appl Skin Physiol 1999;12:211–220
47. Dickerson C, Undem B, Bullock B, Winchurch RA. Neuropeptide regulation of proinflammatory cytokine response. J Leukoc Biol 1998;63:602–602.
48. Liu J, Chen M, Wang X. Calcitonin gene-related peptide inhibits lipopolysaccharide-induced interleukin-12. Immunol 2000;101:61–67.
49. Nong YH, Titus RG, Ribeiro JM, Remold HG. Peptides encoded by the calcitonin gene inhibit macrophage function. J Immunol 1989;143:45–49.
50. Torii H, Hosoi J, Beissert S, Xu S, Fox FE, Asahina A, Takashima A, Rook AH, Granstein RD. Regulation of cytokine expression in macrophages and the Langerhans cell-like line XS52 by calcitonin gene-related peptide. J Leukoc Biol 1997;61:216–223.
51. Karalis K, Mastorakos G, Chrousos GP, Tolis G. Somatostatin analogues suppress the inflammatory reaction *in vivo*. J Clin Invest 1994;93:2000–2006.
52. LeVasseur SA, Gibson SJ, Helme RD. The measurement of capsaicin-sensitive sensory nerve fiber function in elderly patients with pain. Pain 1990;41:19–25.
53. Helme RD, LeVasseur SA, Gibson SJ. RSI revisisted: evidence for psychological and physiological differences from an age, sex and occupation matched control group. Aust N Z J Med 1992;22:23–29.
54. Basile S, Khalil Z, Helme RD. Skin vascular reactivity to the neuropeptide substance P in rats with peripheral neuropathy. Pain 1993;52:217–222.
55. Rebhun J, Botvin J. Histamine flare, a neurovascular response. Ann Allergy 1980;45: 59–62.
56. Tang Y, Kohm AP, Sanders VM, Jones SB. Soluble protein antigen immunization increases efferent sympathetic activation in immune organs. Soc Neurosci 1999;25:942.
57. Yang H, Wang L, Huang CS, Ju G. Plasticity of GAP-43 innervation of the spleen during immune response in the mouse. Neuroimmunomodulation 1998;5:53–60.
58. Leo NA, Callahan TA, Bonneau RH. Peripheral sympathetic denervation alters both the primary and memory cellular immune responses to herpes simplex virus infection.

Neuroimmunomodulation 1998;5:22–35.
59. Kalinichenko VV, Mokyr MB, Graf LH, Jr., Cohen RL, Chambers DA. Norepinephrine-mediated inhibition of antitumor cytotoxic T lymphocyte generation involves a beta adrenergic receptor mechanism and decreased TNF-alpha gene expression. J Immunol 1999;163:2492–2499.
60. Panina Bordignon P, Mazzeo D, Lucia PD, D'Ambrosio D, Lang R, Fabbri L, Self C, Sinigaglia F. Beta$_2$-agonists prevent Th1 development by selective inhibition of interleukin-12. J Clin Invest 1997;100:1513–1519.
61. Kasahara K, Tanaka S, Hamashima Y. Suppressed immune response to T-cell dependent antigen in chemically sympathectomized mice. Res Commun Chem Pathol Pharmacol 1977;18:533–542.
62. Stenzel-Poore MP, Vale WW, Rivier C. Relationship between antigen-induced immune stimulation and activation of the hypothalamic-pituitary-adrenal axis in the rat. Endocrinology 1993;132:1313–1318.
63. Delrue Perollet C, Li KS, Neveu PJ. No change in plasma corticosterone and splenic norepinephrine during humoral and cellular immune responses to sheep erythrocytes in C3H mice. Neuroimmunomodulation 1995;2:36–43.
64. Shanks N, Francis D, Zalcman S, Meaney MJ, Anisman H. Alterations in central catecholamines associated with immune responding in adult and aged mice. Brain Res 1994;666:77–87.
65. Hall NR, McClure JE, Hu SK, Tare NS, Seals CM, Goldstein AL. Effects of 6-hydroxydopamine upon primary and secondary thymus dependent immune responses. Immunopharmacol 1982;5:39–48.
66. Zalcman S, Green-Johnson JM, Murray L, Wan W, Nance DM, Greenberg AH. Interleukin-2-induced enhancement of an antigen-specific IgM plaque-forming cell response is mediated by the sympathetic nervous system. J Pharmacol Exp Ther 1994;271:977–982.
67. Kouassi E, Li YS, Boukhris W, Millet I, Revillard JP. Opposite effect of the catecholamines dopamine and norepinephrine on murine polyclonal B cell activation. Immunopharmacol 1988;16:125–137.
68. Li YS, Kouassi E, Revillard JP. Differential regulation of mouse B cell activation by beta-adrenoceptor stimulation depending on type of mitogens. Immunol 1990;69:367–372.
69. Alaniz RC, Thomas SA, Perez-Melgosa M, Mueller K, Farr AG, Palmiter RD, Wilson CB. Dopamine ß-hydroxylase deficiency impairs cellular immunity. Proc Natl Acad Sci 2001;96:2274–2278.
70. Sanders VM, Munson AE. Beta adrenoceptor mediation of the enhancing effect of norepinephrine on the murine primary antibody response *in vitro*. J Pharmacol Exp Ther 1984;230:183–192.
71. Sanders VM, Munson AE. Kinetics of the enhancing effect produced by norepinephrine and terbutaline on the murine primary antibody response *in vitro*. J Pharmacol Exp Ther 1984;231:527–531.
72. Sanders VM, Baker RA, Ramer-Quinn DS, Kasprowicz DJ, Fuchs BA, Street NE. Differential expression of the β-2-adrenergic receptor by Th1 and Th2 clones: implications for cytokine production and B cell help. J Immunol 1997;158:4200–4210.
73. Ramer-Quinn DS, Baker RA, Sanders VM. Activated T helper 1 and T helper 2 cells differentially express the β-2-adrenergic receptor. J Immunol 1997;159:4857–4867.
74. Kohm AP, Sanders VM. Suppression of antigen-specific Th2 cell-dependent IgM and IgG1 production following norepinephrine depletion *in vivo*. J Immunol 1999;162:5299–5308.

75. Wan W, Vriend CY, Wetmore L, Gartner JG, Greenberg AH, Nance DM. The effects of stress on splenic immune function are mediated by the splenic nerve. Brain Res Bull 1993;30:101–105.
76. Green-Johnson JM, Zalcman S, Vriend CAY, Dolina S, Nance DM, Greenberg AH. Role of norepinephrine in suppressed IgG production in epilepsy- prone mice. Life Sci 1996;59: 1121–1132.
77. Frohman EM, Vayuvegula B, Gupta S, Van den Noort S. Norepinephrine inhibits gamma-interferon-induced major histocompatibility class II (Ia) antigen expression on cultured astocytes via beta-2-adrenergic signal transduction mechanisms. Proc Natl Acad Sci 1988;85:1292–1296.
78. Frohman EM, Vayuvegula B, Van den Noort S, Gupta S. Norepinephrine inhibits gamma-interferon-induced MHC class II (Ia) antigen expression on cultured brain astrocytes. J Neuroimmunol 1988;17:89–101.
79. Sanders VM, Munson AE. Norepinephrine and the antibody response. Pharmacol Rev 1985;37:229–248.
80. Scicchitano R, Bienenstock J, Stanisz AM. *In vivo* immunomodulation by the neuropeptide substance P. Immunol 1988;63:733–735.
81. Stanisz AM, Befus D, Bienenstock J. Differential effects of vasoactive intestinal peptide, substance P, and somatostatin on immunoglobulin synthesis and proliferations by lymphocytes from Peyer's patches, mesenteric lymph nodes, and spleen. J Immunol 1986;136:152–156.
82. Helme RD, Eglezos A, Dandie GW, Andrews PV, Boyd RL. The effect of substance P on the regional lymph node antibody response to antigenic stimulation in capsaicin-pretreated rats. J Immunol 1987;139:3470–3473.
83. Nilsson G, Ahlstedt S. Increased delayed-type hypersensitivity reaction in rats neuromanipulated with capsaicin. Int Arch Allergy Appl Immunol 1989;90:256–260.
84. Girolomoni G, Tigelaar RE. Capsaicin-sensitive primary sensitive neurons are potent modulators of murine delayed-type hypersensitivity reactions. J Immunol 1990;145: 1105–1112.
85. Veronesi B, Williams WC, Smialowicz RJ, Sailstad DM, Doerfler D, Selgrade MJK. Neuropeptide denervation alters both the elicitation and the induction phases of contact hypersensitivity in mice. Toxicol Appl Immunopharm 1998;153:243–249
86. Geppetti P, Frilli S, Renzi D, Santicioli P, Maggi CA, Theodorsson E, Fanciullacci M. Distribution of calcitonin gene-related peptide-like immunoreactivity in various rat tissues: correlation with supstance P and other tachykinins and sensitivity to capsaicin. Regul Pept 1988;23:289–298.
87. Kawamura N, Tamura H, Obana S, Wenner M, Ishikawa T, Nakata A, Yamamoto H. Differential effects of neuropeptides on cytokine production by mouse helper T cell subsets. Neuroimmunomodulation 1998;5:9–15.
88. Asahina A, Hosoi J, Beissert S, Stratigos A, Granstein RD. Inhibition of the induciton of delayed-type and contact hypersensitivity by calcitonin gene-related peptide. J Immunol 1995;154:3056–3061.
89. Kitazawa T, Streilein JW. Hapten-specific tolerance promoted by calcitonin gene-related peptide. J Invest Dermatol 2000;115:942–948.
90. Maghni K, Taha R, Afif W, Hamid Q, Martin JG. Dichotomy between neurokinin receptor actions in modulating allergic airway responses in an animal model of helper T cell type 2 cytokine-associated inflammation. Am J Resp Crit Care Med 2000;162:1068–1074.

91. Bland JH, Eddy WM. Hemiplegia and rheumatoid hemiarthritis. Arthritis Rheum 1968;11: 72–80.
92. Ueno Y, Sawada K, Imura H. Protective effect of neural lesion on rheumatoid arthritis. Arthritis Rheum 1983;26:118.
93. Kaplan R, Robinson CA, Scavulli JF, Vaughan JH. Propranolol and the treatment of rheumatoid arthritis. Arthritis Rheum 1980;23:253–255.
94. Levine JD, Fye K, Heller PH, Basbaum AI, Whiting-O'Keefe Q. Clinical response to regional intravenous guanethidine in patients with rheumatoid arthritis. J Rheumatol 1986;13:1040–1043.
95. Jancso N, Jancso-Gabor A, Szolcsanyi J. Direct evidence for neurogenic inflammation and its prevention by denervation and by pretreatment with capsaicin. Br J Pharmacol 1967;31: 138–151.
96. Jancso G, Kiraly E, Jancso-Gabor A. Pharmacologically induced selective degeneration of chemosensitive primary sensory neurons. Nature 1977;270:741–743.
97. Levine JD, Dardick SJ, Roizen MF, Helms C, Basbaum AI. Contribution of sensory afferents and sympathetic efferents to joint injury in experimental arthritis. J Neurosci 1986;6:3423–3429.
98. Levine JD, Coderre TJ, Helms C, Basbaum AI. β_2-adrenergic mechanisms in experimental arthritis. Proc Natl Acad Sci 1988;85:4553–4556.
99. Colpaert FC, Donnerer J, Lembeck F. Effects of capsaicin on inflammation and on the substance P content of nervous tissues in rats with adjuvant arthritis. Life Sci 1983;32: 1827–1834.
100. Levine JD, Coderre TJ, Covinsky K, Basbaum AI. Neural influences on synovial mast cell density in the rat. J Neurosci Res 1990;26:301–307
101. Coderre TJ, Basbaum AI, Dallman MF, Helms C, Levine JD. Epinephrine exacerbates arthritis by an action at presynaptic β2-adrenoceptors. Neurosci 1990;34:521–523.
102. Coderre TJ, Basbaum AI, Levine JD. Neural control of vascular permeability: Interactions between primary afferents, mast cells, and sympathetic efferents. J Neurophsyiol 1989;62: 48–58.
103. Green PG, Luo J, Heller PH, Levine JD. Modulation of bradykinin-induced plasma extravasation in the rat knee joint by sympathetic co-transmitters. Neurosci 1993;52: 451–458.
104. Green PG, Basbaum AI, Helms C, Levine JD. Purinergic regulation of bradykinin-induced plasma extravasation and adjuvant-induced arthritis. Proc Natl Acad Sci 1991;88:4162–4165.
105. Green PG, Miao FJP, Jänig W, Levine JD. Negative feedback neuroendocrine control of the inflammatory response in rats. J Neurosci 1995;15:4678–4686.
106. Coderre TJ, Chan AK, Helms C, Basbaum AI, Levine JD. Increasing sympathetic nerve terminal-dependent extravasation correlates with decreased arthritic joint injury in rats. Neurosci 1991;40:185–189.
107. Ohta Y, Yamane M, Sohda T, Makino H. TAK-603 selectively suppresses Th1-type cytokine production and inhibits the progression of adjuvant-induced arthritis. Immunol 1997;92:75–83.
108. Svelander L, Mussener A, Erlandsson-Harris H, Kleinau S. Polyclonal Th1 cells transfer oil induced arthritis. Immunol 1997;91:260–265.
109. Yoshino S, Murata Y, Ohsawa M. Successful induction of adjuvant arthritis in mice by treatment with a monoclonal antibody against IL-4. J Immunol 1998;161:6904–6908.

110. Zhang L, Mia MY, Zheng CL, Hossain MA, Yamasaki F, Tokunaga O, Kohashi O. The preventive effects of incomplete Freund's adjuvant and other vehicles on the development of adjuvant-induced arthritis in Lewis rats. Immunol 1999;98:267–272.
111. Yudoh K, Matsuno H, Nakasawa F, Yonezawa T, Kimura T. Reduced expression of the regulatory CD4+ T cell subset is related to Th1/Th2 balance and disease severity in rheumatoid arthritis. Arthritis Rheum 2000;43:617–627.
112. Gracie JA, Forsey RJ, Chan WL, Gilmour A, Leung BP, Greer MR, Kennedy K, Carter R, Wei XQ, Xu D, Field M, Liew FY, McInnes IB. A proinflammatory role for IL-18 in rheumatoid arthritis. J Clin Invest 1999;104:1393–1401.
113. Yin Z, Siegert S, Neure L, Grolms M, Liu L, Eggens U, Radbruch A, Braun J, Sieper J. The elevated ratio of interferon gamma-/interleukin-4-positive T cells found in synovial fluid and synovial membrane of rheumatoid arthritis patients can be changed by interleukin-4 but not by interleukin-10 or transforming growth factor-beta. Rheumatology 1999;38: 1058–1067.
114. Steiner G, Tohidast-Akrad M, Witzmann G, Vesely M, Studnicka-Benke A, Gal A, Kunaver M, Zenz P, Smolen JS. Cytokine production by synovial T cells in rheumatoid arthritis. Rheumatology 1999;38:202–213.
115. Hollo K, Glant TT, Garzo M, Finnegan A, Mikecz K, Buzas E. Complex pattern of Th1 and Th2 activation with a preferential increase of autoreactive Th1 cells in BALB/c mice with proteoglycan(aggrecan)-induced arthritis. Clin Exp Immunol 2000;120:167–173.
116. Mattsson L, Lorentzen JC, Svelander L, Bucht A, Nyman U, Klareskog L, Larsson P. Immunization with alum-collagen II complex suppresses the development of collagen-induced arthritis in rats by deviating the immune response. Scand J Immunol 1997;46: 619–624.
117. Mussener A, Lorentzen JC, Kleinau S, Klareskog L. Altered Th1/Th2 balance associated with non-major histocompatibility complex genes in collagen-induced arthritis in resistant and non-resistant rat strains. Eur J Immunol 1997;27:695–699.
118. Caspi RR, Silver PB, Chan CC, Sun B, Agarwal RK, Wells J, Oddo S, Fujino Y, Najafian F, Wilder RL. Genetic susceptibility to experimental autoimmune uveoretinitis in the rat is associated with an elevated Th1 response. J Immunol 1996;157:2668–2675.
119. Lorton D, Bellinger DL, Duclos M, Felten SY, Felten DL. Application of 6-hydroxydopamine into the fatpads surrounding the draining lymph nodes exacerbates adjuvant-induced arthritis. J Neuroimmunol 1996;64:103–113.
120. Lorton D, Lubahn C, Klein N, Schaller J, Bellinger DL. Dual role for noradrenergic innervation of lymphoid tissue and arthritic joints in adjuvant-induced arthritis. Brain Behav Immun 1999;13:315–334.
121. Lorton D, Lubahn C, Engan C, Schaller J, Felten DL, Bellinger DL. Local application of capsaicin into the draining lymph nodes attenuates expression of adjuvant-induced arthritis. Neuroimmunomodulation 2000;7:115–125.

B. PATHOPHYSIOLOGY

The Immune-Neuroendocrine Circuitry
Edited by I. Berczi and A. Szentivanyi

Neurogenic inflammation in the airways

PETER J. BARNES

National Heart & Lung Institute, Imperial College, Dovehouse Street, London SW3 SLY, U.K.

ABSTRACT

Release of neuropeptides, including tachykinins and calcitonin gene-related peptide, from sensory nerves via and axon or local reflex may have inflammatory effects in the airways. This neurogenic inflammation may be initiated by activation of sensory nerves by inflammatory mediators and irritants. Neurogenic inflammation is well developed in rodents and may contribute to the inflammatory response to allergens, infections and irritants in animal models. However, the role of neurogenic inflammation in airway inflammatory diseases, such as asthma and COPD is still uncertain, as there is still little direct evidence for the involvement of sensory neuropeptides in human airways. Initial clinical studies using strategies to block neurogenic inflammation have not been encouraging, but it is important to study more severe forms of airway disease in more prolonged studies in the future to explore the role of neurogenic inflammation.

1. INTRODUCTION

There is a close interaction between airway nerves and inflammation. Many inflammatory mediators may modulate cholinergic and sensory nerves in the airways through the activation of receptors on nerve terminals [1]. However, sensory nerves in turn may also amplify inflammation in the airways through the release of neurotransmitters. This neurogenic inflammation is described in many organs, and has been documented in the upper and lower respiratory tract in several species [2,3]. The idea that sensory nerves may amplify and spread the inflammatory response has attracted considerable attention, as it may be contributory to the inflammation in airway disease, such as asthma and chronic obstructive pulmonary disease (COPD). Neurogenic inflammation may also be involved in allergic rhinitis [4].

1.1. Neuropeptides and airway inflammation

Airway sensory nerves express several neuropeptides, which might contribute to the inflammatory response. Although classically neuropeptides are released from autonomic and sensory nerves, there is increasing evidence that these peptides may be synthesised and released from inflammatory and non-neural structural cells, particularly in disease. Inflammatory cytokines may increase the expression of neuropeptide genes in inflammatory cells, so that they become a major source of the neuropeptide at the inflammatory site. Neuropeptides have multiple inflammatory and immune effects on the airways, thereby intensifying the ongoing inflammation. In turn inflammatory mediators may amplify or sometimes dampen neuropeptide effects. Inflammatory mediators may increase the release of neuropeptides from sensory and other nerves, may increase the expression of neuropeptide genes in neural and inflammatory cells, may increase the expression of neuropeptide receptors and may decrease the degradation of neuropeptides.

2. TACHYKININS

Substance P (SP) and neurokinin A (NKA), but not neurokinin B, are localised to unmyelinated sensory nerves (C-fibres) in the airways of several species [5]. SP-immunoreactive nerves are abundant in rodent airways, but are sparse in human airways, however. Tachykinins are also be expressed human macrophages, which also express tachykinin receptors [6].

2.1. Effects on airways

Tachykinins have many effects on airway functions that are mediated via NK_1-receptors (preferentially activated by SP) and NK_2-receptors (activated by NKA). Tachykinins constrict smooth muscle of human airways *in vitro* via NK_2-receptors. The contractile response to NKA is significantly greater in smaller human bronchi than in more proximal airways, indicating that tachykinins may have a more important constrictor effect on more peripheral airways, whereas cholinergic constriction tends to be more pronounced in proximal airways. This is consistent with the autoradiographic distribution of tachykinin receptors, which are distributed to small and large airways. NKA causes bronchoconstriction after both intravenous and inhaled administration in asthmatic subjects. Mechanical removal of airway epithelium potentiates the bronchoconstrictor response to tachykinins, largely because the neutral endopeptidase (NEP), which is a key enzyme in the degradation of tachykinins in airways is strongly expressed on epithelial cells [7].

SP stimulates mucus secretion from submucosal glands in human airways *in vitro* and is a potent stimulant to goblet cell secretion in guinea pig airways via activation of NK_1-receptors. NK_1-receptors also mediate the increased plasma exudation and the vasodilator response to tachykinins. Tachykinins may also interact with inflammatory and immune cells, although whether this is of pathophysiological significance is still uncertain. SP degranulates skin mast cells and eosinophils through a non-receptor-mediated mechanism. Tachykinins also enhance eosinophil chemotaxis. Tachykinins may activate alveolar macrophages and monocytes to release inflammatory cytokines, such as interleukin (IL)-6.

SP stimulates proliferation of blood vessels (angiogenesis) and may therefore be involved in the new vessel formation that is found in asthmatic airways. SP and NKA also stimulate the proliferation and chemotaxis of human lung fibroblasts, suggesting that tachykinins may contribute to the fibrotic process in chronic asthma [8]. Tachykinins also enhance cholinergic neurotransmission by facilitating acetylcholine release at cholinergic nerve terminals and by enhancing ganglionic transmission [9].

A third tachykinin, NKB, has not been detected in lung, but its preferred receptors, NK_3-receptors appear to mediate some airway effects [10,11].

NK_1 and NK_2 receptors are widely expressed in human airways, including submucosal glands, airway vessels, airway smooth muscle and some inflammatory cells [12,13].

2.2. Metabolism

Tachykinins are subject to degradation by at least two enzymes, angiotensin converting enzyme (ACE) and NEP [14]. ACE is predominantly localised to vascular endothelial cells and therefore degrades intravascular peptides, whereas NEP is important for degrading tachykinins in the airways. The activity of NEP may therefore determine tachykinin responsiveness in the airways. Inhibition of NEP by phosphoramidon or thiorphan markedly potentiates bronchoconstriction *in vitro* in animal and human airways and after inhalation in vivo [15,16]. The activity of NEP is reduced by mechanical removal of the epithelium, some virus infections, cigarette smoke and hypertonic saline. Several of the stimuli known to induce bronchoconstrictor responses in asthmatic patients have been found to reduce the activity of airway NEP.

3. CALCITONIN GENE-RELATED PEPTIDE

CGRP-immunoreactive nerves are abundant in the respiratory tract of several species and is co-stored and co-localised with SP in afferent nerves. CGRP is a potent vasodilator, which has long lasting effects and potently dilates bronchial vessels *in vitro* and *in vivo*. Receptor mapping studies have demonstrated that CGRP receptors are localised predominantly to bronchial vessels rather than to smooth muscle or epithelium in human airways [17]. CGRP may be an important mediator of arterial vasodilatation and increased blood flow in response to sensory nerve stimulation in the bronchi. And might mediate hyperaemia in asthma. CGRP has variable effects on airway smooth muscle tone and appears to act indirectly through the release of other constrictors, such as endothelin. Like tachykinins, CGRP is chemotactic for eosinophils [18].

4. OTHER SENSORY NEUROPEPTIDES

In addition to tachykinins and CGRP, airway sensory nerves contain several other neuropeptides that might participate in neurogenic inflammation. These include galanin, cholecystokinin octapeptide, and enkephalins [5]. The recently described peptide secretoneurin, derived from secretogranin II (chromogranin C), is localised to sensory nerves and is a potent attractant of eosinophils [18], but its role in asthma has not been explored.

5. NEUROGENIC INFLAMMATION IN ANIMAL AIRWAYS

In rodents there is now considerable evidence for neurogenic inflammation in the airways due to the antidromic release of neuropeptides from nociceptive nerves or C-fibres via an axon reflex [19,20] and it is possible that it may contribute to the inflammatory response in asthma (Figure 1).

5.1. Excitatory NANC nerves

Excitatory non-adrenergic non-cholinergic (e-NANC) nerves have been demonstrated in the respiratory tract of rodents, particularly guinea pigs, both *in vitro* by electrical field stimulation and *in vivo* by vagus nerve stimulation. The e-NANC bronchoconstrictor response is abolished

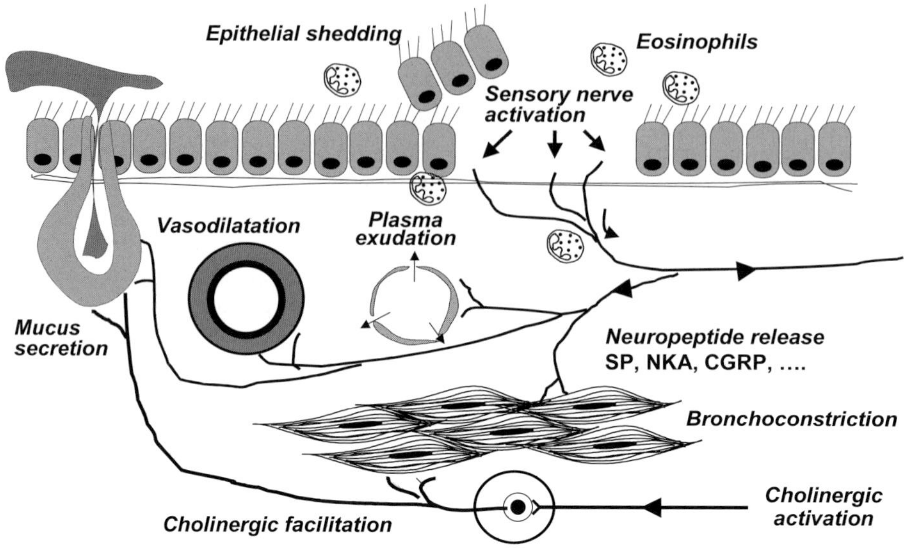

Figure 1. Neurogenic inflammation in asthmatic airways via retrograde release of peptides from sensory nerves via an axon reflex. Substance P (SP) causes vasodilatation, plasma exudation and mucus secretion, whereas neurokinin A (NKA) causes bronchoconstriction and enhanced cholinergic reflexes and calcitonin gene-related peptide (CGRP) vasodilatation.

by capsaicin pre-treatment and by tachykinin antagonists.

5.2. Neurogenic airway inflammation in animal models

There are several lines of evidence that neurogenic inflammation may be important in animal models that may have relevance to asthma. These models have usually been in rodents where tachykinin effects are pronounced and may not be predictive of the role of tachykinins in human airways, however. There are four main experimental approaches which have been used to assess the role of sensory neuropeptides in animal models of asthma; these include studies of depletion with capsaicin depletion, enhancement with inhibitors of NEP, tachykinin receptor antagonists and inhibitors of sensory neuropeptide release.

5.3. Capsaicin depletion studies

Capsaicin pre-treatment to deplete neuropeptides from C-fibres, either in neonatal animals (which results in degeneration of C-fibres) or acute treatment in adult animals (resulting in depletion of sensory neuropeptides). In rat trachea capsaicin pre-treatment inhibits the microvascular leakage induced by irritant gases, such as cigarette smoke and inhibits goblet cell discharge and microvascular leak induced by cigarette smoke in guinea pigs [20]. Capsaicin-sensitive nerves may also contribute to the bronchoconstriction and microvascular leak induced by isocapnic hyperventilation, hypocapnia, inhaled sodium metabisulphite, nebulised hypertonic saline and toluene diisocyanate in rodents. In guinea pigs capsaicin pre-treatment has little or no effect on the acute bronchoconstrictor or plasma exudation response to allergen inhalation in sensitised animals. Administration of capsaicin increases airway responsiveness in guinea pigs to cholinergic agonists, and this effect is prevented by prior treatment with capsaicin, suggesting that capsaicin-sensitive nerves release products which increase airway responsiveness. In a virus model of airway hyperresponsiveness in guinea pigs capsaicin pre-treatment completely blocks the virus-induced hyperresponsiveness [21]. In pigs capsaicin pre-treatment inhibits the vasodilator response to allergen (which may be mediated by the release of CGRP). In allergic sheep capsaicin pre-treatment prevents the airway hyperresponsiveness to both allergen and cholinergic agonists.. In a model of chronic allergen exposure in guinea pigs capsaicin pre-treatment results in complete inhibition of airway hyperresponsiveness, without any change in the eosinophil inflammatory response and also prevents the increased responsiveness to ovalbumin *in vitro*. In rabbits neonatal capsaicin treatment inhibits the airway hyperresponsiveness associated with neonatal allergen sensitisation, although this does not appear to be associated with any change in content of sensory neuropeptides in lung tissue. These studies suggest that in rodents capsaicin-sensitive nerves may play a role in airway hyperresponsiveness following allergen and virus infection. There has been speculation that mast cells in the airways might be activated by capsaicin-sensitive nerves. Histological studies have demonstrated a close proximity between mast cells and sensory nerves and ganglia in airways. Allergen exposure has effects on ion transport in guinea pig airways, which are dependent on capsaicin-sensitive nerves [22].

5.4. Inhibition of neuropeptide metabolism

The activity of NEP may be an important determinant of the extent of neurogenic inflammation in airways and inhibition of NEP in rodent by thiorphan or phosphoramidon has been shown to enhance neurogenic inflammation in various rodent models [14]. NEP is not specific to tachykinins and is also involved in the metabolism of other bronchoactive peptides, including kinins and endothelins. Certain virus infections enhance e-NANC responses in guinea pigs and *Mycoplasma* infection enhances neurogenic microvascular leakage in rats, an effect which is mediated by inhibition of NEP activity. Influenza virus infection of ferret trachea *in vitro* and of guinea pigs *in vivo* inhibits the activity of epithelial NEP and markedly enhances the bronchoconstrictor responses to tachykinins. Similarly Sendai virus infection potentiates neurogenic inflammation in rat trachea. This may explain why respiratory tract virus infections are so deleterious to patients with asthma. Hypertonic saline also impairs epithelial NEP function, leading to exaggerated tachykinin responses, and cigarette smoke exposure has a similar effect which can be explained by an oxidising effect on the enzyme. Inhalation of IL-1β is associated with increased responsiveness to bradykinin and this may be due to inhibition of NEP expression [23]. Thus, many of the agents which lead to exacerbations of asthma appear to reduce the activity of NEP at the airway surface, thus leading to exaggerated responses to tachykinins (and other peptides) and so to increased airway inflammation.

5.5. Tachykinin antagonists

Specific antagonists of tachykinin receptor subtypes provide a more specific tool to investigate the role of tachykinins in animal models. The NK_1-receptor antagonist CP 96,345 blocks the plasma exudation response to vagus nerve stimulation and to cigarette smoke in guinea-pig airways [24], without affecting the bronchoconstrictor response, which is blocked by the NK_2-antagonist SR 48,968 [25]. CP 96,345 also blocks hyperpnoea- and bradykinin-induced plasma exudation in guinea pigs [26,27], but has no effect on the acute plasma exudation induced by allergen in sensitised animals [27].

5.6. Inhibition of sensory neuropeptide release

Several agonists act on pre-junctional receptors on airway sensory nerves to inhibit the release of neuropeptides and neurogenic inflammation [28,29]. Opioids are the most effective inhibitory agonists, acting via pre-junctional μ-receptors and inhibit cigarette smoke-induced discharge from goblet cells in guinea-pig airways *in vivo* and to inhibit ozone-induced hyperreactivity in guinea pigs, which appears to be mediated via sensory nerves. Nociceptin, an endogenous opioid peptide that activates a novel opioid receptor (ORL_1), also modulates excitatory-NANC nerves [30]. Several other agonists are also effective and may act by opening a common calcium-activated large conductance potassium channel in sensory nerves [31]. Openers of other potassium channels, which achieve the same hyperpolarisation of the sensory nerve are also effective in blocking neurogenic inflammation in rodents and block cigarette smoke induced goblet cell secretion in guinea pigs [32]. Local anaesthetics, such as lidocaine, may also inhibit the release of neuropeptides by blocking sensory nerve activation (Barnes et al., 1990).

6. NEUROGENIC INFLAMMATION IN HUMAN AIRWAYS AND ASTHMA

Although, as discussed above, there is extensive evidence for a role for neurogenic inflammation in the lower airways of rodents and some other species, the relevance of neurogenic inflammation is far less certain. Although it was proposed several years ago that neurogenic inflammation and peptides released from sensory nerves might be important as an amplifying mechanism in asthmatic inflammation [33] (Figure 1), there is little direct supportive evidence for this idea. However, this is partly because it has proved difficult to apply the experimental approaches used in animals to human volunteers.

6.1. Sensory nerves in human airways

In comparison with rodent airways, SP- and CGRP-immunoreactive nerves are very sparse in normal human airways. Quantitative studies indicate that SP-immunoreactive fibres constitute only 1% of the total number of intra-epithelial fibres, whereas in guinea pig they make up 60% of the fibres [34]. Neither e-NANC responses nor a contractile response to capsaicin have been observed in human airways *in vitro*, however.

However, there is plasticity in the peripheral nervous system and the pattern of innervation might change in inflammatory disease. Chronic inflammation may lead to changes in the pattern of innervation, through the release of neurotrophins, such as nerve growth factor (NGF), from inflammatory cells. In chronic arthritis and inflammatory bowel disease there is an increase in the density of SP-immunoreactive nerves [35,36]. NGF markedly increases the transcription of the preprotachykinin-A (PPT-A) gene in cultured sensory neurons [37]. Adjuvant-induced inflammation in rat spinal cord increases the gene expression of PPT-A [38]. Allergen challenge is associated with a doubling in PPT-A mRNA-positive neurons in nodose ganglia of guinea pigs and an increase in SP-immunoreactivity in the lungs [39].

A striking increase in SP-like immunoreactive nerves was reported in the airways, particularly in the submucosal region, of patients with fatal asthma [40]. However, other studies have failed to confirm an increase in SP in lungs from patients with asthma [41] or any increase in SP-immunoreactive nerves in bronchial biopsies of asthmatic patients [42]. Elevated concentrations of SP and NKA have been found in bronchoalveolar lavage of patients with asthma, with a further rise after allergen challenge [43,44]. Similarly, SP has been detected in the sputum of asthmatic patients after hypertonic saline inhalation.

6.2. Sensory nerve activation

Sensory nerves in asthmatic airways may become "hyperalgesic" as a result of exposure to inflammatory mediators, such as prostaglandins and certain cytokines (IL-1β, TNFα and neuroptrophins). Other mediators, such as kinins, may then activate hyperalgesic nerves more readily.

Capsaicin induces bronchoconstriction and plasma exudation in guinea pigs and increases airway blood flow in pigs. In humans capsaicin inhalation causes cough and a *transient* bronchoconstriction which is inhibited by cholinergic blockade and is probably due to a laryngeal reflex [46,47]. This suggests that neuropeptide release does not occur in human airways, although it is possible that insufficient capsaicin reaches the lower respiratory tract because the dose is limited by coughing. In patients with asthma, there is no evidence that capsaicin induces a greater degree of bronchoconstriction than in normal individuals [46].

Bradykinin is a potent bronchoconstrictor in asthmatic patients and also induces coughing and a sensation of chest tightness, which closely mimics a naturally occurring asthma. Yet it is a weak constrictor of human airways *in vitro*, suggesting that its potent constrictor effect is mediated indirectly. Bradykinin is a potent activator of bronchial C-fibres in guinea pig airways and sensitises these nerves to other irritants [48]. In asthmatic patients an inhaled non-selective tachykinin antagonist FK-224 reduces the bronchoconstrictor response to inhaled bradykinin and also to block the cough response in those subjects that coughed in response to bradykinin [49], although this was not confirmed in a later study [50].

6.3. Studies with NEP inhibitors

In rodents inhibition of NEP with thiorphan or phosphoramidon results in striking potentiation of tachykinin and sensory nerve induced effects, and has been used as an approach to explore the potential for neurogenic inflammation in disease [14]. Intravenous acetorphan, which is hydrolysed to thiorphan, while potentiating the weal and flare response to intradermal SP, had no effect on baseline airway calibre or on bronchoconstriction induced by a "neurogenic" trigger sodium metabisulphite in asthmatic patients [51]. The lack of effect could be due to inadequate inhibition of NEP in the airways, particularly at the level of the epithelium. Nebulised thiorphan potentiates the bronchoconstrictor response to inhaled NKA in normal and asthmatic subjects, but there is no effect on baseline lung function in asthmatic patients [16,52], indicating that there is unlikely to be any basal release of tachykinins. NEP is strongly expressed in the human airways, but there is no evidence based on immunocytochemical staining or *in situ* hybridisation that it is defective in asthmatic airways [53]. After inhaled thiorphan the bronchoconstrictor response to inhaled NKA is further enhanced in asthmatic subjects provides supportive functional data that NEP function may not be impaired, at least in mild asthma [52]. Of course, it is possible that NEP may become dysfunctional after viral infections or exposure to oxidants and thus contribute to asthma exacerbations.

6.4. Tachykinin responsiveness

In inflammatory bowel disease there is evidence for a marked up-regulation of tachykinin receptors, particularly in the vasculature, suggesting that chronic inflammation may lead to changes in tachykinin receptor expression. NK_1-receptor gene expression is increased in the lungs of asthmatic patients [54]. This might be due to increased transcription in response to activation of transcription factors, such as AP-1, which are activated in human lung by cytokines such as TNFα. A consensus sequence both for AP-1 binding has been identified upstream of the NK_1-receptor gene. Corticosteroids conversely reduce NK_1-receptor gene expression, presumably via an inhibitory effect on AP-1 activation. Abnormal expression of NK_2-receptors has also been described in asthma [55], although the distribution of NK1 and NK2 receptors is apparently normal in asthma and COPD [13]. Recent studies suggest that β2-agonists may increase the expression of NK_2 receptors on airway smooth muscle [56].

6.5. Modulation of neurogenic inflammation

Apart from tachykinin receptor antagonists, neurogenic inflammation may be modulated by either preventing activation of sensory nerves, or preventing the release of neuropeptides. Both approaches may be tried in asthmatic patients, using currently available drugs, although these approaches are not as specific as tachykinin antagonists, as the drugs used have additional effects.

Activation of sensory nerves may be inhibited by local anaesthetics, but it has proved to be very difficult to achieve adequate local anaesthesia of the respiratory tract. Inhalation of local anaesthetics, such as lidocaine, does not have a consistent inhibitory effect on airway challenges, and indeed may even promote bronchoconstriction in some patients with asthma [57]. This paradoxical bronchoconstriction may be due to the greater anaesthesia of laryngeal afferents which are linked to a tonic non-adrenergic bronchodilator reflex [58]. Other drugs may inhibit the activation of airway sensory nerves. Sodium cromoglycate and nedocromil sodium may have direct effects on airway C-fibres [59] and this might contribute to their anti-asthma effect. Nedocromil sodium is highly effective against bradykinin-induced and sulphur dioxide-induced bronchoconstriction in asthmatic patients [60,61] , which are believed to be mediated by activation of sensory nerves in the airways. In addition, cromones inhibit the e-NANC neural bronchoconstriction due to tachykinin release form sensory nerves in guinea-pig bronchi *in vitro*, indicating an effect on release of sensory neuropeptides as well as on activation [62]. The diuretic frusemide (furosemide) given by nebulisation, behaves in a similar fashion to nedocromil sodium and inhibits metabisulphite-induced bronchoconstriction in asthmatic patients [63] and also e-NANC and cholinergic bronchoconstriction in guinea pig airways *in vitro*[64]. In addition, nebulized furosemide also inhibits certain types of cough [65], providing further evidence for an effect on sensory nerves.

Many drugs act on pre-junctional receptors to inhibit the release of neuropeptides.. Opioids are the most effective inhibitors, but an inhaled μ-opioid agonist, the pentapeptide BW 443C, is ineffective in inhibiting metabisulphite-induced bronchoconstriction, which is believed to act via neural mechanisms [66]. However, BW443C may be degraded by NEP in the airway epithelium and may not therefore reach a high enough concentration in the vicinity of the airway sensory nerves.

Some clinical studies in asthma have been undertaken with tachykinin antagonists. The relatively weak non-selective tachykinin antagonist FK-224 has an inhibitory effect on bradykinin-induced bronchoconstriction in asthma [49]. A more potent NK_1-receptor antagonist reduces the duration of exercise-induced asthma, but has no effect on the maximal bronchoconstriction, suggesting an effect on blood vessels rather than airway smooth muscle [50]. However another potent NK_1-receptor antagonist CP 99,994 had no effect on hypertonic saline-induced bronchoconstriction or on cough [68]. Further studies of tachykinin antagonists in chronic asthma and COPD are needed.

7. CONCLUSIONS

While there is convincing evidence that neuropeptides released from sensory nerves contribute to airway inflammation in rodents and some other species, there is relatively little evidence that neurogenic inflammation (axon reflex) is important in asthma. Sensory neuropeptides are not prominent in human airways and the initial studies that showed an apparent increase in SP-immunoreactive nerves in asthmatic airways have not been confirmed. However, SP ands NKA are released in asthmatic airways and may have some effect, particularly if their metabolism is impaired through defective expression or function of NEP, or if there is an increased expression or sensitivity of tachykinin receptors. Capsaicin depletion studies are not feasible in the lower airways of humans, but the recent availability of tachykinin antagonists and drugs that block neuropeptide release from sensory nerves will be a more definitive means of studying neurogenic inflammation in asthma. So far the results of clinical studies with potent tachykinin antagonists have shown little effect in challenge studies. However, it is possible that sensory neuropeptides may only be involved in more severe asthma or in exacerbations, so that further studies with tachykinin antagonists are indicated. There is little doubt that afferent nerves are sensitised in asthma, resulting is symptoms of cough and chest tightness, but this does not necessarily result in neurogenic inflammation. The role of neurogenic inflammation in COPD also requires further study. Cigarette smoke activates sensory nerves and this may result in mucus hypersecretion via a local reflex and the release of sensory neuropeptides. In the future, research may focus on the mechanisms of afferent nerve sensitisation in the airways and the role of neurotrophins in modulating sensory nerve activation.

REFERENCES

1. Barnes PJ. Modulation of neurotransmission in airways. Physiol Rev 1992;72: 699–729.
2. Maggi CA, Giachetti A, Dey RD, Said SI. Neuropeptides as regulators of airway function: vasoactive intestinal peptide and the tachykinins. Physiol Rev 1995;75: 277–322.
3. Barnes PJ. Neurogenic inflammation. Respirat Physiol 2001 (in press)
4. Sanico AM, Atsuta S, Proud D, Togias A. Dose-dependent effects of capsaicin nasal challenge: in vivo evidence of human airway neurogenic inflammation. J Allergy Clin Immunol 1997;100: 632–641.
5. Uddman R, Hakanson R, Luts A, Sundler F. Distribution of neuropeptides in airways. In: Barnes PJ, ed. Autonomic Control of the Respiratory System. London: Harwood Academic, 1997: 21–37.
6. Ho WZ, Lai JP, Zhu XH, Uvaydova M, Douglas SD. Human monocytes and macrophages express substance P and neurokinin-1 receptor. J Immunol 1997;159: 5654–5660.
7. Frossard N, Rhoden KJ, Barnes PJ. Influence of epithelium on guinea pig airway

responses to tachykinins: role of endopeptidase and cyclooxygenase. J Pharmacol Exp Ther 1989;248: 292–298.
8. Harrison NK, Dawes KE, Kwon OJ, Barnes PJ, Laurent GJ, Chung KF. Effects of neuropeptides in human lung fibroblast proliferation and chemotaxis. Am J Physiol 1995;12:L 278–283.
9. Watson N, Maclagan J, Barnes PJ. Endogenous tachykinins facilitate transmission through parasympathetic ganglia in guinea-pig trachea. Br J Pharmacol 1993;109: 751–759.
10. Daoui S, Cognon C, Naline E, Emonds-Alt X, Advenier C. Involvement of tachykinin NK3 receptors in citric acid-induced cough and bronchial responses in guinea pigs. Am J Respir Crit Care Med 1998;158: 42–48.
11. Daoui S, Cui YY, Lagente V, Emonds-Alt X, Advenier C. A tachykinin NK3 receptor antagonist, SR 142801 (osanetant), prevents substance P-induced bronchial hyperreactivity in guinea-pigs. Pulm Pharmacol Ther 1997;10: 261–270.
12. Carstairs JR, Barnes PJ. Autoradiographic mapping of substance P receptors in lung. Eur J Pharmacol 1986;127: 295–296.
13. Mapp CE, Miotto D, Braccioni F, Saetta M, Turato G, Maestrelli P, Krause JE, Karpitskiy V, Boyd N, Geppetti P, Fabbri LM. The distribution of neurokinin-1 and neurokinin-2 receptors in human central airways. Am J Respir Crit Care Med 2000;161: 207–215.
14. Di Maria GU, Bellofiore S, Geppetti P. Regulation of airway neurogenic inflammation by neutral endopeptidase [In Process Citation]. Eur.Respir.J. 1998;12: 1454–1462.
15. Lotvall JO, Elwood W, Tokuyama K, Barnes PJ, Chung KF. Differential effects of phosphoramidon on neurokinin A- and substance P-induced airflow obstruction and airway mkcrovascular leakage in guinea pig. Br J Pharmacol 1991;104: 945–949.
16. Cheung D, Bel EH, den Hartigh J, Dijkman JH, Sterk PJ. An effect of an inhaled neutral endopeptidase inhibitor, thiorphan, on airway responses to neurokinin A in normal humans in vivo. Am Rev Respir Dis 1992;145: 1275–1280.
17. Mak JCW, Barnes PJ. Autoradiographic localization of calcitonin gene-related peptide binding sites in human and guinea pig lung. Peptides 1988;9: 957–964.
18. Dunzendorfer S, Meierhofer C, Wiedermann CJ. Signaling in neuropeptide-induced migration of human eosinophils. J Leukoc Biol 1998;64: 828–834.
19. McDonald DM. Neurogenic inflammation in the respiratory tract: actions of sensory nerve mediators on blood vessels and epithelium of the airway mucosa. Am Rev Respir Dis 1987;136:S65–S72
20. Solway J, Leff AR. Sensory neuropeptides and airway function. J Appl Physiol 1991;71: 2077–2087.
21. Ladenius ARC, Folkerts G, van der Linde HJ, Nijkamp FP. Potentiation by viral respiratory infection of ovalbumin-induced guinea-pig tracheal hyperresponsiveness: role for tachykinins. Br J Pharmacol 1995;115: 1048–1052.
22. Sestini P, Bienenstock J, Crowe SE, Marshall JS, Stead RM, Kakuta Y, Perdue MH. Ion transport in rat tracheal ganglion in vitro. Role of capsaicin-sensitive nerves in allergic reactions. Am Rev Respir Dis 1990;141: 393–3397.
23. Tsukagoshi H, Sakamoto T, Xu W, Barnes PJ, Chung KF. Effect of interleukin-1b on airway hyperresponsiveness and inflammation in sensitized and non-sensitized Brown-Norway rats. J Allergy Clin Immunol 1994;93: 464–469.
24. Lei Y-H, Barnes PJ, Rogers DF. Inhibition of neurogenic plasma exudation in guinea pig airways by CP-96,345, a new non-peptide NK_1-receptor antagonist. Br J Pharmacol 1992;105: 261–262.

25. Advenier C, Naline E, Toty L, Bakdach H, Emors-Alt X, Vilain P, Breliere J-C, Le Fur G. Effects on the isolated human bronchus of SR 48968, a potent and selective nonpeptide antagonist of the neurokinin A (NK_2) receptors. Am Rev Respir Dis 1992;146: 1177–1181.
26. Solway J, Kao BM, Jordan JE, Gitter B, Rodger IW, Howbert JJ, Alger LE, Necheles J, Leff AR, Garland A. Tachykinin receptor antagonists inhibit hypernea-induced bronchoconstriction in guinea pigs. J Clin Invest 1993;92: 315–323.
27. Sakamoto T, Barnes PJ, Chung KF. Effect of CP-96,345, a non-peptide NK_1-receptor antagonist against substance P-, bradykinin-, and allergen-induced airway microvascular leak and bronchoconstriction in the guinea pig. Eur J Pharmacol 1993;231: 31–38.
28. Barnes PJ, Belvisi MG, Rogers DF. Modulation of neurogenic inflammation: novel approaches to inflammatory diseases. Trends Pharmacol Sci 1990;11: 185–189.
29. Spina D. Modulation of sensory nerve function in the airways. Pulm Pharmacol Ther 1998;11: 319–330.
30. Giuliani S, Maggi C. Inhibition of tachykinin release from peripheral endings of sensory nerves by nociceptin, a novel opioid peptide. Br J Pharmacol 1996;118: 1567–1570.
31. Stretton CD, Miura M, Belvisi MG, Barnes PJ. Calcium-activated potassium channels mediate prejunctional inhibition of peripheral sensory nerves. Proc Natl Acad Sci USA 1992;89: 1325–1329.
32. Kuo H-P, Rohde JAL, Barnes PJ, Rogers DF. K+ channel activator inhibition of neurogenic goblet cell secretion in guinea pig trachea. Eur J Pharmacol 1992;221: 385–388.
33. Barnes PJ. Asthma as an axon reflex. Lancet 1986;i: 242–245.
34. Bowden J, Gibbins IL. Relative density of substance P-immunoreactive nerve fibres in the tracheal epithelium of a range of species. FASEB J 1992;6:A1276
35. Levine JD, Dardick SJ, Roizan MF, Helms C, Basbaum AI. Contribution of sensory afferents and sympathetic efferents to joint injury in experimental arthritis. J Neurosci 1986;6: 3423–3429.
36. Holzer P. Local effector functions of capsaicin-senstive sensory nerve endings: involvement of tachykinins, calcitonin gene related peptide, and other neuropeptides. Neuroscience 1988;24: 739–768.
37. Lindsay RM, Harmar AJ. Nerve growth factor regulates expression of neuropeptide genes in sensory neurons. Nature 1989;337: 362–364.
38. Minami M, Kuraishi Y, Kawamura M, Yamaguchi T, Masu Y, Nakanishi S. Enhancement of preprotachykinin A gene expression by adjuvant-induced inflammation in the rat spinal cord: possible inducement of substance P-containing spinal neurons in nociceptor. Neurosci Lett 1989;98: 105–110.
39. Fischer A, McGregor GP, Saria A, Philippin B, Kummer W. Induction of tachykinin gene and peptide expression in guinea pig nodose primary afferent neurons by allergic airway inflammation. J Clin Invest 1996;98: 2284–2291.
40. Ollerenshaw SL, Jarvis D, Sullivan CE, Woolcock AJ. Substance P immunoreactive nerves in airways from asthmatics and non-asthmatics. Eur Resp J 1991;4: 673–682.
41. Lilly CM, Bai TR, Shore SA, Hall AE, Drazen JM. Neuropeptide content of lungs from asthmatic and nonasthmatic patients. Am J Respir Crit Care Med 1995;151: 548–553.
42. Howarth PH, Djukanovic R, Wilson JW, Holgate ST, Springall DR, Polak JM. Mucosal nerves in endobronchial biopsies in asthma and non-asthma. Int Arch Allergy Appl Immunol 1991;94: 330–333.
43. Nieber K, Baumgarten CR, Rathsack R, Furkert J, Oehame P, Kunkel G. Substance P and

b-endorphin-like immunoreactivity in lavage fluids of subjects with and without asthma. J Allergy Clin Immunol 1992;90: 646–652.
44. Heaney LG, Cross LJ, McGarvey LP, Buchanan KD, Ennis M, Shaw C. Neurokinin A is the predominant tachykinin in human bronchoalveolar lavage fluid in normal and asthmatic subjects. Thorax 1998;53: 357–362.
45. Tomaki M, Ichinose M, Miura M, Hirayama Y, Yamauchi H, Nakajima N, Shirato K. Elevated substance P content in induced sputum from patients with asthma and patients with chronic bronchitis. Am J Respir Crit Care Med 1995;151: 613–617.
46. Fuller RW, Dixon CMS, Barnes PJ. The bronchoconstrictor response to inhaled capsaicin in humans. J Appl Physiol 1985;85: 1080–1084.
47. Midgren B, Hansson L, Karlsson JA, Simonsson BG, Persson CGA. Capsaicin-induced cough in humans. Am Rev Respir Dis 1992;146: 347–351.
48. Fox AJ, Lalloo UG, Belvisi MG, Bernareggi M, Chung KF, Barnes PJ. Bradykinin-evoked sensitization of airway sensory nerves: a mechanism for ACE-inhibitor cough. Nature Med 1996;2: 814–817.
49. Ichinose M, Nakajima N, Takahashi T, Yamauchi H, Inoue H, TAKISHIMA T. Protection against bradykinin-induced bronchoconstriction in asthmatic patients by a neurokinin receptor antagonist. Lancet 1992;340: 1248–1251.
50. Schmidt D, Jorres RA, Rabe KF, Magnussen H. Reproducibility of airway response to inhaled bradykinin and effect of the neurokinin receptor antagonist FK-224 in asthmatic subjects. Eur J Clin Pharmacol 1996;50: 269–273.
51. Nichol GM, O'Connor BJ, Le Compte JM, Chung KF, Barnes PJ. Effect of neutral endopeptidase inhibitor on airway function and bronchial responsiveness in asthmatic subjects. Eur J Clin Pharmacol 1992;42: 495–498.
52. Cheung D, Timmers MC, Bel EH, den Hartigh J, Dijuman JH, Sterk PJ. An isolated neutral endopeptidase inhibitor, thiorphan, enhances airway narrowing to neurokin A in asthmatic subjects in vivo. Am Rev Respir Dis 1992;195:A682.
53. Baraniuk JN, Ohkubo O, Kwon OJ, Mak JCW, Davies R, Twort C, Kaliner M, Letarte M, Barnes PJ. Localization of neutral endopeptidase (NEP) mRNA in human bronchi. Eur Respir J 1995;8: 1458–1464.
54. Adcock IM, Peters M, Gelder C, Shirasaki H, Brown CR, Barnes PJ. Increased tachykinin receptor gene expression in asthmatic lung and its modulation by steroids. J Mol Endocrinol 1993;11: 1–7.
55. Bai TR, Bramley AM. Effect of an inhibitor of nitric oxide synthase on neural relaxation of human bronchi. Am J Physiol. 1993;264:L425-L430
56. Katsunuma T, Roffel AF, Elzinga CR, Zaagsma J, Barnes PJ, Mak JC. b_2-Adrenoceptor agonist-induced upregulation of tachykinin NK_2 receptor expression and function in airway smooth muscle. Am J Respir Cell Mol Biol 1999;21: 409–417.
57. McAlpine LG, Thomson NC. Lidocaine-induced bronchoconstriction in asthmatic patients. Relation to histamine airway responsiveness and effect of preservative. Chest 1989;96: 1012–1015.
58. Lammers J-WJ, Minette P, McCusker M, Chung KF, Barnes PJ. Capsaicin-induced bronchodilatation in mild asthmatic subjects: possible role of nonadrenergic inhibitory system. J Appl Physiol 1989;67: 856–861.
59. Jackson DM, Norris AA, Eady RP. Nedocromil sodium and sensory nerves in the dog lung. Pulm Pharmacol 1989;2: 179–184.
60. Dixon CMS, Fuller RW, Barnes PJ. The effect of nedocromil sodium on sulphur dioxide

induced bronchoconstriction. Thorax 1987;42: 462–465.

61. Dixon N, Jackson DM, Richards IM. The effect of sodium cromoglycate on lung irritant receptors and left ventricular receptors in anasthetized dogs. Br J Pharmacol 1979;67: 569–574.
62. Verleden GM, Belvisi MG, Stretton CD, Barnes PJ. Nedocromil sodium modulates non-adrenergic non-cholinergic bronchoconstrictor nerves in guinea-pig airways in vitro. Am Rev Respir Dis 1991;143: 114–118.
63. Nichol GM, Alton EWFW, Nix A, Geddes DM, Chung KF, Barnes PJ. Effect of inhaled furosemide on metabisulfite- and methacholine induced bronchoconstriction and nasal potential difference in asthmatic subjects. Am Rev Respir Dis 1990;142: 576–580.
64. Elwood W, Lotvall JO, Barnes PJ, Chung KF. Loop diuretics inhibit cholinergic and non-cholinergic nerves in guinea pig airways. Am Rev Respir Dis 1991;143: 1340–1344.
65. Ventresca GP, Nichol GM, Barnes PJ, Chung KF. Inhaled furosemide inhibits cough induced by low chloride content solutions but not by capsaicin. Am Rev Respir Dis 1990;142: 143–146.
66. O'Connor BJ, Chen-Wordsell M, Barnes PJ, Chung KF. Effect of an inhaled opioid peptide on airway responses to sodium metabisulphite in asthma. Thorax 1991;46:294P.
67. Ichinose M, Miura M, Yamauchi H, Kageyama N, Tomaki M, Oyake T, Ohuchi Y, Hida W, Miki H, Tamura G, Shirato K. A neurokinin 1-receptor antagonist improves exercise-induced airway narrowing in asthmatic patients. Am J Respir Crit Care Med 1996;153: 936–941.
68. Fahy J, Wong HH, Geppetti P, Reiss JM, Harris SC, MacLean DB, Nadel JA, Boushey HA. Effect of an NK_1 receptor antagonist (CP-99,994) on hypertonic saline-induced bronchoconstriction and cough in male asthmatic subjects. Am J Resp Crit Care Med 1995;152: 879–884.

The Immune-Neuroendocrine Circuitry
Edited by I. Berczi and A. Szentivanyi

Defensins: antimicrobial peptides with a broad spectrum of biological activity

E.A. KORNEVA and V.N. KOKRYAKOV

Department of General Pathology and Pathophysiology Institute of Experimental Medicine, 197376, St. Petersburg, Russia

ABSTRACT

Defensins are antimicrobial cationic peptides with a cystine-stabilized amphipathic structure. These substances are normally localized in phagocytes (neutrophil, monocyte/macrophage) and in the epithelial cells of mucous membranes and skin. Some defensins are released into the blood during the course of infection, inflammation and stress. Defensins function not only as endogenous animal antibiotic molecules, killing microbial cells and enveloped viruses, but also as physiological regulators. Defensins are implicated in the regulation of endocytosis, chemotaxis, mast cell degranulation and inflammation. Moreover, these molecules are modulators of hemostasis and neuroendocrine-immune interaction. Defensins lower the stress-induced elevation of corticosteroid levels in the blood, and abolish the stress-induced inhibition of the humoral immune response. These facts support the hypothesis that defensins are antibacterial peptides with a broad spectrum of biological activity.

1. INTRODUCTION

Defensins were discovered as low-molecular-weight lysosomal cationic proteins in the neutrophils of guinea pigs [1]. They are the biologically active substances of neutrophils and of some macrophages that contribute to the killing of phagocytosed microorganisms [1,2]. Therefore, the functional role of defensins in phagocytosis and inflammation is most investigated. In the last few years, defensins have been identified in many other cells and tissues in humans and animals [3].

Antibiotic peptides that participate in innate immunity against infectious agents are widespread in nature. It is now generally accepted that the antimicrobial peptides produced in higher organisms are analogous to endogenous antibiotics [3,4]. Low concentrations of defensins are also present in the blood of normal individuals. For example, in human the normal plasma level is 10–40 ng/ml [5]. In pathological situations, such as infection or inflammation, the blood level of these peptides is profoundly increased and ranges from 100 to 900 ng/ml [5]. During severe inflammation, the concentration of defensins in the blood and in the focus of inflammation may be extremely high, up to several μg/ml. In this situation, defensins may act on hemostasis and on endocrine-immune interaction [6].

Human defensins

1 2 3 4 5 6

HNP-1	ACYCRIP-ACIAGERRYGTCIYQGRL-WAFCC
HNP-2	CYCRIP-ACIAGERRYGTCIYQGRL-WAFCC
HNP-3	DCYCRIP-AClAGERRYGTCIYQGRL-WAFCC
hBD-1	DHYNCVSSGGQCLYSACPIFTKIQGTCYRGKAKCCK
hBD-2	TCLKSGAICHPVFCPRRYKQIGTCGLPGTKCCKKP

Rabbit defensins

NP-1	VVCACRRAL-CLPRERRAGFCRIRGRI-HPLCCRR
NP-2	VVCACRRAL-CLPLERRAGFCRIRGRI-HPLCCRR
NP-3a	GICACRRR-FCPNSERFSGYCRVNGAR-YVRCCSRR
NP-3b	GRCVCRKQLLCSYRERRIGDCKIRGVR-FPFCCPR
NP-4	VSCTCRRF-SCGFGERASGSCTVNGVR-HTLCCRR
NP-5	VFCTCRGFL-CGSGERASGSCTINGVR-HTLCCRR
NP-6	GICACRRR-FCLNFEQFSGYCRVNGAR-YVRCCSRR
desG1-NP-6	ICACRRR-FCLNFEQFSGYCRVNGAR-YVRCCSRR

Rat defensins

RtNP-1	VTCYCRRT-RCGFRERLSGACGYRGRI-YRLCCR
RtNP-2	VTCYCRST-RCGFRERLSGACGYRGRI-YRLCCR
RtNP-3	CSCRTS-SCRFGERLSGACRLNGRI-YRLCC
RtNP-4	ACYCRIG-ACVSGERLTGACGLNGRI-YRLCCR

Figure 1. The primary amino acid sequences of defensins.
The single-letter amino acid codes: A-alanine, C-cysteine, D-aspartic acid, E-glutamic acid, F-phenylalanine, G-glycine, H-histidine, I-isoleucine, L-leucine, M-methionine, N-asparagine, P-proline, Q-glutamine, R-arginine, S-serine, T-threonine, V-valine, W-tryptophane, Y-tyrosine.
1, 2, 3, 4, 5, 6 – number of cysteine residues.
The cysteine bonds of human defensins are shown schematically as solid lines. The peptide alignments are indexed to the most highly conserved residues, which are in bold.

A discussion of the well-known properties of defensins (e.g. antibacterial activity, killing activity in phagocytes) as well as their possible role as regulatory mediators in physiological and pathophysiological processes follows. Defensins are peptides with a broad spectrum of biological activity [4].

2. THE STRUCTURE AND ANTIMICROBIAL ACTIVITY OF DEFENSINS

Antimicrobial peptides are widespread in nature. On the basis of structural homology, several groups of antimicrobial peptides have been described: defensins [3,7], protegrins [8], cecropins [9], magainin, and bactenecin, etc. Their basic function is the maintenance of innate immunity in animals [3,4,9]. In 1963, peptides of the defensin family were isolated from the neutrophils of guinea pigs and were identified as substances of high antibacterial and fungicidal activity [1]. The primary structure of defensins, from the neutrophils of rabbits, guinea pigs, rats and humans was determined by Lehrer and co-workers [10]. Today, the structural and functional properties of numerous defensins are known in various cells and tissues of many species of insects [11], birds [12], and mammals, including humans [10]. Typical structural features of defensins include a high basic amino acid content (e.g. arginine, lysine) and the presence of six conservatively located residues of cysteine that form three intramolecular disulfide bonds (Figure 1). It is reasonable to assume that disulfide bridges increase the stability of defensins against the digesting action of numerous proteinases that are present in leukocytes and produced by microbes. This resistance prolongs their antimicrobial action during phagocytosis and inflammation.

The family of defensins is divided into four subgroups: α-, β-, θ- and insect defensins. θ-defensins from leukocytes of monkeys, *Macaca rhesus,* unlike other types of defensins are macrocyclic peptides, whose cycle structure is formed by the connection of N- and C-terminal portions of the molecule by peptide linkage [13].

Defensins are simultaneously cationic and amphipathic molecules with spatially separated hydrophilic positively charged and hydrophobic side groups. The electrostatic forces of attraction between the positively charged groups of defensins and negatively charged components of the microbial envelope (e.g. lipopolysaccharides, teichoic acids, phosphatidylglycerol, cardiolipin) determine the absorption of these peptides onto the surface of microorganisms. Defensins can penetrate into the cell cytoplasmic membrane due to the hydrophobic interaction between lipophilic side groups of peptides and the apolar tails of fatty acids of phospholipids. During the electrostatic and hydrophobic interactions of defensins with lipids, the membranes are perforated. This results in a K+ efflux, disturbance of ionic balance, dissipation of the membrane potential and uncoupling from oxidation and phosphorylation, resulting in a decreased intracellular quantity of ATP. These changes in the structure and functional activity of microbes lead to death (killing effect of defensin).

Today, defensins have been identified in phagocytes (neutrophils, monocytes/macrophages and hemocytes of invertebrates), mucosal epithelial cells of the gastrointestinal, respiratory and urinogenital tract [14] and in skin cells [15]. The localization of defensins in the phagocytes and epithelial cells make it possible to suggest the participation of these peptides in innate immunity as universal antibiotics [4,16].

3. DEFENSINS AND INFLAMMATION

Defensins play various roles in inflammation. Firstly, they are microbicidal agents acting during phagocytosis by neutrophils and monocyte/macrophages, and they are partly responsible for the protective effect of the inflammatory process [4,10]. Secondly, neutrophils secrete defensins and other antimicrobial substances. Consequently, leukocyte defensins, together with epithelial defensins, can influence the development of inflammation. Their ability to directly increase the permeability of blood vessels [17] as well as through the degranulation of mast cells [18,19] is well known.

The mobilization of phagocytes and lymphocytes during inflammation is dependent on defensins as well. In recent years it has been shown that neutrophils are a source of chemotactic factors for neutrophils and mononuclear cells (monocytes/macrophages, T-lymphocytes). The *in vitro* chemotactic activity of human α-defensins (HNP-1, HNP-2) for monocytes/macrophages was shown earlier [20]. The same defensins were active as chemotactic molecules for neutrophils, monocytes/macrophages and the T-lymphocytes of BALB/c mice in the current experiments [21]. Therefore, defensins can play an important role in a change of cellular populations during the inflammatory focus, as described in the classical works by E. Metchnikoff [2].

It should be stressed that human β-defensins from epithelial can also attract dendritic cells and memory T-lymphocytes into the site of inflammation. This mobilization is necessary for the initiation of an acquired immune response. In recent *in vitro* experiments the chemotactic activity of human β-defensin 2 (hBD2) was demonstrated in dendritic cells and in a population of

T-memory-lymphocytes [22]. It was found that β-defensin is a putative ligand selectively interacting with the chemokine receptor CCR6. As a result, β-defensin 2 produced by epithelial cells has been assumed to promote the acquired immune response by recruiting dendritic cells and T-memory-cells into the site of pathogen penetration. On the basis of these data, defensins can be considered as mediators between innate and acquired immune mechanisms. There is evidence to indicate that human α-defensins (HNP-1, -2, -3) act as adjuvants, reinforcing the humoral immune response to intranasally applied ovalbumin [23]. Clearly, it is indicated that neutrophils are a dominant cell type during the first stage of inflammation and may produce peptides which initiate the acquired immune response to foreign protein antigens.

Due to their electrochemical properties, defensin molecules and related peptides may effectively interact with bacterial and fungal surfaces. As a result, a reduction in the negative charge of microbial cell surface can be observed, which facilitates the endocytic activity of phagocytes. This ability of defensins to reinforce the endocytotic activity of neutrophils and monocytes/macrophages can be considered as opsonic in nature during phagocytosis and inflammation [24]. Taken together, the data considered here testify to the plural role of defensins in inflammation and immunity.

4. DEFENSINS AND HEMOSTASIS

At sites of inflammation, the formation and maintenance of hemostasis at the level of microcirculation is important. Cationic peptides, which are secreted from neutrophils under the influence of inflammatory mediators, participate in this process. Many factors that initiate inflammation may cause injury to blood vessels. Under these circumstances keeping blood loss to a minimum is important for survival. Defensins from leukocytes are involved in this process. The first observation on the procoagulant activity of defensins from rabbits was made in an *in vitro* system [25]. Defensins as polycations have the ability to electrostatically interact with heparin and to neutralize its anticoagulant properties.

In collaboration with the scientists in the Department of Physiology at the Moscow State University, we investigated the influence of total rabbit defensin on the system of non-enzyme fibrinolysis in rat [6]. The basic conclusion of these experiments is that defensins exert a procoagulant activity by binding and neutralizing blood heparin (heparin is a major component of the non-enzyme fibrinolytic system). In addition, other authors showed the ability of human defensin to inhibit the cascade of enzyme mediated fibrinolysis. In particular, it was shown that human defensins are inhibitors of tissue-type plasminogen activator (tPA)-mediated and plasminogen-depended reactions [27]. This effect is similar to the action of lipoprotein(a) on the system of enzyme fibrinolysis [28]. It is possible, that the anti-fibrinolytic activity of defensins in inflammation prevents the spread of the infection by immobilizing bacteria in fibrin clots.

Platelets, and the physiologically active mediators produced by them, play an important role in the maintenance of hemostasis. In this context, the action of defensins on the functional activity and metabolism of platelets are of interest. It was shown that human defensins (HNP-1, -2, -3) have a dose-dependent effect on the aggregation and functional activity of platelets *in vitro* [29]. At a high concentration (50–100 μg/ml) (microgram/ml), which may occur at the site of inflammation, defensins exert a pro-aggregational effect on platelets. At a low concentration (0. 1–40 μg/ml), which occurs in the blood of normal animals and humans, defensins have an anti-aggregation and anti-secretory effect on platelets. It is necessary to emphasize that in the presence of an infectious pathology the concentration of defensins in the blood can be as high

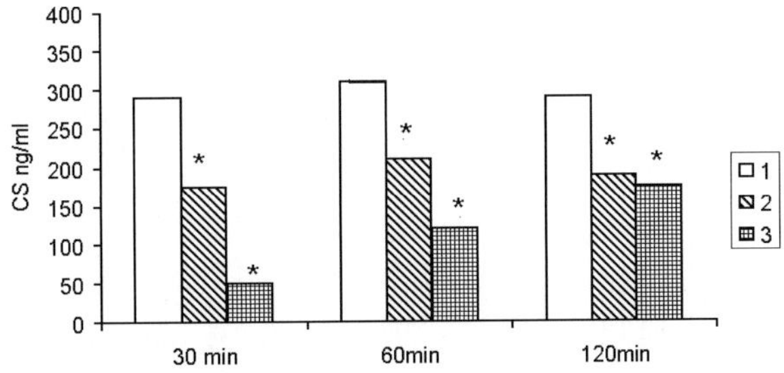

Figure 2. Effect of defensins on blood level of corticosterone in mice subjected to rotation stress.

1 control animals injected with physiological saline 10 min. before stress.

2 animals which received 20 ng/g total rabbit defensin 10 min. before stress.

3 animals which received 2 μg/g total rabbit defensin 10 min. before stress.

Abscissa: time after stress ending. Ordinate: concentration of corticosterone (CS) in samples (ng/ml).

Asterisk: $p<0.05$ vs. 1.

as one or more μg/ml [5]. At this concentration the anti-aggregation property of defensins is expressed. It is probable that defensins secreted from neutrophils can act as mediators of leukocyte-platelet interactions, which are directed towards maintenance of hemostasis during infectious diseases and stress. All these situations are accompanied by neutrophil mobilization from the bone marrow.

5. THE EFFECT OF DEFENSINS ON CORTICOSTEROID PRODUCTION AND STRESS

Of special interest is the possible participation of defensins in the regulation of the steroidogenic activity of the adrenals. A group of Canadian scientists have isolated peptides from rabbit lung, which suppressed the ACTH-induced production of corticosteroids by adrenal cells *in vitro* [30]. It was established that these substances were defensins and they were named corticostatins [6]. Subsequently, defensins with corticostatic activity were detected in human neutrophils and in the bone marrow cells of guinea pigs and rats [31]. Studies on the molecular mechanisms of the corticostatic activity of defensins revealed that this effect is due to the competitive binding of defensin peptides to the ACTH receptors of adrenal cells in culture [32]. This binding interrupted the signal transduction cascade induced by ACTH, which resulted in the inhibition of synthesis and secretion of glucocorticoids. The authors suggested that defensins play a neuroimmune regulatory role by modulating the activity of ACTH-adrenal axis [33]. The corticostatic action of defensins was also confirmed by Japanese scientists [34] as well as in our research [35].

We investigated the effect of exogenous defensins obtained from rabbit neutrophils on the endocrine system and immune activity of stressed animals. The induction of adrenocortical secretion by ACTH was also studied. These models are suitable for the investigation of corticostatic activity of defensins *in vivo,* since the elevation of glucocorticoid production takes place under these conditions.

During rotational stress (10 minutes at 78 rpm), it was established that rabbit defensins, injected i.p. 10 minutes prior to the beginning of stress, caused a decrease in blood corticosterone levels in mice. The depressive action of defensin on the stress-induced glucocorticoid eleva-

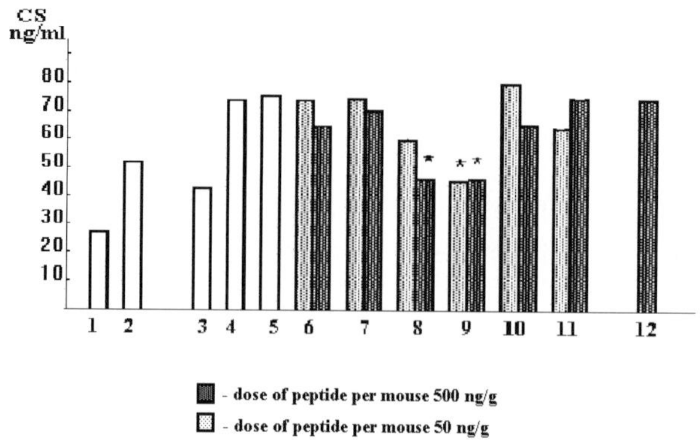

Figure 3. Effect of isolated fractions of rabbit defensins on the ACTH –induced elevation of corticosterone level in the blood of mice.

1 intact animals.
2 animals injected with physiological saline.
3 animals injected with NP-1 defensin.
4 animals injected with ACTH in concentration 0.02 IU/g.
5 animals injected with physiological saline 30 min. before ACTH injection.
6 animals injected with NP-1 defensin 30 min. before ACTH injection.
7 animals injected with NP-2 defensin 30 min. before ACTH injection.
8 animals injected with NP-3a defensin 30 min. before ACTH injection.
9 animals injected with NP-3b defensin 30 min. before ACTH injection.
10 animals injected with NP-4 defensin 30 min. before ACTH injection.
11 animals injected with NP-5 defensin 30 min. before ACTH injection.
12 animals injected with protamine sulfate (clupein) 30 min. before ACTH injection.
Asterisk: $p<0.05$ vs. group 4, 5.

tion was most pronounced after 30 minutes of rotation (Figure 2). The injection of defensins one hour before, or 30 minutes after, stress had no effect. Similar changes in corticosterone blood levels were observed after the i.p. injection of rabbit defensins in rats prior to severe cooling stress (-20^{0}C, 20 minutes). Corticosterone in such animals was approximately half of the level present in animals subjected to stress alone. (62–72 ng/ml vs. 128–145 ng/ml, respectively) [35]. The corticostatic activity of rat defensin (RTNP 1) was shown later by other investigators in rats, using soft electroshock stress [36]. The authors pointed out that defensin only suppressed corticosterone production in stressed animals.

The hormonal shifts in stressed animals after defensin treatment reveals a novel and uniform physiological mechanism, which plays a role in host defense. As suggested by Zhu et al., [30], defensins may influence corticosteroid levels in the blood by binding to ACTH receptors. After binding to these receptors defensins inhibit, in a competitive manner, the production of corticosterone by adrenal cells. Our experiments on the corticostatic activity of rabbit α-defensins (NP-1, NP-2, NP-3a, NP-3b, NP-4, NP-5) in a murine model of ACTH induced glucocorticoid elevation, supports this hypothesis (Figure 3). NP-3a and NP-3b were most active as corticostatins *in vivo*. These data correlate with the results of testing individual corticostatic activity of defensins *in vitro* [37].

Of special interest is the investigation of corticostatic properties of antimicrobial peptides

Rabbit defensins

NP-1	VVCACRRAL-CLPRERRAGFCRIRGRI-HPLCCRR
NP-2	VVCACRRAL-CLPLERRAGFCRIRGRI-HPLCCRR
NP-3a	GICACRRR-FCPNSERFSGYCRVNGAR-YVRCCSRR
NP-3b	GRCVCRKQLLCSYRERRIGDCKIRGVR-FPFCCPR
NP-4	VSCTCRRF-SCGFGERASGSCTVNGVR-HTLCCRR
NP-5	VFCTCRGFL-CGSGERASGSCTINGVR-HTLCCRR
NP-6	GICACRRR-FCLNFEQFSGYCRVNGAR-YVRCCSRR
desG1-NP-6	ICACRRR-FCLNFEQFSGYCRVNGAR-YVRCCSRR

Porcine protegrins

PG-1	RGGRLCYCRRRFCVCVGR
PG-2	RGGRLCYCRRRFCICV
PG-3	RGGGLCYCRRRFCVCVGR

Figure 4. The primary amino acid sequences of antimicrobial peptides. The single-letter amino acid codes are the same as for Figure 1.

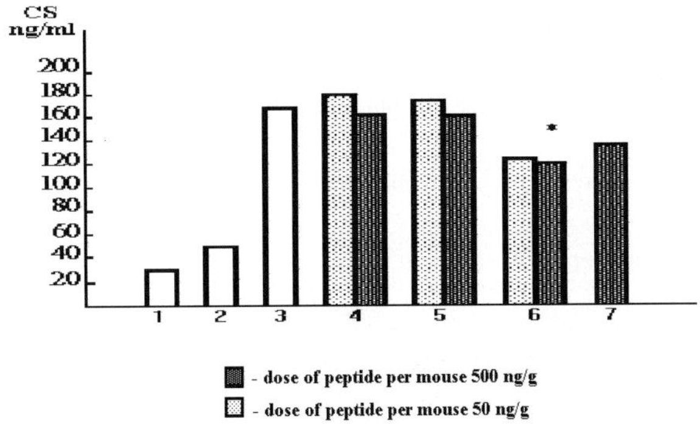

Figure 5. Effect of isolated fractions of porcine protegrins on the ACTH-induced elevation of corticosterone blood level in mice.

1 intact animals.

2 animals injected with physiological saline.

3 animals injected with ACTH in concentration 0.02 IU/g.

4 animals injected with PG-1 30 min. before ACTH injection.

5 animals injected with PG-2 30 min. before ACTH injection.

6 animals injected with PG-3 30 min. before ACTH injection.

7 animals injected with protamine sulfate (clupein) 30 min. before ACTH injection.

Asterisk: $p<0.05$ vs. group 3.

from porcine leukocytes, called protegrins, discovered by our lab in a collaboration with our American colleagues [8]. The primary structure of protegrins is similar to defensins NP-3a (Figure 4). The corticostatic activity of three protegrins (PG-1, PG-2, PG-3) was studied in mice during ACTH induced glucocorticoid elevation. It was established that only PG-3 reduced the corticosteroid blood level (Figure 5). It is interesting to note, however, that PG-3 has the most structural similarity to defensin NP-3a. It is reasonable to assume that glycine, which is at N-position in NP-3a and occupies the 4th position in protegrin PG-3 (it is absent in protegrin PG-1 and PG-2) plays a role in the corticostatic effect of defensins and protegrins in addition to the three arginine residues (RRR). The importance of glycine residues in the corticostatic action of defensins has also been demonstrated by other investigators [38]. They isolated two new defensins (NP-6, desG1-NP-6) from rabbit spleen. The primary structure of desG1-NP-6

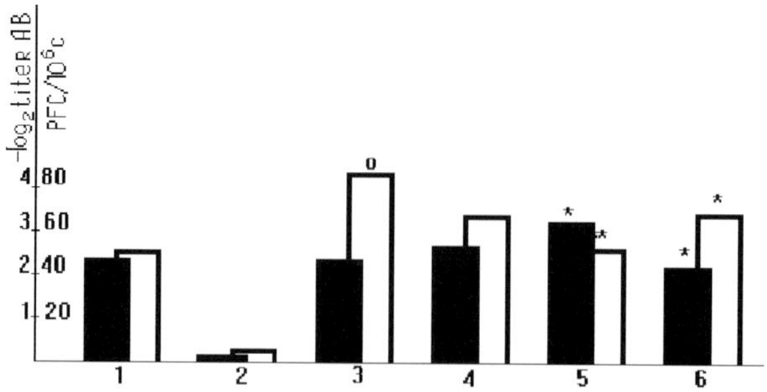

Figure 6. Effect of intraperitoneally injected total rabbit defensin on stress induced immunosuppression in mice.
1 immunized animals not subjected to stress.
2 animals subjected to combined stress before immunization.
3 unstressed animals which received rabbit defensins in dose of 20 μg/g.
4 unstressed animals which received rabbit defensins in dose 2 μg/g.
5 animals subjected to combined stress before immunization, which received 20 μg/g of rabbit defensins 10 min. before stress.
6 animals subjected to combined stress before immunization, which received 2 μg/g of rabbit defensins 10 min. before stress.
Circle: $p<0.02$ vs. group 1. Asterisk: $p<0.05$ vs. group 2. Hatched columns: $-\log_2$ titer AB (antibody). Unhatched columns: PFC/10^6 splenocytes.

defensin was very similar to that of NP-3a (three replacements in a position 11, 13 and 15) only it lacks an N-end glycine (Figure 4). This new variant of defensins (desG1-NP-6) had very weak corticostatic activity *in vitro*. These data favor the existence of a natural relationship between the primary structure of defensins and protegrins and their functional activity.

6. THE EFFECT OF DEFENSINS ON IMMUNE FUNCTION DURING STRESS

The corticostatic action of defensins and protegrins may play a physiological role. Glucocorticoid hormones influence host defense, including the immune response, by various mechanisms. The hypersecretion of glucocorticoids by the adrenals may be one of the possible reasons for the immune system being compromised [33]. In light of the data concerning the corticostatic activity of antimicrobial peptides, one may assume that the immunosuppressive influence of severe stress can be prevented by defensins. During combined stress in mice (immobilization for one hour at +4°C and then for 20 hours at room temperature) a profound inhibition of the antibody response to sheep erythrocytes was shown. The intraperitoneal injection of total rabbit defensins before stress canceled this stress-induced immunosuppression (Figure 6). We also demonstrated that total rabbit defensin prevented the immunosuppression of stressed rats (Figure 7).

It should be emphasized that corticostatic influence is not solely responsible for the immunoprotective action of defensins. It is necessary to take into account that defensins have chemotactic and opsonic properties as well, which mobilize immune cells and enhance their functional activity [22].

The evidence available supports the conclusion that the functional importance of neutrophil mobilization during stress (neutrophilia) is not limited to a preventive reinforcement of the anti-

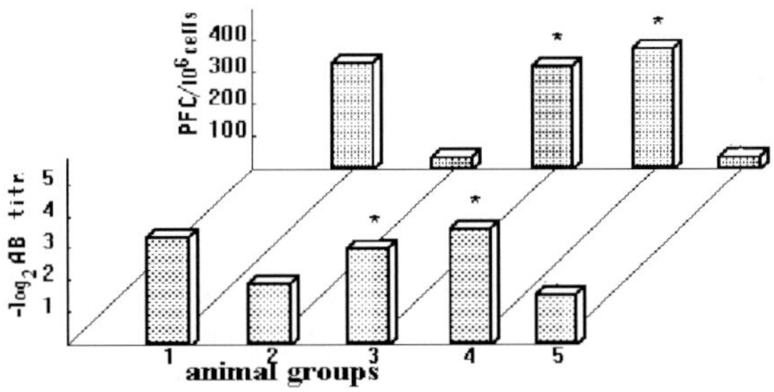

Figure 7. Effect of intraperitoneally injected total rabbit defensin on the humoral immune response of stressed rats.

1 immunized animals not subjected to stress.

2 animals subjected to severe cold stress before immunization.

3 animals subjected to severe cold stress before immunization which received rabbit defensins at the dose of 150 ng/g 10 min. before stress.

4 animals subjected to severe cold stress before immunization which received rabbit defensins at the dose of 150 ng/g 10 min. after stress.

5 animals subjected to severe cold stress before immunization which received protamine sulfate (clupein) at the dose of 150 ng/.g

Asterisk: $p<0.05$ vs. group 2.

microbial barrier of the organism. During mobilization of neutrophils the permanent secretion of defensins and other physiologically active substances takes place. These humoral factors exert corticostatic and immunoprotective effects, and influence the interaction of the immune and neuroendocrine systems for the optimization of host defense during stress [33,35].

ACKNOWLEDGEMENT

This work was supported by Russia Fund of Fundamental Investigations grants # 99-04-49508, # 00-04-48996 and by National Institute of Health by Fogarty Award TW00355.

We thank Olga Shamova, Galina Aleshina, Elena Fomicheva, Dmitriy Orlov, and Viktoria Ignatieva for their expert technical assistance.

NOTE: LEGENDS TO FIGURES

Experiments were carried out on male (CBAxC57Bl/6) F_1 hybrid mice weighing 18 to 20 grams and Wistar rats weighing 180 to 200 grams. Several models of stress were used in the study. In the mouse experiments rotation stress was used. The mice were rotated for 10 minutes at 78 rpm. Total rabbit defensin (obtained from neutrophils [7]) in doses of 20 ng/g and 2 µg/g was administered intraperitoneally 10 minutes before stress. Blood was collected for corticosterone determination by decapitation of the animals at 30, 60, and 120 minutes after stress. Each experimental group contained 5 animals. Combined stress was used in mice for studying the effects of defensins on the antibody response. The animals were restrained in special stands and placed in a refrigerator (+ 4–5^{0}C) for 1 hour. The mice were then moved in the same stands to premises at room temperature (18–20^{0}C) for 24 hours. During the 24 hour period the animals received no food or water. Rabbit defensins at doses of 2 μg/g and 20 μg/g body weight were injected i.p. 10 minutes before combined stress. Sheep erythrocytes at a dose of $5x10^7$ cells were administered immediately after the stress ended. On the 6th day after immunization the animals were decapitated and blood was taken for an antibody titer ($-\log_2$ titer) determination by the method of direct hemagglutination. Spleens were also collected for plaque forming cell (PFC) assay.

Rats were subjected to stress by 20 minute cooling in a freezer at a temperature of -20^0C. Ten minutes before stress, the animals were injected with rabbit defensins at a dose of 150 ng/g. After 30 minutes, the animals were decapitated, and blood was taken for corticosterone determination.

The effect of defensins on the ACTH-induced increase in the blood levels of glucocorticoid hormones was studied in mice. For this purpose, ACTH ("Organon", Netherlands) at a dose of 0.02 IU/g was injected 30 minutes after intraperitoneal administration of individual fractions of rabbit defensins or porcine protegrins (isolated from neutrophils [8]). Control mice received equal volumes of physiological saline. The level of serum corticosterone was determined by radioimmunoassay, using a standard kit (RIA Kit ^{3}H corticosterone, Diagnostic product, California, USA).

REFERENCES

1. Zeya HI, Spitznagel JK. Antibacterial and enzymatic basic proteins from leukocyte lysosomes: separation and identification. Science 1963; 142: 1085–1087.
2. Metchnikoff E. Immunity in infective diseases. Cambridge University Press. Cambridge 1905.
3. Lehrer RI, Bevins CL, Ganz T. In: Mucosal Immunology Ogra PL et al editors. Defensins and other antimicrobial peptides. San Diego et al. Academic Press, 1999: 89–99.
4. Kokryakov VN. The biology of antibiotics from animal sources (in Russian). St. Petersburg: Nauka, 1999.
5. Panyutich AV, Panyutich EA, Krapivin VA, Baturevich EA, Ganz T. Plasma defensin concentrations are elevated in patients with septicemia or bacterial meningitis. J Lab Clin Med 1993; 122: 202–207.
6. Zhu Q, Hu J, Mulay S, Esch F, Shimasaki S, Solomon S. Isolation and structure of corticostatic peptides from rabbit fetal and adult lung. Proc Natl Acad Sci USA 1988; 85: 592–596.
7. Selsted ME, Szklarek D, Lehrer RI. Purification and antibacterial activity of antimicrobial peptides of rabbit granulocytes. Infect Immun 1984; 45: 150–154.
8. Kokryakov VN, Harwig SS, Panyutich EA et al. Protegrins: leukocyte antimicrobial peptides that combine features of corticostatic defensins and tachyplesins. FEBS Lett 1993; 327: 231–236.
9. Boman HG. Peptide antibiotics and their role in innate immunity. Annu Rev Immunol 1995; 13: 61–92.
10. Lehrer RI, Lichtenstein AK, Ganz T. Defensins: antimicrobial and cytotoxic peptides of mammalian cells. Annu Rev Immunol 1993; 11: 105–128.
11. Hoffmann JA. Innate immunity of insects. Curr Opin Immunol 1995; 7: 4–10.
12. Harwig SS, Swiderek KM, Kokryakov VN et al. Gallinacins: Cysteine - rich antimicrobial peptides of chicken leukocytes. FEBS Lett 1994; 342: 281–285.
13. Tang YQ, Yuan J, Ösapay G et al. A cyclic antimicrobial peptide produced in primate

leukocytes by the legation of the two truncated -defensins. Science 1999; 286: 498–502.
14. Huttner KM, Bevins CL. Antimicrobial peptides as mediators of epithelial host defense. Pediatr Res 1999; 45: 785–794.
15. Harder J, Bartels J, Christophers E, Schroder JM. A peptide antibiotic from human skin. Nature 1997; 387: 861.
16. Lehrer RI, Ganz T. Antimicrobial peptides in mammalian and insect host defence. Curr Opin Immunol 1999; 11: 23–27
17. Ranadive NS, Cochrane CG. Isolation and characterization of permeability factors from rabbit neutrophils. J Exp Med 1968; 128: 605–622.
18. Yamashita T, Saito K. Purification, primary structure and biological activity of guinea pig neutrophil cationic peptides. Infect Immun 1989;57: 2405–2409.
19. Befus AD, Mowat C, Gilchrist M, Hu J, Solomon S, Bateman A. Neutrophil defensins induce histamine secretion from mast cells: mechanisms of action. J Immunol 1999; 163: 947–953.
20. Territo MC, Ganz T, Selsted ME, Lehrer RI. Monocyte chemotactic activity of defensins from human neutrophils. J Clin Invest 1989; 84: 2017–2020.
21. Chertov 0, Michiel DF, Xu L. et al. Identification of defensin-1, defensin-2, and CFP37/ azurocidin as T-cell chemoattractant protein released from interleukin-8-stimulated neutrophils. J Biol Chem 1996; 271: 2935–2940.
22. Yang D, Chertov O, Bykovskaia SN et al. B-defensins: linking innate and adaptive immunity through dendritic and T cell CCR6. Science 1999; 286: 525–528.
23. Lillard JW, Boyaka PN, Chertov 0, Oppenheim JJ, McGhee JR. Mechanisms for induction of acquired host immunity by neutrophil peptide defensins. Proc Natl Acad Sci USA 1999; 96: 651–656.
24. Fleischmann J, Selsted ME, Lehrer RI. Opsonic activity of MCP-I and MCP-2 cationic peptides from rabbit alveolar macrophages. Diagn Microbiol Dis 1985; 3: 233–242.
25. Hawiger J, Horn RG, Koening WG, Collins RD. Activation and release of lysosomal enzymes from isolated leukocytic granules by liposomes. Jale J Biol Med 1969; 42: 57–69.
26. Kudriashov BA, Ashmarin IP, Liapina LA et al. Functional state of the animal anticoagulating system after injection of defensins in blood (in Russian). Sechenov Physiol J 1988; 74: 1759–1764.
27. Higazi AA, Barghouti I I, Abu-Much R. Identification of an inhibitor of tissue-type plasminogen activator-mediated fibrinolysis in human neutrophils. A role for defensin. J Biol Chem 1995; 270: 9472–9477.
28. Lin J, Harper PC, Gurewich V. Fibrin-bound lipoprotein (a) promotes plasminogen binding but inhibits fibrin degradation by plasmin. Biochemistry 1994; 33: 2554–2557.
29. Tkatchenko SB, Ashmarin IP, Rudko IA, Korneva EA, Kokryakov VN, Kubatiev AA. Influence of defensin on functional activity of platelets (in Russian). Bull Exp Biol Med 1993; 112: 23–25.
30. Zhu Q, Singh A, Bateman A, Esch F, Solomon S. The corticostatin anti-ACTH and cytotoxic activity of peptides isolated from fetal, adult and tumor-bearing lung. J Steroid Biochem 1987; 27: 1017–1423.
31. Solomon S. Corticostatins. TEM 1993; 4: 260–264.
32. Zhu Q, Bateman A, Singh A, Solomon S. Isolation and biological activity of corticostatic peptides (ANTI-ACTH). Endocr Res 1989; 15: 129–149.
33. Bateman A., Singh A., Kral Th., Solomon S. The immune-hypothalamic-pituitary-adrenal

axis. Endocrine Revs 1989; 10; 92–112.
34. Tominaga T, Fukata J, Naito Y et al. Effects of corticostatin-1 on rat adrenal cells in vitro. J Endocrinol 1990;125: 287–292.
35. Shamova OV, Lesnikova MP, Kokryakov VN, Shkhinek EK, Korneva EA. Effect of defensins on the blood level of corticosterone and the immune response during stress. Bull Exp Biol Med 1993; 115: 728–731.
36. Cervini LA, Gray WR, Kaiser R et al. Rat corticostatin R4: synthesis, disulfide bridge assignment, and in vivo activity. Peptides 1995; 16: 837–842.
37. Zhu Q, Solomon S. Isolation and mode of action of rabbit corticostatic (antiadrenocortico tropin) peptides. Endocrinology 1992; 130: 1413–1423.
38. Fuse N, Hayashi Y, Fukata J et al. Purification and characterization of new antiadrenocotropin rabbit neutrophil peptides (defensins). Eur J Biochem 1993; 216: 653–659.

The Immune-Neuroendocrine Circuitry
Edited by I. Berczi and A. Szentivanyi

The Acute Phase Response

ISTVAN BERCZI and ANDOR SZENTIVANYI

Department of Immunology, Faculty of Medicine, The University of Manitoba, Winnipeg, Manitoba R3E OW3, Canada; and Department of Internal Medicine, Faculty of Medicine, The University of South Florida, Tampa, Florida 33612, USA

ABSTRACT

Mild infection or sublethal dose of endotoxin elicit a brief elevation of growth hormone (GH) and prolactin (PRL) in the serum. These hormones are pro-inflammatory and immunostimulatory. In severe trauma, sepsis and shock, GH and PRL are suppressed, whereas glucocorticoids and catecholamines are elevated. Under these conditions an acute phase response is induced by immune-derived cytokines, primarily IL-1, IL-6, TNFα, which elicit a neuroendocrine response and initiate major metabolic alterations. Fever and catabolism prevails, whereas the synthesis of acute phase proteins in the liver, cell proliferation in the bone marrow, and protein synthesis by leukocytes is elevated. This is an emergency reaction to save the organism after the adaptive immune system has failed to contain and eliminate the infectious agent. During sepsis and endotoxin shock the systemic activation of the complement system and of leukocyte-derived releasing enzymes, tissue-derived brake-down products and highly toxic cytokines seriously threaten survival. Glucocorticoids regulate pro-inflammatory cytokine production and potentiate the secretion of acute phase proteins into the serum. Some of these proteins, such as C reactive protein, LPS binding protein and mannose binding protein are designed to combine with microorganisms and trigger their destruction by the activation of the complement system and of phagocytes. The increased production of some complement components also helps host resistance. The rise in serum fibrinogen promotes blood clotting, which can serve to isolate the invading agent by triggering thrombosis in infected tissues. A number of enzyme inhibitors are produced as acute phase proteins, which are likely to serve to curb the nonspecific damage inflicted by enzymes released from activated phagocytes and from damaged cells. Catecholamines are elevated, which serve to inhibit inflammatory responses and to promote, even initiate, the acute phase response. Serum leptin is also increased, which governs energy metabolism and it is a major stimulator of the immune system.

If the acute phase reaction fails to protect the host, shock will develop. Patients with subclinical adrenal insufficiency succumb to septic shock almost invariably if glucocorticoid therapy is not given. However, glucocorticoid treatment of septic patients with normal adrenal function has not been helpful. During the acute phase response the T-cell regulated adaptive immune response is switched off and natural immune mechanisms are amplified several hundred to a thousand times within 24–48 hours. This phenomenon has been designated as immunoconversion. Immunoconversion is initiated by immune derived cytokines, and involves profound neuroendocrine and metabolic changes, all in the interest of host defence. Thus, natural immunity is essential for a first and last line of defence and the neuroendocrine system is an important promoter and regulator of this fundamental form of immune defence.

1. INTRODUCTION

Fever was regarded as a beneficial healing process in Egypt, Greece and in the Roman and Persian empires. This appreciation continued until the first half of the 19th century, when "fever therapy" was introduced to treat diverse diseases, such as rheumatoid arthritis, manic depression, headaches, neurosyphilis, herpes zoster, and various eye and skin disorders [1]. Initially, fever therapy was done with whole Gram-negative microorganisms until Boivin et al [2] purified *endotoxin* from such bacteria. Endotoxin was first tested in animals and later on it was used for the treatment of asymptomatic neurosyphilis in human beings by inducing therapeutic febrile reactions [3]. Today, numerous beneficial effects can be induced by sublethal doses of endotoxin in animals. These include increased resistance against infectious agents and to injury [4], protective effect against X-irradiation [5], a powerful adjuvant effect to enhance the antibody response [6,7] a powerful antitumor effect [8], stimulation of bone marrow function [5,9] the activation of blood clotting and complement system [10], and promotion of regeneration and healing in the liver [11], and possibly even in the central nervous system [12].

At the end of the 19th century it was observed that glucosuria occurs during infectious disease. This was followed by the demonstration of hyperglycemia, insulin resistance, the detection of endogenous pyrogen and identification of a leukocyte derived endogenous mediator (LEM). LEM was capable of inducing acute phase reactants in the liver during severe infectious disease. Pyrogens were also shown to induce ACTH release. By 1975, it was known that there was a neuroendocrine response to infectious disease, which involved glucocorticoids, mineralcorticoids, insulin, glucagons, growth hormone, and a metabolic response [13,14,15]. During the early 1980s, it was found that interleukin-1 (IL-1) is an endogenous pyrogen. Endogenous pyrogen was known to release ACTH at that time. On this basis the hypothesis was proposed in 1986 that IL-1 is one of the immune derived mediators that acts on the pituitary gland [16]. Indeed, IL-1 was found to induce pituitary hormone release in a number of laboratories [17,18,19,20,21]. The thymic peptide, thymosin was shown to release ACTH in 1983 [22]. Today it is clear that immune derived cytokines, including IL-1, are important inducers of the neuroendocrine and metabolic response to infection, which is now known as the *acute phase response* (APR).

Hans Selye [23] discovered that a variety of noxious agents cause a profound involution of the thymus, spleen and lymph nodes, and the enlargement of the adrenal gland. In subsequent experiments Selye showed that these changes were mediated by the activation of the pituitary-adrenal axis and that glucocorticoids were responsible for the lymphoid involution observed [24]. Physical, chemical, biological agents or even emotional factors could evoke this neuroendocrine response, which were designated by Selye as *stressors* and the their effect was defined as *stress*. He termed the body's reaction to stress as the *general adaptation syndrome*. Stressed animals and individuals show an initial alarm reaction, which is followed by a period of adaptation when the organism is resistant towards the stressor and the endocrine and other parameters return to normal. Eventually, with lasting stress, a breakdown due to exhaustion will occur, which can lead to death [25,26]. Selye [27] discovered the inhibitory effects of glucocorticoids on inflammation and made observations indicting that immune reactions are also subject to stress induced alterations [28]. He demonstrated the influence of sex steroid hormones on lymphoid organs [29]. It is now apparent that the general adaptation syndrome of Selye is analogous to the intensive and highly coordinated emergency defence reaction in man and in higher animals, now designated as the *acute phase response* [30,31,32,33].

2. THE RESPONSE TO INJURY

In a broad sense the agents capable of causing injury may be classified into physical, chemical and biological categories. Without exception injured cells release chemokines and cytokines, which in turn attract and activate leukocytes that lead to inflammation and in most instances to regeneration and healing.

2.1. Physical injury

Of the physical agents, ultraviolet irradiation of the skin activates keratinocytes to release cytokines and other mediators, which leads to an inflammatory response and to immunosuppression. X-irradiation causes apoptosis; cold injury causes mast cell degranulation that leads to inflammation. Burn injury leads to a major neuroendocrine response characteristic of APR and to a profound suppression of adaptive immunity, reduced serum immunoglobulin, complement and properdin levels. Cytokines such as IL-1, IL-6 and TNFα are elevated in the blood. Trauma leading to tissue damage and bleeding activates the coagulation, kallikrein-kinin, fibrinolytic and complement systems. The complement split products, C3a and C5a, activate mast cells and fibroblasts and attract neutrophilic granulocytes and macrophages. Cytokines and other mediators released by platelets, mast cells, macrophages, neutrophils and activated endothelial cells play a major role in the local inflammatory response to injury. Severe injury and bleeding may cause elevated cytokine levels in the blood, such as TNFα, IL-1 and IL-6, which in turn initiate a profound neuroendocrine and metabolic response characteristic of APR [34].

Major surgery evokes an APR, transient immunosuppression and alterations in gastrointestinal function. Normal function is usually restored after a few days; however, in a subgroup of patients homeostasis may be lost and the development of systemic inflammatory response syndrome ensues. Recent clinical trials suggest that immunomodulatory nutrients, including glutamine, arginine, n-3 polyunsaturated fatty acids and dietary nucleotides, may promote restoration of normal tissue function post-operatively and prevent the occurrence of inflammatory syndrome [35]. In patients undergoing elective surgery a significant partial correlation was found between plasma leptin and serum IL-6 concentration. Patients with APR had a significantly elevated level of plasma leptin per unit fat mass [36]. Statistically significant elevation of plasma leptin concentration occurred in patients 24 h after surgery. Leptin levels quickly returned to preoperative levels 48–72 h after operation. Regression analysis revealed the highest coefficient of leptin levels with TNF-alpha and for IL-6 24 h after surgery ($r = 0.29$, $p < 0.05$). There was no significant correlation between leptin and IL-1 and between leptin and soluble IL-2 receptor, respectively [37].

2.2. Chemical injury

A vast amount of chemicals have the potential to cause inflammation by irritation, by inducing necrosis, apoptosis, or even to trigger immune mechanisms by altering self proteins. The cytokine response to chemical insults could be mediated by neurogenic pathways, by those outlined under trauma, and by antigen mediated activation of the immune system [38,39,40]. A toxicity experiment in 25 beagle dogs showed that about half of the animals exhibited increases of fibrinogen and/or WBC counts compared to the pre-dosing values. This response was dose-dependent and transient lasting for 1–2 days. These changes were most remarkable after intravenous application of the toxic agents [41].

2.3. Biological injury

2.3.1. Infection and parasitic infestation

Infections frequently occur in immunocompromised patients, which include the elderly, patients on immunosuppressive therapy, cancer patients and those suffering from severe trauma or shock. The adaptive immune system is designed to control infections locally. The penetration of infectious agents or their products into the bloodstream frequently elicits febrile illness, which in actual fact amounts to APR. Many pathogenic microorganisms are capable of polyclonal lymphocyte activation [42], which leads to massive cytokine production.

Glucocorticoids (GC) are increased shortly before the onset of fever. The circadian rhythm is lost if the infection becomes subacute or chronic, and glucocorticoid levels generally fall to values below normal [15]. Patients with subclinical GC deficiency succumb to infection sooner than those with a normal GC response [43,44]. *Insulin*, *glucagon* (GLN), *epinephrine* (EP), and *norepinephrine* (NEP) are also elevated during infection [45,46,47,48].

In sheep, bacteremia caused a surge of plasma βEND, followed by increases in PRL and GH and a depression of plasma LH. Only the GH response was inhibited by the opioid antagonist drug, naloxone (NAL) [49]. Septic patients with suppressed basal IN levels and elevated GLN did not respond to treatment with human GH by elevated serum levels of insulin-like growth factor I (IGF-I) [50].

2.3.2. Bacterial endotoxin

Lipopolysaccharide (LPS), also known as *endotoxin,* is a constant component of the outer cell membrane of Gram-negative bacteria. Biochemically it can be divided into polysaccharide and lipid regions, the latter serving as anchor to the outer membrane of bacteria. The core glycolipid, *lipid A,* plus the core polysaccharide, is an *obligatory component* of the bacterial cell wall (Figure 1), as mutants lacking this component have not been described. In contrast, there is great variation of the polysaccharide chain sugars, which differ from one bacterial strain to the other and provide epitopes for specific immune reactions and for the serological classification of Gram-negative bacteria [51]. The capsular polysaccharides, which are present in bacteria forming "smooth" colonies and provide the so-called K antigens, and the O-specific polysaccharides of LPS, contain *heterologous epitopes.* These are recognized specifically by the adaptive immune system, according to the individual bacterial strains. On the other hand, lipid A is evolutionarily highly conserved and shows extensive cross-reactivity amongst all Gram negative bacterial strains whether they are pathogenic or saprophyte. This antigenic determinant may be classified as a *homologous epitope*, or *homotope* for short, which is present in all Gram-negative bacteria, regardless of strain and pathogenecity. Lipid A is recognized by both the natural and the adaptive immune system. It is possible to produce anti-lipid A antibodies, including monoclonal antibodies, with deliberate immunization [52,30].

For a long time lipid A was known, and still is, as the "toxic moiety" of endotoxin, which is capable of a massive activation of the immune system, and in turn, immune derived cytokines may induce shock and death [53]. Lipopolysaccharide binding proteins (LBP) are present in the serum, which are produced by the liver. LBP has been identified in the serum of multiple spe-

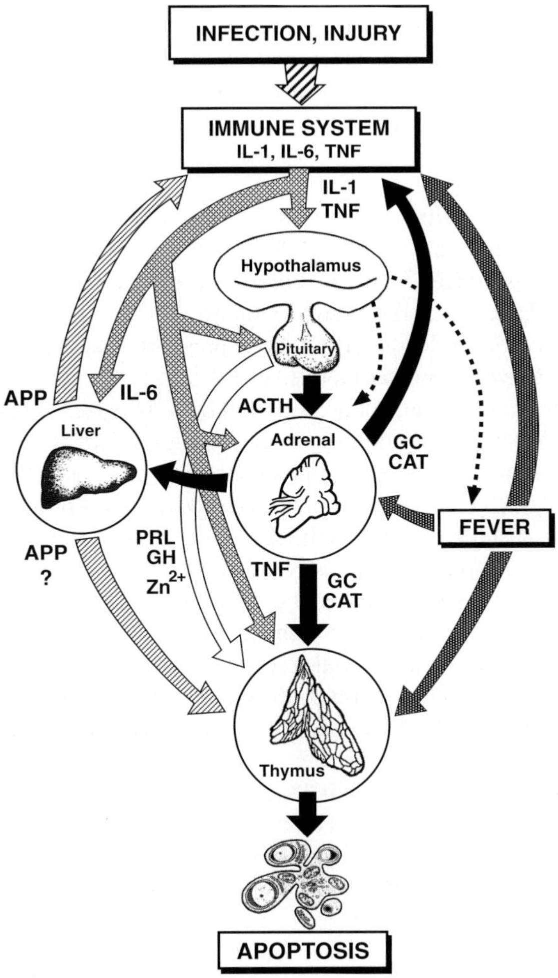

Figure 1. Major regulatory pathways in acute phase response-associated thymic involution.
Proinflammatory cytokines IL-1, -6 and TNFα typically produced in response to a systemic infection and/or tissue damage. These cytokines in turn stimulate the hypothalamus-pituitary-adrenal (HPA) axis. Increased levels of glucocorticoids together with TNFα are of prime importance in the induction of thymic involution via apoptosis of cortical double positive ($CD4^+8^+$) thymocytes. Among other factors/changes induced during the APR, which are likely to contribute to the transient involution of the thymus are fever, catecholamines, downregulation of growth hormone and prolactin synthesis or action and hypozincaemia. Adapted from reference [32].
IL = interleukin, TNF = tumor necrosis factor, ACTH = adrenocorticotropic hormone, APP = acute phase proteins, GH = growth hormone, PRL = prolactin, GC = glucocorticoids, CAT = catecholamines.

cies, including rabbits, rats, mice, pigs, cattle, non-human primates, and humans. LBP combines with lipid A after Gram-negative infection or LPS injection [54]. LBP is a 60 kDa glycoprotein present in normal serum at concentrations of 0.5–10 μg/ml and its level may surpass 200 μg/ml during APR [55]. Its specificity is directed towards the hydrophobic lipid A portion of LPS with an affinity of 10^{-9} M for both smooth and rough forms of LPS [56]. LBP acts as an opsonin for

LPS bearing particles by enhancing its interaction with the CD14 cell surface molecule, which is present on the surface of monocytes, macrophages and neutrophilic granulocytes. LBP serves as a lipid transfer protein that facilitates the rate of transfer of LPS to CD14. This enables the cell to respond to extremely low levels of LPS, which otherwise are unable to elicit a biological response [57]. Another serum protein which mediates LPS recognition by CD14 is septerin [58]. This protein has not been studied extensively to date.

CD14 is a 55 kDa glycoprotein present on the surface of monocytes, macrophages and neutrophils. It lacks a transmembrane sequence [59]. CD14 plays an important role in mediating the induction of cytokines by LPS, such as tumor necrosis factor-α (TNFα), IL-6 and IL-8, in monocytes and macrophages. 1-25-Dehydroxyvitamin D3 (VD3) induced CD14 expression in a premonocytic cell line. IL-4 decreased CD14 expression, and IL-4 or interferon-γ (IFN-γ) inhibited soluble CD14 release by monocytes. TNFα and LPS enhanced CD14 expression by monocyte cell lines. In human neutrophils, TNFα, GM-CSF, G-CSF and formyl peptide have all increased CD14 expression. In addition to mediating the LPS response, CD14 appears to participate in diverse cellular responses that involve cell-to-cell contact [60].

Soluble CD14 is of 50–53 kD molecular weight single chain protein present in the serum that contains intrachain disulphide bonds. The serum concentration in normal human plasma is 6 μg/ml and increases in hospitalized patients, especially in those suffering from autoimmune disease. LPS and TNFα increase the release of CD14, whereas IL-4 decreases its release *in vitro* from normal peripheral blood monocytes. Soluble CD14 inhibits the biological activities of LPS and is assumed to present LPS to endothelial cells [60]. The recently identified Toll-like receptor 4 may be essential for signal transduction by complexes of LPS and membrane bound CD14, leading to NFκB activation [61].

There is evidence to indicate that molecules other than CD14 play a role in cell signalling by LPS. One such molecule is CD18, which belongs to the integrin family of adhesion molecules and exists in three forms, CD11a/CD18, CD11b/CD18, and CD11c/CD18. All three are capable of binding LPS and participate in the phagocytosis of bacteria. CD18 is not essential for mediating cellular responses to LPS. LPS up-regulated CD11b/CD18 expression on human neutrophils [61].

The sensitivity to LPS toxicity differs considerably among various species of mammals. Lower vertebrates, such as frog and fish, show extreme resistance [62]. Nevertheless some observations suggest that at least some species of fish (Tilapia oreochnomis mossanbicus, Teleosti) react to LPS and show an integumental and a cortisol response [63]. In contrast, the horseshoe crab (Limulus polyphemus) responds to LPS with fatal intravascular coagulation. This is due to the activation by endotoxin of a clottable protein of Limulus blood, which is produced by circulating amoebacytes [64,65,66].

It seems clear on the basis of the above presented information, that lipid A is not a toxin at all for animal cells, but rather, it is a highly conserved homologous epitope, which has been singled out during evolution by vertebrate and some invertebrate animals for defence purposes against Gram-negative bacterial infections. Lipid-A, which is the specific homotope involved, is recognized by serum proteins and cell surface receptors which are capable of activating the coagulation and complement systems and various members of the leukocyte series. This enables the animal to mount a rapid and effective immune defence reaction against all Gram-negative pathogens [30].

Table I Major neuroendocrine changes induced by endotoxin.

The HPA Axis	Response	Immunostimulatory Hormones	Response	Gonadotropins & Steroid Hormones	Response
CRF	↑	TRH	↓	LH	↑↓
AVP	↑	TSH	0↓	FSH	↓
ACTH	↑	T4	↓	E2	↑↓
GC	↑	T3	↓	TS	↑↓
α-MSH	↑	PRL	↑↓	DHEA	↓?
β-END	↑	GH	↑↓	PS	↑↓
CAT	↑	IGF-I	↓		
		INS	↑		
		GLU	↑		
		leptin	↑		

NOTE: ↑ = increase, ↓ = decrease, ↑↓ = variable effect, 0 = no effect, ? = predicted change.
Adapted from Ref. [33].

3.2.3. Cytokine response to LPS

After systemic administration of LPS to mice, TNFα is significantly increased in the blood at 1–2 hours, which is followed by a decline, the TNF level returning to normal at around 4 hours. In adrenalectomized (ADX) animals an exaggerated TNF response occurs, whereby TNF levels in the plasma rise approximately 60 times, and sensitivity to the lethal effects of TNF increases approximately 500 times, over that of normal controls. This excessive response and increased mortality can be prevented if the animals are pre-treated with dexamethasone. The inhibition of cortisone synthesis in the adrenals by metyrapone also leads to an increased susceptibility (~15 times) to LPS [67,68,69]. Similar kinetics of TNF release was found in man after LPS infusion [70].

Interleukin-1 also rises in the blood of mice after endotoxin administration reaching the maximum at about 4 hours and elevated levels remain up to 24 hrs [69]. Circulating IL-6 increases significantly after LPS administration. In man IL-6 peaked at 120 minutes after LPS administration, which was not inhibited by glucocorticoids or by repeated LPS administration [53,71,72].

Leukemia inhibitory factor (LIF) rose moderately in mice after a sub-lethal injection of LPS and rose progressively during lethal septic shock induced by E-coli. LIF administration to mice induced catabolism and had a protective effect if given prior to the administration of bacteria [73].

Additional cytokines and mediators also play a role in endotoxin shock that include interleukin-8 [74], interleukin-10 [75], interferon-γ [76], TNF synthesis inhibitor [77], interleukin receptor antagonist [78], platelet activating factor, colony stimulating factor, prostaglandins and thromboxanes [53,34,79].

2.3.4. Neuroendocrine response to endotoxin

Wexler et al [80] observed for the first time the stimulation of ACTH in rats by endotoxin as detected by the depletion of ascorbic acid and cholesterol in the adrenal glands. LPS was ineffective in causing these changes in the adrenal glands of hypophysectomized rats.

Endotoxin, infectious disease, and various forms of injury all elicit a neuroendocrine response via the stimulation of cytokines [53,34,79]. Profound changes occur in serum hormone levels

(Table I). These changes are likely to be more complicated than are indicated in the table. It is rapidly becoming clear that dynamic and diurnal changes of hormones should be taken into consideration. Much remains to be elucidated about the significance of these hormonal alterations. Nevertheless, it is very clear that the hypothalamus-pituitary-adrenal (HPA) axis exerts a powerful suppressive effect on the adaptive immune system and controls the level of inflammatory cytokines. Through the activation of this axis, specific immune reactions are profoundly suppressed, whereas the induction of acute phase proteins in the liver and natural antibodies are stimulated by glucocorticoids and catecholamines, which are also elevated. Therefore, the conversion of the immune system from the adaptive mode of reactivity to the amplification of natural immunity is largely due to the activation of this axis. [53,81,82,83]. Prolactin and growth hormone are immunostimulatory and usually rise within the first hour after endotoxin injection, which is followed by a decline and the level may become low normal to subnormal in serious cases of endotoxin shock. Luteinizing hormone (LH), follicle stimulating hormone (FSH), estrogens, androgens, progesterone, and thyroid hormones all decline during infection and endotoxin shock, as a rule. Insulin, glucagon, α-melanocyte stimulating hormone (MSH), endorphin, and arginine vasopressin are increased during endotoxemia [53,34,79].

2.3.5. *Endotoxin shock*

Geller et al [84] observed that the administration of cortisone to mice prior to, or simultaneously with, a lethal dose of LPS protected the majority of the animals from death, but there was no protection when cortisone was administered after LPS. The protective effect of GC against endotoxin shock has been since observed repeatedly in various species [84,85,86].

No correlation was found between TNF serum levels and the lethal effects induced by different types of LPS in rats [87]. The co-treatment of rats with low doses of TNF and LPS resulted in the rapid demise of the animals leading to 100% mortality within 4 hours [88]. Pre-treatment of rats with a single low i.v. dose of TNF prevented subsequent death from a lethal dose of TNF 24 hours later and also provided protection against the lethal effects of LPS applied 24 hours later [90]. The lethal effect of endotoxin could be inhibited in various animal models by the opioid antagonist, naloxone, by indomethacin, by the IL-1 receptor antagonist, by monoclonal antibodies to TNFα and by the pharmacological inhibition of platelet activating factor (PAF) [34].

2.3.6. *Immunological injury*

Substances that are otherwise harmless to the body may cause severe injury if there is an abnormal immune response (hypersensitivity) directed against them. Tissue injury may be mediated by IgE antibodies (immediate hypersensitivity, allergy, asthma, anaphylaxis), by sensitized T lymphocytes (delayed hypersensitivity, contact dermatitis, etc.), by immune complexes and by phagocytic cells activated either by antibodies or cytokines. Humoral or cell mediated immune reactions may also be directed against self antigens that could lead to autoimmune disease. A detailed discussion of these mechanisms is beyond the scope of this Chapter. Instead, the reader is referred to other chapters in this volume and to some key references [34,38,39,53,91–96].

3. THE ACUTE PHASE RESPONSE

Wannemacher et al [97] isolated a protein from leukocytes and named it leukocytic endogenous mediator (LEM), which stimulated the uptake of amino acids by the liver in ADX, Hypox, thyroidectomized or diabetic rats. Similar stimulation could not be duplicated by pharmacological doses of a large variety of hormones. LEM augmented RNA synthesis and enhanced the hepatic production of a number of acute phase plasma globulins.

The APR is the reaction of the body to injury caused by physical, chemical and biological agents, including infectious disease. Clinically, APR is characterized by fever, loss of appetite, inactivity and sleepiness. The reaction is mediated by cytokines which appear in the circulation and function as acute phase hormones affecting the central nervous system (CNS), the neuroendocrine system and virtually the function of every other tissue and organ in the body. IL-1, IL-6 and TNFα have been identified as major mediators of the endocrine and metabolic changes characteristic of APR. However, several other cytokines have been found to be inducers of APP [98,99,100,101]. ACTH and GC, leptin, EP, NEP, GLN, AVP, and ALD are elevated during the APR, whereas GH, estrogens, androgens, IN, and thyroid hormones may be either elevated or suppressed, depending on the severity of the condition (Table I) [102,30,31,32,33,103].

During APR (eg. sepsis) the serum level of leptin (LEP) rises rapidly. Cytokines, especially TNF-α, causes this elevation. LEP inhibits glucocorticoid and IL-6 production. The levels of LEP in serum correlate positively with the survival of patients with septicaemia. LEP stimulates the production of IL-1ra, which protects against LPS toxicity in mice. LEP stimulated the production of IL-1δ in murine glial cells. Exogenous LEP upregulated both phagocytosis and the production of proinflammatory cytokines in animals. Leptin is also involved in wound healing and angiogenesis [103–112].

An alteration of protein synthesis by the liver is most characteristic for APR. The synthesis of acute phase proteins (APP) is initiated, whereas the synthesis of some normal serum constituents such as albumin and transferrin is decreased. The concentration of APP increase dramatically in the serum. For example, in man, C reactive protein (CRP) and serum amyloid A (SAA) may increase over 1,000-fold. Fibrinogen, α1-antitrypsin and certain complement and properdin components (factor B and C3) show a more moderate increase [30,33].

C reactive protein binds to C-type pneumococcal cell walls in the presence of Ca^{2+}. CRP has been identified in multiple species, including mammals, chicken, fish and crab. In the serum CRP consists of 5 identical subunits, which form a ring-shaped molecule named *pentraxin*. This term now also stands for a protein family. In man, serum amyloid P also belongs to this family. The mature subunits consist of 206 amino acids with 23 kDa molecular weight [113].

CRP recognizes specifically several homotopes, such as phosphocholine, with polysaccharides containing galactose, with some biologic polycations, such as protamine, poly-*L-lysin* and with myelin basic protein. CRP is in its pentameric form, if Ca^{2+} is present, and binds phosphocholine and galactans, whereas in the absence of Ca^{2+}, it becomes monomeric and binds various polycations. These homotope determinants are frequently present on the surface of bacteria, fungi, parasites and damaged cells and tissues. After combination with the specific ligand, CRP activates complement by the classical pathway, induces chemotaxis and enhances phagocytosis by neutrophilic leukocytes and monocytes and elicits tumoricidal activity in macrophages, all of which are complement dependent. In addition, CRP stimulates the synthesis of IL-1, TNFα, and potentiates the cytotoxic activity of T lymphocytes, natural killer (NK) cells and platelets. CRP localizes *in vivo* at sites of inflammation. It binds platelet activating factor (PAF) and blocks its activity. The clinical determination of CRP is diagnostic for the presence of infectious and inflammatory disease [113–115].

Upon denaturation or after attachment onto polystyrol plates, free CRP subunits attain a third conformation referred to as neoCRP. NeoCRP is a membrane protein on NK cells and macrophages and functions as galactose-specific receptors. It also accumulates at injured sites of tissue. Monocyte/macrophages express a specific CRP receptor. Proteolytic fragments of CRP activate macrophages and neutrophils [113]. Human CRP protected mice from an otherwise lethal S. pneumoniae infection [116].

The LBP shows a 100-fold increase (from 0.5–50 μg/ml) in the serum during an APR. LBP is capable of opsonizing LPS bearing particles, and thus may be required for the activation of complement by endotoxin through the alternate pathway. LBP-LPS complexes are also potent stimulators of cytokines from monocytes and macrophages after combining with CD14 on the surface of these cells [54].

Other APP are *proteinase inhibitors*, such as α2-macroglobulin, α1-acid glycoprotein, antithrombin III, α1-acute phase globulin, and α1-proteinase inhibitor, which are abundant in the rat. Kupffer cells stimulate α2-macroglobulin synthesis by hepatocytes *in vitro* in the presence of 10^{-9} M DEX [117]. Fibrinogen is also and APP with an important role in blood clotting and healing. Alpha-macrofetoprotein (α-MFP) is a strong inhibitor of inflammatory mediators, such as histamine, bradykinin, serotonin, PGE2 and to inhibit polymorphonuclear chemotaxis. Catecholamines and GC induce α-MFP in normal rats. The α-MFP level induced by CAT was very high, comparable to those observed in the post-injury phase, whereas the effect of GC was moderate. In ADX rats the effect of CAT on α-MFP synthesis was greatly diminished, whereas the moderate effect of GC remained. The combination of GC and CAT induced extremely high α-MFP levels in ADX animals. Some other APP, such as haptoglobin and α1-major acute phase protein, were affected differently by these hormones [118]. Haptoglobin is an acute phase protein. It binds haemoglobin, thus preventing iron loss and renal damage. Haptoglobin is an antioxidant, has antibacterial activity and plays a role in modulating many aspects of the acute phase response [119]. The acute phase proteins alpha(1)-acid glycoprotein and alpha(1)-antitrypsin exert antiapoptotic and anti-inflammatory effects and contribute to the delayed type of protection associated with ischemic preconditioning in the kidney and in other insults [120].

Mouse mannan-binding lectins (MBL)-A and MBL-C were studied in various strains and in APR. The MBL-A and MBL-C levels in 10 laboratory mice strains were found to vary between 4 μg/ml to 12 μg/ml, and 16 μg/ml to 118 μg/ml, respectively. After the induction of APR by the i.p. injection of casein or LPS, MBL-A was found to increase approximately two-fold, with a maximum after 32 h, while MBL-C did not increase significantly. Serum amyloid A peaked at 15 h with an approximate 100-fold increase [121]. Mannose-binding lectin (MBL) is a serum protein characterized by both collagenous regions and lectin domains, which plays an important role in innate immune defence. It binds to the repeating sugar arrays on many microbial surfaces through multiple lectin domains. Following binding, MBL is able to activate the complement system via an associated serum protease, MASP-2. There is an increased incidence of infections in individuals with mutations of MBL and an association with the autoimmune disorders SLE and rheumatoid arthritis [122].

IL-1 and TNF induced a full range of APP *in vivo*, but only a limited number of APP were induced by these cytokines in cultured liver cells compared with crude cytokine preparations from macrophages. This led to the discovery of IL-6 as a major inducer of APP synthesis. Additional cytokines, namely IFN-γ, leukemia inhibitory factor, TGF-β, and oncostatin M, were found to be active as direct inducers of APP from the liver. IL-6 activates the genes of APP through the DNA binding protein called NF-IL-6. NF-IL-6 is a pleiotropic mediator of many inducible genes involved in the acute-, immune- and inflammatory responses, similarly to NFκB. Both NF-IL-6 and NFκB binding sites are present in the inducible genes, such as IL-6, IL-8 and several acute phase genes [101].

Adrenalin evokes a high level of IL-6 in rats, which can be antagonized by propranolol. When IL-6 release is blocked, the fast reacting APP, α2-macroglobulin and cysteine protease inhibitor are strongly depressed. Isoprenalin, a β2-adrenergic receptor agonist, also causes very high levels of IL-6, indicating that β2 receptors are involved [123].

The relationship between spontaneously occurring activation of the acute phase response and leptin levels were examined in 29 chronic hemodialysis patients. When CRP was elevated, leptin levels were significantly reduced, as were the negative acute phase proteins albumin and transferrin. Serum amyloid A, ceruloplasmin, alpha-1 acid glycoprotein, and IL-6 were all significantly increased at the maximum CRP level, compatible with general activation of the acute phase response. The change in leptin correlated negatively with the change in CRP, as did changes in albumin [124]. Pro-inflammatory cytokines, especially TNFα, induce inflammatory hyper-leptinemia. This is an integral part of APR and necessary for comprehensive immunocompetence. This hints at the existence of an integrated communication network to co-ordinate the energy status of the animal with the ability to fight pathogens [125].

The thymus is severely affected by stress and APR, which rapidly leads to the loss of thymocytes manifesting in profound involution. The thymus is a central primary lymphoid organ responsible for the generation of mature, functional thymus-derived (T) lymphocytes. In turn T lymphocytes govern adaptive immune reactions through their regulatory function. In addition, the thymus exerts endocrine functions, the significance of which is not yet fully appreciated. A large body of evidence attests for the existence of a complex neuroendocrine control of thymus physiology. Long standing observations indicate that the thymus becomes involuted during the stress response. During stress or more forcefully, during APR, the HPA axis is activated. This results in the suppression of the T-cell-dependent adaptive immune response, which is supported further by the suppression of the hormones that are essential for the maintenance of the thymus and of T lymphocytes (e.g., PRL, GH, IGF-I). Catecholamines and glucocorticoids, which are released in large quantities during APR, induce apoptosis in the thymus with a striking efficiency. The elevated level of TNFα and zinc deficiency, that develop during APR, also contribute to thymic involution and to the suppression of the adaptive immune system [126].

In patients with APR the GH-IGF-I axis is suppressed. Growth hormone action is attenuated. TNFα, IL-1β, and IL-6 inhibit GH-induced Spi 2.1 mRNA expression in hepatocytes. Thus the inhibition of GH signalling pathways during APR results in the reduced expression of GH-responsive genes [127]. These observations prompted clinical trials with GH treatment, which were aimed at preventing the severe catabolic state and improving immunocompetence. However, so far the results are not encouraging. In fact a controlled clinical trial showed that after GH treatment the proportion of patients that did not survive was significantly elevated [128]. Deaths attributed to "septic shock or uncontrolled infection" occurred nearly four times more commonly in GH treated patients compared to placebo receiving controls. Although no data were given regarding immune parameters, the authors suggested that alterations in immune functions have contributed to the fatalities.

Although there is no clear-cut consensus concerning the interpretation of this important clinical trial, one must keep in mind that patients with acute critical illness (i.e. APR) are involved. It is now becoming apparent that the acute phase response is a massive neuroimmune and metabolic response that mobilizes all the resources of the body in the interest of host defence and survival. The findings of Takala and co-workers [128] suggest strongly that the suppression of the GH – IGF-I axis in APR is required for intense catabolism to take place. A rapid release of nutrients and of energy is necessary under these conditions in order to support maximally the defence system of the body, that includes the HPA axis, the sympathetic nervous system, the bone marrow, CD5+ B lymphocytes, leukocytes and the liver [30,32,126]. The adaptive immune system is controlled by thymus-derived (T) lymphocytes and it needs several days to a week for an effective response. During APR no time is available for an adaptive immune reaction, and therefore, this system is shut down, primarily by the cytokine and endocrine alterations that take place. The thymus and T cell function is heavily dependent on the GH/PRL – IGF-I axis and it is suppressed profoundly by the elevated levels of glucocorticoids and catecholamines [30,32,126]. Recent observations showed that GH inhibits the production of acute phase proteins in rats with burn injury and in human hepatocytes [32,129,130]. One may argue that the most efficient way to fuel the intensive systemic effort for survival in APR is by the rapid breakdown of bodily tissues. GH is a powerful anabolic hormone, which supports the T lymphocyte dependent immune system, and acts as antagonist of the HPA axis that promotes APR [16,30,32,126,131,132]. The results of this controlled trial supports the hypothesis that the inhibition of the HPA axis and of catabolism by GH treatment in APR hampers the bodie's defence mechanisms, which may have fatal consequences. This is further supported by the results of GH treatment in burn injury discussed below.

GH is a potent anabolic agent that can be administered with minimal risk of untoward side-effects to children with burn injury. GH reduces the catabolic effects of trauma through stimulation of protein synthesis. Accelerated wound healing and reduction in tissue-wasting are clear benefits that reduce the overall morbidity associated with burn injury in children [133]. Twenty-eight thermally injured children received either 0.2 mg/kg/day of recombinant human growth hormone (rHGH) or saline within 3 days of admission and for at least 25 days. IGF-I and IGFBP-3 increased with rHGH treatment, whereas IGFBP-1 decreased in serum compared with placebo ($p < .05$). Burned children treated with rHGH required significantly less albumin substitution and demonstrated a decrease in serum CRP, amyloid-A and an increase in retinol-binding protein compared with placebo ($p < .05$). rHGH decreased serum TNFα and IL-1β, whereas no changes were found for serum IL-1α, IL-6, and IL-10 compared with placebo ($p < .05$). Free fatty acids were elevated in burned children who received rHGH ($p < .05$) [134]. Severely burned children receiving IGF-I/BP-3 showed a decrease in IL-1β and TNFα, followed by a decrease in type I acute phase proteins. This was associated with a concomitant increase in constitutive hepatic proteins [137].

In rats with burn injury rhIGF-I and IGFBP-3 decreased the proinflammatory cytokines IL-1β and TNFα, followed by a decrease in type I acute phase proteins. IGF-I/BP-3 had no effect on IL-6 and type II acute phase proteins. Decreases in acute phase protein and proinflammatory cytokine synthesis were associated with increases in constitutive hepatic proteins, total liver protein content, and hepatocyte proliferation. IGF-I/BP-3 attenuates the hypermetabolic response after thermal injury and may improve the clinical outcome [135].

In rats with burn injury treatment with hepatocyte growth factor (HGF) increased serum albumin, transferrin, α2-macroglobulin, IL-6 and TNFα, compared with controls ($p<.05$). Serum IGF-I decreased in rats receiving HGF, whereas liver IGF-I concentrations were higher, compared with controls ($p<.05$). Hepatic protein concentrations were higher in the HGF group with a concomitant increase in total liver weight and a strong mitogenic effect on hepatocytes, compared with controls ($p<.05$) [136].

Panhypopituitary patients had higher concentrations of oromucoid, fibrinogen, soluble E-selectin, P-selectin, ICAM-1 and lower levels of transferrin (TRF) than did controls. After 12-month rhGH replacement the patients showed an increase of TRF and decrease of soluble adhesion molecules similar to control levels [138].

Fever has been associated with improved survival and shortened disease duration in non-life-threatening infections. However, the influence of fever and the effects of antipyresis in patients with sepsis has not been prospectively studied [139]. Nine healthy elderly volunteers (median,

66 years) and eight young controls (median, 24 years;) were given endotoxin i.v. (2 ng/kg). The elderly subjects showed more prolonged fever compared to the young individuals. TNFα, soluble TNF receptors (sTNFR-I), IL-6, IL-8, IL-10, and IL-1ra in plasma increased markedly in both groups. The elderly group showed larger initial increases in TNFα and sTNFR-I levels and prolonged increased levels of sTNFR-I. Monocyte concentrations decreased in both groups, with the elderly group showing a more rapid decrease and a slower subsequent increase. The elderly group had a more rapid increase in CRP than did the young group [140].

Cachexia is the clinical consequence of a chronic, systemic inflammatory response. The changes seen are multidimensional and highly coordinated. Most obvious is a redistribution of the body's protein content, with preferential depletion of skeletal muscle and an increase in the synthesis of APP involved in APR. The physiologic, metabolic, and behavioral changes of cachexia are tightly regulated by cytokines, which signal the synthesis of APP as well as changes in intermediary metabolism that provide substrate and energy. The increase in the rate of protein degradation limit the ability of hypercaloric feeding to reverse the depletion of lean mass [141]. The *anorexia of infection* is part of the host's APR and is beneficial in the beginning, but deleterious, if long lasting. Bacterial cell wall compounds (e.g., LPS, peptidoglycans), microbial nucleic acids and viral glycoproteins trigger the APR and anorexia by stimulating the production of proinflammatory cytokines (e.g., interleukins , TNFα, interferons), which serve as endogenous mediators. Cytokines increase leptin and it may contribute to, but is not essential for, the anorectic effects of microbial products and cytokines. Cytokines can reach the CNS through the circumventricular organs and through active or passive transport mechanisms or they can act through receptors on endothelial cells of the brain vasculature and stimulate the release of subsequent mediators such as eicosanoids. De novo CNS cytokine synthesis also occurs in response to peripheral infections. The central mediators of the anorexia during infection appear to be neurochemicals involved in the normal control of feeding, such as serotonin, dopamine, histamine, CRF, neuropeptide Y, and αMSH. Reciprocal, synergistic, and antagonistic interactions between various pleiotropic cytokines, and between cytokines and neurochemicals, form a complex network that mediates the anorexia during infection [142].

The APR was studied in well-nourished (WN) rats; (2) undernourished (UN) rats as a consequence of chronic food restriction, and (3) UN rats re-fed to restoration of their body weights. UN rats developed an arrest in body weight gain and exerted basal hypoglycemia, hypotriglyceridemia, hypoleptinemia, hypercorticosteronemia and enhanced adrenal glucocorticoid content; however, no changes in basal total protein, ACTH and TNF plasma levels and in anterior pituitary ACTH concentrations were found. LPS-induced hypoglycemia developed in WN rats was abolished in UN animals, and both ACTH and TNF plasma concentrations, albeit significantly ($p < 0.05$) higher than the respective basal values, were significantly ($p < 0.05$) lower in UN than in WN control rats. The LPS-induced glucocorticoid release was similar in WN and UN animals. LPS treatment did not modify basal plasma leptin levels, regardless of the group. UN rats fed ad libitum for 15 days restored their body weight, and the various metabolic dysfunctions seen in UN rats in both basal and post-LPS conditions were fully normalized [104].

In patients following cardiopulmonary resuscitation, type-1 APP (CRP, α1-acidglycoprotein, serum amyloid A) and type-2 APP (haptoglobin, α1-antitrypsin) increased consistently within 1–2 days and transferrin was downregulated. This APP response occurred irrespective of the cause of arrest, the estimated time of anoxia, clinical course or patient outcome and was not different in patients with and without infectious complications. Exposure of healthy volunteers to less severe but more prolonged hypoxia did not induce APP, although a time dependent increase of serum erythropoietin (EPO) was measurable under these conditions [143].

In acute coronary syndromes elevated concentrations of acute phase reactants, such as CRP, are found. The presence of APP predict future risk in apparently healthy subjects. APR is associated with elevated levels of fibrinogen, a strong risk factor for coronary heart disease, with autocrine and paracrine activation of monocytes by IL-6 in the vessel wall contributing to the deposition of fibrinogen. There is increased blood viscosity, platelet number and activity. The raised serum amyloid A lowers HDL-cholesterol levels. IL-6 decreases lipoprotein lipase (LPL) activity and monomeric LPL levels in plasma, which increases macrophage uptake of lipids. In fatty streaks and in the atheromatous 'cap' and 'shoulder' regions, macrophage foam cells and smooth muscle cells (SMC) express IL-6. This suggests a role for IL-6, along with IL-1 and TNFα, in the progression of atherosclerosis. Both IL-1 and TNFα induce the release of IL-6 from several cell types, including SMC. During vascular injury SMC are exposed to platelet products, and cytokine production by SMC further contributes to vascular damage. In addition, circulating IL-6 stimulates the hypothalamic-pituitary-adrenal (HPA) axis, the activation of which is associated with central obesity, hypertension and insulin resistance [144].

A minimal acute-phase response observed in the elderly does not necessarily establish the existence of an inflammatory process but may also reflect a variety of noninflammatory states, including obesity, sleep disturbance, depression, chronic fatigue, and low levels of physical activity [145]. *Aging* is associated with increased inflammatory activity, increased circulating levels of TNFα, IL-6, cytokine antagonists and acute phase proteins. Chronic low-grade inflammation in aging promotes an atherogenic profile and is related to age-associated disorders (eg, Alzheimer disease, atherosclerosis, type 2 diabetes, etc.) and enhanced mortality risk. A dysregulated production of inflammatory cytokines, delayed termination of inflammatory activity and a prolonged fever response suggest that the acute phase response is altered in aging [146].

The *metabolic syndrome* is characterized by cardiovascular and diabetes risk factors generally linked to insulin resistance, and obesity. Cross-sectional analysis demonstrated that markers of inflammation and endothelial dysfunction predict the development of type 2 diabetes and weight gain in adults. There is biological evidence to suggest that chronic activation of the innate immune system may underlie the metabolic syndrome [147].

The association between circulating IL-6 levels and C-reactive protein on *insulin resistance* and blood pressure was studied in 228 healthy volunteers. The plasma IL-6 concentration was significantly associated with systolic (SBP) and diastolic (DBP) blood pressure, fasting insulin, and the fasting insulin resistance index (FIRI) in all subjects. When smokers were excluded from the analysis, plasma IL-6 levels correlated with percent fat mass, absolute fat mass, SBP, DBP, fasting insulin levels, and FIRI. The latter associations persisted after controlling for body mass index, for fasting insulin and FIRI, respectively. Although IL-6 levels were significantly associated with fasting insulin and FIRI in men, these significant correlations were not observed in women. Conversely, although IL-6 levels were significantly associated with SBP and DBP in women, these coefficients were not statistically significant in men. All of these associations were lost among smokers and remained significant in nonsmokers. Serum CRP levels correlated significantly with IL-6 in all the subjects, but mainly in nonsmokers and men. This significant relationship was lost among smokers. CRP was associated with fasting insulin and FIRI , but not with SBP or DBP (P = NS), in all subjects. The associations between CRP and these parameters were similar in men and women and in smokers and nonsmokers. For insulin and FIRI they were stronger in women and in nonsmokers. In conclusion, circulating IL-6 levels, by inducing either hypertension in women or insulin resistance in men, constitute a significant pro-atherogenic cytokine [148].

In a group of 30 older diabetic patients with poor glycemic control, serum levels of lipid peroxides, measured as thiobarbituric acid-reacting substances (TBARS); CRP; IL-6 and the soluble form of its receptor (sIL-6R), were evaluated at baseline and after 2 and 3 months of therapy. Thirty asymptomatic, untreated individuals with abnormal fasting glycemia, but otherwise healthy status, were matched for age, sex, and weight served as controls. At baseline, glycemia, HbA1C, TBARS, CRP, IL-6 and sIL-6R were significantly higher in diabetic patients than in controls ($p<0.001$). After treatment, glycemia significantly decreased with respect to baseline values, as did serum levels of TBARS, CRP, IL-6 and sIL-6R, respectively. In diabetic patients and in asymptomatic hyperglycemic subjects TBARS and IL-6 were independently correlated with CRP [149]. Plasma IL-6 and TNFα were significantly increased in type 2 diabetes compared to normal subjects. Basal production of IL-6 and TNFα in cultured diabetic blood was markedly depressed in comparison with non-diabetic samples. IL-6 and TNFα production was increased in blood in response to LPS in both groups, IL-6 being slightly but significantly higher in controls [150]. Polymorphism of the IL-6 gene influences the relationship among insulin sensitivity, postload glucose levels, and peripheral white blood cell count [151].

Serum *leptin* concentrations in type 2 diabetes are partly related to an acute-phase response, independent of body mass index and sex. The association of hyperleptinaemia with elevated serum cortisol provides a mechanism for leptin resistance in type 2 diabetes, as glucocorticoids inhibit the central action of leptin [152].

In *allergy* patients increased IL-6 levels correlated with greater erythema extent, lower mean arterial blood pressure, and a longer duration of symptoms. There was an inverse relationship between CRP and histamine levels [153].

In patients with *rheumatoid arthritis* (RA) IL-6 correlated closely with CRP and with erythrocyte sedimentation rate (ESR). These three parameters correlated well with serum cortisol, which is increased in active RA [154]. Neuropeptides and hormones not only affect a systemic acute phase response of RA patients, but actually produced by, and their receptors are expressed on, cells within the inflammatory joint. Therefore these mediators also modulate local inflammation directly in the joints [155]. Acute-phase serum amyloid A (A-SAA) is a major component of the acute-phase response. A sustained acute-phase response in rheumatoid arthritis (RA) is associated with increased joint damage. A-SAA is produced by inflamed RA synovial tissue [156].

CRP and C3a levels were significantly higher in 56 patients with *IgA nephropathy* as compared with 55 healthy controls and with 42 patients with hypertension or nonimmune renal diseases. Mean CRP but not C3a levels were significantly higher in IgA nephropathy patients with disease progression than in those with stable renal function [157].

4.1. *The acute phase response induced by endotoxin*

The biological response to LPS has been studied extensively and will be discussed below as a typical example of APR, the basic principle of which will apply to APR in general. Moreover, LPS is frequently involved in APR even if the initial causative agent does not involve infection with Gram-negative bacteria, as LPS is known to absorb from the intestinal tract during severe trauma and shock [34].

High-mobility group-1 (HMG-1) is an abundant nuclear DNA-binding protein that stabilizes nucleosome formation, facilitates gene transcription, and regulates the activity of steroid hormone receptors. HMG-1 is released by endotoxin-activated monocytes, it activates downstream cytokine release and may cause delayed endotoxin lethality [158].

The vasopressor, angiotensin II (AII), activates transcriptional expression of its precursor, angiotensinogen. This "positive feedback loop" occurs through an angiotensin receptor-coupled pathway that activates a multihormone-responsive enhancer of the angiotensinogen promoter, termed the acute-phase response element (APRE). AII stimulates the NFκB transcription factor pathway by activating latent DNA-binding activity of NFκB subunits through a phorbol ester-sensitive (PKC-dependent) mechanism [159].

Animals exposed to LPS will produce less cytokines to a second dose of LPS which may be observed already at 1 hr after the first exposure. This is known as *endotoxin tolerance* and it is manifested in decreased cytokine production, reduced pathological changes, and in a significantly increased resistance to endotoxin, infectious agents, and to toxic and other noxious insults [74,160,161]. The induction of both endotoxin toxicity and tolerance is mediated by macrophages [162]. The elevated secretion of hormones of the HPA axis, which inhibit cytokine production and antagonize the proinflammatory and neurotoxic effect of cytokines, such as TNFα after LPS challenge, plays an important role in the development of endotoxin tolerance [67,163,164].

The i.p. injection of LPS to rats dose-dependently increased IL-1β protein levels in the hypothalamus, hippocampus, dorsal vagal complex, cerebellum, posterior cortex, and pituitary 2 hrs after the injection. There were no significant differences in brain IL-1β levels between sham-operated and vagotomized rats [165].

In human CRP transgenic mice deficient in C3 or C5, there was a diminished induction by LPS of the acute-phase proteins CRP and serum amyloid P-component (SAP). LPS could induce IL-6, but not IL-1β in complement-deficient mice. Human C5a induced IL-1beta and caused significant elevation of both serum CRP and SAP in human CRP transgenic mice. However, in human CRP transgenic IL-6-deficient mice, recombinant human C5a was ineffective [166].

The newly synthesized acute phase proteins are also essential for the development of endotoxin tolerance and these proteins exert anti-microbial and immunoregulatory functions [34]. The importance of the liver in the development of endotoxin tolerance is emphasized by the observation that D-galactosamine ((GalN)), which increases profoundly endotoxin sensitivity, intoxicates the liver [167]. Moreover, GalN- treated mice cannot develop endotoxin tolerance [163]. Metallothionein (MT), which is a low-molecular weight, cysteine-rich, metal-binding protein, is also induced in the acute-phase reaction. MT-null mice were more sensitive to LPS/GalN-induced lethality than wild-type mice. Messenger RNA level of AGP in response to LPS/GalN was decreased in MT-null mice compared to wild-type mice [168].

Treatment of rats with the cyclooxygenase inhibitors either attenuated (meloxicam) or abolished (diclofenac) LPS-induced fever, but had no effect on the LPS-induced rise of plasma cortisol or IL-6. The serum TNF response to LPS was enhanced by both drugs, reflecting the negative feedback control exerted by prostaglandins on cytokine (specifically TNF) formation. These results provide the first evidence that the prostaglandin-dependent inflammatory pathway for fever induction is distinct from the pathway of HPA axis activation [169].

IL-6 was identified as the principal mediator of APR. Via its receptor, gp130, IL-6 induces DNA-binding of STAT transcription factors on regulatory elements of target genes. TNF-alpha was shown to be involved in several models of liver failure as a mediator for both cytotoxicity and cell proliferation. It activates NFκB, thereby triggering inflammatory processes [170] IL-1β concomitantly induces NFκB activation and dephosphorylates IL-6-activated STAT1. The latter mechanism might account for the inhibition by IL-1β of the IL-6-dependent induction of type II acute-phase genes [171]. Soluble gp130 is the natural inhibitor of IL-6 responses dependent on soluble IL-6 receptor [172].

The expression of the G(s)-coupled prostanoid receptors EP2-R, EP4-R, and DP-R, but not the IP-R, was up-regulated by treatment of hepatocytes in vitro or rats with IL-6. In IL-6-treated hepatocytes PGE2 in turn attenuated the IL-6-induced alpha-2-macroglobulin formation via a cyclic AMP-dependent signal chain [173].

The nuclear hormone receptors, *peroxisome proliferator-activated receptor alpha* (PPARα) and liver *retinoid X receptor* (RXR) play key roles in regulation of hepatic lipid metabolism. LPS induces a rapid, dose-dependent decrease in RXRα, RXRβ and RXRγ proteins in hamster liver. Maximum inhibition was observed at 4 h for RXRα and RXRβ and at 2 h for RXRγ. LPS markedly decreased both basal and Wy-14,643-induced expression of acyl-CoA synthetase, a well characterized PPARα target [174]. LPS elicits a dramatic increase in the synthesis and secretion of triglyceride (TG)-rich lipoproteins by the liver and the inhibition of lipoprotein lipase. This cytokine-induced "lipemia of sepsis" was customarily thought to represent the mobilization of lipid stores to fuel the host response to infection. However, since lipoproteins can also bind and neutralize LPS, it is hypothesized that TG-rich lipoproteins (VLDL and chylomicrons) are also components of an innate immune response to infection [175].

Elevation of IL-6, soluble TNF-alpha receptors and soluble IL-6 receptor levels were detected during liver regeneration after partial hepatectomy [176]. In a mouse model of acute liver failure induced by GalN administration, a single low dose of a hyper-IL-6-encoding adenoviral vector maintained liver function, prevented the progression of liver necrosis, and induced liver regeneration, leading to dramatically enhanced survival [177]. In mice liver IL-1 receptor antagonist (IL-1Ra) was up-regulated after systemic LPS administration and local turpentine injection. After LPS stimulation, the hepatic production of sIL-1Ra correlated with the increase in plasma IL-1Ra levels. The total amount of IL-1Ra present in the liver induced by LPS was six- and tenfold higher than in the lung and spleen. Soluble IL-1Ra, but not intracellular IL-1Ra mRNA was induced in hepatocytes in vivo by LPS. In IL-6(-/-) mice IL-6 mediated the turpentine-induced production of IL-1Ra mRNA by the liver [178].

In bone marrow donors treated with G-CSF, IL-6 is a major physiological effector molecule that induces bone metabolism and an acute-phase reaction along with mobilization of CD34+ cells in the peripheral blood. IL-6 may be responsible for osteopenia observed during short- and long-term G-CSF treatment [179]. IL-6 treatment of rabbits caused an accelerated release of polymorphonuclear cells from the bone marrow. The newly released PMN cells were less deformable, and preferentially sequestered in lung microvessels [180].

Serum MIF levels were significantly elevated on day 1 in patients with septic shock, as opposed to trauma patients and control individuals. In septic patients MIF levels were parallel to cortisol, but in contrast to ACTH, showed persistent elevation. Non-survivors of septic shock showed significantly higher MIF levels than did survivors. Patients with septic adult respiratory distress syndrome (ARDS) showed higher MIF levels than those without ARDS. Multiple logistic regression analysis revealed that both MIF and ARDS were independent predictors of adverse outcome. In septic patients, significant correlations were established between MIF and cortisol and MIF and IL-6 levels and disease severity scores. No relation was found between MIF and markers of the acute phase response (procalcitonin, CRP, and LBP). In multitrauma patients, MIF levels were not elevated at any time point. During immune-mediated inflammation (such as septic shock) MIF is an important neuroendocrine mediator, which acts as a contraregulator of the immunosuppressive effects of glucocorticoids [181].

Anti-inflammatory cytokines, such as the IL-1 receptor antagonist [78], the TNF synthesis inhibitor [77], IL-10 [75], leukemia inhibitory factor [73], all participate in the down regulation of the noxious effects of endotoxin. Interferon-γ antagonizes the development of endotoxin tolerance [76]. *In vitro* observations revealed that the direct exposure of macrophages to LPS also leads to decreased activation and cytokine production upon re-exposure [182,183].

Previte et al [184] discovered that the exposure of endotoxin to ionizing radiation resulted in significant loss of toxicity. Subsequently Bertok and coworkers [185] demonstrated over a number of years that radiodetoxified endotoxin is capable of boosting host resistance against infectious agents, radiation, septic shock, tourniquet shock, intestinal ischemic shock, hemorrhagic shock, X-irradiation and even against immunosuppression by anti-lymphocytic serum [186]. Radiodetoxified endotoxin has been tested clinically for the treatment of infectious disease in man [187]. Monophosphoryl lipid A preparations of low toxicity are also studied in animals and man for boosting host resistance to infections and traumatic events [188,189].

Fever is an unmistakable indicator of elevated cytokine levels in the blood stream and it is the hallmark of APR. Various insults, such as irritation, trauma or toxic agents, are also capable of immune activation by nonspecific pathways, which can lead to APR [34]. One may think of many reasons for the protective effect of acute immune activation during APR. In this context one should remember that pathophysiological immune activation, regardless of whether it is specific or nonspecific, will lead to shared host defence reactions. These reactions include inflammation, phagocytosis, cytotoxicity, neutralization. In the case of systemic reactions the immunoeffector function is supported by neuroendocrine and metabolic responses. Once the pathogenic agent has been eliminated, the immune system contributes to the healing process [34,190,191,192,32,33].

Procalcitonin and CRP levels were evaluated in 38 blood samples of pediatric patients with unexplained fever or with sepsis. Procalcitonin and CRP elevations were found to correlate significantly ($P < 0.01$) [193]. Serum *retinol* decreases transiently during the acute phase response and can thus result in misclassification of vitamin A status. Serum retinol was lowest in subjects aged <10 yrs and increased with age. Concentrations were higher in males than in females aged 20–59 y. CRP concentration (>/=10 mg/L) was lowest in subjects aged <20 yrs (</=4%) and increased with age to a maximum of nearly 15%. An elevated CRP concentration was 2.4-fold greater in females than in males aged 20–59 y. Serum retinol was lower in subjects with elevated CRP concentrations [194].

Sera of patients with severe sepsis/septic shock inhibited the bioactivity of LPS including TNFα secretion compared with control sera. The depletion of LBP from sepsis sera and addition of rhLBP at concentrations found in severe sepsis confirmed that LBP was the major serum component responsible for inhibitory effects. This may represent a novel defense mechanism of the host in severe sepsis and during bacterial infections [195].

4. THE BIOLOGICAL SIGNIFICANCE OF THE ACUTE PHASE RESPONSE

Although it is host defence that comes into mind when one is considering the acute phase response it is clear that the cytokines and the acute phase proteins are produced under physiological conditions and therefore, are expected to fulfil important functions in healthy animals/humans. Some examples that illustrate this point are given below.

Proinflammatory cytokines, such as IL-1β or IL-6, have been implicated as mediators in both pre-term and term labor, particularly in association with intrauterine infection. The oxytocin (OT) receptor gene promoter displays constitutive basal activity and is negatively regulated by both IL-1β and IL-6. This suppression by IL-1β and IL-6 depends on the -1203/-722 region of the OT receptor promoter, which contains binding sites for NF-IL6, acute phase response element, and NF-κB [196].

In newborn babies there was an increase in IL-6 during the first day, followed by an increase in CRP, serum amyloid A and procalcitonin on the second postnatal day. The level of prealbumin fell after birth reaching its lowest value at 3 days of age. Interleukin-1β remained unchanged. The duration of the sleeping period was longer during the first day. There was an increase in body temperature during the first day and maximal weight loss was observed during the first 2 days. These findings indicate that the normal birth process and extra-uterine adaptation stimulates an acute phase reaction in the newborn infant [197].

Strenuous exercise increased plasma levels of TNFα, IL-1, IL-6, IL-1 receptor antagonist, TNF receptors, IL-10, IL-8 and macrophage inflammatory protein-1. IL-6 increased up to 100-fold after a marathon race and the increase was tightly related to the duration and intensity of the exercise. IL-6 is produced in the skeletal muscle in response to exercise and it has growth factor abilities. It is likely to be beneficial and may be involved in mediating exercise-related metabolic changes [198]. Exercise induces immune changes and also alters neuroendocrinological factors including catecholamines, growth hormone, cortisol, beta-endorphin, and sex steroids. Exercise-associated muscle damage initiates the inflammatory cytokine cascade [199]. LPS enters the circulation in athletes after ultra-endurance exercise and may, together with muscle damage, be responsible for the increased cytokine response and hence GI complaints [200].

Three months of estrogen replacement therapy in healthy postmenopausal women decreased low-density lipoprotein with simultaneous decreases in tissue factor pathway inhibitor, a major inhibitor of the extrinsic coagulation pathway. C-reactive protein, a component of the acute phase response, was increased [201].

In 574 nondiabetic elderly men and women insulin strongly and significantly ($P < 0.001$) correlated with the markers of inflammation, C-reactive protein, alpha-1-antichymotrypsin, and IL-6, when adjusted for age and gender. Insulin also correlated with the soluble ICAM-1. This indicates that low-grade inflammation and the cellular adhesion molecule soluble ICAM-1 are an integral part of insulin resistance in nondiabetic elderly [202].

Traditionally it is held that endotoxin is harmful and causes disease. However, LPS itself is devoid of toxicity in some lower animal species. The enormous differences in toxicity for various species are due to the ability of LPS to activate the immune system of the animal in question. Therefore, it is the immune system and not LPS that kills the host. Many other microbial pathogens are able to activate the immune system nonspecifically, which is often referred to as "polyclonal lymphocyte activation" [203]. Some of these immunogens are known as "*superantigens*" [204,205,206]. The immunobiology of superantigens is complex and is beyond the scope of this chapter. However, many of them are pyrogenic substances [207], and therefore, capable of activating APR.

It should be clear to the reader by now that acute immune activation is beneficial to the host in a natural setting. It represents a rapid and efficient way of mounting a massive immune defence against the pathogens right at the site of invasion. The target is promptly identified by the immune system locally at the infection site, where complement, blood clotting or various subsets of white blood cells are activated. Thus, the coagulation of the blood at a focus of infection may provide local defence by the isolation of microbes, whereas disseminated intravascular coagulation may lead to disaster [30,32,33,126].

When the immune system fails to control the infection/insult locally, APR will develop. APR is an emergency reaction, which represents a switch of the immune system from the specific, adaptive mode of reactivity, which is commanded by the thymus derived T lymphocyte system to a less specific, but very rapid and intense reaction. Natural antibodies and newly synthesized APP by the liver play a major role in the identification of the target and the activation and regulation of immune effector mechanisms during APR. A severe febrile illness represents the mobilization of all resources of the host in the interest of defeating/eliminating the pathogen and achieving survival/recovery. This ancient defence mechanism is very demanding on the body, but by and large, it is very successful, as in the overwhelming majority of cases febrile illness will lead to recovery and healing. On this basis, one may suggest that APR is truly beneficial and only in rare and extreme cases will it result in severe disease, shock and death [30,32,33,126].

LPS exerts an effect on the central nervous system, on other endocrine organs and on many other tissues and organs in the body [208,209,210,211]. The various organs and tissues contain "resident macrophages" or related cells such as the glia cells in the central nervous system, the Kupffer cells in the liver, Langerhans cells in the skin, etc., which have the capacity to react to LPS with cytokine production [212,213]. Therefore, systemically applied LPS has the capacity to activate the immune and the neuro-endocrine systems, and also the liver via locally induced cytokines. Thus, indirect activation by blood borne cytokines is supplemented by direct activation by LPS of the key organs necessary to execute an acute phase response. This is complemented also by neural communication. For instance, the vagus nerve plays a role in the activation of the ACTH-adrenal axis and the initiation of a behavioural response after the intraperitoneal injection of LPS [214,215].

Endotoxin is always present in the gastrointestinal tract, even in germ free animals [9]. However, endotoxin does not absorb from the gut of rats even when mucosal damage is inflicted [216]. The reason for this is that bile acids destroy the integrity of lipid A, which leads to loss of biological activity. If bile is diverted from the gastrointestinal tract, oral intoxication with LPS can readily be achieved [217]. The liver is an important clearance organ of LPS via bile secretion [218]. Bertok suggested that bile functions as a physico-chemical defence barrier against endotoxin toxicity in vertebrate animals [219].

LPS has an enormous potential to boost host resistance and the acute phase response is capable of protecting the host efficiently. On this basis it is possible to suggest that throughout evolution a pathophysiological mechanism has evolved which utilizes intestinal LPS for the rapid conversion of the immune system from adaptive reactivity to acute phase host defence by the amplification of natural immunity. This *immunoconversion* may be achieved simply by the control of bile secretion. Indeed, there is ample evidence in the literature to illustrate that intestinal endotoxin readily absorbs after X-irradiation, trauma or even after stressful situations [34,9]. It has been shown already that liver regeneration was stimulated by intestinal endotoxin as was granulopoiesis [9,11]. There are even experiments to suggest that healing in the central nervous system is promoted by LPS [12]. However, under extreme cases intestinal endotoxin has been blamed for fatality in trauma patients [34].

The adaptive immune system fails gradually in many people due to aging, and may also fail during disease or as the result of various insults to the body. Stressful insults initially mobilize the adaptive immune system to enhance immune reactivity in peripheral tissues, which are likely sites of penetration by infectious agents [41]. Should the pathogenic insult continue, it means that the adaptive immune defense system has failed. The only alternative that remains for host defence is to rely fully on the natural immune system for survival. The neuroendocrine and metabolic alterations during the APR lead to a complete switch of the entire immune system to the natural mode of reactivity. This is achieved by the sudden and gross elevation in the serum of polyspecific natural antibodies and liver-derived APP. At the same time the adaptive immune response is rendered non-functional, primarily by the profound inhibition of the thymus/T cell immunoregulatory system. These changes lead to *immunoconversion from adaptive to natural immune host defence* [53 30,34,32,33].

One may conclude on the basis of evidence reviewed here that the acute phase response makes use of evolutionarily preserved elements of ancient defence mechanisms of lower animals. These elements got integrated into a rapid, highly coordinated and very efficient acute immune response, during which the body's physiology undergoes profound alterations in support of host defence. There can be little doubt that a detailed understanding of febrile illness will lead to new insights to human disease and to a better patient care.

REFERENCES

1. Nowotny, A. In search of active sites in endotoxins. In: Nowotny A, editor Beneficial Effects of Endotoxins. New York: Plenum Press, 1983;1–55.
2. Boivin A, Mesrobeanu J, Mesrobeanu L. Technique pour la preparation des polysaccharides microbiens specifiques. CR Soc Biol 1933;113:490.
3. Favorite GO, Morgan HR. Effects produced by the intravenous injection in man of a toxic antigenic material derived from *Eberthella typhosa*: Clinical, hematological, chemical and serological studies. J Clin Invest 1942;21:589.
4. Greisman SE. Induction of endotoxin tolerance. In: Nowotny A, editor. Beneficial Effects of Endotoxins. New York: Plenum Press, 1983;149–178.
5. Behling UH. The radioprotective effect of bacterial endotoxin. In: Nowotny A, editor. Beneficial Effects of Endotoxins. New York: Plenum Press, 1983;127–148.
6. Parant M. Effect of LPS on nonspecific resistance to bacterial infections. In: Nowotny A, editor. Beneficial Effects of Endotoxins. New York: Plenum Press, 1983;179–196.
7. Johnson AG. Adjuvant action of bacterial endotoxins on antibody formation. A historical perspective. In: Nowotny A, editor. Beneficial Effects of Endotoxins. New York: Plenum Press,1983;249–253.
8. Butler RC. Enhancement of nonspecific resistance to tumor by endotoxin. In: Nowotny A, editor. Beneficial Effects of Endotoxins. New York: Plenum Press, 1983;497–512.
9. Walker RI. Possibilities for use of endotoxin in compromised subjects. In: A. Nowotny, editor. Beneficial Effects of Endotoxins. New York: Plenum Press, 1983;197–212.
10. Abdelnoor AM. Effect of LPS on the clotting system. In: Nowotny A, editor. Beneficial Effects of Endotoxins. New York: Plenum Press,1983;75–90.
11. Cornell RP. Restriction of gut-derived endotoxin impairs DNA synthesis for liver regeneration. Am J Physiol. 1985;249(5 Pt 2):R563–9.
12. Windle WF, Chambers WW. Regeneration in spinal cord of cat and dog. J Comp Neurol. 1950;93:241–258.
13. Williams JL, Dick GF. Decreased dextrose tolerance in acute infectious disease. Arch Intern Med 1932;50:801.
14. Fox MJ, Kuzma JF, Washam WT. Transitory diabetic syndromes associated with meningococcic meningitis Arch Intern Med 1947;79:614.
15. Beisel WR. Metabolic response to infection. Ann Rev Med. 1975;26:9–20.
16. Berczi I. The influence of pituitary-adrenal axis on the immune system. In: Berczi I, editor. Pituitary function and immunity. Boca Raton, FL: CRC Press, 1986;49–132.
17. Berkenbosch F, van Oers J, del Rey A, Tilders F, Besedovsky H. Corticotropin-releasing factor-producing neurons in the rat activated by interleukin-1. Science. 1987;238:524–6.
18. Sapolsky R, Rivier C, Yamamoto G, Plotsky P, Vale W. Interleukin-1 stimulates the secretion of hypothalamic corticotropin-releasing factor. Science. 1987;238:522–4.

19. Uehara A, Gottschall PE, Dahl RR, Arimura A. Interleukin-1 stimulates ACTH release by an indirect action which requires endogenous corticotropin releasing factor. Endocrinology. 1987 ;121:1580–2.
20. Bernton EW, Beach JE, Holaday JW, Smallridge RC, Fein HG. Release of multiple hormones by a direct action of interleukin-1 on pituitary cells. Science. 1987;238:519–21.
21. Nakamura H, Motoyoshi S, Kadokawa T. Anti-inflammatory action of interleukin 1 through the pituitary-adrenal axis in rats. Eur J Pharmacol. 1988;151:67–73.
22. Healy DL, Hodgen GD, Schulte HM, Chrousos GP, Loriaux DL, Hall NR, Goldstein AL. The thymus-adrenal connection: thymosin has corticotropin-releasing activity in primates. Science. 1983;222:1353–5.
23. Selye H. A syndrome produced by diverse nocuous agents. Nature 1936;138:32.
24. Selye H. Thymus and the adrenals in response of the organism to injuries and intoxications. Br J Exp Pathol 1936;17:234–248.
25. Selye H. The general adaptation syndrome and the diseases of adaptation. J Clin Endocrinol 1946;6:117–230
26. Selye H. Stress and disease. Science 1955;122:625.
27. Selye H. Effect of ACTH and cortisone upon an "anaphylactoid reaction". Can Med Assoc J 1949;61:553–556.
28. Karady S, Selye H, Brownie JSL. The influence of the alarm reaction on the development of anaphylactic shock. J Immunol 1938;35:335.
29. Selye H. Morphological changes in the fowl following chronic overdosage with various steroids. J Morphol 1943;73:401–421.
30. Berczi I. Neurohormonal host defence in endotoxin shock. Ann NY Acad Sci 1998;840: 787–802.
31. Berczi I. The stress concept and neuroimmunoregulation in modern biology. Ann NY Acad Sci 1998;851:3–12.
32. Haeryfar SMM, Berczi I. The thymus and the acute phase response. Cell Mol Biol 2001;47:145–156.
33. Berczi I, Bertok L, Chow D. Natural immunity and neuroimmune host defence. Ann NY Acad Sci 2001;917:248–257.
34. Berczi I, Nagy E. Neurohormonal control of cytokines during injury. In: Rothwell NJ, Berkenbosch F, editors. Brain Control of Responses to Trauma. Cambridge University Press, 1994;32–107.
35. O'Flaherty L, Bouchier-Hayes DJ. Immunonutrition and surgical practice. Proc Nutr Soc 1999;58:831–7.
36. Moses AG, Dowidar N, Holloway B, Waddell I, Fearon KC, Ross JA. Leptin and its relation to weight loss, ob gene expression and the acute-phase response in surgical patients. Br J Surg 2001;88:588–93.
37. Maruna P, Gurlich R, Fried M, Frasko R, Chachkhiani I, Haluzik M. Leptin as an acute phase reactant after non-adjustable laparoscopic gastric banding. Obes Surg 2001;11: 609–14.
38. Barnetson StC, Gawkroger D. Hypersensitivy - type IV. In: Roitt I, Brostoff J, Male D, editors. Immunology, 2nd edition. London: Gower Med Pub, 1989;22.1–22.
39. Sinha AA, Lopez MT, McDevitt HO. Autoimmune disease. The failure of self tolerance. Science 1990;248:1380–1388
40. Eastman A, Barry MA. The origins of DNA breaks - A consequence of DNA damage,

DNA repair, or apoptosis. Cancer Invest 1992;10:229–240.
41. Hoshiya T, Watanabe D, Akagi K, Mizoguchi Y, Kamiya K, Mizuguchi H, Kumahara M,Toya H, Nagashima Y, Okaniwa A. Acute phase response in toxicity studies. I. Survey of beagle dogs subjected to single-dose toxicity studies. J Toxicol Sci 2001;26:95–102.
42. Banck G, Forsgren A. Many bacterial species are mitogenic for human blood lymphocytes. Scand J Immunol 1978;8:347–354.
43. Sibbald WJ, Short A, Chohen MP, Wilson RF. Variations in adrenocortical responsiveness during severe bacterial infections. Unrecognized adrenocortical insufficiency in severe bacterial infections. Ann Surg 1977;186:29–33
44. Rothwell PM, Udwadia ZF, Lawler PG. Cortisol response to corticotropin and survival in septic shock. Lancet 1991;337:582–583.
45. Curnow RT, Rayfield EJ, George DT, Zenser TV, De Rubertis FR. Altered hepatic glycogen metabolism and glucoregulatory hormones during sepsis. Am J Physiol 1976;230:1296–1301.
46. Iochida LC, Tominaga M, Matsumoto M, Sekikawa A, Sasaki H. Insulin resistance in septic rats - A study by the euglycemic clamp technique. Life Sci 1989;45:1567–1573.
47. Lang CH, Dobrescu C. Sepsis-induced changes in in vivo insulin action in diabetic rats. Amer J Physiol 1989;257:E301–308.
48. Lang CH, Dobrescu C. Gram-negative infection increases noninsulin-mediated glucose disposal. Endocrinology 1991;128:645–653.
49. Leshin LS, Malven PV. Bacteremia-induced changes in pituitary hormone release and effect of naloxone. Am J Physiol 1984;247:E585–591.
50. Dahn MS, Lange MP, Jacobs LA. Insulinlike growth factor 1 production is inhibited in human sepsis. Arch Surg 1988;123:1409–1414.
51. Westphal O, Jann K, Himmelspach K. Chemistry and immunochemistry of bacterial lipopolysaccharides as cell wall antigens and endotoxins. Prog Allergy. 1983;33:9–39.
52. Ramachandra RN, Berczi A, Sehon AH, Berczi I. Inhibition of lipid A and lipopolysaccharide (LPS) induced cytokine secretion, B cell mitogenesis and lethal shock by lipid A specific murine monoclonal antibodies. J Infect Dis 1993;167:1151–1159.
53. Berczi I. Neuroendocrine defence in endotoxin shock. (A review) Acta Microbiol Hung 1993;40:265–302.
54. Raetz CRH, Ulevitch RJ, Wright SD, Sibley CH, Ding A, Nathan CF. Gram-negative endotoxin: An extraordinary lipid with profound effects on eukaryotic signal transduction. FASEB J 1991;5:2652–2660.
55. Tobias PS, Mathison J, Mintz D, Lee JD, Kravchenko V, Kato K, Pugin J, Ulevitch RJ. Participation of lipopolysaccharide-binding protein in lipopolysaccharide-dependent macrophage activation. Am J Respir Cell Mol Biol. 1992;7:239–45.
56. Schumann RR. Function of lipopolysaccharide (LPS)-binding protein (LBP) and CD14, the receptor for LPS/LBP complexes: a short review. Res Immunol. 1992;143:11–5.
57. Mathison JC, Tobias PS, Wolfson E, Ulevitch RJ. Plasma lipopolysaccharide (LPS)-binding protein. A key component in macrophage recognition of gram-negative LPS. J Immunol. 1992;149:200–6.
58. Wright SD, Ramos RA, Patel M, Miller DS. Septin: a factor in plasma that opsonizes lipopolysaccharide-bearing particles for recognition by CD14 on phagocytes. J Exp Med. 1992;176:719–27.
59. Haziot A, Chen S, Ferrero E, Low MG, Silber R, Goyert SM. The monocyte differentiation antigen, CD14, is anchored to the cell membrane by a phosphatidylinositol linkage. J

Immunol. 1988;141:547–52.
60. Kielian TL, Blecha F. CD14 and other recognition molecules for lipopolysaccharide: a review. Immunopharmacology. 1995;29:187–205.
61. Rivest S, Nadeau S, Lacroix S, Laflamme N. Proinflammatory signal transduction pathways in the CNS during systemic immune response. In New Foundation of Biology, Berczi I, Gorczynski RM, Editors, Neuroimmune Biology, Volume I. Berczi I, Szentivanyi A, Series editors, Elsevier 2001, pp. 163–173.
62. Berczi I, Bertok L, Bereznay T. Comparative studies on the toxicity of Escherichia coli lipopolysaccharide endotoxin in various animal species. Can J Microbiol 1966;12:1070–71.
63. Balm PH, van Lieshout E, Lokate J, Wendelaar Bonga SE. Bacterial lipopolysaccharide (LPS) and interleukin 1 (IL-1) exert multiple physiological effects in the tilapia Oreochromis mossambicus (Teleostei). J Comp Physiol [B]. 1995;165:85–92.
64. Bang FB. A bacterial disease of Limulus polyphemus. Bull Johns Hopkins Hosp 1956; 98: 325–351.
65. Levin J, Bang FB. A description of cellular coagulation in the Limulus. Bull Johns Hopkins Hosp 1964;115:337–345.
66. Levin J, Bang FB. The role of endotoxin in the extracellular coagulation of limulus blood. Bull Johns Hopkins Hosp 1964;115:265–274.
67. Ramachandra RN, Sehon AH, Berczi I. Neuro-hormonal host defence in endotoxin shock. Brain Behav Immun 1992;6:157–169.
68. Parant M, Le Contel C, Parant F, Chedid L. Influence of endogenous glucocorticoid on endotoxin-induced production of circulating TNF-alpha. Lymphokine Cytokine Res. 1991;10:265–71.
69. Zuckerman SH, Shellhaas J, Butler LD. Differential regulation of lipopolysaccharide-induced interleukin 1 and tumor necrosis factor synthesis: effects of endogenous and exogenous glucocorticoids and the role of the pituitary-adrenal axis. Eur J Immunol. 1989;19:301–5.
70. Richardson RP, Rhyne CD, Fong Y, Hesse DG, Tracey KJ, Marano MA, Lowry SF, Antonacci AC, Calvano SE. Peripheral blood leukocyte kinetics following in vivo lipopolysaccharide (LPS) administration to normal human subjects. Influence of elicited hormones and cytokines. Ann Surg. 1989;210:239–45.
71. Flohe S, Heinrich PC, Schneider J, Wendel A, Flohe L. Time course of IL-6 and TNF alpha release during endotoxin-induced endotoxin tolerance in rats. Biochem Pharmacol. 1991;41:1607–14.
72. Barber AE, Coyle SM, Marano MA, Fischer E, Calvano SE, Fong Y, Moldawer LL, Lowry SF. Glucocorticoid therapy alters hormonal and cytokine responses to endotoxin in man. J Immunol. 1993;150:1999–2006.
73. Waring PM, Waring LJ, Billington T, Metcalf D. Leukemia inhibitory factor protects against experimental lethal Escherichia coli septic shock in mice. Proc Natl Acad Sci U S A. 1995;92:1337–41.
74. Granowitz EV, Porat R, Mier JW, Orencole SF, Kaplanski G, Lynch EA, Ye K, Vannier E, Wolff SM, Dinarello CA. Intravenous endotoxin suppresses the cytokine response of peripheral blood mononuclear cells of healthy humans. J Immunol. 1993;151:1637–45.
75. Berg DJ, Kuhn R, Rajewsky K, Muller W, Menon S, Davidson N, Grunig G, Rennick D. Interleukin-10 is a central regulator of the response to LPS in murine models of endotoxic shock and the Shwartzman reaction but not endotoxin tolerance. J Clin Invest. 1995;96:

2339–47.
76. Matic M, Simon SR. Effects of gamma interferon on release of tumor necrosis factor alpha from lipopolysaccharide-tolerant human monocyte-derived macrophages. Infect Immun. 1992;60:3756–62.
77. Schade FU, Franke C, Schlegel J, Rietschel ET. Formation of a TNF synthesis inhibitor in endotoxin tolerance. Prog Clin Biol Res. 1995;392:513–21.
78. Henricson BE, Neta R, Vogel SN. An interleukin-1 receptor antagonist blocks lipopolysaccharide-induced colony-stimulating factor production and early endotoxin tolerance. Infect Immun. 1991;59:1188–91.
79. Berczi I, Szentivanyi A. The pituitary gland, psychoneuroimmunology and infectious disease. In: Friedman H, Klein T, Friedman AL, editors Psychoneuroimmunology, stress and infectious disease. Boca Raton, FL: CRC Press, 1996;79–109.
80. Wexler BC, Dolgin AE, Tryczynski EW. Effects of bacterial polysaccharide (Piromen) on the pituitary-adrenal axis: Adrenal ascorbic acid, cholesterol and histological alterations. Endocrinology 1957;61:300–308.
81. Wilder RL. Neuroendocrine-immune system interactions and autoimmunity. Annu Rev Immunol. 1995;13:307–38.
82. Torpy DJ, Chrousos GP. In: Chikanza IC, editor. Neuroendocrine Immune Mechanisms of Rheumatic Diseases. Bailliere's Clin Rheumatol London: Bailliere Tindall 1996;10: 181–198.
83. Berczi I, Chalmers IM, Nagy E, Warrington RJ. The immune effects of neuropeptides. In: Chikanza IC Editor. Neuroendocrine Immune Mechanisms of Rheumatic Diseases. Bailliere's Clin Rheumatol, London: Bailliere Tindall,1996;10:227–257.
84. Geller P, Merrill ER, Jawetz E. Effects of cortisone and antibiotics on lethal action of endotoxins in mice. Proc Soc Exp Biol Med 1954;86:716–719.
85. Jansen NJG, Vanoeveren W, Hoiting BH, Wildevuur CRH. Methylprednisolone prophylaxis protects against endotoxin-induced death in rabbits. Inflammation 1991;15: 91–101.
86. Izumi T, Bakhle YS. Modification by steroids of pulmonary oedema and prostaglandin E2 pharmacokinetics induced by endotoxin in rats. Brit J Pharmacol 1988;93:955–963.
87. Sanchez-Cantu L, Rode HN, Yun TJ, Christou NV. Tumor necrosis factor alone does not explain the lethal effect of lipopolysaccharide. Arch Surg , 1991;126:231–235.
88. Ciancio MJ, Hunt J, Jones SB, Filkins JP. Comparative and interactive in vivo effects of tumor necrosis factor " and endotoxin. Circ Shock1991;33:108–120.
89. Mier JW, Vachino G, Klempner MS, Aronson FR, Noring R, Smith S, Brandon EP, Laird EP, Atkins MB. Inhibition of interleukin-2-induced tumor necrosis factor release by dexamethasone - Prevention of an acquired neutrophil chemotaxis defect and differential suppression of interleukin-2-associated side effects. Blood 1990;76:1933–1940.
90. Sheppard BC, Fraker DL, Norton JA. Prevention and treatment of endotoxin and sepsis lethality with recombinant human tumor necrosis factor. Surgery 1989;106:156–161.
91. Paul WE. The immune system: An introduction. In: W.E. Paul, editor. Fundamental Immunology, 4th edition. New York:Lippincott-Raven, 1999;1–18.
92. Brostoff J, Hall T. Hypersensitivity - Type I. In: Roitt IM, Brostoff J, Male DK, editors. Immunology, 2nd edition. London: Gower Med Pub, 1989;19.1–19.
93. Hay F. Hypersensitivity - Type III. In: Roitt IM, Brostoff J, Male DK, editors. Immunology, 2nd edition. London: Gower Med Pub, 1989,21.1–21.
94. Talal N, editor. Autoimmunity. Genetic, Immunologic, Virologic and Clinical Aspects.

Academic Press, 1977.

95. Derijk R, Berkenbosch F. The immune-hypothalamo-pituitary-adrenal axis and autoimmunity. Intern J Neuroscience 1991;59:91–100.
96. Homo-Delarche F, Fitzpatrick F, Christeff N, Nunez EA, Bach JF, Dardenne M. Sex steroids, glucocorticoids, stress and autoimmunity. J Steroid Biochem Mol Biol 1991; 40: 619–637.
97. Wannemacher RW, Jr, Pekarek RS, Thompson WL, Curnow RT, Beall FA, Zenser TV, deRubertis RF, Beisel WR. A protein from polymorphonuclear leukocytes (LEM) which affects the rate of hepatic amino acid transport and synthesis of acute-phase globulins. Endocrinology 1975;96:651–661.
98. Jamieson JC, Lammers G, Janzen R, Woloski BMRNJ. The acute phase response to inflammation: The role of monokines in changes in liver glycoproteins and enzymes of glycoprotein metabolism. Comp Biochem Physiol 1987;87B:11–15.
99. Berczi I. Neurohormonal immunoregulation. Endocr Pathol 1990;1:197–219.
100. Spangelo BL, Macleod RM. Immunopeptide regulation of anterior pituitary function. Int J Neurosci. 1990;51:369–70.
101. Akira S, Kishimoto T. IL-6 and NF-IL6 in acute-phase response and viral infection. Immunol Rev 1992;127:26–50.
102. Baue AF. Neuroendocrine response to severe trauma and sepsis. In: Faist E, Ninnemann J, Green D, editors. Immune consequences of trauma, shock and sepsis. Springer Verlag, 1989;17–32.
103. Bornstein SR, Licinio J, Tauchnitz R, Engelmann L, Negrao AB, Gold P, Chrousos GP. Plasma leptin levels are increased in survivors of acute sepsis: associated loss of diurnal rhythm, in cortisol and leptin secretion. J Clin Endocrinol Metab 1998;83:280–3.
104. Giovambattista A, Chisari AN, Corro L, Gaillard RC, Spinedi E. Metabolic, neuroendocrine and immune functions in basal conditions and during the acute-phase response to endotoxic shock in undernourished rats. Neuroimmunomodulation 2000;7:92–8.
105. Okamoto S, Irie Y, Ishikawa I, Kimura K, Masayuki Saito. Central leptin suppresses splenic lymphocyte functions through activation of the corticotropin-releasing hormone-sympathetic nervous system. Brain Res 2000;855:192–97.
106. Chautard T, Spinedi E, Voirol M, Pralong FP, Gaillard RC. Role of glucocorticoids in the response of the hypothalamo-corticotrope, immune and adipose systems to repeated endotoxin administration. Neuroendocrinology 1999;69:360–69.
107. Torpy DJ, Bornstein SR, Chrousos GP. Leptin and interleukin-6 in sepsis. Horm Metab Res 1998;30:726–9.
108. Faggioni R, Fantuzzi G, Gabay C, Moser A, Dinarello CA, Feingold KR, Grunfeld C. Leptin deficiency enhances sensitivity to endotoxin-induced lethality. Am J Physiol 1999;276:R136–42.
109. Takahashi N, Waelput W, Guisez Y. Leptin is an endogenous protective protein against the toxicity exerted by tumor necrosis factor. J Exp Med 1999;189:207–12.
110. Finck BN, Kelley KW, Dantzer R, Johnson RW. *In vivo* and *in vitro* evidence for the involvement of tumor necrosis factor-alpha in the induction of leptin by lipopolysaccharide. Endocrinology 1998;139:2278–83.
111. Loffreda S, Yang SQ, Lin HZ, Karp CL, Brengman ML, Wang DJ, Klein AS, Bulkley GB, Bao C, Noble PW, Lane MD, Diehl AM. Leptin regulates proinflammatory immune responses. FASEB J 1998;12:57–65.
112. Fantuzzi G, Faggioni R. Leptin in the regulation of immunity, inflammation, and

hematopoiesis. J Leukoc Biol 2000;68:437–46.
113. Kolb-Bachofen V. A review on the biological properties of C-reactive protein. Immunobiology. 1991;183:133–45.
114. Ballou SP, Kushner I. C-reactive protein and the acute phase response. Adv Intern Med. 1992;37:313–36.
115. Young B, Gleeson M, Cripps AW. C-reactive protein: a critical review. Pathology. 1991 ;23:118–24.
116. Mold C, Nakayama S, Holzer TJ, Gewurz H, Du Clos TW. C-reactive protein is protective against Streptococcus pneumoniae infection in mice. J Exp Med. 1981 ;154:1703–8.
117. Bauer J, Birmelin M, Northoff GH, Northemann W, Tran-Thi TA, Ueberberg H, Decker K, Heinrich P. Induction of rat "2-macroglobulin in vivo and in heaptocyte primary cultures: synergistic action of glucocorticoids and a Kupffer cell derived factor. FEBS Lett 1984;177:89–94.
118. van Gool J, Boers W, Sala M, Ladiges NC. Glucocorticoids and catecholamines as mediators of acute-phase proteins, especially rat alpha-macrofoetoprotein. Biochem J. 1984;220:125–32.
119. Wassell J. Haptoglobin: function and polymorphism. Clin Lab 2000;46(11–12):547–52.
120. Daemen MA, Heemskerk VH, van't Veer C, Denecker G, Wolfs TG, Vandenabeele P, Buurman WA. Functional protection by acute phase proteins alpha(1)-acid glycoprotein and alpha(1)-antitrypsin against ischemia/reperfusion injury by preventing apoptosis and inflammation. Circulation 2000;102:1420–6.
121. Liu H, Jensen L, Hansen S, Petersen SV, Takahashi K, Ezekowitz AB, Hansen FD, Jensenius JC, Thiel S. Characterization and quantification of mouse mannan-binding lectins (MBL-A and MBL-C) and study of acute phase responses. Scand J Immunol 2001;53:489–97.
122. Turner MW, Hamvas RM. Mannose-binding lectin: structure, function, genetics and disease associations. Rev Immunogenet 2000;2:305–22.
123. Van Gool J, Van Vugt H, Helle M, Aarden LA. The relation among stress, adrenalin, interleukin 6, and acute phase proteins in the rat. Clin Immunol Immunopathol 1990;57: 200–210.
124. Don BR, Rosales LM, Levine NW, Mitch W, Kaysen GA. Leptin is a negative acute phase protein in chronic hemodialysis patients. Kidney Int 2001;59:1114–20.
125. Finck BN, Johnson RW. Tumor necrosis factor-alpha regulates secretion of the adipocyte-derived cytokine, leptin. Microsc Res Tech 2000;50:209–15.
126. Haeryfar SM, Berczi I. The thymus and the acute phase response. Cell Mol Biol (Noisy-le-grand) 2001;47:145–56.
127. Bergad PL, Schwarzenberg SJ, Humbert JT, Morrison M, Amarasinghe S, Towle HC, Berry SA. Inhibition of growth hormone action in models of inflammation. Am J Physiol Cell Physiol 2000;279(6):C1906–17.
128. Takala J, Ruokonen E, Webster NR, et al. Increased mortality associated with growth hormone treatment in critically ill adults. N Engl J Med 1999;341:785–92.
129. Jaschke MG, Herndon DN, Wolf SE, et al. Recombinant human growth hormone alters acute phase reactant proteins, cytokine expression, and liver morphology in burned rats. J Surg Res 1999;83:122–29.
130. Derfalvi B, Igaz P, Fulop KA, Szalai C, Falus A. Interleukin-6-induced production of type II acute phase proteins and expression of junB gene are downregulated by human recombinant growth hormone in vitro. Cell Biol Int 2000;24:109–14.

131. Berczi I, Nagy E, Asa SL, Kovacs K. Pituitary hormones and contact sensitivity in rats. Allergy 1983;38:325–330.
132. Berczi I, Nagy E, Asa SL, Kovacs K. The influence of pituitary hormones on adjuvant arthritis. Arthritis Rheum 1984;27:682–688.
133. Ramirez RJ, Wolf SE, Herndon DN. Is there a role for growth hormone in the clinical management of burn injuries? Growth Horm IGF Res 1998 ;8 Suppl B:99–105.
134. Jeschke MG, Barrow RE, Herndon DN. Recombinant human growth hormone treatment in pediatric burn patients and its role during the hepatic acute phase response. Crit Care Med 2000;28:1578–84.
135. Jeschke MG, Herndon DN, Barrow RE. Insulin-like growth factor I in combination with insulin-like growth factor binding protein 3 affects the hepatic acute phase response and hepatic morphology in thermally injured rats. Ann Surg 2000;231:408–16.
136. Jeschke MG, Herndon DN, Wolf SE, DebRoy MA, Rai J, Thompson JC, Barrow RE. Hepatocyte growth factor modulates the hepatic acute-phase response in thermally injured rats. Crit Care Med 2000;28:504–10.
137. Jeschke MG, Barrow RE, Herndon DN. Insulinlike growth factor I plus insulinlike growth factor binding protein 3 attenuates the proinflammatory acute phase response in severely burned children. Ann Surg 2000;231:246–52.
138. Kvasnicka J, Marek J, Kvasnicka T, Weiss V, Markova M, Stepan J, Umlaufova A. Increase of adhesion molecules, fibrinogen, type-1 plasminogen activator inhibitor and orosomucoid in growth hormone (GH) deficient adults and their modulation by recombinant human GH replacement. Clin Endocrinol (Oxf) 2000;52:543–8.
139. Hasday JD, Fairchild KD, Shanholtz C. The role of fever in the infected host. Microbes Infect 2000;2:1891–904.
140. Krabbe KS, Bruunsgaard H, Hansen CM, Moller K, Fonsmark L, Qvist J, Madsen PL, Kronborg G, Andersen HO, Skinhoj P, Pedersen BK. Ageing is associated with a prolonged fever response in human endotoxemia. Clin Diagn Lab Immunol 2001;8:333–8.
141. Kotler DP. Cachexia. Ann Intern Med 2000;133:622–34.
142. Langhans W. Anorexia of infection: current prospects. Nutrition 2000;16:996–1005.
143. Oppert M, Gleiter CH, Muller C, Reinicke A, von Ahsen N, Frei U, Eckardt KU. Kinetics and characteristics of an acute phase response following cardiac arrest. Intensive Care Med 1999;25:1386–94.
144. Yudkin JS, Kumari M, Humphries SE, Mohamed-Ali V. Inflammation, obesity, stress and coronary heart disease: is interleukin-6 the link? Atherosclerosis 2000;148:209–14.
145. Kushner I. C-reactive protein elevation can be caused by conditions other than inflammationand may reflect biologic aging. Cleve Clin J Med 2001;68:535–7.
146. Bruunsgaard H, Pedersen M, Pedersen BK. Aging and proinflammatory cytokines. Curr Opin Hematol 2001;8:131–6.
147. Duncan BB, Schmidt MI. Chronic activation of the innate immune system may underlie the metabolic syndrome. Sao Paulo Med J 2001;119:122–7.
148. Fernandez-Real JM, Vayreda M, Richart C, Gutierrez C, Broch M, Vendrell J, Ricart W. Circulating interleukin 6 levels, blood pressure, and insulin sensitivity in apparently healthy men and women. J Clin Endocrinol Metab 2001;86:1154–9.
149. Arnalich F, Hernanz A, Lopez-Maderuelo D, Pena JM, Camacho J, Madero R, Vazquez JJ, Montiel C. Enhanced acute-phase response and oxidative stress in older adults with type II diabetes. Horm Metab Res 2000;32:407–12.
150. Pickup JC, Chusney GD, Thomas SM, Burt D. Plasma interleukin-6, tumour necrosis

factor alpha and blood cytokine production in type 2 diabetes. Life Sci 2000;67:291–300.
151. Fernandez-Real JM, Broch M, Vendrell J, Gutierrez C, Casamitjana R, Pugeat M, Richart C, Ricart W. Interleukin-6 gene polymorphism and insulin sensitivity. Diabetes 2000;49: 517–20.
152. Pickup JC, Chusney GD, Mattock MB. The innate immune response and type 2 diabetes: evidence that leptin is associated with a stress-related (acute-phase) reaction. Clin Endocrinol (Oxf) 2000;52:107–12.
153. Lin RY, Trivino MR, Curry A, Pesola GR, Knight RJ, Lee HS, Bakalchuk L, Tenenbaum C, Westfal RE. Interleukin 6 and C-reactive protein levels in patients with acute allergic reactions: an emergency department-based study. Ann Allergy Asthma Immunol 2001;87: 412–6.
154. Boss B, Neeck G. Correlation of IL-6 with the classical humoral disease activity parameters ESR and CRP and with serum cortisol, reflecting the activity of the HPA axis in active rheumatoid arthritis. Z Rheumatol 2000;59 Suppl 2:II/62–4.
155. Sakane T, Suzuki N. Neuro-endocrine-immune axis in human rheumatoid arthritis. Arch Immunol Ther Exp (Warsz) 2000;48:417–27.
156. O'Hara R, Murphy EP, Whitehead AS, FitzGerald O, Bresnihan B. Acute-phase serum amyloid A production by rheumatoid arthritis synovial tissue. Arthritis Res 2000;2:142–4.
157. Janssen U, Bahlmann F, Kohl J, Zwirner J, Haubitz M, Floege J. Activation of the acute phase response and complement C3 in patients with IgA nephropathy. Am J Kidney Dis 2000;35:21–8.
158. Yang H, Wang H, Tracey KJ. HMG-1 rediscovered as a cytokine. Shock 2001;15:247–53.
159. Jamaluddin M, Meng T, Sun J, Boldogh I, Han Y, Brasier AR. Angiotensin II induces nuclear factor (NF)-kappaB1 isoforms to bind the angiotensinogen gene acute-phase response element: a stimulus-specific pathway for NF-kappaB activation. Mol Endocrinol 2000;14:99–113.
160. Cavaillon JM. The nonspecific nature of endotoxin tolerance. Trends Microbiol. 1995;3: 320–4.
161. Mengozzi M, Ghezzi P. Cytokine down-regulation in endotoxin tolerance. Eur Cytokine Netw. 1993;4:89–98.
162. Freudenberg MA, Galanos C. Induction of tolerance to lipopolysaccharide (LPS)-D-galactosamine lethality by pretreatment with LPS is mediated by macrophages. Infect Immun. 1988;56:1352–7.
163. Evans GF, Zuckerman SH. Glucocorticoid-dependent and -independent mechanisms involved in lipopolysaccharide tolerance. Eur J Immunol. 1991;21:973–9.
164. Szabo C, Thiemermann C, Wu CC, Perretti M, Vane JR. Attenuation of the induction of nitric oxide synthase by endogenous glucocorticoids accounts for endotoxin tolerance in vivo. Proc Natl Acad Sci U S A. 1994;91:271–5.
165. Hansen MK, Nguyen KT, Goehler LE, Gaykema RP, Fleshner M, Maier SF, Watkins LR.Effects of vagotomy on lipopolysaccharide-induced brain interleukin-1beta protein in rats. Auton Neurosci 2000;85:119–26.
166. Szalai AJ, van Ginkel FW, Wang Y, McGhee JR, Volanakis JE. Complement-dependent acute-phase expression of C-reactive protein and serum amyloid P-component. J Immunol 2000;165:1030–5.
167. Galanos C, Freudenberg MA, Reutter W. Galactosamine-induced sensitization to the lethal

effects of endotoxin. Proc Natl Acad Sci U S A. 1979;76:5939–43.
168. Kimura T, Itoh N, Takehara M, Oguro I, Ishizaki JI, Nakanishi T, Tanaka K. Sensitivity of metallothionein-null mice to LPS/D-galactosamine-induced lethality. Biochem Biophys Res Commun 2001;280:358–62.
169. Roth J, Hubschle T, Pehl U, Ross G, Gerstberger R. Influence of systemic treatment with cyclooxygenase inhibitors on lipopolysaccharide-induced fever and circulating levels of cytokines and cortisol in guinea-pigs. Pflugers Arch 2002;443:411–417.
170. Streetz KL, Wustefeld T, Klein C, Manns MP, Trautwein C.Mediators of inflammation and acute phase response in the liver. Cell Mol Biol (Noisy-le-grand) 2001;47:661–73.
171. Shen X, Tian Z, Holtzman MJ, Gao B. Cross-talk between interleukin 1beta (IL-1beta) and IL-6 signalling pathways: IL-1beta selectively inhibits IL-6-activated signal transducer and activator of transcription factor 1 (STAT1) by a proteasome-dependent mechanism. Biochem J 2000;352 Pt 3:913–9.
172. Jostock T, Mullberg J, Ozbek S, Atreya R, Blinn G, Voltz N, Fischer M, Neurath MF, Rose-John S. Soluble gp130 is the natural inhibitor of soluble interleukin-6 receptor transsignaling responses. Eur J Biochem 2001;268:160–7.
173. Fennekohl A, Lucas M, Puschel GP. Induction by interleukin 6 of G(s)-coupled prostaglandin E(2) receptors in rat hepatocytes mediating a prostaglandin E(2)-dependent inhibition of the hepatocyte's acute phase response. Hepatology 2000;31:1128–34.
174. Beigneux AP, Moser AH, Shigenaga JK, Grunfeld C, Feingold KR. The acute phase response is associated with retinoid X receptor repression in rodent liver. J Biol Chem 2000;275:16390–9.
175. Harris HW, Gosnell JE, Kumwenda ZL. The lipemia of sepsis: triglyceride-rich lipoproteins as agents of innate immunity. J Endotoxin Res 2000;6:421–30.
176. Fulop AK, Pocsik E, Brozik M, Karabelyos C, Kiss A, Novak I, Szalai C, Dobozy O, Falus A. Hepatic regeneration induces transient acute phase reaction: systemic elevation of acute phase reactants and soluble cytokine receptors. Cell Biol Int 2001;25:585–92.
177. Hecht N, Pappo O, Shouval D, Rose-John S, Galun E, Axelrod JH. Hyper-IL-6 gene therapy reverses fulminant hepatic failure. Mol Ther 2001;3(5 Pt 1):683–7.
178. Gabay C, Gigley J, Sipe J, Arend WP, Fantuzzi G. Production of IL-1 receptor antagonist by hepatocytes is regulated as an acute-phase protein in vivo. Eur J Immunol 2001;31: 490–9.
179. Carstanjen D, Regenfus M, Muller C, Salama A. Interleukin-6 is a major effector molecule of short-term G-CSF treatment inducing bone metabolism and an acute-phase response. Exp Hematol 2001;29:812–21.
180. Suwa T, Hogg JC, Klut ME, Hards J, van Eeden SF. Interleukin-6 changes deformability of neutrophils and induces their sequestration in the lung. Am J Respir Crit Care Med 2001;163:970–6.
181. Beishuizen A, Thijs LG, Haanen C, Vermes I. Macrophage migration inhibitory factor and hypothalamo-pituitary-adrenal function during critical illness. J Clin Endocrinol Metab 2001;86:2811–6.
182. Seatter SC, Bennet T, Li MH, Bubrick MP, West MA. Macrophage endotoxin tolerance. Tumor necrosis factor and interleukin-1 regulation by lipopolysaccharide pretreatment. Arch Surg. 1994;129:1263–9.
183. Li MH, Seatter SC, Manthei R, Bubrick M, West MA. Macrophage endotoxin tolerance: effect of TNF or endotoxin pretreatment. J Surg Res. 1994;57:85–92.
184. Previte JJ, Chang Y, El-Bisi HM. Detoxification of *Salmonella typhimurium*

lipopolysaccharide by ionizing radiation. J Bacteriol 1967;93:1607.
185. Fust G, Bertok L, Juhasz-Nagy S. Interactions of radio-detoxified Escherichia coli endotoxin preparations with the complement system. Infect Immun. 1977;16:26–31.
186. Bertok L. Stimulation of nonspecific resistance by radiation-detoxified endotoxin. In:. Nowotny A, editor. Beneficial Effects of Endotoxin. New York: Plenum Press, 1983;213–226.
187. Banhegyi D, Varnai F, Horvath A, Bertok L. Treatment of severe immune deficiency with radio-detoxified endotoxin. J Med. 1992;23:154–6.
188. Gustafson GL, Rhodes MJ, Hegel T. Monophosphoryl lipid A as a prophylactic for sepsis and septic shock. Prog Clin Biol Res. 1995;392:567–79.
189. Astiz ME, Rackow EC, Still JG, Howell ST, Cato A, Von Eschen KB, Ulrich JT, Rudbach JA, McMahon G, Vargas R, et al. Pretreatment of normal humans with monophosphoryl lipid A induces tolerance to endotoxin: a prospective, double-blind, randomized, controlled trial. Crit Care Med. 1995;23:9–17.
190. Anisman H, Baines MG, Berczi I, et al. Neuroimmune mechanisms in health and disease: 1. Health. Can Med Assoc J 1996;155:867–874.
191. Anisman H, Baines MG, Berczi I, et al. Neuroimmune mechanisms in health and disease: 2. Disease. Can Med Assoc J 1996;155:1075–1082
192. Berczi I, Chow DA, Sabbadini ER. Neuroimmunoregulation and Natural Immunity. Domestic Anim Endocrin 1998;15:273–281.
193. Somech R, Zakuth V, Assia A, Jurgenson U, Spirer Z. Procalcitonin correlates with C-reactive protein as an acute-phase reactant in pediatric patients. Isr Med Assoc J 2000;2: 147–50.
194. Stephensen CB, Gildengorin G. Serum retinol, the acute phase response, and the apparent misclassification of vitamin A status in the third National Health and Nutrition Examination Survey. Am J Clin Nutr 2000;72:1170–8.
195. Zweigner J, Gramm HJ, Singer OC, Wegscheider K, Schumann RR. High concentrations of lipopolysaccharide-binding protein in serum of patients with severe sepsis or septic shock inhibit the lipopolysaccharide response in human monocytes. Blood 2001;98: 3800–8.
196. Schmid B, Wong S, Mitchell BF. Transcriptional regulation of oxytocin receptor by interleukin-1beta and interleukin-6. Endocrinology 2001;142:1380–5.
197. Marchini G, Berggren V, Djilali-Merzoug R, Hansson LO. The birth process initiates an acute phase reaction in the fetus-newborn infant. Acta Paediatr 2000;89:1082–6.
198. Pedersen BK. Special feature for the Olympics: effects of exercise on the immune system: exercise and cytokines. Immunol Cell Biol 2000;78:532–5
199. Pedersen BK, Hoffman-Goetz L. Exercise and the immune system: regulation, integration, and adaptation. Physiol Rev 2000;80:1055–81
200. Jeukendrup AE, Vet-Joop K, Sturk A, Stegen JH, Senden J, Saris WH, Wagenmakers AJ. Relationship between gastro-intestinal complaints and endotoxaemia, cytokine release and the acute-phase reaction during and after a long-distance triathlon in highly trained men. Clin Sci (Lond) 2000;98:47–55.
201. Luyer MD, Khosla S, Owen WG, Miller VM. Prospective randomized study of effects of unopposed estrogen replacement therapy on markers of coagulation and inflammation in postmenopausal women. J Clin Endocrinol Metab 2001;86:3629–34.
202. Hak AE, Pols HA, Stehouwer CD, Meijer J, Kiliaan AJ, Hofman A, Breteler MM, Witteman JC. Markers of inflammation and cellular adhesion molecules in relation to

insulin resistance in nondiabetic elderly: the Rotterdam study. J Clin Endocrinol Metab 2001;86:4398–405.
203. Banck G, Forsgren A. Many bacterial species are mitogenic for human blood B lymphocytes. Scand J Immunol. 1978;8:347–54.
204. Fleischer B, Gerlach D, Fuhrmann A, Schmidt KH. Superantigens and pseudosuperantigens of gram-positive cocci. Med Microbiol Immunol (Berl). 1995;184:1–8.
205. Levinson AI, Kozlowski L, Zheng Y, Wheatley L. B-cell superantigens: definition and potential impact on the immune response. J Clin Immunol. 1995;15(6 Suppl):26S–36S.
206. Zumla A. Superantigens, T cells and microbes. Clin Infect Dis 1992;15:313–320.
207. Kotb M. Bacterial pyrogenic exotoxins as superantigens. Clin Microbiol Rev. 1995;8: 411–26.
208. Spangelo BL, deHoll PD, Kalabay L, Bond BR, Arnaud P. Neurointermediate pituitary lobe cells synthesize and release interleukin-6 in vitro: effects of lipopolysaccharide and interleukin-1 beta. Endocrinology. 1994;135:556–63.
209. Chao HS, Poisner AM, Poisner R, Handwerger S. Lipopolysaccharides inhibit prolactin and renin release from human decidual cells. Biol Reprod. 1994;50:210–4.
210. Di Santo E, Sironi M, Pozzi P, Gnocchi P, Isetta AM, Delvaux A, Goldman M, Marchant A, Ghezzi P. Interleukin-10 inhibits lipopolysaccharide-induced tumor necrosis factor and interleukin-1 beta production in the brain without affecting the activation of the hypothalamus-pituitary-adrenal axis. Neuroimmunomodulation. 1995;2:149–54.
211. Faggioni R, Fantuzzi G, Villa P, Buurman W, van Tits LJ, Ghezzi P. Independent down-regulation of central and peripheral tumor necrosis factor production as a result of lipopolysaccharide tolerance in mice. Infect Immun. 1995;63:1473–7.
212. Das KP, McMillian MK, Bing G, Hong JS. Modulatory effects of [Met5]-enkephalin on interleukin-1 beta secretion from microglia in mixed brain cell cultures. J Neuroimmunol. 1995;62:9–17.
213. Scotte M, Hiron M, Masson S, Lyoumi S, Banine F, Teniere P, Lebreton JP, Daveau M. Differential expression of cytokine genes in monocytes, peritoneal macrophages and liver following endotoxin- or turpentine-induced inflammation in rat. Cytokine. 1996;8: 115–20.
214. Gaykema RP, Dijkstra I, Tilders FJ. Subdiaphragmatic vagotomy suppresses endotoxin-induced activation of hypothalamic corticotropin-releasing hormone neurons and ACTH secretion. Endocrinology. 1995;136:4717–20.
215. Laye S, Bluthe RM, Kent S, Combe C, Medina C, Parnet P, Kelley K, Dantzer R. Subdiaphragmatic vagotomy blocks induction of IL-1 beta mRNA in mice brain in response to peripheral LPS. Am J Physiol. 1995;268(5 Pt 2):R1327–31.
216. Berczi I, Bertok L, Baintner K, Jr, Veress B. Failure of oral Escherichia coli endotoxin to induce either specific tolerance or toxic syndromes in rats. J Path Bact 1968;96:481–486.
217. Kocsar LT, Bertok L, Varteresz V. Effect of bile acids on the intestinal absorption of endotoxin in rats. J Bacteriol. 1969;100:220–3.
218. Mimura Y, Sakisaka S, Harada M, Sata M, Tanikawa K. Role of hepatocytes in direct clearance of lipopolysaccharide in rats. Gastroenterology. 1995;109:1969–76.
219. Bertok L. Physico-chemical defence of vertebrate organisms: the role of bile acids in defence against endotoxins. Persp Biol Med 1977;21:70–76.

Edited by I. Berczi and A. Szentivanyi

Autoimmune Disease

ISTVAN BERCZI and ANDOR SZENTIVANYI

Department of Immunology, Faculty of Medicine, The University of Manitoba, Winnipeg, Manitoba R3E OW3, Canada; and Department of Internal Medicine, Faculty of Medicine, The University of South Florida, Tampa, Florida 33612, USA

ABSTRACT

Auto-reactive lymphocytes and natural antibodies do exist physiologically and participate in the homeostatic regulation of bodily functions. Mature T and B lymphocytes with autoimmune effector potential do develop under normal conditions but are in equilibrium with suppressor cells and with other inhibitory elements, the end result of which is self-tolerance. Immune homeostasis does not rely exclusively on internal suppressive mechanisms of the immune system. Immunocompetence is dependent on, and adaptive and natural immune responses and inflammation are controlled by neuroendocrine regulatory mechanisms. By definition autoimmune disease develops upon the loss of self-tolerance. An autoimmune reaction is due to de-regulated lymphocyte proliferation and maturation into effector cells directed towards self-antigens. Deregulated growth is also characteristic of cancer cells and it is clear that multiple genes must be de-regulated for the development of uncontrolled cell growth. Evidence is emerging that the de-regulation of multiple genes are required also for the development of autoimmune disease. Current evidence supports the hypothesis that de-regulation at the cellular level as well as an altered neuroendocrine milieu are necessary for autoimmmune disease to develop. Autoimmune reactions may be specific for self-molecules, cells, organs or tissues. These reactions are due to de-regulated cells of the adaptive immune system. Rheumatoid diseases are characterized by polyclonal lymphocyte activation, inflammation and acute phase reactants, which are indicative of a disorder of the innate immune system. Abnormalities of the hypothalamus – growth and lactogenic hormone – insulin-like growth factor axis, of the hypothalamus – pituitary – adrenal axis and of the hypothalamus – pituitary – gonadal axis are frequently observed in autoimmune disease. An imbalance of these major immunoregulatory axes of the pituitary gland is very likely to play an etiological role in the pathogenesis of autoimmune disease. It is also becoming apparent that defective regulation by neurotransmitters and neuropeptides contribute significantly to the pathogenesis of autoimmune and inflammatory conditions.

1. INTRODUCTION

Higher organisms must defend themselves constantly against a multitude of pathogenic microorganisms and parasites. Because of their short generation time, pathogenic organisms can evolve much faster through inherited mutations than their vertebrate hosts could develop resistance genes that are passed to subsequent generations. Talmage [1] and Burnet [2,3] predicted that the immune system must generate a diversity of lymphocyte antigen receptors through rapid somatic mutations in order to develop responsiveness to an enormous variety of antigens that has relevance to host defence. Burnet also proposed that autoimmune lymphocyte clones are produced during somatic mutation, which have to be eliminated in order to prevent autoimmune disease. Today this process is known as the thymic selection of T lymphocytes. Until recently it was held that self- tolerant T lymphocytes are selected in the thymus and then are released to the periphery. Evidence was produced that autoreactive T cells are killed *in situ* by the activation of the suicide pathway. Autoreactive B cell clones are also eliminated during development in the bone marrow [4,5]. In addition, there are peripheral mechanisms that contribute to the maintenance of self tolerance. Antigen induced anergy [6], apoptosis [5], the activation of suppressor cells [7] and cytotoxic mechanisms [8] all seem to contribute to the peripheral maintenance of self-tolerance. In recent years much emphasis was given to peripheral suppressor cells that mediate to a large extent self tolerance and anergy. According to recent evidence the thymus produces both *self-reactive effector* and *suppressor cells*, which reach an equilibrium peripherally in the tissues, where the autoantigen is located. Therefore, self tolerance is not due to the lack of outoreactive cells as originally thought, but rather it is a dynamic balance between effector and suppressor lymphocytes that maintain *immune-homeostasis* throughout the organism and significantly contributes the homeostatic control of the entire organism [9]. The evidence presented in this chapter supports fully the concept that the immune system forms a regulatory network with the endocrine and the nervous systems, and that this network is fundamental to the function of higher organism in health and disease.

Autoimmune disease is explained most frequently with the breakdown of self-tolerance [10]. Certain serotypes of MHC antigens are associated with autoimmunity. MHC molecules present self-antigens to T cell precursors maturing in the thymus. Thymocytes that are activated by self-antigen are eliminated through apoptosis. Therefore, MHC antigens determine the shaping of the T cell repertoire (e.g. effector and suppressor cells) released to the periphery. A failure of this thymic selection could result in autoimmune disease. MHC bearing APC also present antigen in the periphery. The induction by cytokines (e.g. IFNγ) of class II MHC antigens in tissues where it is normally not present is viewed by some as a trigger mechanism for autoimmune disease [11]. However, the association of autoimmune disease with MHC is never absolute, which indicates that factors other than antigen presentation play important roles in the development of autoimmune disease. The role of the T cell receptor (TCR) in the etiology of autoimmune disease has also been considered at length with similar conclusions [10]. It is generally accepted that the development of autoimmune disease involves T cells of the helper type as B cells and killer/suppressor T cells all require T cell help to develop autoreactivity. It is now apparent that suppressor CD25+ cells in the CD4+ T-cell population exert a dominant control of self-reactive T-cells and contribute to the maintenance of immunologic self-tolerance. In normal naive animals CD25+ suppressor cells are produced by the thymus and they play critical roles not only in preventing autoimmunity but also in controlling tumor immunity and transplantation tolerance [12,13]. The *"dominant tolerance" models* provide explanations for natural autoimmune activities, notably in the regulation of inflammatory reactions and, more generally, in the "quality control" of immune responses [14]. Because *self tolerance* is now explained by *dominant equilibrium* between effector and suppressor cells, autoreactivity must come from *dis-equilibrium*. Recent studies as well as clinical observations indicate that normal immunoglobulin also protects against autoimmune disease. Self-reactive natural antibodies are normal serum components and likely to fulfil important regulatory function as well as provide host defence against infections, cancer and autoimmune disease [15,16].

2. PROLACTIN AND AUTOIMMUNE DISEASE

2.1. Experimental autoimmune disease

We observed that hypophysectomized (hypox) rats did not develop polyarthritis after challenge with complete Freund's adjuvant. The treatment of intact animals with bromocriptine (BRC) had a similar inhibitory effect. Adjuvant arthritis (AA) developed normally in Hypox or BRC suppressed animals if the animals were also given replacement doses of either PRL or GH. Additional treatment with ACTH inhibited this restoration. The treatment of Hypox rats with FSH, LH and TSH was without any effect [17,18]. Others observed that the postpartum exacerbation of collagen induced arthritis was suppressed in mice by BRC treatment. Approximately 50% reduction in disease severity was achieved [19]. However, Mattsson et al [20] found that PRL aggravated collagen induced arthritis in mice if given during the immunization phase, whereas at a later stage treatment with BRC caused exacerbation, which suggests a pathogenic role of hypoprolactinemia.

Experimental allergic encephalitis (EAE) induced in Lewis rats by immunization with rat spinal cord in Freund's complete adjuvant could be inhibited significantly by BRC treatment [21]. Similarly, treatment with BRC alone decreased the incidence of experimental autoimmune uveitis (EAU) in female rats only. However, BRC treatment significantly decreased the antibody titers in both sexes of rats, when immunized with the S antigen to elicit the disease. Low dose of cyclosporin-A (CSA) (2 mg/kg/day) partially decreased the incidence of EAU in female rats and had no effect on the antibody response. When BRC and CSA were applied together at the same dosage the incidence of EAU was markedly inhibited. Both anti-S antigen antibody titers and lymphocyte proliferative responses were markedly decreased [22].

Mice of the B/W strain spontaneously develop an autoimmune disease similar to systemic lupus erythematosus (SLE) in man. The treatment of female B/W animals with BRC inhibited the production of anti-DNA IgG antibodies in serum and increased survival. In contrast, if female B/W mice were grafted with syngeneic pituitaries, which caused a prolonged hyperprolactinemia, a premature occurrence of albuminuria was observed. The concentration of immune complexes containing murine retroviral gp70 and IgG were elevated in the serum and mortality was accelerated [23]. Pregnancy with or without suckling and pseudopregnancy also accelerated the autoimmune disease in B/W mice. The immunological changes were most severe in the pseudopregnant animals that were hyperprolactinemic for 18 weeks [24]. It was also established that the frequencies of PRL receptor bearing T cells and receptor density per cell was increased with age in the autoimmune NZB mice, but not in C57/BL6 mice [25]. An impairment of PRL release was also observed in male Sprague-Dawley rats after the induction of autoimmune diabetes by streptozotocin [26].

In multiple studies, elevated serum PRL stimulated the appearance or progression of murine lupus. Disease acceleration is augmented by estrogen stimulation of PRL secretion and is independent of immunosuppressive effects of androgens such as testosterone. The suppression of serum PRL concentrations by BRC inhibited the development of murine SLE [27]. Estrogen and PRL have been shown to aggravate autoimmunity in the NZB/NZW F1 (B/W) mouse model of SLE. However, estrogen stimulates prolactin secretion. Hyperprolactinemic mice with either low or high serum estrogen levels had accelerated development of albuminuria at 24 and 32 weeks of age, when compared to normal and hypoprolatinemic mice. High estrogen/high PRL mice also had a higher percentage of anti-DNA antibodies compared to mice in the low estrogen/low PRL and the high estrogen/low PRL groups. IgG levels were not significantly different between groups. Mean survival was shortest in the high estrogen/high PRL group (34+/-1.0 weeks) and longest in the high estrogen/low PRL group (42+/-1.2 weeks; $P < 0.05$). *In vitro* the proliferation and IL-2 production by lymphocytes of mice with high levels of serum estrogen were depressed [28].

Estrogen and PRL have a reciprocal endocrinologic relationship and both hormones have pleiotropic effects on the immune system and both hormones modulate autoimmunity. Estrogen appears to suppress cell-mediated and to augment humoral immunity, whereas PRL appears to stimulate both forms of immunity. Both hormones modulate IFNγ secretion. In the NZB x NZW F1 mouse model of SLE, PRL accelerates disease expression, whereas estrogen, when devoid of its PRL stimulating properties, is immunosuppressive and inhibits IL-2 production. Therefore, estrogen, because of its endocrinological and immune effects, may directly or indirectly stimulate or inhibit immune responses. These dichotomous effects have limited the success of therapeutic attempts by the pharmacologic manipulation of estrogens in human autoimmune disease. In contrast, the reduction of immunostimulatory concentrations of PRL with BRC has successfully suppressed development or expression of murine and human autoimmune disease [29]. Treatment with PRL protected mice against the induction of diabetes by low dose of streptozotocin [30].

2.2. Prolactin and autoimmune disease in man

Originally hyperprolactinemia was considered to cause decreased libido, amenorrhea, galactorrhea and infertility in the human. It is now clear that hyperprolactinemia is associated also with autoimmune disorders. At screening the British population aged 38 years or older, 0.7% of the men and 2.5% of the women had serum PRL levels greater than 600 mU/l. For men, 2.5% were above 400 mU/l. Hyperprolactinaemia was defined as greater than 400 mU/l in men and greater than 600 mU/l in women. On repeat testing, hyperprolactinaemia was found in 1.4% of men and 1.2% of women. No significant relationship was found in either sex between hyperprolactinaemia and age or evidence of autoimmune thyroid disease. In women, there was no association with age, distance beyond the menopause or duration of reproductive years, but PRL levels were slightly higher in those on oestrogen therapy [31]. Elevated micro and/or macrovascular anti-endothelial cell antibodies were found in sera from 19/25 hyperprolactinemic women (76%). Titers of autoantibodies did not correlate with PRL level, or with the use or duration of anti-PRL treatment. None of the individuals examined presented clinical manifestations of autoimmune disease [32].

High basal serum PRL levels were found in 8 male patients suffering from SLE. Estrone levels were also increased, whereas testosterone and dehydrotesterone levels were decreased in serum. There was a decreased pituitary gonadotropin response to LHRH stimulation [33]. Similarly, PRL levels were elevated in 9 pregnant SLE patients. Five patients having active SLE showed the highest PRL levels and one of them had fetal wastage. In active SLE cases serum testosterone and estradiol (E2) levels were decreased significantly from weeks 10 to 30 [34]. Serum PRL levels were significantly higher in 29 male SLE patients, when compared to normal controls, regardless of treatment with corticosteroids [35]. Moderate hyperprolactinaemia was observed in a subset of patients with SLE and serum PRL correlated with clinical and serological disease activity [36].

In several clinical studies conducted in patients with juvenile SLE and in adult male and female SLE patients there was no correlation between the moderate hyperprolactinemia found and clinical manifestations, thyroid abnormalities or the production of autoantibodies [37,38,39,40]. Increased plasma PRL concentration (> 20 ng/ml) was recorded in 11 of 35 patients (31%) with lupus. No correlation was found between plasma PRL and GH, IL-6, cortisol, C-reactive protein or with the European Consensus Lupus Activity Measurement score. Patients with SLE selected by the attending physician for prednisone therapy in doses > or = 10 mg/day were more frequently hyperprolactinemic [41].

In a 31-year-old female patient with SLE accompanied by prolactinoma the disease flared when the plasma levels of PRL increased, and subsided when these levels decreased following the administration of bromocriptine, irrespective of the glucocorticoid dosage. In the 29 plasma samples obtained during the clinical course of this patient, PRL levels significantly correlated to the serum anti-DNA antibody titers ($r = 0.55$, $p < 0.05$) and inversely to the serum complement activity ($r = -0.33$, $p < 0.05$) [42].

The published clinical results concerning the relationship between PRL and SLE activity are contradictory. This is due, in part, to the statistical power of the studies. A combined analysis of these studies showed that PRL is related to lupus activity, without establishing a formal causal relationship [43].

Serum PRL, testosterone (T), E2, FSH and LH levels were obtained from 35 males with SLE and compared with 33 age-matched normal controls. No significant differences in serum T, E2, PRL levels and E2/T ratio were observed between patients and controls. Patients had significantly higher levels of FSH and LH. Five (14%) patients, but none of the controls, had both low T and elevated LH. Although PRL or T levels did not correlate with disease activity, the ratio of PRL to T showed a significant correlation with SLEDAI scores ($p = 0.42$. $P = 0.01$) [44].

Plasma PRL levels in SLE patients were significantly greater than in the normal controls. Prolactin biopotency was not altered with the disease, however, the PRL detected in plasma was associated with immunoreactive proteins of 130 and 23 kD. The concentration of the 130-kD PRL-like species was 10-fold higher in inactive SLE versus active SLE patients [45].

Plasma concentrations of PRL, GH and cortisol were determined in venous blood during insulin-induced hypoglycaemia (0.1 U/kg b.w., i.v.) in ten SLE patients and in 12 age-, gender- and weight-matched healthy subjects. Basal PRL concentrations were higher in patients vs healthy controls (12 vs 6 ng/ml, $P < 0.01$), though still within the physiological range. Insulin-induced plasma PRL and GH increases were similar, but the plasma cortisol response showed moderate attenuation in patients. PRL release in response to TRH administration (200 μg i.v.) was the same in patients and control subjects. This was true also for patients with low activity of the disease [46]. The response of cortisol to insulin-induced hypoglycemia and of PRL to TRH was measured in seven women with active untreated SLE and in ten paired control subjects. The basal serum levels of cortisol and PRL in the SLE group were not significantly different from those of the control group. The cortisol response after hypoglycemia was significantly lower in SLE patients, however the integrated cortisol response to hypoglycemia expressed as area under the response curve (AUC) did not differ significantly in either group. Although the peak of PRL response and AUC after TRH were no different the interval-specific 'delta' response was significantly higher in SLE than the control group at 0–60 min ($P=0.02$) and 0–90 min ($P = 0.01$) after TRH injection [47].

Anti PRL antibodies were found in 5% (13/259) of SLE patients, and all of them had hyperprolactinemia. The presence of antibodies was independent of lupus activity, but high level of serum PRL was associated with lupus activity independently of other studied variables. Patients with antibodies had higher PRL levels than did anti-PRL negative patients [48,49]. In a woman with SLE there was a marked increases in circulating 150-kDa PRL during pregnancy. The predominant form in serum from healthy pregnant women was little (or monomeric) PRL. In this patient big big PRL was due to the presence of autoantibodies forming an IgG-23 kDa PRL complex. This IgG-PRL complex was fully bioactive in the Nb2 rat lymphoma cell assay. When injected into the rats the IgG-PRL complex was cleared more slowly than serum containing predominantly monomeric PRL [50]. SLE patients without big, big PRL had significantly higher levels of disease activity, cause-proven HPRL, and menstrual disturbances compared with patients with big big PRL. The big big PRL fractions contained a single 23-kd nonglycosylated PRL. The bioactivity of big big PRL in the Nb2 cells was very similar to that of 23-kd nonglycosylated PRL. In rats the PRL-IgG complex was eliminated more slowly than monomeric PRL [51,52]. In hyperprolactinemic SLE patients high levels of PRL in the serum and urine could be related to severe renal disease [53].

Prolactin and IL-6 were elevated in the cerebrospinal fluid of SLE patients with central nervous system involvement, when compared to patients with SLE without CNS involvement and with patients suffering from neurocysticercosis, or with healthy women ($p<0.001$). Four of seven patients had higher levels of IL-6 and PRL in CSF than in serum. A positive correlation between PRL and IL-6 in CSF of SLE was established ($r=0.88$, $p<0.001$). The mean serum PRL concentrations were not significantly different between any of the groups studied [54].

Functionally significant polymorphism was found that alters PRL promoter activity and mRNA levels in the lymphocytes of SLE patients. Disease association studies in a cohort of SLE patients demonstrated an increased frequency of the prolactin -1149 G allele compared with control subjects. Altered local PRL production by immune cells may contribute to disease progression by affecting T cell function [55]. In the presence or absence of Concanavalin-A, peripheral mononuclear cells from SLE patients secreted significantly higher ($P < 0.001$) amounts of bioactive PRL-like species than normal cells. A 60-kDa PRL-like molecule was isolated from both normal and SLE mononuclear cell extracts [56].

After PRL stimulation, the IgG concentrations of supernatants from SLE peripheral blood mononuclear cells (PBMC) were significantly higher than those of unstimulated PBMC (median 394 ng/ml). Physiological concentration of PRL (20 ng/ml) induced IgG production more effectively (median 1139 ng/ml) than PRL at 100 ng/ml (median 1029 ng/ml). PRL did not stimulate IgG production in normal PBMC. A significant correlation between PRL induced IgG production and the disease activity (ECLAM) of the patients with SLE was seen. The maximum amount of IgG induced by PRL *in vitro* depended on the serum PRL concentrations of the patients [57].

Homovanillic acid (HVA), the principal metabolite of dopamine (DA), is removed from the brain into the blood. It is assumed that the turnover DA in the brain can be determined through an analysis of its metabolites. Lupus patients had significantly decreased levels of HVA compared to controls ($P < 0.0001$). The average level of PRL in lupus patients was higher than in healthy controls ($P < 0.001$). For the 3 month duration of this study, a significant variation was observed in the levels of HVA in the clinically active patients ($P < 0.05$) compared to inactive patients [58].

In patients with Sjögren's syndrome (SS) low basal activity of the HPA axis was associated with pituitary hyporesponsiveness to exogenous CRH, as well as hyporesponsiveness of the adrenal glands to endogenous ACTH. There was an overall attenuated and delayed LH and FSH response to LHRH stimulation, indicative of central dysfunction of the gonadal axis. TSH levels were elevated with evidence of mild hypothyroidism. Basal PRL concentrations were also elevated. Both TSH and PRL showed relatively increased responses to TRH stimulation. Adrenal and gondadal steroid hormone deficiency and elevated PRL levels, probably affect immune function greatly in SS patients [59]. Patients with primary Sjögren's syndrome have moderately increased levels of serum PRL. The elevation of PRL is especially evident in patients diagnosed at a young age with active immunological disease. Serum PRL is correlated with the index for internal organ disease. Primary SS may be preceded by hyperprolactinemia for many years [60].

PRL-like proteins are overexpressed in glandular epithelial cells of labial salivary glands from Sjögren's syndrome patients and correlate with the aggressiveness of the disease [61]. Prolactin is synthesized in the acinar cells of salivary glands from patients with SS. Cathepsin D immunoreactivity was considerably greater ($p < 0.0001$) in both the acini and the ducts of patients with SS as compared with control subjects. Cathepsin B, but not D, was expressed in about 20% of infiltrating mononuclear cells of SS patients. Treatment of both healthy and SS minor salivary glands with PRL significantly (p< 0.05 top < 0.0001) enhanced cathepsin B and D expression in acinar and ductal cells. PRL produced locally in salivary glands of SS patients could play a role in the pathogenesis of Sjogren's syndrome through the activation of proteolytic activity by cathepsins B and D [62].

Twelve patients with multiple sclerosis (MS) and twelve healthy controls were studied at baseline and with TRH stimulation for PRL secretion. Patients with MS had significantly higher PRL levels at baseline than controls (10.2+/-1.6 vs 6.4 4+/-0.57 ng/ml, P=0.042), however, values were within the normal range. After TRH stimulation peak PRL levels were significantly higher in patients with MS (57.08+/-6.144 vs 32.94+/-4.92 ng/ml, P=0.006) [63]. In women, serum PRL was found to be significantly higher only in Asian type MS (mean=23.1 ng/ml, n=25) while it did not change significantly in men. Hyperprolactinemia was significantly associated with acute relapse involving the optic nerves. All MS patients with hyperprolactinemia (7 women with Asian type MS and 2 women with Western type MS) showed recurrent opticomyelitis, and also had a higher age of onset, a higher Expanded Disability Status Scale score, a greater visual impairment, and higher cell counts and protein contents in the cerebrospinal fluid than did the normoprolactinemic patients [64].

Seven euthyroid male patients with Addison's disease were studied four times, with an interval of 1–3 months, as follows: 1) during normocortisolism (hydrocortisone infusion, 0.4 mg/kg, over 19.5 hr) ; 2) normocortisolism with coadministration of naloxone (at 25 μg/kg x hr during the last 6.5 hr); 3) hypocortisolism (24 hr withdrawal of hydrocortisone, followed by 19.5 hr saline infusion); and 4) hypocortisolism plus naloxone administration. Mean PRL levels increased significantly during hypocortisolism, compared with normocortisolism (5.8 +/- 0.4 vs. 3.6 +/- 0.2 μg/L; $P < 0.001$), and naloxone induced an increase in PRL levels both during hypo- and normocortisolism (7.1 +/- 0.7 vs. 4.7 +/- 0.5 microμg/L, respectively; $P < 0.01$). PRL responses to TRH were similar during hypo- and normocortisolism and without any change during opioid receptor blockade [65]. Hyperprolactinemia was found in 4 patients with Addison's disease. Two females had galactorrhea. In 3 cases a complete resolution of hyperprolactinemia occurred after corticosteroid replacement therapy, but in the fourth patient remission was only partial [66].

Autoimmune factors play a role in liver cirrhosis. Basal serum PRL concentrations were significantly higher in cirrhotic than in control individuals. After treatment with metergoline, PRL levels were decreased significantly in patients, but the decrease was less than what was observed in normal individuals. Serum GH levels were also elevated in cirrhotic individuals which was not influenced by metergoline treatment [67].

Autoimmune lymphocytic hypophysitis occurs most frequently in young women in association with pregnancy and lactation. It often mimics prolactin producing pituitary adenomas clinically and may be accompanied by mild hyperprolactinemia. Histological examination shows a massive lymphocytic infiltration and widespread adenohypophyseal cell damage. The interdigitation of activated lymphocytes with adenohypophyseal cells has been observed by the electron microscope. No immune complex deposits were found. Hyperpituitarism is present in patients with lymphocytic hypophysitis, which is not restricted to the secretion of PRL [68]. A 54-year old man with hyperprolactinemia had a large chromophobe pituitary adenoma that concurred with a granulomatous hypophysitis [69].

In 14 children with insulin dependent diabetes basal PRL levels did not differ from values of controls when examined after diagnosis. However, in 6 children there an exaggerated TRH-induced PRL response was observed. Basal and TRH-induced TSH secretions were normal, but plasma T3 and T4 concentrations were significantly lower in patients than in controls [70].

Thyroid autoimmunity was studied in 92 hyperprolactinemic patients, 4 of whom had acromegaly. High titers of thyroglobulin and microsomal antibodies were found in 1 acromegalic woman and 12 female patients with either adenomatose or idiopathic hyperprolactinemia. Low titers of one or both antibodies were found in 9 other euthyroid women. In 185 control individuals the prevalence of thyroglobulin and microsomal antibodies was 3.3% in females and 2.5% in males. These findings indicate that autoimmune thyroid disorders, especially asymptomatic autoimmune thyroiditis, occur in hyperprolactinemic women with a prevalence far exceeding that in the general population [71]. Patients with Hashimoto's thyroiditis exhibited significantly higher prolactin and lower cortisol values than did control subjects [72]. Fourteen of 74 female patients with various thyroid conditions had increased serum PRL. In the overt primary hypothyroid group the incidence of hyperprolactinemia was 42.4%. This was significantly higher than in any other group having normal serum thyroxine. There was a close association between the increment in serum PRL and of free triiodothyronine above the basal level after TRH administration [73].

The serum levels of immunoreactive and bioactive PRL and GH were determined in 84 patients (64 females and 20 males) suffering from RA and in 73 normal individuals (42 females and 31 males). Circulating PRL from RA patients exhibited a highly significant deficiency in bioactivity by the Nb2 lymphoma proliferation assay, whereas immunoreactive PRL (determined by RIA) was near normal. Age matched male patients showed a significantly lower serum prolactin levels by RIA than did controls. Within the patient group, those having anemia and high reticulocyte count had elevated PRL bioactivity when compared with patients with anemia and low reticulocyte count, who had a decreased bioactivity of PRL when compared to patients without anemia. PRL isolated from sera of 5 RA patients showed decreased bioactivity in comparison with PRL separated from 5 sex matched controls. Serum factors capable of enhancing or inhibiting the response of Nb2 cells to ovine PRL were also detected in sera of RA patients [74,75]. Others also observed normal serum PRL levels in RA patients by RIA [35,34] although Jara-Quezada and coworkers [34] found 2 patients out of 30 with hyperprolactinemia. Patients with RA showed an exaggerated PRL response to TRH challenge [76].

Neeck et al [77] studied the circadian rhythm of corticosteroids PRL and GH in 26 RA patients. None of the patients had been treated with corticosteroids or with disease modifying drugs. Hormone levels were determined from blood samples taken every 2 hours for a 24-hour period. An increased adrenal corticosterol secretion coupled with altered circadian rhythm has been observed. The serum levels of PRL and GH were in the low normal range. In patients with high inflammatory activity there was a tendency to lose the nocturnal increase of PRL and GH in the serum.

Juvenile rheumatoid arthritis (JRA) and juvenile systemic lupus erythematosus (JSLE) are more common in girls. Normal to low levels of cortisol have been observed despite the high serum levels of IL-6, IL-1β, and TNFα, which activate the HPA axis. PRL levels were normal in children with JRA, most of whom were not active. Higher PRL levels were present in children with active anti-nuclear antibody (ANA) positive JRA complicated by uveitis. A trend toward high PRL levels was seen in 33 children with JSLE. High serum PRL levels are present in patients with active juvenile ankylosing spondylitis (JAS) only. Growth retardation is a feature of JRA. Patients with JRA have low to normal levels of GH and low levels of IGF-1. High levels of FSH and LH are found in children with JSLE while the levels in JRA tend to be normal. Testosterone levels are low in patients with JRA. No significant differences in estrogen levels have been found in patients with JRA or JSLE and matched controls [78].

Mean serum concentrations of PRL were significantly increased (8.9 +/- 4.0 ng/ml) in anti-nuclear antibody positive girls suffering from juvenile chronic arthritis and in patients with spondyloarthropathy (7.8 +/- 2.4 ng/ml), compared to those with previous acute postinfectious arthritis in remission (4.4 +/- 0.3 ng/ml) ($p = 0.01$ and $p = 0.025$, respectively) and to controls. A significant correlation between PRL and IL-6 concentrations ($r = 0.604$, $p = 0.002$) was observed from the whole series [79]. In 4 patients suffering from severe endogenous iridocyclitis, treatment with BRC for hyperprolactinemia, galactorrhea and Parkinson's disease, alleviated the autoimmune condition [80].

Serum PRL levels were significantly higher in men with RA than in controls ($p=0.0015$). High PRL levels correlated with the duration of RA ($r=0.23$; $p=0.01$) and with functional stage according to the Steinbrocker classification ($r=0.24$; $p=0.01$). PRL concentrations did not correlate with the low levels of androgens observed [81].

The chemokine, macrophage inflammatory protein-1alpha (MIP-1α) was twice as high in active than in inactive RA than in controls. Parallel to acute phase response and joint count and high levels of MIP-1α, there were remarkable elevations of serum PRL [82].

PRL was exclusively expressed on fibroblast-like synovial cells and lymphocytes infiltrating into the synovium in patients with RA. Inhibitors of PRL release, such as BRC, inhibited proliferation and the production of proinflammatory cytokines and collagenases by RA synovial cells [83].

2.3. Therapeutic suppression of PRL in autoimmune disease

Disease activity was suppressed in six of seven SLE patients treated with BRC. All had elevated disease activity and five became hyperprolactinemic after treatment stopped. Manipulating serum prolactin provides a means of treating clinical SLE activity [84].

A double blind, placebo controlled study was performed on 66 SLE patients with BRC treatment at a fixed daily dosage of 2.5 mg. At entry, serum PRL was (mean+/-s.d.) 24.8 ng/ml+/-18.4 in the BRC treatment group and fell to 5.8+/-9.0 after 12 months of treatment. The corresponding PRL values in controls were 23.7+/-22.1 pretreatment and 20.3+/-14 after 12 months. The SLEDAI score on the fifth visit was decreased significantly in the BRC group vs controls: 0.9+/-1.4 vs 2.6+/-4.5 ($P < 0.05$). The absolute number of flares in each group was similar. However, the mean number of flares/patient/month was decreased significantly in the BRC group compared to the control group (0.08+/-0.1 vs 0.18+/-0.2, $P = 0.03$). Long term treatment with a low dose of BRC appears to be a safe and effective means of decreasing SLE flares in SLE patients [85].

Prolactin stimulated disease in autoimmune NZB/NZW mice and treatment with BRC suppressed anti-DNA antibodies and prolonged life spans. In patients with SLE in an open-label study, a double blind study, and a study comparing BRC to hydroxychloroquine provided evidence that BRC therapy reduced flares and suppressed disease activity in SLE [86,87,88].

A number of clinical therapeutic trials using 2.5 to 30 mg of BRC per day in a single or divided dose, have shown efficacy with minimal side effects in the treatment of rheumatic and other autoimmune diseases. In RA, BRC has induced immunosuppression of several immune parameters and has been associated with improvements in morning stiffness, grip strength, numbers of swollen/painful joints, and the Health Assessment Questionnaire disability index. In two double blind studies, BRC reduced the number of SLE flares and was as effective as hydroxychloroquine in reducing lupus disease activity indices, respectively. In case reports, BRC has been used successfully in the treatment of Reiter's syndrome enthesopathy and psoriatic arthritis. The efficacy of BRC in the treatment of uveitis and multiple sclerosis is suggested but remains to be verified [89].

2.4. Pregnancy and autoimmunity

McMurray et al [24] performed elegant animal experiments for the examination of the role of PRL, pregnancy, suckling and pseudopregnancy on the spontaneous autoimmune SLE-like disease of B/W mice. Pregnancy with or without suckling and pseudopregnancy induced long term hyperprolactinemia, which accelerated the appearance of anti-DNA antibodies and antibodies to the viral protein, GP-70. In pseudopregnant animals, which were hyperprolactinemic, significant elevation occurred in titers of both antibodies at 18 and 24 weeks and to GP-70 also at 30 weeks. The two other hyperprolactinemic groups (pregnant non suckling, and pregnant suckling) also showed accelerated antibody production with similar kinetics to the pseudopregnant group though the magnitude of these elevations was somewhat less and thus not quite significant when compared to virgin females. It was concluded on the basis of these observations that it was the elevated levels of PRL, and not some other pregnancy related immunoregulatory factor(s) that led to the acceleration of autoantibody production. These investigators showed already in an earlier paper [23] that in female B/W mice treated with BRC, the elevation of anti-DNA antibodies and serum IgG was delayed and survival was increased. If, however, female B/W mice were grafted with SPG, which caused prolonged hyperprolactinemia, a premature occurrence of albuminuria was observed, the concentration of complexes GP-70 and IgG were elevated in the serum and mortality was accelerated.

These results leave little doubt that in this autoimmune disease model PRL does indeed contribute to the pathogenesis of the disease. However, the relationship between elevated PRL levels and the autoimmune disease in these animals appears to be complex. For instance, the formation of autoantibodies occurred only after 12 weeks of elevated PRL levels, which suggests that PRL did not stimulate directly autoantibody production by mature B lymphocytes in which case one would have expected the rise of antibodies much sooner. It is possible that PRL affected the autoreactive cells (e.g. B, T cells or both) at an earlier stage of their development. Alternatively, the elevated PRL level stimulated antibody formation and suppressor immunoregulatory mechanisms parallel and hence the kinetics of autoantibody production has been altered. The activity of the lymphotropic retrovirus, which has a pathogenic role in this system could have been influenced by PRL as well, which in turn led to the kinetic changes of autoantibody production. PRL was found to enhance the production of retrovirus by lymphoid and other tissues [90].

The observations that SLE in man and a similar disease in animals is aggravated by hyperprolactinemia and pregnancy [91,92], whereas rheumatoid arthritis goes into partial or complete remission during pregnancy in the majority of patients [93,94] is intriguing. It is certain that hyperprolactinemia due to physiological or pathological causes does not lead to autoimmune disease in the majority of individuals. This suggests that additional immunoregulatory defects are necessary for the promotion of the autoimmune process by PRL. This could be an uneven stimulation of effector and suppressor immune mechanisms in the case of SLE. Such abnormality may be caused by the retrovirus present in B/W mice, for instance [95,96]. Infectious agents are known to trigger autoimmune disease also in man [97]. Hyperprolactinemia also activates the HPA axis as demonstrated in pituitary grafted animals [98). Thus, under normal conditions, the neuroendocrine system may act to balance out the immunostimulatory effect of PRL by the increased secretion of glucocorticoids, which are immunosuppressive. However, in autoimmune disease the HPA axis may not react normally to such a stimulus [99].

Current evidence suggests that the beneficial effect of pregnancy on RA is due in fact to the elevated levels of PRL. If this is correct, then the conclusion is almost inevitable that PRL is capable of reversing the inflammatory process by the stimulation of suppressor immune- and/or endocrine mechanisms. Therefore, in RA the immunosuppressor elements must function normally at a higher serum level of PRL. Indeed the findings of Mattsson et al. [20] in mice, suggest that during the initial (immunization) phase of arthritis, PRL supports the pathogenesis, whereas at later stages, when suppression is to be generated, PRL may be beneficial by supporting this response.

Estriol treatment reduced the severity of EAE significantly compared with placebo treatment. Serum estriol levels that approximated estriol levels during late pregnancy were capable of ameliorating the disease. Major basic protein-specific T-lymphocyte responses from estriol-treated EAE significantly increased production of IL-10 [100].

SLE may flare during any trimester of pregnancy, as well as in the puerperium. Flares are usually mild, affecting skin and joints, and, unless affecting the kidney, do not confer any adverse prognosis on pregnancy outcome. Recent data link sex hormones, particularly prolactin, to SLE activity, which may be one explanation for the high frequency of SLE flares during pregnancy. No data support the thesis that corticosteroids prevent SLE flares during pregnancy [101].

Recently lactation and prolactin have been considered as predictors of onset, flare, or relapse of arthritis. Among genetically susceptible women, breastfeeding is associated with an increased risk of rheumatoid arthritis, particularly after the first pregnancy [102]. The striking increase in corticosteroids, estrogen, and progesterone during pregnancy may suppress RA onset or activity through the regulation of cytokines such as TNF alpha, IL-1, IL-6, IL-12, and IL-10. The relative adrenal- and gonadal-deficiency in the postpartum period further supports the view that hormonal deficiencies predispose to the development or increased activity of RA. Additional evidence indicates that adrenal dysfunction in RA patients results in dysregulation of corticosteroid and DHEA production. Adrenal hypofunction develops prior to the development of RA, or within the first year of disease activity. This suggests that adrenal deficiency is a primary abnormality in RA [103].

Disease development of RA is mitigated by estrogen and pregnancy whereas SLE tends to flare during pregnancy and in response to estrogen. RA is regarded as a TH1 immune response-driven disease. Pregnancy induces a shift from TH1 to TH2 immune response, increasing the anti-inflammatory cytokines IL-4 and IL-10, which may contribute to the gestational amelioration of RA. The shift to a TH2 immune response may trigger SLE manifestations that are dependent on humoral immune responses such as lupus nephritis. Prolactin is elevated in SLE patients of both sexes and correlated to disease activity in several studies. The hyperprolactinemia of lactation seems to influence postpartum behavior of SLE as well as RA [104].

The excessive production of IL-12 and TNFα has been causally linked to rheumatoid arthritis and multiple sclerosis. In 18 women with normal pregnancies in their third trimester *ex vivo* monocytic IL-12 production was about 3-fold and TNFα production was approximately 40% lower than postpartum values. Urinary cortisol and norepinephrine excretion and serum levels of 1,25-dihydroxyvitamin D3 were 2- to 3-fold higher than postpartum values. These hormones can directly suppress IL-12 and TNFα production by monocytes/macrophages and could mediate the remission of RA and MS during pregnancy [105].

3. THE HPA AXIS AND AUTOIMMUNE DISEASE

A major role of the HPA axis is to control the immune system and to prevent tissue damage. The interaction of the HPA axis and immune system constitutes a new endocrine feedback loop [106,107,108]. Cytokines, such as IL-1, IL-2, IL-6, IL-11, IL-12, TNFα, and INFγ can activate the HPA axis. The resulting increase in glucocorticoid blood levels can affect immune cell activity directly or by controlling the production of cytokines. This indicates the existence of an immunoregulatory cytokine-HPA axis circuit. There is evidence to indicate that this circuit plays a role in controlling excessive inflammatory reactions and the non-specific expansion of immune cells. Defects in HPA axis are implicated in susceptibility to, and severity of, animal models of rheumatoid arthritis, and are hypothesized to contribute to the human disease [109,110].

In adjuvant-induced arthritis of rats CRF is no longer the major ACTH-releasing factor. This is also true for experimental allergic encephalomyelitis, eosinophilia myalgia syndrome, systemic lupus erythematosus, and leishmaniasis. Arginine vasopressin (AVP) takes over as the major stimulator of the axis. The arthritic rat is unable to mount a response to acute stressors, but is responsive to an acute immune challenge. This change presumably reflects an adaptation to the development of chronic inflammation. Serotonin given at the time of the development of the inflammatory episode reduces the severity of the inflammation. [111]. Corticotropin-releasing hormone secreted from leukocytes and neuronal terminals in peripheral tissues also plays a role in regulating inflammation [112].

3.1. Animal models of autoimmune disease

We observed that hypophysectomized or BRC treated rats developed adjuvant arthritis (AA) only if given replacement doses of either PRL or GH. If the restored animals were also given replacement doses of ACTH, the development of AA could be prevented. However, this could be reversed towards disease manifestation with higher doses of PRL or GH and AA development could be suppressed again with elevated doses of ACTH. In the case of EAE Hypox rats died invariably within a few days after immunization and for this reason we could not conduct similar experiments, as originally intended. However, we were able to show that BRC also suppressed the development of EAE [17,113,18]. These experiments clearly indicated that AA was stimulated by PRL and GH and that the ACTH-adrenal axis acted antagonistically to the development of this experimental inflammatory disease. It was also indicated that immunization was lethal for hypox animals. Our suspicion was that immune derived cytokines kill the animals in the absence of a functional HPA axis. This point was clarified in adrenalectomized (ADX) mice given LPS. Serum TNF rose some 60 times over the level of normal mice in ADX animals 2 hrs after LPS injection. The sensitivity of ADX animals to the lethal effect of LPS was increased some 500 times. Dexamethasone treatment reversed this excessive sensitivity. Moreover, the inhibition of glucocorticoid synthesis in the adrenal gland of normal mice by methyrapone increased their sensitivity to LPS over 10 times [114]. Taken together, these experiments demonstrated the vital importance of the pituitary gland in the regulation of cytokine production and of autoimmune reactions.

That the HPA axis plays a role in the development of autoimmune disease was also suggested by observations in chicken of the obese strain, which develop spontaneous autoimmune thyroid disease. These birds did not respond to immunization with elevated glucocorticoid levels [115]. The observation that in adrenalectomized (ADX) rats experimental autoimmune encephalomyelitis (EAE) took a much more severe course, leading to death, compared with controls supported further the role of the HPA axis in autoimmunity [106,116].

Inbred Lewis rats develop a variety of autoimmune conditions after deliberate immunization. Sternberg and coworkers [117,118] observed that Lewis rats had a markedly impaired ACTH and glucocorticoid response after immunization with group A streptococcal cell wall peptidoglycan polysaccharide. After such immunization Lewis rats developed arthritis, which was similar to rheumatoid arthritis (RA). Most importantly, Fisher rats, that have identical MHC antigens, exhibited normal ACTH and glucocorticoid responses and did not develop arthritis after immunization with the same antigen. Ineffective CRH production by hypothalamic neurons was found to be the underlying abnormality responsible for the development of disease in Lewis rats. These observations indicate that the ACTH adrenal axis plays an important role in the termination of autoimmune reactions.

In addition to the obese chicken, impaired surges of glucocorticoid hormones were found after stimulation of the HPA axis by antigens or certain cytokines in murine SLE and avian scleroderma. It is also clear that the classical pathways of glucocorticoid-T-cell interactions can take place in the thymus itself, which has been shown to be a site of steroid hormone production [119]. Autoimmune disease (AID)-prone mice produced nearly the same levels of plasma corticosterone after injection of recombinant human IL-1α as normal mice, but had baseline corticosterone levels consistently higher, which led to significantly lower corticosterone increase ratios. ACTH levels increased after IL-1α injection, but there was no difference in the ratios of AID-prone and normal strains. Corticosterone-binding globulin levels showed no difference [120].

Analysis of several inbred rat strains led to the hypothesis that HPA axis abnormalities may contribute, in part, to susceptibility to both autoimmune disease and addiction [121]. Genetic linkage and segregation studies in inflammatory susceptible and resistant rat strains indicate that this susceptibility is multigenic and polygenic. Multiple loci on multiple chromosomes, each with a weak effect, control this trait, and there is a large environmental contribution to its variability [122].

In a murine model of SLE (MRL +/+.MRL lpr/lpr) circulating autoantibodies increase with age, there is a decrease in hypothalamic CRH mRNA expression and an increase in AVP mRNA expression. Disease progression is associated with increased autoantibody levels, and there is altered hypothalamic peptide expression. Circulating corticosterone levels are increased and there is a trend for levels of corticosterone binding globulin to decrease [123].

The recovery of animals from experimental autoimmune encephalomyelitis (EAE) is clearly dependent on an increase in endogenous glucocorticoid levels. In this case immune derived IL-1 stimulated by the encephalitogenic antigen triggered, at least in part, the endocrine response during the disease [124].

Murine relapsing EAE can be profoundly suppressed by restraint stress (RST) when initiated prior to neuroantigen immunization. Nadolol (NAD), a β2-adrenergic antagonist, did not reverse the RST-induced suppression of EAE. RU486 or aminoglutethimide, which block the action of peripheral glucocorticoids, partially reversed EAE suppression. Exogenous corticosterone mimicked the effects of RST, in terms of suppression of EAE, decrease in lymphoid cell numbers and decrease in Th1 cytokine production [125].

The hypothalamic-pituitary-gonadal and -adrenal axes (HPG and HPA) play important roles in various animal models of autoimmune disease. These include systemic lupus erythematosus in mice and collagen-induced arthritis (CIA) in mice and rats, and the streptococcal cell wall, adjuvant and avridine arthritis models in rats. In Lewis and Fischer F344 rats CRF and corticosteroids of the HPA axis play important roles in the regulation of inflammation and the predisposition to autoimmune diseases. The HPA axis is intimately linked to the HPG axis and is sexually dimorphic. Estrogens stimulate higher corticosteroid responses in females [126].

3.2. Autoimmune conditions in man

Hench and coworkers [127] proposed the first theory for the role of pituitary hormones in the pathogenesis of rheumatoid arthritis. These investigators observed clinical improvement in arthritis patients after treatment with either glucocorticoids or ACTH. These observations led to a widespread use of glucocorticoid analogues for the treatment of RA, as well as of a variety of other inflammatory conditions. However, the serious side effects associated with such treatment became widely recognized. Eventually the consensus evolved that these hormone analogues control the disease as pharmacological agents, rather than by the replacement of a hormonal deficiency as originally proposed. This conclusion was supported by repeated observations that glucocorticoid serum levels were within the normal range in patients with rheumatoid arthritis. Eventually most investigators rejected the neuroendocrine theory of the pathogenesis of RA. However, the observation that RA patients show a blunted glucocorticoid response after ACTH administration supports this possibility [128,129].

Glucocorticoids, which are anti-inflammatory, rise during the day and decrease during the night, whereas the level of GH and PRL, which are pro-inflammatory, are decreased at daytime and elevated at night. Chikanza and Panayi [130] proposed that the perpetuation of inflammation in RA may be due to the abnormality of the HPA axis, which leads to an imbalance of anti-inflammatory and pro-inflammatory hormones. They proposed that morning stiffness is the result of the exaggeration of this imbalance at night. The PRL deficiency observed by us, and the blunted nocturnal release of GH and PRL in patients with severe RA, observed by Neeck [77], do not seem to support this hypothesis. Nevertheless, it is still possible that a hormonal imbalance exists, which is primarily due to the malfunction of the ACTH-adrenal axis. The decreased bioactivity of PRL, the blunted nocturnal release of PRL and GH, and the lack of growth in Still's disease, may all represent the suppression of pro-inflammatory hormones by inflammatory mediators [129].

Mean ACTH and cortisol levels did not differ between RA patients and matched controls. Patients tended to have a less pronounced ACTH response ($p = 0.08$) and had a significantly smaller cortisol response ($p < 0.01$) than healthy controls in response to the stressors. This difference in cortisol response was reduced, but remained significant when ACTH responsiveness was accounted for by entering it as a covariate ($p = 0.03$). ACTH and cortisol levels and responses were not influenced by basal IFNγ and IL-4 as reflections of Th1 and Th2 cell activity, respectively. The HPA axis activity was not linked to current disease activity [131].

Female RA patients seem to depend more on the HPA axis, whereas male patients seem to depend more on the hypothalamus-pituitary-gonadal (HPG) axis for immunoregulation.

Hypoandrogenism may play a pathogenetic role in male RA patients because adrenal and gonadal androgens, both products of the HPA and HPG axes, are natural immunosuppressors. The production of the adrenal androgens, DHEAS and DHEA is altered in nonglucocorticoid-treated premenopausal RA patients. The menopausal peak of RA suggests that estrogens and/or progesterone deficiency also play a role in the disease [132].

Plasma IL-6 levels showed a circadian variation and were significantly increased in RA patients, compared to controls. There is a positive temporal correlation between plasma IL-6 and ACTH/cortisol, with IL-6 preceding ACTH and cortisol by 1 and 2 hrs, respectively. In the same patients, a negative effect of cortisol upon IL-6 was detected with a delay of 5 hrs. In patients with early untreated RA the overall secretory activity of the HPA axis remains "inappropriately" normal despite elevated plasma IL-6 levels [133]. In 64 RA patients with active disease previously untreated with glucocorticoids, IL-6 correlated with CRP ($p < 0.001$), with ESR ($p < 0.001$) and with serum cortisol ($p = 0.019$). ESR and CRP correlated with each other, and cortisol also correlated with ESR ($p = 0.002$) and CRP ($p < 0.001$). Thus, IL-6 is an important disease activity parameter in RA that is closely related to both the classical humoral disease activity and the HPA axis [134].

In chronic inflammatory diseases, such as RA, an alteration of the HPA stress response results in inappropriately low cortisol secretion in relation to ACTH secretion. DHEA was also significantly lower after ACTH stimulation in patients with RA with no prior corticosteroid treatment than in healthy controls. DHEA inhibits cytokine production by the inhibition of NF-κB activation. Therefore, low levels of this hormone may be deleterious in chronic inflammatory diseases. DHEA is a potent inhibitor of IL-6 and of other cytokines, such as TNF. This may be a significant risk factor in RA. The serum level of DHEAS was significantly lower in patients with chronic inflammatory diseases. DHEAS is the precursor for sex steroid hormones, such as testosterone and 17 beta-estradiol. A lack of this hormone leads to a significant sex hormone deficiency in the periphery. A combined therapy is suggested with corticosteroids plus DHEA in chronic inflammatory diseases [135].

Plasma ACTH and CRH levels were significantly lower in RA patients than in osteoarthritis patients. In the synovial fluid CRH and IL-6 concentrations were higher in RA patients than in controls [136].

RA patients showed no difference in plasma concentrations of resting or stimulated ACTH, cortisol, PRL and TSH, when compared to healthy controls. In contrast, the GH response to GHRH was blunted in RA patients. However, if one considers the presence of chronic inflammation, normal plasma ACTH and cortisol concentrations in patients are actually inappropriately low [137]. In 20 patients with RA cortisol, ACTH and GH responses to insulin-induced hypoglycaemia were impaired in 65%, 85% and 30% of the patients, respectively. The basal and peak hormone levels were similar between patients with active RA and patients in remission [138].

Normal to low levels of cortisol have been observed in children with active JRA despite the high serum levels of IL-6, IL-1β and TNFα, which activate the HPA axis. In inflammatory disease cortisol levels should be elevated due to the activation of the HPA axis by pro-inflammatory cytokines [78].

Patients with Sjogren's syndrome (SS) were characterized by significantly lower ACTH levels and cortisol levels compared to controls. A blunted pituitary and adrenal response to ovine CRH was observed in comparison with controls. Basal TSH levels were significantly elevated in patients [139]. The functional integrity of the adrenal, gonadal, and thyroid axes was assessed in Sjögren's syndrome. Low basal activity of the HPA axis was associated with pituitary hyporesponsiveness to exogenous CRH, and of the adrenal glands to endogenous ACTH, indicating a central deficiency of the adrenal axis. Adrenal and gondadal steroid hormone deficiency, and elevated PRL levels, probably greatly affect immune function in SS patients [59].

Patients with primary Sjogren's syndrome had significantly lower basal mean DHEA-S values compared with healthy controls ($p < 0.05$) and significantly lower DHEA-S values after stimulation. Cortisol was normal, but he cortisol/DHEA-S ratio in the patient group was higher than in controls ($p < 0.05$). A correlation was found between basal ACTH and DHEA-S values in the patients ($r = 0.650$; $p = 0.05$). No correlation was seen between disease activity or age and the serum concentration of DHEA-S [140].

In female patient with Sjogren's syndrome a central deficiency was indicated in all three neuroendocrine axes: adrenal, gonadal, and thyroid. It is likely that the pathological effect of these changes involves the synergistic and antagonistic effects of multiple hormones, making the specific effects of individual hormones difficult to discern [141].

Urinary excretion of free cortisol (UFC), its nyctohemeral variation and its relation to indices of diabetic angiopathy was investigated in 130 subjects with insulin dependent diabetes (IDDM). The mean age of patients was 15.2+/-4.8 years and diabetes duration 7.3+/-5 years. 48 controls of comparable age were also examined. The total 24-hour UFC excretion was greater in diabetics than in controls ($p=0.002$) and also greater in diabetic males than in females ($p=0.006$), while no sex difference was detected in the controls. UFC correlated to carotid intimal plus medial thickness ($p=0.002$), urinary albumin, diastolic, systolic and mean blood pressure, serum creatinine, age and diabetes duration ($p<0.001$). Diabetic patients had significantly greater 24-h UFC outputs than controls. There was a significant positive correlation between 24-h UFC outputs and duration of diabetes, diabetic retinopathy or with diabetic cardiovascular complications. These results further show that diabetic patients have mild dysregulation of the HPA axis [142].

Thirteen patients with MS were studied at baseline and with provocative tests of HPA axis function (e.g. ovine CRH, AVP, and ACTH stimulation). MS patients had significantly higher plasma cortisol levels at baseline than did matched controls. Patients with MS showed normal plasma ACTH responses to ovine CRH, a blunted ACTH responses to AVP and normal cortisol responses to high and low dose ACTH stimulation. These data do not support a role for hypocortisolism in fully established MS [143]. Patients with MS had an exaggerated rise in plasma cortisol concentrations in the Dex-CRH test ($p< 0.05$), indicating hyperactivity of the HPA system. The degree of hyperactivity was moderate in relapsing-remitting MS patients, intermediate in secondary progressive MS patients, and marked in primary progressive MS patients. Differences were significant between the three patient groups ($p < 0.005$), and between control subjects and each patient group. Indicators of HPA axis activation correlated with neurologic disability (Kurtzke's Expanded Disability Status Scale), but not with the duration of the disease, number of previous relapses, previous corticosteroid treatments, or depressed mood (Hamilton Depression Scale). The ACTH test was normal in 31 of the 33 patients studied [144]. The time-integrated cortisol response to human CRH stimulation was lower in the patients with secondary progressive MS, than in patients with primary progressive MS and in healthy subjects. The time-integrated beta-endorphin response to CRH was greater in patients with relapsing-remitting MS compared with the others. Feedback regulation was normal as assessed by the dexamethasone suppression test [145].

A double blind, placebo-controlled trial was conducted to evaluate the neuroendocrine effect of cotreatment with the antidepressant, moclobemide, as an adjunct to oral corticosteroids in multiple sclerosis. The patients were evaluated using the combined dexamethasone-CRH test and clinically on days 1, 30, and 75. Moclobemide co-treatment resulted in normalization of the HPA axis response, whereas the HPA system hyperresponse was maintained with steroids alone [146].

4. SEX HORMONES AND AUTOIMMUNE DISEASE

Gonadal hormones directly, and indirectly via the HPA axis, alter the tone of the immune system and of the inflammatory responses. Abnormalities in the complex interactions between the HPA axis, immune and the gonadal systems may prove to be fundamental in the genesis and perpetuation of autoimmune disease [106,108]. Sex hormone levels in both humans and experimental animals correlate with the activity of their cytokine-secreting cells. Sex hormones influence the cytokine milieu in autoimmune patients and contribute to the skewed cytokine milieu characteristic of SLE. While sex hormones alone cannot cause autoimmune disease, abnormal hormone levels may set the stage for other factors (genetic, infectious) to trigger disease [147]. Several members of the sex hormone binding globulin family are implicated in the regulation of hormonal and inflammatory processes, which may be pertinent to the accelerated atherosclerosis seen in systemic lupus [148].

In synovial tissues and cells from RA patients, androgen and/or estrogen receptors, high concentrations of biologically active steroids, key enzymes of steroid metabolism, and significant changes of estrogen to androgen ratio, are all present. This strongly suggests that individual immune cells, including synovial macrophages, behave as steroid-sensitive cells [149].

More than 12,000 genes were screened by a microarray system from estradiol (E2)-treated mice protected from EAE vs. control mice with severe EAE. E2 treatment affected about 10% of the genes tested, but only 18 cytokine, chemokine/receptor, adhesion molecule, or activation genes were up- or down-regulated more than 2.4-fold. Down-regulated genes included TNFα; peptidoglycan recognition proteins (Pgrp); RANTES; and neural cell adhesion molecule (MCP-1). Up-regulated genes included CTLA-4, TGFβ3, IL-18, and two IFNγ-induced genes, the chemokines: monocyte chemoattractant protein-1 (MCP-1) and macrophage inflammatory protein-1beta (MIP-1β), vascular cell adhesion molecule (VCAM), and disintegrin metalloprotease [150].

Estrogen effectively and equally suppressed the development of EAE in IL-4 and IL-10 knockout mice and in wild type littermates. Significant disease suppression was also seen in IFNγ-deficient mice. A concomitant reduction was observed in the number of proinflammatory cytokine- and chemokine-producing cells in the CNS. There was a profound decrease in the frequency of TNFα-producing cells in the CNS and the periphery. This indicates that one mechanism by which estrogen protects females from the development of cell-mediated autoimmunity is through a hormone-dependent regulation of TNFα production [151].

Diestrus (<100 pg/ml in serum) levels of E2 significantly reduced the clinical manifestations of active EAE in both male and female mice. Estriol was also effective. The reduction in disease severity was accompanied by a coincidental reduction in the number and size of inflammatory foci in the CNS. Recipients of encephalitogenic T cells from low-dose estrogen-treated mice developed less severe paralysis than mice receiving T cells from placebo-treated mice. Estrogen treatment at the onset of active EAE failed to reduce disease severity. These results challenge the idea that increased susceptibility to autoimmunity in females is dependent on physiological levels of estrogen [152].

In transgenic mice carrying the heavy chain gene of pathogenic anti-DNA antibodies a sustained increase in estradiol altered B cell development, leading to the survival, expansion, and activation of autoreactive marginal zone B cells. This implicates that this B cell subset is involved in autoimmunity [153].

In the NZB x NZW F1 mouse model of SLE, PRL accelerates disease expression, whereas estrogen, devoid of its PRL stimulating properties, is immunosuppressive and inhibits IL-2 production. Estrogen, because of its action on both the endocrine and immune systems, may directly or indirectly stimulate or inhibit immune responses [154].

Calcineurin steady-state mRNA levels and phosphatase activity increase when estrogen was cultured with T cells from SLE patients. This increase was dose-dependent, hormone-specific and temporally regulated. The estrogen receptor antagonist, ICI 182,780, inhibited the increase in calcineurin mRNA and phosphatase activity, while cycloheximide had no effect. Both ERα and ERβ were shown to be expressed in human T cells. However, calcineurin did not respond to estrogen stimulation in T cells from normal females, males and lupus males [155].

Clinical research indicates that estrogens are elevated in patients with SLE of both sexes. A preferential hydroxylation of estrone and increased oxidation of testosterone in SLE patients maximizes the effects of estrogens on T cell functions [156]. Two healthy young women developed full-blown Sjogren's syndrome 3 years after starting estrogen-increasing therapy [157].

Bone marrow transplantation in an androgen-resistant mouse strain permits the creation of chimeric animals with androgen receptor-defective lymphoid or epithelial/stromal cellular compartments. Hormonal manipulation in these animals have indicated that thymic epithelial cells and bone marrow stromal cells are mediators of androgenic effects on immature lymphocytes [158].

T cells from female mice transferred more severe EAE than T cells from male animals. Androgen-selected T cell lines secreted less IFNγ and more IL-10 than untreated cell lines. Clinical EAE induced by androgen-selected T cell lines was less severe than the disease induced by untreated T cell lines. Androgen treatment of naive T cells, during their first encounter with Ag, resulted in a shift in the balance of Th1/Th2 cytokines. This phenotype was maintained during subsequent stimulations in the absence of androgen [159]. The higher susceptibility of female animals to EAE is due primarily to a protective effect of testosterone in male animals. The decrease in disease severity during late pregnancy appears to be due, at least in part, to high levels of estriol, which is present during this period [160].

Hypoandrogenic male SLE patients did not have overt features of hypogonadism but had a higher prevalence of CNS disease than those with normal androgen levels. Although PRL or testosterone (T) levels did not correlate with disease activity, the ratio of PRL to T showed a significant correlation with SLEDAI scores ($p = 0.42$. $P = 0.01$) [44].

In cultures of myelin basic protein (MBP)-primed splenocytes DHEA significantly decreased T cell proliferation and the secretion of IFN-gamma, IL-12, TNF-alpha and NO in response to myelin basic protein. This was associated with a decrease in activation and translocation of NF-κB. *In vivo* DHEA significantly reduced the severity and incidence of acute EAE, along with a decrease in demyelination/inflammation and expressions of (pro)inflammatory cytokines in the CNS [161].

DHEAS concentrations were significantly decreased in both women and men with inflammatory arthritis (IA) ($P < 0.001$). After taking into consideration the erythrocyte sedimentation rate, previous glucocorticoid usage, current treatment with nonsteroidal anti-inflammatory drugs, duration of disease and homeostasis-model assessment, the differences in patients and controls were markedly attenuated in women ($P = 0.050$) and were no longer present in men ($P = 0.133$) [162]. In young premenopausal women, significantly low serum DHEAS levels were found 4 to 20 years prior to onset of RA. Dissociation of serum cortisol and DHEAS levels was also found in the youngest, but not in older, pre-RA subjects [163].

In premenopausal RA patients, who never received glucocorticoid therapy, DHEA and DHEAS basal concentrations were found to be significantly lower ($p<0.05$) than in controls. Significantly higher basal levels of IL-6 and IL-12 ($p<0.05$) were found in patients. After low dose ACTH challenge, the DHEA area under the curve value was significantly lower ($p<0.01$) in patients than in controls. Basal DHEA levels showed a significant negative correlation with the erythrocyte sedimentation rate and platelet count, as well as with the Steinbrocker class of the disease ($p<0.05$). Normal plasma cortisol response was present after oCRH and ACTH challenge in patients in spite of their inflammatory condition. After ACTH challenge, IL-6 levels decreased significantly ($p<0.05$), whereas IL-12 levels were unchanged. No significant changes in IL-6 and IL-12 levels were found after oCRH challenge [164].

Low cortisol and DHEAS responses to testing in patients with RA patients not treated with corticosteroids, should be clearly regarded as a "relative adrenal insufficiency" in the setting of a sustained inflammatory process, as shown by high serum IL-6 levels. Androgens seem implicated in the pathophysiology of autoimmune disorders as natural immunosuppressors. Low plasma DHEAS levels are mainly observed in female RA patients. The menopausal peak of RA suggests that estrogens and/or progesterone deficiency also play a role in the disease. Acquired changes of sex steroid metabolism seem also modulate peripheral sex hormone levels [165]. The adrenal androgen, DHEA, was significantly lower after ACTH stimulation in patients with RA without prior corticosteroids than in healthy controls. DHEA, is an inhibitor of cytokines due to an inhibition of NFκB activation. Low levels of this hormone may be deleterious in chronic inflammatory diseases [135].

Current evidence indicates that: 1) Inappropriately low secretion of cortisol in relation to inflammation is a typical feature of the disease in patients with RA. 2) The secretion of adrenal androgens is significantly reduced. This is a problem in postmenopausal women and elderly men due to a lack of downstream sex hormones. 3) Serum levels of testosterone are markedly reduced in RA. 4) Sympathetic nerve fibers are markedly reduced in the synovial tissue of patients with RA, whereas proinflammatory sensory fibers (substance P) are present. 5) Substance P serves to continuously sense painful stimuli in the periphery. The nociceptive input from the inflamed joint shows a large amplification in the spinal cord. SP is released continuously into the joint, which leads to constant pain and inflammation. Alterations of the systemic anti-inflammatory feedback systems contribute significantly to the pathogenesis of RA [166].

In 10 patients with primary Sjogren's syndrome significantly lower basal and stimulated mean DHEA-S values were found, compared with healthy controls. Correlation was found between basal ACTH and DHEA-S but no correlation was seen with disease activity or age [167].

The available experimental and clinical data suggest that low androgens and abnormalities in the prolactin-gonadal axis are the most consistent hormonal aberrations found in human SLE [168]. New immunomodulating therapies are emerging for SLE. It is hoped that DHEA (prasterone), will help reduce the need for steroid treatment among patients with mild-to-moderate disease [169].

Serum testosterone levels are not elevated in patients with ankylosing spondylitis (AS). Elevated serum levels of the adrenal steroids 17α-hydroxyprogesterone and DHEAS have been found in patients with AS. Oral estrogen therapy in female patients and human chorionic gonadotrophin injections in male patients with AS, increased the 17 beta-estradiol/testosterone ratio and resulted in a moderate clinical improvement [170].

Androgens influence pancreatic islets and autoimmune infiltration in NOD and NODscid mice. This indicates that the gender difference in diabetes incidence in NOD mice is related to target organ-specific androgen effects [171]. In female patients with insulin dependent diabetes serum leptin was increased during late puberty. This was associated with low IGF-1 and hyperandrogenemia, and correlated with increased urinary albumin excretion [172].

5. AUTOIMMUNE ENDOCRINE DISORDERS

Autoantibodies are known to occur against hormones, such as insulin, prolactin, thyroxin, triiodothyronin and thyroid stimulating hormone. Receptors for hormones and neurotransmitters are also targeted by autoantibodies, including the acethylcholin receptor in myoneural junctions, the insulin receptor, the TSH receptor. Frequently such autoantibodies lead to clinical symptoms, such as diabetes, hyper- or hypothyroidism and myasthenia gravis. Antibodies or cell-mediated immune reactions may be directed against the tissues of endocrine organs. Lymphocytic hypophysistsis, Grave's disease and Hashimoto's thyroiditis, adrenal disorders, disorders of pancreatic islets (diabetes), ovarian and testicular disorders can all be induced by autoimmune reactions [173].

In addition to the organ specific diseases caused by autoimmunity, polyendocrine autoimmune disorders also exist. Two distinct groups have been identified: (i) *type I polyendocrine autoimmunity*, or the "candida endocrinopathy syndrome", is a familial disorder characterized by hypoparathyroidism, primary adrenal insufficiency and mucocutaneous candidiasis, which occurs in 50% of affected individuals. The associated conditions include pernicious anemia, thyroiditis, gonadal failure, which is due to antibodies to cytochrome P_{450} and there is defective dental enamel formation. This disorder is inherited in an autosomal recessive pattern and is equally prevalent in both sexes. (ii) *type II polyendocrine autoimmunity* is more common and is also known as *Schmidt's syndrome*. The associated conditions are autoimmune adrenal and thyroid disorders and IDDM. A variant form, which is sometimes classified as type III, consists of autoimmune thyroid disease and IDDM (IIIa), pernicious anemia (IIIb), or vitiligo, alopecia, IDDM and pernicious anemia (IIIc). Hypophysitis is occasionally associated with and ovarian failure, which may precede the adrenal insufficiency. The letter is attributed to antibodies to ovarian cells. Antibodies to melanocytes can cause vitiligo, and to *tyrosinase*, the rate-limiting enzyme in melatonin formation has been suggested to be the major antigen. This syndrome has its peak incidence between 20 and 60 years of age. It is common to multiple generations in a family, and the affected individuals show more than one disease. In a given individual several diseases may arise at once or years may separate the onset of different disorders. The development of autoantibodies to target tissues may precede the manifestation of clinical symptoms by several years and may not be manifested in specific target organ effects. The inheritance pattern is complex. In some families the susceptibility is inherited in an autosomal dominant fashion and is associated with specific HLA haplotypes [173].

Some tumors metastasize preferentially to tissues that produce paracrine growth factors capable of stimulating their growth [174]. It is possible that a similar mechanism operates for deregulated autoimmune lymphocytes in endocrine glands that produce immunostimulatory hormones. Such paracrine stimulation of autoimmune lymphocytes could take place in the pituitary gland, in the thyroid and in the panceatic islets of Langerhans. Some of these organs may support the primary autoimmune response until memory cells develop. For instance an immune response to cytochrome P_{450} could evolve this way. Once the effector cells generated are fully mature they could home to other tissues and organs, such as the adrenal, for instance, that displays the same antigen and cause autoimmune damage. A decreased adrenal cell function would then enhance the recruitment of autoimmune lymphocytes and acceleration/widening of the autoimmune process [173]. Alternatively, endocrine glands that secrete immunosuppressive hormones (adrenals, gonads) may be protected against the development of a primary autoimmune response against them, but still would be susceptible to attack by memory effector cells, that resist immunosuppression by glucocorticoids. CD4+ killer cells actually require estradiol to exert cell mediated cytotoxicity. Estradiol also has the capacity to sensitise target cells for killer cell induced lysis [10,173].

Yet another scenario, which should be considered is, that at least some of the immune responses directed against hormones, hormone receptors and endocrine cells are actually regulatory in nature. It is now clear that immune-derived cytokines regulate hormone secretion in the pituitary gland and elsewhere and that stromal immunocytes in endocrine organs may exert such regulation [10]. It remains to be seen how important are immune-derived signals for normal endocrine function.

6. THE NERVOUS SYTEM AND AUTOIMMUNITY

Complex models of autoimmune disease, such as inflammatory arthritis in rats and thyroiditis in chickens, indicate a role for blunted HPA axis and dysregulated sympathoneuronal responses in susceptibility to autoimmune disease [122].

The interaction between the autonomic nervous system and HPA axis with the immune system is disturbed in chronic RA. Neuropeptides play a role in the regulation of cytokines that modulate immunity. The blocking of catecholamine actions can ameliorate the symptoms of RA [175]. The HPA axis plays an important role in regulating and controlling immune responses and it's dysfunction has been implicated in the pathogenesis of RA. CRH, one of the main hormones of the axis, is also released peripherally at the site of inflammation and may modulate inflammatory responses locally [176].

7. DISCUSSION

Traditionally autoimmune disease has been defined as the *loss of self-tolerance*. In other words the lymphocytes are activated inadvertently by self-antigens and secrete autoantibodies or differentiate into effector cells causing damage to the target organ/tissue. This process requires a partial de-regulation of the lymphocytes involved, because they are activated and will proliferate and differentiate into the "forbidden" clones of Burnet, eventually causing autoimmune disease. The elaborate control of cell proliferation is well illustrated by the process of oncogenesis. It is clear that the *deregulation* of a single gene is not sufficient for the creation of a malignant cell, but, rather, a whole complement of growth promoting and suppressor genes need to be deregulated [177]. The *autoimmune lymphocytes* may be regarded as *partially deregulated cells,* which are activated by a stimulus (i.e. self antigen) towards which they should be refractory. The possible relationship between neoplasia and autoimmune disease has been reviewed [178,179].

Below we discuss briefly the mechanisms that appear to be relevant to the pathogenesis of autoimmune disease. The 3 signal model for cell proliferation, which is presented in detail in the chapter on immunocompetence, will serve as a basis for our deliberations. It is assumed that lymphocytes require hormonal competence signals (e.g. GLH or equivalent), cell-to-cell adherence signals (such as antigen presentation) and cytokine signals for the maintenance of self-tolerance. Lymphocyte activation is always initiated at the adherence signal level (e.g. antigen, mitogen), which needs to be followed up by specific cytokines in order to achieve cell proliferation and differentiation [180,10].

The growth pattern of the Nb2 thymic lymphoma, which is dependent on PRL for proliferation, illustrates the dominance of the competence signal. These cells require the PRL signal, which makes them competent to respond to growth factor(s) present in the serum [181]. That the proliferation of lymphoid cells is regulated by GH and PRL is well demonstrated [181–183]. The profound immunosuppressive effect of anti-proliferative drugs [184] illustrates the potential for immunoregulation by hormones that affect lymphocyte growth. On this basis we proposed earlier that the principal mechanism by which the pituitary gland regulates immune function is growth control [185].

7.1. Immunocompetence and self-reactivity

A compelling body of evidence suggests that growth and lactogenic hormones (PRL, GH, PL) have the capacity to maintain the immune system in a functional, competent state [186,10]. The hypothalamus-pituitary-adrenal axis functions as an antagonist of the GLH-IGF axis and serves to control excessive immune/inflammatory reactions. The HPA axis consists of CRH/AVP, ACTH, α-MSH, βEND, glucocorticoids and DHEA-S. Every peptide belonging to this axis is known to be involved with the regulation of acute immune/inflammatory reactions, including fever and of other aspects of CNS function, cytokine and immune effector activity. Glucocorticoids are amongst the most potent immunosuppressive hormones, whereas DHEA-S serves as a precursor steroid, to be metabolized into bioactive steroid hormones by immune and inflammatory cells [106,187,10]. During severe acute illness AVP may take over the function of CRH [188], although the significance of this is not fully understood at this time.

There is compelling evidence to indicate that the GLH-IGF axis in equilibrium with the HPA maintain immunocompetence as well as are capable of regulating the thresholds of stimuli required for immune activation. The gonadal axis joins in with immunomodulatory influence. PRL interacts intensively with estradiol and testosterone of the gonadal axis [106,107].

It is apparent from the above survey that autoimmune diseases frequently show abnormalities of PRL or GH, of the HPA axis, including DHEA and DHEA-S, estrogens and androgens. At least some of the endocrine changes observed may be the consequence of the systemic immune/inflammatory reaction. However, females with JRA may have low DHEA serum levels for 4–20 years prior to disease onset, and long standing hyperprolactinemia was observed also in some SLE patients and in those with Sjogren's syndrome and there was adrenal hypofunction prior to the onset of RA [163,84,60,103]. Even when a defect is not pre-existing, laboratory tests for challenging PRL, GH or HPA secretory activity frequently reveal abnormalities. Unfortunately most of the studies so far tend to examine only one or a few hormones of the hypothalamic-pituitary regulatory system. The correct approach would be to assess all the known stimulatory and inhibitory hormones in relation to each other. Such assessment should give a better idea about the possible maladjustment of the neuroendocrine system. Another problem is that cytokines that are most relevant to autoimmune disease are seldom used as yet for testing. Clearly, hyperprolactinemia with a normal HPA axis may have a similar effect than normal PRL levels in association with a blunted HPA axis. The studies of Mok and Lau [44] are pioneering from the point of view of assessing the PRL/testosterone ratio in SLE, which corrrelated with clinical status.

The consensus is that the selection of the T cell effector/suppressor repertoire in the thymus is based on the establishment of a threshold mitogenic signal for *self reactivity*. It seems now clear that the size of the signal is not dependent solely on MHC-TCR interaction, but rather on the entire hormonal/cytokine environment (milieu). Cell proliferation in the thymus is dependent on GH or PRL and there is evidence that thymic function is also influenced by steroid and thyroid hormones [181,189]. Therefore, the threshold may change with the change of *endocrine milieu*, which could result in autoreactivity of properly selected T cell subsets, or alternately, to deficient immune response, depending on the direction of the change. This mechanism may operate during T cell maturation in the thymus and also on mature T cells in the periphery. There is evidence that the thymus continues to generate T lymphocytes throughout postnatal life, despite the onset of thymic involution [190–192]. Therefore, hormonal changes could affect thymic selection for the entire lifetime of an individual [180]. Some observations that are compatible with this hypothesis are discussed below.

In rats, man and in some other species, if thymic regeneration is inhibited by cyclosporin-A (CSA) after lethal irradiation, a serious autoimmune condition develops once the drug has been discontinued, even if syngeneic or autologous bone marrow cells were used for reconstitution [193]. CSA inhibits intracellular signal transduction pathways in T cell precursors [194], and thus may be considered as an agent creating the equivalent of stimulatory hormone deficiency or inhibitory hormone excess. These observations indicate that thymus has a remarkable capacity to adjust to this new milieu with selection. However, once the drug is removed the sudden return of signal transduction to a normal level will now allow the clones released to the periphery during the suppressed state as self-tolerant, to become autoaggressive. Although CSA is known to affect a number of other organs and tissues, and to induce endocrine changes [195,196] it is still reasonable to consider altered signal transduction of thymocytes as the cause of autoimmune disease.

Endocrine changes are numerous and occur continuously. For instance, one may refer to the daytime elevation of glucocorticoids and the nocturnal release of growth GH and PRL [77]. Other daily activities, such as food intake, exercise, emotional stimuli, etc., elicit endocrine changes [197]. Endocrine changes are associated with puberty, the sex of the individual, and with aging. Pregnancy is also accompanied by profound changes in the endocrine milieu [198,199]. Adaptation to the seasons also elicits endocrine changes [200,201]. Infection, trauma, and immunization have a well recognized effect on the neuroendocrine system, which may lead to profound functional and metabolic alterations [202,114]. Excessive stimuli, commonly defined as “stress”, which may be physical or mental, also lead to neuroendocrine changes [203,204].

In normal individuals the endocrine response to physiological and most pathological stimuli does not lead to autoimmune disease. However, there is good evidence that stress modulates immune function in animals via neuroendocrine pathways [205,206]. It has been demonstrated that stress situations also alter the immune response in man [207]. Moreover, there is evidence to indicate that emotional stress situations play a role in the pathogenesis of at least some autoimmune conditions, such as RA or Grave’s disease [208]. Therefore, it is conceivable that the neuroendocrine imbalance which is present during stress or during some mental conditions such as depression, contributes to the development of autoimmune disease by altering the thymic selection process, peripheral regulation, or both in *predisposed individuals*.

Recent reviews on the role of endocrine alterations in autoimmune disease conclude the following: Hypoprolactinemia results in immunodeficiency, while hyperprolactinemia enhances active systemic lupus erythematosus, Reiter’s disease, juvenile and adult rheumatoid arthritis (RA), autoimmune thyroiditis, multiple sclerosis, and cardiac allograft rejection. PRL gene polymorphisms have now been shown to be linked to RA [209]. Elevated serum levels of PRL in adjuvant arthritis, collagen type II-induced arthritis in animals, in murine and human SLE, and in murine and rat autoimmune type I diabetes may influence the outcome of the disease. It is suggested that mild hyperprolactinemia is a risk factor for the development of autoimmunity. This occurs in adrenocortical deficiency or postpartum. In human SLE, PRL appears to favor the production of anti-double stranded DNA [210]. Lymphocytes from patients with active SLE produce increased amounts of PRL, and this extrapituitary PRL may participate in aberrant immune processes in SLE. In addition cytokine stimulation of pituitary PRL secretion and defects in dopamine metabolism have been described in patients with SLE [211]. In SLE there is a strong sexual predilection for females. Clinical research indicates the presence of a more estrogenic environment in patients with SLE of both sexes. A preferential hydroxylation of estrone and increased oxidation of testosterone in patients with SLE maximizes the effects of estrogens on T cell functions. Prolactin functions as stimulant of immune functions when elevated in SLE [156]. The available experimental and clinical data suggests that low androgens and abnormalities in the prolactin-gonadal axis are the most consistent hormonal aberrations found in human SLE [212].

Elevated levels of glucocorticoids cause thymic involution by increasing the rate of apoptosis [213]. Thus adrenal steroid hormones are capable of influencing both thymic selection and peripheral lymphocyte function, which makes these hormones very powerful in immunosuppression. The observation that some sex hormone analogues show strong immunomodulatory action without having major effects on reproductive organs [214] suggested the possibility that endogenous immuno-regulatory steroid hormones may exist. Indeed it is now emerging that DHEAS and its derivatives play a major role in immunoregulation. Moreover, the thymus is a steroidogenic organ, which emphasizes further the role of steroid hormones in immunoregulation [10].

One must remember that autoimmune conditions are associated with a whole host of neuroendocrine and metabolic alterations and that there is heterogeneity even with the same clinical diagnosis. The neuroendocrine abnormalities that have been observed in patients with RA are as follows: (i) adrenal deficiency prior to the onset of RA [103]; (ii) abnormal diurnal rhythm of cortisol [77,216]; (iii) a blunted glucocorticoid response after ACTH challenge [215]; (iv) elevated plasma cortisol 21-acetate [216]; (v) normal or decreased cortisol and normal catecholamines in urine [77,217]; (vi) normal cortisol but elevated serum IL-1, IL-6, and TNF in juvenile rheumatoid arthritis [209]; (vii) reduced serum androgen levels and impaired testosterone response to HCG challenge [218,219,220]; (viii) estrogen and progesterone deficiency [165]; (ix) decreased serum DHEA and DHEA-S levels with negative correlation with disease activity [162,163,164,135]; (x) abnormal diurnal rhythm of GH and PRL [77]; (xi) decreased bioactivity of serum PRL [74,75]; (xii) elevated PRL in anti-nuclear antibody positive juvenile rheumatoid arthritis [79,78]; (xiii) elevated serum PRL levels in man with RA that correlates with the Steinbrocker index [82]; (xiv) a blunted TSH and exaggerated PRL response to TRH stimulation [53]; (xv) PRL secretion by synovial T cells that perpetuate inflammation [83]; (xvi) the presence of GH and IGF-I in synovial fluid [221]; (xvii) blunted GH response to GHRH [137]; (xviii) paradoxical elevation of GH during glucose infusion [77]; (xix) insulin resistance [222,223]; (xx) elevated basal insulin and exaggerated insulin response to glucose loading [77]; (xxi) decreased plasma glucagon [77]; (xxii) IGF-I deficiency in juvenile rheumatoid arthritis [224]; (xiii) a blunted erythropoietin response to anemia [225,226]; (xxiv) diminished serum endorphin levels [227]; (xxv) impairment of parathyroid hormone response to the infusion of ethylenediamine tetraacetic acid [228]; (xxvi) decreased salivary concentration of epidermal growth factor [229]; and (xxvii) elevated serum gastrin levels [230,231]. With this kind of complexity of neuoroendocrine alterations, which is further complicated by neurological end immunological changes, it should not come as a surprise that autoimmune diseases in general and rheumatoid arthritis in particular are multi-genic and heterogeneous.

7.2. Cell-to-cell signaling and autoimmunity

A malignant cell must de-regulate itself first from "contact inhibiton" by neighbouring cells in order to become a tumor cell. In an analogous way a lymphocyte must loose "self tolerance" i.e. get de-regulated from inhibiton by suppressor immune mechanisms. Evidence is rapidly increasing that self-antigen recognizing suppressor T lymphocytes have a dominant role in restricting potentially self reactive T cells from mounting an autoimmune response. This interaction takes place in an antigen specific manner within target tissues. Therefore, self-tolerance does not mean the lack of self-reactivity, but rather it is a dynamic equilibrium between the effector and suppressor T cells [13,14]. The situation is very similar indeed, if not identical, to the contact inhibition of tumor cells, as the suppressor cells and other suppressing mechanisms maintaining the immune system in homeostasis. Without suppression autoimmune reactions are free to develop. At this point no data are available for the possible hormonal regulation of tissue bound suppressor T cells [10].

Other mechanisms that permit autoimmune reactions to develop are: defective lymphocyte selection in the thymus or bone marrow, defects in the generation of suppressor cells in mucosa associated lymphoid tissue, defects in programmed cell death, in antigen presentation due to the presence of certain MHC phenotypes or mutant antigen receptors, immunoglobulin deficiency, deficient handling of immune complexes, unidentified genetic defects, the breakdown of vascular barrier to privileged sites. Drugs and infectious agents, especially viruses, are also capable of triggering autoimmunity [232,173,10].

7.3. Cytokines and autoimmune disease

Autoimmune reactions may be mediated by organ/tissue specific antibodies or cells, which are analogous reactions to adaptive immune reactions. It is now emerging that nucleated target cells do have defence mechanisms against autoaggressive cells by decreasing their susceptibility to killer cell-induced lysis or even killing CTL by the Fas/FasL pathway. Some tumor cells and cells of the CNS have been shown to have effective self-defence by this mechanism against autoaggressive lymphocytes [233–235]. However, autoimmune reactions frequently cause organ or tissue damage. Organ/tissue specific autoimmune reactions could kill out completely the target cell population (e.g. type I diabetes) or may cause inhibition/stimulation/tissue destruction, depending on the type of immune reaction directed towards the organ (thyroiditis) [173,10].

The cytokines that play a role in specific autoimmune reactions are the ones participating normally in cell mediated (IL-2, IFNγ) and humoral (IL-4, -5, -6, -10) adaptive immunity. In thyroiditis the inappropriate expression of MHC-II antigens by the target organ has been proposed to have a causal relationship [12]. However, such expression is induced by IFNγ, which is likely to be produced by auto-aggressive T lymphocytes that migrated into the target tissue prior to MHC expression. Nevertheless this mechanism could play an important role in the exacerbation of the disease. De-regulated cytokine production would likely to induce autoimmune reactions. This is indicated by the occurrence of autoimmunity in patients after cytokine therapy [236–238]. Interestingly, autoimmune reactions appear to be part of host immune defence against cancer, and thus very beneficial to the host [239–241]. Suppressor T lymphocytes also exert specific autoreactivity, which is fundamental for the maintenance of immune homeostasis and of homeostasis of the organism. It is conceivable that the antigen excess caused by cancer upsets the suppressor/effector immune balance and is sufficient to trigger tumor-directed autoimmune reactions, which have protective value. Alternatively, in progressing tumors suppressor cells are known to inhibit tumor rejection [242].

Rheumatoid diseases are systemic disorders, which show the characteristics of de-regulated innate immune reactions. These disorders may be initiated by infection at times, but frequently there is no identifiable cause. SLE is characterized by the production of "polyclonal" autoantibodies. In healthy individuals CD5+ B lymphocytes produce *poly-reactive natural antibodies* with autoreactivity. T lymphocytes are not required for natural antibody production. These natural antibodies are germ-line coded and do not show the generation of diversity and affinity maturation as adaptive antibodies do. It is now apparent that autoreactive natural antibodies fulfil important physiological regulatory functions [15,243,16] the full potential of which has not yet been realized. The important cytokines that play a causative role in rheumatoid diseases are identical to those participating in the acute phase response, namely IL-6, TNFα and IL-1. In several studies the serum level of IL-6 and of TNF was in correlation with neuroendocrine alterations and with clinical symptoms. Here there is a definite potential for de-regulated cytokine production to initiate the disease process. However, so far there is little if any evidence on this point. The neuroendocrine system is fundamental to the regulation of natural immunity [244–246]. On this basis it comes natural to suggest that neuroendocrine imbalance would play an etiological role in rheumatoid diseases. Indeed the evidence available to date is compatible with this hypothesis.

7.4. The evolution of autoimmune disease

It is very well established that immune reactions have the capacity to adapt to the environment by genetically altering their antigen receptors that recognize pathogens and aberrant cells in the body. In B lymphocytes this adaptation goes one step further to progressively increase antibody avidity, which is known as "affinity maturation". This leads to a significant increase with time in the affinity of the antibody produced [10]. The accumulation of autoantibodies with increasing avidity may then cause more and more damage to the target tissue. In principle, class switch of the autoantibody may also have a deleterious effect, if the damaging potential of the new class is increased.

Another phenomenon that has been observed in T cell-mediated anti-viral immunity is "*epitope spreading*". Here a virus infection within the nervous system may eventually lead to the development of autoimmunity towards the CNS, for instance [246]. Although the mechanism of this phenomenon is not entirely clear, it is likely that viral antigens play a carrier function and stimulate autoreactivity against autoantigens that functions as "haptens" in the system. It is well established that haptens coupled to carriers are capable of inducing autoimmune reactions. Other intracellular pathogens may also induce autoimmune reactions (e.g. tuberculosis, leprosy) by a similar mechanism. Infectious agents are also able to stimulate autoimmune reactions through crossreaction with host antigens, or by polyclonal lymphocyte activation, some of which is mediated by "superantigens" [6,11,247–249]. Polyclonal lymphocyte activation leads to immunosuppression by excess cytokine production, which in turn activates the ACTH-adrenal axis. Superantigens may also cause alterations in thymic selection, which could further impair immune function [248].

During adaptive immune responses memory cells will evolve, which are the toughest cells of the immune system. They have the capacity for long term survival, they are resistant to glucocorticoids and also survive hypohysectomy, react instantaneously to the specific antigen by activation and cytokine production and [10]. It should be obvious to the reader that the development of memory cells against autoantigens will make the disease induced very difficult to control indeed.

Autoimmune reactions may evolve from inducing a local lesion to the development of a full-blown acute phase response. In APR immune-derived cytokines trigger profound neuroendocrine and metabolic changes the end result of which is immunoconversion. The thymus dependent immune system is suppressed, whereas innate immune mechanisms, such as natural antibodies, bone marrow and leukocyte activity and the production of acute phase proteins by the liver, are rapidly amplified. This alternate activation of the immune system is happening at the expense of other bodily functions and organs, most of which are inhibited and undergo intense catabolism. The APR creates a *pro-inflammatory milieu*, which could have fatal consequences, if the inflammatory process is not terminated on time [243–245]. The liver has PRL receptors and it is known to play a role in the development of tolerance to orally induced antigens [250]. The liver also plays an important role in the acute phase response by the synthesis of many of the acute phase proteins and it also produces some of the complement components [203].

Finally, sex-related differences and age- and life cycle related changes in the neuroendocrine and immune systems definitely have an effect on the development of autoimmune disease. For instance the onset of SLE peaks in young females at about the time of reproductive age. On the other hand the incidence of rheumatoid arthritis is highest 2 years after the onset of menopause. Pregnancy aggravates SLE and ameliorates RA. There are examples of stress-related onset of autoimmune disease as well. During aging the adaptive immune system gradually declines, which necessitates more and more reliance on natural immune defence. This leads to the creation of a *pro-inflammatory milieu* with elevated blood levels of IL-6 and TNF. Such milieu leads to insulin resistance and may culminate in type II diabetes eventually. In susceptible individuals this milieu also supports the development and evolution of inflammatory diseases [129,243,10].

Immune complexes occur frequently in the blood and the kidney may play a physiological role in the removal and elimination of these complexes. Glomerulonephritis frequently occurs in animals and patients with an excessive level of immune complexes in the circulation. The kidney has receptors for PRL [251], and kidney damage or nephrectomy leads to hyperprolactinemia [252]. This suggests a regulatory interaction between the kidney and PRL. The interaction of the kidney with the immune system and with PRL in relation to immune complex and autoimmune disease is yet to be investigated. This is especially important because renal lesions contribute to morbidity and mortality very significantly in immune complex disease. The fact that phagocyte function is also stimulated by PRL [253,254] may also be relevant to the development of kidney damage in immune complex disease.

Defective neural regulation of inflammation plays a pathogenic role in allergy and asthma, in the symmetrical form of rheumatoid arthritis and in gastrointestinal inflammatory disease and very likely in other forms of chronic inflammation [256,257].

The challenge now is not to prove, whether or not neuroendocrine factors are involved in autoimmune disease, but to understand the significance of neuroendocrine and immunological changes involved in these conditions in their entire complexity. A better understanding of neuroimmunoregulation holds the promise of new approaches to the treatment of immune and inflammatory diseases with the use of hormones, neurotransmitters, neuropeptides and drugs that modulate these newly recognized immune regulators.

ACKNOWLEDGEMENTS

The work discussed in this paper has been supported in part by the Arthritis Society of Canada, the Medical Research Council of Canada and the Manitoba Health Research Council.

REFERENCES

1. Talmage DW. Allergy and immunology. Ann Rev Med 1957;8:239–256.
2. Burnet FM. Auto-immune disease I. Modern immunological concepts. Br Med J 1959; 5: 645–650.
3. Burnet FM. Autoimmune disease II. Pathology of the immune response. Br Med J 1959;5: 720–725.
4. Boehmer H von, Kisielow P. Self-nonself discrimination by T cells. Science 1990; 248: 1368–73.
5. Adelstein S, Pritchard-Briscoe H, Anderson TA, Crosbie J, Gammon G, Loblay RH, et al. Induction of self-tolerance in T cells but not B cells of transgenic mice expressing little self antigen. Science 1991; 251:1223–5.
6. Honjo T. Seppuku and autoimmunity. Science 1992; 258:591–2.
7. Gauer A, Wiers B, Liu A, Rothbard J, Fathman CG. Amelioration of autoimmune encephalomyelitis by myelin basic protein synthetic peptide-induced anergy. Science 1992; 258:1491–4.
8. Qin S, Cobbold SP, Pope H, Elliott J, Kioussis D, Davies J, Waldmann H. "Infectious" transplantation tolerance. Science 1993; 259:974–977.
9. Suhrbier A, Burrows SR, Fernan A, Lavin MF, Baxter GD, Moss DJ. Peptide epitope induced apoptosis of human cytotoxic T lymphocytes. Implications for peripheral T cell deletion and peptide vaccination. J Immunol 1993; 150:2169–78.
10. Berczi I, Szentivanyi A. Immune reactions. In: Berczi I, Szentivanyi A, series editors. The immune-neuroendocrine circuitry – History and progress. Neuroimmone Biology Vol. 3. Amsterdam: Elsevier, 2003;315–377.
11. Sinha AA, Lopez MT, McDevitt HO. Autoimmune diseases: the failure of self tolerance. Science 1990; 248:1380–8.
12. Bottazzo GF, Mirakian R, De Lazzari F, Mauerhoff T, Todd I, Pujol-Borrell R. Autoimmune endocrine/organ-specific disorders: clinical diagnostic relevance and novel approaches to pathogenesis. In: Berczi I, Kovacs K, editors. Hormones and immunity. Lancaster, UK: MTP Press Ltd, 1987; 296–311.
13. Sakaguchi S, Sakaguchi N, Shimizu J, Yamazaki S, Sakihama T, Itoh M, Kuniyasu Y, Nomura T, Toda M, Takahashi T. Immunologic tolerance maintained by CD25+ CD4+ regulatory T cells: their common role in controlling autoimmunity, tumor immunity, and transplantation tolerance. Immunol Rev 2001;182:18–32.
14. Coutinho A, Hori S, Carvalho T, Caramalho I, Demengeot J. Regulatory T cells: the physiology of autoreactivity in dominant tolerance and "quality control" of immune responses. Immunol Rev 2001;182:89–98.
15. Hardy RR, Li YS, Allman D, Asano M, Gui M, Hayakawa K. B-cell commitment, development and selection. Immunol Rev 2000;175:23–32.
16. Chow D, Kraut R, Wang X. Natural immune regulation of activated cells. In: Berczi I, Gorcynski R, Editors. New Foundations of Biology, Neuroimmune Biology. Volume 1,

Berczi I, Szentivany A. series editors, Amsterdam: Elsevier, 2001;331–345.

17. Berczi I, Nagy E. A possible role of prolactin in adjuvant arthritis. Arthritis Rheum 1982;25:591–594.
18. Berczi I, Nagy E, Asa SL, Kovacs K. The influence of pituitary hormones on adjuvant arthritis. Arthritis Rheum 1984;27:682–688.
19. Whyte A, Williams RO. Bromocriptine suppresses postpartum exacerbation of collagen-induced arthritis. Arthritis Rheum , 1988;31:927–840.
20. Mattsson R, Mattsson A, Hansson I, Holmdahl R, Rook GAW, Whyte A. Increased levels of prolactin during, but not after, the immunisation with rat collagen II enhances the course of arthritis in DBA/1 mice. Autoimmunity 1992; 11:163–70.
21. Nagy E, Berczi I, Wren GE, Asa SL, Kovacs K. Immunomodulation by bromocriptine. Immunopharmacology 1983;6:231–43.
22. Palestine AG, Muellenberg-Coulombre CG, Kim MK, Gelato, Nussenblatt RB. Bromocriptine and low dose cyclosporine in the treatment of experimental autoimmune uveitis in the rat. J Clin Invest , 1987; 79:1078–81.
23. McMurray R, Keisler D, Kanuckel K, Izui S, Walker SE. Prolactin influences autoimmune disease activity in the female B/W mouse. J Immunol 1991;147:3780–7.
24. McMurray R, Keisler D, Izui S, Walker SE. Effects of parturition, suckling, and pseudopregnancy on parameters of disease activity in the B/W mouse model of systemic lupus erythematosus. J Rheumatol 1993;20:1143–51.
25. Gagnerault MC, Touraine P, Savino W, Kelly PA, Dardenne M. Expression of prolactin receptors in murine lymphoid cells in normal and autoimmune situations. J Immunol, 1993;150:5673–81.
26. Kampa IS, Rosenberg JM, Sideman M, Kampa LL. Prolactin levels in streptozotocin-induced diabetic rats. Horm Metab Res 1986; 18:419–20.
27. McMurray RW. Prolactin in murine systemic lupus erythematosus. Lupus 2001;10:742–7.
28. Elbourne KB, Keisler D, McMurray RW. Differential effects of estrogen and prolactin on autoimmune disease in the NZB/NZW F1 mouse model of systemic lupus erythematosus. Lupus 1998;7:420–7.
29. McMurray RW. Estrogen, prolactin, and autoimmunity: actions and interactions. Int Immunopharmacol 2001;1:995–1008.
30. Holstad M, Sandler S. Prolactin protects against diabetes induced by multiple low doses of streptozotocin in mice. J Endocrinol 1999;163:229–34.
31. Vanderpump MP, French JM, Appleton D, Tunbridge WM, Kendall-Taylor P. The prevalence of hyperprolactinaemia and association with markers of autoimmune thyroid disease in survivors of the Whickham Survey cohort. Clin Endocrinol (Oxf) 1998;48: 39–44.
32. Krause I, Blumenfeld Z, Malchinsky M, Cohen S, Blank M, Eldor A, Weksler B, Schweitzer K, Shoenfeld Y. Anti-endothelial cell antibodies in the sera of hyperprolactinemic women. Lupus 1998;7:377–82.
33. Lavalle C, Loyo E, Paniagua R, Bermudez JA, Herrera J, Graef A, Gonzalez-Barcena D, Fraga A. Correlation study between prolactin and androgens in male patients with systemic lupus erythematosus. J Rheumatol 1987; 14:268–72.
34. Jara-Quezada L, Graef A, Lavalle C. Prolactin and gonadal hormones during pregnancy in systemic lupus erythematosus. J Rheumatol 1991;18:349–53.
35. Folomeev M, Prokaeva T, Nassonova V, Nassonov E, Masenko V, Ovtraht N. Prolactin

levels in men with SLE and RA. J Rheumatol 1990; 17:1569–70.

36. Jacobi AM, Rohde W, Ventz M, Riemekasten G, Burmester GR, Hiepe F. Enhanced serum prolactin (PRL) in patients with systemic lupus erythematosus: PRL levels are related to the disease activity. Lupus 2001;10:554–61.
37. Ronchezel MV, Len CA, Spinola e Castro A, Sacchetti S, Lourenzi VM, Ajzen S, Tufik S, Hilario MO. Thyroid function and serum prolactin levels in patients with juvenile systemic lupus erythematosus. J Pediatr Endocrinol Metab 2001;14:165–9.
38. Kozakova D, Rovensky J, Cebecauer L, Bosak V, Jahnova E, Vigas M. Prolactin levels and autoantibodies in female patients with systemic lupus erythematosus. Z Rheumatol 2000;59 Suppl 2:II/80–4.
39. Mok CC, Lau CS, Tam SC. Prolactin profile in a cohort of Chinese systemic lupus erythematosus patients. Br J Rheumatol 1997;36:986–9.
40. Mok CC, Lau CS, Lee KW, Wong RW. Hyperprolactinemia in males with systemic lupus erythematosus. J Rheumatol 1998;25:2357–63.
41. Rovensky J, Blazickova S, Rauova L, Jezova D, Koska J, Lukac J, Vigas M. The hypothalamic-pituitary response in SLE. Regulation of prolactin, growth hormone and cortisol release. Lupus 1998;7:409–13.
42. Funauchi M, Ikoma S, Enomoto H, Sugiyama M, Ohno M, Hamada K, Kanamaru A. Prolactin modulates the disease activity of systemic lupus erythematosus accompanied by prolactinoma. Clin Exp Rheumatol 1998;16:479–82.
43. Blanco-Favela F, Quintal-Alvarez G, Leanos-Miranda A. Association between prolactin and disease activity in systemic lupus erythematosus. Influence of statistical power. J Rheumatol 1999;26:55–9.
44. Mok CC, Lau CS. Profile of sex hormones in male patients with systemic lupus erythematosus. Lupus 2000;9:252–7.
45. Cruz J, Avina-Zubieta A, Martinez de la Escalera G, Clapp C, Lavalle C. Molecular heterogeneity of prolactin in the plasma of patients with systemic lupus erythematosus. Arthritis Rheum 2001;44:1331–5.
46. Rovensky J, Jurankova E, Rauova L, Blazickova S, Lukac J, Veselkova Z, Jezova D, Vigas M. Relationship between endocrine, immune, and clinical variables in patients with systemic lupus erythematosus. J Rheumatol 1997;24:2330–4.
47. Gutierrez MA, Garcia ME, Rodriguez JA, Rivero S, Jacobelli S. Hypothalamic-pituitary-adrenal axis function and prolactin secretion in systemic lupus erythematosus. Lupus 1998;7:404–8.
48. Leanos-Miranda A, Pascoe-Lira D, Chavez-Rueda KA, Blanco-Favela F. Antiprolactin autoantibodies in systemic lupus erythematosus: frequency and correlation with prolactinemia and disease activity. J Rheumatol 2001;28:1546–53.
49. Leanos-Miranda A Pascoe-Lira D, Chavez-Rueda KA, Blanco-Favela F. Detection of macroprolactinemia with the polyethylene glycol precipitation test in systemic lupus erythematosus patients with hyperprolactinemia. Lupus 2001;10:340–5.
50. Leanos-Miranda A, Pascoe-Lira D, Chavez-Rueda KA, Blanco-Favela F. Persistence of macroprolactinemia due to antiprolactin autoantibody before, during, and after pregnancy in a woman with systemic lupus erythematosus. J Clin Endocrinol Metab 2001;86:2619–24.
51. Leanos A, Pascoe D, Fraga A, Blanco-Favela F. Anti-prolactin autoantibodies in systemic lupus erythematosus patients with associated hyperprolactinemia. Lupus 1998;7:398–403.

52. Leanos-Miranda A, Chavez-Rueda KA, Blanco-Favela F. Biologic activity and plasma clearance of prolactin-IgG complex in patients with systemic lupus erythematosus. Arthritis Rheum 2001;44:866–75.
53. Miranda JM, Prieto RE, Paniagua R, Garcia G, Amato D, Barile L, Jara LJ. Clinical significance of serum and urine prolactin levels in lupus glomerulonephritis. Lupus 1998;7:387–91.
54. Jara LJ, Irigoyen L, Ortiz MJ, Zazueta B, Bravo G, Espinoza LR. Prolactin and interleukin-6 in neuropsychiatric lupus erythematosus. Clin Rheumatol 1998;17:110–4
55. Stevens A, Ray D, Alansari A, Hajeer A, Thomson W, Donn R, Ollier WE, Worthington J, Davis JR. Characterization of a prolactin gene polymorphism and its associations with systemic lupus erythematosus. Arthritis Rheum 2001;44:2358–66.
56. Larrea F, Martinez-Castillo A, Cabrera V, Alcocer-Varela J, Queipo G, Carino C, Alarcon-Segovia D. A bioactive 60-kilodalton prolactin species is preferentially secreted in cultures of mitogen-stimulated and nonstimulated peripheral blood mononuclear cells from subjects with systemic lupus erythematosus. J Clin Endocrinol Metab 1997;82:3664–9.
57. Jacobi AM, Rohde W, Volk HD, Dorner T, Burmester GR, Hiepe F. Prolactin enhances the in vitro production of IgG in peripheral blood mononuclear cells from patients with systemic lupus erythematosus but not from healthy controls. Ann Rheum Dis 2001;60: 242–7.
58. Ferreira C, Paes M, Gouveia A, Ferreira E, Padua F, Fiuza T. Plasma homovanillic acid and prolactin in systemic lupus erythematosus. Lupus 1998;7:392–7.
59. Johnson EO, Moutsopoulos HM. Neuroendocrine manifestations in Sjogren's syndrome. Relation to the neurobiology of stress. Ann N Y Acad Sci 2000;917:797–808.
60. Haga HJ, Rygh T. The prevalence of hyperprolactinemia in patients with primary Sjogren's syndrome. J Rheumatol 1999;26:1291–5.
61. Steinfeld S, Rommes S, Francois C, Decaestecker C, Maho A, Appelboom T, Heizmann CW, Kiss R, Pochet R. Big prolactin 60 kDa is overexpressed in salivary glandular epithelial cells from patients with Sjogren's syndrome. Lab Invest 2000;80:239–47.
62. Steinfeld S, Maho A, Chaboteaux C, Daelemans P, Pochet R, Appelboom T, Kiss R.Prolactin up-regulates cathepsin B and D expression in minor salivary glands of patients with Sjogren's syndrome. Lab Invest 2000 Nov;80(11):1711–20.
63. Azar ST, Yamout B. Prolactin secretion is increased in patients with multiple sclerosis. Endocr Res 1999;25:207–14.
64. Yamasaki K, Horiuchi I, Minohara M, Osoegawa M, Kawano Y, Ohyagi Y, Yamada T, Kira J. Hyperprolactinemia in optico-spinal multiple sclerosis. Intern Med 2000;39:296–9.
65. Hangaard J, Andersen M, Grodum E, Koldkjaer O, Hagen C. The effects of endogenous opioids and cortisol on thyrotropin and prolactin secretion in patients with Addison's disease. J Clin Endocrinol Metab 1999;84:1595–601.
66. Lever EG, McKerron CG. Auto-immune Addison's disease associated with hyperprolactinaemia. Clin Endocrinol 21:451–7, 1984.
67. Borzio M, Caldara R, Ferrari C, Borzio F, Piepoli V, Romussi M. Prolactin and growth hormone responses to the serotonin antagonist, metergoline, in liver cirrhosis: further evidence for serotoninergic dysfunction. Horm Metab Res 1982;14:277–8.
68. Asa SL, Bilbao JM, Kovacs K, Josse RG, Kreines K. Lymphocytic hypophysitis of pregnancy resulting in hypopituitarism. A distinct clinico-pathologic entity. Ann Intern Med 1981;95:166–71.
69. Holck S, Laursen H. Prolactinoma coexistent with granulomatous hypophysitis. Acta

Neuropathol 1983;61:253–7.
70. Abe K, Matsuura N, Fukushima N, et al. Plasma prolactin response to thyrotropin releasing hormone in children with newly diagnosed insulin dependent diabetes. Tohoku J Exp Med 1983;140:29–34.
71. Ferrari C, Boghen M, Paracchi A, Rampini P, Raiteri F, Benco R, Romussi M, Codecasa F, Mucci M, Bianco M. Thyroid autoimmunity in hyperprolactinemic disorders. Acta Endocrinol 1983;104:35–41.
72. Legakis I, Petroyianni V, Saramantis A, Tolis G. Elevated prolactin to cortisol ratio and polyclonal autoimmune activation in Hashimoto's thyroiditis. Horm Metab Res 2001;33: 585–9.
73. Notsu K, Ito Y, Furuya H, Ohguni S, Kato Y. Incidence of hyperprolactinemia in patients with Hashimoto's thyroiditis. Endocr J 1997;44:89–94.
74. Berczi I, Cosby H, Hunter T, Baragar F, McNeilly AS, Friesen HG. Decreased bioactivity of circulating prolactin in patients with rheumatoid arthritis. Br J Rheum 26:433–6, 1987.
75. Nagy E, Chalmers IM, Baragar FD, Friesen HG, Berczi I. Prolactin deficiency in rheumatoid arthritis. J Rheumatol 1991;18:1662–8.
76. Ferraccioli G, Cavalieri F, Salaffi F, Fontana S, Scita F, Nolli M, Maestri D. Neuroendocrinologic findings in primary fibromialgia (soft tissue chronic pain syndrome) and in other chronic rheumatic conditions (rheumatoid arthritis, low back pain). J Rheumatol 1990;17:869–73.
77. Neeck G, Federlin K, Graef V, Schatz H, Rusch D, Schmidt KL. Circadian variations of cortisol, prolactin, human growth hormone in patients with rheumatoid arthritis. About interactions between endocrine and immune system. Akt Endokr Stoffw 1988;9:57–63.
78. Chikanza IC. Prolactin and neuroimmunomodulation: in vitro and in vivo observations. Ann N Y Acad Sci 1999;876:119–30.
79. Picco P, Gattorno M, Buoncompagni A, Facchetti P, Rossi G, Pistoia V. Prolactin and interleukin 6 in prepubertal girls with juvenile chronic arthritis. J Rheumatol 1998;25: 347–51.
80. Hedner LP, Bynke G. Endogenous iridocyclitis relieved during treatment with bromocriptine. Am J Ophthalmol , 1985;100:618–9
81. Mateo L, Nolla JM, Bonnin MR, Navarro MA, Roig-Escofet D. High serum prolactin levels in men with rheumatoid arthritis. J Rheumatol 1998;25:2077–82.
82. Kullich WC, Klein G. High levels of macrophage inflammatory protein-1alpha correlate with prolactin in female patients with active rheumatoid arthritis. Clin Rheumatol 1998;17: 263–4.
83. Nagafuchi H, Suzuki N, Kaneko A, Asai T, Sakane T. Prolactin locally produced by synovium infiltrating T lymphocytes induces excessive synovial cell functions in patients with rheumatoid arthritis. J Rheumatol 1999;26:1890–900.
84. Walker SE, McMurray RW, Houri JM, Allen SH, Keisler D, Sharp GC, Schlechte JA. Effects of prolactin in stimulating disease activity in systemic lupus erythematosus. Ann N Y Acad Sci 1998;840:762–72.
85. Alvarez-Nemegyei J, Cobarrubias-Cobos A, Escalante-Triay F, Sosa-Munoz J, Miranda JM, Jara LJ. Bromocriptine in systemic lupus erythematosus: a double-blind, randomized, placebo-controlled study. Lupus 1998;7:414–9.
86. Walker SE, Jacobson JD. Roles of prolactin and gonadotropin-releasing hormone in rheumatic diseases. Rheum Dis Clin North Am 2000;26:713–36.
87. Walker SE. Bromocriptine treatment of systemic lupus erythematosus. Lupus 2001;10:

762–8.
88. Walker SE. Treatment of systemic lupus erythematosus with bromocriptine. Lupus 2001;10:197–202.
89. McMurray RW. Bromocriptine in rheumatic and autoimmune diseases. Semin Arthritis Rheum 2001;31:21–32.
90. Chen HW, Meier H, Heininger HJ, Huebner RJ. Tumorigenesis in strain DW/J mice and induction by prolactin of the group-specific antigen of C-type RNA tumor virus. J Natl Cancer Inst 1972; 49:1145–53.
91. Petri M, Howard D, Repke J. Frequency of lupus flare in pregnancy. The Hopkins Lupus Pregnancy Center experience. Arthritis Rheum 1991;34:1538–45.
92. Jara LJ, Lavalle C, Espinoza LR. Does prolactin have a role in the pathogenesis of systemic lupus erythematosus. J Rheumatol 1992;19:1333–6.
93. Oka M. Effect of pregnancy on the onset and course of rheumatoid arthritis. Ann Rheum Dis 1953;12:227–32.
94. Kaplan D, Diamond H. Rheumatoid arthritis and pregnancy. Clin Obstet Gynecol 1965; 8: 286–303.
95. East J, de Sousa MAB, Prosser PR, Jaquet H. Malignant changes in New Zealand black mice. Clin Exp Immunol 1967; 2:427–43.
96. Izui S, McConahey PJ, Clark JP, Hang LM, Hara I, Dixon FJ. Retroviral gp70 immune complexes in NZB x NZW F2 mice with murine lupus nephritis. J Exp Med 1981; 154: 517–28.
97. Sinha AA, Lopez MT, McDevitt HO. Autoimmune diseases: The failure of self tolerance. Science 1990; 248:1380–8.
98. Adler RA. The anterior pituitary-grafted rat: a valid model of chronic hyperprolactinemia. Endocr Rev 1986;7:302–13.
99. Berczi I. Prolactin, pregnancy and autoimmune disease. (editorial) J. Rheumatol. 1993;20: 1095–1100.
100. Kim S, Liva SM, Dalal MA, Verity MA, Voskuhl RR. Estriol ameliorates autoimmune demyelinating disease: implications for multiple sclerosis. Neurology 1999;52:1230–8.
101. Khamashta MA, Ruiz-Irastorza G, Hughes GR. Systemic lupus erythematosus flares during pregnancy. Rheum Dis Clin North Am 1997;23:15–30.
102. Hampl JS, Papa DJ. Breastfeeding-related onset, flare, and relapse of rheumatoid arthritis. Nutr Rev 2001;59:264–8.
103. Kanik KS, Wilder RL. Hormonal alterations in rheumatoid arthritis, including the effects of pregnancy. Rheum Dis Clin North Am 2000;26:805–23.
104. Ostensen M. Sex hormones and pregnancy in rheumatoid arthritis and systemic lupus erythematosus. Ann N Y Acad Sci 1999;876:131–43.
105. Elenkov IJ, Wilder RL, Bakalov VK, Link AA, Dimitrov MA, Fisher S, Crane M, Kanik KS, Chrousos GP. IL-12, TNF-alpha, and hormonal changes during late pregnancy and early postpartum: implications for autoimmune disease activity during these times. J Clin Endocrinol Metab 2001;86:4933–8.
106. Berczi I. The influence of pituitary-adrenal axis on the immune system. In: Berczi I, editor Pituitary function and immunity. Boca Raton, FL: CRC Press, 1986, 49–132.
107. Berczi I. Immunoregulation by pituitary hormones. In pituitary function and immunity. Berczi I. Editor. Boca Raton FL, CRC Press, 1986, pp. 227–240.
108. Torpy DJ, Chrousos GP. The three-way interactions between the hypothalamic-pituitary-adrenal and gonadal axes and the immune system. Baillieres Clin Rheumatol 1996;10:

181–98.
109. Besedovsky HO, del Rey A. The cytokine-HPA axis feed-back circuit. Z Rheumatol 2000;59 Suppl 2:II/26–30.
110. Morand EF, Leech M. Hypothalamic-pituitary-adrenal axis regulation of inflammation in rheumatoid arthritis. Immunol Cell Biol 2001;79:395–9.
111. Harbuz MS, Conde GL, Marti O, Lightman SL, Jessop DS.The hypothalamic-pituitary-adrenal axis in autoimmunity. Ann N Y Acad Sci 1997;823:214–24.
112. Jessop DS, Harbuz MS, Lightman SL. CRH in chronic inflammatory stress. Peptides 2001;22:803–7.
113. Nagy E, Berczi I, Wren GE, Asa SL, Kovacs K. Immunomodulation by bromocriptine. Immunopharmacology 1983;6:231–243.
114. Ramachandra, RN, Sehon AH, Berczi I. Neuro-hormonal host defence in endotoxin shock. Brain Behav Immun 1992; 6:157–69.
115. Schauenstein K, Fassler R, Dietrich H, Schwartz S, Kromer G, Wick G. Disturbed immune-endocrine communication in autoimmune disease. Lack of corticosterone response to immune signals in obese strain chickens with spontaneous autoimmune thyroiditis. J Immunol 1987; 139:1830–3.
116. MacPhee IAM, Antoni FA, Mason DW. Spontaneous recovery of rats from experimental allergic encephalomyelitis is dependent on regulation of the immune system by endogenous adrenal corticosteroids. J Exp Med 1989; 169:431–45.
117. Sternberg EM, Hill JM, Chorousos GP, Kamilaris T, Listwak SJ, Gold PW, et al. Inflammatory mediator-induced hypothalamic-pituitary-adrenal axis activation is defective in streptococcal cell wall arthritis-susceptible Lewis rats. Proc Natl Acad Sci USA 1989; 86: 2374–8.
118. Sternberg EM, Young WS III, Bernardini R, Calogero AE, Chrousos GP, Gold PW, et al. A central nervous system defect in biosynthesis of corticotropin-releasing hormone is associated with susceptibility to streptococcal cell wall-induced arthritis in Lewis rats. Proc Natl Acad Sci USA 1989; 86:4771–5.
119. Wick G, Sgonc R, Lechner O. Neuroendocrine-immune disturbances in animal models with spontaneous autoimmune diseases. Ann N Y Acad Sci 1998;840:591–8.
120. Lechner O, Hu Y, Jafarian-Tehrani M, Dietrich H, Schwarz S, Herold M, Haour F, Wick G. Disturbed immunoendocrine communication via the hypothalamo-pituitary-adrenal axis in murine lupus. Brain Behav Immun 1996;10:337–50.
121. Wilder RL, Griffiths MM, Cannon GW, Caspi R, Gulko PS, Remmers EF. Genetic actors involved in central nervous system/immune interactions. Adv Exp Med Biol 2001;493: 59–67.
122. Jafarian-Tehrani M, Sternberg EM. Neuroendocrine and other factors in the regulation of inflammation. Animal models. Ann N Y Acad Sci 2000;917:819–24.
123. Shanks N, Moore PM, Perks P, Lightman SL. Alterations in hypothalamic-pituitary-adrenal function correlated with the onset of murine SLE in MRL +/+ and lpr/lpr mice. Brain Behav Immun 1999;13:348–60.
124. del Rey A, Besedovsky HO. The cytokine-HPA axis circuit contributes to prevent or moderate autoimmune processes. Z Rheumatol 2000;59 Suppl 2:II/31–5.
125. Dowdell KC, Gienapp IE, Stuckman S, Wardrop RM, Whitacre CC. Neuroendocrine modulation of chronic relapsing experimental autoimmune encephalomyelitis: a critical role for the hypothalamic-pituitary-adrenal axis. J Neuroimmunol 1999;100:243–51.
126. Wilder RL. Hormones and autoimmunity: animal models of arthritis. Baillieres Clin

Rheumatol 1996;10:259–71.
127. Hench PS, Kendall EC, Slocumb CH, Polley HF. The effect of a hormone of the adrenal cortex (17-hydroxyl-11-dehydrocorticosterone: compound E) and of pituitary adrenocorticotrophic hormone on rheumatoid arthritis. Ann Rhem Dis 1949; 8:97–104.
128. Hill SR Jr, Ulloa A, Starnes WR, Holley HL. Corticosteroids in rheumatoid arthritis. Studies on Aldosterone, 17-hydroxycorticosteroid, and 17-ketosteroid response to corticotropin in male patients with rheumatoid arthritis. Arch Intern Med 1963; 112:603–11.
129. Berczi I, Baragar FD, Chalmers IM, Keystone EC, Nagy E, Warrington RJ. Hormones in self-tolerance and autoimmunity: a role in the pathogenesis of rheumatoid arthritis? Autoimmunity, 1993;16:45–56.
130. Chikanza IC, Panayi GS. Hypothalamic-pituitary mediated modulation of immune function: prolactin as a neuroimmune peptide. Brit J Rheumatol 1991; 30:203–7.
131. Dekkers JC, Geenen R, Godaert GL, Glaudemans KA, Lafeber FP, van Doornen LJ, Bijlsma JW. Experimentally challenged reactivity of the hypothalamic pituitary adrenal axis in patients with recently diagnosed rheumatoid arthritis. J Rheumatol 2001;28:1496–504.
132. Cutolo M, Villaggio B, Foppiani L, Briata M, Sulli A, Pizzorni C, Faelli F, Prete C, Felli L, Seriolo B, Giusti M. The hypothalamic-pituitary-adrenal and gonadal axes in rheumatoid arthritis. Ann N Y Acad Sci 2000;917:835–43.
133. Mastorakos G, Ilias I. Relationship between interleukin-6 (IL-6) and hypothalamic-pituitary-adrenal axis hormones in rheumatoid arthritis. Z Rheumatol 2000;59 Suppl 2: II/75–9.
134. Boss B, Neeck G. Correlation of IL-6 with the classical humoral disease activity parameters ESR and CRP and with serum cortisol, reflecting the activity of the HPA axis in active rheumatoid arthritis. Z Rheumatol 2000;59 Suppl 2:II/62–4.
135. Straub RH, Scholmerich J, Zietz B. Replacement therapy with DHEA plus corticosteroids in patients with chronic inflammatory diseases--substitutes of adrenal and sex hormones. Z Rheumatol 2000;59 Suppl 2:II/108–18.
136. Nishioka T, Kurokawa H, Takao T, Kumon Y, Nishiya K, Hashimoto K.Differential changes of corticotropin releasing hormone (CRH) concentrations in plasma and synovial fluids of patients with rheumatoid arthritis (RA). Endocr J 1996;43:241–7.
137. Templ E, Koeller M, Riedl M, Wagner O, Graninger W, Luger A. Anterior pituitary function in patients with newly diagnosed rheumatoid arthritis. Br J Rheumatol 1996;35: 350–6.
138. Demir H, Kelestimur F, Tunc M, Kirnap M, Ozugul Y. Hypothalamo-pituitary-adrenal axis and growth hormone axis in patients with rheumatoid arthritis. Scand J Rheumatol 1999;28:41–6.
139. Johnson EO, Vlachoyiannopoulos PG, Skopouli FN, Tzioufas AG, Moutsopoulos HM. Hypofunction of the stress axis in Sjogren's syndrome. J Rheumatol 1998;25:1508–14.
140. Valtysdottir ST, Wide L, Hallgren R. Low serum dehydroepiandrosterone sulfate in women with primary Sjogren's syndrome as an isolated sign of impaired HPA axis function. J Rheumatol 2001;28:1259–65.
141. Johnson EO, Skopouli FN, Moutsopoulos HM. Neuroendocrine manifestations in Sjogren's syndrome. Rheum Dis Clin North Am 2000;26:927–49.
142. Roy MS, Roy A, Brown S. Increased urinary-free cortisol outputs in diabetic patients. J Diabetes Complications 1998;12:24–7.
143. Michelson D, Stone L, Galliven E, Magiakou MA, Chrousos GP, Sternberg EM, Gold PW.

Multiple sclerosis is associated with alterations in hypothalamic-pituitary-adrenal axis function. J Clin Endocrinol Metab 1994;79:848–53.
144. Then Bergh F, Kumpfel T, Trenkwalder C, Rupprecht R, Holsboer F. Dysregulation of the hypothalamo-pituitary-adrenal axis is related to the clinical course of MS. Neurology 1999;53:772–7.
145. Wei T, Lightman SL. The neuroendocrine axis in patients with multiple sclerosis. Brain 1997;120:1067–76.
146. Then Bergh F, Kumpfel T, Grasser A, Rupprecht R, Holsboer F, Trenkwalder C. Combined treatment with corticosteroids and moclobemide favors normalization of hypothalamo-pituitary-adrenal axis dysregulation in relapsing-remitting multiple sclerosis: a randomized, double blind trial. J Clin Endocrinol Metab 2001;86:1610–5.
147. Verthelyi D. Sex hormones as immunomodulators in health and disease. Int Immunopharmacol 2001;1:983–93.
148. Merrill JT, Lahita RG. Sex hormone binding globulins and atherosclerotic risk in systemic lupus. Lupus 2000;9(3):217–22.
149. Castagnetta L, Cutolo M, Granata OM, Di Falco M, Bellavia V, Carruba G. Endocrine end-points in rheumatoid arthritis. Ann N Y Acad Sci 1999;876:180–91.
150. Matejuk A, Dwyer J, Zamora A, Vandenbark AA, Offner H. Evaluation of the effects of 17beta-estradiol (17beta-e2) on gene expression in experimental autoimmune encephalomyelitis using DNA microarray. Endocrinology 2002;143:313–9.
151. Ito A, Bebo BF Jr, Matejuk A, Zamora A, Silverman M, Fyfe-Johnson A, Offner H. Estrogen treatment down-regulates TNF-alpha production and reduces the severity of experimental autoimmune encephalomyelitis in cytokine knockout mice. J Immunol 2001;167:542–52.
152. Bebo BF Jr, Fyfe-Johnson A, Adlard K, Beam AG, Vandenbark AA, Offner H. Low-dose estrogen therapy ameliorates experimental autoimmune encephalomyelitis in two different inbred mouse strains. J Immunol 2001;166:2080–9.
153. Grimaldi CM, Michael DJ, Diamond B. Cutting edge: expansion and activation of a population of autoreactive marginal zone B cells in a model of estrogen-induced lupus. J Immunol 2001;167:1886–90.
154. McMurray RW. Sex hormones in the pathogenesis of systemic lupus erythematosus. Front Biosci 2001;6:E193–206.
155. Rider V, Abdou NI. Gender differences in autoimmunity: molecular basis for estrogen effects in systemic lupus erythematosus. Int Immunopharmacol 2001;1:1009–24.
156. Lahita RG. The role of sex hormones in systemic lupus erythematosus. Curr Opin Rheumatol 1999;11:352–6.
157. Nagler RM, Pollack S. Sjogren's syndrome induced by estrogen therapy. Semin Arthritis Rheum 2000;30:209–14.
158. Olsen NJ, Kovacs WJ. Effects of androgens on T and B lymphocyte development. Immunol Res 2001;23:281–8.
159. Bebo BF Jr, Schuster JC, Vandenbark AA, Offner H. Androgens alter the cytokine profile and reduce encephalitogenicity of myelin-reactive T cells. J Immunol 1999;162:35–40.
160. Voskuhl RR, Palaszynski K. Sex hormones in experimental autoimmune encephalomyelitis: implications for multiple sclerosis. Neuroscientist 2001;7:258–70.
161. Du C, Khalil MW, Sriram S. Administration of dehydroepiandrosterone suppresses experimental allergic encephalomyelitis in SJL/J mice. J Immunol 2001;167:7094–101.
162. Dessein PH, Joffe BI, Stanwix AE, Moomal Z. Hyposecretion of the adrenal androgen

dehydroepiandrosterone sulfate and its relation to clinical variables in inflammatory arthritis. Arthritis Res 2001;3:183–8.
163. Masi AT, Chatterton RT, Aldag JC. Perturbations of hypothalamic-pituitary-gonadal axis and adrenal androgen functions in rheumatoid arthritis: an odyssey of hormonal relationships to the disease. Ann N Y Acad Sci 1999;876:53–62.
164. Cutolo M, Foppiani L, Prete C, Ballarino P, Sulli A, Villaggio B, Seriolo B, Giusti M, Accardo S. Hypothalamic-pituitary-adrenocortical axis function in premenopausal women with rheumatoid arthritis not treated with glucocorticoids. J Rheumatol 1999;26:282–8.
165. Cutolo M, Giusti M, Foppiani L, Seriolo B, Briata M, Bisso A, Faelli F, Felli L, Prete C, Pizzorni C, Sulli A. The hypothalamic-pituitary-adrenocortical and gonadal axis function in rheumatoid arthritis. Z Rheumatol 2000;59 Suppl 2:II/65–9.
166. Straub RH, Cutolo M. Involvement of the hypothalamic--pituitary--adrenal/gonadal axis and the peripheral nervous system in rheumatoid arthritis: viewpoint based on a systemic pathogenetic role. Arthritis Rheum 2001;44:493–507.
167. Valtysdottir ST, Wide L, Hallgren R. Low serum dehydroepiandrosterone sulfate in women with primary Sjogren's syndrome as an isolated sign of impaired HPA axis function. J Rheumatol 2001;28:1259–65.
168. McMurray RW. Estrogen, prolactin, and autoimmunity: actions and interactions. Int Immunopharmacol 2001;1:995–1008.
169. Wallace D. Current and emerging lupus treatments. Am J Manag Care 2001;7(16 Suppl): S490–5.
170. Giltay EJ, van Schaardenburg D, Gooren LJ, Popp-Snijders C, Dijkmans BA. Androgens and ankylosing spondylitis: a role in the pathogenesis? Ann N Y Acad Sci 1999;876: 340–64.
171. Rosmalen JG, Pigmans MJ, Kersseboom R, Drexhage HA, Leenen PJ, Homo-Delarche F. Sex steroids influence pancreatic islet hypertrophy and subsequent autoimmune infiltration in nonobese diabetic (NOD) and NODscid mice. Lab Invest 2001;81:231–9.
172. Rudberg S, Persson B. Serum leptin levels in young females with insulin-dependent diabetes and the relationship to hyperandrogenicity and microalbuminuria. Horm Res 1998;50:297–302.
173. Berczi I, Asa S. Endocrine-immune interactions. In: Kovacs K, Asa S, editors, Functional Endocrine Pathology, 2nd edition, Boston, MA:Blackwell Scientific Publishers, 1998;1050–1091.
174. Nicolson GL. Paracrine and autocrine growth mechanisms in tumor metastasis to specific sites with particular emphasis on brain and lung metastasis. Cancer Metastasis Rev 1993;12:325–343.
175. Baerwald CG, Burmester GR, Krause A. Interactions of autonomic nervous, neuroendocrine, and immune systems in rheumatoid arthritis. Rheum Dis Clin North Am 2000;26:841–57.
176. Eijsbouts AM, Murphy EP. The role of the hypothalamic-pituitary-adrenal axis in rheumatoid arthritis. Baillieres Best Pract Res Clin Rheumatol 1999;13:599–613.
177. Weinberg RA. Tumor suppressor genes. Science 1991; 254:1138–46.
178. Santana V, Rose NR. Neoplastic lymphoproliferation in autoimmune disease - an updated review. Clin Immunol Immunopathol 1992; 63:205.
179. Berczi I, Chow DA, Baral E, Nagy E. Neuroimmunoregulation and cancer. Int. J. Oncol. 13:1049 1051,1998.
180. Berczi I. The role of the growth and lactogenic hormone family in immune function.

Neuroimmunomodulation 1994;1:201–216.
181. Berczi I, Nagy E, Matusik RJ, Friesen HG. Pituitary hormones regulate c-myc and DNA synthesis in lymphoid tissue. J Immunol 1991; 146:2201–6.
182. Hooghe R, Devos S, Dogusan Z, Hooghe-Peeters EL. Signal transduction and modulation of gene expression by prolactin in human leukocytes. In: Matera L, Rapaport R, editors. Growth and Lactogenic Hormones, Neuroimmune Biology, Vol. 2. Berczi I, Szentivanyi A, series editors. Amsterdam: Elsevier 2002;123–136.
183. Weigent D. Growth Hormone and Insulin-Like Growth Factor-1 Production by Cells of the Immune System. In Growth and Lactogenic Hormones, Matera L, Rapaort R. Eds. Neuroimmune Biology, Vol. 2. Berczi I, Szentivanyi A. Series Ed. Elsevier 2002. Pp. 87–100.
184. Calabresi P, Parks RE Jr. Antiproliferative agents and drugs used for immunosuppression. In: Gilman AG, Goodman LS, Gilman A, editors. The pharmacological basis of therapeutics. New York :Macmillan, 1980;1256–1313.
185. Berczi I, Nagy E. The effect of prolactin and growth hormone on hemolymphopoietic tissue and immune function. In: Berczi E, Kovacs K, editors. Hormones and immunity. Lancaster, UK: MTP Press Ltd, 1987; 145–71.
186. Matera L, Rapaport R. Editors, Growth and lactogenic hormones. Neuroimune Biology series, volume 2 Berczi I, Szentivanyi A. Series editors. Amsterdam, Elsevier, 2002.
187. Berczi I, Nagy E. Hormones as immune modulating agents. In: Krezina T, editor. "Handbook of Immune Modulating Agents". Marcel Dekker, 1997;75–120.
188. Aubry JM, Bartanusz V, Jezova D, Belin D, Kiss JZ. Single stress induces long-lasting elevations in vasopressin mRNA levels in CRF hypophysiotrophic neurones, but repeated stress is required to modify AVP immunoreactivity. J Neuroendocrinol 1999;11:377–84.
189. Dardenne M, Savino W. Neuroendocrine control of the thymic epithelium: modulation of thymic endocrine function, cytokeratin expression and cell proliferation by hormones and neuropeptides. Progr Neuroendocrinimmun 1990; 3:18–25.
190. Stutman O. Ontogeny of T cells. Clin Immunol Allergy 1985; 5:191–234.
191. Scollay RG, Butcher EC, Weissman IL. Thymus migration: quantitative studies on the rate of migration of cells from the thymus to the periphery in mice. Eur J Immunol 1980; 10: 210–18.
192. Donskoy E, Goldschneider I. Thymocytopoiesis is maintained by blood-borne precursors throughout postnatal life. J Immunol 1992; 148:1604–1612.
193. Hess AD, Fischer AC. Immune mechanisms in cyclosporine-induced syngeneic graft-versus-host disease. Transplantation 1989; 48:895–900.
194. O'Keefe SJ, Tamura J, Kincaid RL, Tocci MJ, O'Neill EA. FK-506- and CsA-sensitive activation of the interleukin-2 promoter by calcineurin. Nature 1992; 357:692–4.
195. Wahlstrom HE, Lavelle-Jones M, Endres D, Akimoto R, Kolterman O, Moossa AR. Inhibition of insulin release by cyclosporine and production of peripheral insulin resistance in the dog. Transplantation 1990; 49:600–4.
196. Villanua MA, Steger RW, Szary A, Esquifino AI, Bartke A. Cyclosporine treatment in hyperprolactinemic male rats. 2. Effects at the hypothalamic level. Prog Neuroendocrinimmunol 1992; 5:52–61.
197. Silverman AJ. Rheumatoid arthritis. In: Freedman A, Kaplanj H, Sadock B, editors. Comprehensive Textbook of Psychiatry. Vol. 2, 2nd ed, Williams & Wilkins Co, 1975; 1694.
198. Muller E, Giuseppe G, Scapginini. Neurotransmitters and Anterior Pituitary Function. New

York: Academic Press, 1977.
199. Germain G, Ferre F. Hormones et parturition chez les primates. Ann Endocrinol (Paris) 1987; 48:311–21.
200. Curlewis JD. Seasonal prolactin secretion and its role in seasonal reproduction: a review. Reprod Fert Develop 1992; 4:1–23.
201. Tuominen RK, Mannisto PT, Mattila J. Studies on the site and mechanism of inhibitory action of intracerebral histamine on the cold-stimulated thyrotropic secretion in male rats. Brain Res 1985; 343:329–35.
202. Berczi I, Nagy E. Neurohormonal control of cytokines during injury. In: Rothwell NJ, Berkenbosch F, editors. Brain Control of the Response to Injury. Cambridge: University Press, 1993; pp. 32–107.
203. Dantzer R, Kelley KW. Minireview. Stress and immunity: an integrated view of relationships between the brain and the immune system. Life Sci 1989; 44:1995–2008.
204. Jasmin G, Cantin M, editors. Stress Revisited 1: Neuroendocrinology of Stress. Methods and Achievements in Experimental Pathology, vol 14. Basel: Karger, 1991.
205. Keller SE, Schleifer SJ, Demetrikopoulos MK. Stress-induced changes in immune function in animals: hypothalamo-pituitary-adrenal influences. In: Ader R, Felten DL, Cohen N, eds. Psychoneuroimmunology, 2nd ed. New York: Academic Press, 1991; 771–87.
206. Shavit Y. Stress-induced immune modulation in animals: opiates and endogenous opioid peptides. In: Ader R, Felten DL, Cohen N, editors. Psychoneuroimmunology, 2nd ed. New York: Academic Press, 1991; 789–806.
207. Kiecolt-Glaser JK, Glaser R. Stress and immune function in humans. In: Psychoneuroimmunology, 2nd ed. New York: Academic Press, 1991; 849–67.
208. Winsa B, Adami HO, Bergstrom R, Gamstedt A, Dahlberg PA, Adamson U, et al. Stressful life events and Graves' disease. Lancet 1991; 338:1475–9.
209. Chikanza IC. Neuroendocrine immune features of pediatric inflammatory rheumatic diseases. Ann N Y Acad Sci 1999;876:71–80.
210. Neidhart M. Prolactin in autoimmune diseases. Proc Soc Exp Biol Med 1998;217:408–19.
211. Jara LJ, Vera-Lastra O, Miranda JM, Alcala M, Alvarez-Nemegyei J. Prolactin in human systemic lupus erythematosus. Lupus 2001;10:748–56.
212. Lahita RG. Sex hormones and systemic lupus erythematosus. Rheum Dis Clin North Am 2000;26:951–68.
213. Compton MM, Cidlowski JA. Thymocyte apoptosis. A model of programmed cell death. Trends Endocrinol Metab 1992; 3:17–23.
214. Schuurs AHWM, Deckers GHJ, Schot LPC, Verheul HAM. Lack of correlation between endocrinological and immunological effects of androgen-derived steroids. In: Berczi I, Kovacs K, eds. Hormones and immunity. Lancaster, UK: MTP Press Ltd, 1987; 120–9.
215. Warren JE. Diurnal plasma corticoid studies and their relation to morning stiffness in rheumatoid arthritis. Ann Rheumatic Dis 1956; 15:392–3.
216. Margraf H, Wise L, Ballinger WF. Cortisol 21-acetate: apparently abnormal plasma levels in patients with rheumatoid arthritis. Clin Chim Acta 1972; 39:247–51.
217. Pal SB. Pattern of the excretion of urinary cortisol, tetrahydro-cortisone, allo-tetrahydrocortisol and tetrahydrocortisol in normal human individuals and in patients with rheumatic disorders. Clin Chim Acta 1967; 15:57–67.
218. Feher KG, Feher T, Merétey K. Interrelationship between the immunological and steroid hormone parameters in rheumatoid arthritis. Exp Clin Endocrinol 1986; 87:38–42.

219. Cutolo M, Balleari E, Giusti M, Monachesi M, Accardo S. Sex hormone status of male patients with rheumatoid arthritis - evidence of low serum concentrations of testosterone at baseline and after human chorionic gonadotropin stimulation. Arthritis Rheum 1988; 31: 1314–7.
220. Sambrook PN, Eisman JA, Champion GD, Pocock NA. Sex hormone status and osteoporosis in postmenopausal women with rheumatoid arthritis. Arthritis Rheum 1988; 31:973–8.
221. Denko CW, Boja B, Moskowitz RW. Growth factors, insulin-like growth factor-1 and growth hormone, in synovial fluid and serum of patients with rheumatic disorders. Osteoarthritis Cartilage 1996;4:245–9.
222. Svenson KL, Lundqvist G, Wide L, Hallgren R. Impaired glucose handling in active rheumatoid arthritis: relationship to the secretion of insulin and counter-regulatory hormones. Metabolism 1987; 36:940–3.
223. Svenson KL, Pollare T, Lithell H, Hallgren R. Impaired glucose handling in active rheumatoid arthritis: relationship to peripheral insulin resistance. Metabolism 1988; 37: 125–30.
224. Allen RC, Jimenez M, Cowell CT. Insulin-like growth factor and growth hormone secretion in juvenile chronic arthritis. Ann Rheum Dis 1991; 50:602–6.
225. Hochberg MC, Arnold CM, Hogans BB, Spivak JL. Serum immunoreactive erythropoietin in rheumatoid arthritis - impaired response to anemia. Arthritis Rheum 1988; 31:1318–21.
226. Boyd HK, Lappin TRJ. Erythropoietin deficiency in the anemia of chronic disease. Eur J Haematol 1991; 46:198–201.
227. Denko CW, Aponte J, Gabriel P, Petricevic M. Serum beta-endorphin in rheumatic disorders. J Rheumatol 1982; 9:827–33.
228. Ekenstam EA, Benson L, Hallgren R, Wide L, Ljunghall S. Impaired secretion of parathyroid hormone in patients with rheumatoid arthritis: relationship to inflammatory activity. Clin Endocrinol 1990; 32:323–8.
229. Kelly SM, Crampton J, Hunter JO. Decreased salivary epidermal growth factor in rheumatoid disease: a possible mechanism for increased susceptibility to gastric ulceration. Brit Med J 1990; 301:422–3.
230. Rooney PJ, Grennan DM, Sturrock RD, Brooks PM, Dick WC. Serum immunoreactive gastrin: specificity for rheumatoid arthritis, bimodality of distribution, and failure of effect of anti-inflammatory drugs. Ann Rheum Dis 1976; 35:40–45.
231. De Witte TJ, Geerdink PJ, Lamers CB, Boerbooms AM, Van der Korst JK. Hypochlorhydria and hypergastrinaemia in rheumatoid arthritis. Ann Rheum Dis 1979; 38:14–17.
232. Sabbadini E, Berczi I. The submandibular gland: a key organ in the neuro-immunoregulatory network? (Review). Neuroimmunomodulation 1995;2:184–202.
233. Badie B, Schartner J, Prabakaran S, Paul J, Vorpahl J. Expression of Fas ligand by microglia: possible role in glioma immune evasion. J Neuroimmunol 2001;120:19.
234. Riffkin CD, Gray AZ, Hawkins CJ, Chow CW, Ashley DM. Ex vivo pediatric brain tumors express Fas (CD95) and FasL (CD95L) and are resistant to apoptosis induction. Neuro-oncol 2001;3:229–40.
235. Ichinose M, Masuoka J, Shiraishi T, Mineta T, Tabuchi K. Fas ligand expression and depletion of T-cell infiltration in astrocytic tumors. Brain Tumor Pathol 2001;18:37.
236. Abdel-Raheem MM, Potti A, Kobrinsky N. Severe Evans's syndrome secondary to interleukin-2 therapy: treatment with chimeric monoclonal anti-CD20 antibody. Ann Hematol 2001;80:543–5.

237. Passos de Souza E, Evangelista Segundo PT, Jose FF, Lemaire D, Santiago M. Rheumatoid arthritis induced by alpha-interferon therapy. Clin Rheumatol 2001;20:297–9.
238. Stavroyianni N, Stamatopoulos K, Viniou N, Vaiopoulos G, Yataganas X. Autoimmune hemolytic anemia during alpha-interferon treatment in a patient with chronic myelogenous leukemia. Leuk Res 2001;25:1097–8.
239. Chan C, O'Day J. Melanoma-associated retinopathy: does autoimmunity prolong survival? Clin Experiment Ophthalmol 2001;29:235.
240. van Elsas A, Sutmuller RP, Hurwitz AA, Ziskin J, Villasenor J, Medema JP, Overwijk WW, Restifo NP, Melief CJ, Offringa R, Allison JP. Elucidating the autoimmune and antitumor effector mechanisms of a treatment based on cytotoxic T lymphocyte antigen-4 blockade in combination with a B16 melanoma vaccine: comparison of prophylaxis and therapy. J Exp Med 2001;194:481–489.
241. Hansen MH, Nielsen H, Ditzel HJ. The tumor-infiltrating B cell response in medullary breast cancer is oligoclonal and directed against the autoantigen actin exposed on the surface of apoptotic cancer cells. Proc Natl Acad Sci U S A 2001;98:12659–64.
242. Fu T, Shen Y, Fujimoto S. Tumor-specific CD4(+) suppressor T-cell clone capable of inhibiting rejection of syngeneic sarcoma in A/J mice. Int J Cancer 2000;87:680–7.
243. Berczi I, Bertok L, Chow, D. Natural immunity and neuroimmune host defence. Ann NY Acad Sci 2001;917 :248–257.
244. Berczi I, Chalmers IM, Nagy E, Warrington RJ. The immune effects of neuropeptides. Balliere's Clinical Rheumatology, I. Chicanza, editor, 1996;10:227–257.
245. I. Berczi, D.A. Chow, E. Baral, E. Nagy. Neuroimmunoregulation and cancer. Int. J. Oncol. 13:1049–1051,1998.
246. Owens T, Tran EH, Hassan-Zahracc M, Babcock A, Krakowski ML, Fournier S, Jensen MB, Finsen B. The pathogenesis of encephalitis. In: Berczi I, Gorcynski R, editors. New Foundations of Biology, Neuroimmune Biology. Volume, Berczi I, Szentivany A. Series Editors, Amsterdam: Elsevier, 2001; 387–397.
247. Black G, Forsgren A. Many bacterial species are mitogenic for human blood lymphocytes. Scand J Immunol 1978; 8:347–54.
248. Schwab JH, Brown RR, Anderle SK, Schlievert PM. Superantigen can reactivate bacterial cell wall-induced arthritis. J Immunol 1993; 150:4151–9.
249. Blackman M, Kappler J, Marrack P. The role of T cell receptor in positive and negative selection of T cells. Science 1990; 248:1335–41.
250. Weiner HL, Mackin GA, Matsui M, Orav, EJ, Khoury SJ, Dawson DM, Hafler DA. Double-blind pilot trial of oral tolerization with myelin antigens in multiple sclerosis. Science 1993;259:1321–4.
251. Posner BI, Kelly PA, Shiu RPC, Friesen HG. Studies of insulin, growth hormone and prolactin binding: Tissue distribution, species variation and characterization. Endocrinology 1974; 95:521–31.
252. Biasioli S, Mazzali A, Foroni R, et al. Chronobiological variations of prolactin (PRL) in chronic renal failure (CRF). Clin Nephrol 1988; 30:86–92.
253. Davila DR, Edwards III CK, Arkins S, Simon J, Kelley KW. Interferon-γ-induced priming for secretion of superoxide anion and tumor necrosis factor-α declines in macrophages from aged rats. FASEB J 1990; 4:2906–11.
254. Rovensky J, Vigas M, Marek J, Blazickova S, Vydetelkova L, Takac A: Evidence for immunomodulatory properties of prolactin in selected invitro and invivo situations. Int J Immunopharmacol 1991; 13:267–72.

255. Anisman H, Baines MG, Berczi I, Bernstein CN, Blennerhassett MG, Gorczynski RM, Greenberg AH, Kisil FT, Mathison RD, Nagy E, Nance DM, Perdue MH, Pomerantz DK, Sabbadini ER, Stanisz A, Warrington RJ. Neuroimmune mechanisms in health and disease: 2. Disease. CMAJ 1996;155:1075–82.

The Immune-Neuroendocrine Circuitry
Edited by I. Berczi and A. Szentivanyi

Immunodeficiency

ISTVAN BERCZI and ANDOR SZENTIVANYI

Department of Immunology, Faculty of Medicine, The University of Manitoba, Winnipeg, Manitoba RE3 0W3, Canada; and Department of Internal Medicine, Faculty of Medicine, The University of South Florida, Tampa, Florida 33612, USA

ABSTRACT

Traditionally immunodeficiency is attributed to internal structural and/or regulatory defects of the immune system, although it is recognized that some of the deficiencies are complex. Animal experiments and some observations in man clearly indicate that abnormalities of the neuroendocrine system are also able to cause immunodeficiency.

Hypophysectomized (Hypox) rats are immunodeficient. If the residual serum PRL present in Hypox animals is neutralized, the thymus, spleen, adrenals and gonads become atrophic; severe combined immunodeficiency, anemia, wasting and death will occur in 6–8 weeks. The spontaneous occurrence of joint and complete deficiency of pituitary GH and PRL has never been demonstrated in man or in animals, indicating that this condition must be lethal. Immunological memory cells survive hypophysectomy and resist glucocorticoids. The pituitary GH/PRL-IGF-I axis supports the adaptive T cell dependent immune system. This axis is suppressed during febrile illness (acute phase response, APR), where natural immune mechanisms are amplified.

The hormones of the hypothalamus-pituitary-adrenal axis, especially glucocorticoids and dehydroepyandrosterone-sulphate (DHEA-S), are required for normal immune function. DHEA-S serves as the precursor for internal steroidogenesis by the immune system. The importance of steroid hormones is emphasised by the fact that the thymus is a steroidogenic organ. During enhanced immune/inflammatory activity the HPA axis exerts a suppressive control over immune and inflammatory reactions and is essential for the maintenance of immune homeostasis. Animals lacking adrenal function may be killed by excess cytokine production after simple immunization or after minute amounts of endotoxin, which are harmless to normal animals. The HPA axis is also the prime neuroendocrine mediator of APR, which is fundamental for survival in emergency situations (e.g. sepsis), where adaptive immune mechanisms are ineffective. Cytokines (IL-1, -6, TNFα) initiate these neuroendocrine, metabolic and immune alterations in APR. The HPA axis coordinates the nervous, endocrine and immune systems during APR by regulating fever and CNS activity in general, and contributes to the conversion of the immune system from the adaptive mode of reactivity to boosting of natural immune mechanisms and to metabolic alterations (catabolism in most tissues). The fat cells-derived hormone, leptin, is elevated during APR and regulates energy metabolism in conjunction with TNF.

Growth hormone (GH) treatment of patients with severe acute illness worsened their survival. One possible reason for this is that GH interfered with the HPA axis and with catabolism that is fundamental to boosting natural immune defence during APR.

Estradiol is required for the cytotoxic activity of CD4+ T lymphocytes, which fulfill major immunoregulatory function and is also important in thymic physiology. Androgens and progesterone are immunosuppressive steroids. Sex hormones are fundamental for reproductive function, where the immune system is an active participant. Disturbances of sex hormone function due to ageing or to other factors contribute to deficient immune function and to the development of disease.

Whenever the adaptive immune system becomes deficient (e.g. in aging individuals, AIDS patients, or due to malnutrition or to various insults) the body must resort to natural immune defence, including APR, at major metabolic costs, which may lead to malnutrition, wasting and eventually to death. Proper nutrition and hormone supplements (e. g. GH, IGF-I, anabolic steroids) proved useful in at least some of these conditions.

1. INTRODUCTION

At present immunodeficiency is considered to be due to internal defects of the immune system and the possible role of neuroendocrine abnormalities that may play a role is seldom taken into consideration. It is now established that *severe combined immunodeficiency (SCID)* is due to mutations in the common signal transducing chain (γ_c) of type I cytokines (IL-2, -4, -7, -9 and –15), which render this group of cytokines non-functional. Autosomal recessive SCID may be due to adenosine deaminase deficiency, Janus kinase 3 deficiency or recombinase (RAG1 and RAG2) deficiency. Combined immunodeficiency and purine nucleoside phosphorylase deficiency are also combined disorders. In some children the thymus may not develop (diGeorge's syndrome), which leads to the lack of the T-cell dependent adaptive immune system. Defects of T and B lymphocytes, and their subsets, problems with phagocytosis, antigen presentation (e.g. defective MHC expression) and with leukocyte adhesion, immunoglobulin (x-linked agammaglobulinemia, IgA and IgG subclass deficiency, X-linked lymphoproliferative disorder) and complement deficiency all cause partial immunodeficiencies. Interferon γ receptor 1 mutation, lymphocytopenia due to ZAP-70 tyrosine kinase deficiency, defective expression of the TCR-CD3 complex and defective cytokine production also lead to immunodeficiency. An intriguing disease of uncertain etiology is common variable immunodeficiency (acquired hypogammaglobulinemia). Complex disorders are cartilage-hair hypoplasia, Wiscott-Aldrich syndrome, ataxia telangiectasia, Nijman breakage syndrome Chediak-Higashi syndrome, Omenn's syndrome and hyperimmunoglobulinemia E syndrome. Severe deficiencies are lethal, if not corrected [1–3].

2. HORMONES AND IMMUNODEFICIENCY

2.1. Growth and lactogenic hormones (GLH)

Combined immunodeficiency was observed in young hypophysectomized (Hypox) rats. Hypophysectomized rats did not develop cell mediated contact sensitivity reactions and did not produce antibodies after primary immunization. However, the secondary antibody response was only partially inhibited by Hypox. Adjuvant arthritis was also inhibited by Hypox. Bromocriptine (BRC), a dopamine agonist drug, which inhibits pituitary PRL secretion, was able to inhibit immune reactions similarly to Hypox. The immune reactivity of Hypox or BRC treated animals could be fully restored by replacement doses of either PRL or GH. No other pituitary hormones (ACTH, FSH, LH, TSH) were required for immune restoration. Adrenocorticotropic hormone (ACTH) antagonized restoration and replacement doses of other pituitary hormones had no effect on immune function. In Hypox animals the thymus and spleen underwent a profound involution. RNA and DNA synthesis, as established by the incorporation of H^3-thymidine and H^3-uridine, respectively, fell to near background levels, indicating that cell proliferation was at halt in these organs. The situation was identical in the bone marrow. Again, only PRL and GH restored cell proliferation and immunocompetence at replacement doses. No other pituitary hormones (ACTH, FSH, LH, TSH) would influence cell proliferation in these lymphoid organs (Figures 1 and 2). The immune reactivity of rats was in direct and highly significant correlation with their capacity to synthesize DNA and RNA in their lymphoid organs. These data demonstrate clearly that lymphocyte proliferation in lymphoid organs is dependent on pituitary GH and PRL and support strongly the long established fact that immune reactions are dependent on lymphocyte proliferation (i.e. clonal expansion). The only exception to this seems to be that memory lymphocytes are capable of surviving hypophysectomy and to mount a secondary response in the absence of pituitary hormones [4–14]. The immunoregulatory role of GH and PRL has been firmly established during the past two decades in numerous laboratories [15,16].

There seemed to be a contradiction between the long-term survival of Hypox rats (6–8 months) and the profound immunodeficiency that was observed shortly after Hypox. Therefore, the possible presence of residual hormones was investigated in Hypox rats. Indeed, residual PRL (10–20 % of normal) was found in the serum of Hypox rats. If Hypox animals were given daily s.c. injections of a rabbit anti-rat PRL serum, their serum lactogenic activity diminished further. Severe anemia, wasting and immunological anergy developed and death occurred within 8 weeks. Untreated Hypox animals gradually increased their serum lactogenic activity, starting on the 7th week after pituitary removal, which rose up to 50% of control levels by week 9. Hypox animals showed normochromic-normocytic anemia, a grossly reduced immunocompetence, decreased body, thymus, spleen, adrenal, and ovary weights, and decreased DNA and RNA synthesis in the thymus, spleen, and bone marrow. The condition of Hypox animals did not deteriorate further over the 9-week experimental period. All the hematological deficiencies and decrease in organ weights observed in Hypox rats were normalized after grafting with syngeneic pituitaries (SPG) that secrete PRL. The restoring effect of SPG could be inhibited by additional treatment with anti-rat PRL serum. Treatment of Hypox animals with ovine PRL had a restoring effect similar to SPG, which was not inhibited by additional anti-rat PRL serum treatment. Rat and ovine PRL and GH and human placental lactogen all stimulated the incorporation of 3H-thymidine by rat bone marrow cells in vitro. These results demonstrate that PRL alone is capable of supporting multiple organs and of maintaining vital bodily functions for long periods of time [14].

GH has long been shown in animals to stimulate immune and inflammatory reactions. In contrast, clinicians did not find immune abnormalities in pituitary dwarf individuals. On this basis serious doubts were raised about the role of GH in immune function [17–19]. However, pituitary dwarf humans have normal serum PRL levels [20], which explains their normal immune function. Even with modern laboratory approaches it is difficult to demonstrate immune alterations in children after GH therapy, although transient responses can be demonstrated. However, *in vitro* experiments with human lymphocytes indicate the role of GH in immunoregulation [19].

Although much less studied, the evidence available clearly indicates that placental lactogenic hormones (PL) also have the potential of regulating the immune system [7,13,21].

Snell dwarf mice are deficient of the pituitary transcription factor, Pit-1, which controls the production of GH, PRL and of thyroid stimulating hormone (TSH) secretion. These animals have often been presented as completely lacking pituitary GH and PRL secretion. Nevertheless, they maintain near normal immune function. However, low serum level of GH and PRL was detectable in these animals by radioimmunoassay [22–24]. Furthermore, humans with Pit-1 mutations also have subnormal levels of GH, PRL and TSH, and do not show total deficiency [25]. Snell dwarf mice produced less lymphocyte-derived PRL (LPRL) than did their normal littermates. LPRL could be restored to normal levels by thyroxin treatment of Snell lymphocytes *in vitro*. The production of placental lactogen was normal in dwarf mice [26]. Therefore, it appears that Pit-1 deficient mice and humans do in fact, have sufficient pituitary hormone levels, which permit the maintenance of immune function and the function of other organs and tissues at adequate levels, so that survival is possible. The joint and complete deficiency of pituitary GH and PRL has never been demonstrated to date in man or in animals, which is compatible with the experimental observations that joint and complete deficiency is lethal.

Genetically altered (knockout) mice, lacking either PRL or its receptor (PRLR), or IGF-I are immunocompetent. It was concluded on the basis of these observations that these hormones are not obligate immunoregulators, but rather, affect immune reactions as anabolic and stress modulating agents [27–29]. In fact, the data obtained in knockout mice fully confirm, while using a very powerful methodology, that GLH show redundancy in the maintenance of immunocompetence [7–11]. Today a compelling body of experimental evidence indicates that indeed this is the case [16]. Clearly, immune function, as many other functions in the body, are maintained by multiple genes that show redundancy [30].

Growth hormone and PRL share important features with the type-I cytokine family [31]. Functional overlap and redundancy is the norm for type I cytokines (and for other cytokines as well) in the immune system. Type I cytokines receptors consist of a ligand-specific chain and of a shared signal-transducing chain. For instance, in the first group, where IL-2, -4, -7, -9 and -15 belong, there is a common gamma chain (γ_c), for the second group (IL3, -5 and GM-CSF) a common beta chain (β_c) and for the third group (IL-6, -11, oncostatin M, leukemia inhibitory factor, ciliary neurotropic factor and cardiotrophin-1) the common chain is glycoprotein 130. For signal transduction the ligand binding and the signal transducing chains are crosslinked by the specific cytokine. Experiments with knockout mice in this system showed that the elimination of specific cytokines or their specific receptor chains produced minimal if any abnormalities. However, knocking out the shared signal transducing γ_c chain resulted in severe combined immunodeficiency [1–3]. These observations collectively demonstrate that *type I cytokines are indispensable as a group* for normal immune function. Apparently there is enough redundancy in this cytokine group to compensate for the lack of any particular cytokine.

Prolactin and GH do not share receptor chains with any of the above cytokines. However, human GH and other primate GH are known to act on PRL receptors and to exert lactogenic activity [30]. Similarly, IGF-I, IGF-II and insulin show functional overlap in the maintenance of immunocompetence and of other bodily functions [15]. These facts strongly indicate that *functional redundancy* exists within GLH hormones, which explains why the disabling of a single gene is of no consequence for immune function.

The major signal transduction pathway, which involves the Janus kinase (JAK) and signal transducers and activators of transcription (STAT) nuclear regulatory factors, is shared between cytokines and GLH. STAT knockout mice show severe developmental and immune deficiencies [15,31–33]. These facts illustrate that redundancy is present at the receptor-ligand level of the JAK/STAT nuclear regulatory pathway, which includes GLH, type I cytokines and their receptors. The JAK/STAT nuclear regulatory system is of major biological significance not only for normal immune function, but also for the development and normal function of the entire organism [14,15,21,28,31,33–35]. Because the JAK-STAT transcription pathway of PRL and GH is shared with interleukins and hemopoietic growth factors [1,3,15], some regard these hormones as members of the hemopoietic cytokine family. However, GLH have a much wider spectrum of biological activity than any of the type I cytokines. A functional overlap with these cytokines could simply indicate the capacity of GLH hormones to maintain the hemopoietic and immune systems at times when cytokines are in short supply as well as to boost immune activity in situations of emergency, when immune host defence is vital for survival.

In seriously ill patients an acute phase response (APR) develops, which leads to the suppression of the GH-IGF-I axis. This observation prompted several clinical trials with GH treatment, which were aimed at reverting the severe catabolic state and improving immunocompetence in the interest of increased survival. However, so far this hope has not materialized. On the contrary, a controlled clinical trial showed that GH treatment of severely ill patients significantly elevated the proportion of patients that did not survive [36]. Deaths attributed to "septic shock or uncontrolled infection" occurred nearly four times more commonly in GH treated patients compared to placebo receiving patients. Although immune parameters were not followed, the authors suggested that alterations in immune function may have contributed to these fatalities.

During APR the immune system is switched from the adaptive mode of reactivity to the amplification of natural immune mechanisms. The serum levels of IL-1, IL-6 and TNFα are profoundly elevated, which induce complex neuroendocrine and metabolic alterations. The hypothalamic-pituitary-adrenal axis is activated, whereas the serum levels of GH, IGF-I and PRL are suppressed. Tri-iodothyronine is also diminished (sick euthyroid syndrome). The cytokines cause fever, and alter neuroendocrine function, which is followed by catabolism and by the suppression of the T lymphocyte-dependent adaptive immune system. In contrast, the nervous system, the bone marrow, leukocytes and the liver are hyper-active, fueling the natural immune system. There is a rapid rise in serum natural antibodies and liver-derived acute-phase proteins such as endotoxin-binding protein and C-reactive protein. These antibodies and acute phase proteins recognize homologous crossreactive epitopes (homotopes) on microbes and on altered self components in a polyspecific fashion and activate immune defense mechanisms after combining with the respective homotope. Host defense against toxins and other noxious agents are also increased during the APR [37–39].

The findings of Takala and co-workers [36] strongly suggest that the suppression of the GH – IGF-I axis in acutely ill patients is in fact required for the intense catabolism to take place. Under these conditions the defense system of the body, that includes the hypothalamus – pituitary – adrenal axis, the sympathetic nerves system, the bone marrow, CD5+ B lymphocytes, leukocytes and the liver [37–39], must be maximally supported by nutrients and energy. It is suggested that this requirement can only be satisfied by massive catabolic activity in muscle and in other organs and tissues, and not by alimentation. Anabolism is antagonistic to APR and catabolism.

The adaptive immune system is controlled by thymus-derived (T) lymphocytes and needs several days to a week for an effective response of host defence. During APR (e.g. sepsis) no time is available for an adaptive immune reaction, and therefore, this system becomes useless and is shut down, primarily by the cytokine and endocrine alterations that take place. The thymus and T cell function is heavily dependent on the GH/PRL – IGF-I axis and it is suppressed profoundly by the elevated levels of glucocorticoids and catecholamines. TNF and possibly zinc deficiency, contribute to this suppression [37–39]. Recent observations showed that GH inhibits the production of acute phase proteins in rats with burn injury and in human hepatocytes [40,41]. These findings are compatible with the hypothesis that in acute illness catabolism is the only way to cover the nutrient and energy requirements of the innate immune system, which must be maximally activated. GH is a powerful anabolic hormone, which supports the T lymphocyte dependent immune system, and acts as antagonist of the HPA axis that promotes APR [6,8,11,37–39,42]. The results of the controlled trial of Takala et al [36] support the hypothesis that the inhibition of the HPA axis and of catabolism by GH treatment in APR hampers the bodies defence mechanisms, which may have fatal consequences.

The atrophic thymus in old rats may be partially restored by GH [44]. Both GH and PRL maintain adaptive immunocompetence, which is antagonized by the hypothalamus-pituitary adrenal axis [42]. In addition, the pituitary gland is capable of fine tuning the level of immune reactivity and plays a fundamental role in the induction of immunoconversion from adaptive to enhanced natural immune reactivity in APR [5,6,9,37–39].

A compelling body of evidence indicates that after activation by antigen or mitogen lymphocytes produce their own PRL and/or GH. There is evidence to suggest, that this local production makes the rapid proliferation required for an immune response feasible [45–50]. The situation is similar to the development of the embryo, where placental lactogenic hormones make it possible for the embryo to grow at a very rapid rate. The production of placental GLH is independent from the pituitary gland and is controlled by "placental" promoters. This allows placental lactogens to override the regulatory power of the maternal pituitary gland during pregnancy in the interest of assuring the proper development of the fetus. Therefore, while conception is clearly dependent on normal pituitary function, the fetus becomes independent from such influence [35].

The placental promoter is associated with the lymphocyte PRL gene [51]. However, Pit-1 was detected in lymphocytes as well [52]. Therefore, it is possible that once the lymphocyte PRL gene is activated, pituitary PRL is no longer required to support lymphocyte growth or function. When the immune response is over, most of the activated lymphocytes will undergo apoptosis. This is governed by a complex mechanism that involves the delivery of death signals, primarily by the Fas-FasL system. However, a specialized subset called memory cells, will survive [53].

The secondary antibody response shows partial dependence on the pituitary gland. In these experiments the rats were immunized first, hypophysectomized and immunized again. The antibody response to the second stimulus was of similar magnitude to the primary response [6]. This suggests that memory cells maintained their reactivity after Hypox, but the rapid recruitment of naïve lymphocytes, which occurs in normal animals, could not take place. It was demonstrated by Cho and co-workers [54] that in recombinase deficient (RAG-1-/-) mice, which are lymphopenic, naïve T lymphocytes undergo *"homeostasis-stimulated" proliferation*, which is MHC restricted, and develop into memory cells in the absence of antigenic stimulation. These cells acquire the phenotypic and functional characteristics of antigen-induced memory CD8+ T cells. They lyse target cells directly and respond to lower doses of antigen than do naïve cells and secrete IFN faster upon restimulation. Interleukin-2, or co-stimulation by CD28, do not participate and effector cells are not formed during this homeostatic differentiation. These findings indicate that memory T cells may be generated and maintained under the influence of *physiological immunoregulatory mechanisms*, in the complete absence of immune stimulation by antigen, adhesion molecules and cytokines.

Immature thymocytes of rodents are killed by glucocorticoids, whereas mature thymocytes are resistant. The helper, suppressor and killer functions of T lymphocytes and the production of interleukins by them are all inhibited by glucocorticoids. In contrast, memory cells and of cells mediating the graft-versus-host reaction are not inhibited by glucocorticoids [42,55]. Prolactin and GH antagonize the immunosuppressive effects of the ACTH-adrenal axis [6,8,11]. These original observations are now generally accepted by the scientific community [56]. One may suggest on the basis of the above evidence, that naive T cells are maintained in the absence of antigenic stimulation by pituitary GLH, whereas memory T cells are autonomous. Memory cells survive and resist glucocorticoids, most likely because they produce autocrine GLH that enable these cells to survive and to resist adverse conditions, including the APR.

This pattern of obtaining gradual independence by memory cells assures maximal host defence while the maturation and selection process of lymphocytes in the thymus and bone marrow is tightly controlled by the neuroimmune regulatory system. Pituitary GLH is important for the development of lymphocytes and for the maintenance of mature naïve cells in a state of immunocompetence. It is likely that during an immune response lymphocyte-derived GLH assumes a prominent role in the maintenance of lymphocyte proliferation. Finally, it is likely that memory cells rely on autocrine GLH for long term survival and function. This is to be substantiated further experimentally.

2.2. The hypothalamus-pituitary-adrenal axis

The predominant function of the HPA axis is thought to be to restrict immune and inflammatory reactions to certain levels, which are compatible with the safety and survival of the organism. However, there are also indications that physiological levels of glucocorticoids (GC) are in fact required for normal immune function. This suggests that the role of the HPA axis is much wider than having a mere suppressive effect [56].

The thymic epithelium is capable of synthesizing GC, which plays a significant role in the normal development of thymocytes [57,58]. Thymus-derived GC antagonize T lymphocyte antigen receptor (TCR) - mediated cell deletion, thus promoting the survival of thymocytes. The inhibition of thymus GC biosynthesis causes an increase in thymocyte apoptosis. The expression of CD5 on CD4+CD8+ thymocytes (an indicator of TCR-mediated activation) increased in a TCR- and MHC-dependent manner when corticosteroid production or responsiveness was decreased [59].

Physiological concentrations of corticosterone (CON) (the principal GC of the rat), accelerated the anti-T cell receptor antibody-induced growth of rat lymphocytes after 2–3 days in culture, which was followed by inhibited cell growth after 5–7 days. CON increased the expression of the IL-2 receptor α chain on $TCR\alpha\beta^{+}$ on $CD4^{+}$ T cells after 48–72 hours in culture [60]. In human T cells GC induced apoptosis in a Fas-independent, but caspase-dependent manner. On the other hand, GC protected T cells from Fas-mediated apoptosis, which is also caspase-dependent [61,62].

Glucocorticoid treatment of dendritic cells (DC) enhanced the secretion of the anti-inflammatory cytokine IL-10 and the type 2 cytokine IL-5 by T cells [63]. GC increased the endocytic activity of human dendritic cells *in vitro*. This favoured the scavenging of antigen from the external milieu, decreased antigen concentration and availability, and simultaneously inhibited the capacity of DC to stimulate T cells [64].

Glucocorticoids increased the expression of HLA antigens and IFN-γ receptors by monocytes [65,66], induced high affinity IL-1 receptors on human blood lymphocytes [67], upregulated high affinity IL-6 receptors on human epithelial cells [68] and potentiated the induction of Fc-γ receptors in human monocytes by IFN-γ [69]. Low concentrations of GC induced the production of MIF from macrophages. In turn, MIF is capable of overriding the GC-mediated inhibition of cytokine secretion by LPS activated monocytes and to antagonize the protective effect of GC in lethal endotoxemia [70].

Cutaneous application of GC (50 μg diflucortolone valerate in ethanol) to mice on the flank seven times on alternate days, augmented passive cutaneous anaphylactic reaction on the ear skin, which was induced by intravenous applications of monoclonal IgE anti-DNP antibody and local DNP challenge. In contrast, GC inhibited the reactions when applied to the challenge sites. GC induced a late phase but not an early phase cutaneous reaction in mast cell deficient WBB6F1 v/v mice by IgE anti-DNP antibody [71].

Hydrocortisone treatment of mice daily for 2 weeks (0.5 or 1.0 mg/day/mouse) induced severe leukocytopenia in all immune organs. Granulocytes (Gr-1^+ Mac-1^+) were increased in number in the bone marrow and peripheral blood, which was due to an elevated level of myelopoiesis in the bone marrow. The granulocytes were functionally activated as estimated by the Ca^{2+} influx and superoxide production. The proportion of primordial T cells ($CD3^{int}$IL-$2R^+$) in the thymus and in the bone marrow was also increased. Mice that were administered hydrocortisone became susceptible to stress and developed gastric ulcers when they were exposed to restraint stress for 12 h. [72].

GC also contributes to the inhibition of immune and inflammatory reactions by the aqueous humor of the anterior chamber of the eye, which is an immunologically privileged site. In the aqueous humor of mice, rats and humans significant concentrations of GC and little or no cortisol binding globulin were found [73]. GC also contributed to strain specific variations in the immune response and, along with other steroid hormones, to the age-related abnormalities of immune inflammatory reactions [74].

These observations would indicate that the consequences of GC deficiency should be much wider than the development of autoimmune and inflammatory diseases. However, complete GC deficiency is lethal, if not corrected, and for this reason there is no chance for the development of immune deficiency conditions. If produced experimentally by Hypox or adrenalectomy (ADX), such animals show extreme sensitivity to the lethal effect of LPS (500 times increase) or may even die after immune activation by adjuvants, because cytokine production, especially of TNF, rise to very high levels in the absence of HPA function, which has lethal consequences [75,76]

The glucocorticoid response peaks 36 hours following murine cytomegalovirus (MCMV) infection, which is coincident with circulating levels of the cytokines IL-12, IFN-γ, TNF, and IL-6, and is dependent on IL-6 for maximal release. Adrenalectomized mice were more susceptible than intact mice to MCMV-induced lethality, and the increased sensitivity was reversed by hormone replacement. Lack of endogenous GC resulted in increases in IL-12, IFN-γ, TNF, and IL-6 production as well as in mRNA expression for a wider range of cytokines, including IL-1α and IL-1β. TNF, but not IFN-γ, was required for increased lethality in the absence of endogenous hormone [77].

Glucocorticoids are also important participants in the acute phase response, where the adaptive, thymus-dependent immune system is suppressed and the capacity of the innate immune system is amplified. Glucocorticoids, TNF, catecholamines, and possibly zinc deficiency, which is required for GH action, participate in the paralysis of the thymus under these conditions. At the same time interleukin-6, glucocorticoids and catecholamines convert the liver from the synthesis of ordinary serum proteins to the synthesis of acute phase proteins. During APR the serum level of C-reactive protein, which is an important defence molecule, may rise up to 1000 times within 24–48 hrs. [78–80]. Again, HPA axis deficiency should lead to an inefficient host defence in febrile illness and to elevated susceptibility to the lethal consequences of infections.

3.3. Sex hormones

Estradiol, testosterone, the steroid hormone precursor, dehydroepiandrosterone-sulphate (DHEA-S) and progesterone are major immunoregulatory steroid hormones. DHEA-S is of special importance, because it is converted by the immune system to specific bioactive steroids that are required for normal immune function [81]. Deficiencies of sex steroids lead to immune disorder and deficiency as outlined briefly below.

Murine CD4^{+} T cells are dependent on *estrogen* for cytotoxic activity. CD4^{+} T cells fulfil important immunoregulatory function as well as they provide defence against viral infections [53,82]. Sex hormone metabolism, including androgen synthesis, occurs in human lymphocytes [83]. In aging rats the regeneration of the thymus after castration was inhibited by testosterone treatment. Testosterone was converted to estradiol by aromatase and estradiol induced thymic involution. Aromatase inhibitors restored the thymus in aging rats with normal histological appearance. These drugs also enlarged the thymus in young intact animals [84].

Trauma, hemorrhage and shock leads to profound immunosuppression in male animals and in men. This is mediated by *testosterone* and may be corrected by anti-androgenic drugs. The testosterone receptor antagonist, flutamide, restores the depressed immune function in males after trauma and hemorrhage. Flutamide improved plasma corticosterone content, and the diminished ACTH-induced corticosterone release and adrenal cAMP response [85,86]. The sexually dimorphic immune response in young males and females to trauma-hemorrhage reverses as sex hormone levels decline with age [87]. Testosterone-treated female mice that had experienced hemorrhage showed significant depression in splenic and peritoneal macrophage IL-1 and IL-6 production, comparable with the values seen in male mice [88].

Amongst 84 patients that sustained blunt injuries a sex difference in the early posttraumatic cytokine release by peripheral blood mononuclear cells was not detectable. However, in males, the development of severe sepsis later was found to coincide with significantly decreased testosterone and increased estradiol serum levels, whereas in females, such differences in sex hormone levels were not detectable [89]. Estrogen prevented the androgen-induced immunodepression in castrated male mice [87]. Short-term therapy of male mice with the androgen receptor blocker, flutamide at 25 mg/kg BW, following trauma-hemorrhage had protective effects on immune functions [90].

The weak androgen, *dehydroepiandrosterone*, and its metabolites are major immunoregulatory hormones. Cholesterol is converted to pregnenolone (PREG) in the adrenal gland, which serves as a precursor for the synthesis of DHEA. DHEA is regarded as a "mother steroid" from which other hormones of high biological activity, such as *testosterone* (TS), *androstendiol* (AED), and *dehydrotestosterone* (DHT) are derived [91]. Both PREG and DHEA are metabolized by the spleen and that 7α-hydroxy derivatives of DHEA and PREG are 160 times more active in stimulating the antibody response in mice than are the precursor hormones [92]. Androstendiol and *androstentriol* (AET), which are also derived from DHEA, are more potent immunoregulators than DHEA itself. DHEA suppressed Con-A or LPS-driven lymphocyte proliferation, AED had little influence and AET potentiated these responses. The secretion of IL-2 and IL-3 by Con-A activated lymphocytes was depressed by DHEA and potentiated by AET. The suppression of Con A induced lymphocyte proliferation by hydrocortisone was strongly antagonized by AET, AED showed a weak anti-glucocorticoid effect and DHEA had no effect [93].

DHEA levels decrease during chronic illness and dexamethasone also suppresses plasma levels of DHEA-sulphate (DHEA-S), androstendiol and testosterone in patients. ACTH treatment corrects this suppressive effect [91]. DHEA-S levels are depressed in postmenopausal women with rheumatoid arthritis [94].

Serum DHEA declines progressively after the second decade by approximately 20% per decade and will decrease to 10–15% of maximum by age 85–90. The immune system is gradually disregulated in ageing animals and man, which is referred to as *immune senescence*. The proportion of $CD5^+$ B cells is increased with age, which produce natural antibodies with frequent autoreactivity. Th1 type helper cell activity and the production of IL-2 decline, whereas the activity of Th2 helper cells and the production of IL-4, -5 and -6 increase. Treatment of aged animals with DHEA-S led to the reversal of immunosenescence and restored the reactivity of old animals [74 , 95–97].

The activity of IFNγ-secreting cells correlated with serum DHEA-S levels in men and pre-menopausal women. Post-menopausal women had fewer cells secreting cytokines and the activity of these cells did not correlate with sex hormone levels [98]. Postmenopausal women received physiologic doses of DHEA for 3 weeks in a prospective randomized double blind crossover study, separated by a 2 week washout period. A decrease was observed in $CD4^+$ T cells and increase in $CD8^+/CD56^+$ (natural killer) cells in treated individuals. T cell mitogenesis and IL-6 responses were inhibited, whereas NK cell cytotoxicity was increased dramatically by DHEA [99].

Progesterone (PS) plays a fundamental role in the protection of the fetus against maternal immune reactions. Amongst infertile women treated with the fertility drug, clomiphene citrate, the only women who conceived were those in whom an immunosuppressive effect of the drug was present, whereas women with no immunosuppressive effect did not achieve pregnancy [100]. In women suffering from primary habitual abortions the incidence of pregnancies after immunotherapy was increased when PS treatment was also given, suggesting a synergistic effect [101].

In female patients with thyroid and ovarian autoimmune disease and rheumatoid arthritis progesterone deficiency is present [102]. In three patients the severe pre-menstrual exacerbation of asthma was reported which did not respond to treatment with high dose glucocorticoids. Treatment by intravenous PS corrected the decreased airflow and led to the reduction of daily doses of prednisone [103].

3. NUTRITION

Undernutrition in rats during the very critical periods of gestation and lactation, induces in their male pups at weaning: (a) reduced circulating leptin levels and body weight values; (b) metabolic adaptation to normal carbohydrate metabolism; (c) hyporesponsiveness of the HPA and immune (TNFα) axes during endotoxemia; and (d) leptin resistance at the adrenocortical level [104]. Leptin deficiency increases susceptibility to infectious and inflammatory stimuli and is associated with dysregulation of cytokine production and of hematopoiesis. Leptin promotes Th1 mediated cellular immunity. Low leptin levels regulate the neuroendocrine and immune alterations during starvation [105].

4. ACQUIRED IMMUNODEFICIENCY SYNDROME

The functional derangement of every endocrine organ system has been reported in association with HIV infection. Some of the abnormalities are similar to the ones seen in other systemic illness, whereas others are unique to HIV infection. The clinical significance of many of these abnormalities is not well understood [106].

In HIV infected men fasting bioactive (but not immunoreactive) PRL was higher ($P = 0.03$), but the stimulated PRL (both immunoreactive and bioactive) was lower than in healthy men ($P <$ or $= 0.01$). Fasting serum TSH was normal while its response to metoclopramide was absent in patients but not in controls ($P = 0.049$). A 23.5-kD PRL was the predominant circulating isoform both in patients and healthy controls. CD4+ T lymphocytes correlated negatively with fasting bioactive PRL ($P = 0.008$) and positively with the area under the PRL (both immunoreactive and bioactive) curves ($P < 0.001$). The viral load negatively correlated with the area under the curve of the bioactive/immunoreactive PRL ratio ($P = 0.008$). These results suggest the presence of a decreased, but not absent dopaminergic tone in patients [107,108].

In women with acquired immune deficiency wasting syndrome a significant shift of adrenal steroid metabolism away from androgenic pathways and toward cortisol production was observed. The ovarian androgen response to hCG stimulation was intact [109]. Of 63 very ill AIDS patients 12 (19%) had adrenal insufficiency [110]. Adrenocortical insufficiency occurs in approximately 20% of the cases with HIV disease [111]. Serum cortisol and ACTH values and CD4 cell count were evaluated in 25 perinatally HIV-1-infected children. Only one of eight severely ill children had primary adrenal insufficiency. Mean CD4 cell count of each group declined but no correlations were found between cortisol or ACTH values and CD4 cells. These data argue against the hypothesis of a cortisol-induced shift from T-helper-1 (Th1) to Th2 cytokine production profile as the pathogenetic mechanism of progression to AIDS [112].

Androgen deficiency is common among men and women with HIV infection. Low testosterone concentrations are associated with lower CD4 cell count, advanced stage of illness, medication use, and weight loss. Serum bioavailable (free) testosterone concentration is the most useful laboratory indicator. A number of testosterone preparations for treatment of androgen deficiency in HIV-infected men now exist. Preliminary studies suggest that use of physiological testosterone administration is of benefit in HIV-infected women [113].

In 11 children with HIV infection hypothyroidism indicated by elevated basal TSH and abnormal TRH response was common. This may have contributed to failure of growth. Replacement therapy resulted in correction of abnormal TSH and improvement of height z scores [114].

Polyunsaturated fats, fiber, and alcohol intake are strongly associated with insulin resistance and hyperlipidemia in human immunodeficiency virus (HIV)-infected patients with fat redistribution and may be important targets for dietary modification [115]. Central diabetes insipidus caused by viral infections has been reported in immunosuppressed patients, such as those with AIDS. Herpes simplex virus caused diabetes insipidus in a patient immunosuppressed by Cushing's syndrome [116].

Malnutrition is a leading cause of morbidity in HIV disease. Early nutritional interventions are necessary to prevent or ease its impact [117]. GH treatment of HIV-infected patients with fat accumulation at 3 mg/d resulted in a decrease in total body fat and an increase in lean body mass. While insulin sensitivity and glucose tolerance initially worsened, they subsequently improved toward baseline [118]. Toxic epidermal necrolysis (TEN) is a serious drug reaction that is occurring with increased frequency among those affected with HIV and AIDS Nd. It has a grave prognosis if advanced. It may be managed with wound care, intravenous nutrition, human GH, and G-CSF [119]. The specific therapies recommended for HIV-associated wasting include testosterone replacement, other anabolic steroids, and recombinant human growth hormone. Other adjunctive measures, such as progressive resistance exercise and cytokine modulation, may also be utilized [120].

Androgenic anabolic steroids (AAS) are used for the treatment of male hypogonadism. AAS therapy have a favourable anabolic effect on patients with chronic diseases and muscle catabolism and has been recommended for the treatment of patients with acquired immunodeficiency syndrome wasting and in severely catabolic patients with severe burns and possibly also in renal failure-associated wasting [121].

5. AGEING

Studies in mice established a relationship between immune function and longevity [122]. The mechanisms underlying age-related changes in immunity are likely to be multi-factorial, with both genetic and environmental factors playing a significant role [123]. *Immunosenescence* is reflected by a sum of disregulations of the immune system and its interaction with other systems, many of which would appear to implicate age-related deficiencies of the immune responses. Several studies suggest an association between immune function and individual longevity. There are observations, especially in very old healthy people that several immune functions are unexpectedly well preserved and substantially comparable to those observed in young subjects. These findings raise the question of whether the alterations that can be observed in the immune parameters of the elderly are a cause or a result of underlying disease processes. Studies on centenarians revealed a remodelling of the immune system rather than deterioration, suggesting that the changes observed during immunosenescence do not correspond to immunodeficiency. Age-related T-cell immune deficiency is preceded by thymic involution and is influenced by several intrinsic and extrinsic factors. The role of monocytes and macrophages in T-cell activation changes with advancing age [124].

Immune ageing manifests in the progressive appearance of immune dysregulations, mainly of acquired immunity leading to gradual increases in TH2:TH1 cell ratio. Decreased thymic function, and later accumulative antigen pressure lead to these changes over the lifespan. Innate immunity (macrophage function) is preserved during the ageing process and in the elderly this leads to macrophage-lymphocyte dysequilibrium, which is particularly critical during on-going disease. Indeed, any disease may induce long-lasting acute-phase reactions in aged patients. This leads to body nutritional reserve (mainly protein) losses. Episodes of disease in the aged patient progressively deplete body nutritional reserves and lead to protein-energy malnutrition, undernutrition-associated immunodeficiency, and finally cachexia. Protein-energy malnutrition is found in more than 50% of hospitalized elderly patients and in most elderly diseased subjects. Micronutrient deficit or low levels are common in home-living self-sufficient apparently-healthy elderly subjects [125].

Gastrointestinal mucosal immunosenescence reflects deficits in: (i) the differentiation and/or migration (homing) of immunoglobulin A immunoblasts to the intestinal lamina propria, and (ii) the initiation and/or regulation of local antibody production. (iii) There is a significant age-related increase in the incidence and severity of gastrointestinal infectious diseases [126].

Prolactin and GH function as lymphoid growth and differentiation factors. Long-standing hyper- or hyposecretion of PRL or GH does not induce major alterations in the immune system. Subclinical changes in chemotaxis or phagocytosis by granulocytes or macrophages or natural killer cell activity have been reported in some of these conditions. The GH-IGF-I axis is dysregulated in ageing, in catabolic states and in critical illness. Hyperprolactinaemia may play an aggravating role in systemic lupus erythematosus, in autoimmune thyroiditis and in other autoimmune diseases [127].

In healthy nonagenarians and centenarians the highest number of NK cells and the best preserved cytolytic function was found in individuals with a preserved metabolism of thyroid hormones and vitamin D and integrity of muscle mass. The NK cell number and/or cytolytic activity of healthy subjects over 90 years old was positively correlated with serum levels of vitamin D, while T3, FT4, i-PTH hormones and lean body mass were associated only with NK cell number [128]. In postmenopausal patients with rheumatoid arthritis IL-6 positively correlated with increased bone resorption and with high disease activity. Serum levels of VD3 inversely correlated with T-cell activation [129].

Thymectomy in mice not only reduces the immune response, but also deteriorates learning performances. Cytokines such as IL-1, IL-6 and FNF and corticosterone affect the induction of hippocampal long-term potentiation, which is a synaptic model of memory. Alzheimer's patients show disordered immune function in addition to cognitive deficit. The brain lesions of Alzheimer's patients may be associated with abnormal immune reactions occurring in the brain [130]. It is emerging that the human thymus is functional well into the sixth decade and plays a role throughout life to optimize immune function [131]. The human thymus may be considered as a chimeric organ comprised of both central and peripheral lymphoid tissues. It was postulated that the thymic epithelial atrophy and decrease in thymopoiesis that occurs in myasthenia gravis, HIV-1 infection, and thymic involution may in part derive from cytokines or other factors produced by peripheral immune cells within the thymic perivascular space [132]. Age-related changes in the thymus are quantitative, not qualitative. The increases in naive T-cell numbers in patients receiving antiretroviral therapy for AIDS are largely derived from the thymus. This provides direct evidence for the functional capacity of the adult thymus [133].

T cells show the largest age-related differences in distribution and function. The antibody production capacity of B cells also shows an age-related decline. Habitual physical activity may enhance NK cell activity and check certain aspects of the age-related decline in T cell function [134]. Long-term memory to viruses that establish latency and to repeatedly encountered viruses such as influenza may be severely impaired by "replicative senescence", a genetically programmed process affecting most somatic cell types of human origin. Memory CD8+ T cells with hallmarks of replicative senescence have been identified in vivo. Such cells may contribute to compromised viral immunity and response to vaccines. Furthermore, the presence of such cells may negatively influence homeostatic mechanisms that control the size of the memory T cell pool in elderly persons [135]

Mutation makes essentially no contribution to the diversity of endogenous serum IgM, IgG, or IgA in young mice. The titer of mutated immunoglobulin in T cell-proficient mice rises strikingly with age, such that the major proportion of serum Ig in adults is somatically mutated, with the mutation load in IgG being some 10-fold greater than in IgM [136]. The analysis of circulating T lymphocytes from 120 healthy subjects ranging in age from 18 to 105 years revealed that naive T cells decreased sharply with age. The exhaustion of the naive CD8(+) T-cell reservoir suggests that this T-cell pool is a major target of the aging process and may define a parameter possibly related to the life span of humans [137].

The Fas/FasL pathway plays an important role in regulating the life and death of the immune cells and has been implicated in aging, human immunodeficiency virus infection, drug abuse, stress, and cancer development. Deregulated Fas/FasL interaction leads to diseases [138].

6. CONCLUSIONS

It is clear from this chapter that immunodeficiency develops frequently under the influence of disease, starvation and of other noxious insults as well as during ageing. In most cases initially there is normal neuroimmune function, which may grow gradually deficient as time goes on. It is also possible that inherited abnormalities of the neuroendocrine system would cause immunodeficiency, as it was shown in experimental animals. However, this is not observed very frequently in clinical practice either because such mutations are lethal (e.g. babies with the lack of pituitary gland are not viable; this should apply for joint GH/PRL deficiency as well), or the deficiency must be compensated (adrenal, thyroid) for other reasons and the immunological effect of such treatment is never recognized.

It is now well established that the GH-IGF-I axis gradually deteriorates with ageing, although it is not entirely sure whether this is inevitable with aging, or if it is due to disease and to the frequent appearance of APR. The energy and metabolic requirements of APR can only be satisfied by the rapid brake-down of body components in order to fuel the nervous system, the bone marrow, liver and of leukocytes, which are major participants in this emergency defence reaction. The frequent occurrence of APR will then lead to malnutrition, which may develop into cachexia. It is entirely feasible that the deficiency of the GH/IGF-I axis is disease related in the first place because this axis is suppressed during APR. This can lead to the decline of the T cell compartment and of thymic atrophy, creating a vicious circle that leads to more and more severe problems in the long run. More information is needed on these problems.

Current evidence clearly suggests that the age related decline of DHEA-S creates a shortage of this steroid precursor, which is deleterious for immune function and indeed this should create problems for the entire organism. However, it is also clear that the relationship between DHEA and immune function is complex. For one thing the immune system must be able to be in proper condition for the synthesis of bioactive derivatives from this precursor hormone. If this capability is missing or if it is impaired, no favourable results can be hoped from DHEA treatment, as it has been observed repeatedly. Again, more information is required on this point.

The estrogen deficit that occurs after menopause clearly creates immunological abnormalities. This may also have to do with the gradual decline of T cell function as estradiol was shown to be required for CD4+ T cell mediated cytotoxicity and also for normal thymus function. Again information is missing on this problem.

Lymphocytes are capable of producing both GH and PRL. The PRL gene was shown to have placental regulatory sequences in B cells. The production of placental lactogenic hormones is independent of maternal regulatory influences, which secures optimal conditions for the development of the fetus, even against the mother's interests [35]. It is possible that memory lymphocytes similarly grow independent from the pituitary gland during their "homeostatic" differentiation by expressing the placental-type PRL gene. Such autonomous memory cells will than be very effective in providing host defence. It is now established that the percentage of genetically diversified immunoglobulin and of memory T lymphocytes increase in the serum with age, which necessitates that germ-line immunoglobulin and naïve T cells should decline [136,137]. As the percentage of memory T and B lymphocytes increase, host defence becomes very efficient, and the immunological role of the thymus and of the bone marrow decreases. This could be the true cause of thymic atrophy. Clearly, the above outlined changes need not be looked at as pathological, rather, they should be regarded as the age-related physiological re-modeling of the immune system [124]. Indeed, long-lived human beings maintain their metabolic and immune function along with muscle mass. Thyroid function and vitamin D levels are preserved. The preservation of lean mass attest for normal anabolic hormone levels (e.g. GH-IGF-I and steroids) as well [129]. These observations, together with the findings that the atrophic thymus may be functionally re-activated at any age [131–133], suggest indeed that the age related decline of immunity is not a physiological necessity, but rather, it is due to pathological conditions. If this is the case, then it must be preventable. Indeed, patients with wasting syndrome are treated already with some success by sophisticated alimentation and with anabolic hormones, such as GH and steroids.

In general the neuroendocrine and metabolic alterations in AIDS patients comply with the changes observed in APR, although it is clear that HIV itself is capable of causing specific alterations. Infectious agents can cause immunodeficiency and immunodeficiency may lead to more infectious disease, which lead to frequent APR, wasting cachexia and death. This is the usual fate of patients with AIDS. The current treatment of AIDS patients with nutrients and with anabolic hormones has only palliative value.

That thymectomy in mice seriously affected the learning process and memory [130] is most interesting. This finding is highly relevant to the ageing process and truly enlightens again the inter-related function and significance of the neuroimmune regulatory network, which governs the function of the entire organism for life.

REFERENCES

1. Warren JL. Type I cytokines, interferons and their receptors. In: Paul WE, editor. Fundamental Immunology 4th edition. New York: Lippincott-Raven, 1999;741–74.
2. Buckley RH. Primary immunodeficiency diseases. In: Fundamental Immunology, 4th edition, Paul WE, editor, Lippincott-Raven, Philadelphia, 1999;1427–153.
3. Notarangelo LD, Gilliani S, Mazza C, Mella P, Savoldi G, Rodriguez-Perez C, Mazzolari E, Fiorini M, Duse M, Plebani A, Ugazio AG, Vihinen M, Candotti F, Schumacher RF. Of genes and phenotypes: the immunological and molecular spectrum of combined immune deficiency. Defects of the gamma (c) -JAK3 signaling pathway as a model. Immunol Rev 2000;178:39–48.
4. Nagy E, Berczi I. Immunodeficiency in hypophysectomized rats. Acta Endocrinol. 1978;89:530–537.
5. Nagy E, Berczi I. Prolactin and contact sensitivity. Allergy 1981;36:429–431.
6. Berczi I, Nagy E, Kovacs K, Horvath E. Regulation of humoral immunity in rats by pituitary hormones. Acta Endocrinol 1981;98:506–513.
7. Nagy E, Berczi I, Friesen HG. Regulation of immunity in rats by lactogenic and growth hormones. Acta Endocrinol 1983;102:351–357.
8. Berczi I, Nagy E, Asa SL, Kovacs K. Pituitary hormones and contact sensitivity in rats. Allergy 1983;38:325–330.
9. Nagy E, Berczi I, Wren GE, Asa SL, Kovacs K. Immunomodulation by bromocriptine. Immunopharmacology 1983;6:231–243.
10. Nagy E, Berczi I. Pituitary dependence of bone marrow function. Brit J Haematol. 1989;71:457–462.
11. Berczi, I., E. Nagy, S.L. Asa, K. Kovacs. The influence of pituitary hormones on adjuvant arthritis. Arthritis Rheum. 27:682–688, 1984.
12. Nagy E, Berczi I. Immunodulation by tamoxifen and pergolide. Immunopharmacology 1986;12:145–153.

13. Berczi I, Nagy E. Placental lactogen is a hemopoietic hormone. Brit J Haematol 79: 355–358, 1991.
14. Nagy E, Berczi I. Hypophysectomized rats depend on residual prolactin for survival. Endocrinology 1991;128:2776–84.
15. Berczi I, Szentivanyi A. Growth and lactogenic hormones, insulin-like growth factor, insulin. In: Szentivanyi A, Berczi I, editors. The Immune-Neuroendocrine Circuitry, Neuroimmune Biology, Vol. 3, Amsterdam:Elsevier, 2003;129–153.
16. Matera L, Rapaport R, editors. Growth and lactogenic hormones. In: Berczi I, Szentivanyi A, series editors. Neuroimmune Biology, Volume 3. Amsterdam: Elsevier, 2002.
17. Selye H. Stress and disease. Science, 1955; 122:625.
18. Berczi I. The effect of growth hormone and related hormones on the immune sytem. In: Berczi I, editor. Pituitary function and immunity. Boca Raton, FL: CRC Press, 1986; 133–159.
19. Rapaport R. Growth hormone therapy and immune function. In: Berczi I, Gorczynsky R, editors. New foundation of Biology. NeuroImmune Biology Vol. 1, Amsterdam:Elsevier, 2001; 403–408.
20. Kaplan SL, Grumbach MM, Friesen HG, Costom BH. Thyrotropin-releasing factor (TRF) effect on secretion of human pituitary prolactin and thyrotropin in children and in idiopathic hypopituitary dwarfism: further evidence for hypophysiotropic hormone deficiencies. Journal of Clinical Endocrinology and Metabolism 1972, 35:825–830.
21. Soares MJ. PRL-like protein (PLP-A) as an embryonic-maternal linker between trophoblast cells and uterine NK lymphocytes.
22. Sinha YN, Salocks CB, Vanderlaan WP. Pituitary and serum concentrations of prolactin and GH in Snell dwarf mice. Proc Soc Exp Biol Med 1975;150:207–10.
23. Barkley MS, Bartke A, Gross DS, Sinha YN. Prolactin status of hereditary dwarf mice. Endocrinology 1982;110:2088–96.
24. Soares MJ, Bartke A, Colosi P, Talamantes F. Identification of a placental lactogen in pregnant Snell and Ames dwarf mice. Proc Soc Exp Biol Med 1984;175:106–8.
25. Pfaffle RW, Blankenstein O, Wuller S, Kentrup H. Combined pituitary hormone deficiency: role of Pit-1 and Prop-1. Acta Paediatr Suppl 1999;88:33–41.
26. Gala RR. The influence of thyroxine, growth hormone and prolactin alone and in combination on the production of prolactin-like activity by splenocytes from Snell dwarf mice. Life Sci 1995;57:113–22.
27. Horseman ND, Zhao W. Montecino-Rodrigez E, Tanaka M, Nakashima K, Engle SJ, Smith F, Markoff E, Dorshkind K. Defective mammopoiesis, but normal hematopoiesis, in mice with a targeted disruption of the prolactin gene. EMBO J 1997;16:6926–35.
28. Bole-Feysot C, Goffin V, Edery M, Binart N and Kelly PA. Prolactin (PRL) and its receptor: Actions, signal transduction pathways and phenotypes observed in PRL-receptor knockout mice. Endocrine Rev 1998; 19:225–68.
29. Dorshkind K, Horseman ND. The roles of prolactin, growth hormone, insulin-like growth factor-I, and Thyroid hormones in lymphocyte development and function. Insights from genetic models of hormone and hormone receptor deficiency. Endocrine Rev 21:292–312, 2000.
30. Berczi, I. Neuroimmune Biology – An introduction. In: Berczi I, Gorczynski R, editors. New Foundation of Biology; Neuroimmune Biology Series, vol. 1, Elsevier, 2001;3–45.
31. Freeman ME, Kanyicska B, Lerant A, Nagy G. Prolactin: structure, function, and regulation of secretion. Physiol Rev 2000;80:1523–631.

32. Ivashkiv LB. Jak-STAT signaling pathways in cells of the immune system. Rev Immunogenet 2000;2:220–30.
33. Berczi I. The role of the growth and lactogenic hormone family in immune function. Neuroimmunomodulation 1994;1:201–216.
34. Baines MG. Olfactory stimuli and allo-recognition. In: Berczi I, Gorczynski R, editors. New Foundation of Biology, Neuroimmune Biology, Vol. 1. 2001; 313–329.
35. Handwerger S, Freemark M. The roles of placental growth hormone and placental lactogen in the regulation of human fetal growth and development. J Pediatr Endocrinol Metab 2000;13:343–56.
36. Takala J, Ruokonen E, Webster NR, et al. Increased mortality associated with growth hormone treatment in critically ill adults. N Engl J Med 1999;341:785–92.
37. Berczi I. Neurohormonal host defence in endotoxin shock. Ann NY Acad Sci 1998;840: 787–802.
38. Haeryfar SMM, Berczi I. The thymus and the acute phase response. Cell Mol Biol 2001;47:145–156.
39. Berczi I, Bertok L, Chow D. Natural immunity and neuroimmune host defence. Ann NY Acad Sci 2001;917:248–257.
40. Jeschke MG, Herndon DN, Wolf SE, et al., Recombinant human growth hormone alters acute phase reactant proteins, cytokine expression, and liver morphology in burned rats. J Surg Res 1999; 83:122–29.
41. Derfalvi B, Igaz P, Fulop KA, Szalai C, Falus A. Interleukin-6-induced production of type II acute phase proteins and expression of junB gene are downregulated by human recombinant growth hormone in vitro. Cell Biol Int 2000;24:109–14.
42. Berczi I. The influence of pituitary-adrenal axis on the immune system. In: Berczi I, ed. Pituitary function and immunity. Boca Raton, FL: CRC Press, 1986; 49–132.
43. Berczi I, Nagy E, Matusik RJ, Friesen HG. Pituitary hormones regulate c-myc and DNA synthesis in lymphoid tissue. J Immunol 1991;146:2201–2206.
44. Kelley KW, Brief S. Westly HJ. GH_3 pituitary adenoma cells can reverse thymic aging in rats. Proc Natl Acad Sci USA 1986;83:5663–5667.
45. Montgomery DW, Zukoski CF, Shah GN, Buckley AR, Pacholczyk T, Russell DH: Concanavalin A-stimulated murine splenocytes produce a factor with prolactin-like bioactivity and immunoreactivity. Biochem Biophys Res Commun 1987;145:692–698.
46. Hartmann DP, Holaday JW, Bernton DW. Inhibition of lymphocyte proliferation by antibodies to prolactin. FASEB J 1989;3:2194–2202.
47. Clevenger CV, Russell DH, Appasamy PM, Prystowsky MB. Regulation of interleukin-2 driven T-lymphocyte proliferation by prolactin. Proc Natl Acad Sci USA 1990;87:6460–6464.
48. Weigent D. GH-IGFI production by cells of the immune system. In: Matera L, Rapaport R, editors. Growth and lactogenic hormones. Neuroimmune Biology, Vol. 2, Berczi I, Szentivanyi A, Series editors. Amsterdam:Elsevier, 2002;87–100.
49. de Vos S, Dogusan Z, Hooghe-Peeters E. Signal transduction and modulation of gene expression by prolactin in human leukocytes. In: Matera L, Rapaport R editors. Growth and lactogenic hormones. Neuroimmune Biology, Vol. 2, Berczi I, Szentivanyi A, series editors. Amsterdam:Elsevier, 2002;123–136.
50. Sabharwal P, Glaser R, Lafuse W, Varma S, Liu Q, Arkins S, Kooijman R, Kutz L,Kelley KW, Malarkey WB. Prolactin synthesized and secreted by human peripheral blood mononuclear cells:an autocrine growth factor for lymphoproliferation. Proc Natl Acad Sci

U S A 1992;89:7713–6.
51. DiMattia GE, Gellersen B, Duckworth ML, Friesen HG. Human prolactin gene expression: The use of an alternative noncoding exon in decidua and the IM-9-P3 lymphoblastoid cell line. J Biol Chem 1990;265:16412–421.
52. Delhase M, Vergani P, Malur A, Hooghe Peters EL, Hooghe RJ. The transcription factor Pit-1/HGF-1 is expressed in hematopoietic and lymphoid tissues. Eur J Immunol 1993;23: 951–955.
53. Berczi I. Szentivanyi A. Immune reactions. In: Berczi I, Szentivanyi A, editors. The immune-neuroendocrine circuitry – History and progress. Neuroimmune Biology vol. 3. Berczi I, Szentivanyi A, series editors. Amsterdam:Elsevier, 2003;315–377.
54. Cho BK, Rao VP, Ge Q, Eisen HN, Chen J. Homeostasis-stimulated proliferation drives naive T cells to differentiate directly into memory T cells. J Exp Med 2000 21;192:549–556.
55. Claman HN. Corticosteroids-immunologic and anti-inflammatory effects. In: Berczi I, Kovacs K, editors. Hormones and immunity. Lancaster, UK: MTP Press, 1987;38–42.
56. Berczi I, Szentivanyi A. The Hypothalamus-Pituitary-Adrenal Axis and Opioid Peptides. In : Berczi I, Szentivanyi A, editors. The immune-neuroendocrine circuitry – History and progress. Neuroimmune Biology vol. 3. Berczi I, Szentivanyi A, Series Ed. Elsevier, Amsterdam, 2002.
57. Vacchio MS, Papadopoulos V, Ashwell JD, Steroid production in the thymus: implications for thymocyte selection. J Exp Med 1994;179: 1835–46.
58. Tolosa E, King LB, Ashwell JD. Thymocyte glucocorticoid resistance alters positive selection and inhibits autoimmunity and lymphoproliferative disease in MRL-lpr/lpr mice. Immunity 1998;8:67–76.
59. Vacchio MS, Lee JY, Ashwell JD. Thymus-derived glucocorticoids set the thresholds for thymocyte selection by inhibiting TCR-mediated thymocyte activation. J Immunol 1999;163:1327–33.
60. Wiegers GJ, Labeur MS, Stec IEM, Klinkert WEF, Holsboer F, Reul JMHM. Glucocorticoids accelerate anti-T cell receptor-induced T cell growth. J Immunol 1995;155: 1893–1902.
61. Riccardi C, Cifone MG, Migliorati G. Glucocorticoid hormone-induced modulation of gene expression and regulation of T-cell death: role of GITR and GILZ, two dexamethasone-induced genes. Cell Death Differ. 1999;6:1144–5.
62. Zipp F, Wendling U, Beyer M, Grieger U, Waiczies S, Wagenknecht B, Haas J,Weller M. Dual effect of glucocorticoids on apoptosis of human autoreactive and foreign antigen-specific T cells. J Neuroimmunol 2000;110:214–22.
63. Vieira PL, Kalinski P, Wierenga EA, Kapsenberg ML, de Jong EC. Glucocorticoids inhibit bioactive IL-12p70 production by in vitro-generated human dendritic cells without affecting their T cell stimulatory potential. J Immunol 1998;161:5245–51.
64. Piemonti L, Monti P, Allavena P, Leone BE, Caputo A, Di Carlo V. Glucocorticoids increase the endocytic activity of human dendritic cells. Int Immunol 1999;11:1519–26.
65. Gerrard TL, Cupps TR, Jurgensen CH, Fauci AS. Increased expression of HLA-DR antigens in hydrocortisone-treated monocytes. Cell Immunol 1984;84: 311–316.
66. Strickland RW, Wahl LM, Finbloom DS. Corticosteroids enhance the binding of recombinant interferon-gamma to cultured human monocytes. J Immunol 1986;137: 1577–1580.
67. Akahoshi T, Oppenheim JJ, Matsushima K. Induction of high-affinity interleukin 1 receptor on human peripheral blood lymphocytes by glucocorticoid hormones. J Exp Med

1988;167: 924–936.
68. Snyers L, De Wit L, Content J. Glucocorticoid up-regulation of high-affinity interleukin 6 receptors on human epithelial cells. Proc Natl Acad Sci 1990;87: 2838–2842.
69. Girard MT, Hjaltadottir S, Fejes-Toth AN, Guyre PM. Glucocorticoids enhance the - interferon augmentation of human monocyte immunoglobulin G Fc receptor expression, J Immunol 1987;138: 3235–3241.
70. Calandra T, Bernhagen J, Metz CN, Spiegel LA, Bacher M, Donnelly T, Cerami A, Bucala R. MIF as a glucocorticoid-induced modulator of cytokine production. Nature 1995;377: 68.
71. Katayama I, Igawa K, Minatohara K, Nishioka K. Topical glucocorticoid augments IgE-mediated passive cutaneous anaphylaxis in Balb/C mice and mast cell deficient WBB6F1 v/v mice. Clin Exp Allergy 1997;27:1477–83.
72. Maruyama S, Minagawa M, Shimizu T, Oya H, Yamamoto S, Musha N, Abo W, Weerasinghe A, Hatakeyama K, Abo T. Administration of glucocorticoids markedly increases the numbers of granulocytes and extrathymic T cells in the bone marrow. Cell Immunol 1999;194:28–35.
73. Knisely TL, Hosoi J, Nazareno R, Granstein RD. The presence of biologically significant concentrations of glucocorticoids but little or no cortisol binding globulin within aqueous humor: relevance to immune privilege in the anterior chamber of the eye. Invest Ophthalmol Vis Sci 1994;35:3711–23.
74. Daynes RA, Araneo BA, Hennebold J, Enioutina E, Mu HH. Steroids as regulators of the mammalian immune response. J Invest Dermatol 1995;105: 14S–19S.
75. Ramachandra, RN, Sehon, AH, Berczi I. Neuro-hormonal host defence in endotoxin shock. Brain Behav Immun 1992;6:157–169.
76. Berczi I, Szentivanyi A. The acute phase response. In : Berczi I, Szentivanyi A, editors. The immune-neuroendocrine circuitry – History and progress. Neuroimmune Biology vol. 3. Berczi I, Szentivanyi A, series editor. Amsterdam: Elsevier, 2003;463–494.
77. Ruzek MC, Pearce BD, Miller AH, Biron CA. Endogenous glucocorticoids protect against cytokine-mediated lethality during viral infection. J Immunol 1999;162:3527–33.
78. Haeryfar SMM, Berczi I. The thymus and the acute phase response. Cell Mol Biol 2001;47:145–156.
79. Berczi I. The influence of pituitary-adrenal axis on the immune system. In: Berczi I, editor. Pituitary function and immunity. Boca Raton, FL:CRC Press, 1986;49–132.
80. Berczi I, Bertok L, Chow D. Natural immunity and neuroimmune host defence. Ann NY Acad Sci. 2001;917:248–257.
81. Berczi I, Szentivanyi A. Steroid hormones. In: Berczi I, Szentivanyi A, editors. The immune-neuroendocrine circuitry – History and progress. Neuroimmune Biology, Vol. 3. Berczi I, Szentivanyi A, series editors. Amsterdam:Elsevier, 2003;221–270.
82. Muller D, Chen M, Vikingsson A, Hildeman D, Pederson K. Oestrogen influences $CD4^+$ T-lymphocyte activity in vivo and in vitro in 2-microglobulin-deficient mice. Immunology 1995;86: 162–7.
83. Zhou Z, Shackleton CH, Pahwa S, White PC, Speiser PW. Prominent sex steroid metabolism in human lymphocytes. Mol Cell Endocrinol 1998;138:61–9.
84. Greenstein BD, de Bridges EF, Fitzpatrick FTA. Aromatase inhibitors regenerate the thymus in aging male rats. Int J Immunopharm 1992;14: 541–53.
85. Ba ZF, Wang P, Koo DJ, Zhou M, Cioffi WG, Bland KI, Chaudry IH. Testosterone receptor blockade after trauma and hemorrhage attenuates depressed adrenal function. Am J Physiol

Regul Integr Comp Physiol 2000;279:R1841–8.
86. Messingham KA, Shirazi M, Duffner LA, Duffner LA, Emanuele MA, Kovacs EJ. Testosterone receptor blockade restores cellular immunity in male mice after burn injury. J Endocrinol 2001;169:299–308.
87. Kahlke V, Angele MK, Schwacha MG, Ayala A, Cioffi WG, Bland KI, Chaudry IH.Reversal of sexual dimorphism in splenic T lymphocyte responses after trauma-hemorrhage with aging. Am J Physiol Cell Physiol 2000;278:C509–16.
88. Angele MK, Schwacha MG, Ayala A, Chaudry IH. Effect of gender and sex hormones on immune responses following shock. Shock 2000;14:81–90.
89. Majetschak M, Christensen B, Obertacke U, Waydhas C, Schindler AE, Nast-Kolb D, Schade FU. Sex differences in posttraumatic cytokine release of endotoxin-stimulated whole blood: relationship to the development of severe sepsis. J Trauma 2000;48:832–9; discussion 839–40.
90. Wichmann MW, Ayala A, Chaudry IH. Male sex steroids are responsible for depressing macrophage immune function after trauma-hemorrhage. Am J Physiol 1997;273(4 Pt 1): C1335–40.
91. Regelson W, Loria R, Kalimi M. Dehydroepiandrosterone (DHEA) – the "mother steroid." I. Immunologic action. Ann NY Acad Sci 1994;719: 553–63.
92. Morfin R, Courchay G. Pregnenolone and dehydroepiandrosterone as precursors of native 7-hydroxylated metabolites which increase the immune response in mice. J Steroid Biochem Molec Biol 1994;50: 91–100.
93. Padgett DA, Loria RM. In vitro potentiation of lymphocyte activation by dehydroepiandrosterone, androstenediol, and androstenetriol. J Immunol 1994;153: 1544.
94. Hall GM, Perry LA, Spector TD. Depressed levels of dehydroepiandrosterone sulphate in postmenopausal women with rheumatoid arthritis but no relation with axial bone density, Ann Rheum Dis 1993;52: 211–14.
95. Weksler ME. Immune senescence and adrenal steroids: immune dysregulation and the action of dehydroepiandrosterone (DHEA) in old animals. Eur J Clin Pharmacol 1993;45: S21–3.
96. Araneo B, Dowell T, Woods ML, Daynes R, Judd M, Evans T. DHEAS as an effective vaccine adjuvant in elderly humans. Ann NY Acad Sci 1995;774: 232–48.
97. Danenberg HD, Ben-Yehuda A, Zakay-Rones Z, Friedman G. Dehydroepiandrosterone (DHEA) treatment reverses the impaired immune response of old mice to influenza vaccination and protects from influenza infection. Vaccine 1995;13: 1445–8.
98. Verthelyi D, Klinman DM. Sex hormone levels correlate with the activity of cytokine-secreting cells in vivo. Immunology 2000;100:384–90.
99. Casson PR, Andersen RN, Herrod HG, Stentz FB, Straughn AB, Abraham GE Buster JE. Oral dehydroepiandrosterone in physiologic doses modulates immune function in postmenopausal women. Am J Obstet Gynecol 1993;169: 1563–9.
100. Scarpellini F, Scarpellini L, Dino N, Benvenuto P. Progesterone immunosuppressive levels and luteal steroid profiles in the cycles induced with clomiphene citrate. Clin Exp Obstet Gynecol 1993;20:182–8.
101. Check JH, Tarquini P, Gandy P, Lauer C. A randomized study comparing the efficacy of reducing the spontaneous abortion rate following lymphocyte immunotherapy and progesterone treatment versus progesterone alone in primary habitual aborters. Gynecol Obstet Invest 1995;39: 257–61.
102. Valentino R, Savastano S, Tommaselli AP, Riccio A, Mariniello P, Pronesti G, DeDivitiis

PM, Lombardi G. Hormonal pattern in women affected by rheumatoid arthritis. J Endocrinol Invest 1993;16: 619–24.
103. Benyon HLC, Garbett ND, Barnes PJ. Severe premenstrual exacerbations of asthma: effect of intramuscular progesterone. Lancet 1988;2: 370–2.
104. Chisari AN, Giovambattista A, Perello M, Gaillard RC, Spinedi ES. Maternal undernutrition induces neuroendocrine immune dysfunction in male pups at weaning. Neuroimmunomodulation 2001;9:41–8.
105. Faggioni R, Feingold KR, Grunfeld C. Leptin regulation of the immune response and the immunodeficiency of malnutrition. FASEB J 2001;15:2565–71.
106. Bhasin S, Singh AB, Javanbakht M. Neuroendocrine abnormalities associated with HIV infection. Endocrinol Metab Clin North Am 2001;30:749–64.
107. Parra A, Ramirez-Peredo J, Larrea F, Cabrera V, Coutino B, Torres I, Angeles A, Perez-Romano B, Ruiz-Arguelles G, Ruiz-Arguelles A. Decreased dopaminergic tone and increased basal bioactive prolactin in men with human immunodeficiency virus infection. Clin Endocrinol (Oxf) 2001;54:731–8.
108. Parra A, Ramirez-Peredo J, Larrea F, Perez-Romano B, Cabrera V, Torres I, Reyes-Nunez V, Ruiz-Arguelles G, Ruiz-Arguelles A. Serum prolactin is associated with apoptosis in men with human immunodeficiency virus infection. Immunol Cell Biol 2001;79:285–90.
109. Grinspoon S, Corcoran C, Stanley T, Rabe J, Wilkie S. Mechanisms of androgen deficiency in human immunodeficiency virus-infected women with the wasting syndrome. J Clin Endocrinol Metab 2001;86:4120–6.
110. Wolff FH, Nhuch C, Cadore LP, Glitz CL, Lhullier F, Furlanetto TW. Low-dose adrenocorticotropin test in patients with the acquired immunodeficiency syndrome. Braz J Infect Dis 2001;5:53–9.
111. Gonzalez-Gonzalez JG, de la Garza-Hernandez NE, Garza-Moran RA, Rivera-Morales IM, Montes-Villarreal J, Valenzuela-Rendon J, Villarreal-Perez JZ. Prevalence of abnormal adrenocortical function in human immunodeficiency virus infection by low-dose cosyntropin test. Int J STD AIDS 2001;12:804–10.
112. Lala R, Palomba E, Matarazzo P, Altare F, Tovo PA. ACTH and cortisol secretions in children with perinatal HIV-1 infection. Pediatr AIDS HIV Infect 1996;7:243–5.
113. Mylonakis E, Koutkia P, Grinspoon S. Diagnosis and treatment of androgen deficiency in human immunodeficiency virus-infected men and women. Clin Infect Dis 2001;33: 857–64.
114. Rana S, Nunlee-Bland G, Valyasevi R, Iqbal M. Thyroid dysfunction in HIV-infected children: is L-thyroxine therapy beneficial? Pediatr AIDS HIV Infect 1996;7:424–8.
115. Hadigan C, Jeste S, Anderson EJ, Tsay R, Cyr H, Grinspoon S. Modifiable dietary habits and their relation to metabolic abnormalities in men and women with human immunodeficiency virus infection and fat redistribution. Clin Infect Dis 2001;33:710–7.
116. Torres AM, Kazee AM, Simon H, Knudson PE, Weinstock RS. Central diabetes insipidus due to herpes simplex in a patient immunosuppressed by Cushing's syndrome. Endocr Pract 2000;6:26–8.
117. Berger DS. Combating malnutrition in HIV infection. AIDS Patient Care STDS 1996;10: 94–100.
118. Lo JC, Mulligan K, Noor MA, Schwarz JM, Halvorsen RA, Grunfeld C, Schambelan M.The effects of recombinant human growth hormone on body composition and glucose metabolism in HIV-infected patients with fat accumulation. J Clin Endocrinol Metab 2001;86:3480–7.

119. Bae RJ, Orgill DP, DeBiasse MA, Demling R. Management of a patient with advanced AIDS and toxic epidermal necrolysis using human growth hormone and G-CSF. AIDS Patient Care STDS 1997;11:125–9.
120. Polsky B, Kotler D, Steinhart C. HIV-associated wasting in the HAART era: guidelines for assessment, diagnosis,and treatment. AIDS Patient Care STDS 2001;15:411–23.
121. Basaria S, Wahlstrom JT, Dobs AS. Clinical review 138: Anabolic-androgenic steroid therapy in the treatment of chronic diseases. J Clin Endocrinol Metab 2001;86:5108–17.
122. Rothschild H, Jazwinski SM. Human longevity determinant genes. J La State Med Soc 1998;150:272–4.
123. Burns EA, Goodwin JS. Immunodeficiency of aging. Drugs Aging 1997;11:374–97.
124. Chakravarti B, Abraham GN. Aging and T-cell-mediated immunity. Mech Ageing Dev 1999;108:183–206.
125. Lesourd B. Immune response during disease and recovery in the elderly. Proc Nutr Soc 1999;58:85–98.
126. Schmucker DL, Heyworth MF, Owen RL, Daniels CK. Impact of aging on gastrointestinal mucosal immunity. Dig Dis Sci 1996;41:1183–93.
127. Velkeniers B, Dogusan Z, Naessens F, Hooghe R, Hooghe-Peters EL. Prolactin, growth hormone and the immune system in humans. Cell Mol Life Sci 1998;54:1102–8.
128. Mariani E, Ravaglia G, Forti P, Meneghetti A, Tarozzi A, Maioli F, Boschi F, Pratelli L, Pizzoferrato A, Piras F, Facchini A. Vitamin D, thyroid hormones and muscle mass influence natural killer (NK) innate immunity in healthy nonagenarians and centenarians. Clin Exp Immunol 1999;117:206.
129. Oelzner P, Franke S, Muller A, Hein G, Stein G. Relationship between soluble markers of immune activation and bone turnover in post-menopausal women with rheumatoid arthritis. Rheumatology (Oxford) 1999;38:841–7.
130. Nishiyama N. Thymectomy-induced deterioration of learning and memory. Cell Mol Biol (Noisy-le-grand) 2001;47:161–5.
131. Haynes BF, Markert ML, Sempowski GD, Patel DD, Hale LP. The role of the thymus in immune reconstitution in aging, bone marrow transplantation, and HIV-1 infection. Annu Rev Immunol 2000;18:529–60
132. Haynes BF, Hale LP. The human thymus. A chimeric organ comprised of central and peripheral lymphoid components. Immunol Res 1998;18:175–92.
133. Douek DC, Koup RA. Evidence for thymic function in the elderly. Vaccine 2000;18: 1638–41.
134. Shinkai S, Konishi M, Shephard RJ. Aging, exercise, training, and the immune system. Exerc Immunol Rev 1997;3:68–95.
135. Effros RB. Long-term immunological memory against viruses. Mech Ageing Dev 2000;121:161–71.
136. Williams GT, Jolly CJ, Kohler J, Neuberger MS. The contribution of somatic hypermutation to the diversity of serum immunoglobulin: dramatic increase with age. Immunity 2000;13: 409–17.
137. Fagnoni FF, Vescovini R, Passeri G, Bologna G, Pedrazzoni M, Lavagetto G, Casti A, Franceschi C, Passeri M, Sansoni P. Shortage of circulating naive CD8(+) T cells provides new insights on immunodeficiency in aging. Blood 2000;95:2860–8.
138. Sharma K, Wang RX, Zhang LY, Yin DL, Luo XY, Solomon JC, Jiang RF, Markos K, Davidson W, Scott DW, Shi YF. Death the Fas way: regulation and pathophysiology of CD95 and its ligand. Pharmacol Ther 2000;88:333–47.

V. IMMUNE-NEUROENDOCRINE CIRCUITRY

Edited by I. Berczi and A. Szentivanyi

The Immune-Neuroendocrine Circuitry

ISTVAN BERCZI and ANDOR SZENTIVANYI

Department of Immunology, Faculty of Medicine, The University of Manitoba, Winnipeg, Manitoba R3E OW3, Canada; and Department of Internal Medicine, Faculty of Medicine, The University of South Florida, Tampa, Florida 33612, USA

ABSTRACT.

That a healthy mind is fundamental to general well being has been recognized since prehistoric times and proverbs analogous to "Healthy body - healthy mind" exist in many languages. Scientific inquiries with regards to mind-body interactions commenced over a century ago. Hans Selye discovered that stress activates the hypothalamus-pituitary-adrenal-thymus axis, which results in the development of the 'general adaptation syndrome'. This syndrome is characterised by elevated resistance of stressed animals to diverse insults [1]. Andor Szentivanyi and colleagues discovered [2–4] that hypothalamic lesions prevent anaphylactic death in guinea pigs. This was the first experimental evidence for the sweeping regulatory power of the nervous system over violent, life threatening immune reactions. Miklos Jancso and coworkers [5]. described that the nervous system also controls the inflammatory response.

In the seventies and eighties a handful of laboratories started to re-examine various aspects of neuroimmune-interaction. It was established that pituitary hormones have the capacity to stimulate, inhibit and modulate immune responses. Placental and pituitary hormones are also involved in the development of the immune system and maintenance of immunocompetence. It was also described that lymphoid organs are innervated and that neurotransmitters and neuropeptides are important regulators of immune and inflammatory reactions. It became gradually apparent that immune derived cytokines and nerve impulses serve as feedback signals towards the neuroendocrine system. Compelling evidence was produced, indicating that immune reactions may be conditioned in the classical pavlovian sense and that emotions affect immune function. Evidence is increasing rapidly for the physiological role of cytokines and of immunocytes in the function of various organs and tissues, and in reproduction. It is also becoming obvious that Selye's general adaptation syndrome really corresponds to the acute phase response. This is a multi-faceted and highly co-ordinated systemic defence reaction, which involves the *conversion* of the immune system from a specific, adaptive mode of reactivity to a rapidly amplifiable, poly-specific reaction mediated by natural immune mechanisms. Immunological (poly)specificity is assured by profoundly elevated levels of natural antibodies and liver-derived acute phase proteins.

Much has been learned about the regulation of cell activation, growth and function from immunological studies. Burnet's clonal selectional theory designated the antigen as the sole activator [6]. Bretcher and Cohn recognised first that at least 2 signals are required [7]. This was followed by numerous studies on cell-to-cell interaction within the immune system and led to our current understanding of the importance of cell adhesion molecules and cytokines in cell activation and proliferation. This, coupled with the available information about the mechanisms of action of hormones and neurotransmitters, of signal transduction and nuclear regulatory pathways paves the way to understanding how higher organisms function in their entire complexity. It is now apparent that the Nervous- Endocrine- and Immune-Systems form a systemic regulatory network, which is capable of regulating all aspects of bodily functions in health and disease. Thus, Neuroimmune Biology provides new foundations to Biology.

1. HISTORY

The first paper on the role of the nervous system in immunological reactions was published in 1898 by Salomonsen and Madsen [8]. This was followed by a book on vagotomy in 1910 [9] that explained anaphylaxis and allergy with a pathological increase in the firing of the vagus nerve. Hyperergic inflammatory reactions were observed in denervated tissue during the 1930s. However, such tissue shows hyperreactivity to neurotransmitters that are present in serum or may reach the tissue through local axonal reflexes. Today it is clear that the sympathetic and the parasympathetic nerveus systems are in a dynamic equilibrium, which changes continuously according to the local regulatory demands of the organs and tissues innervated. Pavlov's students applied successfully the methodology of conditioning to the phenomenon of anaphylaxis. Unfortunately contemporary scientists rejected these findings.

Szentivanyi and co-workers during 1949–1952 [2–4] obtained definitive scientific evidence for the dominant regulatory role of the hypothalamus (tuber cinereum) over anaphylactic reactions in guinea pigs. Other investigators confirmed these original observations by producing evidence that both the hypothalamus and the pituitary gland are key regulators of the immune system. Today much evidence is available in the literature to support the role of the neuroendocrine system in immunoregulation as is obvious from the material presented in this book.

2. THEORETICAL PROGRESS.

The recognition of the immunoregulatory role of the hypothalamus in 1951 coincided with the advent of knowledge about some basic aspects of immunology. The nature of antibody diversity occupied the focus of interest in the light of new genetic studies completed. The "instructionist" theories of immune activation have been abandoned on the basis of new knowledge in favour of "selection theories" as advanced by Talmage in 1957 [10] and Burnet [6]. These theories served as the basis for the elucidation of the genetic and molecular basis of lymphocyte function during immune responses, as we know it today. The recognition that the immune system is highly adaptable through receptor mutation, possesses memory and is capable of responding under *in vitro* conditions led to the conclusion that it is a virtually autonomous system that defends the body from foreign pathogens. This view is still maintained by most immunologists who see no need or use for additional systemic regulatory intervention.

Under these conditions the scientific community at large has ignored the evidence that accumulated slowly with regards to the interaction of the Neuroendocrine and Immune systems. Neverteless, the significance of this relationship has been re-emphasised in 1966 [11] and it was proposed that immune homeostasis and the homeostasis of the organism are closely linked. Much evidence has accumulated since then to support this hypothesis.

3. THE CELLULAR FOUNDATION OF NERVOUS TISSUE AND ITS ORGANIZATION

The nervous system and the central organ of the adaptive immune sytem, the thymus originate from the neural crest, and these two systems share numerous adhesion molecules and soluble mediators. All neural tissue consists of nerve cells called *neurons*, and supporting cells. Neurons have long fibers called *axons* and short ones named *dendrites*. Supporting elements are the *Schwann cells* and *oligodendrocytes* that provide the nerve axons with *myelin* sheaths. *Microglia* are bone marrow derived and macrophage-related. *Astrocytes* are plentiful but ill defined in their function. *Ependymal cells* line the internal cavities of the brain. The blood vessels in the brain are lined with endothelial cells forming tight junctions that form the *blood-brain-barrier*.

4. THE AUTONOMIC NERVOUS SYSTEM

The autonomic nervous system may be divided into *sympathetic* (adrenergic) and *parasympathetic* (cholinergic) compartments, with *norepinephrine* and *acethylcholine* as neurotransmitters. Anatomically this system is often referred to as the *thoracolumbar* or sympathetic and *craniosacral* or parasympathetic division. *Sympathetic chromaffin tissues* are the *adrenal medullary cells, paraganglia* and the *carotid body* and homologous tissues of the great vessels in the thorax.

The nervous system uses ionic currents and neurotransmitters to communicate, which are transmitted from one cell to the other through *synapses*. Synapses may be either electrical (gap junctions) or chemical (neuroeffector junctions). *Neuroeffector junctions* are formed between nerve cells and non-nervous effector cells. A junctional contact consists of a pre-synaptic membrane of the axon, a post-synaptic membrane of the dendrite or perycaryon and an inter-membrane synaptic cleft (~200–300 Å). Because of lack of anatomical continuity, the nerve impulse is transmitted from the first to the second cell by specific mediators called *neurotransmitters*. The transmitters are secreted into the cleft where they reach their receptors on the target cell membrane.

Adrenergic receptors may be subdivided into α and β types which contain several subtypes, and cholinergic receptors may be *muscarinic* or *nicotinic*. Biogenic amines (histamine, serotonin, acethylcholine), amino acids (γ-aminobutyric acid – GABA, glycine), peptides (such as substance-P, neurotensin, calcitonin-gene related peptide, somatostatin, vasoactive intestinal peptide, the endogenous opioid peptides, vasopressin and oxytocin and thymic peptides, etc), non-conventional neurotransmitters (nitric oxide, carbon monoxide) and growth factors (nerve growth factor, neurotrophins), kinins (kallikrein, bradykinin, leukokinin) serve as neurotransmitters.

5. THE ENDOCRINE HYPOTHALAMUS AND CYTOKINES

The hypothalamus contains neurons that coordinate and integrate distant organ system activities through the effector functions of the endocrine and autonomic nervous systems ("endocrine"' hypothalamus). Peptide neurotransmitters produced by hypophysiotropic neurons in the hypothalamus control anterior pituitary hormone secretion. Nearly 40 peptides are present in the median eminence (ME)* of the hypothalamus, which are produced in different hypothalamic nuclei: corticotropin releasing factor (CRF) in the paraventricular nucleus (PVN); growth hormone releasing hormone (GHRH) in the arcuate nucleus; somatostatin (SRIF) in the anterior periventricular area; thyrotropin-releasing hormone (TRH) in the PVN, and dopamine (DA) in the A12 region of the hypothalamus and the A2 and A4 regions of the brain stem. The neurons containing luteinizing hormone-releasing hormone (LHRH) are scattered in the diagonal band of Broca, the medial septum, the medial preoptic and suprachiasmatic areas, and the lateral basal hypothalamus.

Several cytokines, including IL-1α, IL-1β, IL-2, IL-6, TNF-α, and IFN-γ are now known to affect the release of pituitary hormones by an action on the hypothalamus and/or the pituitary gland. Their predominant effects are to stimulate the hypothalamic -pituitary-adrenal axis and to suppress the hypothalamic-pituitary-thyroid and gonadal axes. IFN-γ inhibits the HPA axis. The main effect of cytokines is to stimulate the release of hypothalamic CRF. This effect is then augmented by direct pituitary actions of IL-1 and IL-6 from hypothalamic and/or pituitary sources (possibly from both) on basal and CRF-stimulated ACTH secretion. Conversely, glucocorticoids have been known for some time to have a negative feedback on interleukin synthesis in cells of the immune system. The predominant effect of IL-1 on the hypothalamic-pituitary-gonadal axis is inhibitory via a central action on gonadotropin-releasing hormone (GnRH) secretion. IL-1, TNF-α, and IFN-γ have an inhibitory effect on thyroid hormone secretion acting directly at the hypothalamic-pituitary level. IL-1, IL-6, TNF-α, and the interferons (α, β, γ) all stimulate prolactin release from anterior pituitary cells. IL-6 release is stimulated from the anterior pituitary by IL-1, vasoactive intestinal peptide, pituitary adenylate cyclase-activating polypeptide (PACAP), and calcitonin gene-related peptide. IL-1 predominantly stimulates somatostatin release and this may result in an overall suppression of GH secretion.

6. ALTERED EFFECTOR RESPONSES

During neurotransmission, whether synaptic or neuroeffector, the second unit should always be regarded as an *effector cell*, regardless of its nervous or non-nervous nature. In the *autonomic system* there are three types of effector cells: the *neurons*, the *smooth muscle*, and *exocrine gland cells*. These are the principal target cells of the pharmacologic mediators of allergic responses. The mediators of antigen-antibody responses, when viewed from the standpoint of their physiologic function, are among the natural chemical organizers of autonomic action. Denervation or mediator antagonist drugs cause hypersensitivity in the effector cells towards the transmitter in short supply.

The hypothalamus has two reciprocally antagonistic divisions: *the anterior hypothalamus,* which mediates primarily *cholinergic* responses, and the *posterior hypothalamus*, the stimulation of which results largely in *adrenergic* responses. A balance between these divisions is thought to be important in maintaining normal autonomic functions (e.g., blood pressure). The posterior hypothalamus is normally suppressed by inhibitory impulses transmitted from the sinoaortic barorecaptors, which, among others, keep the posterior hypothalamus in check. When histamine or acethylcholine is given, blood pressure falls, the sinoaortic tension decreases, causing a shift to sympathetic activity. This leads to the release of catechols, which tends to correct and limit the blood pressure drop. Catecholamines increase blood pressure, which shifts the hypothalamic

*The ME constitutes the final common hypothalamic pathway for signals from the brain to the pituitary.

balance to the cholinergic side. It is possible to produce imbalance of the hypothalamus by applying histamine, acethylcholine, or catecholamines directly on hypothalamic structures: by the electrolytic removal or electrical stimulation of one of the divisions of the hypothalamus; and by many other stimuli.

7. SOME EVOLUTIONARY MORPHOREGULATORY AND FUNCTINAL ASPECTS OF THE IMMUNE-NEUROENDOCRINE CIRCUITRY

The immune system is distributed throughout the body and its basic function is local disease prevention and regulation. The nervous system and the thymus epithelium are both develop from the neural crest. Neuronal immunoglobulin gene superfamilies (Ig SFs) and cytokines and their receptors play important roles during the embryonic development of both systems. The neural cell adhesion molecule (N-CAM) belongs to IgSF and plays an important role in the development of the central nervous system by cell-to-cell and cell-to-matrix adhesion and promotion of synapse formation. Extracellular matrix (ECM) proteins have neurite-promoting activity. These ECM proteins are: type IV collagen, fibronectin, laminin, and a newly discovered protein, *neurite outgrowth factor* (NOF), which promotes neurite outgrowth from sympathetic, parasympathetic, sensory, motor, and central neurons. Cytokines, that are shared by the immune and nervous systems also play a role in neural development and function. They include the cilliary neurotropic factor, the leukemia inhibitory factor, oncostatin M, nerve growth factor, brain-derived neurotropic factor, neurotropin-3 and neurotropins 4/5. Factors known to regulate NGF secretion in astrocytes include IL-1, IL-4 and IL-5. The origin of cytokine receptors also represents a classical example of molecular co-evolution in the immune and nervous systems.

Immunomodulation through the T-cell lineage in rodents can be achived by the manipulation of hemispheric lateralization in the brain. It appears that the right hemisphere controls the inductive influence on T cells of signal emitted by the left hemisphere.

Two phenomena are associated with learning and memory in the nervous sytem: habituation and potentiation. Habituation defines decreased responsiveness to a stimulus presented repetitively over time and potentiation refers to increased responsiveness to repetitive stimuli. The hippocampus plays a special role in learning. Experiments on hippocampal slices indicate that the biochemical changes in the synapse represent the molecular bases for long-term memory. They are: (1) a monoamine (serotonin) and a glutamate receptor (NMDA receptor) are involved; (2) binding of the neurotransmitter (serotonin, glutamate) by these receptors initiates a cascade of enzymatic reactions; (3) the first step in this cascade is the activation of a G-protein mediated signal transduction that ultimately affects the efficacy of synaptic transmission. In B memory cells the affinity maturation of the antigen receptor takes place that potentiates signal transduction after antigenic stimulation. Other possible connections between neuronal and immunological memory are: the survival gene, bcl-2, plays a significant role in both immune and neuronal memory; so far IL-1α has been found to be present only in a subset of T-memory cells and hippocampal neurons; transgenic mice overproducing bcl-2 have a long-term persistence of immunoglobulin-secreting cells and an extended lifetime for memory B cells. This suggests that the biochemical basis of all memory-forming processes are highly similar.

Lymphocytic IL-1α is a cell- and species-specific factor, which is capable of increasing β-adrenoceptor concentration and induction of its gene. These observations are highly important both for normal airway physiology as well as for the possible nature of the β-adrenergic dysregulation in asthma by adding an entirely new dimension to the β-adrenergic theory of the atopic abnormality in bronchial asthma.

Stress alters immune reactivity through specific interactions at every level of the neuroimmune regulatory system. This is based on four critical features shared by both systems: (1) they are composed of extraordinarily large numbers of phenotypically distinct cells organized into intricate networks. (2) Cells of both systems synthesize, secrete, and/or release shared effector molecules. (3) Recognition of these effector molecules is realized by the same cellular receptors and second messenger mechanisms of both cell systems. (4) These cellular and molecular determinants make a continuous, multi-lateral flow of information, which is the *sine qua non* of the unique interactions within the immune-neuroendocrine circuitry.

Neuropeptides take on added significance as immunomodulators, since it is now known that lymphoid organs are directly innervated with nerves secreting these agents. From the standpoint of the integration of information in the immune - neuroendocrine circuitry, future studies will have to examine these parallel signaling pathways in isolation.

8. VIRUS ASSOCIATED IMMUNE AND PHARMACOLOGIC MECHANISMS IN DISORDERS OF RESPIRATORY AND CUTANEOUS ATOPY

8.1. Infection associated immune and pharmacologic mechanisms in atopy

The prototype of the non-atopic immediate hypersensitivity is localized or generalized anaphylaxis, whereas manifestations of *atopic immediate hypersensitivity* include *bronchial asthma, hay fever, allergic rhinitis, chronic urticaria*, and *atopic dermatitis*. Only a minority of the population shows some form of atopic disease in spite of the fact that, by and large, identical conditions of antigens must be presumed to exist for all members of the population. The nature of the constitutional basis of atopy, that is, of the underlying determinant for the development of atopic disease, is as yet unexplained. Two possible explanations for the pathogenesis of these conditions have survived: 1) atopy is a *primary disorder of the immune system* with sequelae in the various effector tissues; and 2) a concept of atopy as a *primary autonomic imbalance*, essentially beta adrenergic in character, with sequelae in effector cells, including those engaged in the production of antibodies. The autonomic imbalance is perceived as caused not by some disorder of the autonomic nervous system itself but by a defector functioning of its effector cells. These two concepts are not mutually exclusive. The IgE antibody, which mediates allergic reactions, is essentially identical with atopic reagin in various animal species.

In anaphylaxis we are dealing with a normal (physiologic) antibody response to an unnatural exposure to antigen, whereas in atopic allergic an "abnormal" antibody response to natural antigenic exposure seems to be involved. Anaphylactic reactivity of the sensitized individual depends on the release of an amount of pharmacologically active effector molecules sufficient to be toxic for most members of the same species. In contrast, individuals with atopic disease possess a quantitatively and qualitatively abnormal reactivity to otherwise nontoxic concentrations of endogenously released or exogenously administered pharmacologic mediators. The quantitative change consistently is in the direction of a decreased response when beta-adrenergic agents are the agonists and in the direction of an increased response when any one of the other pharmacologically active effector molecules are involved.

Infection and a number of unrelated stimuli may play major contributory roles in atopy. Atopic disease with its spontaneous pattern of *familial occurrence* cannot be induced at will. The essential difference between immediate hypersensitivities of the non atopic and atopic varieties is that the former conditions are mediated by normal immune and pharmacologic mechanisms, whereas atopy is based on abnormal immune and pharmacologic mechanisms.

Filip, Szentivanyi and Mess [2–4] used hypothalamic "imbalanced" anaphylactic guinea pigs as models for atopic disease. By electrolytic removal of one hypothalamic division or electric stimulation of the antagonistic division, it was possible to alter profoundly the anaphylactic reactivity of guinea pigs both immunologically and pharmacologically. From both the immunologic and pharmacologic standpoints, the conditions so produced more closely approximated those of the human atopic state than does anaphylaxis. Later Szentivanyi and Fishel [12] found that the Bordetella Pertussis- induced hypersensitive state served as a more appropriate model. The hypersensitivity of the pertussis sensitized mouse to pharmacologic mediators was found to be due to an acquired or genetically determined autonomic imbalance caused primarily by a reduced functioning of the adenylate cyclase coupled beta adrenergic receptors and the associated cyclic AMP system.

8.2. The beta-adrenergic theory of atopic disease

This theory regards atopic disorders (i.e., perennial and seasonal allergic rhinitis, bronchial asthma, and atopic dermatitis) not as immunologic diseases but as unique patterns of altered reactivities to a broad spectrum of immunologic, psychic, infectious, chemical and physical stimuli. The antigen-antibody interaction is given the same role as that of a broad category of nonspecific stimuli that function only to trigger the same defective homeostatic mechanism in the various effector cells involved in immediate hypersensitivities [13]. Regardless of the immunologic or non-immunologic nature of the triggering event, the chemical realization of hypersensitivity would be expected to be brought about by the same mediators. The theory postulated that the constitutional basis of atopy lies in a relatively unresponsive beta-adrenergic effector system and the resultant autonomic imbalance deprives the effector tissues of their normal counter regulatory adjustment. This leads to a unique pattern of quantitatively and qualitatively altered reactivity to the chemical organizers of autonomic action, mostly in response to trivial trauma. This alters immune reactions, such as chemotaxis, phagocytosis and cytotoxicity. T cell mediated immunity is suppressed, whereas IgE production and Fc receptor expression is stimulated, mast cell mediator release is exaggerated, bronchial smooth muscle constriction is augmented, mucus secretion is increased and eosiniphilia will develop. All these alterations point to the malfunctioning of the *adenylate cyclase-coupled beta-adrenergic receptor* and the associated cyclic nucleotide complex. Current evidence favors the possibility that there are inherited and/or acquired multiple abnormalities in the receptor – adenylate cyclase – cyclic AMP system of all effector cells that are critical in the organization of immune reactivities.

Atopic abnormality may be 1) acquired by functional receptor regulatory shifts caused by hormonal changes, infection (viral, bacterial, etc) allergic tissue injury or other event; 2) genetically determined; or 3) caused by autoimmune disease. One, two or all three of these effector mechanisms may be operative in a particular disease.

There is an important relationship between asthma and viral respiratory infection. A history of childhood viral respiratory illness is a risk factor for the development of *chronic obstructive airway syndromes* in later life. If such infection leads to obstructive airway disease, the resultant manifestation is likely to be a "wheezy" or asthmatic type of obstructive airway disease. Respiratory viruses commonly induce exacerbations of bronchospasm in the older child, and in adults with known asthma. However, not all viruses can be implicated in this phenomenon. This is a finding that is difficult to reconcile with the damaged epithelium hypothesis (assuming equivalent of infection) and raises the question of other possible mechanisms related to the biochemical properties of the virus. An endotoxin-like action of influenza vaccine has been described and vaccination with purified LPS from E. Coli resulted in decreased number of beta-adrenergic receptors in guinea pig lung. It was also observed that a simple colonization of the respiratory tract by virus was not sufficient to provoke asthma: such attacks occurred only when the infection produced symptoms of fever, malaise, cough or coryza. The *dominant role of fever* in these episodes immediately suggests the profound involvement of adrenergic effector mechanisms.

In patients with atopic dermatitis the increased susceptibility to viral infections is not restricted to dermatotropic viruses but rather reflects an abnormal host response to viral infections in general. Defective *cytotoxic T cells* and also the association of abnormally functioning macrophages and natural killer cells, appear to have an important role in the impaired host defense against viral infections in atopic dermatitis. A reduced production of *IFN*α in children with atopic dermatitis as well as of *IFN*γ in atopic patients with food allergy has recently been demonstrated. Lymphocytic *cyclic AMP-phosphodiesterase*, that destroys cyclic AMP, is increased in atopic dermatitis and in allergic respiratory disease of adults, and this increased activity correlated closely with histamine release from basophils. A significant elevation of phosphodiesterase activity was reconfirmed in newborns with a positive atopic history in first-degree relatives, compared to newborns with a negative history. However, there was no correlation between phosphodiesterase activity and histamine release. This strongly suggests that increased cyclic AMP phosphodiesterase activity plays a primary role in the pathogenesis of atopic disease. Studies of peripheral blood leukocytes and lymphocytes in atopic dermatitis have frequently demonstrated impaired beta adrenergic reactivity.

8.3. The allergic tissue injury as a developmental mechanism of beta adrenergic subsensitivity

Allergic tissue injury may be initiated by antigen-specific *IgE antibodies* that combine with Fcε receptors on various cell types and trigger mediator release upon encounter with the antigen. Various *noxious agents* that are capable of triggering *asthma* are capable of releasing inflammatory mediators from the same target cells. Accounting only for those pharmacologic mediators where the cell-type has been identified, the spectrum of mediator-storing, synthesizing, or transporting cells, includes the *neutrophil leucocytes* that produce and release slow-reacting substance of anaphylaxis (SRS-A), eosinophil chemotactic factor of anaphylaxis (ECF-A), enzymes, vascular permeability factors, kinin-generating substances, a complement- activating factor, histamine-releasers, and a neutrophil inhibitory factor (NIF); *basophilic leucocytes* that release histamine, SRS-A, ECF-A, neutrophil chemo- tactic factor (NCF) and platelet-activating factor (PAF); *the murine basophilic leucocyte* releasing histamine, SRS-A, ECF-A, PAF, and serotonin; *eosinophilic leucocytes* that secrete histamine, PAF, and possibly SRS-A; the *mast cell* with histamine, SRS-A, ECF-A, NCF and PAF content, the *murine mast cell* that contains histamine, SRS-A, ECF-A, PAF, NCF and serotonin, the *"chromaffin-positive" mast cells* with dopamine content in ruminants – in other mammals possibly norepinephrine; the *enterochromaffin cell* containing serotonin; the *chromaffin cell* containg catecholamines; the *platelet,* - depending on species, may contain histamine, serotonin, catecholamines, and prostaglandins; the *neurosecretory cell* with histamine, serotonin, catecholamines, acetylcholine, and prostaglandin cotent, and *nerve cells* that potentially produce all amine-mediators as well as prostaglandins and kinins. Collectively, these pharmacologically active agents produce an increase in blood flow, in capillary permeability, constriction of smooth muscles, and secretion of mucus by exocrine glands. These are the manifestations that dominate the clinical picture of immediate hypersensitivities and the associated inflammatory responses. Pharmacologically active adrenergic agents can both release as well as inhibit the release of allergic mediators. Such non-immunologic histamine release renders sensitized mast cells incapable of responding to antigen challenge with histamine release.

Methylxanthines inhibit histamine release, possibly because they are competitive inhibitors of the specific phosphodiesterase that inactivates cyclic 3', 5'AMP and thereby induce "adrenergic action" by increasing the intracellular concentration of the compound. The adenylate cyclase system therefore must be considered as a critical regulatory system in allergic histamine release. The allergic release of histamine or of other pharmacologic mediators of immediate hypersensitivity is also inhibited by prostaglandins of the E series, prostacycline, adenosine, and histamine. These are substances that interact with cell membrane receptors that activate adenylate cyclase. Current evidence suggests that cyclic AMP acts early in the release process, that it is linked to the obligatory inward flux of calcium, and that it is related to microtubule function.

8.4. Allergic tissue injury, beta-adrenergic subsensitivity and asthma

Recent studies in patients with extrinsic asthma and in animal models of experimental asthma indicated that the allergic tissue injury itself may result in the development of some forms of beta-adrenergic subsensitivity. Beta-adrenergic subsensitivity of airway smooth muscle that was associated with airways hyper-reactivity in a canine asthma model which was due to a deficiency of beta-adrenoceptors. All post receptor beta-adrenergic responses that were measured (cAMP, protein kinase, relaxation) tended to be depressed in animals with airways hyper-reactivity. These observations suggest that the allergic process can cause beta-adrenergic subsensitivity. However, beta-adrenergic subsensitivity in asthma can be shown to occur in the absence of allergic symptoms or beta-adrenergic medication, and under circumstances in which prior or concurrent beta adrenergic medication can be only one contributing factor to defective beta-adrenergic function. This is also reflected by the presence of beta-adrenergic subsensitivity in atopic dermatitis in which beta-adrenergic medication is not used as a therapeutic modality. Nevertheless, endogenous release of catecholamines in response to the allergic tissue injury may contribute to the development of beta-adrenergic subsensitivity through *homologous desensitization* of the beta-adrenergic receptors. At the same time the endogenous release of other pharmacologic mediators (i.e., histamine) in response to the allergic tissue injury may contribute to the beta-adrenergic subsensitivity through *heterologous desensitization*.

Enhanced "releasability" of histamine from basophils and mast cells has been shown to occur in atopic dermatitis and in bronchial asthma. There is also evidence of increased spontaneous in vitro IgE-secretion by lymphocytes from patients with atopic dermatitis. The regulation of IgE secretion in man has not been elucidated in full detail. Furthermore, the exact pathogenetic role of IgE-mediated reactions in atopic dermatitis is still controversial.

8.5. Autoimmunity as a developmental mechanism of beta-adrenergic subsensitivity

Myasthenia gravis is caused by autoantibodies directed at nicotinic acetylcholine receptors at the neuromuscular end-plates, Graves' disease involves autoantibodies to the thyrotropin receptor, and the severe insulin resistance in Type II insulin-resistant diabetes that has been ascribed to autoantibodies to the insulin receptor. It has been recently described that autoantibodies to beta adrenoceptors can be identified in the plasma of some subjects with atopic allergy. Such autoantibodies bound to calf lung membranes, and inhibited stereospecific beta-adrenergic radioligand binding to calf lung beta adrenoceptors, and also precipitated solubilized calf lung beta-adrenergic receptors. The presence of such autoantibodies to beta-adrenoceptors in patients correlated well with a reduced beta - and an increased alpha-adrenergic responsiveness. Virus infections can elicit autoantibody formation.

9. RECEPTORS, LIGANDS AND SIGNAL TRANSDUCTION

9.1. Adhesion molecules

Adhesion molecules are cell membrane receptors that mediate cell-to-cell and cell-to-matrix communication and are involved in organ, tissue and cell specific regulation under physiological and pathophysiological conditions. These molecules are present in all cells of the body, some may be unique to the tissue/cell type, while others are shared amongst numerous tisssues/organs and cells. This regulatory system has the power to alter/modify/tune systemic regulatory signals and to generate signals locally in order to match tissue/cell function with local demands in health and disease. Moreover adhesion molecules are involved in immune-activation and function as well as in immune neuroendocrine communication. Four major families of adhesion molecules are distinguished. (i) The *immunoglobulin superfamily* (IgSF): Fc receptors for immunoglobulins, the CD4 and CD8 co-receptors of T cells, major histocompatibility antigens, and the T and B lymphocyte antigen receptor are members of this family. (ii) *Integrins* are transmembrane glycoprotein receptors that play critical roles in matrix-to-cell and cell-to-cell signaling and play key roles in the early development of invertebrates and vertebrates, in inflammation and in tumor metastasis. Beta-1 integrins play crucial roles in lymphocyte adhesion, migration, proliferation and differentiation. Beta-2 integrins mediate leukocyte cell-to-cell and cell-to-matrix adhesions during inflammation and immune responses. (iii) *Selectins* recognize carbohydrate molecules on pathogenic microorganisms as well as on self-components. They play a role in cellular interactions, function as cell surface receptors that mediate leukocyte recirculation and natural killer cell activity as well as play a role in inflammation. Patients with neutrophils deficient in selectin ligand suffer from recurrent infections, persistent leukocytosis and severe growth and mental retardation. (iv) *Cadherins* are abundant in the central nervous system (CNS), but also present in many other tissues. Classical cadherins, desmosomal cadherins, seven-pass transmembrane cadherins and protocadherins are major families belonging to this superfamily. Cell-to-cell signaling by cadherins affects cytoskeletal proteins and the organization of adherens junctions. Seven-pass transmembrane cadherins signal through G-protein coupled receptors, regulate cell polarity and thus determine neuronal morphology. Protocadherins are numerous and several of them display an immunoglobulin-like genetic structure. The level of cadherin expression by cells determines the strength of adhesions and the type determines the specificity of cell interactions. Cadherins regulate epithelial cell shape and differentiation, contribute to central nervous system (CNS) regionalization, morphogenesis, fiber tract formation and to the maintenance of functional structures. In reproductive structures, cadherins are responsive to steroid hormones.

9.2. Immunoglobulins (Ig)

Immunoglobulins are serum proteins produced by bone marrow-derived (B) lymphocytes and function as specific antibodies. Five classes are known in man and higher animals: IgA, IgD,

IgE, IgG and IgM. Human IgA is further divided into 2, and IgG into 4 subclasses. The dimeric heavy (H) chains, that are unique to each class are designated as α-, δ-, ε-, γ- and μ-chains, respectively. Each Ig molecule contains two identical light (L) chains, which may belong to 2 major classes: κ or λ. The variable (V) regions of one H and L chain form one antigen combining site, giving rise to 2 sites per Ig molecule. Both H and L chains have constant (C) regions. The C region of the H chains mediates biological functions, such as complement fixation, phagocytosis, cytotoxicity, secretion, regulation of cellular functions and transportation across membranes. IgA is secreted to mucosal surfaces as a dimer and IgM exists both as pentamer and hexamer in serum and secretions. The mammalian genome contains multiple genes coding for variable, diversity (D), joining (J) and constant regions of immunoglobulins. The germline genes are spliced and joined together to achieve VDJC configuration for H chains and VJC for L chains. During the joining process mutations also occur, which give rise to affinity maturation of antibodies during the immune response. Natural antibodies are produced by $CD5^+$ B lymphocytes and preserve their germ-line cofiguration, without mutations. These antibodies recognize highly conserved homologous epitopes (homotopes) in microbes and self components and provide wide sprectrum, polyspecific protection against pathogens. It is now emerging that autoreactive natural antibodies fulfill important regulatory functions. Protection against cancer and against autoimmune disease are amongst the functions fulfilled by natural antibodies.

9.3. Growth and lactogenic hormones (GLH)

Prolactin (PRL), growth hormone (GH) and placental lactogenic hormones (PL) belong to this family. These hormones show molecular heterogeneity and the variant hormones differ in their biological activity. GLH receptors are also heterogeneous and share common features with type I cytokine receptors. GLH commonly use signal transducers and activators of transcription (STAT) family transcription factors in the early phase of induction. However, they also act through other signaling systems.

Growth hormone and prolactin induce insulin-like growth factor-I (IGF-I) in cells of the immune system and in other cells in the body. IGF receptors belong to the transmembrane tyrosine kinase family which signal through phospholipase C-protein kinase C pathway.

The embryonic development of the immune system is supported by placental lactogenic hormones. After parturition pituitary GH and PRL in conjunction with IGF-I maintain the bone marrow, thymus and the secondary lymphoid tissue (spleen, lymph nodes, mucosal and cutaneous lymphoid system) in a functional state. In the joint and complete absence of pituitary GH and PRL, the immune system loses rapidly cellularity, which is associated with a profound decrease in immunocompetence. Bone marrow function is also decreased. Full restoration is possible with replacement doses of either PRL or GH. Pituitary GLH in conjunction with IGF-I support all immune functions that involve humoral and cell mediated immune responses and also the internal immunoregulatory pathways. On this basis, pituitary GLH have been designated as the hormones of *immunocompetence.*

Insulin (INS) is a pleiotropic growth factor, which interacts with both GH and IGF-I. Insulin is required for normal immune development and function. In general, INS stimulates immune reactions but inhibitory effects have also been observed. Insulin appears to be a key regulator of immune function in acute illnesses, which are also characterized by insulin resistance.

9.4. The hypothalamus-pituitary-adrenal (HPA) axis

Corticotropin releasing factor (CRF) regulates *adrenal corticotropic hormone* (ACTH) release, integrates the stress response in the CNS and acts centrally as an immunosuppressive agent by causing sympathetic outflow. Lymphocytes also produce CRF, and it functions within the immune system as an internal regulator. *Eurocortin* (UCN) is a novel endogenous ligand for Type II CRF receptors which is produced within the immune system.

ACTH exerts an immunosuppressive and anti-inflammatory effect mainly by the stimulation of glucocorticoid production by the adrenal glands. Lymphocytes express receptors for ACTH. ACTH is also produced within the immune system and has a direct regulatory effect on immune function. In addition to ACTH, β-*endorphin* (END) and α-*melanocyte stimulating hormone* (αMSH) are derived from the pro-opiomelanocortin molecule in the pituitary gland and in other tissues including the immune system. Kappa, delta and mu-type opioid receptors are present in immune sites. Endogenous opioids have a variable effect on immune function, which may be the result of direct action on the immune system or the modulation of immune function through the CNS. Alpha-MSH is a major inhibitor of fever and inflammation and an antagonist of proinflammatory cytokines. Alpha-MSH is produced within the immune system, in the skin and in many other tissues and functions locally as an immunoregulator.

9.5. The pituitary-thyroid axis

Thyrotropin releasing hormone (TRH) is produced in the hypothalamus and also within the immune system. In the pituitary gland, TRH regulates the release of thyroid stimulating hormone (TSH), prolactin and growth hormone. Within the immune system, TRH has an effect on the development of intestinal T lymphocytes, on the production of interleukin-2 and interferon-γ. TSH also effects immunocytes directly and receptors are detectable at low concentration on B and T lymphocytes, monocytes and NK cells; whereas, high levels were detected on dendritic cells. Human monocytes and lymphocytes synthesize TSH. TSH stimulated IL-2 production by lymphocytes, NK cell cytotoxicity and the phagocytic activity and the production of IL-1β and IL-12 by dendritic cells. Bone marrow cells also express TSH receptors and respond to TSH by increased cytokine production. Thyroxin (T4) and tri-iodothyronine (T3) directly effect lymphocytes and monocytes through nuclear receptors. Lymphocytes are able to convert T4 to T3 which regulates their metabolism, maturation and function. The immunological effects of thyroid hormones are diverse and somewhat controversial. While thyroid deficiency is usually, but not always, associated with immune deficiency, supplemental treatment with T3 yielded mostly negative results. Primary B lymphocyte development is defective in mice deficient in the TSH receptor and thyroid receptor α-negative mice. Other hematopoietic lineages and the development of secondary lymphocytes were normal in these mice.

9.6. Nerve growth factor and leptin and neuropeptides

Nerve growth factor (NGF) shares tyrosine kinase receptors with other neurotropic factors. These receptors belong to the TNF receptor superfamily and are present in lymphoid cells and cells of the nervous system. In general, NGF stimulates immune responses with the exception of IgE production which is inhibited. Nerve growth factor is stimulatory for mast cells and neutrophilic leukocytes and locally exerts a proinflammatory effect. However, systematically applied NGF suppresses inflammation. During inflammatory reactions, lymphocyte-derived

NGF protects the nervous system and other tissues from damage.

Leptin (LEP) is produced primarily by adipocytes and structurally is related to the GLH cytokine family. LEP signals by a class I cytokine receptor of which two isoforms are known. Leptin regulates energy metabolism, reproductive function and lymphocyte development and function. It stimulates TH-1 mediated immune responses and upregulates phagocytosis and the production of proinflammatory cytokines. During the acute phase response, LEP rises rapidly in response to elevated TNF levels. LEP appears to contribute significantly to survival in sepsis by moderating glucocorticoid and IL-6 production, stimulating IL-1 receptor antagonist and potentiating the immune response.

Arginine vasopressin (AVP) exerts an immunostimulatory effect through V_1 type receptors. AVP is produced by lymphocytes.

Substance P (SP) is produced in both the nervous system and in the immune system. SP is a major mediator of neurogenic inflammation and is capable of causing mast cell discharge, increased capillary permeability and smooth muscle construction. SP has a stimulatory effect on immune phenomena in general. Its effect on immunoglobulin secretion is variable. SP stimulates bone marrow cytokines and the formation of granuloma tissue.

Calcitonin gene related peptide (CGRP) stimulates mast cell discharge and inflammatory response whereas it has an inhibitory effect on immune reactions. T lymphocytes synthesize CGRP. *Somatostatin* (SOM) is an antagonist of SP and inhibits inflammatory and immune reactions. *Vasoactive intestinal peptide* (VIP) is produced within the immune system and the receptors are present on monocytes and lymphocytes. In general, VIP functions as an immunosuppressive neuropeptide and is involved in the immunosuppressive properties of aqueous humour in the eye. Its effect on immunoglobulin secretion and NK cell activity is variable. VIP downregulated the inflammatory response. Pituitary adenylate cyclase (PACAP) activating peptide has a similar immunosuppressive and antiinflammatory effect.

9.7. Cytokines

These are locally acting tissue hormones that function as autocrine and paracrine regulators. They have a regulatory influence on growth, differentiation, function, inhibition and apoptosis of cells in various tissues.

Type I Cytokines: Interleukin (IL)-2, -4, -7, -9 and -15 are immunoregulatory cytokines that are recognized by type I membrane bound glycoprotein receptors. These receptors are heterodiamers of a cytokine specific chain and a common γ_c chain, which is involved in signal transduction. Interleukin-3, -5 and granulocyte-macrophage colony stimulating factor (GM-CSF) are hemopoietic cytokines with similar receptors that share a signaling β_c chain. Interleukin –6, IL-11, oncostatin M (OSM), leukemia inhibitory factor (LIF), ciliary neurotropic factor (CNTF) and cardiotrophin-1 have diverse targets that range from the hematopoietic and immune systems to the cardiovascular and central nervous systems. Receptors for these cytokines share glycoprotein (gp) 130. Interleukin-12 is a covalent dimer of a 35 and 50 kDa peptide, the latter is analogous to the soluble IL-6Rα chain. Interleukin-12 promotes cellular immunity and hemopoiesis. IL-21 and 23 are new members of this cytokine family.

Type II Cytokines: Type I interferons (IFNα, β, ω) and the Type II IFNγ belong to this group. Type I interferons share receptors and induce the same signals. They exert anti-proliferative and anti-viral effects and stimulate cytolysis by lymphocytes, NK cells and macrophages. Interferon-γ forms a homodimer (34kDa) and its receptor consists of 2 chains (IFNGR-1 and IFNGR-2). The IFNγ receptor is ubiquitously expressed even by platelets.

Interleukin-10 is an inhibitory cytokine regulating the growth and differentiation of B cells, T cells, mast cells, granulocytes, dendritic cells, keratinocytes and endothelial cells. The IL-10 receptor is related to the IFN receptor. IL-19, 20 and 21 are recently discovered cytokines related to IL-10.

Proinflammatory Cytokines: 1) *Tumor necrosis factor family* – Tumor necrosis factor (TNF), lymphotoxins (LTα, β) belong here. TNF and LTα share receptors and LTβ has a specific receptor as well. These receptors, the p75 nerve growth factor receptor and a number of other related receptors share a "death domain", which is involved in the induction of apoptosis. LTα and other TNF ligands bind to their receptors as trimers. TNF exists both as a soluble and membrane bound protein and LTβ is membrane bound. Many cells can produce TNF in response to noxious agents, infections or immune stimuli. These cytokines regulate the growth of some cells while inhibit others, promote inflammation and exert a neuroendocrine and a metabolic effect during acute phase reactions.

The interleukin-1 family – IL-1α, IL-1β and the IL-1 receptor antagonist (IL-1ra) belong in this family. These cytokines coordinate the host response to infection and injury. IL-1ra is the only known natural cytokine receptor antagonist. Pro-IL-1 is stored in the cytoplasm of various cells and needs to be enzymatically activated before it can be secreted. Various enzymes, including the IL-1β converting enzyme (ICE) activate pro-IL-1. Interleukin-1 is secreted in response to diverse stimuli, including infection, immunological and inflammatory processes and endogenous mediators of cell and tissue injury. Two receptors, IL-1RI and IL-1RII and a receptor associated protein, IL-1RacP, with putative signal transducing function, are known. Type II receptor serves as a "decoy" receptor with no signaling capacity. Immune and inflammatory cells express receptors. IL-1 is a pleiotropic cytokine that coordinates host response to infection and injury and is a major messenger between the neuroendocrine and the immune systems. Interleukin-18 is a new member of the IL-1 family of cytokines.

Chemokines (CEN) – Chemokines are the largest cytokine group. They show chemotactic and proinflammatory properties. They play a major role in acute and chronic inflammation. Chemokines also promote the immune responses, contribute to lympho- and hematopoiesis, vascularization, cell-to-cell adhesion, leukocyte recirculation and homing. Injury, polyclonal lymphocyte stimulation, antigens and endogenous cytokines stimulate CEM production. Four groups of CEM may be distinguished on the basis of their receptor specificity: CXC(α), CC(β), C(γ) and CX3C. These receptors are seven pass transmembrane proteins and are coupled with G-proteins. Chemokines are stored intracellularly as inactive precursors and have to be enzymatically cleaved before secretion in their active form. Interleukin-16 and –17 are novel members of this cytokine family.

Transforming growth factor-β (TGF-β) – This superfamily includes over 30 members, all of which are multifunctional regulating growth, differentiation, inflammation and immune function. Three isotypes are present in mammals: TGF-β2 and TGF-β3 are important in embryogenesis and cellular differentiation, whereas TGF-β1 is an immunoregulator. TGF-β is produced as an inactive protein and it is bound to cell surfaces or to matrix components before it is activated enzymatically. Three receptors exist. Type I and type II receptors are transmembrane serine/threonine kinases that are involved in signal transduction; type III receptors are beta glycans and present TGF-β to receptors I and II. TGF-β is a powerful immunosuppressor and has a systemic anti-inflammatory effect. It promotes the induction of oral immunological tolerance, promotes tissue repair and wound healing and regulates cell proliferation and differentiation. Macrophage inhibitory protein-1 (MIC-1) is a divergent member of the TGF-β superfamily.

Migration inhibitory factor (MIF) – MIF is a pleiotropic cytokine that is produced within the immune and neuroendocrine systems, including the pituitary gland. MIF is secreted by the pituitary gland during infection and injury and regulates glucocorticoid action systemically and locally and it is a major regulator of immune function. MIF promotes cytokine production, angiogenesis, and the immune/inflammatory process.

9.8. Steroid hormones

Glucocorticoids, aldosterone, estrogens, androgens and vitamin D play major roles in the neuroimmune regulatory network are. The receptors for these hormones are nuclear transcription factors that directly regulate gene expression. *Glucocorticoids* are produced in the adrenal gland and in the thymus and are fundamental to immune function. Physiological levels are required for the normal development and functioning of the immune system. Under pathophysiological conditions the serum level of glucocorticoids is elevated, which plays an important role in the suppression of the adoptive immune response, which is T cell-dependent. Glucocorticoids exert a powerful anti-inflammatory effect. Elevated glucocorticoids during acute phase reactions augment the production of natural antibodies and of acute phase proteins by the liver.

Luteinizing hormone releasing hormone and *gonadotropins* exert a direct regulatory influence on the immune system, in addition to the regulation of sex steroid hormones. In turn immune-derived cytokines regulate the production of gonadotropins. These mechanisms insure the coordination of reproduction with health status and prevent inopportune conception.

Estrogens regulate thymus function and suppress cell-mediated immune reactions. The antibody response and natural immunity (NK cytotoxicity, phagocytosis) are augmented by estradiol. *Testosterone* is immunosuppressive during trauma and shock. Many of the immunological effects of testosterone are due to its conversion to estradiol by aromatase in the thymus and in other lymphoid organs. The adrenal androgen, *dehydroepyandrosterone* (DHEA), stimulates immune reactions in experimental animals and in man and antagonizes the immunosuppressive effect of glucocorticoids. The age-related decline of DHEA may contribute to the immunodeficiency that develops in elderly individuals. *Progesterone* is a powerful immunosupressive hormone and plays a major role in the protection of the fetus during mammalian reproduction. Progesterone also contributes to the generation of self tolerance and protects against the excessive activation of the immune system.

1, 25-dihydroxy vitamin D3 (VD3) is a powerful immunosuppressive steroid hormone. Deficiency of VD3 plays an important role in the development of autoimmune disease. It is also required for normal immune function and for proper defense against infectious disease and cancer.

9.9. Enzymes

Members of the serine protease family induce a wide range of biological effects by activating several hormones and growth factors. They also participate in enzyme cascades for inflammation, blood coagulation, complement activation, and other reactions. The action of these proteases is regulated by the endocrine system and by several inhibitors, which are also under endocrine control. Many inhibitors of these enzymes belong to a single family of proteins called serine protease inhibitors or *serpins*.

10. NEUROIMMUNE FUNCTION

10.1. Physiology

10.1.1. Immunocompetence

Immune reactions, which are based on lymphocyte proliferation, are regulated by mechanisms that are involved in the growth control of all cells in higher animals. Growth and lactogenic hormones (GLH) are required for the development and function of the immune system. GLH deliver the first signal to cells, including lymphocytes, that prepares them for proliferation, differentiation and function. This signal is designated as the *competence signal,* which is required for lymphocyte growth and is obligatory for the maintenance of immunocompetence. The second group of signals, that control cell growth, is delivered by cell-to-cell and cell-to-matrix signaling and are designated as *stromal or adherence signals.* Adhesion molecules, tissue bound hormones, cytokines and matrix components deliver these signals. Within the immune system antigen presentation represents a dominant signal for which cell-to-cell interaction is obligatory. Adhesion molecules are fundamental to the organization of multi-cellular organisms and the signals delivered by them serve the basis of species, organ and tissue specific growth and development. The adhesion molecule system has been perfected during evolution from self-recognition to individually specific antigen recognition. Adhesion molecules also play a role in the elimination of degenerated and neoplastic cells. Cell-to-cell signaling has a dominant power over other signals to commit the cell to proliferation. The cell cycle is then completed after the delivery of *cytokine signals.* The nature and combination of these three groups of signals will determine the fate of each cell, which may be survival, proliferation, differentiation and function or alternately apoptosis. Hormones and neurotransmitters, that alter signal delivery, modulate further this basic pattern of animal cell growth. Current evidence that GLH maintain immunocompetence, which enables the immune system to respond to specific antigenic and mitogenic stimuli.

10.1.2. Antigen presentation

Antigen presentation takes place by cell-to-cell interaction whereby a complex signaling process via cell surface adhesion molecules initiates the adaptive (antigen specific) immune responses. Without exception the macromolecular antigen undergoes proteolytic breakdown (processing) and peptide fractions (~ 9–24 residues) are presented by antigen presenting cells (APC) in association with surface MHC antigens. External antigens are engulfed by phagocytic mononuclear cells and are digested in endosomes. The peptide fragments are then joined with MHC-II molecules in endocytic vesicles (endocytic pathway of processing) prior to expression on the cell surface. The APC of this pathway are resposible for the induction of the antibody response and of delayed type hypersensitivity reactions. Endogenous antigens are processed in the cytoplasm, in enzyme-containing proteosomes (cytosolic pathway) and the peptides generated are associated with MHC-I in the endoplsmic reticulum, which in turn is expressed on the surface of all nucleated cells in the body. Cytotoxic T lymphocytes recognize MHC-I-peptide complexes on the cell surface and destroy infected and cancer cells.

Because the antigen presenting system requires the digestion of the antigen after phagocytosis, it protects against extracellular and intracellular pathogens, it also exposes hidden antigenic determinents, decreases the impact of mutations by employing short peptides and allows for self-non-self discrimination. Non-classical MHC antigens are also involved in antigen presentation. Specialized surface molecules (CD1) mediate carbohydrate and lipid presentation. Heat shock proteins normally serve to eliminate dead cells from tissues but also form complexes with antigen, which are taken up by APC through a specific receptor (CD91). Such complexes are taken up by macrophages and dendritic cells, digested and are presented by MHC-I. The dominant role of antigen presentation in the adaptive immune system illustrates the fundamental regulatory function of adherence signals in multi-cellular organisms.

10.1.3. Immune reactions

Under physiological conditions the *immune system* provides continuous defence against infectious agents and cancer, and is part of the homeostatic neuroimmune regulatory circuit that co-ordinates the normal function of the entire organism. The immune system provides local and mobile defence and regulation and it has enormous capacity to deliver defense and regulatory molecules to sites that are in need. Every organ and tissue possesses stromal lymphoid elements that intervene locally to control autoimmune reactions, inflammation, and in general, participate in the physiological processes.

Adaptive cell mediated and humoral immunity and immunological memory are reactions exerted by T and B lymphocytes in concert with members of the leukocyte series. *Innate* host defense relies on non-immune mechanisms, on specialized immune cells, such as natural killer cells, $\gamma\delta$T lymphcytes and $CD5^+$ B cells. The complement system and of T and B lymphocytes activated by alternate pathways are also part of the natural immune system. The natural immune system relies on germ-line coded receptors that recognizes evolutionarily highly preserved homologous epitopes (homotopes) on microbes and also on self components.

Cell-to-cell interactions by the antigen receptor and of MHC molecules and by other adhesion molecules are fundamental to immune activation as well as to stromal regulation. Cytokines are an essential part of this regulatory system. In addition, the immune system interacts with the neural elements and mediators, parenchymal and stromal cells as part of the local neuroimme circuits that govern the organ/tissue under physiological circumstances.

Immunoregulation by the HPA axis. Glucocorticoid hormones exert a multitude of functions that affect virtually every cell in the body. The physiological significance of glucocorticoids is most remarkable at times of stress, when the hypothalamic-pituitary-adrenal axis (HPA) is fully mobilized. The same hypothalamic hormone that stimulates the HPA axis, CRF also mediates behavioral, autonomic and neuroendocrine responses to stress in rodents and primates. Hyperfunction of CRF neurons, therefore, appears to underlie a variety of psychiatric, gastrointestinal, cardiovascular, metabolic and reproductive illnesses attributable to stress. Nonpeptide CRF type-1 receptor antagonists are novel tools to moderate CRH activity in the brain after systemic administration [14]

10.2. Pathophysiology

10.2.1. Neurogenic inflammation

The release of neuropeptides, including tachykinins and calcitonin gene-related peptide, from sensory nerves via an axon or local reflex may have inflammatory effects in the airways. This neurogenic inflammation may be initiated by the activation of sensory nerves by inflammatory mediators and irritants. The phenomenon of neurogenic inflammation is well developed in rodents and may contribute to the inflammatory response to allergens, infections and irritants in animal models. However, the role of neurogenic inflammation in airway inflammatory diseases, such as asthma and chronic obstructive pulmonary disease is still uncertain. There is still little direct evidence for the involvement of sensory neuropeptides in human airways. Initial clinical studies using strategies to block neurogenic inflammation have not been encouraging. In order to clarify the situation, it is necessary to perform prolonged studies of more severe forms of airway disease in the future to explore the role of neurogenic inflammation [15].

10.2.2. Defensins (DEF)

Defensins are antimicrobial cationic peptides with a cysteine-stabilized amphipathic structure. Defensins are normally localized in phagocytes (neutrophil, monocyte/macrophage) and in the epithelial cells of mucous membranes and skin. Some DEF are released into the blood during the course of infection, inflammation or stress. DEF function not only as endogenous animal antibiotic molecules, killing microbial cells and enveloped viruses, but also as physiological regulators. Defensins are implicated in the regulation of endocytosis, chemotaxis, mast cell degranulation and inflammation. Moreover, these molecules are modulators of hemostasis and neuroendocrine-immune interaction. Defensins lower the stress-induced elevation of corticosteroid levels in the blood, and abolish the stress-induced inhibition of the humoral immune response. These facts support the hypothesis that DEF are antibacterial peptides with a broad spectrum of biological activity [16].

10.2.3. The acute phase response

Mild infection or sublethal dose of endotoxin elicits a brief elevation of GH and PRL in the serum, which are proinflammatory and immunostimulatory. In severe trauma, sepsis and shock, GH and PRL are suppressed, whereas glucocorticoids and catecholamines are elevated. Under these conditions an acute phase response (APR) is initiated by immune-derived cytokines, primarily IL-1, IL-6, TNFα. These cytokines elicit a profound neuroendocrine and metabolic response. Fever and catabolism prevails, whereas the synthesis of acute phase proteins (APP) in the liver, cell proliferation in the bone marrow, and protein synthesis by leukocytes is elevated. This is an emergency reaction to save the organism after the adaptive immune sytem has failed to protect the host. During sepsis and endotoxin shock the systemic activation of the complement system and of leukocyte-derived enzymes, tissue-derived brake-down products and highly toxic cytokines seriously threaten survival. The hypothalamus-pituitary-adrenal axis is activated. Glucocorticoids play a major role in the regulation of proinflammatory cytokine production and potentiate the secretion of acute phase proteins. Some APP, such as C reactive protein, LPS binding protein and mannose binding protein in the serum are designed to combine with microorganisms and trigger their destruction by the activation of complement system and of phagocytes. The increased production of some complement components also helps host resistance. The rise in serum fibrinogen promotes blood clotting which can serve to isolate the invading agent by triggering thrombosis in infected tissues. A number of enzyme inhibitors are produced as APP, which are likely to serve to curb the nonspecific damage inflicted by enzymes. *Catecholamines* are also elevated, which serve to inhibit inflammatory responses and to promote, even initiate, the acute phase response. Serum *leptin* is also increased, which governs energy metabolism as well as it has an immunostimulatory effect. If the acute phase reaction fails to protect the host, shock will develop. Patients with subclinical *adrenal insufficiency* succumb to septic shock almost invariably if glucocorticoid therapy is not given. However, glucocorticoid treatment of septic patients with normal adrenal function has not been helpful.

During the acute phase response the T-cell regulated adaptive immune response is switched off and natural immune mechanisms are amplified several hundred to a thousand times within 24–48 hours. This phenomenon has been designated as *immunoconversion,* which is initiated by immune derived cytokines, and involves profound neuroendocrine and metabolic changes, all in the interest of host defense. Therefore, *natural immunity* is essential for a first and last line of defense and the neuroendocrine system is an important promoter and regulator of this fundamental form of immune defence.

10.2.4. Autoimmunity

Auto-reactive lymphocytes and natural antibodies do exist physiologically and participate in the homeostatic regulation of bodily functions. Under normal conditions mature T lymphocytes with autoimmune effector potential are in equlibrium with suppressor cells and with other inhibitory elements, the end result of which is *self-tolerance*. *Immune homeostasis* does not rely exclusively on internal suppressive mechanisms of the immune system. Immunocompetence is dependent on growth and lactogenic hormones. Adaptive and natural immune responses and inflammation are controlled by neuroendocrine regulatory mechanisms. By definition autoimmune disease develops upon the loss of self-tolerance. An autoimmune reactions are due to *de-regulated lymphocyte proliferation* and maturation into effector cells directed towards self antigens. There is evidence to indicate that the de-regulation of multiple genes are required for the development of autoimmune disease. It appears that that de-regulation at the cellular level as well as an *altered neuroendocrine milieu* are necessary for autoimmmune disease to develop.

Autoimmune reactions may be specific for self-molecules, cells, organs or tissues. These specific reactions are due to de-regulated cells of the adaptive immune system. Rheumatoid diseases are characterized by polyclonal lymphocyte activation, inflammation and acute phase reactants, which are indicative of a disorder of the innate immune system. Abnormalities of the hypothalamus – growth and lactogenic hormone – insulin like growth factor axis, of the hypothalamus – pituitary – adrenal axis and of the hypothalamus – pituitary – gonadal axis are freequently observed in autoimmune disease. An *imbalance* of these major immunoregulatory axes of the pituitary gland is very likely to play an etiological role in the pathogenesis of autoimmune disease. It is also becoming apparent that defective regulation by neurotransmitters and neuropeptides contribute significantly to the pathogenesis of autoimmune and inflammatory conditions.

10.2.5. Immunodeficiency

According to current consensus, immunodeficiency is attributed to internal structural and/or regulatory defects, although it is recognized that some of the deficiencies are complex. Animal experiments and some observations in man clearly indicate that abnormalities of neuroendocrine sytem are also able to cause immunodeficiency.

In hypohysectomized (Hypox) rats immune function is inhibited. If the residual serum PRL present in such animals is neutralized, the thymus, spleen, adrenals and gonads become atrophic, severe combined immunodeficiency, anemia, wasting and death will occur in 6–8 weeks. The spontaneous occurrence of joint and complete deficiency of pituitary GH and PRL has never been demonstrateed in man or in animals. This suggests that this condition must be lethal. Memory cells survive hypophysectomy and resist glucocorticoids. Pituitary PRL and GH support the adaptive T cell dependent immune system and are suppressed during febrile illness (acute phase response), where natural immune mechanisms are amplified.

The hormones of the HPA axis, especially glucocorticoids and DHEA-S, are required for normal immune function. Animals lacking adrenal function may be killed by excess cytokine production after simple immunization or after minute amounts of endotoxin, which are harmless to normal animals. The HPA axis is also the prime neuroendocrine mediator of APR, which is fundamental for survival in emergency situations (e.g. sepsis), where adaptive immune mechanisms are ineffective. Cytokines (IL-1, -6, TNFα) initiate these neuroendocrine, metabolic and immune alterations in APR. The HPA axis coordinates the nervous, endocrine and immune systems during APR by regulating fever and CNS activity in general. This axis contributes to the conversion of the immune system from the adaptive mode of reactivity to boosting of natural immune mechanisms and to metabolic alterations (catabolism in most tissues).

Growth hormone treatment of patients with severe acute illness worsened their survival. One possible reason for this is that GH interfered with boosting natural immune defence by inhibiting the HPA axis and catabolism, which are fundamental to APR.

Estradiol is required for the cytotoxic activity of CD4+ T lymphocytes, which fulfill major immunoregulatory function. *Androgens* and *progesterone* are immunosuppressive steroids. Sex hormones are fundamental for reproductive function, where the immune system is an active participant. Disturbances of sex hormone function due to ageing or to other factors contribute to deficient immune function and to the development of disease.

Whenever the adaptive immune system becomes deficient (e.g. in aging individuals, AIDS patients, or as a consequence of malnutrition or of various insults) the body must resort to natural immune defence, including APR. APR develops at major metabolic costs to the host, and may lead to malnutrition, wasting and eventually to death. Proper nutrition and hormone supplements (e. g. GH, IGF-I, anabolic steroids) proved useful in at least some of these conditions.

11. THE NEUROIMMUNE REGULATORY NETWORK

11.1. Basic concepts and principles in neuroimmune regulation

The rapid accumulation of experimental data in diverse systems that has relevance to neuroimmune regulation has led to contradictions, misconceptions and confusion, which need to be addressed and clarified. Some of the key issues are addressed below both from the theoretical and practical points o view.

11.1.1. Hierarchy

It is clear that the *hypothalamus is the highest regulatory organ* of immune and inflammatory reactions. The hypothalamus communicates with all tissues and organs in the body, rather than having a “bi-directional” affair with the immune system, which is commonly held today. Indeed, there is ample evidence to support the proposal that the Nervous-, Endocrine- and Immune systems form a regulatory network that involves the entire organism and governs all functions from conception till death. Network communication is achieved by innervation, by hormones and cytokines distributed through the blood, and by re-circulating cells of the immune system (Figure 1). Innervation is *bi-directional*, as the CNS sends signals to each organ and tissue in the body and receives feedback signals through sensory nerves and by the vagus nerve. Communication by hormones and cytokines is of *multi-directional* network type. The immune system is mobile and is capable of “patrolling” the body by T cell recirculation. Other immune cells (B cells, macrophages), together with T cells, are taking up residence in various organs and tissues as *stromal regulatory elements* and also home to sites of immune/inflammatory reactions. Immune- and tissue-derived *cytokines* provide *feedback* towards the hypothalamus. Shared mediators assure efficient communication between members of the regulatory triad as well as with the tissues and organs of the entire organism [18].

11.1.2. Competence

It is known for over a century that GH is capable of promoting the proportional growth of all tissues and organs in the body. Naturally, this includes the immune organs, such as the spleen, thymus and lymph nodes. Receptors for GLH are expressed on every tissue and organ in the body. Current indications are that all cells require some members of this hormone family for normal growth and function. GLH in conjunction with IGF-I exert a growth stimulatory and an anti-apoptotic effect on all cells and thus assure maintenance in a functional state, i.e. the cells are *competent* to respond to additional stimuli. On this basis GLH may be defined as *competence hormones*. The *competence signal is the first in the signaling hierarchy* of cell activation (No. 1 in Figure 1), as it is the pre-requisite for the survival, growth, differentiation and function of each cell in the body. There is evidence to indicate that the immune system is dependent on GLH for the maintenance of *immunocompetence*. This is related to the general function of these hormones to promote growth and development, as lymphocyte growth is a prerequisite of adaptive immune reactions [19].

Tissue specific growth factors, *cytokines* and *adhesion molecules* play important roles in cell to cell signaling and the regulation of growth and differentiation of various cells in the body. It is apparent that proportional growth, although ultimately controlled by the systemic level of GH, is achieved through the coordinated interaction of systemic and local growth regulatory signals, much of which is tissue/organ specific and also function-related. GH in conjunction with IGF-I promote the survival and growth of all cells in the body and maintain them in a state of *competence*. Prolactin, which may be regarded as a modified growth hormone, is also capable of providing competence to most tissues and organs, with the possible exception of the skeleton where its growth promoting effect is limited. Like GH, PRL also induces IGF-I in its target cells. The IGF-I signal may be regarded as the ubiquitous cytokine signal that is needed for the survival of competent cells. On this basis one may suggest that *competence hormones maintain their target cells by a direct stimulatory effect on the genome and by the stimulation of IGF-I or equivalent cytokine.*

Prolactin synthesis is detectable in numerous tissues, including lymphoid tissue. There is compelling evidence to suggest that tissue-derived PRL fulfills autocrine/paracrine regulatory functions. Within the immune system, small lymphocytes are in a quiescent state and do not synthesize significant amounts of mediators. They need to be activated in order to do so. Based on current experimental evidence, one may propose that inactive small lymphocytes are dependent for survival on pituitary GH/PRL and IGF-I. The dependence of cell survival in the thymus, spleen and bone marrow on GH/PRL supports this hypothesis. Moreover, pituitary PRL and GH maintain vital bodily functions and thus must act as survival hormones for the entire organism [19–21]. Animals with joint and total deficiency of GH/PRL do not exist, and such deficiency has not been convincingly demonstrated in man to date [22].

11.1.3. Redundancy

Failure of the neuroimmune regulatory system invariably leads to the death of the organism. In order to avoid frequent failures, the system must have multiple and overlapping regulatory pathways with a high degree of flexibility and plasticity. This is achieved by isologous forms of regulatory molecules, multiple forms of receptors, and by the existence of functionally overlapping or interchangeable regulatory pathways. This is true for growth and lactogenic hormones for the IGF/insulin system for steroid hormones, neuropeptides, cytokines, chemokines and for the various immunoglobin classes. It is quite common in immunology that unsuspected redundancy is revealed in the system by knocking out a particular gene. Similarly, the disabling of prolactin or growth hormone, or even IGF-I, would not paralyze immune function. These experiments and clinical observations indicate clearly that the functional integrity of the neuroimmune regulatory network is maintained even after very severe insults/deficiencies. Elaborate physiological and pathophysiological mechanisms assure that this systemic regulatory network remains functional under the most adverse conditions.

11.1.4. Homeostasis

Healthy individuals and animals maintain their body temperature, blood pressure heart rate metabolism and the concentration of various ingredients in the serum and in tissue fluids within standard physiological ranges, which is characteristic of the species. This has been termed by Claude Bernard over a century ago as "*milieu interieaux*" [23], now designated as *homeostasis*. Under physiological, or homeostatic conditions two basic forms of immune reactivity can be observed, e.g., *innate or natural immunity* and *adaptive immunity*. These function hand-in-hand, providing efficient host defense quietly in total harmony with the organism.

Emotional and physical stress activates the HPA axis, which has a suppressive effect on the adaptive immune system by the inhibition of the thymus and of T lymphocytes. Severe trauma or infection will induce *febrile illness*, which is now termed *the acute phase response* (APR). During APR the innate immune system is amplified and the adaptive, T-cell-dependent system is suppressed. This *immunoconversion* is a very intensive and highly co-ordinated process, that requires the catabolic breakdown of body constituents in order to fuel the neuroendocrine and immune systems (bone marrow, leukocytes) and the intensive production of liver-derived acute phase defense molecules [24]. Because the immune-, neuroendocrine- and metabolic alterations are strictly regulated, the use of the term *allostasis* [25,26], in comparison with homeostasis, is justified. The *allostatic milieu* assures the promotion of host defense at the expense of other tissues and organs. Immunodeficient individuals rely more and more frequently on APR for host defece, which may create a negative energy balance leading to cachexia and eventually to death.

11.2. Cell-to-cell regulation

Traditionally the cells of all tissues and organs have been divided to stromal cells, which were thought to provide for the structure of organs and serve as a frame for holding together the parenchymal or functioning cell. It is now apparent that stromal cells interact actively with parenchymal cells and this interaction is fundamental to the functional regulation of the tissue/organ. Invariably the stroma of all tissues and organs contain immune derived elements such as lymphocytes, macrophages or more specialized cells that include the glia cells in the nervous system, Kuppfer cells in the liver, the Langerhans cells in the skin, etc. These lympoid cells play important functional roles both in health and disease.

Blood vessels and endothelial cells lining the blood vessels are also active participants in lymphocyte recirculation and in local immune/inflammatory reactions. These cells interact both with the circulatory elements of the immune system and locally with elements of the tissue/organ. Cell to cell regulation in tissues is mediated by adhesion molecules that have complimentary binding sites. These molecules are capable of delivering activation or inhibitory signals [26–28].

Upon lymphocyte activation adhesion molecules and other cell membrane receptors have the capacity to co-aggregate within the semi-fluid cell membrane (capping). This brings the molecules/receptors in close proximity and allows the interaction of immuunoreceptor thyroid based activation motives (ITAM) and -inhibitory motives (ITIM). These motifs promote activation by phosphorylation, or inhibit activation by dephosphorylation of signal transuding molecules, respectively. According to the phosphorylation/dephosphorylation balance amongst the capped receptors, the cell may be activated or inhibited as the final outcome. The relevance of this phenomenon to cell function is especially well established for NK cells and lymphocytes. However evidence is increasing that this "higher order" interaction may apply to many other cell types and receptors. Also, the phenomena "receptor crosstalk" has been observed in several systems [28–31]. Adhesion signals have a dominant power to regulate cell activation, proliferation, differentiation and function, or alternately inhibition or commit cells to the suicide pathway. Adhesion signals are in *second* place in the hierarchy of cell signaling (Figure 1).

The immune system consists of mobile cells that are able to home readily to specific target tissues and also to sites of infection, injury, regeneration and healing. T lymphocytes extravasate through postcapillary venules, enter the tissue compartment and then, after having "patrolled" the tissues, gather through the lymphoid system into the thoracic duct and ultimately into the blood. This T cell recirculation is a continuous phenomenon and is thought to be the basis of "immunological surveillance" to control cancer, and to remove infected and degenerated cells from tissues. Stromal lymphoid cells play important physiological roles and are fundamental to host defense, regeneration and repair. Adhesion molecules mediate immunocyte homing and lymphocyte re-circulation. Blood vessels also provide important barrier function in some tissues and organs that are known as immunologically privileged sites. The blood brain barrier is very important from the point of view of neuroimmune interaction.

Ultimately, adhesion molecules, that are capable of delivering non-diffusible cell-to-cell or cell-to-matrix signals, determine the proliferation and functional activation of all cells in the body. Adherence signals are capable of regulating cell function specifically on an individual basis. For example the antigen presenting cell delivers regulatory signals to the antigen specific T lymphocyte [18].

11.3. Innervation

All the immune organs are innervated and functional neurotransmitter receptors are present on immune cells. These facts create the foundation of the research field of neuroimmunology, which examines the homeostatic interactions between the nervous and immune systems in the process of host resistance. Much evidence is available which shows the effects of neurotransmitters on various *in vitro* measures of immune cells. However, the complex interactions created by neural-immune interactions at sites of antigen challenge and within immune organs cannot be accurately modeled in tissue culture systems. This fact created a need for focusing attention on *in vivo* model systems, which maintain the local microenvironment created when antigen is encountered in skin or mucosal membranes. Such experiments showed that nerves are important regulators of lymphid organs and also of immune/inflammatory reactions [32–34].

The central nervous system has the capacity to deliver neurotransmitters and neuropeptides to all tissues and cells in the body. Interestingly the spleen contains only sympathetic efferent nerve fibers. Tissue mast cells are form synapses with nerve fibers. Neurogenic inflammation is the direct result of the discharge of inflammatory mediators from mast cells after stimulation by neurotransmitters (primarily by substance P) released from sensory nerve terminals. Neurotransmitters and neuropeptides, (e.g. cathecolamines, substance P, somatostatin) play major roles in the regulation of immune/inflammatory responses.

During the acute phase response there is a massive release of catecholamines into the circula-

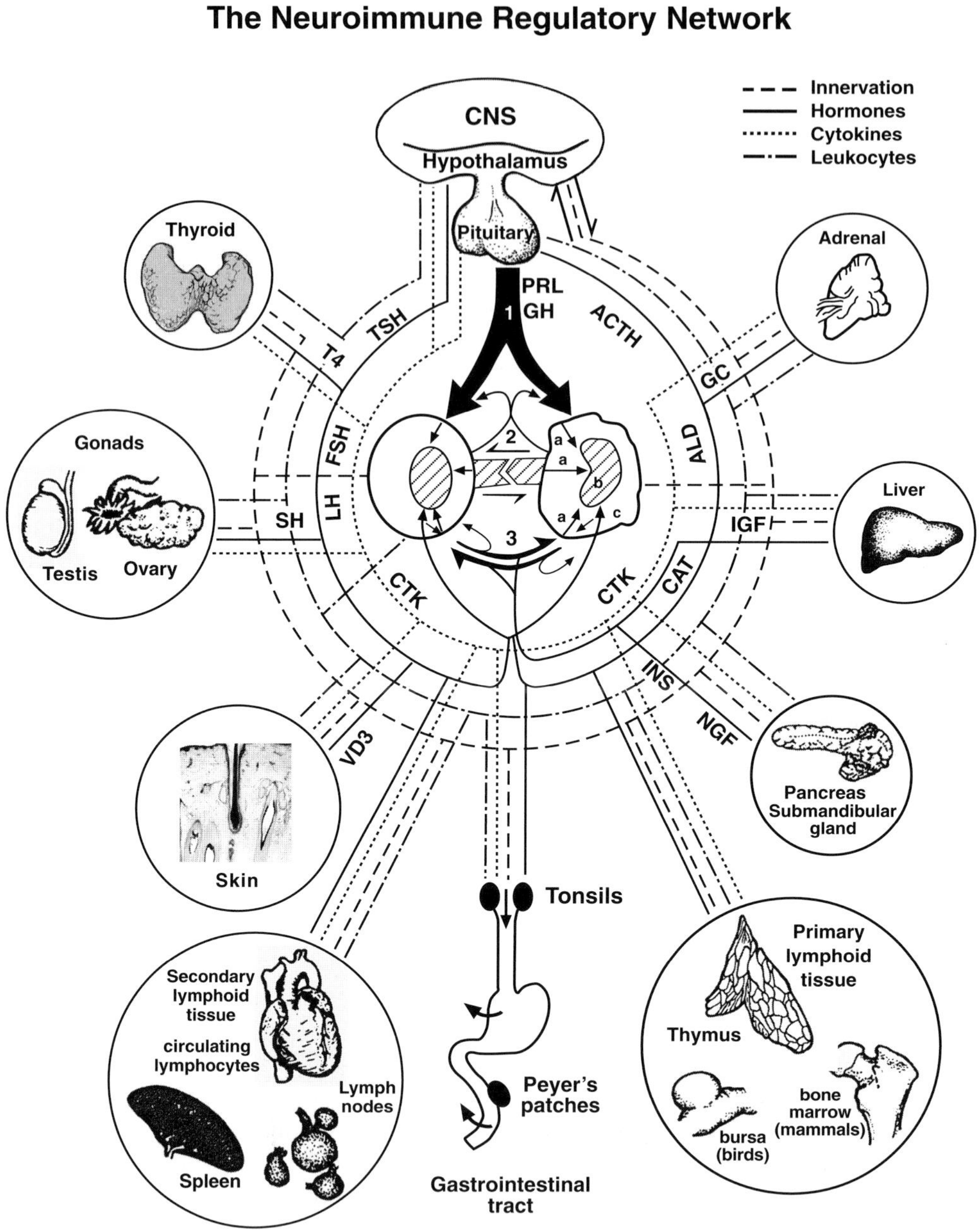

Figure 1. The Neuroimune Regulatory Network.

This figure shows the interaction of major systemic neuroimmune regulatory pathways with local paracrine-autocrine regulatory circuits. Some key organs are also shown with their major regulatory input. Each organ/tissue has its local circuits, which interact with the systemic network via innervation, hormones and cytokines. Mobile elements of the immune system home to all organs and tissues and participate in physiological and pathophysiological processes. Neuroimmune regulation is fundamental to the development and function of all cells in the body from conception till death.

Immunoregulation is used as an example for detailed explanation.

It is proposed that the development and maintenance of lymphocytes and other cells in the body in a functional state is dependent on competence hormones. Additional signals are required for fine tuned functional regulation that includes hormones, adhesion molecules, neurotransmitters, neuropeptides and cytokines. There is a hierarchy for the 3 major groups of signals as given below:

1. Competence signal. During post-natal life pituitary GH and/or PRL deliver to lymphocytes and to all other cells in the body this signal. Many tissues also produce competence hormones ectopically, including the immune system. Ectopic PRL/GH fulfil a local autocrine/paracrine regulatory function during immune reactions. This local regulatory circuit makes rapid lymphocyte proliferation possible, which is a prerequisite of immune reactions. After lymphocyte activation some cytokines may be able to initiate/potentiate the competence signals within the immune system. This remains to be clarified.

2. Stromal/adherence signals. Antigen presentation by specialized cells is an adherence signal and a dominant lymphocyte activator. This is accompanied by additional co-stimulatory adherence signals, eventually leading to lymphocyte activation. Adherence signals also play a role in the induction of immunological tolerance, in lymphocyte survival, and in the induction of programmed cell death (PCD). These signals fulfil tissue-, site- and cell-specific regulation in the body, i.e. the function/fate of individual cells are determined at this level.

3. Cytokine signals. Lymphocyte activation is completed by cytokine signals, which lead to cell proliferation, differentiation, and functional activation. Cytokines may also perform inhibitory function (e.g. TGF-β, interferon-γ) or cause PCD (e.g. tumor necrosis factors).

a. ***Signal modulation***. Some steroid hormones, catecholamines, endorphins/ enkephalins and chemokines are capable of modulating the process of signal delivery from the cell membrane to the nucleus by regulating Ca^{2+} influx, cAMP and cGMP.

b. ***Signal regulation***. Glucocorticoids, sex and other steroid hormones (SH), tri-iodothyronin (T3) and vitamin D3 control lymphocyte signalling by the regulation of nuclear transcription factors. These steroid hormones and T3 play a regulatory role also in cell differentiation and in the elimination of unwanted cells via the induction of PCD.

c. ***Local hormone activation***. Bioactive thyroid and steroid hormones are locally generated from inactive precursors by immunocytes (e.g. T3, E2, androstenediol, androstenetriol, and vitamin D3) while the primary function of others (corticosteroids, estradiol, progesterone, aldosterone) is systemic immunoregulation. The thymus itself is also a steroidogenic organ.

Quiescent lymphocytes do not synthesize DNA and exert minimum metabolic activity. Pituitary GH/PRL, some adhesion signals and serum IGF-I play a key role in the maintenance of these lymphocytes until functional activation occurs. Neurotransmitters and neuropeptides are locally acting functional regulators, basically acting as signal modulators and cytokines. Some neuropeptides may be able to deliver competence signals.

The primary and secondary lymphoid organs, the mucosal and cutaneous lymphoid systems contribute leukocytes to the circulation that mediate systemic immunity. Circulating leukocytes also penetrate tissues and become “stromal cells”. These sessile lymphoid elements fulfil important local regulatory function. The adrenal, thyroid and gonads produce important immunoregulatory hormones. The liver contributes to immune function by producing serum IGF-I, by promoting the induction of oral immunological tolerance and it plays a major role in the acute phase response. The pancreas produces insulin and the submndibular gland nerve growth factor, which are major immunoregulators. These glands also produce glandular kallikrein, which have a major immunoregulatory function. All other tissues and organs have an input into the neuroimmune regulatory network via nerve impulses and by cytokine production.

Abbreviations: ACTH = adrenocorticotropic hormone; ALD = aldosterone; CAT = cathecolamines; CNS = central nervous system; CTK = cytokines; FSH = follicle stimulating hormone; GC = glucocorticoids; GH= growth hormone; IGF-I insulin-like growth factor-I, INS = insulin; LH = luteinizing hormone; NGF = nerve growth factor; PRL = prolactin; SH = sex hormones; TSH= thyroid stimulating hormone; T4 = thyroxin; VD3 = vitamon D.

tion, which is known as "sympathetic outflow". Catecholamines are important regulators in the acute phase response, which is an emergency defense reaction. Sensory nerves provide feedback signals towards the CNS from sites of injury, inflammation, and infections. The vagus nerve carries feedback signals to the CNS from visceral organs.

11.4. Humoral communication

Historically the humoral mediators of cell communication have been classified as hormones that act at distant targets, neurotransmitters and neuropeptides, and locally produced hormone-like mediators, now called cytokines. One may also include here immunoglobulins, which originate from B lymphocytes. Immunoglobulins have evolved from adhesion molecules. In addition, virtually every cell membrane bound molecule is present in the serum, which includes MHC molecules and receptor-like-binding proteins.

By now it is clear that "classical" hormones, neurotransmitters and neuropeptides are widely synthesized at various other sites, including the immune system. Moreover, cytokines, which have been originally discovered within the immune system, are now known to be synthesized in other tissues and organs, including the neuroendocrine system. Therefore, the historical definition of hormones neurotransmitters and neuropeptides no longer applies. Rather, systemic and locally produced mediators complement each other, so that optimal function is assured both under physiological and pathophysiological conditions.

Cytokine signals are third in the hierarchy of signals required for cell activation (Figure 1). In the mitotic cycle, third signals are delivered during the late G_1 phase. *Insulin-like growth factor 1* and *insulin* are the classical hormones that act at this stage and initiate the synthetic (S) phase, which will then progress to mitosis [35]. It is clear, however, that at least some of the cytokines produced locally in the bone marrow, e.g. IL-3 and GM-CSF are actually capable of stimulating IGF-I in their target cells [36]. Therefore, these cytokines should be considered as competence hormones, delivering signal 1 and not signal 3. It has been also observed on B-cell hybridomas in culture that PRL potentiated the effect of interleukins for the stimulation of proliferation and antibody formation [37]. Moreover, PRL and GH, which are the prototypes of competence hormones, are also produced locally. In cell biology epidermal growth factor, platelet-derived growth factor and fibroblast growth factor have been designated as competence hormones [35]. In addition estradiol is capable of inducing IGF-I in mammary tissue [38]. These observations indicate that competence may be inducible by hormones and cytokines other than GLH. However, *in vivo* observation in GLH deficient animals strongly indicate that the total lack of GLH leads to wasting disease and death. Therefore, GLH is fundamental to the maintenance *of vital bodily functions*, including immune function. A simple explanation for this could be that the production of all the other factors that are capable of delivering signal 1 in the mitotic cycle requires competent cells to begin with and the production of such cells rests with the availability of GLH. Indeed, we have observed that PRL disappeared from lymphoid tissue of rats after hypophysectomy [17]. Clearly, GH, PRL, IGF-I and insulin are readily available in significant quatities in the serum at all times, and thus provide competence to all cells in the body whether they are active or in a quiescent state. Further studies are required to elucidate these questions in more detail.

Numerous receptors in immunology and several hormone receptors need to be cross-linked by the ligand in order to deliver an activation signal to the target cell. This mechanism provides an important regulatory function in that cross-linking may take place only at an optimal receptor/ligand ratio, whereas low or high ligand concentrations would not be able to activate the receptor. When more than one receptor isotype is available, the homo- and hetero-diamers formed by the specific ligand may have different biological functions. This is the case with steroid hormones. In addition, cross-linking may be one of the important mechanisms that promote capping of the receptors prior to activation.

12. THE SIGNIFICANCE OF NEUROIMMUNE BIOLOGY

The fundamental significance of the central nervous system for normal development and for the maintenance of health throughout the life cycle of higher organisms is generally accepted by the scientific community and by the general public. These organisms rely continuously on their central nervous system for the regulation of all bodily functions. Some time ago it was realized that the Endocrine system is an essential part of this systemic regulatory circuitry. Now a case can be made for the Immune system, which also belongs to this systemic regulatory network of higher organisms that play a fundamental coordinating role form conception till death. Some important aspects of this regulation are discussed below [17].

Placental hormones play a major role in the development of the fetus. Indeed, now there is decisive evidence to indicate that placental GLH gradually overrule the maternal pituitary gland and regulate metabolism and the level of developmental hormones even in the maternal organism in the interest of normal fetal development. Virtually all other hormones required for fetal development are produced in the placenta in a self sufficient and autonomous manner [39,40]. These hormones govern fetal development, which include the development of the immune system. In this context it is interesting to note, that fetuses with the congenital lack of the pituitary gland develop normally *in utero*, but are not viable after parturition [41]. This indicates that placental and perhaps maternal hormones fully support fetal development and that the pituitary gland takes over the regulatory role of the placenta after parturition. Sinha and Vanderlaan treated newborn mice with anti-PRL antibody, which induced wasting disease and death in about 30% of the animals [42]. Mice are born with an immature immune system and probably also, with an immature pituitary gland. It is likely that during the transition period of immune and pituitary maturation the newborn pups rely for survival on PRL received via placental transfer and in the colostrum. If this PRL is neutralized, there is no other competence hormone to take over, which could result in wasting and death. This situation is analogous to the case of hypophysectomized rats, where there is residual PRL in the serum of long-term survivors. If this PRL is neutralized by antibodies, immune and bone marrow function ceases, the thymus, spleen, adrenals and gonads become very athrophic and a rapid wasting disease and death occurs within 6 weeks [19]. The vital importance of GLH as a group is further emphasized by the fact that joint and complete deficiency of PRL and GH has never been described. Pit-1 deficient mice and human beings, which have been claimed to be free of pituitary GLH, do in fact have detectable serum hormone levels [43–46]. This explains their survival.

It is known that old people may be deficient in GH. However, PRL deficiency is virtually unknown. Again, it is possible that those individuals that become deficient in both hormones pass away before their condition would be recognized. Preliminary clinical evidence indicates that the treatment of immunodeficiency-related wasting disease with GH and with anabolic steroids is beneficial. These crucial observations, along with much additional evidence presented in this volume, indicate that *GLH maintain vital bodily functions*, including immune function, in the entire organism from conception till death. These hormones provide all cells in the body with *competence* to develop and function. The fundamental importance of these hormones in the biology of higher organisms cannot be over-emphasized.

In contrast with GLH, which have an overall stimulatory effect on the organism, the hormones of hypothalamus – pituitary – adrenal (HPA) axis act both synergistically with GLH under physiological conditions and antagonistically during pathophysiological events by inhibiting growth, metabolism, immune responses and inflammation. Glucocorticoid and other steroid hormone receptors are nuclear transcription factors and are necessary for normal signal transcription by membrane bound receptors. Thus they are integrated into the signal transduction sequence under physiological conditions. The immune system and other tissues have the capacity to generate bioactive steroid hormones from the adrenal steroid precursor-hormone, *dehydroepyandrosterone-sulphate* (DHEA-S).

There is evidence to indicate that the HPA axis begins to function already within the fetus, prior to the activation of pituitary GLH secretion [41]. This suggests that it is of vital importance to have the nuclear control of signaling in place for normal development and function of the organism, including immune function. Animals that lack HPA function respond with exaggerated cytokine production after immunization or if given minute doses of LPS. Such animals die easily, due to cytokine toxicity. The HPA axis plays a fundamental regulatory role also during the acute phase response by promoting the production of acute phase proteins and keeping cytokine production and immune activation at a level, which is compatible with survival. In rheumatoid arthritis and in other chronic inflammatory conditions the HPA axis responds at a subnormal level to inflammatory cytokines, which is likely to play an etiological role in the prolongation of the disease process. DHEA declines with ageing and also during chronic illness, which contributes to the deficiency of adaptive immunity under these circumstances.

The immunomodulatory role of estradiol and progesterone is fundamental to the success of mammalian reproduction. Thymus function is regulated by estradiol, some of which is testosterone derived in males. Estrogens suppress cellular immunity and enhance the humoral response and the activity of phagocytes. Testosterone has an overall suppressive effect, which appears to be significant for protection against autoimmune reactions, but it is a definite disadvantage in trauma-relatd immunosuppression. Because castration does not create life threatening immune disturbances, these hormones my be designated as facultative immunomodulators that help to fine tune immune reactions to accommodate some physiological events such as the reproductive cycle in females.

By now it is clear that the immune system is integrated with the neuroendocrine systems to form a systemic regulatory network in higher organisms. Immune function is required for reproductive success. All tissues and organs contain immune derived stromal cells, which participate in normal development and function as well as provide protection under pathophysiological conditions. Immune derived cells (such as the glia cells in the CNS, the Langerhans cells in the skin, Kupffer cells in the liver and similar monocyte/macrophage related cells in other tissues) interact with parenchymal cells. They play major roles as cytokine producing regulatory cells that are also capable of phagocytosis and antigen presentation. Parenchymal cells, such as keratinocytes in the skin, neurons in the brain, enterocytes in the gut and liver cells form important local regulatory circuits with immune-derived stromal cells [47–51]. According to recent evidence MHC-specific suppressor and effector T lymphcytes are present in all tissues, balancing against each other during homeostasis. The effector T cells rapidly mount immunity against foreign invaders and against cancer, and also participate in the elimination of degenerated cells. The immune system is fundamental also to regeneration and healing [52].

Immunological studies in ageing animals revealed a correlation between longevity and the preservation of immune function. In healthy 90 and 100 year old individuals the level of NK cell-mediated cytotoxicity was similar to that of young subjects. Thyroid hormone and vitamin D levels were maintained and lean body mass was preserved [53]. These observations point again to the fundamental importance of the *neuroimmune regulatory triad* in the maintenance of long and healthy lifespan. Clearly, this systemic regulatory network is fundamental to the development and normal function of higher organisms throughout the entire life cycle.

13. FUTURE PERSPECTIVES

The human genome has been sequenced and functional studies of the entire genome are underway (Genomics) with modern methodology. The power of these approaches is unprecedented and is expected to produce major advances to our understanding at the cellular level. The need for the integration of the accumulated knowledge has also been realized and is promoted by several groups world-wide. These efforts run under the term of Neuroimmunology, NeuroImmunoModulation, PsychoNeuroImmunology, Integrative Physiology and the all-inclusive term, *Neuroimmune Biology*. There is no doubt that these efforts will lead to major advances in our understanding of higher organisms in their entire complexity, as well as yield abundant practical information for more rational approaches to the prevention and management of diseases.

Major advances will be made in our understanding of the physiology and pathophysiology of higher organisms, including man, at the molecular, cellular and organizational level. The conditions and requirements of health will be recognized, which will make *Preventive Medicine* the first choice for the management of public health problems. The etiology of most diseases that are not curable today will be clearly delineated. If prevention is not feasible, it will be possible to correct the deficiencies either by replacing the key regulatory substances that are missing or even better, by providing the patient with good genes, healthy cells, tissues or organs that will perform the missing/defective function. Recent developments in the field of molecular science, genetics, immunology, stem cell science and medicine give us legitimate hope that most, if not all the human diseases that cause much suffering today, will be prevented and or eradicated in the brave new world of *Neuroimmune Biology*. The implications are similar for all other areas of Biology, including Animal Science and Animal Husbandry.

REFERENCES

1. Selye H. The general adaptation syndrome and the diseases of adaptation. J Clin Endocrinol 1946; 6: 117–230.
2. Filipp G, Szentivanyi A, Mess B. Anaphylaxis and nervous system.Part I. Acta Med Hung 1952; 2: 163–73.
3. Szentivanyi A, Filipp G. Anaphylaxis and the nervous system. Part II. Ann Allergy 1958; 16: 143–151.
4. Filipp G, Szentivanyi A. Anaphylaxis and the nervous system. Part III. Allergy 1958; 16: 306–311.
5. Jancso N. Role of the nerve terminals in the mechanism of inflammatory reactions. Bull. Millard Fillmore Hosp (Buffalo NY) 1960; 7–53.
6. Burnet FM: The Clonal Selection Theory of Acquired Immunity. Cambridge University Press, 1959.
7. Bretscher P, Cohn M: A theory of self-nonself discrimination. Science 1970;169:1042–

1049.
8. Salomonsen and Mandsen. Compt. rend. de l'Academie des Sciences. 1898;125: 1229.
9. Eppinger and Hess. Die Vagotonie. Hirschwald. Berlin. 1910.
10. Talamge DW. Allergy and immunology. Ann Rev Med 1957;8:239–256.
11. Szentivanyi A, Fishel CW. Die Amin-mediatorstoffe der allergischen Reaktion und die reaktionfahiegheit ihrer Erfolgeszellen. In Pathogese und Therapie allergisher Reaktionen. Filipp G, Szentivanyi A, editors, Ferdinand Enke Verlag, Stutgart, 1966; pp.588–683.
12. Szentivanyi A, Fishel CW. Neurohumoral concepts of bronchial asthma. Acta Allergologica 1974; 29 (Suppl.11):26.
13. Szentivanyi A. The beta adrenergic theory of the atopic abnormality in bronchial asthma. J. Allergy 1968; 42: 203.
14. Gabry KE, Chrousos G, Gold PW. The Hypothalamic-Pituitary-Adrenal (HPA) Axis: A Major Mediator of the Adaptive Responses to Stress. In: Berczi I, Szentivanyi A, series editors. The immune-neuroendocrine circuitry. History and Progress. Neuroimmune Biology, Vol. 3. Amsterdam: Elsevier, 2003;379–414.
15. Barnes PJ. Neurogenic inflammation in the airway. In: Berczi I, Szentivanyi A, series editors. The immune-neuroendocrine circuitry. History and Progress. Neuroimmune Biology, Vol. 3. Amsterdam: Elsevier, 2003;437–450.
16. Korneva EA, Kokryakov VN. Defensins: antimicrobial peptides with a broad spectrum of biological activity. In: Berczi I, Szentivanyi A, series editors. The immune-neuroendocrine circuitry. History and Progress. Neuroimmune Biology, Vol. 3. Amsterdam: Elsevier, 2003;451–463.
17. Berczi I. Neuroimmune Biology: An introduction. In "New foundation of Biology" I. Berczi I, Gorczynski R. Editors, Neuroimmune Biology Volume 1, Berczi I, Szentivanyi A, Series Editors, Elsevier, 2001. Pp. 3–45, 2001.
18. Berczi I. The role of the growth and lactogenic hormone family in immune function. Neuroimmunomodulation 1994;1:201–216.
19. Nagy E, Berczi I. Pituitary dependence of bone marrow function. Brit. J. Haematol. 1989;71:457–462.
20. Berczi I, Nagy E, Matusik RJ, Friesen HG. Pituitary hormones regulate c-myc and DNA synthesis in lymphoid tissue.J. Immunol. 1991;146:2201–2206.
21. Nagy E, Berczi I. Hypophysectomized rats depend on residual prolactin for survival. Endocrinology 1991;128:2776–2784.
22. Berczi I. Foreword: The neuroimmune biology of growth and lactogenic hormones. In "Growth and lactogenic hormones" Matera L, Rapaport R. Editors, Neuroimune Biology, Volume 2, Berczi I, Szentivanyi A. Series Editors. Elsevier, 2002.
23. Bernard C. Lecons sur le phenomenes de la vie communs aux animaux et au vegetaux. Balliere, Paris, 1878.
24. Berczi I. Bertok L. Chow D. Natural immunity and neuroimmune host defence. Ann NY Acad Sci 2001;917:248–257.
25. McEwen BS. Allostasis, allostatic load, and the aging nervous system: role of excitatory amino acids and excitotoxicity. Neurochem Res 2000;25:1219–31.
26. Berczi I, Nagy E. Neurohormonal control of cytokines during injury. In: "Brain Control of Responses to Trauma," N.J. Rothwell, F. Berkenbosch (Eds.), Cambridge University Press, 1994, pp. 32–107.
27. Tanaka Y. Integrin activation by chemokines: relevance to inflammatory adhesion cascade during T cell migration. Histol Histopathol 2000;15:1169–76.

28. Renard V, Cambiaggi A, Vely F, Blery M, Olcese L, Olivero S, Bouchet M, Vivier E. Transduction of cytotoxic signals in natural killer cells: a general model of fine tuning between activatory and inhibitory pathways in lymphocytes. Transplant Rev 1997; 155: 205–221.
29. Bourguignon LY, Jy W, Majercik MH, Bourguignon GJ. Lymphocyte activation and capping of hormone receptors. J Cell Biochem 1988;37:131–50.
30. Jans DA. The mobile receptor hypothesis revisited: a mechanistic role for hormone receptor lateral mobility in signal transduction. Biochim Biophys Acta 1992;1113:271–6.
31. Horvat RD, Nelson S, Clay CM, Barisas BG, Roess DA. Intrinsically fluorescent luteinizing hormone receptor demonstrates hormone-driven aggregation. Biochem Biophys Res Commun 1999;25:382–5.
32. Pletsityi DF, Averi'anova LL. The influence of vagotomy on the development of experimental endocarditis. Dokl Akad Nauk SSSR 1966;167:238–40. Russian
33. Williams JM, Petersen RG, Shea PA, Schemdtje JF, Bauer DC, Felten DL. Sympathetic innervation of murine thymus and spleen: evidence for a functional link between the nervous and immune systems. Brain Res Bull 1981;6:83–94.
34. Nance DM. Innervation of the spleen in the rat: evidence for absence of afferent innervation. Burns J Brain Behav Immun 1989;3:281–90.
35. Aaronson SA: Growth factors and cancer. Science 1991;254:1146–1153.
36. Kelley KW, Arkins S, Minshall C, Liu Q, Dantzer R. Growth hormone, growth factors and hematopoiesis. Horm Res 1996;45(1–2):38–45.
37. Richards SM, Garman RD, Keyes L, Kavanagh B, McPherson JM. Prolactin is an antagonist of TGF-beta activity and promotes proliferation of murine B cell hybridomas. Cell Immunol 1998;184:85–91.
38. Fournier B, Gutzwiller S, Dittmar T, Matthias G, Steenbergh P, Matthias P. Estrogen receptor (ER)-alpha, but not ER-beta, mediates regulation of the insulin-like growth factor I gene by antiestrogens. J Biol Chem 2001;276:35444–9.
39. Handwerger S, Freemark M. The roles of placental growth hormone and placental lactogen in the regulation of human fetal growth and development. J Pediatr Endocrinol Metab 2000;13:343–56.
40. Reis FM, Florio P, Cobellis L, Luisi S, Severi FM, Bocchi C, Picciolini E, Centini G, Petraglia F. Human placenta as a source of neuroendocrine factors. Biol Neonate 2001;79(3–4):150–6.
41. Potter EL, Craig JM: Pathology of the fetus and the infant. 3rd edition. Chicago, Year Book, 1975.
42. Sinha YN, Vanderlaan WP. Effect on growth of prolactin deficiency induced in infant mice. Endocrinology 1982;110:1871–8.
43. Sinha YN, Salocks CB, Vanderlaan WP. Pituitary and serum concentrations of prolactin and GH in Snell dwarf mice. Proc Soc Exp Biol Med 1975;150:207–10.
44. Barkley MS, Bartke A, Gross DS, Sinha YN. Prolactin status of hereditary dwarf mice. Endocrinology 1982;110:2088–96.
45. Soares MJ, Bartke A, Colosi P, Talamantes F. Identification of a placental lactogen in pregnant Snell and Ames dwarf mice. Proc Soc Exp Biol Med 1984;175:106–8.
46. Pfaffle RW, Blankenstein O, Wuller S, Kentrup H, Combined pituitary hormone deficiency: role of Pit-1 and Prop-1. Acta Paediatr Suppl 1999;88:33–41.
47. Yu Y, Tang L, Wang J, Liu S, Wang W, An H, Qi R, Zhang M, Cao X. Psoriatic lesional keratinocytes promote the maturation of human monocyte-derived langerhans cells.

Dermatology 2002;204:94–9.
48. Heins N, Malatesta P, Cecconi F, Nakafuku M, Tucker KL, Hack MA, Chapouton P, Barde YA, Gotz M. Glial cells generate neurons: the role of the transcription factor Pax6. Nat Neurosci 2002;5:308–15.
49. Dionne S, Ruemmele FM, Seidman EG. Immunopathogenesis of inflammatory bowel disease: role of cytokines and immune cell-enterocyte interactions. Nestle Nutr Workshop Ser Clin Perform Programme 1999;2:41–57; discussion 58–61.
50. Vodovotz Y, Liu S, McCloskey C, Shapiro R, Green A, Billiar TR. The hepatocyte as a microbial product-responsive cell. J Endotoxin Res 2001;7:365–73.
51. Kmiec Z. Cooperation of liver cells in health and disease. Adv Anat Embryol Cell Biol 2001;161:1–151.
52. Berczi I, Szentivanyi A. Immune reactions. In: Sventivanyi A, Berczi I, editors. The Immune-Neuroendrocrine Biology, Vol. 3, Amsterdam, Elsevier 2003;315–377.
53. Mariani E, Ravaglia G, Forti P, Meneghetti A, Tarozzi A, Maioli F, Boschi F, Pratelli L, Pizzoferrato A, Piras F, Facchini A. Vitamin D, thyroid hormones and muscle mass influence natural killer (NK) innate immunity in healthy nonagenarians and centenarians. Clin Exp Immunol 1999;116:19–27.

Keyword index